AF538439

A System Engineering Approach to Imaging

A System Engineering Approach to Imaging

Norman S. Kopeika

SPIE Optical Engineering Press

A Publication of SPIE—The International Society for Optical Engineering
Bellingham, Washington USA

Library of Congress Cataloging-in-Publication Data

Kopeika, Norman S.
A system engineering approach to imaging / Norman S. Kopeika.
p. cm.
Includes bibliographical references and index.
ISBN 0-8194-2377-7 (hardcover)
1. Imaging systems. 2. Systems engineering. I. Title.
TK8315.K66 1998
621.36'7—dc21 96-37101
CIP

Published by

SPIE—The International Society for Optical Engineering
P.O. Box 10
Bellingham, Washington 98227-0010
Phone: 360/676-3290
Fax: 360/647-1445
Email: spie@spie.org
WWW: http://www.spie.org/

Printed in the United States of America.

Second Printing

Dedication

This book is dedicated to my father, to my mother of Blessed Memory, to our children Avi, Chedvah, and Yaakov, and particularly to my dear wife Miriam who provided the sweet homelife that made this effort possible.

The author is also grateful to his parents for supporting his university studies, and to them and his father-in-law of Blessed Memory and his mother-in-law for their encouragement during the course of his career.

Contents

Introduction

The science of imaging by human beings begins with the human visual system and thus dates back to the initial existence of mankind. The human visual system is limited to the "visible" spectrum, which extends from violet wavelengths on the order of 400 nm (750-THz frequency) to red wavelengths on the order of 700 nanometers (429-THz frequency). However, during the last century or so, mankind has succeeded in extending his imaging capability to beyond the visible spectrum toward both shorter (ultraviolet and x-ray images, for example) and longer (infrared, millimeter wave, and microwave, for example) wavelengths.

A plot of the electromagnetic (EM) spectrum is shown in Fig. I.1. The visible spectrum is defined by the spectral response limits of the human visual system, which peaks at green wavelengths—usually the color most common in areas of human habitation. The visible is also characterized by very little atmospheric absorption of radiation, albeit with lots of light scattering in the atmosphere, thus causing "the sky" to exist. Another characteristic of the visible is the transparency of most types of glass, thus making it suitable to fashion from glass lenses and other optical components intended to transmit light.

If one branches out from the visible, at the shorter wavelength end is the ultraviolet (UV) region. The *near-UV* region extends from the visible down to about the 200-nm wavelength. Below that is the *far-UV* region, also known as the *vacuum UV* because the extremely strong atmospheric absorption of such radiation permits radiation transmission essentially under vacuum conditions only. Most of this absorption is by oxygen and ozone, which is high up in the atmosphere. This absorption is essential for human existence down below, because vacuum UV and shorter wavelength radiation is ionizing; that is, the photon energies are so high that they cause permanent changes in media on which they are incident because they give rise to ionization in them (hence, the danger implied in the recent discovery of the ozone hole over Antarctica). In the near UV, only special UV-transmitting glasses can be used in optical components. Most ordinary types of glass become opaque at wavelengths on the order of 350 nm or below.

If one proceeds from the visible toward the infrared (IR), nature behaves almost as in the visible except for the human visual system. Glasses that are transparent in the visible are also transparent in the near IR out to close to the 3 μm wavelength. Therefore, lenses and other optical components useful in the visible are also generally quite applicable in the near IR. From about 3 μm wavelength out to about 14 μm wavelength is the middle or thermal infrared. This wavelength region is most appropriate for thermal imaging. Special materials that are transparent to such radiation are required for lenses and optical components.

Wavelengths from about 15 to about 100 μm comprise the *far-IR* region. Here, there is essentially negligible atmospheric transmission because of strong atmospheric absorption properties.

The spectral region from about the 100 μm wavelength to 1 mm is known as the *near millimeter wave*, or *submillimeter wave*, region. Strong atmospheric absorption

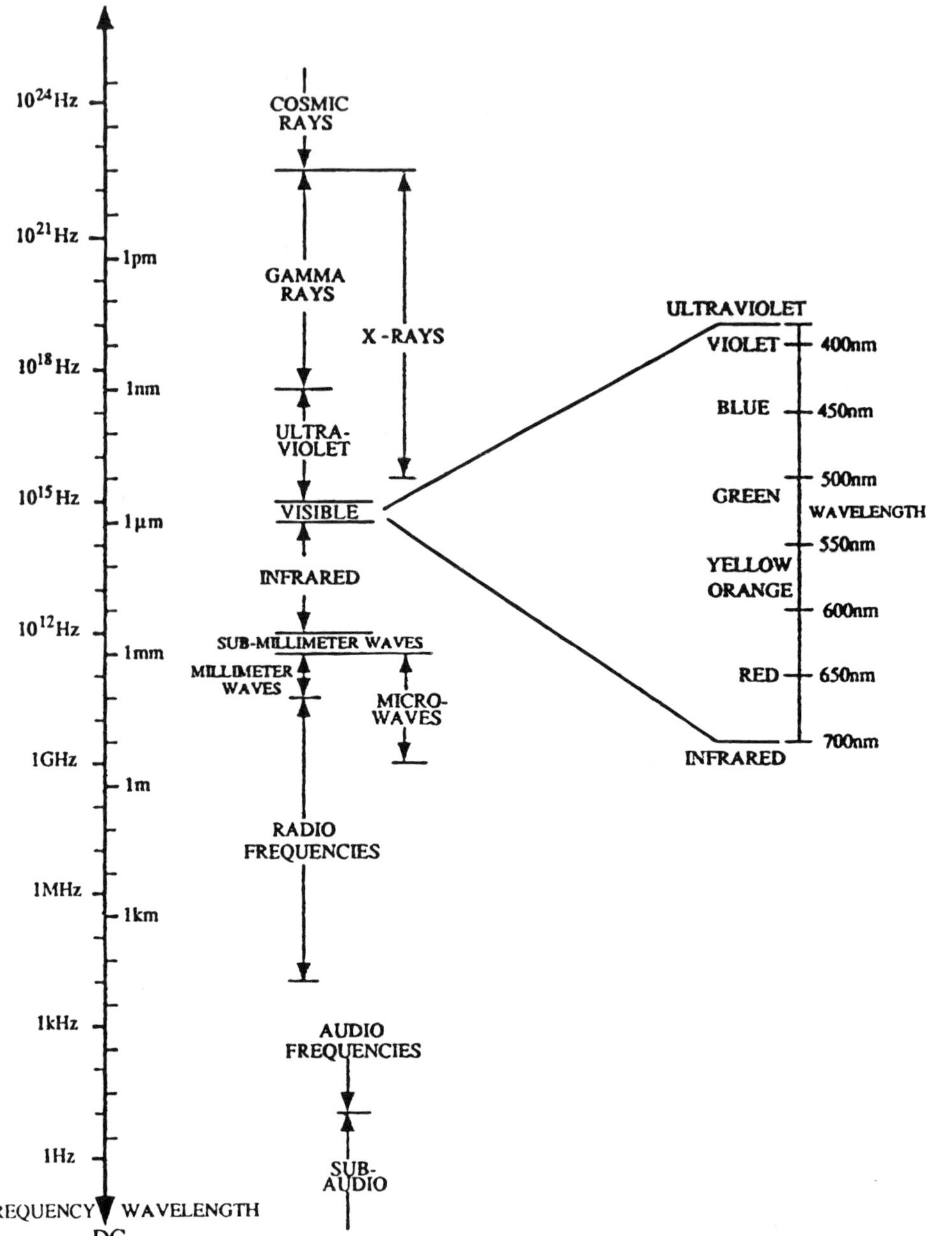

Fig. I.1 Plot of the electromagnetic spectrum.

characteristics of the far IR extend down to about the 1-mm wavelength, and much of it is caused by water vapor absorption.

Millimeter waves (from about 1- to 10-mm wavelength) form a bridge between optical and microwaves. Techniques and concepts from both of the latter are readily employed for millimeter waves. Atmospheric transmission at these and longer wavelengths is generally much better than at optical wavelengths.

In general, the longer the wavelength, the poorer the resolution or detail in image quality. This limitation stems from the diffraction effects discussed in Part 3.

This poor resolution can only be compensated for by increasing the antenna aperture according to the increase in wavelength. This is extremely difficult to do at wavelengths as long as microwaves because optical-type resolution would require apertures on the order of kilometers. Therefore, with the exception of synthetic apertures, imaging is seldom employed at such long wavelengths because of the exceedingly poor image quality. Radars generally just present blips rather than image shapes. However, microwave imaging can be obtained with near-optical quality by employing aperture dimension-to-wavelength ratios that are similar to each other in both spectral regions. This has been shown experimentally at the University of Pennsylvania using arrays of smaller antennas that are equivalent to kilometer-order apertures.

I.1 SYSTEM POINT OF VIEW

This textbook is about imaging, particularly from the system engineering point of view. The approach is relatively broad, being applicable all across the spectrum, with advantages and disadvantages of imaging in various spectral regions being considered. The system engineering approach is important in system design and system analysis so as to determine final image quality or to design for a given required image quality. This requires quantitative criteria with which to define image quality. It also requires analytical tools with which to determine the weak link in the system, the link that has the chief role in limiting the image quality. If either mechanical vibrations or atmospheric effects are the weak link, for example, it makes no sense in noise-limited imaging to use a high-resolution imager, with the ability to provide detail 10 times smaller than that permitted by the vibrational or atmospheric blurring, because the high-resolution imager will only permit the blur itself to be viewed clearly without providing the small detail required in the final image. In such cases a much less expensive low-resolution sensor will provide essentially the same final image quality as the high-resolution sensor. However, for contrast-limited imaging it is possible to restore the image and get rid of essentially all the corrupting blur if the image degradation can be characterized correctly. In such cases, higher resolution sensors can be advantageous. Thus, good engineering design must also relate to system analysis and to cost effectiveness.

An engineering tool very appropriate for system design and analysis is the concept of the *transfer function*, which relates output and input to imaging system. Network transfer functions are an essential tool in electrical engineering. Similarly, optical transfer functions are an essential tool in image systems. Much of this book, therefore, is devoted to the concept and applications of optical transfer functions, including their relationship to target acquisition probabilities. In most imaging systems, the weakest link is often vibrational motion or atmospheric blurring. Consequently, Parts 4 and 5 are added so as to present specific optical transfer functions with which to describe such phenomena. They are then applied in Part 6 to image restoration and system analysis. First, however, the concepts basic to geometrical optics, transfer functions, devices, and imaging in general are presented in tutorial fashion in the earlier chapters.

I.2 TYPICAL IMAGING SYSTEM

A block diagram for a typical imaging system is presented in Fig. I.2. An important parameter to consider in the object plane scene is contrast between the object and its background. This varies, often very strongly, with wavelength. In the open atmosphere it varies also with weather. The propagation channel, such as atmosphere,

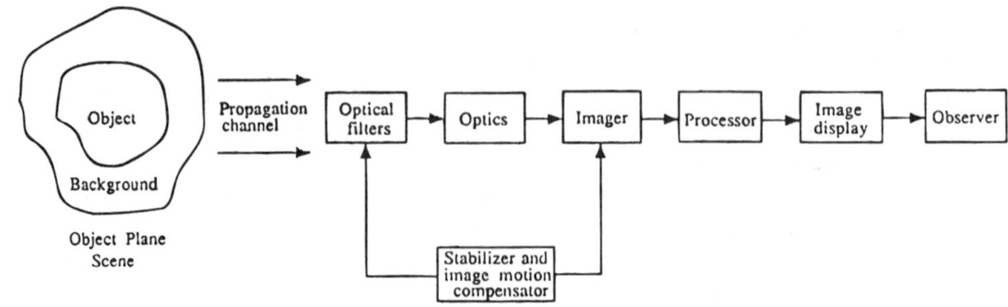

Fig. I.2 Block diagram of typical imaging system.

space, water, or optical fibers, for example, affects not only the transmission of the object plane scene radiation to the receiver, but may also introduce distortion, image blur, noise, and background radiation, all of which vary with wavelength and magnification.

At the receiver end, optical filters can be selected to pass those wavelengths at which contrast and the signal-to-noise ratio are best. The optics are used to collect the received radiation and to form the image that is projected onto the imager, which must be situated in the image plane. The imager may be a film camera, or a TV system, or a thermal imaging system, or a holographic system, etc. The image may be processed for image restoration to improve resolution, target acquisition times and probabilities, etc. The image display should be appropriate to the observer's requirements. For imaging systems involving image motion, either because the receiver or the object or both are moving, image motion compensation is usually essential so as to provide a relatively still picture. When mechanical vibrations are involved, such as in robotics and in moving reconnaissance systems, a stabilizer is almost always necessary so as to limit image blur. Knowledge as to the transfer function characterizing image blur can be used in restoration in order to remove the effects of image blur.

I.3 ORGANIZATION OF BOOK

Chapter 1 provides the theoretical background with which to consider image propagation through the propagation channel and the conceptual foundation for geometric optics. Chapter 2 uses geometrical optics to explain image formation.

Part 2 is concerned with transfer of radiant power in order to permit determination of the image plane signal-to-noise ratio (SNR). In some cases, the radiant power of interest originates in the object plane. In others, the object plane may be illuminated by some external source. The manner in which this radiation power is transferred determines the brightness of the final image.

The image quality may be limited by *contrast* with regard to background radiation. Such an image is also called *blur limited*. In other cases image quality may be limited by fluctuations in photon arrival rate or, in photoelectronic imaging systems, by electronic noise. To characterize both contrast- or blur-limited and noise-limited situations, it is necessary to evaluate each according to relevant image SNRs.

We begin this part first with concepts of radiant power transfer in Chapter 3 in order to be able to calculate later on the signal power in the final image. Light sources in Chapter 4 are also presented from the standpoint of radiometry to extend this capability. Finally, photon and electronic noise mechanisms are presented in

Chapter 5 so as to permit SNR calculations immediately afterwards in Chapter 5. These are for detection primarily and are extended to imaging in Chapter 11.

Part 3 deals with imaging from an engineering point of view. Whereas Chapter 2 considered imaging from the standpoint of geometrical or ray optics, this part of the book considers imaging from an overall system engineering point of view. Hence, wave properties of light must also be included. Inclusion of such properties is referred to as *physical optics*, rather than geometrical optics, which describes ray rather than wave phenomena. First, diffraction effects, which are neglected in geometrical optics, must be included.

Chapter 7 considers Fraunhofer diffraction in particular, and the concepts of optical Fourier transforms and spatial rather than temporal frequencies are introduced. Chapter 8 then considers the effects of diffraction on image quality, and the concept of an optical transfer function involving two-dimensional spatial rather than single-dimensional temporal frequencies is developed. This tool is basic to engineering optics, and forms the main thrust of the book. Chapter 9 relates the mathematical concept of the transfer function to its physical implications concerning contrast. The application of optical transfer function concepts to image system design is introduced in Chapter 10, much of which is also based on the related concept of contrast. Various degrees of image quality according to the Johnson chart are introduced there and the methods and probabilities of "acquiring" targets under such resolution conditions are described. Numerical examples are presented. While Chapter 10 considers contrast-limited imaging, noise-limited imaging is considered in Chapter 11. Chapter 12 considers the transfer function and threshold contrast of the human visual system, which is normally at the output stage of most imaging systems. Finally, Chapter 13 describes various types of imaging hardware, using imaging criteria such as transfer function and contrast function described in the preceding chapters.

Perhaps the strongest tie between optics and electronics lies in the similar mathematics that can be used to describe the respective systems of interest—the mathematics of Fourier analysis and "system" theory. The fundamental reason for the similar mathematics is not merely the common interest in "information," but rather certain basic properties that electronic and imaging systems share. For example, many electronic networks and imaging devices share the properties of *linearity* and *invariance*. Spatial invariance or spatial stationarity is also called *isoplanaticism*. It implies that if an input complex field amplitude (CFA) distribution $U_i(x)$ yields an output $U_0(x_1')$, then an input $U_i(X - x_1)$ yields an output $U_0(x' - x_1')$, where x_1' is related to x_1 by system magnification. Any network or device (electronic, optical, or otherwise) that possesses these two properties can be described mathematically with considerable ease using the techniques of *frequency analysis*. Thus, just as it is convenient to describe an audio amplifier in terms of its (temporal) frequency response, so too it is often convenient to describe an imaging system in terms of its (spatial) frequency response.

The similarities do not end when the linearity and invariance properties are absent. Certain nonlinear optical elements (in particular photographic film) have input output relationships that are directly analogous to the corresponding characteristics of nonlinear electronic components (diodes, vacuum tubes, etc.), and similar mathematics of analysis can be applied in both cases.

It is particularly important to recognize that the similarity of the mathematical structures can be exploited not only for analysis purposes, but also for *synthesis* purposes. Thus, just as the spectrum of a temporal function can be intentionally manipulated in a prescribed fashion, so too can the spectrum of a spatial function be modified in various desired ways. Consequently, the frequency domain in both electronic and optical systems is the basis for most data processing techniques. Most optical data processing techniques find application in the fields of imaging, image enhancement, pattern recognition and correlation, memories, data recording, etc. The recent history of electro-optics is rich with examples of such important advances achieved by the application of Fourier synthesis techniques. The future will undoubtedly see many more benefits result from the merger of the two disciplines, benefits that will enrich both optics and the communication sciences.

Parts 4 and 5, respectively, discuss MTFs for image motion and vibrations and the atmosphere. Part 6 considers image restoration based on the MTF limiting image resolution. Restoration of images from blur can improve image quality very considerably, as shown in the pictorial examples of Chapter 18. Fine, small detail hidden previously by the blur is "unmasked" and easily resolvable. Mathematically, this results from a considerable broadening of overall system MTF brought about by the restoration. A result of such image quality and resolution improvements is improved target acquisition probabilities and longer ranges. This pertains primarily to contrast-limited imaging.

Although noise is also enhanced, particularly at high spatial frequencies, if the signal-to-noise ratio is high to begin with, the image restoration process "corrects" the overall system MTF in both the vertical and horizontal directions. The former correction improves contrast, particularly at high spatial frequencies; the latter increases spatial frequency bandwidth, thereby improving both resolution and target acquisition probabilities and ranges. An example is shown in Fig. 18.10.

Chapter 18 describes several methods of such restoration, based on knowledge of the MTF limiting the image quality. In real life, such limitations usually stem not from the electronics or optics, but rather from atmospheric distortions (over long paths) or mechanical vibrations and motion. These are emphasized in the pictorial demonstrations in Chapter 18. Chapter 19 then considers implications of such restoration from atmospheric blur on target acquisition ranges and probabilities.

Numerous examples with solutions are presented in almost every chapter to illustrate the mathematical concepts and to give them physical meaning. Exercises are presented at the ends of chapters to supplement the texts. Some of the exercises integrate material from earlier chapters. They actually form an important and integral part of the teaching text, and for this reason a Solutions Manual with complete solutions is provided.

The general approach utilized in the text is on imaging principles rather than devices, with an emphasis on *system concepts which relate image perception and understanding at the output to image sensing at the input*. Many of the exercises, beginning in Chapter 10, are oriented in this direction.

The earlier chapters are intended as a senior-year undergraduate text, whereas the later ones are intended as a graduate-level engineering text.

ACKNOWLEDGMENTS

The author benefited greatly from the graduate study program in electro-optics developed by Professors N. Farhat and J. Bordogna at The Moore School of Electrical Engineering, University of Pennsylvania, when he was a student there. Portions of this book are based on these lectures.

In addition, the author benefited greatly from discussions with colleagues through the years and particularly, in more recent times, with Prof. Stan Rotman of our Electrical and Computer Engineering Department at Ben-Gurion University of the

Negev. The suggestions and encouragement of Prof. R. Barry Johnson in the planning of this book, and the comments and suggestions of the various reviewers are deeply appreciated.

The author greatly appreciates the timely and careful typing of Ms. Phillipe Lambert, and the meticulous editing of Ms. Sue Price and the SPIE staff.

Most of all, the author is indebted to his students who have interacted with him in the classroom over the years and in particular to those who have carried out the research whose results are incorporated in this book, especially in Parts 4, 5, and 6, to which Dr. Danny Sadot, and soon to be Drs. Itai Dror and Ofer Hadar have contributed greatly.

". . . for R. Chanina said, I have learned much from my teachers, even more from my colleagues, but from my students most of all" (Babylonian Talmud, Taanith 7a)

תושלב״ע

"... for it is not as man sees, for what man sees is visible to the eyes..."

(Samuel 1, 16:7)

Part One

GEOMETRICAL OPTICS

CHAPTER

1

Electromagnetic Waves and Rays

Light wave propagation is an electromagnetic phenomenon. It is governed by the same equations that are used for microwaves and other electromagnetic waves. In this chapter, we review the wave propagation equations, which are derived from Maxwell's equations. A question basic to imaging is whether or not we can justify considering electromagnetic waves to be rays, for this implies that we must ignore the diffraction phenomena associated with electromagnetic wave propagation that limit image quality. Nevertheless, many concepts fundamental to imaging derive from geometrical optics in which diffraction effects are excluded. Geometrical optics, by definition, deals only with rays. The Eikonal equation defines situations in which there is justification for not considering diffraction effects. If diffraction can be excluded, then electromagnetic propagation can be considered from the standpoint of ray propagation, which simplifies the mathematics. This is the thrust of Part 1 of this book. Effects of diffraction are added later in Part 3.

Discrete changes in the refractive index or dielectric constant of propagation channels involve changes in ray direction at such interfaces as a result of reflection, refraction, or diffraction. Geometrical optics considers only the first two. The medium in which the image propagates can involve attenuation as well as image blur resulting from undesired changes in medium refractive index and light scattering by particulates, which give rise to random changes in the light's angle of arrival at the imager. The image formed at the receiver output is a result of refraction at *discrete* interfaces if the image propagates through ordinary lenses, or reflection at *discrete* interfaces if the image is reflected from mirrors. The fundamentals of such processes are treated here and utilized later for imaging.

In nonhomogeneous media, the refractive index may change on a *continual* basis, thereby giving rise to continual changes in ray direction. Ray propagation in such media can be described via the ray equation. Chapter 1 concludes with an example involving use of the ray equation in a graded-index optical fiber. Such fibers are also used as lenses and are considered from that standpoint in Chapter 2.

1.1 THE NATURE OF LIGHT

Attempts to describe the nature of light involve an apparent contradiction. On the one hand, light waves are known to behave like any other electromagnetic wave;

the physical laws describing propagation, reflection and refraction, and attenuation accurately describe the behavior of light waves. On the other hand, some optical phenomena are defined and studied in terms of photons rather than waves. The discovery of the photoelectric effect showed beyond any doubt that the electromagnetic wave model was not adequate for describing the absorption of light by a photosensitive material.

In some cases, researchers have rationalized that when there are lots of photons, they can no longer be distinguished as individual particles but behave collectively as a continuum—an electromagnetic wave. This point of view is acceptable for studying propagation phenomena but cannot explain phenomena such as the interaction of light waves, or photon beams, with the semiconductor material in a photodiode.

The wave versus particle dilemma can be addressed in a more formal and productive way using techniques beyond the scope of this introductory book. Quantum mechanics explains some of the properties of photons in terms of wave packets. Much progress has been made toward reconciling the dilemma of the wave versus the particle. Quantum mechanics also calls on the uncertainty principle to define some questions as unanswerable.

In this book, both electromagnetic wave and photon concepts will be used each in the places best suited to the phenomena studied. It is comforting to know that although these two models for light wave phenomena are different, we can rest assured that they are not really in conflict [1.1].

Although photon theory can adequately describe large-scale optical effects such as reflection and refraction, it cannot explain finer scale phenomena such as interference and diffraction.

The presently held qualitative concept of diffraction was given by Fresnel in 1815. He showed that the approximately rectilinear character of light could be interpreted on the assumption that light is a wave motion, and that the diffraction fringes could thus be accounted for in detail. Later the work of Maxwell in 1864 theorized that light waves must be electromagnetic in nature. Furthermore, observation of polarization effects indicated that light waves are transverse (that is, the wave motion is perpendicular to the direction in which the wave travels). In this *wave* or *physical optics viewpoint* the electromagnetic waves radiated by a small optical source can be represented by a train of *spherical* wavefronts with the source at the center as shown in Fig. 1.1. A *wavefront* is defined as the locus of all points in the wave train that have the same phase.

When the wavelength of the light is much smaller than the object (or opening) that it encounters, the spherical wavefronts appear to this object or opening as straight lines. This is true for small portions of the spherical wavefront. In this case the light wave can be represented as a *plane wave*, and its direction of travel can be indicated by a light ray that is drawn perpendicular to the phase front. Thus large-scale optical effects such as reflection and refraction can be analyzed by the simple geometrical process of *ray tracing*. This view of optics is referred to as *ray* or *geometrical optics* and forms the thrust of Part 1 of this book. The concept of light rays is very useful because the rays show the direction of energy flow in the light beam.

Propagation of electromagnetic waves is characterized by both diffraction and ray phenomena, all of which are based on Maxwell's equations.

1.2 MAXWELL'S EQUATIONS

Maxwell's equations in differential form are:

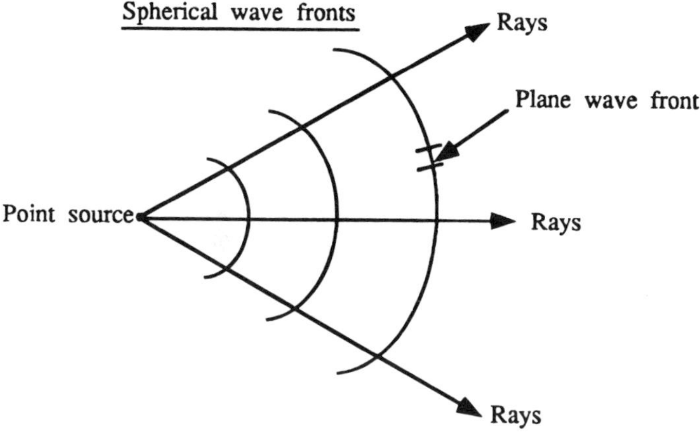

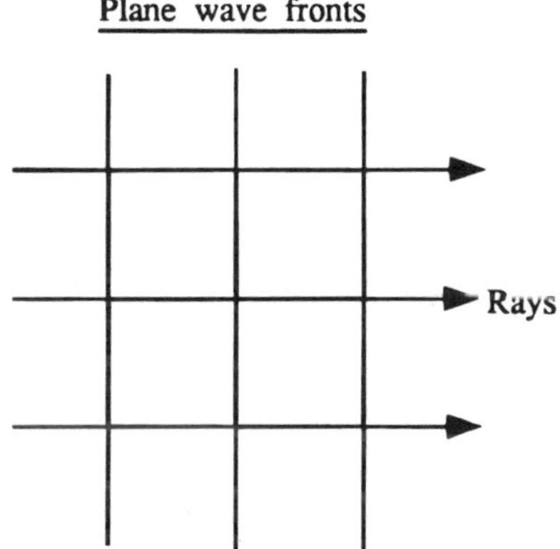

Fig. 1.1 Spherical and plane wave fronts.

$$\nabla \times \overline{E} = -\frac{\partial B}{\partial t} \tag{1.2.1a}$$

$$\nabla \times \overline{H} = \overline{J} + \frac{\partial \overline{D}}{\partial t} \tag{1.2.1b}$$

$$\nabla \cdot B = 0 \tag{1.2.1c}$$

$$\nabla \cdot D = \rho \tag{1.2.1d}$$

There are, in addition, the three constitutive relationships:

$$\overline{D} = \epsilon\overline{E}; \qquad \overline{B} = \mu\overline{H} \qquad \overline{J} = \sigma\overline{E} \tag{1.2.2}$$

where ϵ, μ, and σ are the permittivity, permeability, and conductivity, respectively, of the medium. It is usually assumed that ϵ, μ, and σ are not functions of time and that the medium is linear, homogeneous, and isotropic; ϵ, μ, and σ are therefore constant and uniform throughout the medium. However, there are some circumstances under which these assumptions are not valid, such as in nonhomogeneous media, which are considered later in this chapter.

When E and H are sinusoidal functions of time t, the time dependence can be represented with the exponential $\exp(j\omega t)$, where ω is radian frequency. Equations (1.2.1a) and (1.2.1b) then become

$$\nabla \times \overline{E} = -\mu \frac{\partial \overline{H}}{\partial t} = -j\omega\mu\overline{H} \tag{1.2.3a}$$

$$\nabla \times \overline{H} = (\sigma + j\omega\epsilon)\overline{E} \tag{1.2.3b}$$

1.3 THE WAVE EQUATION

The wave equation, developed from the time-dependent Maxwell's equations, describes the propagation of an electromagnetic wave through a uniform medium. Taking the curl of Eq. (1.2.3a) and then substituting for $\nabla \times \overline{H}$ from Eq. (1.2.3b) results in

$$\begin{aligned}\nabla \times \nabla \times \overline{E} &= -j\omega\mu\nabla \times \overline{H} \\ &= -j\omega\mu(\sigma + j\omega\epsilon)\overline{E} \\ &= (\omega^2\mu\epsilon - j\omega\mu\sigma)\overline{E}\end{aligned} \tag{1.3.1}$$

This is further reduced, using the vector identity

$$\nabla \times \nabla \times \overline{E} = [\nabla(\nabla \cdot \overline{E}) - \nabla^2\overline{E}] \tag{1.3.2}$$

and the assumption of a linear, homogeneous, isotropic, charge-free region. In such a region, according to Eq. (1.2.1d) $\nabla \cdot \overline{E} = 0$ and

$$\nabla \times \nabla \times \overline{E} = -\nabla^2\overline{E} \tag{1.3.3}$$

Substituting this into Eq. (1.3.1) one obtains

$$\begin{aligned}\nabla^2\overline{E} &= -(\omega^2\mu\epsilon - j\omega\mu\sigma)\overline{E} \\ &= \gamma^2\overline{E}\end{aligned} \tag{1.3.4a}$$

where $\gamma^2 = -\omega^2\mu\epsilon + j\omega\mu\sigma$. Similarly, by taking the curl of Eq. (1.2.3b) and by making use of Eqs. (1.2.3a) and (1.3.2) it is easy to show that

$$\nabla^2\overline{H} - \gamma^2\overline{H} = 0 \tag{1.3.4b}$$

Equations (1.3.4) are the wave equations, otherwise known as Helmholtz's equations. They can be used in any orthogonal coordinate system. Each can be expressed as three scalar equations. For example,

$$\frac{\partial^2 E_y}{\partial x^2} + \frac{\partial^2 E_y}{\partial y^2} + \frac{\partial^2 E_y}{\partial z^2} - \gamma^2 E_y = 0 \tag{1.3.5}$$

1.3.1 Plane Wave Propagation

A uniform unbounded plane wave, such as that shown in the lower part of Fig. 1.1, is assumed to be propagating in the z direction. Loci of constant phase are planes perpendicular to the z axis. Without loss of generality one can orient a rectangular coordinate system so that $E_x = E_z = 0$. Thus $\overline{E} = \hat{y}E_y$, and the vector equation [Eq. (1.3.4a)] reduces to one scalar equation, Eq. (1.3.5). The plane of the wave is then the x-y plane. Since the wave is uniform and unbounded in the x-y plane, its partial derivatives with respect to x and y are both zero. Equation (1.3.5) then reduces to

$$\frac{\partial^2 E_y}{\partial z^2} - \gamma^2 E_y = 0 \tag{1.3.6}$$

The solution to this equation has the form

$$E_y = A_1 \exp(\gamma z) + A_2 \exp(-\gamma z) \tag{1.3.7}$$

The two solutions represent waves traveling in the $-z$ and $+z$ directions, respectively. In some cases, both of these waves are present. For our purposes, however, only the positive traveling wave will be considered at the moment. Thus

$$E_y = A_y \exp(-\gamma z) \tag{1.3.8}$$

Let $\gamma = (-\omega^2\mu\epsilon + j\omega\mu\sigma)^{1/2}$. Then

$$E_y = A_y \exp[-(\alpha + j\beta)]z = A_y \exp(-\alpha z)\exp(-j\beta z) \tag{1.3.9}$$

By restoring the time-dependence term, $\exp(j\omega t)$, the complete expression for E as a function of time and distance is obtained:

$$E_y(t, z) = A_y \exp(-\alpha z)\exp[j(\omega t - \beta z)] \tag{1.3.10}$$

The amplitude A_y is found from the power or from other constraints imposed for specific situations.

Poynting's theorem describes power flow in a region containing electric and magnetic fields. The form of Poynting's theorem used here is

$$\overline{S} = \overline{E} \times \overline{H} \tag{1.3.11a}$$

where $\overline{S}$ is a vector representing the *instantaneous* power density with magnitude and direction defined by the cross product of $\overline{E}$ and $\overline{H}$; the direction will be normal to the plane in which the $\overline{E}$ and $\overline{H}$ vectors lie. Its units are $\mathrm{W \cdot m^{-2}}$.

For fields having a sinusoidal variation with time, the *average* power density is

$$\overline{S}_{\text{avg}} = \frac{1}{2}\,\text{Re}\{\overline{E} \times \overline{H}^*\} \tag{1.3.11b}$$

where the magnitudes of $\overline{E}$ and $\overline{H}$ are the peak values of the sinusoids and the phase angle between the sinusoids is assumed to be zero; the asterisk indicates the time-domain complex conjugate. If the phase angle between $\overline{E}$ and $\overline{H}$ is ϕ, the average power density given by Eq. (1.3.11b) will include an $\exp(j\phi)$ term. If this phase angle is $\pi/2$, the power is imaginary. Imaginary power means that the instantaneous power, given by Eq. (1.3.11a), alternates in direction with net power flow equal to zero.

Poynting's theorem is not limited to plane waves, but it applies to spherical waves and electrical and magnetic fields in general.

The magnitudes of the $\overline{E}$ and $\overline{H}$ field vectors are related through the properties of the medium. For the plane wave considered earlier, Eq. (1.2.3a) has only the x component:

$$-\frac{\partial E_y}{\partial z} = \gamma E_y = -j\omega\mu H_x \tag{1.3.12}$$

If the conductivity $\sigma = 0$, then $\gamma = j\omega(\mu\epsilon)^{1/2}$. The ratio of electric to magnetic field vectors is then $-j\omega\mu/\gamma$ or

$$\frac{E_y}{H_x} = -\left(\frac{\mu}{\epsilon}\right)^{1/2} \tag{1.3.13}$$

This ratio is the intrinsic impedance η of the given propagation channel. The negative sign indicates that the H vector is directed in the negative x direction.

The intrinsic impedance of space is $(\mu_0/\epsilon_0)^{1/2} \approx 377\ \Omega$. In dielectric media, $\eta = 377/\epsilon_r^{1/2} = 377/n$, where n is the index of refraction and is defined later in Eq. (1.3.18).

The power density in the plane wave can be found using Poynting's theorem, Eq. (1.3.11b). In the plane wave, with $\overline{E} = \hat{y}E_y$ and $\overline{H} = \hat{x}H_x$, the power flow is in the z direction and its average magnitude is $E_yH_x/2$. Since $E_y = \eta H_x$, $|S_{avg}| = |E_y|^2/2\eta$.

The $E_y(t, z)$ term of Eq. (1.3.10) has magnitude $A_y \exp(-\alpha z)$ and phase $(\omega t - \beta z)$. The propagation constant γ is a complex number with units of m^{-1}. The real part α is the attenuation constant and the imaginary part β is the phase or wave propagation constant. Together they describe the behavior of the electric field vector as it moves through space. In the case represented here, propagation is in the z direction.

Propagation through space can be examined by tracing a point of constant phase as t and z change:

$$(\omega t - \beta z) = \text{constant} \tag{1.3.14}$$

The time derivative of this *constant* phase angle is

$$\frac{d}{dt}(\omega t - \beta z) = \omega - \beta \frac{dz}{dt} = 0 \tag{1.3.15}$$

The derivative dz/dt is interpreted as the phase velocity, v, of the wavefront,

$$v = \frac{dz}{dt} = \frac{\omega}{\beta} \tag{1.3.16}$$

and

$$v = \frac{\omega}{\beta} = \frac{\omega}{\omega(\mu\epsilon)^{1/2}} = \frac{1}{(\mu\epsilon)^{1/2}} \tag{1.3.17}$$

In free space $\mu = \mu_0$, $\epsilon = \epsilon_0$, and

$$v = c = \frac{1}{(\mu_0\epsilon_0)^{1/2}} \approx 3 \cdot 10^8 \ (\text{m}\cdot\text{s}^{-1})$$

In other media $\mu = \mu_0\mu_r$ and $\epsilon = \epsilon_0\epsilon_r$, where μ_r and ϵ_r are permeability and dielectric constants, respectively. Consequently,

$$\begin{aligned} v &= \frac{1}{(\mu_0\epsilon_0)^{1/2}(\mu_r\epsilon_r)^{1/2}} \\ &= \frac{c}{(\mu_r\epsilon_r)^{1/2}} = \frac{c}{n} \end{aligned} \tag{1.3.18}$$

where $n = (\mu_r\epsilon_r)^{1/2}$ is the index of refraction. In nonferromagnetic materials $\mu_r \approx 1$ and $n \approx \epsilon_r^{1/2}$.

As a phase front propagates with velocity v, the time variable appears in the form $\omega(t - z/v)$. The field vectors at z are referred to as *retarded* with respect to those at $z = 0$ because variations at z are delayed in time, that is, retarded by the propagation time z/v. Similarly, (z/v) can be called the *retardation time.*

The factor $(\omega t - \beta z)$ represents the variation of the phase angle of the field vectors with time and distance. If z is fixed at some arbitrary value $z = z_1$, then βz_1 is a *constant* phase difference between the field at $z = z_1$ and the field at $z = 0$. Thus, $z = z_1$ defines a *plane* of constant phase as expected with regard to a plane wave, shown in Fig. 1.1. Now, $(\omega t - \beta z_1)$ gives the phase as a function of time. The period T is the time required for the phase angle at z_1 to go through one complete cycle, or 2π radians; $\omega T = 2\pi$ and $T = 2\pi/\omega$. Similarly, $(\omega t_1 - \beta z)$ is the phase angle as a function of distance at a fixed time t_1. The wavelength λ is the distance along the direction of propagation required for the phase angle at time t_1 to go through 2π radians; $\beta\lambda = 2\pi$ and $\lambda = 2\pi/\beta$.

Since

$$\beta = \frac{2\pi}{\lambda} = \frac{\omega}{v} = \frac{2\pi f}{v}$$

then

$$v = f\lambda = \frac{c}{n}$$

and

$$\lambda = \frac{v}{f} = \frac{c}{nf} = \frac{\lambda_0}{n}. \tag{1.3.19}$$

The phase velocity of propagation in free space is c, and the phase velocity in a medium having index of refraction n is c/n. If the wavelength in free space is $\lambda_0 = c/f$, the wavelength in a medium of index n is λ_0/n. It is assumed in Eq. (1.3.19) that f is independent of propagation medium and that the change of phase velocity with refractive index is manifested totally as a change in wavelength with no change in frequency. The basis for this assumption is conservation of energy. Because photon energy is $h \cdot f$, where h is Planck's constant ($6.6256 \cdot 10^{-34}$ J·s), a change in frequency with refractive index would imply a change in photon energy or a change in power of the electromagnetic wave. For lossless media, experience indicates conservation of energy and power when electromagnetic waves, including light, propagate from one medium to another. This is shown later in Example 1.7.

1.3.2 Propagation Channel

From the object plane, the image must propagate to the image plane at the output of the receiver optics. Between object and receiver the image may propagate through the atmosphere, water, optical fibers, receiver optics, etc., all of which can form the propagation channel.

In photoelectronic imaging systems, in which a TV camera tube or charge-coupled device (CCD), for example, is in the image plane, the image must propagate from the image plane through the photosensitive material itself to the layer in which the image is actually detected (converted to an electronic signal).

Propagation through the various refracting and/or reflecting optics causes an image to be formed in the image plane. Propagation through the part of the propagation channel between the object and image planes can result in attenuation, loss of contrast, and image blur. Attenuation is considered first. The following mathematical treatment is general. It is followed in later chapters by detailed treatments on various propagation channels as well as on the image-forming process.

As shown in Eq. (1.3.9) the propagation factor γ is composed of two terms, α and $j\beta$. The former represents attenuation and the latter propagation. By definition, from Eq. (1.3.4a)

$$\begin{aligned}\gamma^2 &= -\omega^2\mu\epsilon + j\omega\mu\sigma \\ &= -\omega^2\mu\epsilon\left(1 - j\frac{\sigma}{\omega\epsilon}\right) \\ &= -\omega^2\mu\epsilon_{eq}\end{aligned} \tag{1.3.20}$$

where

$$\epsilon_{eq} = \epsilon - j\sigma/\omega \tag{1.3.21}$$

The latter is derived directly from Maxwell's curl equation

$$\nabla \times \overline{H} = \overline{J} + j\omega\overline{D} = (\sigma + j\omega\epsilon)\overline{E} = j\omega\epsilon_{eq}\overline{E} \tag{1.3.22}$$

for time-varying fields. This implies that conducting media can be treated mathematically as dielectrics with an equivalent permittivity ϵ_{eq}. In general, because ϵ itself may be complex and is equal to $\epsilon' - j\epsilon''$, then

$$\epsilon_{eq} = \epsilon' - j(\sigma/\omega + \epsilon'') \tag{1.3.23}$$

Accordingly, Eq. (1.3.20) becomes

$$\gamma^2 = -\omega^2\mu\epsilon' + j\omega\mu(\sigma + \omega\epsilon'') \tag{1.3.24}$$

The imaginary terms will be seen to indicate attenuation. The propagation factor may be separated into real and imaginary parts:

$$\gamma = \alpha + j\beta = [j\omega\mu(\sigma + \omega\epsilon'') - \omega^2\mu\epsilon']^{1/2} \tag{1.3.25}$$

When $\sigma + \omega\epsilon'' < \omega\epsilon'$, Eq. (1.3.25) reduces to

$$\gamma = j\beta = j\omega(\mu\epsilon')^{1/2} = j\frac{\omega}{v} \tag{1.3.26}$$

which is consistent with Eq. (1.3.17).

By comparing Eq. (1.3.26) with Eq. (1.3.25), we can see that attenuation derives from the imaginary components of ϵ_{eq}. A general solution of Eq. (1.3.25) is [1.2]

$$\alpha = \omega\left(\frac{\mu\epsilon'}{2}\right)^{1/2}\left[\left(1 + \left(\frac{\sigma + \omega\epsilon''}{\omega\epsilon'}\right)\right)^{2\,1/2} - 1\right]^{1/2} \tag{1.3.27a}$$

$$\beta = \omega\left(\frac{\mu\epsilon'}{2}\right)^{1/2}\left[\left(1 + \left(\frac{\sigma + \omega\epsilon''}{\omega\epsilon'}\right)\right)^{2\,1/2} + 1\right]^{1/2} \tag{1.3.27b}$$

Thus, if the imaginary component of ϵ_{eq} is zero, Eq. (1.3.27a) reduces to zero and Eq. (1.3.27b) to Eq. (1.3.26). However, the greater the imaginary component of

ϵ_{eq}, the greater the wave attenuation. The conductivity losses are manifested as electromagnetic or light wave "heating" of free-charge carriers, that is, the free-charge carriers acquire energy as a result of absorption of the electromagnetic wave. This can affect the performance of detectors. Although in quantum detectors such as photoemissive cathode devices or semiconductor photodiodes (both discussed in chapter 6), the actual detection mechanism derives from quantum effects involving photon absorption, nevertheless, conductivity of the device itself gives rise to losses as light propagates through the detector until the light wave actually reaches the depth at which photon absorption takes place. The losses decrease with decreasing ratio of $(\sigma + \omega\epsilon'')$ to $\omega\epsilon'$.

In addition to conductivity losses, there are those that derive from ϵ''. These losses are a result of quantum absorption. In a detector, such "losses" are desired in the detection region itself for they give rise to the detected electronic signal resulting from conversion of absorbed photons to current carriers, but they are undesirable in the various layers of the detector through which photons must propagate to reach the detection region. If $\sigma << \omega\epsilon''$, then the absorption losses depend on the ratio ϵ''/ϵ', otherwise known as the *loss tangent* of the given medium. However, the significance of ϵ'' and absorption losses in general far transcends effects in detectors. They are generally of much greater importance in terms of the media through which the image propagates to the receiver, which often includes the atmosphere or other media such as water or optical fibers. In the atmosphere, generally $\sigma \approx 0$, but in water the conductivity is not negligible.

The imaginary component of permittivity, ϵ'', is essentially a result of quantum absorption of electromagnetic radiation by the given medium. At visible wavelengths, such absorption generally gives rise to an increase in energy of an electron to a higher energy state. The electron involved is generally the outermost electron of the atom absorbing the photon. At such wavelengths absorption of electromagnetic radiation is therefore generally atomic. At longer wavelengths, such as the infrared, photons have less energy and their effect is generally on the molecule rather than on the atom. That is, the bonds connecting atoms in a given molecule are changed without changing the internal energy of any of the atoms. Such molecular absorption is termed vibrational. At even longer wavelengths, such as in the millimeter wave and microwave range, absorption involves changes in electron spin direction. These effects require much less energy than do atomic or vibrational absorption.

As a consequence of these characteristics, ϵ' generally does not change much with frequency. However, ϵ'' is strongly frequency dependent because such absorptions are resonant to quantum transitions, particularly in dielectrics such as propagation media. Because it is difficult to image through high-loss media, low-loss conditions are emphasized here. For a low-loss or "imperfect" dielectric (see Exercise 1.1),

$$\alpha \approx \omega\left(\frac{\mu}{\epsilon'}\right)^{1/2} \epsilon''/2 \tag{1.3.28}$$

and

$$\beta \approx \omega(\mu\epsilon')^{1/2}\left[1 + \frac{1}{8}\left(\frac{\epsilon''}{\epsilon'}\right)^2\right] \tag{1.3.29}$$

if $\sigma = 0$. Because of absorption, β increases from its value in Eq. (1.3.26). Since

$\beta = 2\pi\lambda$, the implication is a decrease in wavelength in lossy dielectrics. Furthermore, because, as shown following Eq. (1.3.18), the refractive index is equal to $\epsilon_r^{1/2}$, then it can be shown (see Exercise 1.2a) that

$$\alpha = \omega n''/c = \beta_0 n'' \tag{1.3.30}$$

where n'' is the imaginary component of the refractive index and β_0 is the free-space propagation constant whose value is $\omega\sqrt{\mu_0\epsilon_0}$. In general,

$$n = n' - jn'' \tag{1.3.31}$$

For a good conductor, defined as a medium in which $\sigma >> \omega\epsilon'$,

$$n'' \approx \left[\frac{\sigma}{2\omega\epsilon_0}\right]^{1/2} \tag{1.3.32a}$$

(see Exercise 1.3), whereas for a low-loss dielectric (see Exercise 1.2b):

$$n'' = n' \frac{\epsilon''}{2\epsilon'} \tag{1.3.32b}$$

where $n' = \sqrt{\epsilon/\epsilon_0}$.

In semiconductor detectors, Eq. (1.3.32b) is more relevant than Eq. (1.3.32a) because for wavelengths shorter than the detection cutoff the quantum effects are much more dominant than electron heating. However, in various thermal detectors, Eq. (1.3.32a) is more relevant.

In view of Eqs. (1.3.30) and (1.3.32), it is clear that an imaginary permittivity or refractive index component implies attenuation as a result of dielectric absorption or conductor "heating."

As a result of Eqs. (1.3.32),

$$\alpha = \frac{\omega}{c} n'' \approx \left[\frac{\sigma\omega\mu_0}{2}\right]^{1/2} \tag{1.3.33a}$$

and

$$\alpha = \frac{\omega}{c} n'' \approx \frac{\omega\epsilon''}{2}\sqrt{\frac{\mu_0}{\epsilon'}} \tag{1.3.33b}$$

for good conductors (and thermal detectors) and low-loss dielectrics (and quantum detectors), respectively. As a propagation channel example, typical attenuation in seawater as a function of electromagnetic frequency is shown in Fig. 1.2, although transmission does vary with conductivity, salinity, turbidity, depth, etc. It is useful to note that transmission is best at visible wavelengths. More detailed tabulated data are also available [1.4].

The atmosphere, which is the propagation channel in most applications, is treated in detail in Part 5 of this book.

Example 1.1

With regard to Fig. 1.2, consider the applicability of Eq. (1.3.33a). It is known that for frequencies less than 10^{11} Hz, σ is approximately constant in seawater at about 4 mho·m^{-1} and that $\epsilon' \approx 81\epsilon_0$ [1.2]. Assume a frequency of 10^8 Hz and compare the theoretical expression for α with that in Fig. 1.2.

Solution

Since

$$\frac{\sigma}{\omega\epsilon'} = \frac{4}{2\pi 10^8 \cdot 81 \cdot 10^{-9}/36\pi} \approx 9$$

at this frequency, seawater behaves like a conductor rather than a dielectric. Using Eq. (1.3.33a)

$$\begin{aligned} \alpha &\approx \left[\frac{\sigma\omega\mu_0}{2}\right]^{1/2} \\ &\approx \left[\frac{4 \cdot 2\pi \cdot 10^8 \cdot 4\pi \times 10^{-7}}{2}\right]^{1/2} \\ &\approx 39.7 \text{ m}^{-1}. \end{aligned}$$

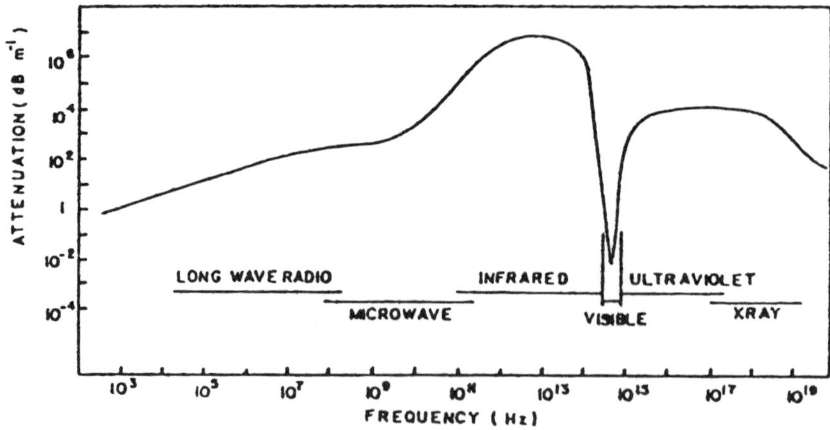

Fig. 1.2 Spectral attenuation of sea water [1.3].

To convert to attenuation A_{tt} in dB·m^{-1}, consider that transmission $\tau = \exp(-\alpha z)$

$$\therefore A_{tt} = \frac{1}{z}\,10 \log \tau = \frac{1}{z}\,10\alpha z \log e = 4.34\alpha$$

$$\therefore A_{tt} = 4.34(39.7) = 172.5 \text{ dB·m}^{-1} \text{ at } 10^8 \text{ Hz}$$

This solution agrees very favorably with Fig. 1.2. Further, attenuation in Fig. 1.2 increases at low frequencies according to $\omega^{1/2}$, as predicted by Eq. (1.3.33a), until at microwave frequencies sea water is no longer a conductor because $\sigma < \omega\epsilon'$. In the infrared attenuation increases because of molecular absorption and in the ultraviolet because of atomic absorption, both of which are expressed by ϵ'' or n''.

Example 1.2

With regard to Fig. 1.2, consider the applicability of Eq. (1.3.30). According to Ref. [1.4], at 10^{13} Hz $n'' \approx 0.328$ for water. Therefore at that frequency

$$\alpha = \frac{\omega}{c}\,n'' = \frac{2\pi 10^{13}}{3 \times 10^8}(0.328) \approx 6.87 \times 10^4 \text{ m}^{-1}$$

$\therefore A_{tt} = 4.34\alpha \approx 2.98 \times 10^5$ dB·m^{-1}. According to Fig. 1.2, A_{tt} should be almost an order of magnitude higher. The discrepancy is attributed to two factors:

1. Reference [1.4] refers to water, whereas Fig. 1.2 refers to seawater, which has a somewhat different chemical composition. In the visible, for example, pure water is transparent, whereas seawater is much less so.
2. Figure 1.2 includes all sources of attenuation, including light scattering by large particulates, which are common in seawater. (Light scattering is addressed in Part 5.)

It should be noted that the drop in attenuation in the visible implies that in this spectral region ϵ'' is very small. Reference [1.4] reports $n'' \approx 10^{-9}$ at such wavelengths for water. This is the only portion of the electromagnetic spectrum that has practical meaning for imaging or even transmission through water or seawater because the attenuation is at its least here.

Example 1.3

The absorption coefficient of calcium fluoride glass at 10 μm wavelength is 1.8 cm^{-1}. How is this related to refractive index n?

Solution

Since the material is glass, the absorption is quantum rather than conducting. Hence,

$$\alpha = \frac{\omega}{c}\,n'' = 1.8 \text{ cm}^{-1} = \frac{2\pi(c/\lambda)n''}{c} = \frac{2\pi}{\lambda}\,n''$$

$$\therefore n'' = \frac{\lambda}{2\pi}(1.8) = 10^{-3}\left(\frac{1.8}{2\pi}\right) = 2.9 \times 10^{-4}$$

where λ has been expressed in centimeters. This attenuation is generally not severe enough to prohibit use of thin lenses of CaF_2 at this wavelength.

1.3.3 Polarization

In paragraph 1.3.1, a transverse electromagnetic (TEM) plane wave propagating in the z direction was considered, in which the electric field was in the y direction and the magnetic field in the x direction. The Poynting vector, resulting from the vector cross product between the two, was in the z direction, which is the direction of propagation.

Propagation in the z direction by a TEM wave can also take place if $\overline{E}$ is in the x direction and $\overline{H}$ in the y direction. For the latter case,

$$E_x(t, z) = A_x \exp(-\alpha z) \exp[j(\omega t - \beta z + \delta)] \tag{1.3.34}$$

where δ is the relative phase difference, if any, between both these waves propagating in the z direction. The direction of the electric field is called polarization. The total electric field is then

$$\overline{E}(t, z) = \hat{x}\overline{E}_x(t, z) + \hat{y}\overline{E}_y(t, z) \tag{1.3.35}$$

If δ is zero or an integer multiple of 2π, the waves are in phase. Equation (1.3.35) refers to a linearized polarized wave with a polarization vector making an angle

$$\theta = \arctan \frac{A_x}{A_y}$$

with respect to E_x and having a magnitude

$$E = (A_x^2 + A_y^2)^{1/2}$$

This case is shown schematically in Fig. 1.3 where $\delta = 0$. Conversely, just as any

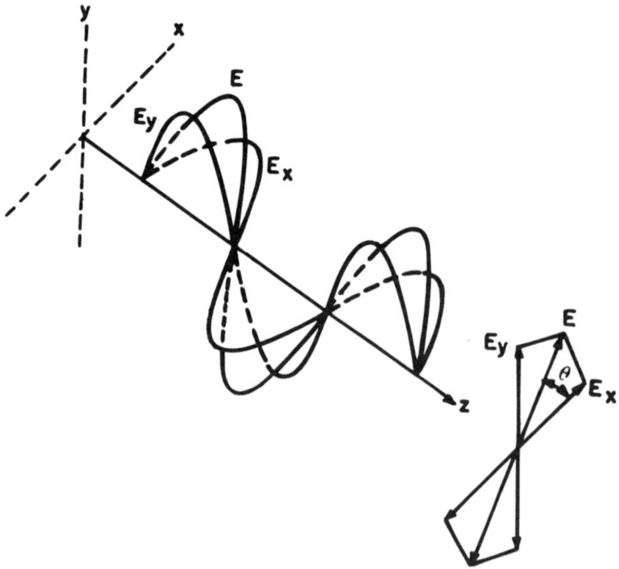

Fig. 1.3 Linearly polarized wave.

two orthogonal plane waves can be combined into a linearly polarized wave, an arbitrary linearly polarized wave can be resolved into two independent orthogonal plane waves that are in phase.

For general values of δ, the wave whose electric field is given by Eq. (1.3.35) is elliptically polarized. The resultant field vector $\overline{E}$ will both rotate and change its magnitude as a function of the angular frequency ω. As shown in Fig. 1.4, the endpoint of E will trace an ellipse at a given point in space with the axis of the ellipse related to the x axis by an angle α given by (see Exercise 1.5)

$$\tan 2\alpha = \frac{2A_x A_y \cos \delta}{A_x^2 - A_y^2} \tag{1.3.36}$$

where the angle α should not be confused with the attenuation constant α. When $A_x = A_y$, the wave is circularly polarized.

1.3.4 Spherical Waves

To utilize the wave (Helmholtz's) equation, Eq. (1.3.4), for spherical wave propagation, it is convenient to express ∇^2 in spherical coordinates. Assuming no directional dependencies ($\partial/\partial\theta = \partial/\phi = 0$), the radial variation of the Laplacian operator is

$$\nabla^2\psi = \frac{1}{r}\frac{\partial^2}{\partial r^2}(r\psi) \tag{1.3.37}$$

where ψ is the complex field amplitude (CFA) of either the electric or magnetic field. The difference between the CFA and the actual field is the time dependence $\exp(j\omega t)$. Because the wave equation applies to both electric and magnetic fields, it is useful to discuss wave propagation in terms of the phasor ψ without specifying electric or

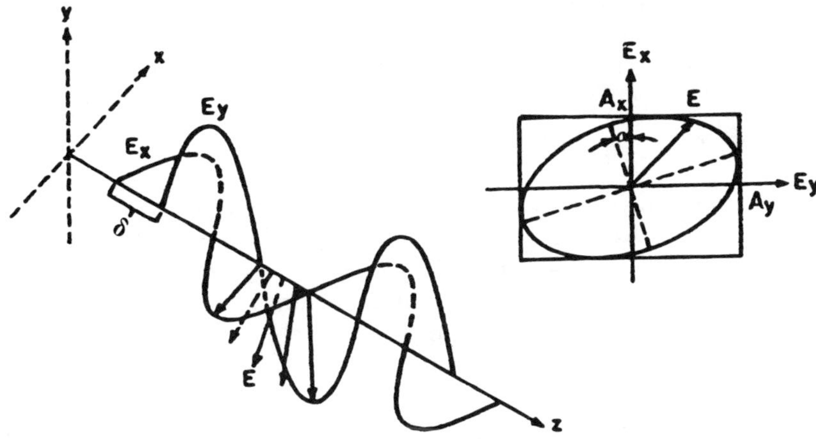

Fig. 1.4 Elliptical polarization.

magnetic field. After substituting this Laplacian into Eqs. (1.3.4), then in terms of the CFA, the wave equation becomes

$$\frac{1}{r}\frac{\partial^2}{\partial r^2}(r\psi) - \gamma^2\psi = 0 \tag{1.3.38}$$

for spherical coordinates. Multiplying by r,

$$\frac{\partial^2(r\psi)}{\partial r^2} - \gamma^2(r\psi) = 0 \tag{1.3.39}$$

Equation (1.3.39) is identical to Eq. (1.3.4) except that the field is replaced by $r\psi$. The solution of Eq. (1.3.39) therefore is the solution of Eq. (1.3.4) divided by r, or

$$\begin{aligned}\psi &= \frac{A_r}{r}\exp(-\gamma r)\\ &= \frac{A_I}{r}\exp(-\alpha r)\exp(-j\beta r)\end{aligned} \tag{1.3.40}$$

where radial propagation in the direction r instead of the direction z has been considered, and the time dependence $\exp(j\omega t)$ has been dropped in keeping with the phasor nature of ψ. Equation (1.3.40) implies that the Poynting vector of a spherical wave drops off with increasing radial distance according to $1/r^2$ if $\alpha = 0$. This is to be expected since the Poynting vector represents power per cross-sectional area. For a spherical wave propagating to the right, as in Fig. 1.1, the cross-sectional area illuminated by the wave increases according to r^2. If power is conserved, then the same electromagnetic power must be spread over a cross-sectional area increasing according to r^2, thereby causing the power density to decrease according to r^{-2}. Result (1.3.40) is therefore consistent with conservation of power.

The loss mechanisms inherent in the attenuation or extinction constant α for plane waves are properties of the propagation medium, according to the conductivity and imaginary components of permittivity or refractive index of the propagation channel. Accordingly, they hold for spherical wave as well as plane wave propagation.

1.4 INTERFACES

When an electromagnetic (EM) wave propagating in one medium is incident at the boundary separating that medium from another, part of the incident wave may be reflected back into the first medium and part of it may penetrate into the second medium. Mathematically, it is simpler to describe such phenomena with plane wave rather than spherical wave representation, although the same properties of reflection and refraction hold for spherical waves as well.

Consider TEM plane waves incident diagonally at angles θ_1 with the normal to the interfaces as shown in Fig. 1.5 for perpendicular and parallel polarization. Perpendicular polarization considers a wave with electric field perpendicular to the page. Parallel polarization considers an EM wave with electric field parallel to the page.

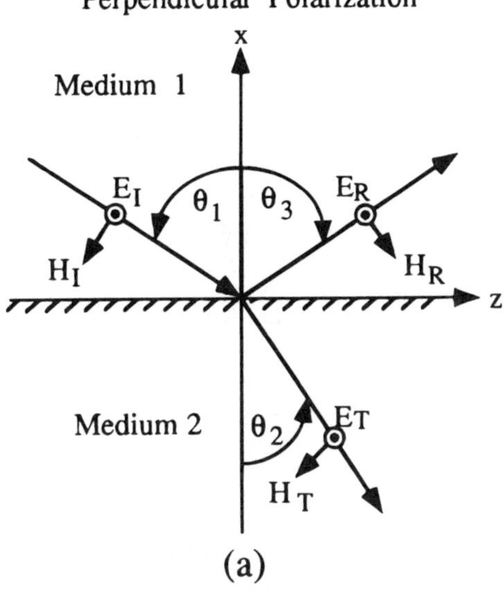

(a)

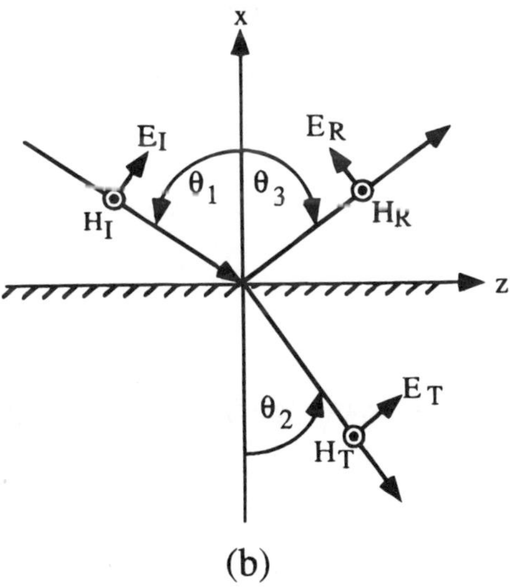

(b)

Fig. 1.5 Incident, reflected, and refracted fields.

In Fig. 1.5(a) for *perpendicular* polarization and with regard to the incident wave direction, the propagation constants in the x and z directions are, respectively,

$$\beta_{1x} = \beta_1 \cos \theta_1 = \omega(\mu_1 \epsilon_1)^{1/2} \cos \theta_1 \tag{1.4.1a}$$

$$\beta_{1z} = \beta_1 \sin \theta_1 = \omega(\mu_1 \epsilon_1)^{1/2} \sin \theta_1 \tag{1.4.1b}$$

where the subscript 1 refers to medium number 1.

Assuming no losses in medium 1, the incident electric field is propagating in the $-x$ and $+z$ directions and is therefore represented as

$$E_{yI}(t, x, z) = A_{yI} \exp(j\omega t) \exp(j\beta_{1x}x) \exp(-j\beta_{1z}z) \tag{1.4.2}$$

where

$$\beta_{1x}^2 + \beta_{1z}^2 = \beta_1^2 = \omega^2\mu_1\epsilon_1 \tag{1.4.3}$$

The magnetic field vector can be calculated from Eq. (1.2.3a) as follows:

$$-\hat{x}\frac{\partial E_y}{\partial z} + \hat{z}\frac{\partial E_y}{\partial x} = -j\omega\mu[\hat{x}H_x + \hat{y}H_y + \hat{z}H_z] \tag{1.4.4}$$

The partial derivatives can be found by using Eq. (1.4.2) for $E_y(t, x, z)$. The vector equation is thus reduced to three scalar equations that give the relationship between E_y and the components of H for the incident wave,

$$\begin{aligned} j\beta_{1x}E_{yI} &= j\omega(\mu_1\epsilon_1)^{1/2}\cos\theta_1 E_{yI} = -j\omega\mu_1 H_{zI} \\ j\beta_{1z}E_{yI} &= j\omega(\mu_1\epsilon_1)^{1/2}\sin\theta_1 E_{yI} = -j\omega\mu_1 H_{xI} \\ 0 &= -j\omega\mu_1 H_{yI} \end{aligned} \tag{1.4.5}$$

from which

$$\begin{aligned} H_{xI} &= -\frac{E_{y1}}{\eta_1}\sin\theta_1 \\ &= -\frac{A_{yI}}{\eta_1}\sin\theta_1 \exp(j\omega t)\exp[j(\beta_1\cos\theta_1)x - j(\beta_1\sin\theta_1)y] \\ H_{yI} &= 0 \\ H_{zI} &= -\frac{E_{yI}}{\eta_1}\cos\theta_1 \\ &= -\frac{A_{yI}}{\eta_1}\cos\theta_1 \exp(j\omega t)\exp[j(\beta_1\cos\theta_1)x - j(\beta_1\sin\theta_1)y] \end{aligned} \tag{1.4.6}$$

where the subscript I refers to incident wave.

The directions for the components of H in these equations are consistent with those shown in Fig. 1.5.

The reflected wave, whose relative direction is θ_3, is propagating in the $+x$ and $+z$ directions in medium 1. Therefore,

$$E_{yR}(t, x, z) = A_{yR} \exp(j\omega t) \exp(-j\beta_{3x}x) \exp(-j\beta_{3z}z) \tag{1.4.7}$$

where

$$\beta_{3x} = \omega(\mu_1\epsilon_1)^{1/2} \cos\theta_3 = \beta_1 \cos\theta_3 \tag{1.4.8a}$$

$$\beta_{3z} = \omega(\mu_1\epsilon_1)^{1/2} \sin\theta_3 = \beta_1 \sin\theta_3 \tag{1.4.8b}$$

Hence, using Eq. (1.2.3a) for the reflected field as was done for the incident field above:

$$\begin{aligned} H_{xR} &= -\frac{E_{yR}}{\eta_1}\sin\theta_3 \\ H_{yR} &= 0 \\ H_{zR} &= \frac{E_{yR}}{\eta_1}\cos\theta_3 \end{aligned} \tag{1.4.9}$$

where the subscript R refers to the reflected wave. The signs of the magnetic field are consistent with the reflected wave in Fig. 1.5 for perpendicular polarization.

The transmitted wave is propagating in the $-x$ and $+z$ directions in medium 2. Hence, using the subscript T for the transmitted wave, we obtain

$$E_{yT}(t, x, z) = A_{yT}\exp(j\omega t)\exp(j\beta_{2x}x)\exp(-j\beta_{2z}z) \tag{1.4.10}$$

The magnetic fields then are

$$\begin{aligned} H_{xT} &= -\frac{E_{yT}}{\eta_2}\sin\theta_2 \\ H_{yT} &= 0 \\ H_{zT} &= \frac{E_{yT}}{\eta_2}\cos\theta_2 \end{aligned} \tag{1.4.11}$$

where

$$\beta_{2x} = \omega(\mu_2\epsilon_2)^{1/2} \cos\theta_2 = \beta_2 \cos\theta_2 \tag{1.4.12a}$$

$$\beta_{2z} = \omega(\mu_2\epsilon_2)^{1/2} \sin\theta_2 = \beta_2 \sin\theta_2 \tag{1.4.12b}$$

and the subscript 2 refers to medium 2.

1.4.1 Angles of Reflection and Refraction

At the interface $x = 0$, boundary conditions for tangential electric fields require for all t and all z that

$$E_{yI}(t, 0, z) + E_{yR}(t, 0, z) = E_{yT}(t, 0, z) \tag{1.4.13a}$$

or

$$A_{yI} \exp(j\omega) \exp(-j\beta_{1z}z) + A_{yR} \exp(j\omega t) \exp(-j\beta_{3z}z) = A_{yT} \exp(j\omega t) \exp(-j\beta_{2z}z) \tag{1.4.13b}$$

The equality of the fields in medium 1 and medium 2 at the interface requires that the time dependence $\exp(j\omega t)$ be the same for all components and that the transmission and reflection parameters be independent of z. The latter condition requires that the z components of the propagation vectors all be equal, or

$$\beta_{1z} = \beta_{2z} = \beta_{3z}$$

or in view of Eqs. (1.4.1b), (1.4.8b), and (1.4.12b)

$$\beta_1 \sin \theta_1 = \beta_2 \sin \theta_2 = \beta_1 \sin \theta_3 \tag{1.4.14}$$

From the first and last parts of Eq. (1.4.14), we can deduce that

$$\sin \theta_1 = \sin \theta_3 \text{ and } \theta_1 = \theta_3 \tag{1.4.15}$$

The angle of reflection is equal to the angle of incidence. From the first two parts of Eq. (1.4.14)

$$\beta_1 \sin \theta_1 = \beta_2 \sin \theta_2$$

or

$$n_1 \sin \theta_1 = n_2 \sin \theta_2 \tag{1.4.16}$$

This is *Snell's law.*

The *angle of reflection law,* Eq. (1.4.15), requires some clarification. If the interface is smooth, that is, inhomogeneities are of dimensions on the order of a fraction of a wavelength, then that boundary acts as a mirror and Eq. (1.4.15) is indeed easily observable since light is reflected essentially at only a single angle, that which is equal to the angle of incidence. Such surfaces are referred to as *specular.* However, if the interface surface is rough, light appears to be scattered in all directions even though, on a microscale, Eq. (1.4.15) still holds, as shown in Fig. 1.6 for a plane wave incident diagonally on the surface. For example, a laser beam reflected by an ideal mirror can be received from only a single angle, as expected from Eq. (1.4.15). However, a laser beam reflected from a rough surface can be seen in every direction

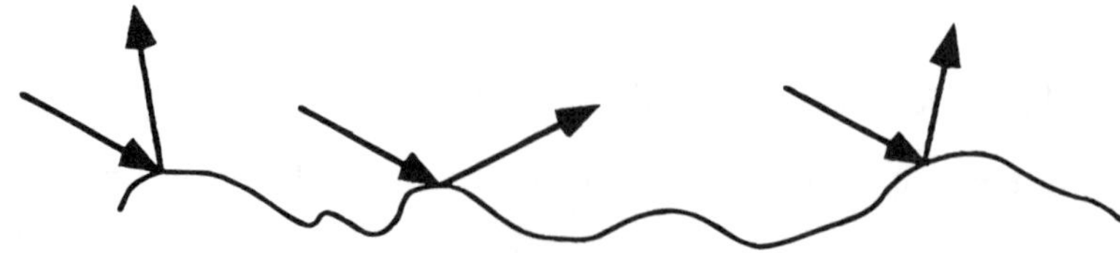

Fig. 1.6 Incident plane wave reflected in many directions by a rough surface.

from the rough surface. Such rough surfaces are termed *diffuse*. An ideally rough surface is one in which the angular dependence of (power) reflection coefficient ρ is

$$\rho = \rho(0) \cos \theta \tag{1.4.17}$$

where $\rho(0)$ is the reflection coefficient at 0° reflection and θ is the angle of reflection, which is always measured with respect to the normal to the surface as in Fig. 1.5. This result is independent of angle of incidence. Equation (1.4.17) is called *Lambert's law*, and such an ideal rough reflecting or scattering surface is called a *Lambertian surface* or a *Lambertian source* of radiation.

Surfaces such as metals that are not smooth enough to act as mirrors but are not nearly as ideally and uniformly rough such that they act as Lambertian sources generally exhibit reflection maxima at both $\theta_3 = 0°$ as expected from Eq. (1.4.17) and at $\theta_3 = \theta_1$ as expected from Eq. (1.4.15). These surfaces reflect fuzzy rather than sharp images. In addition there are indications that light reflected backwards in the direction from which it is incident, that is, $\theta_3 = -\theta_1$, can be of even greater intensity than light scattered in either of the above directions [1.5].

1.4.2 Reflection and Transmission Coefficients (Fields)

Boundary conditions at dielectric interfaces require continuity of not only tangential electric fields, as in Eq. (1.4.13), but also of tangential magnetic fields. As a result

$$H_{zI} + H_{zR} = H_{zT} \tag{1.4.18}$$

By substituting Eqs. (1.4.6), (1.4.9), and (1.4.11) into the preceding equation, we get

$$-\frac{E_{yI}}{\eta_1} \cos \theta_1 + \frac{E_{yR}}{\eta_1} \cos \theta_1 = -\frac{E_{yT}}{\eta_2} \cos \theta_2 \tag{1.4.19}$$

or

$$\left(\frac{\epsilon_1}{\mu}\right)^{1/2} \cos \theta_1 (E_{yI} - E_{yR}) = \left(\frac{\epsilon_2}{\mu}\right)^{1/2} \cos \theta_2 E_{yT} \tag{1.4.20}$$

In terms of indices of refraction

$$n_1 \cos \theta_1 (E_{yI} - E_{yR}) = n_2 (\cos \theta_2) E_{yT} \tag{1.4.21}$$

Since $\overline{E}$ has only a y component, the y subscript can be dropped and the magnitudes of the three $\overline{E}$ fields referred to as E_I, E_R, and E_T. From the previous equation

$$E_I - E_{yR} = \frac{n_2 \cos \theta_2}{n_1 \cos \theta_1} E_T \tag{1.4.22}$$

By using this with Eq. (1.4.13a), equations giving the strengths of the transmitted and reflected fields in terms of the incident field strength can be obtained:

$$E_T = \frac{2n_1 \cos \theta_1}{n_1 \cos \theta_1 + n_2 \cos \theta_2} E_I \tag{1.4.23}$$

$$E_R = \frac{n_1 \cos \theta_1 - n_2 \cos \theta_2}{n_1 \cos \theta_1 + n_2 \cos \theta_2} E_I \tag{1.4.24}$$

These are the *Fresnel equations* for a perpendicularly polarized wave.

A reflection coefficient $r_\perp$, defined as E_r/E_I, is found from Eq. (1.4.24):

$$r_\perp = \frac{n_1 \cos \theta_1 - n_2 \cos \theta_2}{n_1 \cos \theta_1 + n_2 \cos \theta_2} \tag{1.4.25}$$

By defining transmission coefficient as E_T/E_I, Eq. (1.4.23) yields

$$t_\perp = \frac{2n_1 \cos \theta_1}{n_1 \cos \theta_1 + n_2 \cos \theta_2} \tag{1.4.26}$$

For parallel polarization, Fig. 1.5 indicates continuous tangential electric and magnetic field boundary conditions, respectively, such that

$$(E_I - E_R)\cos \theta_1 = E_T \cos \theta_2 \tag{1.4.27a}$$

$$\frac{E_I + E_R}{\eta_1} = \frac{E_T}{\eta_2} \tag{1.4.27b}$$

As a result of these last two equations, the reflection coefficient for parallel polarization is

$$r_{11} = \frac{n_2 \cos \theta_1 - n_1 \cos \theta_2}{n_2 \cos \theta_1 + n_1 \cos \theta_2} \tag{1.4.28}$$

and the transmission coefficient for parallel polarization is

$$t_{11} = \frac{\eta_2}{\eta_1}(1 + r_{11}) = \frac{2n_1 \cos \theta_1}{n_2 \cos \theta_1 + n_1 \cos \theta_2} \tag{1.4.29}$$

for nonferromagnetic media where $\mu_2 \approx \mu_1$. The subscript 11 refers to parallel polarization.

1.4.3 Reflection and Transmission Coefficients (Power)

Poynting vectors indicate flow power per unit *cross-sectional* area. To consider reflection and transmission of EM wave radiant power, Poynting vectors can be used if they relate to the same area of illumination of the interface. In Fig. 1.7, a collimated plane wave illuminates an interface area AC of unit depth into the page. The Poynting vector for the incident wave gives the EM power per cross-sectional area AB also of unit depth. Area AC is larger than area AB by cos θ_1. Therefore, the radiant *power* incident on AC is

$$\frac{E_I^2}{\eta_1} AB = \frac{E_I^2}{\eta_1} AC \cos \theta_1$$

where E_I^2/η_1 is the Poynting vector of the instantaneous incident wave. Part of this power is reflected and part transmitted. The power reflected from the same area is

$$\frac{E_R^2}{\eta_1} AB = \frac{E_R^2}{\eta_1} AC \cos \theta_1$$

since θ_3 in Fig. 1.5 equals θ_1. The total radiant power transmitted through this same area is $(E_T^2/\eta^2)\, AC \cos \theta_2$ since the area AC is larger than the cross-sectional area CD by cos θ_2. Thus, along the area AC, from considerations of power conservation,

$$\frac{E_I^2}{\eta_1} \cos \theta_1 = \frac{E_R^2}{\eta_1} \cos \theta_1 + \frac{E_T^2}{\eta_2} \cos \theta_2 \qquad (1.4.30)$$

The term on the left-hand side is incident radiant power. The first and second terms on the right-hand side are the relative reflected and transmitted radiant powers, respectively. Hence, the power reflection coefficient is

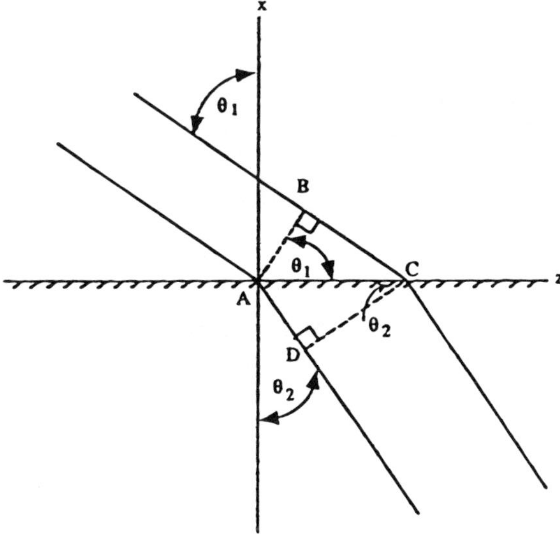

Fig. 1.7 Cross-sectional areas for incident and refracted plane waves.

$$\rho = \frac{(E_R^2/\eta_1)\cos\theta_1}{(E_I^2/\eta_1)\cos\theta_1} = \frac{E_R^2}{E_I^2} \tag{1.4.31}$$

This result is equal to $|r_{11}|^2$ and $|r_\perp|^2$ for parallel and perpendicular polarizations, respectively, where ρ is the coefficient appearing in Eq. (1.4.17) and the field reflection coefficients are described in Eqs. (1.4.28) and (1.4.25), respectively.

The power transmission coefficient is, from Eq. (1.4.30),

$$\begin{aligned}\tau &= \frac{\eta_1 \cos\theta_2}{\eta_2 \cos\theta_1}\left(\frac{E_T}{E_I}\right)^2 \\ &= \left(\frac{n_2}{n_1}\right)\left[\frac{1-(n_1/n_2)^2\sin^2\theta_1}{1-\sin^2\theta_1}\right]^{1/2} |t|^2 \end{aligned} \tag{1.4.32}$$

where $t = t_{11}$ and $t_\perp$ for parallel and perpendicular polarizations, respectively, as defined in Eqs. (1.4.29) and (1.4.26). Note that when Eqs. (1.4.31) and (1.4.32) are substituted into Eq. (1.4.30) the result is

$$1 = \rho + \tau$$

Example 1.4

The refractive index of water at a 3-m wavelength is $1.37 - j0.272$. In trying to image from air ($n = 1$) through water, a light ray reflected by the object traverses upward through the water and is incident at the water/air interface at 30°. What is the angle of refraction in air?

Solution

Snell's law, as seen from the analysis in Section 1.4.1, derives from equality of propagation coefficients β_1 and β_2 in the two media. These depend only on the real part of the refractive index and not the imaginary part, which determines attenuation α. As a result, the *direction* of the light is not related to jn''. Hence, $n_1 \sin\theta_1 = 1.37 \cdot \sin 30° = (1) \sin\theta_2$. Consequently,

$$\theta_2 = \sin^{-1}(1.37 \sin 30°) = 43.2° \text{ or } 0.75 \text{ rad}$$

Example 1.5

What percentage of EM radiation power in Example 1.4 is reflected back into the water?

Solution

From Eq. (1.4.31), $\rho = |r|^2$. For perpendicular polarization

$$r_\perp = \frac{n_1\cos\theta_1 - n_2\cos\theta_2}{n_1\cos\theta_1 + n_2\cos\theta_2}$$

from Eq. (1.4.25). Here,

$$r_{\perp} = \frac{1.37 \cos 30^{\circ} - (1) \cos 43.2^{\circ}}{1.37 \cos 30^{\circ} + (1) \cos 43.2^{\circ}} = 0.24$$

and $\rho = |0.24|^2 = 5.7\%$. For parallel polarization, using Eq. (1.4.28)

$$r_{11} = \frac{n_2 \cos \theta_1 - n_1 \cos \theta_2}{n_2 \cos \theta_1 + n_1 \cos \theta_2} = -0.71$$

and $\rho = |-0.71|^2 = 50.1\%$. If the light is randomly polarized $\rho = |r_{\perp}|^2 + |r_{11}|^2 = 55.8\%$.

Example 1.6

It is well known that when one looks at an object through a plane surface interface between two media, as in an aquarium for example, the objects can be seen clearly but not in their true position. For this reason these are virtual images. In the case of an aquarium, the objects appear closer and therefore larger than they are in reality. Consider Fig. 1.8, where an actual object point in medium 1 is at M. In the case of water as medium 1, the real part of the refractive index $n_1 \approx 1.33$ in the visible, which is greater than that of air ($n_2 \approx 1$). Consider a wave incident at the water/air interface at an angle of incidence of θ_1. Since medium 2, air, for example, is less

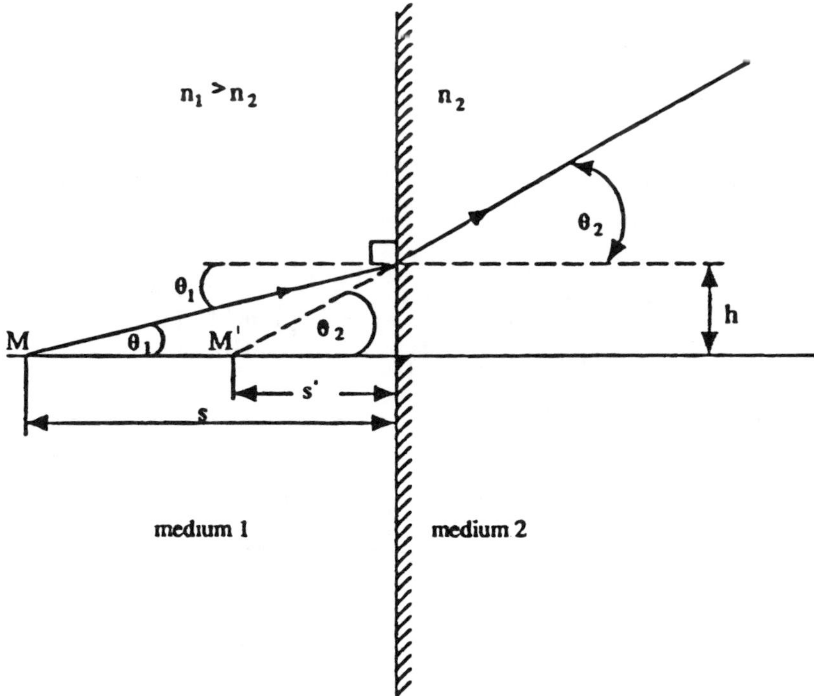

Fig. 1.8 Object (s) and image (s') distances when looking from air (medlum 2) into water (medium 1).

dense optically, the penetrating wave is refracted away from the normal to the surface, that is, $\theta_2 > \theta_1$. As a result, to a viewer in medium 2 the plane wave appears to emanate from point M', which is closer, rather than from the actual point M. Develop an expression for s'/s, where s' is the distance between the interface and M', and s is that between the interface and M.

Solution

$$h = s \tan \theta_1 = s' \tan \theta_2$$

Therefore,

$$\frac{s'}{s} = \frac{\tan \theta_1}{\tan \theta_2} = \frac{\sin \theta_1}{\sin \theta_2} \cdot \frac{\cos \theta_2}{\cos \theta_1} = \frac{n_2}{n_1} \frac{\cos \theta_2}{\cos \theta_1} \tag{1.4.33}$$

For small angles, this reduces to

$$\frac{s'}{s} \approx \frac{n_2}{n_1} \tag{1.4.34}$$

For this reason, when one looks perpendicularly into water, objects in the water appear closer at a distance $(1/1.33) \approx 3/4$ their actual distance.

1.4.4 Brewster and Critical Angles

Example 1.5 indicates a large difference in reflectance according to light polarization. Examination of Eq. (1.4.28) indicates zero reflectance for parallel polarization if $n_2 \cos \theta_1 = n_1 \cos \theta_2$ or

$$\cos \theta_1 = \frac{n_1}{n_2} \cos \theta_2 = \frac{n_1}{n_2} \left[1 - \left(\frac{n_1}{n_2} \right)^2 \sin^2 \theta_1 \right]^{1/2}$$

which implies

$$\cos^2\theta_1 = 1 - \sin^2 \theta_1 = \left(\frac{n_1}{n_2} \right)^2 - \left(\frac{n_1}{n_2} \right)^4 \sin^2 \theta_1$$

or

$$\left[1 - \left(\frac{n_1}{n_2} \right)^4 \right] \sin^2 \theta_1 = 1 - \left(\frac{n_1}{n_2} \right)^2$$

Therefore,

$$\sin^2 \theta_1 = \frac{1 - (n_1/n_2)^2}{1 - (n_1/n_2)^4} = \frac{n_2^2}{n_2^2 + n_1^2} \tag{1.4.35}$$

This angle at which reflectance ceases and $\tau = 1$ is called the Brewster angle. Since this result [Eq. (1.4.35)] is always less than unity, such an angle always exists for parallel polarization. In Example 1.5, it is equal to $\arcsin^2[1/(1 + 1.37^2)] = 36°$. Therefore, at $\theta_1 = 30°$, reflectance in that example is quite small for parallel polarization at that wavelength.

Examination of Eq. (1.4.25) for perpendicular polarization indicates no *real* angle exists at which $r_\perp$ goes to zero. Therefore, there is no Brewster angle for perpendicular polarization.

Snell's law, Eq. (1.4.16), can be written in the form

$$\sin \theta_2 = \frac{n_1}{n_2} \sin \theta_1 \tag{1.4.36}$$

When n_1 is smaller than n_2 (medium 1 optically less dense), θ_2 is less than θ_1. As shown in Fig. 1.9(a) the refracted ray is bent toward the normal. However, as shown in Fig. 1.9(b), when medium 1 is optically denser, the refracted ray in medium 2 is bent away from the normal. In this case, if the angle of incidence θ_1 is increased, a point will eventually be reached where the light ray in medium 2 is parallel to the interface. This point is known as the critical angle of incidence θ_c. When the incident angle θ_1 is greater than the critical angle, the condition for total internal reflection is satisfied; that is, the light is totally reflected back into medium 1. (This is an idealized situation. In practice there is always some tunneling of optical energy through the interface, as is dealt with in Exercise 1.9.) In view of Eq. (1.4.36), $\theta_2 = 90°$ when $\theta_1 = \theta_c = \sin^{-1}(n_2/n_1)$. As long as medium 1 is denser ($n_1 > n_2$), a critical angle always exists.

In examples 1.4 and 1.5, a critical angle exists for light incident from water into air, but not vice versa.

The concept of essentially total reflection from a dielectric, for angles of incidence greater than the critical angle θ_c, is the foundation for step index optical fibers, as shown in Fig. 1.10. As long as $\theta_2 > \theta_c$, the optical rays propagating in the core will remain inside the core. In Fig. 1.10, only for light rays incident to the fiber at angles smaller than ϕ does the fiber act as a waveguide (see Exercise 1.8). The angle ϕ thus defines a core of ray acceptance to the fiber. Numerical aperture is defined as $n_0 \sin\phi$, where n_0 is the refractive index of the medium from which the incident light rays enter the fiber. Referring to Fig. 1.10, the numerical aperture is equal to $(n_1^2 - n_2^2)^{1/2}$ (Exercise 1.8) and thus is a constant depending solely on fiber refractive indices. In water, for example, n_0 is greater than for air, thus causing ϕ to decrease if the fiber is immersed in water. The smaller the difference between core and cladding refractive indices, the greater the angular width of acceptance.

Many imaging devices make use of optical fiber faceplates at their input. Step index fibers should be of narrow diameter so as to permit resolution of small detail in imaging applications.

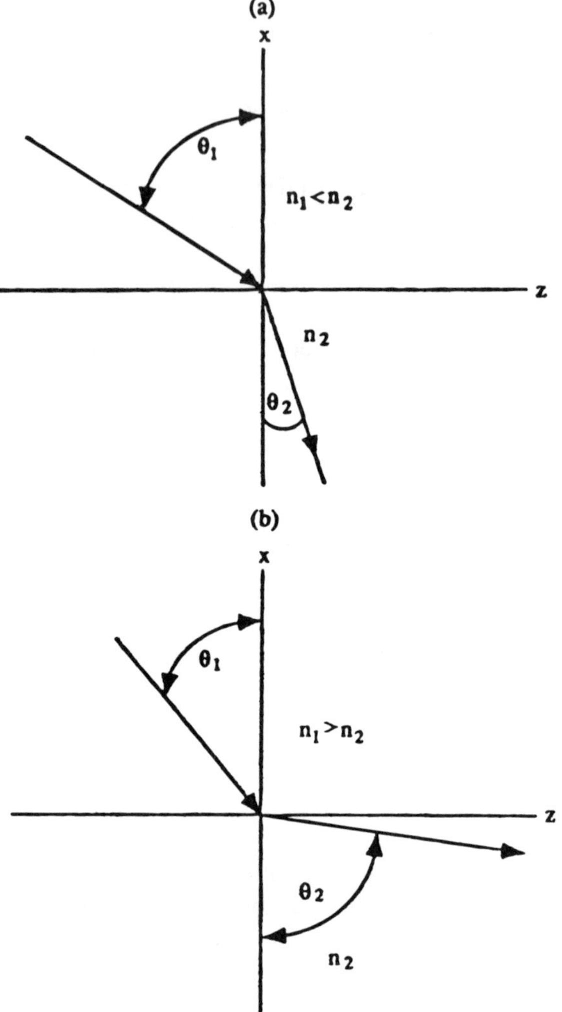

Fig. 1.9 Refraction when medium of incidence is (a) less dense and (b) more dense than medium of refraction.

1.5 INTERFACES IN SERIES

Consider n media in series as in Fig. 1.11. A plane EM wave is incident normally, so that $\theta_1 = \theta_2 = \theta_3 \cdots \theta_n = 0$. The fields in region 1 may be expressed as

$$E_{x1}(z) = A_{i1}\{\exp[-\gamma_1(z + d_1)] + r_{12}\exp[\gamma_1(z + d_1)]\} \tag{1.5.1}$$

$$H_{y1}(z) = \frac{A_{i1}}{\eta_1}\{\exp[-\gamma_1(z + d_1)] - r_{12}\exp[\gamma_1(z + d_1)]\} \tag{1.5.2}$$

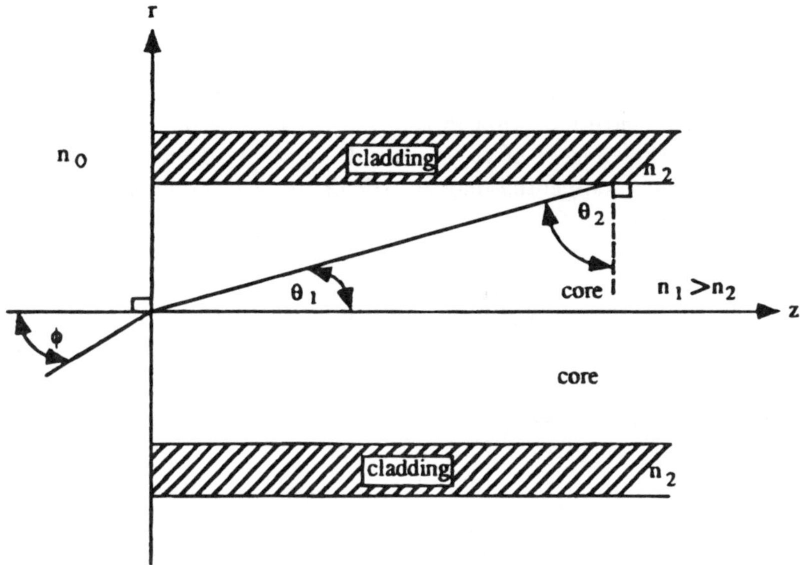

Fig. 1.10 Step-index optical fiber.

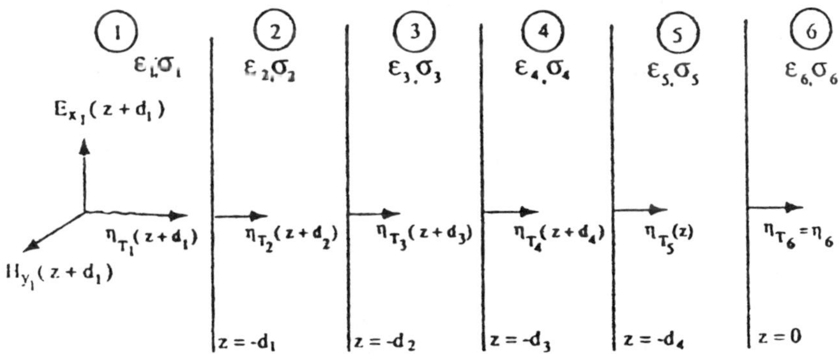

Fig. 1.11 Wave propagation through interfaces in series.

Similarly, in medium 2,

$$E_{x2}(z) = A_{i2}\{\exp[-\gamma_2(z + d_2)] + r_{23}\exp[\gamma_2(z + d_2)]\} \tag{1.5.3}$$

$$H_{y2}(z) = \frac{A_{i2}}{\eta_2}\{\exp[-\gamma_2(z + d_2)] - r_{23}\exp[\gamma_2(z + d_2)]\} \tag{1.5.4}$$

All fields are tangential to relevant interfaces. Reflection coefficients r_{12} and r_{23} cannot be obtained from Eqs. (1.4.25) or (1.4.28), which are two-media situations. The reason derives from the boundary conditions. In the present case, matching electric fields in both media 1 and 2 at the $z = -d_1$ interfaces yields

$$E_{x1}(-d_1) = A_{i1}[1 + r_{12}] = E_{x2}(-d_1) \tag{1.5.5}$$

In the two-media treatment, the r_{23} term did not exist. The reason was that under the latter circumstances there was no interface to the right of the 1-2 interface at $z = -d_1$ to cause a wave reflection in the backward direction (from right to left in Fig. 1.11). Using the concept of ϵ_{eq} to include conductivity in wave impedance, it is permissible to match tangential magnetic fields too as if both media were imperfect dielectrics. Accordingly,

$$A_{i1}\frac{(1 - r_{12})}{\eta_1} = H_{y2}(-d_1) \tag{1.5.6}$$

If Eq. (1.5.5) is divided by Eq. (1.5.6),

$$\eta_1\left(\frac{1 + r_{12}}{1 - r_{12}}\right) = \frac{E_{x2}(-d)}{H_{y2}(-d)} = \eta_{T2}(-d_1) \tag{1.5.7}$$

where η_{T2} is the total wave impedance in medium 2 and is dependent on the electric and magnetic fields of *both* the forward and backward going waves instead of only either wave by itself as intrinsic medium wave impedance η is defined in Eq. (1.3.13). Consequently, η is a constant that is intrinsic to a given medium, whereas η_T varies with position. For example, if Eq. (1.5.3) is divided by Eq. (1.5.4), in medium 2

$$\eta_{T2}(z) = \eta_2\left\{\frac{\exp[-\gamma_2(z + d_2)] + r_{23}\exp[\gamma_2(z + d_2)]}{\exp[-\gamma_2(z + d_2)] - r_{23}\exp[\gamma_2(z + d_2)]}\right\} \tag{1.5.8a}$$

From Eq. (1.5.7)

$$r_{12} = \frac{\eta_{T2}(-d_1) - \eta_1}{\eta_{T2}(-d_1) + \eta_1} \tag{1.5.9}$$

At the $z = -d_2$ interface, matching of tangential fields and treatment similar to Eqs. (1.5.5), (1.5.6), and (1.5.7) leads to

$$\eta_2\left(\frac{1 + r_{23}}{1 - r_{23}}\right) = \frac{E_{x3}(-d)}{H_{y3}(-d)} = \eta_T(-d_2) \tag{1.5.10}$$

Solving for r_{23} yields

$$r_{23} = \frac{\eta_{T3}(-d_2) - \eta_2}{\eta_{T3}(-d_2) + \eta_2} \tag{1.5.11}$$

This can be substituted into Eq. (1.5.8) so that

$$\eta_{T2}(-d_1) = \eta_2 \left\{ \frac{\exp[\gamma_2(d_1 - d_2)] + \dfrac{\eta_{T3}(-d_1) - \eta_1}{\eta_{T3}(-d_1) + \eta_1} \exp[-\gamma_2(d_1 - d_2)]}{\exp[\gamma_2(d_1 - d_2)] - \dfrac{\eta_{T3}(-d_1) - \eta_1}{\eta_{T3}(-d_1) + \eta_1} \exp[-\gamma_2(d_1 - d_2)]} \right\}$$

$$= \eta_2 \left\{ \frac{\eta_{T3}(-d_1) \cosh \gamma_2(d_1 - d_2) + \eta_2 \sinh \gamma_2(d_1 - d_2)}{\eta_2 \cosh \gamma_2(d_1 - d_2) + \eta_{T3}(-d_1) \sinh \gamma_2(d_1 - d_2)} \right\} \tag{1.5.12}$$

In general, relating to Fig. 1.11, the proper procedure would be to start at the right. Since there is no interface to the right of that at $z = 0$, then $r_{67} = 0$. Consequently, (Eq. 1.5.8) applied to this interface yields $\eta_{T6} = \eta_6$, which is a constant intrinsic to medium 6 and therefore does not vary with position. Using this value of wave impedance,

$$r_{56} = \frac{\eta_{T6} - \eta_5}{\eta_{T6} + \eta_5} = \frac{\eta_6 - \eta_5}{\eta_6 + \eta_5} \tag{1.5.13}$$

Since all wave impedances are determined by the permittivity and permeabilities of each medium, which defines and characterizes each medium, r_{56} is known. Next, $\eta_{T5}(-d_4)$ can be determined as

$$\eta_{T5}(-d_4) = \eta_5 \left\{ \frac{\eta_6 \cosh \gamma_5 d_4 + \eta_5 \sinh \gamma_5 d_4}{\eta_5 \cosh \gamma_5 d_4 + \eta_6 \sinh \gamma_5 d_4} \right\} \tag{1.5.14}$$

Next, r_{45} can be calculated

$$r_{45} = \frac{\eta_{T5}(-d_4) - \eta_4}{\eta_{T5}(-d_4) + \eta_4} \tag{1.5.15}$$

as well as $\eta_{T4}(-d_3)$

$$\eta_{T4}(-d_3) = \eta_4 \left\{ \frac{\eta_{T5}(-d_3) \cosh \gamma_4(d_3 - d_4) + \eta_4 \sin \gamma_4(d_3 - d_4)}{\eta_4 \cosh \gamma_4(d_3 - d_4) + \eta_{T5}(-d_3) \sinh \gamma_4(d_3 - d_4)} \right\}$$

Proceeding backwards from right to left, we obtain:

$$r_{34} = \frac{\eta_{T4}(-d_3) - \eta_3}{\eta_{T4}(-d_3) + \eta_3} \tag{1.5.16}$$

$$\eta_{T3}(-d_2) = \eta_3 \left\{ \frac{\eta_{T4}(-d_3) \cosh \gamma_3(d_2 - d_3) + \eta_3 \sinh \gamma_3(d_2 - d_3)}{\eta_3 \cosh \gamma_3(d_2 - d_3) + \eta_{T4}(-d_3) \sinh \gamma_3(d_2 - d_3)} \right\}$$

In this way, values of the fields in each medium can be established.

The underlying physical concept is that the EM wave is assumed to be propagating from left to right and that η_T represents *input* wave impedance. For example, $\eta_{T2}(-d_1)$ represents input wave impedance to medium 2. As such, it includes effects of all media to the right of the $z = -d_1$ boundary as they affect that boundary. That is because $\eta_T(-d_1)$ includes r_{23}, which depends on $\eta_{23}(-d_2)$, etc. Thus, wave solutions derived in this manner are steady-state values that are valid after all transient reflections back and forth in the $+z$ and $-z$ directions between the various interfaces have converged to the steady-state situation.

Example 1.7

An EM wave is incident normally on the dielectric system shown earlier. The EM frequency is 300 THz (1-μm wavelength). Maximum electric field in medium 3 is 1 V$\cdot$m^{-1}. Find the fields and Poynting vector in each medium.

Solution

Starting from the right, $r_{34} = 0$ because there is no interface to the right of $z = 0$.

$$\therefore \eta_{T3}(z) = \eta_3 = \eta_0$$

$$r_{23} = \frac{\eta_{T3}(0) - \eta_2}{\eta_{T3}(z) + \eta_2} = \frac{\eta_3 - \eta_2}{\eta_3 + \eta_2} = \frac{\eta_0 - \eta_0/\sqrt{4}}{\eta_0 + \eta_0/\sqrt{4}} = \frac{1 - 1/2}{1 + 1/2} = \frac{1}{3}$$

where Eq. (1.3.13) has been used.

From Eq. (1.5.12)

$$\eta_{T2}(-d_1) = \eta_2 \left\{ \frac{\eta_{T3}(-d_1)\cos\beta_2 d_1 - j\eta_2 \sin\beta_2 d_1}{\eta_2 \cos\beta_2 d_1 - j\eta_{T3}(-d_1)\sin\beta_2 d_1} \right\}$$

where $\gamma_2 = j\beta_2$ and $d_1 = 10^{-3}$ m. Substituting $\eta_{T3}(-d_1) = \eta_3 = \eta_0$, $\eta_2 = \eta_0/2$, and $\beta_2 d_1 = 2\beta_0 d_1 = 4\pi \cdot 10^6 \cdot 10^{-3} = 4\pi \cdot 10^3$, we obtain:

$$\eta_{T2}(-d_1) = \frac{\eta_0}{2} \left\{ \frac{\eta_0 \cos\beta_2 d_1 - j(\eta_0/2)\sin\beta_2 d_1}{(\eta_0/2)\cos\beta_2 d_1 - j\eta_0 \sin\beta_2 d_1} \right\}$$

$$= \frac{\eta_0}{2} \left\{ \frac{\cos \cdot 4\pi \cdot 10^3 - j(1/2)\sin 4\pi \cdot 10^3}{(1/2)\cos 4\pi \cdot 10^3 - j\sin 4\pi \cdot 10^3} \right\}$$

$$= \frac{\eta_0}{2} \left\{ \frac{1 - 0}{1/2 - 0} \right\} = \eta_0$$

Applying this to Eq. (1.5.9),

$$r_{12} = \frac{\eta_{T2}(-d_1) - \eta_1}{\eta_{T2}(-d_1) + \eta_1} = \frac{\eta_0 - \eta_0}{\eta_0 + \eta_0} = 0$$

Accordingly,

$$\begin{aligned}
E_{x1}(z) &= A_{i1}e^{-j\beta_1(z+d_1)} + r_{12}e^{j\beta_1(z+d_1)} \\
&= A_{i1}e^{-j\beta_0(z+d_1)} \\
H_{y1}(z) &= \frac{A_{i1}e^{-j\beta_1(z+d_1)}}{\eta_0} \\
E_{x2}(z) &= A_{i2}[e^{-j\beta_2 z} + r_{23}e^{j\beta_2 z}] \\
&= A_{i2}[e^{-j2\beta_0 z} + (\tfrac{1}{3})e^{j2\beta_0 z}] \\
H_{y2}(z) &= \frac{A_{i2}}{\eta_2}\left[e^{-j2\beta_0 z} - \left(\frac{1}{3}\right)e^{j2\beta_0 z}\right], \qquad \eta_2 = \eta_0/2 \\
E_{x3}(z) &= A_{13}[e^{-j\beta_3 z} + r_{34}e^{j\beta_3 z}], \qquad r_{34} = 0 \\
&= A_{13}e^{-j\beta_0 z} \\
H_{y3}(z) &= \frac{A_{i3}}{\eta_3}e^{-j\beta_0 z} = \frac{A_{i3}}{\eta_0}e^{-j\beta_0 z} \\
\max[E_{x3}(z)] &= 1 \rightarrow A_{i3} = 1
\end{aligned}$$

where A_{i2} and A_{i3} can be related through boundary conditions at the plane $z = 0$. By matching tangential electric fields,

$$E_{x3}(0) = A_{i3} = 1 = E_{x2}(0) = A_{i2}(1 + \tfrac{1}{3}) \rightarrow A_{i2} = (\tfrac{3}{4})A_{i3} = \tfrac{3}{4}$$

A_{i1} and A_{i2} can be related by matching tangential electric fields at the $z = -d_1$ boundary. There,

$$2\beta_0 \cdot (-d_1) = 4\pi \cdot 10^6 \cdot (-10^{-3}) = -4\pi \cdot 10^3 = -2\pi N$$

where $N = 2 \cdot 10^3$

$$\therefore E_{x1}(-d_1) = A_{i1} = E_{x2}(-d_1) = A_{i2}[e^{j2\pi N} + \tfrac{1}{3}e^{-j2\pi N}] = \tfrac{3}{4}[1 + \tfrac{1}{3}] = 1$$

Since $A_{i1} = A_{i3} = 1$, while $A_{i2} = \frac{3}{4}$,

$$E_{x1}(z) = e^{-j\beta_0(z+d_1)}$$

$$H_{y1}(z) = \frac{e^{-j\beta_0(z+d_1)}}{\eta_0}$$

$$E_{x2}(z) = \tfrac{3}{4}[e^{-j2\beta_0 z} + \tfrac{1}{3}e^{j2\beta_0 z}]$$

$$H_{y2}(z) = \frac{\frac{3}{4}}{\eta_0/2}[e^{-j2\beta_0 z} - \tfrac{1}{3}e^{j2\beta_0 z}]$$

$$E_{x3}(z) = e^{-j\beta_0 z}$$

$$H_{y3}(z) = \frac{e^{-j\beta_0 z}}{\eta_0}$$

Poynting vectors are

$$\overline{P}_1 = \hat{z}\frac{1}{2}\,\mathrm{Re}\,\overline{E}_{x1}(z) \times \overline{H}^*_{y1}(z) = \hat{z}\frac{1}{2\eta_0}\ \mathrm{W\cdot m^{-2}}$$

$$\overline{P}_2 = \hat{z}\frac{1}{2}\,\mathrm{Re}\,\overline{E}_{x2}(z) \times \overline{H}^*_{y2}(z)$$

$$= \hat{z}\,\mathrm{Re}\,\frac{1}{2\eta_2}\left[\left(\frac{3}{4}\right)^2\left(e^{-j2\beta_0 z} + \frac{1}{3}\,e^{j2\beta_0 z}\right)\right]\cdot\left(e^{j2\beta_0 z} - \frac{1}{3}\,e^{-j2\beta_0 z}\right)$$

$$= \hat{z}\,\frac{1}{2\eta_2}\,\mathrm{Re}\left(\frac{3}{4}\right)^2\left[1 - \frac{1}{3}\,e^{-j4\beta_0 z} + \frac{1}{3}\,e^{j4\beta_0 z} - \frac{1}{9}\right]$$

$$= \hat{z}\,\frac{1}{2(\eta_0/2)}\left(\frac{3}{4}\right)^2\left(\frac{8}{9}\right) = \hat{z}\,\frac{1}{2\eta_0}\ \mathrm{W\cdot m^{-2}}$$

$$\overline{P}_3 = \hat{z}\frac{1}{2}\,\mathrm{Re}\,\overline{E}_{x3}(z) \times \hat{H}^*_{y3}(z) = \hat{z}\frac{1}{2\eta_0}\ \mathrm{W\cdot m^{-2}}$$

Note that power densities $\overline{P}_1 = \overline{P}_2 = \overline{P}_3$; this is to be expected from conservation of power since in this example there are no losses and, because the EM wave is a plane wave, the cross-sectional area on which the wave is incident is identical in each medium.

Another interesting point here is that $r_{12} = 0$, which implies that $\rho_{12} = |r_{12}|^2$ also equals zero. Medium 2 is an *antireflection coating* to medium 3 whenever $\eta_1 = \eta_3$. This occurs at wavelengths such that $\beta_2 d_1 = m\pi$, where m is an integer. Both conditions apply in this example, where $\eta_1 = \eta_3 = \eta_0$ and $\beta_2 d_1 = 2\pi N$, the value of N being 2000 in this example. Antireflection coating to lenses or mirrors is very helpful in improving image contrast.

1.6 THE EIKONAL EQUATION

Previously, the incidence of an EM wave to the middle areas of a long continuous interface has been analyzed. At this point, however, what happens to a wave incident at the *edge* of an interface is to be considered. Usually objects to be imaged have edges and boundaries. So do the *optical elements such as lenses and mirrors* that perform the imaging. Radiation incident on such edges becomes diffracted. There is a change in direction of the wave, which is inconsistent with the laws governing refraction and reflection of radiation. The direction of diffraction depends on many factors, such as initial direction, wavelength, dimensions of detail or objects causing the diffraction, etc., and is discussed in Part 3. For the present, we will concern ourselves with the question of whether or not the relative amount of light diffracted is significant. If not, the EM wave can be dealt with from the standpoint of ray optics, which simplifies the mathematics considerably since diffraction is neglected. The criteria for answering this question derive from the Eikonal equation, which lays the foundation for geometrical optics in which waves are treated as rays. The term *Eikonal* is derived from the Greek word for *image*.

The concept of complex field amplitude (CFA) introduced with regard to Eq. (1.3.37) is a convenient tool with which to generalize the wave equation (or Helmholtz's equation) so that it can apply to either the electric or magnetic field of a wave, without bothering to specify which. This is convenient since both fields in an EM wave propagate together with the same propagation properties of phase velocity, propagation and attenuation constants, etc. Accordingly, using Helmholtz's equation

$$\nabla^2\psi + \beta^2\psi = 0 \tag{1.6.1}$$

where is ψ CFA and β is, once again, propagation constant (equal to $\omega^2\mu\epsilon$ or $2\pi/\lambda$). The wave amplitude is comprised of both amplitude and phase. That is,

$$\psi = \psi_0(x, y, z) \exp[\quad j\beta_0 S(x, y, z)] \tag{1.6.2}$$

where S is called an eikonal and represents a surface of constant phase in much the same way that $z = z_1$ represents a plane of constant phase for a plane wave propagating according to $\exp(\pm j\beta z)$. The difference is, however, that Eq. (1.6.2) is much more general. Since $\beta = \beta_0 n$, refractive index n is included in S. The surface of constant phase represented by S is not limited to planes, but includes any and all sorts of indentations and curvatures as long as they are balanced out by refractive index changes in such a way that S, because it includes refractive index as well as spatial distance, indeed represents a surface of constant phase. This is so because for a constant frequency ω or propagation constant β_0 in vacuum, changes in phase of the wave derive only from the integral $\int nds$, where ds is incremental path length. The term S includes, therefore, both path length and refractive index. Its magnitude, deriving from the above integral, determines phase. Accordingly, Eq. (1.6.2) must be substituted into Eq. (1.6.1), yielding

$$\nabla^2\{\psi_0(x, y, z) \exp[-j\beta_0 S(x, y, z)]\} + \beta^2\psi_0(x, y, z) \exp[-j\beta_0 S(x, y, z)] = 0 \tag{1.6.3}$$

This equation involves the Laplacian of a product of functions of position. If $\psi_0(x, y, z)$ is designated $V(x, y, z)$, and $\exp[-j\beta_0 S(x, y, z)]$ is designated $W(x, y, z)$, then from the product property of derivatives

$$\nabla(WV) = V\nabla W + W\nabla V \tag{1.6.4}$$

and

$$\nabla^2(WV) = \nabla[V\nabla W + W\nabla V] = W\nabla^2 V + 2\nabla V\nabla W + V\nabla^2 W \quad (1.6.5)$$

By using the defined V, W definitions for ψ_0 and $\exp(-j\beta_0 S)$, it is easy to show by simple substitution that Eq. (1.6.3) reduces to

$$\beta_0^2\left(\frac{\beta^2}{\beta_0^2} - \nabla S \cdot \nabla S\right)\psi_0 - j\beta_0(2\nabla S \cdot \nabla\psi_0 + \psi_0\nabla^2 S) + \nabla^2\psi_0 = 0 \quad (1.6.6)$$

where $\exp(-j\beta_0 S)$ has been cancelled out of each term. In the preceding equation, $\nabla W = -j\beta_0\nabla S \exp(-j\beta_0 S)$ and $\nabla^2 W = -\beta_0^2(\nabla S)^2 \exp(-j\beta_0 S) - j\beta_0\nabla^2 S \cdot \exp(-j\beta_0 S)$. Now, by definition β/β_0 in Eq. (1.6.6) is equal to refractive index n. Equation (1.6.6) is comprised of both real and imaginary terms, so that it can be decomposed into two separate equations with each being equal to zero. By considering only the real terms and equating their sum to zero,

$$(\nabla S)^2 = n^2 + \frac{\nabla^2\psi_0}{\beta_0^2\psi_0} \quad (1.6.7)$$

Now, $\beta_0 = 2\pi/\lambda_0$. If the finiteness of wavelength is ignored and λ is allowed to reach zero (or equivalently $\beta_0 \rightarrow \infty$), the preceding equation becomes

$$(\nabla S)^2 = n^2 \quad (1.6.8)$$

which is known as the *Eikonal equation*. Waves can be treated as rays when it holds true, that is, if the last term in Eq. (1.6.7) can be ignored. So too can diffraction. The EM wave can then be treated as a ray, behaving according to the laws of reflection and refraction only, with diffraction effects being neglected. This can be understood in the following way.

Physically, the resolution of a wave is essentially on the order of a wavelength. This means that while diffraction is considered to derive from a wave being incident to an edge, the "edge" extends effectively into and out of the object by about one wavelength from the actual physical edge. Therefore, the light diffracted by an edge is essentially that which is incident within about one wavelength from the actual physical edge. If object size is very large compared to wavelength, almost the entire wave is incident at distances further inward than one wavelength from the edge. In this case, only a small percentage of the incident radiation is incident on the edges and diffracted, and the Eikonal equation then holds. In the limit, if λ approaches zero, the edges of the object have zero area and none of the incident light is incident on them and, as a result, none is diffracted. Therefore, the EM wave can be treated solely from the standpoint of rays. However, if object dimensions are not much larger than wavelength, then a high percentage of incident radiation undergoes diffraction because it is incident at the edges according to the above definition of an "edge." In this case, therefore, it is unjustified to ignore diffraction. At microwave wavelengths, which are much larger than optical ones, conditions in which diffraction can be ignored hardly exist. They give rise to sidelobes in the directionality of antennas. Diffraction exists too at optical wavelengths but to a lesser extent. This explanation can be related mathematically to Eqs. (1.6.7) and (1.6.8) in the following simplified fashion.

Consider an aperture whose size is proportional to a radial coordinate, as is demonstrated in the cone shown in Fig. 1.12 for example. The aperture diameter is $2a \sin\theta$ where a is the value of a radial coordinate at the aperture edge in a spherical coordinate system centered at the cone vertex. Consider a point source of light at the cone vertex at the origin of the coordinate system. If there were no diffraction, all light exiting the cone would continue propagating as a light beam in the shape of a pure cone. Diffraction of light by the aperture edge at the cone exit, however, would cause beam widening beyond that of the initial conical shape as the beam propagates further and further away from the cone exit. For a spherical wave, in view of Eq. (1.3.40),

$$\psi_0 = \frac{A_r}{r}$$

$$\nabla\psi_0 = -\frac{A_r}{r^2}$$

$$\nabla^2\psi_0 = \frac{2A_r}{r^3}$$

Therefore, substituting into Eq. (1.6.7),

$$\frac{\nabla^2\psi_0}{\beta_0^2\psi_0} = \frac{2A_r\lambda_0^2}{(2\pi)^2 r^3 A_r/r} = \frac{1}{2\pi^2}\left(\frac{\lambda_0}{r}\right)^2 \tag{1.6.9}$$

Thus, as long as $\lambda_0 << r$, the last term in Eq. (1.6.7) is negligible and the Eikonal equation is valid. At the cone exit, $r = a$, and

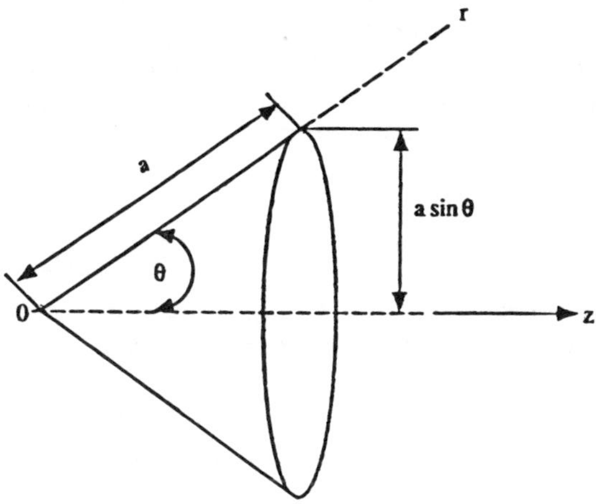

Fig. 1.12 Cone aperture.

$$\frac{\nabla^2\psi_0}{\beta_0^2\psi_0} = \frac{1}{2\pi^2}\left(\frac{\lambda_0}{a}\right)^2 \tag{1.6.10}$$

For aperture or object dimensions much larger than wavelength, Eq. (1.6.10) approaches zero, and the use of ray optics can be justified by the Eikonal equation since the percentage of diffracted light is negligible compared to the total light flux.

Next, we begin to consider the question of what constitutes ray optics.

1.7 THE RAY EQUATION

From Fig. 1.1 it is clear that surfaces of constant phase are perpendicular to the direction of EM wave propagation. Electromagnetic *rays* are defined as the loci of points that form trajectories orthogonal to the constant phase fronts of an EM wave. In an inhomogeneous medium, as constant phase fronts curve because of changes in refractive index, so too do the light rays because they are perpendicular to the constant phase fronts.

It is useful to determine the trajectories of light rays directly without having to construct constant phase fronts from the Eikonal equation, which would be normal to the direction of light ray propagation. This purpose is served by the ray equation, whose usefulness is perhaps most apparent in inhomogeneous media where light rays do *not* travel in straight lines because of continual changes in refractive index. In such media, the usefulness of Snell's law is quite limited because the latter applies primarily to discrete interfaces rather than to continual changes in refractive index. For homogeneous media the ray equation reduces to Snell's law (Exercise 1.13).

Let s be the distance measured along a ray. In Fig. 1.13, a radius vector r is drawn from an origin O to an arbitrary point along a ray. Knowledge of this vector can lead to a mathematical description of the ray path in terms of r, s, and $n(r)$, the latter being the refractive index for a medium assumed in the general case to be inhomogeneous. From Fig. 1.13, the unit vector

$$\hat{s} = \frac{d\bar{r}}{ds} \tag{1.7.1}$$

is tangential to the light ray. Since S in the Eikonal equation defines a surface of constant phase, or wavefront, then ∇S is a vector perpendicular to such a surface. Therefore ∇S is in the direction of the ray. (The fact that ∇S is perpendicular to S $=$ constant is similar to the relationship between equipotential surfaces and electric field lines in electrostatics, where $E = -\nabla V$ and is perpendicular to such equipotential surfaces.) In view of the Eikonal equation and the direction of ∇S,

$$\nabla S = n\hat{s} = n\frac{d\bar{r}}{ds} \tag{1.7.2}$$

The path s is a function of x, y, and z since the light is assumed to not necessarily follow a straight-line path because of the inhomogeneity of refractive index $n(r)$. Consequently,

$$\begin{aligned}\frac{d}{ds} &= \frac{d}{ds}(\hat{s}_x x + \hat{s}_y y + \hat{s}_z z) \cdot \left(\hat{s}_x\frac{d}{dx} + \hat{s}_y\frac{d}{dy} + \hat{s}_z\frac{d}{dz}\right)\\ &= \frac{d\bar{r}}{ds}\cdot\nabla = \hat{s}\cdot\nabla\end{aligned} \tag{1.7.3}$$

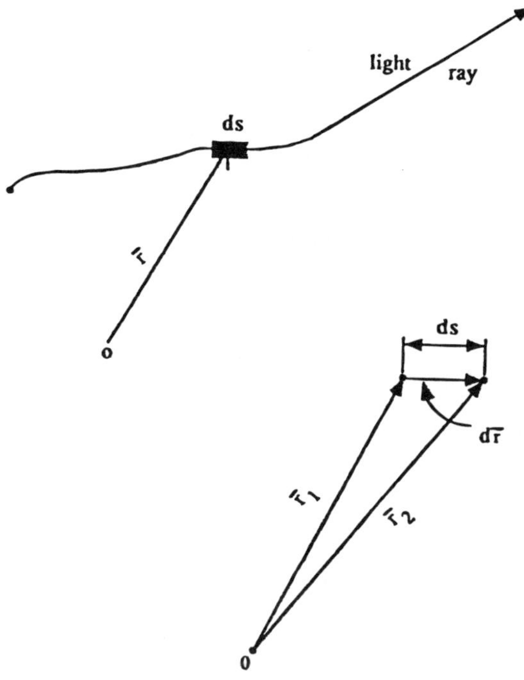

Fig. 1.13 Generalized ray propagation in three-dimensional space (medium not necessarily homogeneous).

where Eq. (1.7.1) has been used on the right-hand side. Using this to evaluate the derivative of Eq. (1.7.2), we get

$$\frac{d}{ds}(\nabla S) = \hat{s} \cdot \nabla(\nabla S) = \hat{s} \cdot (\nabla\nabla S) \tag{1.7.4}$$

where $\nabla\nabla$ is the dyadic tensor operator. By taking the gradient of the Eikonal equation [Eq. (1.6.8)],

$$2\nabla S \cdot \nabla\nabla S = 2n\nabla n \tag{1.7.5}$$

or

$$\frac{\nabla S}{n} \cdot \nabla\nabla S = \nabla n \tag{1.7.6}$$

By virtue of Eq. (1.7.2), $\nabla S/n = \hat{s}$. Substituting into Eq. (1.7.6),

$$\frac{\nabla S}{n} \cdot \nabla\nabla S = \hat{s} \cdot \nabla\nabla S = \nabla n \tag{1.7.7}$$

or, in view of Eq. (1.7.4) and this last result

$$\hat{s} \cdot \nabla\nabla S = \frac{d}{ds}(\nabla S) = \nabla n$$

By substituting Eq. (1.7.2) in the latter

$$\frac{d}{ds}\left(n\frac{d\bar{r}}{ds}\right) = \nabla n \tag{1.7.8}$$

which is the *ray equation* and is expressed in vector form. It shows quantitatively how changes in refractive index alter the location (r) and path length (s) of a light ray propagating in such a medium. When the medium is homogeneous, $\nabla n = 0$ and light rays take straight-line paths (Exercise 1.14). For an inhomogeneous medium, the ray paths are curved.

For Cartesian coordinates, the ray equations decompose to the scalar equations

$$\frac{d}{ds}\left(n\frac{dx}{ds}\right) = \frac{\partial n}{\partial x} \tag{1.7.9}$$

$$\frac{d}{ds}\left(n\frac{dy}{ds}\right) = \frac{\partial n}{\partial y} \tag{1.7.10}$$

$$\frac{d}{ds}\left(n\frac{dz}{ds}\right) = \frac{\partial n}{\partial z} \tag{1.7.11}$$

For paraxial rays that propagate nearly parallel to an optical axis, such as the z axis, for example, $ds \simeq dz$ and the ray equation reduces to the paraxial ray equation

$$\frac{d}{dz}\left(n\frac{d\bar{r}}{dz}\right) = \nabla n \tag{1.7.12}$$

This result is very important for imaging applications because, as will be shown later, paraxial rays produce sharper images. An interesting application of the paraxial ray equation is to graded-index optical fibers, particularly those with a parabolic index of refraction. The relevance to imaging derives from the fact that such fibers can serve also as lenses. This can be advantageous in many applications because of the ruggedness of fibers as compared to ordinary lenses, as well as the ability to insert narrow fibers into narrow locations where the insertion of ordinary lenses is much more difficult. An example of such applications is diagnostic imaging inside the human body, which has become an important medical tool in recent years.

The lens properties of such fibers will be considered in Chapter 2. For the moment, propagation properties are introduced. A parabolic index medium is clearly inhomogeneous ($\nabla n \neq 0$).

Example 1.8

Consider a dielectric waveguide of radius a in which refractive index varies parabolically, as in Fig. 1.14, that is,

$$n = n_0 - \frac{1}{2} n_1 r^2, \qquad n_1 a^2 << n_0 \tag{1.7.13}$$

In this medium, the refractive index is maximum (n_0) along the optical (z) axis, and falls off parabolically in the radial direction with increasing distance from the optical axis. The fall-off rate is very gradual. Substituting into the paraxial ray equation, where it is understood that r is in the radial direction,

$$\frac{d}{dz}\left[\left(n_0 - \frac{1}{2} nr^2\right)\frac{dr}{dz}\right] = \nabla n = -n_1 r \tag{1.7.14}$$

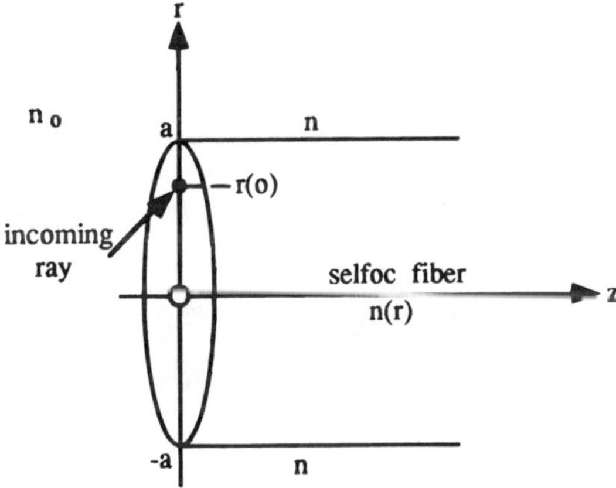

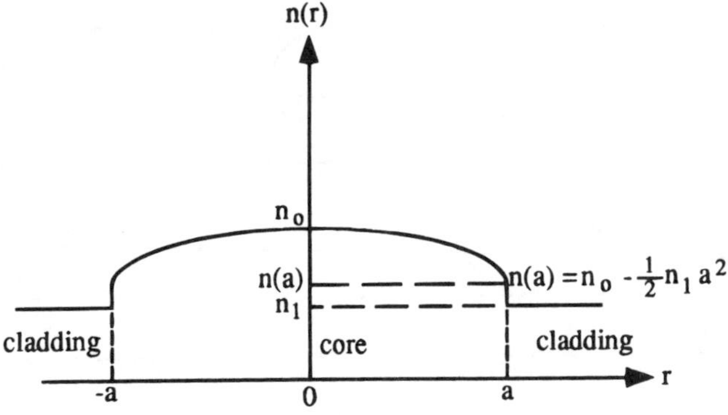

Fig. 1.14 Graded-index (GRIN) selfoc fiber.

or

$$\left(n_0 - \frac{1}{2}\,n_1 r^2\right)\frac{d^2 r}{dz^2} = -n_1 r \tag{1.7.15}$$

Since $n_0 >> \frac{1}{2}n_1 a^2$, this reduces to

$$\frac{d^2 r}{dz^2} = -\frac{n_1}{n_0}\,r \tag{1.7.16}$$

the solution to which, including initial conditions at fiber input, is

$$r(z) = r_0 \cos[(n_1/n_0)^{1/2}z] + (n_0/n_1)^{1/2} r_0' \sin[(n_1/n_0)^{1/2}z] \tag{1.7.17}$$

$$r'(z) = -(n_1/n_0)^{1/2} r_0 \sin[(n_1/n_0)^{1/2}z] + r_0' \cos[(n_1/n_0)^{1/2}z] \tag{1.7.18}$$

where

$$r'(z) = \frac{dr}{dz}, \qquad r_0' = \left.\frac{dr}{dz}\right|_{z=0}, \qquad \text{and } r_0 = r\Big|_{z=0}.$$

Rays thus propagate sinusoidally around the optical axis as they propagate in the z direction. The fact that rays are continually refracted back toward the z axis has led to the description of such media as being *self-focusing* ("selfoc"). If a light ray is incident perfectly parallel to the z axis, that is, $r_0' = 0$, then

$$r(z) = r_0 \cos[(n_1/n_0)^{1/2}z] \tag{1.7.19}$$

$$r'(z) = -(n_1/n_0)^{1/2} r_0 \sin[(n_1/n_0)^{1/2}z] \tag{1.7.20}$$

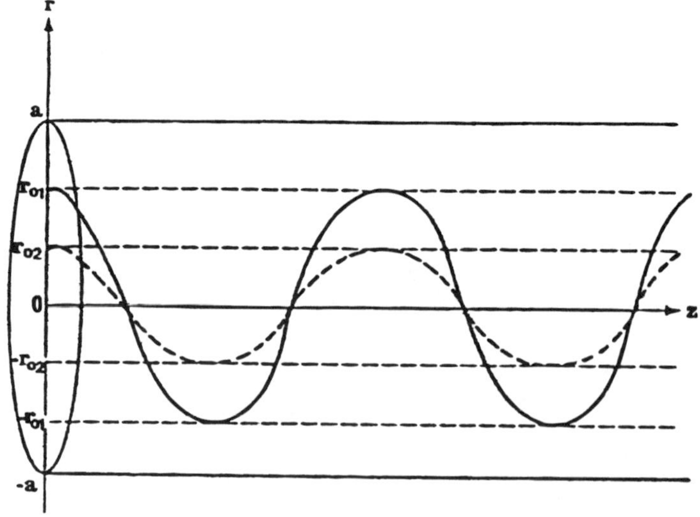

Fig. 1.15 Propagation of initially incident paraxial rays ($r_0' = 0$) at different initial heights (r_{01}, r_{02}) through selfoc fiber.

Such rays enter the fiber with zero degree angles of incidence and refraction in the $z = 0$ plane. Although there is no gradient in the z direction, the refractive index gradient in the r direction causes the wavefront (or location of constant phase) to shift from the radial to the nonradial direction. Since the wavefront is perpendicular to ray direction, the rays are bent or refracted toward the z axis. Such light rays never exceed a distance r_0 away from the axis. This means that while in step-index fibers light rays are "totally" reflected by the core/cladding interface back into the core only for angles of incidence greater than the critical angle, in selfoc fibers rays initially parallel to the fiber axis never even reach the core/cladding interface since $r_0 < a$. Another interesting point is that according to Eq. (1.7.19) zero crossings across the z axis occur at points $z = (2m - 1)(n_0/n_1)^{1/2}\,\pi$, where $m = 0, 1, 2, \ldots$ These locations along the z axis are independent of r_0. This means that initially parallel rays entering at different values of r_0, as in Fig. 1.15, propagate together through the fiber with zero crossings taking place at identical locations on the z axis. This leads to phase changes between input and output being essentially independent of r_0, as is described in Chapter 2.

REFERENCES

1.1. E. Goldin, *Waves and Photons: An Introduction to Quantum Optics*, Wiley, New York, 1982.

1.2. S. Ramo, J. R. Winnery, and T. Van Duzer, *Fields and Waves in Communication Electronics*, Wiley, New York, 1965, p. 337.

1.3. L. R. Lankes, "Optics and the physical parameters of the sea," *Photonic Spectra,* Vol. 4(5), p. 42 (May 1970). © Laurin Publishing Co.

1.4. G. M. Hale and R. Querry, "Optical constants of water in the 200-nm to 200-μm wavelength region," *Appl. Opt.*, Vol. 12, March 1973, pp. 555–563.

1.5. A. Ishimaru, "Experimental and theoretical studies on enhanced backscattering from scatterers and rough surfaces," in *Scattering in Volumes and Surfaces*, M. Nieto-Vesperinas and J. C. Dainty, Eds., Elsevier, North-Holland, 1990, p. 1–15.

EXERCISES

1.1 Show that for low-loss media

$$\alpha \approx \left(\frac{\mu}{\epsilon'}\right)^{1/2}\left(\frac{\sigma + \omega\epsilon''}{2}\right)$$

and

$$\beta \approx \omega(\mu\epsilon')^{1/2}\left[1 + \frac{1}{8}\left(\frac{\sigma + \omega\epsilon''}{\omega\epsilon'}\right)^2\right]$$

1.2 (a) Verify Eq. (1.3.30). (b) Determine how n'' is related to ϵ'' for a low-loss dielectric.

1.3 Determine n' and n'' for a good conductor ($\sigma >> \omega\epsilon'$).

1.4 What is the intrinsic wave impedance η in a low-loss dielectric ($\sigma = 0$) medium?

1.5 Elliptically polarized light can be represented by the two orthogonal waves in Eq. (1.3.35).

Show that

$$\left(\frac{E_x}{A_x}\right)^2 + \left(\frac{E_y}{A_y}\right)^2 = 2\,\frac{E_xE_y}{A_xA_y}\cos\delta$$

which is the equation of an ellipse making an angle α with the x axis, where α is described by Eq. (1.3.36).

1.6 A parallel-polarized plane wave is incident from air on a plane surface. The refractive index of medium 2 is $n_2 = (1 - j)\sqrt{\sigma/(2\omega\epsilon_0}$, where σ is conductivity of medium 2. (a) Develop an expression for Snell's law involving real angles of incidence and refraction. (b) For a frequency of 100 kHz and a conductivity of 40 mho·m^{-1}, what is the angle of refraction?

1.7 A parallel-plane "window" of refractive index n_2 is immersed in a medium of refractive index n_1. A plane wave is incident on the window at its Brewster angle. What percentage of electric field incident to the exit boundary actually exits on the other side?

1.8 Referring to Fig. 1.10, find the maximum value of acceptance angle ϕ permitted so as to ensure that radiation in the core incident on the cladding cannot escape. Show that the numerical aperture is equal to $(n_1^2 - n_2)^{1/2}$.

1.9 An EM wave of perpendicular polarization is incident from the core to the cladding of a step-index optical fiber at an angle of incidence greater than θ_c. Find the total electric and magnetic fields in each medium, as well as the time-average Poynting vectors. Assume a planar interface as in Fig. 1.5(a), and a plane wave from the core incident on the core/cladding interface.

1.10 Example 1.6 considers an air/water interface. It is not exactly appropriate for an aquarium because it neglects the glass interface between air and water. Assume the thickness of this interface, t_1, is small. For an object in water at a distance s from the interface, show that the effect of the glass is to shift the apparent distance even closer to the glass by a distance $s(1 - n_1^{-1})$, where n_1 is the refractive index of the water. Does inclusion of the thin glass window change the result of Example 1.6? Assume paraxial rays, that is, small angles of incidence and refraction.

1.11 When a ray traverses a dielectric plate of refractive index n_2 with parallel plane surfaces, it emerges parallel to its original direction but with a lateral displacement d that increases with angle of incidence. Show that the displacement is given by

$$d = t\sin\theta_1\left(1 - \frac{n_1\cos\theta_1}{n_2\cos\theta_2}\right)$$

where n_1 is the refractive index of the surrounding medium, θ_1 is the angle of incidence in medium 1 on the dielectric and also the angle of refraction in medium 1 upon exit from the dielectric, and θ_2 is the angle of refraction in the dielectric after the entrance plate and also the angle of incidence in the dielectric upon the exit plate, and t is the thickness of the dielectric plate.

1.12 Consider an EM wave of 1 μm wavelength (300-THz frequency) in air incident normally on the network shown in the figure below. Determine the fields and average Poynting vectors in each medium as functions of A_{i1}, where A_{i1} is the electric field amplitude in medium 1 (air).

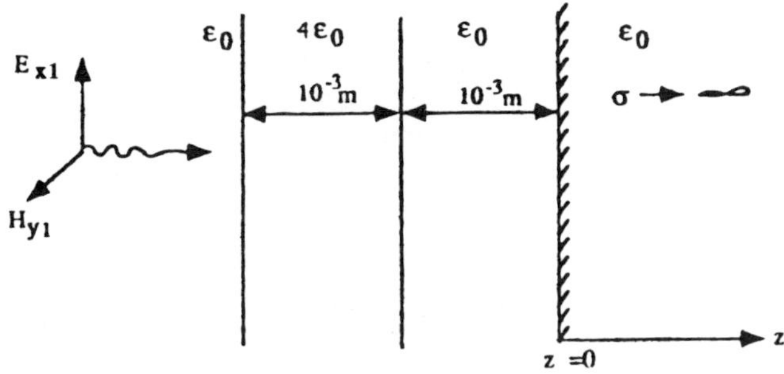

1.13 Apply the ray equation to homogeneous media and show that it reduces to Snell's law. Assume refractive indices are functions of x only in Fig. 1.5.

1.14 Using the ray equation, show that light rays take straight paths in homogeneous media.

CHAPTER

2

Imaging

In Chapter 1 the concept of rays was introduced. When the ratio of wavelength to aperture dimension approaches zero, diffraction effects can be neglected as justified by the Eikonal equation, and electromagnetic waves can be treated as rays propagating in directions perpendicular to locations of constant phase, known as wavefronts.

In this chapter, the concept of electromagnetic (EM) rays is developed further to explain image formation. Central to image formation is the concept of the *optical path*, which determines which paths may be taken by a given ray. These requirements are described by Fermat's principle. The curvature of lens and mirror surfaces optimal to image formation are derived from considerations of Fermat's principle.

After developing conditions required for image formation by a single lens (or mirror), combinations of lenses are considered. The approach is that of system engineering, where the output of a given lens is the input to the following one. Here, output represents the image and input represents the object.

After developing the theoretical requirements for image formation, the selfoc optical fibers, developed in Chapter 1, are shown to be completely analogous to ordinary lenses. Both satisfy these requirements.

Matrix optics are shown to be a useful tool for multielement optical system design.

2.1 FERMAT'S PRINCIPLE

The "Eikonal" S in Eq. (1.6.2) defines the effective path length which includes refractive index. For a homogeneous medium it is equal to the product of geometrical distance traversed by a given light ray and refractive index. The phase difference over this path length is simply $\beta_0 S$, where β_0 is the propagation constant in vacuum. The Eikonal S is thus effective rather than physical path length because S includes refractive index. In general, including the case of propagation in an inhomogeneous medium, the Eikonal

$$S = \int_{P_1}^{P_2} n(x, y, z)\, ds \tag{2.1.1}$$

describes the *optical path* from point P_1 to point P_2. As in the ray equation, the geometrical path length s is also a function of (x, y, z).

Although EM rays traverse straight line paths in homogeneous media, as shown in Exercise 1.14, this is not the case in inhomogeneous media as indicated by the ray equation and the accompanying example there of a parabolic index (Selfoc) optical fiber.

Fermat's principle relates the geometric path selected to the optical path or Eikonal. It is known as the principle of the shortest optical path or the principle of *least* time. In this form, the principle was strong but incomplete. A weaker version, but one that has a wide range of validity, is based on stationarity [2.1]. Stationary values of a function are those for which its total derivative is zero. Because such extrema can be either maxima or minima, Fermat's principle can be stated in the following way. The path chosen by monochromatic light in traveling along an actual ray from point P_1 to point P_2 in a given medium is always such that the time required is stationary with the time required to traverse any other closely adjacent paths. Within the context of this book, Fermat's principle is implemented for least rather than maximum time. The time required for a ray to travel along an actual path from P_1 to P_2 with a phase velocity v is

$$\int_{P_1}^{P_2} \frac{ds}{v} = \frac{1}{c} \int_{P_1}^{P_2} n ds = \frac{S}{c} \tag{2.1.2}$$

The optical path between P_1 and P_2 is recognized as S. If applied to an inhomogeneous medium, such as the atmosphere with its varying density, winds, temperature fluctuations, etc., Eq. (2.1.2) would describe the point-by-point refractions or scatterings of light rays. However, for a homogeneous medium, n is a constant and an immediate consequence of Fermat's principle then is that in such media light travels along straight paths, since the value of the integral must be minimum.

Example 2.1

Derive Snell's law and the law of reflection from Fermat's principle.

Solution

In variational notation Fermat's principle states

$$\delta S = \delta \left[\int_{P_1}^{P_2} n ds \right] = 0 \tag{2.1.3}$$

where δ represents a small variation about the actual path. Consider the general problem of refraction at an interface separating two homogeneous materials of refractive indices n_1 and n_2 as shown in Fig. 2.1. For this situation

$$\delta S = n_1(s_1 + \delta s_1) + n_2(s_2 + \delta s_2) - (n_1 s_1 + n_2 s_2) = 0$$

or

$$n_1 \delta s_1 + n_2 \delta s_2 = 0 \tag{2.1.4}$$

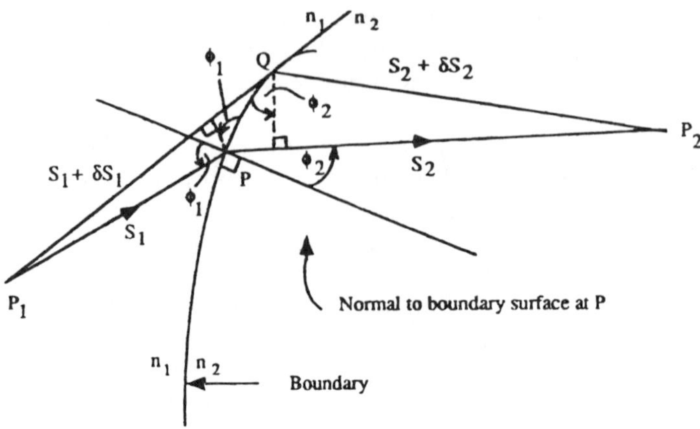

Fig. 2.1 Geometry for derivation of Snell's law of refraction from Fermat's principle.

From Fig. 2.1

$$s_1 + \delta s_1 = s_2 + PQ \sin \phi_1$$

$$s_2 + \delta s_2 = s_2 - PQ \sin \phi_2$$

Substituting into Eq. (2.1.4) yields *Snell's law* of refraction:

$$n_1 \sin \phi_1 = n_2 \sin \phi_2 \tag{2.1.5}$$

which is identical to Eq. (1.4.16). Thus, Fermat's principle describes limitations as to which paths may be taken by a ray traveling from P_1 in medium 1 to P_2 in medium 2, the result being consistent with Snell's law. Paths satisfying Eq. (2.1.5) are those of least propagation time.

Suppose the interface separating n_1 and n_2 is a mirror. Then the ray coming from P_1 is reflected at point P. Snell's law of refraction implies reflection, provided the refractive index after reflection is taken to be the negative of the index before reflection. This is consistent with the original definition of refractive index, which determines the phase velocity in a given medium. A negative refractive index implies a negative velocity or a negative distance traversed in positive time. In Fig. 2.1 a positive distance is implicitly defined from left to right; thus, a negative distance is measured from right to left so that putting $n_1 = -n_2$ in Eq. (2.1.5) indicates the direction in which the ray is reversed after impinging at point$P$ on the mirrored surface.

This is a *very useful concept* since it means any equation associated with refraction can be converted into a corresponding equation for reflection by setting $n_1 = -n_2$. As a result, the various equations derived in this chapter for image formations by means of lenses can be converted to those appropriate for imaging with mirrors with this technique. Here, Snell's law for *reflection* becomes

$$-n_2 \sin \phi_1 = n_2 \sin \phi_2$$

or

$$\sin \phi_1 = -\sin \phi_2 \tag{2.1.6}$$

which indicates that incident and reflected rays lie on opposite sides of the normal to the boundary surface and are at equal angles to it. This is consistent too with the manner in which the angles are drawn in Fig. 1.5 and with the discussion concerning backscattering from rough surfaces.

We can see here that with proper consideration as to the physical meaning of refractive index, Snell's law of *refraction* implies too the law of *reflection*.

2.1.1 Application to Imaging: Thin Lens Formula and Magnification

As an example of a typical lens system, consider the homogeneous, isotropic, round piece of glass of refractive index n centered on an axis 0-0′ (called the *optical axis*) and immersed in air (which has a refractive index close to unity) as shown cross sectionally in Fig. 2.2. Suppose we want to find the light ray connecting point P_1 to point P_2; that is, we want to determine the path light would travel going from P_1 to P_2 through this system. The piece of glass has a refractive index different from that of air. It causes the ray to *bend* or *refract* back toward the system axis. (For this reason it is called a *lens*; further, it is a *thin lens* because its thickness t is much less than its diameter.) Thus, the problem here is simply one of finding the points B and C at which the ray intersects the lens' surfaces (the boundaries between the materials of different refractive indexes), because it is known from Fermat's principle that the ray is a straight line in both the air and glass since each is a homogeneous medium.

It is postulated further in this example that

$$r, s, s' >> x, x', h >> t \tag{2.1.7}$$

Now since r, s, $s' >> x$, x', the angles which the ray P_1ABCP_2 makes with the optical axis are relatively small compared with the size of the whole system. By "relatively small" we mean that the sines and tangents of the angles may be replaced by the angles themselves. Rays such as these which lie close to the optical axis are known as *paraxial rays* and the region through which they travel is called the *paraxial region*.

Since the lens is thin [i.e., by relationship (2.1.7) all system dimensions are much greater than t] the distances of points A, B, and C from the optical axis are very nearly equal to h. Thus, for a thin lens and paraxial rays:

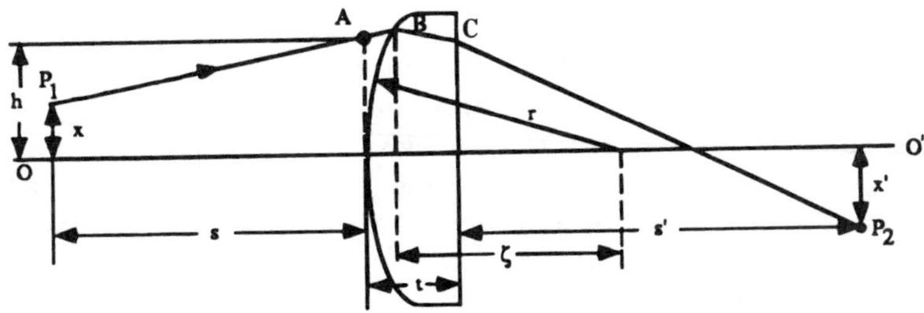

Fig. 2.2 Application of Fermat's principle for imaging of point P_1 at point P_2. For convenience, the radius of curvature of the back surface of the lens is infinite.

$$P_1A = [s_2 + (h - x)^2]^{1/2} \approx s\left[1 + \frac{(h - x)^2}{2s^2}\right] = \frac{(h - x)^2}{2s}$$

$$CP_2 = [(s')^2 + (h - x)^2]^{1/2} \approx s' + \frac{(h - x)^2}{2s'}$$

$$\xi^2 + h^2 \approx r^2 \Rightarrow \xi \approx \sqrt{r^2 - h^2} \approx r - \frac{h^2}{2r} \tag{2.1.8}$$

$$AB \approx r - \xi \approx \frac{h^2}{2r}$$

$$BC \approx t - AB = t - \frac{h^2}{2r}$$

Using Eq. (2.1.8) and taking account of the fact that the path BC is in glass, the optical path length along the ray P_1ABCP_2 can be calculated as a function of h, which is essentially a vertical coordinate in the lens plane:

$$\begin{aligned} S(h) &\approx (1)\left[s + \frac{(h - x)^2}{2s}\right] + (1)\left(\frac{h^2}{2r}\right) + n\left(t - \frac{h^2}{2r}\right) + (1)\left[s' + \frac{(h - x)^2}{2s'}\right] \\ &= s + s' + nt + \frac{1}{2}\left(\frac{x^2}{s} + \frac{x'^2}{s'}\right) - h\left(\frac{x}{s} + \frac{x'}{s'}\right) \\ &\quad + \frac{1}{2}h^2\left(\frac{1}{s} + \frac{1}{s'} - \frac{n - 1}{r}\right) \end{aligned}$$

where n is the refractive index of the lens material and x' can be negative as drawn in Fig. 2.2.

To find the value of h that makes $s(h)$ stationary, $ds(h)/dh$ is set equal to zero, yielding

$$\frac{ds(h)}{dh} = -\left(\frac{x}{s} + \frac{x'}{s'}\right) + h\left(\frac{1}{s} + \frac{1}{s'} - \frac{n - 1}{r}\right) = 0$$

This equation can be solved for h by transposing the constant term to the right-hand side and dividing by the coefficient of h; however, this implies only the possibility of *one unique* ray defined as possibly emanating from P_1 and passing through P_2. We must remember that h is essentially the height at which the ray from P_1 to P_2 is incident on the lens. A unique solution to h implies that there is only one ray path by which an object plane point P_1 can be imaged to the image plane point P_2. This hypothesis defies experiment. It is easy to show that a lens can have a portion of it or even almost all of it covered by an opaque material and yet still form a complete image. The effect of covering part of the lens is to diminish the brightness of the image, but not its spatial extent. This implies that there are really an infinite number of paths through the lens that rays emanating from P_1 can take to get to P_2. In other words, real life experience indicates that there cannot be one unique value of h. Therefore, a better idea is to consider making the solution independent of the value

h. This can be accomplished by setting both the constant term and the coefficient of h equal to zero; hence

$$\frac{1}{s} + \frac{1}{s'} = \frac{n-1}{r} \tag{2.1.9}$$

$$\frac{x'}{x} = -\frac{s'}{s} \tag{2.1.10}$$

If these formulas are satisfied, the optical path length S is not a function of h, which means that every ray departing from P_1 through the lens passes through P_2. This means that the point P_1, the *object*, is *imaged* in P_2.

Equation (2.1.9) is a version of the *thin lens formula*. The ratio x'/x is called the *lateral magnification* and Eq. (2.1.10) shows that object and image heights are proportional. Here, these relationships are seen to be consistent with experiment and with Fermat's principle. Later, they will be derived directly in a more general sense, and implications of a negative value for x' will be presented.

Several interesting physical implications arise out of the preceding analysis:

1. Using Eqs. (2.1.9) and (2.1.10) as solutions to $dS(h)/dh$, it is clear that h can be of any value such that the rays pass through the lens.

2. Since for all rays satisfying Eqs. (2.1.9) and (2.1.10), $dS(h)/dh$ is zero, then *all* such paths are of least time; that is, all such paths are of *identical* optical paths S. This arises from the variation of BC with h—for different heights the physical path through the lens varies. The end result is, therefore, that despite the physical path differences all rays from P_1 to P_2 traversing through the lens are of identical optical path. As such, the phase differences for propagation from P_1 to P_2 through the lens are identical. This is explored further mathematically in Section 2.3.2.

2.2 PARAXIAL RAYS

The preceding application of Fermat's principle considers only the *paraxial* region. This is a thin region about the optical axis that is so narrow that to a first approximation square root terms can be set equal to a binomial series involving only the first few terms, as in Eqs. (2.1.8). It is only under such conditions that the previous analysis is valid. Another convenient advantage of using the paraxial region is that sines of angles of incidence and refraction are approximately equal to the angles (in radians) themselves. In this way, many nonlinear relationships can be conveniently approximated with good accuracy by linear ones. At first glance this concept may seem useless, since the region is infinitesimally narrow and therefore of interest only as a limiting case. However, calculations of the performances of optical systems based on paraxial relationships are of great utility, and their simplicity makes them extremely convenient. Further, experiments with field stops and iris diaphragms (as in Fig. 2.3) which limit possible radial distances from the optical axis indicate that paraxialness contributes noticeably to image sharpness and therefore is a desirable characteristic in imaging systems.

Since most imaging systems of practical value form good images, most of the rays originating at an object point must wind up reasonably close to the paraxial ray point in the image plane in order to minimize image blur. Consequently, paraxial approximations are utilized later in order to describe geometrical optics requirements

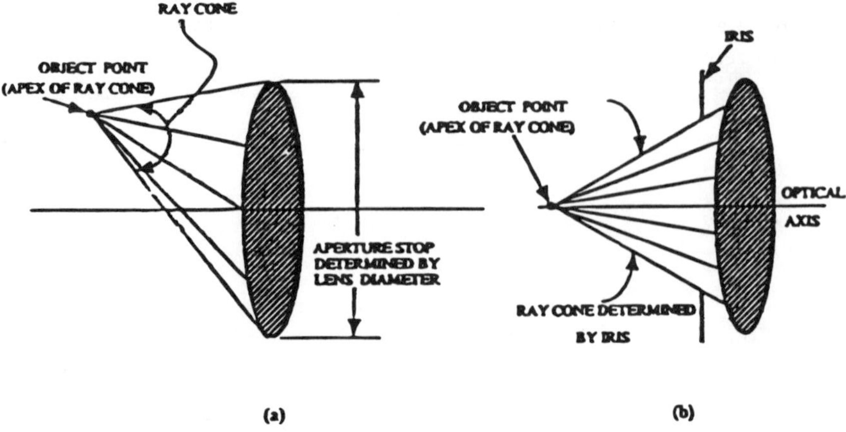

Fig. 2.3 Use of iris or field stop to limit ray cone.

for good image quality. Deviations from paraxialness contribute to aberrations, which are discussed only very briefly at the end of this chapter.

From a geometrical optical point of view a perfect optical system forms a point image of a point object. Graphically, as shown in Fig. 2.4, this implies that two points exist, P_1 and P_2, lying on either size of a lens such that the optical path from P_1 to P_2 is the same by any of the possible routes through the lens. This means that all rays leaving P_1 will intersect at P_2 after passing through the lens. Thus, the geometrical optics interpretation of image formation is that the image point P_2 is located if the point at which any two rays from P_1 can be found.* This implies that in order to attain perfect imaging the optical paths are identical. This is explored further mathematically in Section 2.3.2.

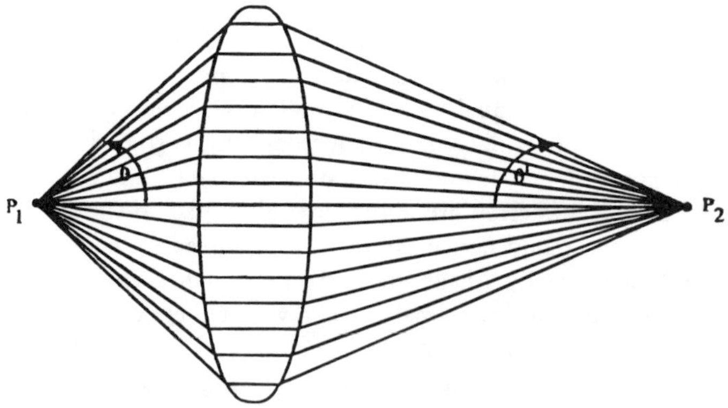

Fig. 2.4 The concept of perfect imaging by a lens: All rays from P_1 cross P_2 after passing through lens.

* In wave optics perfect imaging of the point object means the wave emerging at the output of the lens system must possess spherical wavefronts. This establishes the physical optics definition of a perfect lens as one that converts plane wavefronts to spherical wavefronts. Rays entering such a lens will all be brought to a focus on the other side of the lens.

Paths from an object point to an image point must be independent of the size of the cone (shown in Fig. 2.4) traversed by the rays. This is quite a restriction; yet it is satisfied in most instances by the paraxial ray approximation because here both the height of the object above the optical axis and the angles that the rays make with the optical axis are small. Thus, for example, an optical system of narrow aperture and narrow field of view can accept only paraxial rays emanating from small objects and is therefore capable of perfect imaging as the size of the aperture approaches zero. It is obvious that this type of system is not dealt with in real life; frequently one has to consider systems where the object is of small linear dimensions but is sufficiently close to an optical system with relatively large aperture so that the paraxial ray picture is certainly not applicable. The conditions under which such objects can be imaged accurately are stated in the form of two laws: *Abbe's law* for small objects perpendicular to the optical axis and *Herschel's law* for small linear objects colinear with the optical axis.

If x and x' are object and image heights, respectively, and if θ and θ are ray cone angles as in Fig. 2.4 for the object and image planes respectively, then Abbe's law [2.1, 2.2] states that

$$|nx \sin\theta| = |n''x' \sin\theta'| \tag{2.2.1}$$

where n and n'' are refractive indices in object and image space, respectively, and x' is negative if the image is inverted as in Fig. 2.2. This is also known as the *Sine theorem*. For paraxial rays, this reduces to

$$\left|\frac{x'}{x}\right| = \left|\frac{n\theta}{n''\theta'}\right| \tag{2.2.2}$$

which is referred to as the *Lagrange theorem*. In order for nonparaxial ray image sharpness to be comparable to that for paraxial rays, the two above relationships require

$$\frac{\sin\theta}{\sin\theta'} = \frac{\theta}{\theta'} = \text{constant} \tag{2.2.3}$$

otherwise known as the *sine condition*. Relationship (2.2.2) relates object plane parameters to image plane parameters. Note that if $n = n''$, $|\theta/\theta'| = |s'/s|$ from Eq. (2.1.10). The sine theorem, therefore, implies that the magnitude of lateral magnification, defined in Eq. (2.1.10), is equal to $ns'/n''s$ if the object and image planes are in different media. The sign of the magnification is negative if the image is inverted, as in Fig. 2.2. This is shown in Section 2.6.2. These products comprise system invariants, that is, they are constants. If one changes θ, for example, by varying object distance, then x' and θ' will change too so as to cause each product in Eq. (2.2.1) to remain constant.

Abbe's law plays a major role in the theory of optical instruments in that it must be satisfied for accurate imaging of small linear objects normal to the optical axis by systems of large aperture where the paraxial methods of geometrical optics are not applicable. It is also used to derive the radiance or brightness theorem in Chapter 3.

Herschel's law pertains to a linear element oriented along the optical axis, and states that for object and image plane elements dz and dz' parallel to the optical axis

$$\left|\frac{dz'}{dz}\right| = \frac{n \sin^2\theta/2}{n'' \sin^2\theta'/2}$$

If Eq. (2.2.4) is satisfied, then dz' is a perfect image of dz. For paraxial rays,

$$\left|\frac{dz'}{dz}\right| \approx \frac{n\theta^2}{n''(\theta')^2} = \frac{n}{n''}\left(\frac{s'}{s}\right)^2 \tag{2.2.4}$$

The difference between Eq. (2.2.2) involving lateral magnification (x'/x) and Eq. (2.2.4) involving axial or longitudinal magnification suggests that the magnitude of axial magnification is equal to that of the square of lateral magnification. This is proven in Section 2.6.3.

The Helmholtz law describes a relation among the transverse (or lateral) magnification and the indexes of refraction n and n''. It is expressed as

$$\left(\frac{x'}{x}\right)\left(\frac{\tan\theta'}{\tan\theta}\right) = -\frac{n}{n''} \tag{2.2.5}$$

and is to be proved in Exercise 2.8. Written in the form

$$H = nx\tan\theta = -n''x'\tan\theta' \tag{2.2.6}$$

where the minus sign indicates image inversion, the quantity H represents also an ***invariant*** of the system. This is an important fact because comparing Eq. (2.2.6) with Herschel's law [Eq. (2.2.4)] and Abbè's law [Eq. (2.2.1)], we see that except with in the paraxial region where the sines and tangents may be equated, these laws are contradictory, as are the invariants, Eqs. (2.2.1) and (2.2.6). But Abbe's law must be satisfied in order to achieve a perfect image of a single small line element with a wide aperture system. Since this law contradicts the Helmholtz equation and the two invariants are not equal, no optical system can be perfect. Aberrations deriving from nonparaxial rays are discussed later. The most that can be hoped for is geometrically perfect imaging for one point on the line element. However, this is an academic point for, in practice, the imperfections can be made small.

Note that for paraxial rays Eq. (2.2.5) reduces to Eq. (2.2.2), both of which describe lateral magnification.

2.3 THE THIN LENS

In our study of Fig. 2.2, the concept of a thin lens was introduced. This concept is developed further here because it is useful in a practical sense and because it helps introduce some additional terminology of geometrical optics.

Generally speaking, a lens is an optical component capable of image formation. As seen from the preceding discussion this capability stems basically from the ability of a lens to refract paraxial rays coming from a point of an object on one side of the lens in such a fashion as to make them pass through a common point on the other side of the lens. This common point is called the *focal point*, as shown in Fig. 2.5(a). A *thin lens* is a lens whose thickness is such that a ray traversing it does not change its distance from the optical axis appreciably. This, of course, is equivalent to saying that the lens thickness is small compared to other dimensions of the optical system, such as radii of curvature of the lens surfaces and object and image distances.

The characteristics of a thin lens are determined primarily by the material from which it is made and the shapes of its surfaces. The material properties are explained by a knowledge of the material's refractive index, as explained in Chapter 1. However, little has been said so far about the shape of a lens surface. Recall, for example, that in the example of Fig. 2.1 the surface shape was not discussed in any real detail. Let us examine this parameter now.

2.3.1 The Shape of a Thin Lens Surface for Perfect Focusing

What should be the shape of a thin lens surface in order to really focus incident parallel rays into a single focal point as in Fig. 2.5(a)? To answer this question, consider a paraxial ray incident on a thin lens of refractive index n_2 immersed in a

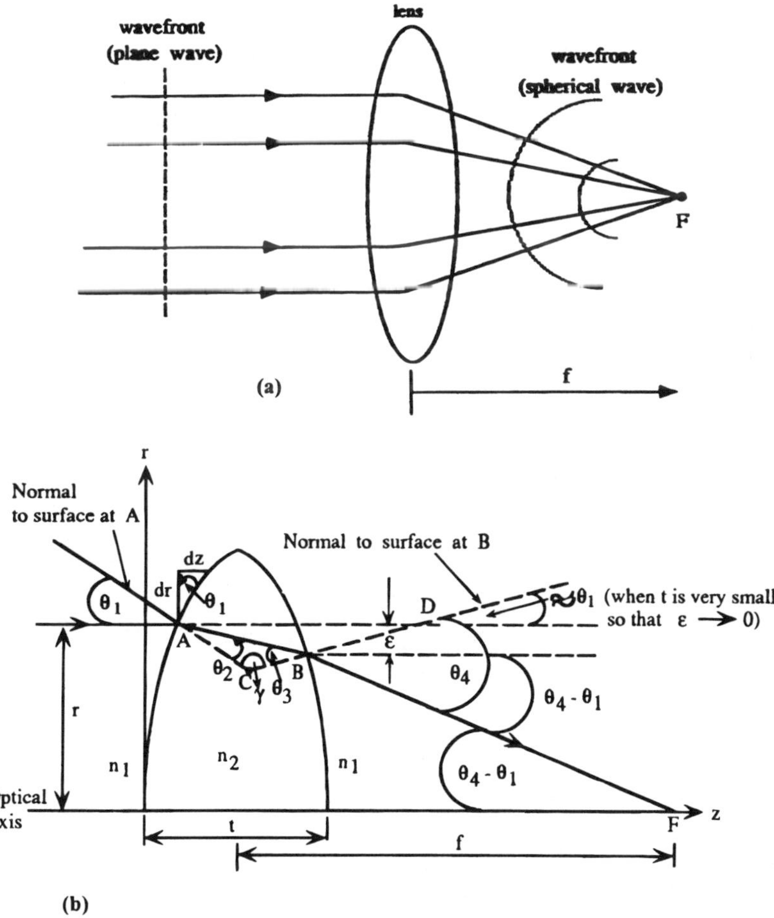

Fig. 2.5 (a) Focusing action of a lens. Plane wave input is converted to spherical wave output. (b) Geometry for determining ideal lens surface.

medium of refractive index n_1 as shown in Fig. 2.5(b). In Fig. 2.5(b) note that when $t \to 0$, the normals to the lens surfaces become essentially parallel so that the output normal is approximately θ_1 with regard to the z axis.

From Fig. 2.5(b), using Snell's law,

$$\frac{\sin\theta_2}{\sin\theta_1} = \frac{n_1}{n_2} = \frac{\sin\theta_3}{\sin\theta_4} \tag{2.3.1}$$

From triangle ABC,

$$\theta_3 = \pi - \gamma - \theta_2$$

and from triangle ADC,

$$\gamma \approx \pi - 2\theta_1$$

Thus,

$$\theta_3 \approx \pi - (\pi - 2\theta_1) - \theta_2 = 2\theta_1 - \theta_2 \tag{2.3.2}$$

We assume that the emerging ray will cross the axis at $z = f >> t$, where f is the *focal length* given by

$$\frac{1}{f} \approx \frac{\tan(\theta_4 - \theta_1)}{x} \tag{2.3.3}$$

Now, since paraxial rays are assumed, the angles in Eqs. (2.3.1) and (2.3.2) are small enough to yield the approximation

$$\frac{\theta_2}{\theta_1} \approx \frac{n_1}{n_2} \approx \frac{\theta_3}{\theta_4} \tag{2.3.4}$$

and

$$\frac{1}{f} \approx \frac{\theta_4 - \theta_1}{r} \tag{2.3.5}$$

From Eqs. (2.3.4) and (2.3.2)

$$\theta_4 \approx \frac{n_2}{n_1}\theta_3 \approx \frac{n_2}{n_1}(2\theta_1 - \theta_2)$$

Substitution in Eq. (2.3.5) yields

$$\frac{1}{f} \approx \frac{1}{r}\left[\frac{n_2}{n_1}(2\theta_1 - \theta_2) - \theta_1\right] = \frac{1}{r}\left[\frac{n_2}{n_1}\theta_1\left(2 - \frac{\theta_2}{\theta_1} - \frac{n_1}{n_2}\right)\right]$$

But by Eq. (2.3.4), $\theta_2/\theta_1 \approx n_1/n_2$, so that

$$\frac{1}{f} \approx \frac{1}{r}\left[\frac{n_2}{n_1}\theta_1 2\left(1 - \frac{n_1}{n_2}\right)\right]$$

$$\approx \frac{2\theta_1}{r}\left(\frac{n_2}{n_1} - 1\right) \tag{2.3.6}$$

To satisfy the definition of "focus" (i.e., parallel rays incident from the left in Figs. 2.5 all pass through focal point F after passing through the lens regardless of the height r at which they enter the lens) the lens surface has to be shaped such that the angle of incidence θ_1 in Eq. (2.3.6) is proportional to r in such a way that f is independent of r. This dependence of θ_1 on r is obtained readily by observing from Fig. 2.5(b) that θ_1 is the slope of the lens surface dz/dr. Thus, from Eq. (2.3.6),

$$\theta_1 \approx \frac{r}{2f(n_2/n_1 - 1)}$$

$$\therefore \quad z(r) = \int \theta_1\, dr = \int_0^r \frac{r\, dr}{2f\left(\frac{n_2}{n_1} - 1\right)} = \frac{r^2}{4f\left(\frac{n_2}{n_1} - 1\right)} \tag{2.3.7}$$

In view of the axial symmetry of the system, Eq. (2.3.7) describes the surface as a paraboloid of revolution.

Equation (2.3.7) indicates that the lens surface required for perfect focusing as a result of refraction is parabolic. When diffraction effects are considered in Part 3, parallel rays incident at lens or aperture edges are seen to be diffracted to points other than the focal point, so another requirement will be that we have to increase the lens diameter by a large amount so as to minimize the relative amount of radiant power incident at lens edges and thus minimize the relative amount of diffraction. However, that would permit nonparaxial rays to enter the image plane. As a result, perfect focusing and imaging are really not possible. A parabolic surface, nevertheless, yields optimum results for a given lens diameter, and these are the types of surfaces utilized for high resolution and image quality requirements, as in astronomy, remote sensing, reconnaissance, etc. Although our analysis here is carried out for a lens, the same result also holds for imaging via mirrors. A parabolic-shaped mirror provides perfect focusing, as shown in Fig. 2.6, if diffraction effects at the mirror edges are neglected.

The geometric optics approach in Fig. 2.6 also explains why parabolic antennas yield maximum directivity or *gain* in millimeter wave or microwave systems. The larger the ratio of aperture size to wavelength, the less the relative intensity of the diffraction effects arising at the aperture edges. Consequently, for relatively large parabolic antennas the gain is very high. For receiving antennas, in accordance with the arrow direction in Fig. 2.6, the high directivity is manifested as narrow field of view. For transmitting antennas, the arrow directions in Fig. 2.6 are reversed, and the high gain is manifested as a narrow transmitted beamwidth.

The same geometrical optics applications apply also to optical wavelengths. Parabolic-shaped mirrors or lenses provide the narrowest beamwidths in transmitters

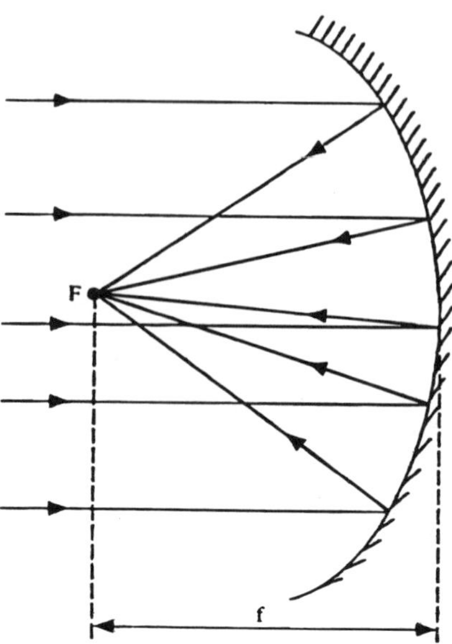

Fig. 2.6 Focusing action of an ideal parabolic mirror where rays incident on mirror edges are not considered.

and narrowest fields of view in receivers. At optical wavelengths, because the wavelengths are small, it is easier to increase the ratio of aperture size to wavelength so as to minimize the relative nonparaxial ray intensity deriving from diffraction at aperture edges.

2.3.2 Phase Implications

Focusing effects cause ray intensities to add at the focal point F in Figs. 2.5 and 2.6. However, from the wave nature of such effects, it is clear that such addition depends on whether or not the various rays intersecting at F are all in phase. For incident plane waves, as in Figs. 2.5 and 2.6, the wavefronts prior to incidence on the lens or mirror are planes perpendicular to the direction of propagation, in accordance with the derivations of the Eikonal and ray equations. However, the lens or mirror converts them to spherical waves converging to the focal point F. The question to be addressed here is whether a parabolic surface acts on incident parallel rays to focus them so that they will all undergo identical phase change, thereby giving rise to intensity addition at the focal point. To consider this question, assume an incident ray parallel to the optical axis at an arbitrary height r and refracted by lens action to pass through F as shown in Fig. 2.7.

The total phase shift ϕ of such a wave is, for wave propagation constant $\beta_0 = \omega/c$,

$$\phi = \frac{\omega}{c}\,[\text{optical path}]$$

Accordingly, for propagation from the r axis to F in Fig. 2.7, the total phase shift of the ray considered is for a thin lens approximately

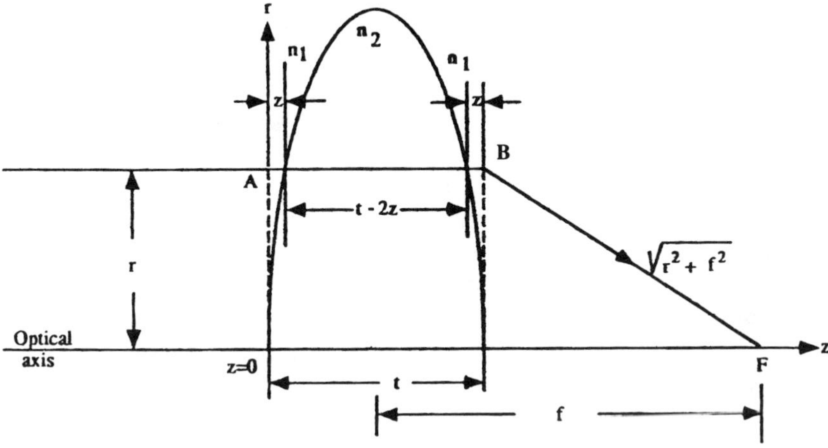

Fig. 2.7 Geometry for considering phase change for a plane wave through a parabolic lens.

$$\phi \approx \frac{\omega}{c}\,[2n_1 z + n_2(t - 2z) + n_1(f^2 + r^2)^{1/2}]$$

and, since for paraxial rays incident on a thin lens, $r^2 << f^2$,

$$\phi \approx \frac{\omega}{c}\left[2n_1 z + n_2(t - 2z) + n_1 f\left(1 + \frac{r^2}{2f^2}\right)\right]$$

$$\approx \frac{\omega}{c}\left[n_2 t + 2z(n_1 - n_2) + n_1 f\left(1 + \frac{r^2}{2f^2}\right)\right]$$

However, the surface $z(r)$ is assumed to be described by Eq. (2.3.7); hence,

$$\phi \approx \frac{\omega}{c}\left[n_2 t + \frac{2r^2 n_1(n_1 - n_2)}{4f(n_2 - n_1)} + n_1 f\left(1 + \frac{r^2}{2f^2}\right)\right]$$

$$\approx \frac{\omega}{c}\left(n_2 t - \frac{n_1 r^2}{2f} + n_1 f + \frac{n_1 r^2}{2f}\right) \tag{2.3.8}$$

$$\approx \frac{\omega}{c}\,(n_2 t + n_1 f)$$

which is independent of r. Therefore, no matter how high off the axis they strike the lens surface (as long as they lie within the paraxial region), all rays arrive at F with the same phase shift ϕ. This analysis, of course, uses the fact that the difference between the actual path and the approximated path is independent of r; therefore, the same error is introduced in the phase shift of Eq. (2.3.8) regardless of the value of r. Also, when one assumes the rays all arrive at the focal point with the same phase shift ϕ, an unreal situation on the lens system is being imposed because infinite energy density is implied by such a constraint. The fact is, however, because of fundamental reasons arising from diffraction as seen in Part 3, a geometrical point object cannot be imaged as a perfect point; at best it can form a pattern of finite size.

Result (2.3.8) indicates that all rays intersecting at F in Fig. 2.5(a) involve an identical optical path from the input vertex plane to the focal point. The same applies to Fig. 2.6.

It is useful to point out that an incident plane wave is equivalent to a spherical wave emanating from a point source a very large distance away, as shown in Fig. 1.1. Therefore it is not surprising that the image is a point image at F according to geometrical optics.

2.3.3 Spherical Surfaces

Although it has been shown that the shape of the lens surface required for perfect focusing is theoretically a paraboloid of revolution, except for high-resolution requirements in practice most lenses have spherical surfaces because these are easier to grind and polish.

However, the imperfections introduced by such a deviation from the theoretically ideal are small when the incident rays are close to the optical axis. To get a feeling for how good this approximation is, consider a paraxial ray impinging on a single spherical surface separating two materials of different refractive indices as shown in Fig. 2.8.

For the coordinate system shown in Fig. 2.8, the equation for the spherical surface is

$$r^2 + (R - z)^2 = R^2$$

or

$$r^2 + R^2\left(1 - \frac{z}{R}\right)^2 = R^2 \tag{2.3.9}$$

When the rays are close to the optical axis, $z << R$ and Eq. (2.3.9) becomes

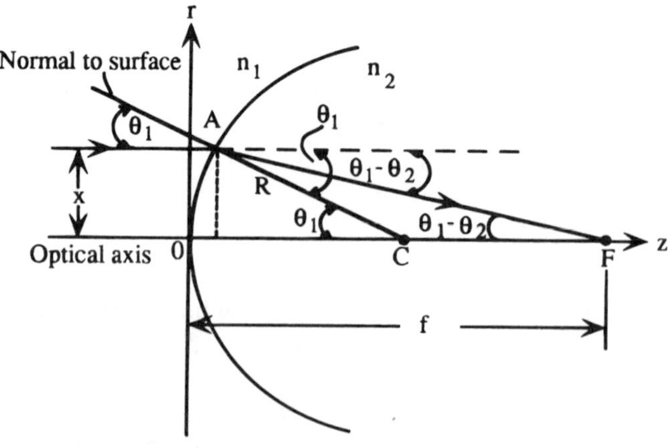

Fig. 2.8 Paraxial rays incident on a spherical surface.

$$r^2 + R^2\left(1 - \frac{2z}{R}\right) \approx R^2$$

or

$$r^2 + R^2 - 2zR \approx R^2$$

so that

$$z(r) \approx \frac{r^2}{2R} \qquad \text{for } z << R \tag{2.3.10}$$

Now in the triangle ACF, the law of sines yields

$$\frac{R}{\sin(\theta_1 - \theta_2)} = \frac{f - R}{\sin \theta_2}$$

For rays close to the optical axis, θ_1 and θ_2 are small, so that

$$\frac{R}{\sin(\theta_1 - \theta_2)} \approx \frac{f - R}{\sin \theta_2}$$

or

$$\theta_2 R \approx f\theta_1 - f\theta_2 - R\theta_1 + R\theta_2$$

or

$$R \approx \frac{f(\theta_1 - \theta_2)}{\theta_1} \approx f\left(1 - \frac{\theta_2}{\theta_1}\right) \tag{2.3.11}$$

But by Snell's law,

$$\frac{\theta_2}{\theta_1} \approx \frac{n_1}{n_2}$$

so that Eq. (2.3.11) becomes

$$R \approx f\left(1 - \frac{n_1}{n_2}\right) \tag{2.3.12}$$

Combining Eqs. (2.3.10) and (2.3.12), we obtain

$$z(r) \approx \frac{r^2}{2f\left(1 - \dfrac{n_1}{n_2}\right)} \tag{2.3.13}$$

which is the equation of the spherical surface of Fig. 2.8 in the region close to the optical axis. Approximation (2.3.13) describes a paraboloid of revolution; hence, over a small arc, a spherical surface of small curvature of large radius R closely approximates a paraboloidal surface. Most common lenses are made with spherical surfaces because they are easier and less expensive than parabolic surfaces; however, for applications requiring high resolution, parabolic surfaces are used.

2.4 DESCRIPTION OF A THIN LENS

We have seen from the preceding discussion that for both theoretical and practical reasons thin lens surfaces are made spherical in shape. Most of them are also circular in shape across their front and back surfaces and are usually set in an optical system with their centers along a common optical axis. Because of these features, most optical systems end up consisting wholly of coaxial surfaces of revolution so that only their cross-sectional view in a plane through the optical axis need be considered. This plane is called a *meridional plane*; a view of such a plane for a single thin lens system is shown in Fig. 2.9. (Thick lenses are discussed in Section 2.8.) Rays that lie in a meridional plane are called *meridional rays*; those that do not are called *skew rays*. Any meridional ray can be defined at a plane normal to the optical axis by two numbers: (1) its height above the optical axis and (2) the angle it makes with the optical axis.

The parameters indicated in Fig. 2.9 are those commonly used to describe a thin lens optical system. Note that *the radii of curvature of the front and back surfaces of the lens are often not the same.*

Associated with each of these surfaces are certain points and distances marked off along the optical axis in Fig. 2.9. These are usually defined as follows:

C_1 = center of curvature of front surface
C_2 = center of curvature of back surface
V_1 = point at which front surface intersects the optical axis; called the *vertex* of the front surface
V_2 = point at which back surface intersects the optical axis; called the *vertex* of the back surface
M = location of a point of the object on the optical axis; called the *object point*
M' = location of a point of the image on the optical axis caused by front surface alone; called the *image point of front* surface; serves also as a *virtual* (imaginary) object for the back surface

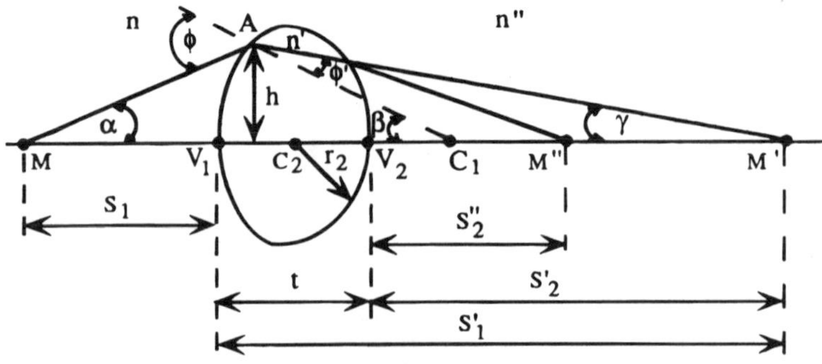

Fig. 2.9 Lens parameters.

M'' = location of a point of the image on the optical axis resulting from refraction by both front *and* back surfaces; called the *image point* of a thin lens
s_1 = distance between object point M and front surface vertex V_1; called the *object distance*
s'_1 = distance between front surface image point M' and front surface vertex V_1; called the *front surface image distance*
s'_2 = distance between front surface image point M' and back surface vertex V_2; called the *back surface object distance*
s''_2 = distance between lens image point M'' and back surface vertex V_2; called the lens *image distance*
F_1 = point on optical axis with property that any ray coming from it or proceeding toward it travels parallel to the optical axis after refraction by the front surface; called the *front surface first focal point*
F'_1 = point on optical axis with property that any ray incident on front surface traveling parallel to the optical axis prior to refraction by the front surface proceeds toward or appears to emanate from it after refraction; called the *front surface second focal point*
F'_2 = same as F_1 defined for back surface
F''_2 = same as F'_1 defined for back surface
f_1 = distance between F_1 and V_1; called *front surface first focal length*
f'_1 = distance between F'_1 and V_1; called *front surface second focal length*
f'_2 = distance between F'_2 and V_2; called *back surface first focal length*
f''_2 = distance between F''_2 and V_2; called *back surface second focal length*.

Planes passing through focal points are called *focal planes*; those passing through image points are called *image planes*, and a plane through the object point is called the *object plane*. Focal points and lengths are to be considered shortly.

Now, in order to develop some analytical and graphical descriptions for the formation of an image by a thin lens, a convention for determining the signs of the distances just defined must be established. This convention, of course, is arbitrary. The sign convention usually used is the following:

1. Unless otherwise stated, a ray of light always propagates from left to right through an optical system.
2. Object and image dimensions are positive when measured upward from the axis and negative when measure downward.
3. All object distances (s) are considered positive when measured to the left of the vertex and negative when measured from the right; all image distances (s') are positive when measured to the right of the vertex and negative when to the left.
4. The *vertex* of a refracting surface is defined as its intersection point with the optical axis (i.e., the axis of symmetry) of the system.
5. A radius of curvature is positive if the direction from the vertex to the center of curvature is from left to right.

With the above definitions and sign conventions in mind and continuing the assumption of paraxial rays, then from Fig. 2.9 and Snell's law,

$$\frac{\phi}{\phi'} = \frac{n'}{n} \tag{2.4.1a}$$

where, as a result of external angles to triangles MAC_1 and C_1AM', respectively,

$$\begin{aligned} \phi &= \alpha + \beta \\ \beta &= \phi' + \gamma \end{aligned} \tag{2.4.1b}$$

Substitution into Eq. (2.4.1a) yields

$$n\alpha + n'\gamma = (n' - n)\beta \tag{2.4.2}$$

For paraxial rays, α, β, and γ are small so that

$$\alpha \approx \frac{h}{s_1} \qquad \beta \approx \frac{h}{r_1} \qquad \gamma \approx \frac{h}{s_1'} \tag{2.4.3}$$

where h is the height above the optical axis at which the ray strikes the front surface of the lens and r_1 is the radius of this surface. Substituting this into Eq. (2.4.2) yields

$$\frac{n}{s_1} + \frac{n'}{s_1'} = \frac{(n' - n)}{r_1} \tag{2.4.4}$$

With $n = 1$, this equation is similar to Eq. (2.1.9) derived from Fermat's principle for the front surface rather than from Snell's law as is done here.

Equation (2.4.4) describes what happens to a paraxial ray as it is refracted at the *front* surface of the lens in Fig. 2.9. When the ray strikes the *back* surface it is refracted once again because $n' \neq n''$. To describe what happens at this second surface, the output or image deriving from the first surface can be considered as the input or *virtual* object for the second surface. Therefore, the distance s_2' is now the object distance for the second surface whereas s_2'' is the image distance for the second surface and for the lens as a whole. Thus, Eq. (2.4.4) if applied to the second surface yields

$$\frac{n'}{s_2'} + \frac{n''}{s_2''} = \frac{n'' - n'}{r_2} \tag{2.4.5}$$

where the sign of r_2 depends on whether the back surface is convex (r_2 negative) or concave (r_2 positive) according to the sign convention. Note that since AM' is a straight line, the virtual object plane medium is seen to be of refractive index n' as if refraction at the back surface did not take place. Since AM'' involves refraction at the back surface and is therefore not a straight line, the image plane refractive index in Eq. (2.4.5) is n''.

For a thin lens, $t << s_1, s_1', s_2', s_2''$. In this case

$$s_1' \approx -s_2'$$

Consequently, Eq. (2.4.5) becomes

$$\frac{n'}{-s_1'} + \frac{n''}{s_2''} = \left(\frac{n' - n}{r_1} + \frac{n'' - n'}{r_2}\right) \tag{2.4.6}$$

Adding Eqs. (2.4.4) and (2.4.6) yields

$$\frac{n}{s_1} + \frac{n''}{s_2''} = \left(\frac{n' - n}{r_1} + \frac{n'' - n'}{r_2}\right) \tag{2.4.7}$$

which is valid for paraxial rays in the thin lens system of Fig. 2.9. Now s_1 is the object distance not only for the front surface but also for the lens as a whole, whereas s_2'' is the image distance not only for the back surface but also for the lens as a whole. This implies that

$$s_1 \triangleq s$$

$$s_2'' \triangleq s'' \tag{2.4.8}$$

Using Eqs. (2.4.8) in Eq. (2.4.7) results in

$$\frac{n}{s} + \frac{n''}{s''} = \left(\frac{n' - n}{r_1} + \frac{n'' - n'}{r_2}\right) \tag{2.4.9}$$

The parameters s and s'' are named the *object* and *image distances*, respectively, of a thin lens.

Equation (2.4.9) is a general image formation formula for a thin lens of refractive index n' having surfaces of different radii of curvature separating two materials of refractive indexes n and n''.

As we shall see later, the quantities $(n' - n)/r_1$ and $(n'' - n')/r_2$ are of importance in their own right. They are called the *powers* of the front and back surfaces, respectively. The geometry for Eq. (2.4.9) is illustrated in Fig. 2.10.

In Fig. 2.10,

F $\triangleq$ front (or primary) focal point

F″ $\triangleq$ back (or secondary) focal point

f $\triangleq$ front (or primary) focal length

f″ $\triangleq$ back (or secondary) focal length

These definitions imply that the front focal point F in Fig. 2.10 is an axial point with the property that any ray coming from it [as in Fig. 2.10(a)] or proceeding toward it [as in Fig. 2.10(b)] travels parallel to the optical axis after refraction by both surfaces of the lens. Examples are shown in Figs. 2.10(a) and (b) for a typical "converging" lens and a typical "diverging" lens where the radii of curvature of the front and back surfaces (i.e., r_1 and r_2 in Fig. 2.9) are assumed equal. The *back focal point* F′ is an axial point with the property that any incident ray travelling parallel

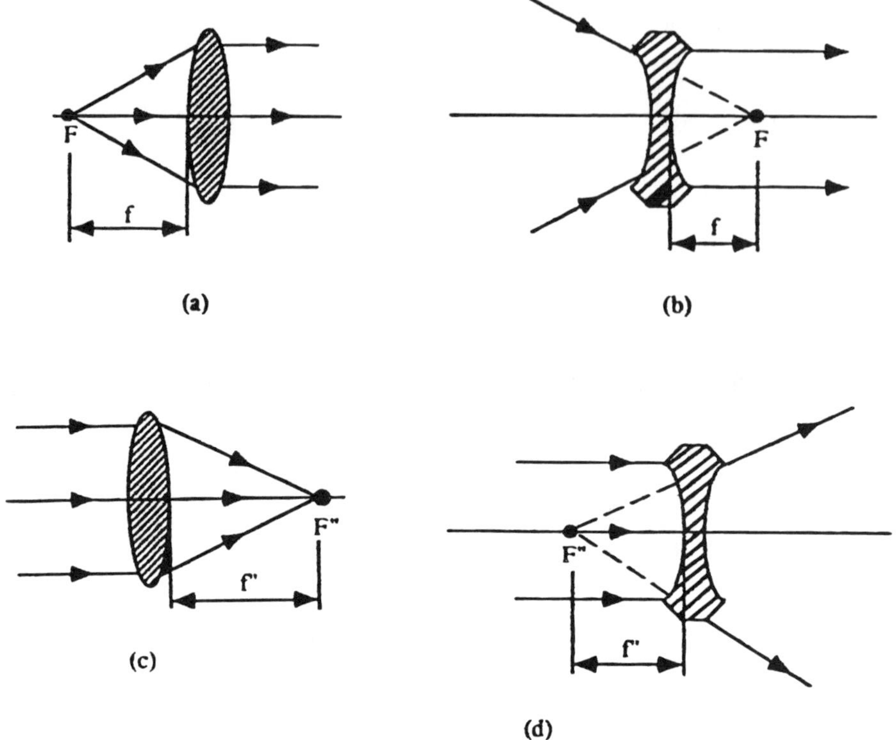

Fig. 2.10 Front and back focal points for convex (positive) lens [parts (a) and (c)] and concave (negative) lens [parts (b) and (d)].

to the optical axis prior to refraction by the lens proceeds toward or appears to emanate from it after refraction. Examples are shown in Figs. 2.10(c) and (d). A plane passing through F normal to the optical axis is called the *front focal plane*; the plane passing through F' is the *back focal plane*.

The *front* and *back focal lengths* f and f' are defined as shown in Fig. 2.10. Note that for a thin lens they are measured to the *center* of the lens. Since f in Fig. 2.10(a) is measured to the left of the front surface vertex, it is a positive quantity according to our sign convention, but f' in Fig. 2.10(b) is negative because it is measured to the right of the vertex.

According to the definition of F an expression for the front focal length f can be developed by setting $s'' \to \infty$ and $s \to f$ in Eq. (2.4.9), yielding

$$\frac{n}{f} + \frac{n''}{\infty} = \left(\frac{n' - n}{r_1} + \frac{n'' - n'}{r_2}\right)$$

To find f' set $s \to \infty$ and $s'' \to f''$ in Eq. (2.4.9) to obtain

$$\frac{n}{\infty} + \frac{n''}{f''} = \left(\frac{n' - n}{r_1} + \frac{n'' - n'}{r_2}\right)$$

Taken together, these two expressions determine front and back focal lengths as follows:

$$\frac{1}{f} = \frac{1}{n}\left[\frac{n' - n}{r_1} + \frac{n'' - n'}{r_2}\right] \tag{2.4.10}$$

$$\frac{1}{f''} = \frac{1}{n''}\left[\frac{n' - n}{r_1} + \frac{n'' - n'}{r_2}\right] \tag{2.4.11}$$

These are related to each other such that

$$\frac{f}{f''} = \frac{n}{n''} \tag{2.4.12}$$

Thus, even for asymmetrical lenses, primary and secondary focal lengths are equal if object and image plane media are identical. If the lens is immersed in one medium only, then $n = n''$ in Fig. 2.9 and Eq. (2.4.9) becomes

$$\frac{n}{s} + \frac{n}{s''} = (n' - 1)\left(\frac{1}{r_1} - \frac{1}{r_2}\right) \tag{2.4.13}$$

If the lens is immersed in air, then $n \approx 1$ and consequently

$$\frac{1}{s} + \frac{1}{s''} = (n' - 1)\left(\frac{1}{r_1} - \frac{1}{r_2}\right) = \frac{1}{f} \tag{2.4.14}$$

where the relation to focal length stems from Eqs. (2.4.10) and (2.4.11). This is the well-known *lens makers'* formula and is the same as Eq. (2.1.9) for $r_2 \to \infty$ (as it does in Fig. 2.2). Equation (2.4.14) is used by lens makers to grind lenses to required focal length.

Lens focal length is seen to derive from lens physical characteristics, that is, material (n') and radii of curvature (r_1, r_2).

Example 2.2

Consider a thin lens with one flat surface as shown in Fig. 2.2. Calculate the radii of curvature of the grinding and polishing tools that must be used to make this lens if the glass has $n' = 1.5$ and the desired focal length is 30 cm.

Solution

For the lens of Fig. 2.2, $r_2 \to \infty$. (Such a lens is called *planoconvex*; thin lenses with $|r_1| = |r_2|$ are called *equiconvex* if r_1 is positive and r_2 is negative, or *equiconcave* if r_1 is negative and r_2 is positive.) Thus, with $s = f$, Eq. (2.4.14) becomes

$$\frac{1}{f} + \frac{1}{\infty} = (n - 1)\left(\frac{1}{r_1} - \frac{1}{\infty}\right)$$

$$\frac{1}{30} = (1.5 - 1)\left(\frac{1}{r_1}\right)$$

$$r_1 = (30)(0.5) = 15 \text{ cm}$$

If the lens is reversed, then $r_1 = \infty$ and

$$\frac{1}{f} + \frac{1}{\infty} = (n - 1)\left(\frac{1}{\infty} - \frac{1}{r_2}\right)$$

$$\frac{1}{30} = (1.5 - 1)\left(-\frac{1}{r_1}\right)$$

$$r_1 = -30\ (0.5) = -15 \text{ cm}$$

According to sign convention, the second surface is now convex and the same positive focal length still results.

Example 2.3

Use the physical implication of refractive index to derive an equation relating object and image distances appropriate for a mirror rather than a lens. Derive an expression for mirror focal length according to the radius of mirror curvature.

Solution

A mirror involves only a single surface since essentially light does not pass through it. Accordingly, Eq. (2.4.4) can be used. For air, $n = 1$. As discussed in the development of Eq. (2.1.6), light reflected from a mirror is of velocity in a direction negative with regard to that incident on a mirror. The reflected light is also in air. Therefore, $n' = -1$. Substituting into Eq. (2.4.4)

$$\frac{1}{s_1} - \frac{1}{s_1'} = \frac{-1 - 1}{r_1} = -\frac{2}{r_1}$$

The subscripts can be dropped since there is only one surface. Because of the reflection properties of a mirror, image and objects planes are both on the same side of a mirror. Therefore, if the object is to the left of the mirror, so too is the image and this must be manifested by changing the sign of s' since rays from the mirror to the image are propagating from right to left. Consequently,

$$\frac{1}{s} + \frac{1}{s'} = -\frac{2}{r}$$

and, for mirrors, sign convention is such that s and s' are positive if they both lie to the left of the vertex.

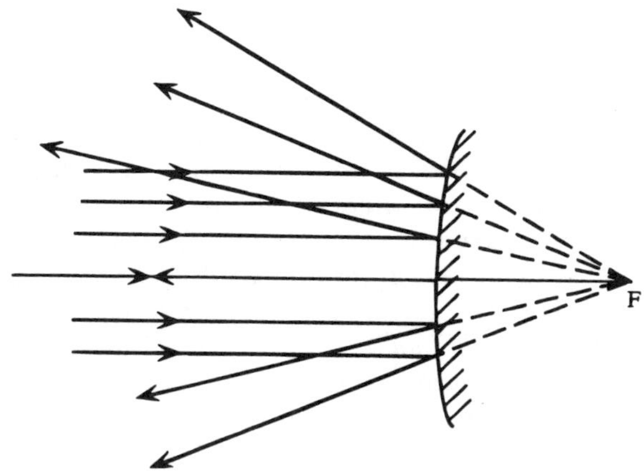

Fig. 2.11 Mirror with negative focal length (to right of mirror instead of to left of it).

Since

$$\frac{1}{s} + \frac{1}{s'} = \frac{1}{f}$$

it is clear that $f = -r/2$ and

$$\frac{1}{s} + \frac{1}{s'} = \frac{1}{f} = -\frac{2}{r} \tag{2.4.15}$$

This equations relates object distance, image distance, and focal length for a concave mirror in Fig. 2.6, for which r is negative according to rule 5 in Section 2.4. If the mirror is convex, and parallel rays are incident on it, then the focal point appears to emanate from the right of the mirror, as in Fig. 2.11. In this case, r is positive and f is negative. The image is virtual since it appears to be from the right of the mirror.

2.5 PHYSICAL EXPLANATION

Consider corresponding object and image plane points M and M', respectively, in Fig. 2.12. Waves propagating from point M toward the lens are spherical waves, as

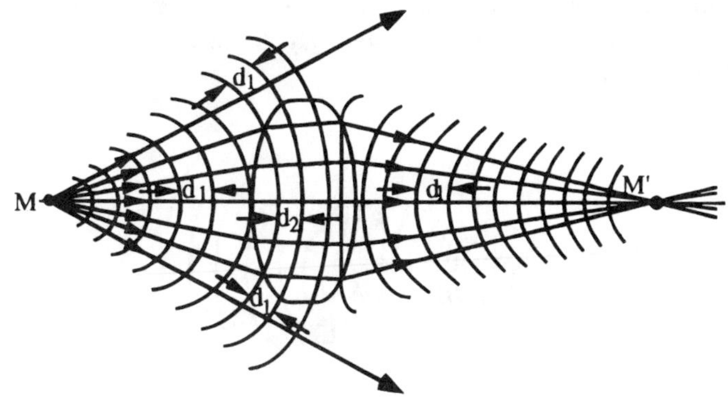

Fig. 2.12 Spherical wave propagation from object to image through a positive lens.

in Fig. 1.1. The object can be decomposed into a series of points such as M. Thus, M can be considered a point source of spherical waves. The actual rays are designated with long arrows in Fig. 2.12 and are perpendicular to the spherical wavefronts. In each interval of time the wavefront in the object medium may be assumed to propagate a distance d_1. The position of the wavefront inside the lens will propagate a smaller distance d_2. As pointed out regarding Eq. (2.3.8), and in Section 2.1.1, the number of possible paths for rays to take when travelling from M to M' is infinite. However, all such permissible paths involve identical phase change, according to Fermat's principle. Therefore in Fig. 2.12 optical path d_1 must equal optical path d_2, or $nd_1 = n'd_2$. Consequently, d_2 equals $(n/n')d_1 = d_1/n'$ for a lens immersed in air and is therefore less than d_1. As the wavefront passes through a positive focal length lens, this effect is repeated in *reverse* at the second surface. The wavefront has been retarded inside the lens since $d_2 < d_1$, and this retardation is greater in the thicker central portion of the lens, causing the curvature of the wavefront to be reversed and converge upon exiting the back surface in Fig. 2.12. To the left of the lens the light from M is diverging. To the right of the lens it is converging. A lens of this type is called a *converging* or *positive* lens. It is convex. A convex lens with refractive index higher than the surrounding media is a converging lens.

A concave lens is thicker at the edges and therefore retards the wavefront more at the edge than at the center and increases the divergence as in Fig. 2.13. After passing through a negative focal length or diverging lens, the wavefront appears, when viewed to the right of the lens, to have originated from M', which is the image of M formed by the lens.

A screen placed at point M' in Fig. 2.12 would cause a concentration of light to be observed there. Therefore, that image is a *real* image. However, a screen placed at M' in Fig. 2.13 would not exhibit a *concentration* of light. All that would be observed would be the general illumination produced by light emanating at M. This type of image is called a *virtual* image. It may be observed directly or it may serve as a source to be reimaged by a subsequent lens, but it cannot be produced on a screen.

In Fig. 2.12, as waves from M strike the vertex they have a radius s and a curvature $1/s$. The latter is consistent with the magnitude of CFA for a spherical wave in Eq. (1.3.40). As the waves leave the lens and converge toward M', they have a radius s' and a curvature $1/s'$. Equations (2.4.4), (2.4.5), and (2.4.7) for each surface and Eq. (2.4.9) for the lens as a whole are known as the *Gaussian formulas*. These, and the lensmakers' formulas, Eqs. (2.4.13) and (2.4.14), may be considered to involve addition and subtraction of quantities proportional to curvatures of spherical surfaces. When these curvatures rather than radii are used the formulas become simpler in form and for some purposes more convenient. It is convenient to introduce at this point the following quantities:

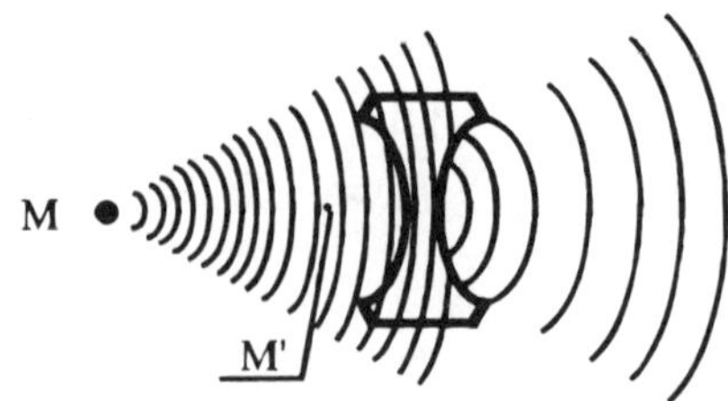

Fig. 2.13 Spherical wave propagation from object to image through a negative lens.

$$V = \frac{n}{s} \quad V' = \frac{n''}{s''}, \quad P_1 = \frac{n' - n}{r_1}, \quad P_2 = \frac{n'' - n'}{r_2} \tag{2.5.1}$$

The first two of these, V and V' are called *reduced vergences* because they are direct measures of the convergence and divergence of the object and image wavefronts, respectively. For a divergent wave from the object s is positive, and so is the vergence V. For a convergent wave, on the other hand, s is negative, and so is its vergence. For a converging wavefront toward the image V' is positive, and for a diverging wavefront V' is negative. Note that in each case the refractive index involved is that of the medium in which the wavefront is located. [Note too that the derivation of the Gaussian formulas are based on Fig. 2.9, which is appropriate for a diverging wave from the object plane and a converging wave to the image plane. For a converging wave from the object plane and/or diverging wave to the image plane angles α and/or γ in Eq. (2.4.3) would be negative. The resulting equations would be consistent with the concepts of positive and negative vergences.] When all distances are measured in meters, the reduced vergences V and V' and the power P are in units called diopters. V may be thought of as the power of the object wavefront just as it touches the front refracting surface and V'' as the power of the corresponding image wavefront that is tangent to the back refracting surface. In these new terms, Eq. (2.4.9) becomes

$$V + V'' = P_1 + P_2 = P \tag{2.5.2}$$

where P is the power of the lens and is equal to the sum of the powers of the two surfaces. For object and image planes in air, $P = 1/f$. Eyeglass prescriptions are given in terms of powers in units of diopters.

Positive focal length or converging lenses exhibit positive powers, whereas negative focal length or diverging lenses exhibit negative powers.

2.6 IMAGE FORMATION BY A THIN LENS VIA RAY TRACING

In our discussions thus far an extended object in an optical system has been treated as though it were made up of individual points, each emanating rays that are collected by the optical system to form into an image. This superposition of images of individual points on the object is valid in most instances because it can be assumed that each object point is an independent radiator of light or, in other words, has little *coherence* with respect to any other point. This is reexamined in Part 3 of this book.

The process of following a ray through a system to determine where it strikes the image plane is called *ray tracing*. Ideally, then, the aim in lens design is to operate on bundles of light rays emanating from some object point to cause them to come together at a single image point and to do this simultaneously for many object points and for light of many wavelengths. However, variations in lens refractive index (or ϵ') with wavelength give rise to different focal points for different wavelengths. This means that the image of a given object point would appear at different locations in the image plane according to wavelength. Such multiple images cause image blur. Thus, deviations or *aberrations* from the perfect image are always evident unless a bundle of *monochromatic* light rays is being traced. Since most optical systems are expected to image with white light, the problem of chromatic aberrations must be dealt with and is considered at the end of this chapter.

With these practical thoughts in mind, we continue to place restrictions on paraxial rays and monochromatic light. With these restrictions ray tracing can be accomplished graphically or analytically by means of a lens formula. In either case, ray tracing is a trigonometric process involving an application of Snell's law at each surface of the optical system.

For convenience, the double prime notation for output of the second surface is dropped since Eq. (2.4.9) makes it unnecessary to consider each surface individually for a thin lens.

Two common techniques for graphically tracing a ray through a thin lens are the *parallel ray method* and the *oblique ray method.* The parallel ray method is illustrated in Fig. 2.14 for a vertical linear object. An image of each object point is formed by drawing two rays based on the definition of focal point: a ray emanating from an object point parallel to the optical axis must pass through the back focal point F' after refraction by the lens; a ray emanating from the same object point and passing through the front focal point F must propagate parallel to the optical axis after refraction. The intersection of these two rays forms an image point in the image plane. Note the ray drawn through the center of the lens. For paraxial rays, if $r_1 = r_2$ this ray passes through two lens surfaces that are approximately plane and parallel to each other. In this case, the ray goes through the lens almost without deviation; such a ray is given the name *chief ray.*

The oblique ray method for paraxial rays and $r_1 = r_2$ is illustrated in Fig. 2.15. Its construction is based on two facts: (1) Parallel rays incident at small angles with respect to the optical axis are focused to a point in the *back focal* plane and (2) a chief ray passes through the lens undeviated. As shown in Fig. 2.15, the chief ray locates a point X in the back focal plane through which a ray (parallel to the chief ray in the object plane) from M must pass after refraction. The continuation of this ray from X locates point M' in the image plane. Similarly, a ray originating from F must proceed parallel to the optical axis upon exiting from the lens. If it is parallel to the chief ray before the lens it must also intersect the chief ray in the back focal plane *since all rays parallel to each other at the lens input must intersect in the back focal plane.* Because these front and parallel rays are oblique to the optical axis, the back focal

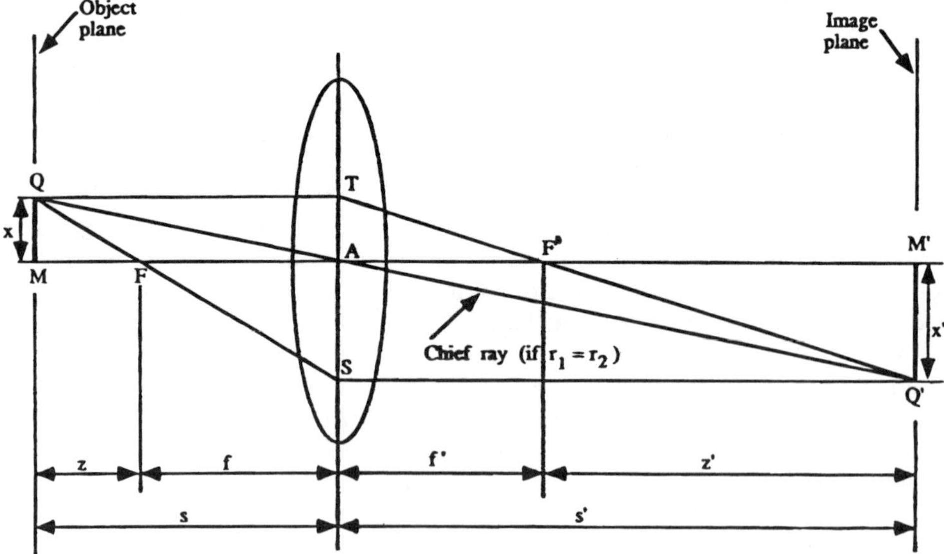

Fig. 2.14 Parallel-ray tracing method.

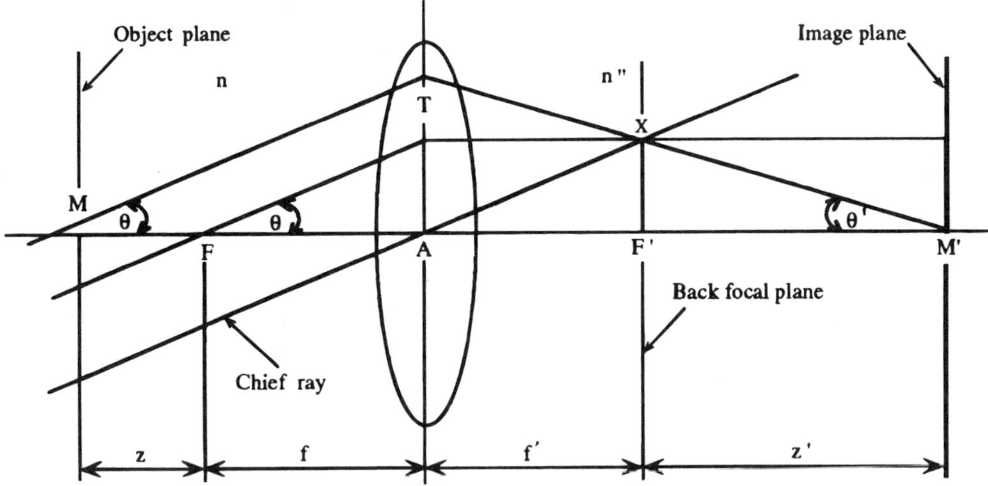

Fig. 2.15 Oblique ray tracing method.

point is off axis. This can be used to prove Helmholtz's law [Eq. (2.2.5)]. See Exercise 2.8.

2.6.1 The Thin Lens Formula

Ray tracing can be carried out analytically by deriving a lens formula based on the graphical procedures previously defined. In general, a lens formula consists of a set of numbers giving the thicknesses and spacings of the various elements and refractive indexes of the materials of which they are made. Thus, for paraxial rays and for the case of thin lenses with $r_1 = r_2$, we can see from the geometry of Fig. 2.14 that

$$\frac{x - x'}{s'} = \frac{x}{f'} \tag{2.6.1}$$

because triangles $Q'TS$ and $F'TA$ are similar. In like manner,

$$\frac{x - x'}{s} = \frac{-x'}{f} \tag{2.6.2}$$

because of the similarity of triangles QTS and FAS. Adding Eqs. (2.6.1) and (2.6.2) yields

$$\frac{x - x'}{s'} + \frac{x - x'}{s} = \frac{x}{f'} - \frac{x'}{f} \tag{2.6.3}$$

But $f = f'$ for a thin lens immersed in a uniform medium so that Eq. (2.6.3) becomes for air

$$\frac{x - x'}{s'} + \frac{x - x'}{s} = \frac{1}{f}(x - x')$$

or

$$\frac{1}{s'} + \frac{1}{s} = \frac{1}{f} \tag{2.6.4}$$

Thus, ray tracing yields the same Gaussian form of the thin lens formula appearing in Eq. (2.4.14).

If similar triangles *QMF* and *FAS* in Fig. 2.14 are considered

$$\frac{x}{z} = \frac{-x'}{f}$$

or

$$z = -xf/x' \tag{2.6.5}$$

In like manner, similar triangles TAF' and $F'M'Q'$ yield

$$\frac{-x'}{z'} = \frac{x}{f'}$$

or

$$z' = -x'f'/x \tag{2.6.6}$$

Multiplication of Eqs. (2.6.5) and (2.6.6) yields, for $n = n''$,

$$zz' = ff' = f^2 \tag{2.6.7}$$

which is known as the *Newtonian form* of the *thin lens* formula.

2.6.2 Lateral Magnification

From similar triangles *QMA* and $Q'M'A$ in Fig. 2.14,

$$\frac{-x'}{x} = \frac{s'}{s} \tag{2.6.8}$$

Since lateral magnification m is defined as x'/x, Eq. (2.6.8) implies

$$m = -\frac{s'}{s} \tag{2.6.9}$$

Another expression for m can be obtained from Eqs. (2.6.5) and (2.6.6):

$$m = \frac{x'}{x} = -\frac{f}{z} = -\frac{z'}{f'} \tag{2.6.10}$$

If m is negative, the image is real and inverted.

Equation (2.6.8) is identical to Eq. (2.1.10) obtained from Fermat's principle. A more general form of lateral magnification is $-ns'/n''s$ as shown following Eq. (2.2.3).

2.6.3 Longitudinal or Axial Magnification

From the Newtonian form of the thin lens formula, Eq. (2.6.7), $zz' = f^2$. Therefore, for small changes along the optical axis $zdz' + z'dz = 0$. Consequently, axial magnification is

$$\frac{dz'}{dz} = -\frac{z'}{z}$$

In view of Eqs. (2.6.5) and (2.6.6),

$$\frac{dz'}{dz} = -\frac{-x'f'/x}{-xf/x'} = -\left(\frac{x'}{x}\right)^2 \frac{f'}{f} = -m^2 \frac{n''}{n} \tag{2.6.11}$$

where Eq. (2.4.12) has been used for f'/f. Note that transversal (lateral) and longitudinal magnifications are different. This is important in three-dimensional imaging. The minus sign is consistent with the z axis. Detail which is "deep" (to the left) in the object domain is also "deep" (to the right) in the image domain, the depth being nonlinear as indicated by the magnitude of the axial magnification. As z (and s) increase for real images, z' (and s') decrease, thus affecting image focus. However, at the same time dz becomes more negative and dz' more positive according to $-m^2$.

Lateral magnification is consistent with Abbé's law and axial magnification with Herschel's law, as described in Section 2.2.

Example 2.4

Two lenses having focal lengths $f_1 = 6$ cm and $f_2 = -9$ cm are placed 4 cm apart. If an object 3 cm high is placed 16 cm in front of the first lens, find (a) the position and (b) the size of the final image.

Solution

The two lenses are analogous to networks in series. Similar to the treatment of two lens surfaces in series, the image or output of the first lens is a *virtual* object or input to the second lens.

(a) For the first lens alone, with s_1' being image distance to it,

$$\frac{1}{s_1} + \frac{1}{s_1'} = \frac{1}{f_1} = \frac{1}{16\text{ cm}} + \frac{1}{s_1'} = \frac{1}{6\text{ cm}}$$

or

$$\frac{1}{s_1'} = \frac{1}{6} - \frac{1}{16} = \frac{10}{96} \rightarrow s_1' = 9.6 \text{ cm}$$

The image produced by the first lens alone is 9.6 cm behind or to the right of it. It is also behind the second lens, because its distance from the first lens is only 4 cm. Nevertheless, this image is a *virtual* object for the second lens and distant from the second lens by $s_2 = 4 \text{ cm} - 9.6 \text{ cm} = -5.6 \text{ cm}$. This object distance is negative because it is behind or to the right of the second lens. The final image position can be determined from

$$\frac{1}{s_2'} = \frac{1}{f_2} - \frac{1}{s_2} = \frac{1}{-9 \text{ cm}} - \frac{1}{-5.6 \text{ cm}}$$

$$s_2' = 14.8 \text{ cm}$$

Since s_2' is positive, the image is 14.8 cm behind or to the right of the second lens.
(b) For the first lens alone lateral magnification is

$$m_1 = -\frac{s_1'}{s_1} = -\left(\frac{9.6 \text{ cm}}{16 \text{ cm}}\right) = -0.6$$

For the second lens alone, it is

$$m_2 = -\frac{s_2'}{s_2} = -\left(\frac{14.8 \text{ cm}}{-5.6 \text{ cm}}\right) = 2.64$$

Overall magnification is $m_1 m_2 = -1.59$. the image is real and inverted. Final image height is

$$x' = x m_1 m_2 = (3 \text{ cm})(-1.59) = -4.77 \text{ cm}$$

where the minus sign again implies image inversion as a result of the negative magnification.

2.7 TWO-LENS COMBINATION

A ray parallel to the optical axis at height h from it is incident on lens L_1 in Fig. 2.16. It is refracted by each lens and intersects the optical axis at F, which represents an equivalent focal point for the two lenses in combination since all rays parallel to the optical axis would intersect at F. If f_1 and f_2 are focal lengths of lenses L_1 and L_2, respectively, our intent is to develop an expression for *equivalent back focal length* f were the two-lens combination to be replaced by a single lens in the plane CH'. The parallel ray AB entering L_1 at height h from the optical axis in Fig. 2.16 is refracted, in the absence of L_2, in the BF_1 *direction to form an image at*

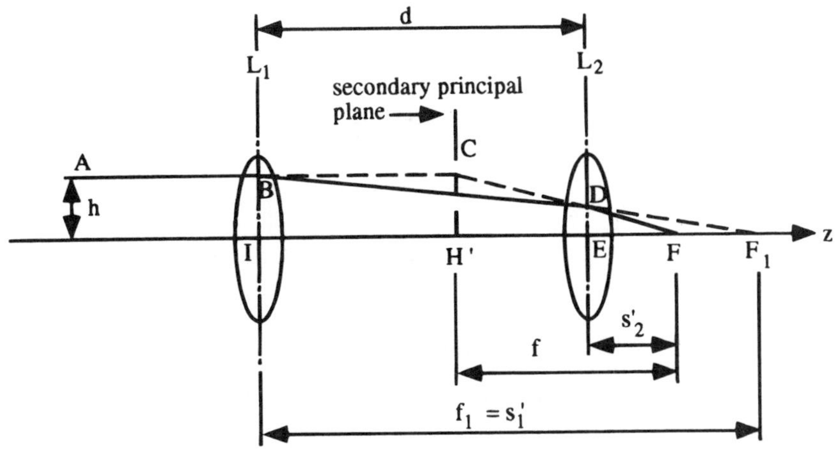

Fig. 2.16 Geometry for determining equivalent back focal length f of two-lens combination.

$$s_1' = (1/f_1 - 1/s)^{-1}$$
$$= \frac{s_1 f_1}{s_1 - f_1}$$

where s and s_1' are, respectively, the distances of the object and its image from lens L_1. Now, since $s_1 = \infty$, then $s_1' = f_1$. The image formed by L_1 is now considered to be a virtual object for L_2. The distance of this object from L_2 is $s_2 = (f_1 - d)$. Applying the thin lens Gaussian formula to L_2, the position s_2' of the image formed by L_2 is obtained:

$$s_2' = \frac{s_2 f_2}{s_2 - f_2} = \frac{-(f_1 - d)f_2}{-(f_1 - d) - f_2} = \frac{(f_1 - d)f_2}{f_1 + f_2 - d} \tag{2.7.1}$$

From Fig. 2.16, the equivalent back focal length of the two-lens combination is given by

$$f = s_2' + EH' \tag{2.7.2}$$

where EH' is obtained by considering similar triangles. Thus, from the similar triangles $CH'F'$ and DEF',

$$\frac{h}{f} = \frac{DE}{f - EH'} \Rightarrow \frac{h}{DE} = \frac{f}{f - EH'}$$

Similarly, from the triangles BIF_1 and DEF_1

$$\frac{h}{f_1} = \frac{DE}{f_1 - d} \Rightarrow \frac{h}{DE} = \frac{f_1}{f_1 - d}$$

Equating the last two, we obtain

$$\frac{f_1}{f_1 - d} = \frac{f}{f - EH^1}$$

$$f_1(f - EH') = f(f_1 - d)$$

$$f_1 f - f_1 EH' = ff_1 - fd \Rightarrow EH' = d\frac{f}{f_1} \tag{2.7.3}$$

Substituting Eqs. (2.7.1) and (2.7.3) into Eq. (2.7.2), we get

$$f = \frac{(f_1 - d)f_2}{f_1 + f_2 - d} + \frac{df}{f_1}$$

$$f\left(1 - \frac{d}{f_1}\right) = \frac{f_1 f_2(1 - d/f_1)}{f_1 + f_2 - d)}$$

$$\frac{1}{f} = \frac{1}{f_1} + \frac{1}{f_2} - \frac{d}{f_1 f_2} \tag{2.7.4}$$

In terms of powers, equivalent power is

$$P = P_1 + P_2 - dP_1P_2 \tag{2.7.5}$$

Equation (2.7.4), derived for rays entering the system from the left, defines the equivalent back or secondary focal length.

It is interesting to consider the equivalent front focal point. Suppose a point source were to be located at F in Fig. 2.17. Refraction occurs at both lenses. A single equivalent lens in plane CH would result in an output beam parallel to the optical axis provided the distance $f = FH$ is equal to the equivalent front or primary focal length of the two lens combination. This equivalent focal length is already given in

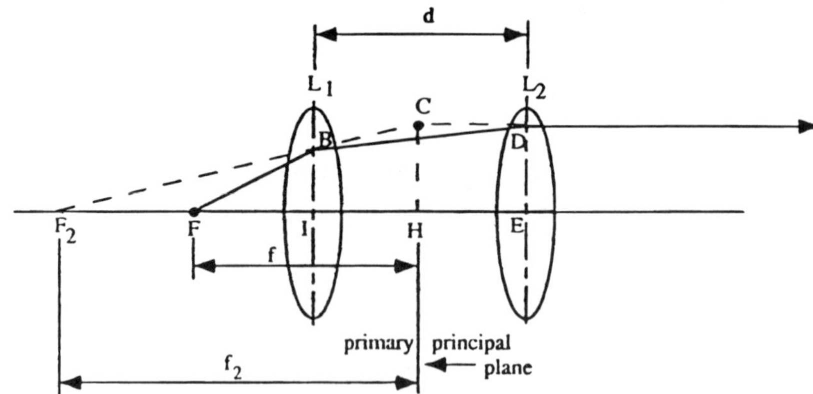

Fig. 2.17 Geometry for determining equivalent front focal length of thin lens combination.

Eq. (2.7.4) for the back focal length. Provided the same medium exists both behind and in front of the optical system ($n = n''$) the front and back focal lengths are equal. However, what is different is the location of the equivalent lens. Equation (2.7.2) indicates the equivalent lens in Fig. 2.16 is located a distance $EH' = df/f_1$ in front of the back lens. A similar analysis for Fig. 2.17 would indicate the equivalent lens is located a distance $HI = df/f_2$ behind the front lens. Note that Fig. 2.16 refers to equivalent back focal length and is therefore relevant to image distance since F is the image point for an object located an infinite distance in front of the first lens. Consequently, if s_2' is the actual image distance with reference to the second lens, image distance with reference to an equivalent lens is, in view of the distance EH' in Fig. 2.16,

$$s'_{\mathrm{eq}} = s'_2 + df/f_1 \tag{2.7.6}$$

where f is equivalent focal length defined in Eq. (2.7.4).

Similarly, Fig. 2.17, which deals with object distance, suggests that if s_1 is object distance with regard to L_1, equivalent object distance is

$$s_{\mathrm{eq}} = s_1 + df/f_2 \tag{2.7.7}$$

Equations (2.7.6) and (2.7.7) imply that the equivalent lens must be envisaged in two planes, one for object distance and the other for image distance calculations. The plane CH' (Fig. 2.16) is called the *secondary principal plane*, and the plane CH (Fig. 2.17) is called the *primary principal plane*. The points H and H', respectively, are called the *primary* and *secondary principal points*. As seen in Fig. 2.16, the secondary principal plane is defined by the intersection of paraxial rays entering the system from the left with those leaving from the right. The primary principal plane is defined as the intersection of paraxial rays leaving the system from the right with those entering from the left.

Example 2.4 has been solved by considering two lenses in series. It can also be solved by using the equivalent lens approach derived here, as in Exercise 2.9.

2.8 THICK LENS

In the development of the Gaussian forms of the thin lens equation in Section 2.4, refractions at each surface of a lens were considered separately. The image or output of the first surface was treated as the input or virtual object for the second surface. The image distance of the first surface was

$$s'_1 = s'_2 + t \tag{2.8.1}$$

Lens thickness t is not negligible for a thick lens. With this clarification, the thick lens can be treated from the standpoint of two surfaces in series separated by the non-negligible distance t. The image of the front surface is the virtual object of the back surface.

Example 2.5

A glass rod 1.8 cm long and of refraction index 1.6 has both ends polished to spherical surfaces with the following radii:

$$r_1 = 2.4 \text{ cm and } r_2 = -2.4 \text{ cm}$$

An object 2 cm high is located on the axis of the rod 8 cm from the first vertex. Assuming air as the surrounding medium, find the following:

(a) the primary and secondary focal lengths for each of the surfaces

(b) the image distance from the first surface

(c) the object distance for the second surface

(d) the final image distance from the second vertex.

Solution

For the first surface, using Eq. (2.4.4) and the focal length definitions at the beginning of Section 2.4,

$$\frac{n}{s_1} + \frac{n'}{s_1'} = \frac{n' - n}{r_1} \text{ at the first surface}$$

The primary focal length is found by setting s_1', the image distance, to infinity, in which case

$$s_1 = f_{\text{primary}} = \frac{nr_1}{n' - n} = \frac{1.0 \times 2.4}{1.6 - 1.0} \text{ cm} = 4.0 \text{ cm} = f_1$$

Similarly, to find $f_{\text{secondary}}$, object distance $s_1 \rightarrow \infty$ and, therefore,

$$s_1' = f_{\text{secondary}} = \frac{n'r_1}{n' - n} = \frac{(1.6)(2.4)}{1.6 - 1.0} \text{ cm} = 6.4 \text{ cm} = f_1'$$

For the second surface, using Eq. (2.4.5)

$$\frac{n'}{s_2'} + \frac{n''}{s_2''} = \frac{n'' - n'}{r_2}$$

Let $s_1'' \rightarrow \infty$, resulting in

$$s_2' = f_{\text{primary}} = \frac{n'r^2}{n'' - n'} = \frac{(1.6)(-2.4)}{1.0 - 1.6} \text{ cm} = +6.4 \text{ cm} = f_2'$$

Let $s_2' \to \infty$. This yields

$$s_2'' = f_{\text{secondary}} = \frac{n'' r_2}{n'' - n'} = \frac{(1.0)(-2.4)}{1.0 - 1.6}\ \text{cm} = +4.0\ \text{cm} = f_2''$$

The image distance from the first surface, s_1' is found from Eq. (2.4.4):

$$\frac{1}{8.0\ \text{cm}} + \frac{1.6}{s_1'} = \frac{1.6 - 1.0}{2.4\ \text{cm}}$$

$$s_1' = 12.8\ \text{cm}$$

The virtual object distance for the second surface, s_2', is the image distance of the first surface referred to the vertex of the second surface. Using Eq. (2.8.1), and sign convention for objects to the right of the imaging surface,

$$s_2' = t - s_1' = 1.8\ \text{cm} - 12.8\ \text{cm} = -10.0\ \text{cm}$$

The final image distance $_2'$ from the second vertex is found from Eq. (2.4.5):

$$\frac{1.6}{-10\ \text{cm}} + \frac{1.0}{s_2''} = \frac{1.0 - 1.6}{-2.4\ \text{cm}}$$

$$s_2'' = 2.53\ \text{cm}$$

This is to the right of the second surface and the entire lens.

It is interesting to note that $f_{\text{secondary}}$ of the first surface (6.4 cm) is larger than the sum of lens thickness and $f_{\text{secondary}}$ of the second surface (1.8 cm + 4.0 cm = 5.8 cm). This means a paraxial ray entering this thick lens will be refracted at each surface and intersect the axis as shown in Fig. 2.18. This is quite similar to Fig. 2.16 for the two-lens combination. In each Figure, CH' defines the secondary principal plane to which image distances are referenced, as in Eq. (2.7.6).

Note too a similar consideration concerning primary focal length, that is, f_{primary} of the second surface (6.4 cm) is greater than the sum of $t + f_1$ (1.8 cm + 4 cm =

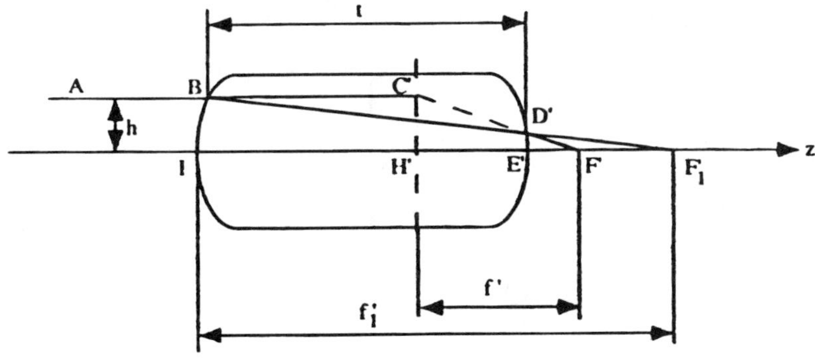

Fig. 2.18 Thick lens for Example 2.5.

5.8 cm). This means a point source at the input to the lens generating a paraxial ray at the output must be positioned as in the two-lens combination of Fig. 2.17. *The thick lens can therefore be modeled as such a two-lens combination with primary and secondary principal planes, which do not coincide because of the lens separation.* However, if the two principal planes coincide, then that lens can be modeled as a thin lens.

By defining an equivalent thick lens focal length f_t according to

$$\frac{n}{f_t} = \frac{n'}{f'_1} + \frac{n''}{f''_2} - \frac{tn''}{f'_1 f'_2} = \frac{n''}{f''} \tag{2.8.2}$$

which reduces for a thick lens immersed in air to

$$\frac{1}{f_t} = \frac{n'}{f'_1} + \frac{1}{f''_2} - \frac{t}{f'_1 f''_2} \tag{2.8.3}$$

the Gaussian form of the thin lens equation can be used for a thick lens provided that object distance s and image distance s' are referenced to primary and secondary principal planes as in Eqs. (2.7.6) and (2.7.7). Because each surface has both primary and secondary focal lengths, these will be restated as

$$s_{eq} = s + f_t t / f'_2 \tag{2.8.4}$$

$$s'_{eq} = s' + f_t t / f'_1 \tag{2.8.5}$$

Imaging with a thick lens can be described with parallel ray tracing as in Fig. 2.19.

Example 2.6

Consider Example 2.5 from the point of view of the Gaussian form of the thin lens equation as applied to a thick lens.

Solution

From Eq. (2.8.3),

$$\frac{1}{f_t} = \frac{1.6}{6.4} + \frac{1}{4} - \frac{1.8}{(6.4)(4)} \rightarrow f_t = 2.33 \text{ cm} = f'' \text{ in air, since } n = n''$$

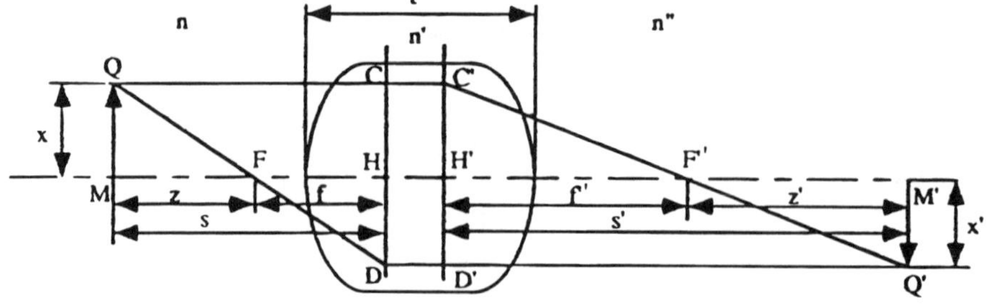

Fig. 2.19 Parallel ray tracing for a thick lens.

From Eq. (2.8.4),

$$s_{eq} = 8 + (2.33)(1.8)/(6.4) = 8.65 \text{ cm}$$

$$\frac{1}{s'_{eq}} = \frac{1}{f_t} - \frac{1}{s_{eq}} = \frac{1}{2.33} - \frac{1}{8.65} \rightarrow s'_{eq} = 3.19 \text{ cm}$$

From Eq. (2.8.5),

$$s' = s'_{eq} - f_t t/f'_1 = 3.19 - \frac{2.33(1.8)}{6.4} = 2.53 \text{ cm}$$

This answer coincides with that obtained using the two-step solution of Example 2.5.

2.9 VIRTUAL IMAGES AND ANGULAR MAGNIFICATION

So far the discussion and geometrical configurations have been for real images for positive lenses. For example, an object point M_1 in Fig. 2.20 will give rise to a real image point M'_1. The wave exiting the lens is converging towards M'_1. The angle of deviation $M_1CM'_1$ produced by the lens is equal to the angle of deviation FCB for a ray starting at the primary focal point and proceeding parallel to the axis upon exiting the lens. Because the lens is positive, the deviation of $M_1CM'_1$ is toward the optical axis. There is no convergence for the ray FCB because CB is parallel to the optical axis. For a real image, object distance must be greater than focal length, as in the case of object point M_1. However, suppose the object distance $s < f$, as for M_2 in the figure. The angular deviation M_2CA is the same angular deviation toward the optical axis as is the case for the other two rays FCB and $M_1CM'_1$. However, since for M_2 object distance is less than focal length, the exiting ray CA is still proceeding away from the axis. For such a case, as depicted in Fig. 2.21, rays will never converge behind the axis so as to form a real image. If one looks toward the left from behind the lens in Fig. 2.21, the point M will *seem* to emanate from M' on the left side of the lens and the point Q will seem to emanate from Q' on the left side of the lens. Such an image is *virtual*. As shown for the object QM in Fig. 2.21, the virtual image $Q'M'$ is erect, contrary to real images, which are inverted.

Real as well as virtual imagery are described by the thin lens equation. Since

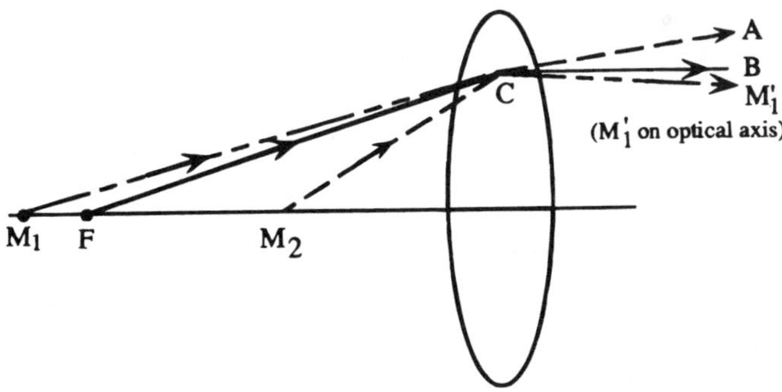

Fig. 2.20 Deviations of exit rays for object distance greater, equal to, and less than focal length.

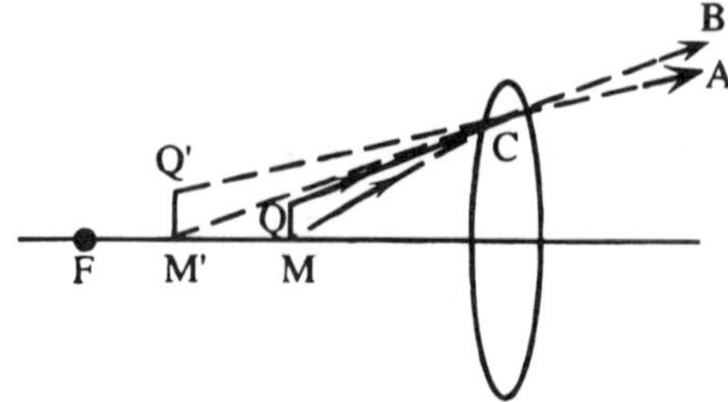

Fig. 2.21 Virtual image $Q'M'$ of object QM. Rays $Q'C$ and $M'C$ forming $Q'M'$ are continuations of rays CA and CB, respectively, exiting lens.

$$\frac{1}{s'} = \frac{1}{f} - \frac{1}{s} = \frac{s-f}{sf}$$

then

$$s' = \frac{sf}{s-f} \tag{2.9.1}$$

For $s > f$, the image is real and to the right of the lens.

However, if $s < f$, then s' in Eq. (2.9.1) is negative. This indicates that virtual images are to the left of a lens. Furthermore, with regard to lateral magnification,

$$m = -\frac{s'}{s} = -\frac{f}{s-f} \tag{2.9.2}$$

Therefore, where $s < f$, m is positive and the image is erect. Equation (2.9.2) indicates also that the image blows up, or infinite magnification occurs, when $s = f$.

This can be demonstrated very easily with a positive lens, preferably of short focal length for convenience, used as a magnifying glass. If the lens is placed *near* a text, the letters appear erect and larger than they really are. This is a virtual image. As the magnifying glass is removed further away from the text, the letters appear to increase in size. This is in accordance with Eq. (2.9.2). Finally, as the glass is removed even further from the text, the letters appear to get even larger in size until they "blow up." This corresponds to the condition $s = f$ in Eq. (2.9.2). If the magnifying glass is removed even further from the text, the text suddenly appears inverted and a real image is formed. As s is increased further, the inverted real image becomes smaller. These observations are very easily verified with any positive lens, conveniently of short focal length.

2.9.1 Angular Magnification

The concept of lateral magnification, as in Eq. (2.9.2), is appropriate for a two-dimensional image projected on a screen or TV monitor, for example. The human visual system also sees depth, and can focus on a continuum of object distances for different targets even within the same scene. For the human visual system, whether it is the output of eyeglasses or microscopes or telescopes, a much more appropriate concept with which to describe magnification is angular magnification.

Consider a simple magnifying glass, as in Fig. 2.21. If $s < f$, the image is virtual and, using Eq. (2.9.2),

$$x' = mx = \frac{f}{f - s}x \tag{2.9.3}$$

As s decreases, *accommodation* (discussed in Chapter 12) permits the eye to change its power and to form larger retinal images. The object looks larger. However, there is a limit as to how close the object can be to the eye and still appear sharp. Usually $s \approx 10$ in. or about 25 cm is taken to be the standard *near point*. At this distance the angle subtended by the object is

$$\tan \alpha = x/25 \approx \alpha \tag{2.9.4}$$

where object dimension x is in centimeters. If a positive lens is placed before the eye, as in Fig. 2.22, the object can be brought much closer to the eye and an image subtending a larger angle

$$\tan \alpha' = x/s \approx \alpha' \tag{2.9.5}$$

is formed on the retina. Since $s < 25$ cm, the object is magnified. Angular magnification is simply the ratio of angle subtended by image to that by object or

$$m_a = \frac{\alpha'}{\alpha} = \frac{x}{s}\frac{25(\text{cm})}{x} = \frac{25(\text{cm})}{s} \tag{2.9.6}$$

The smaller the focal length, the greater the magnification. This can be seen if m_a

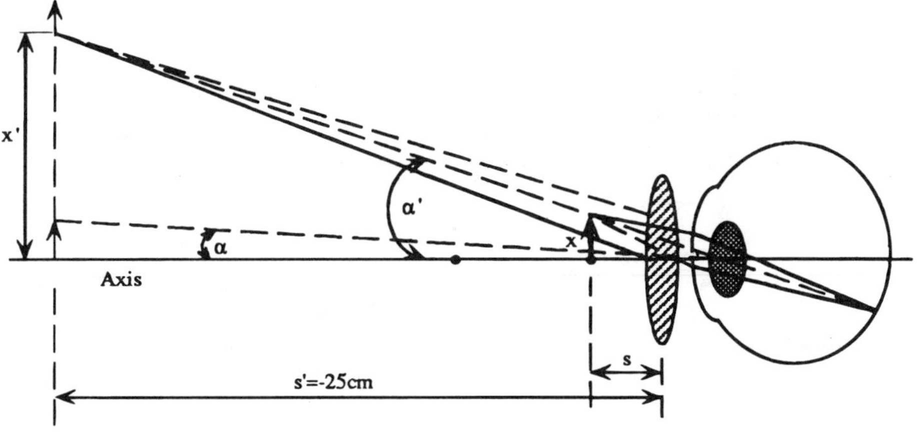

Fig. 2.22 Virtual image angular magnification.

is thought of in terms of image distance rather than object distance. Accordingly, making use of the Gaussian thin lens formula to substitute for $1/s$ in Eq. (2.9.6):

$$m_a = \frac{25(\text{cm})}{f} - \frac{25(\text{cm})}{s'} \tag{2.9.7}$$

Since the image is virtual, s' in Eq. (2.9.7) is negative. For example, if focus is set so that the image appears to be 25 cm away as in Fig. 2.22 ($s' = -25$ cm), then $m_a = 25(\text{cm})/f + 1$. Magnifiers typically have short focal lengths not much different from object distance. Therefore, the expression 25 (cm)/f is commonly used since, practically speaking, s' in Eq. (2.9.7) is usually rather large.

The angular magnification approach is used commonly for magnifiers and microscopes. For eyeglasses as well, shorter focal length or higher powers (in diopters) yield greater magnification.

The concept of angular magnification is also useful to define telescope magnification even though the obtained image is real. Typical astronomical telescopes are shown in Fig. 2.23 where f_o is focal length for the objective (lens or mirror) and f_e is that for the eyepiece. For long-range imaging such that $s \rightarrow \infty$, the separation D between both elements is such that $D = f_e + f_o$.

In the normal state of adjustment the secondary focal plane of the objective coincides with the primary focal plane of the eyepiece, so that incident parallel rays also emerge parallel. The objective forms a real inverted internal image I of height

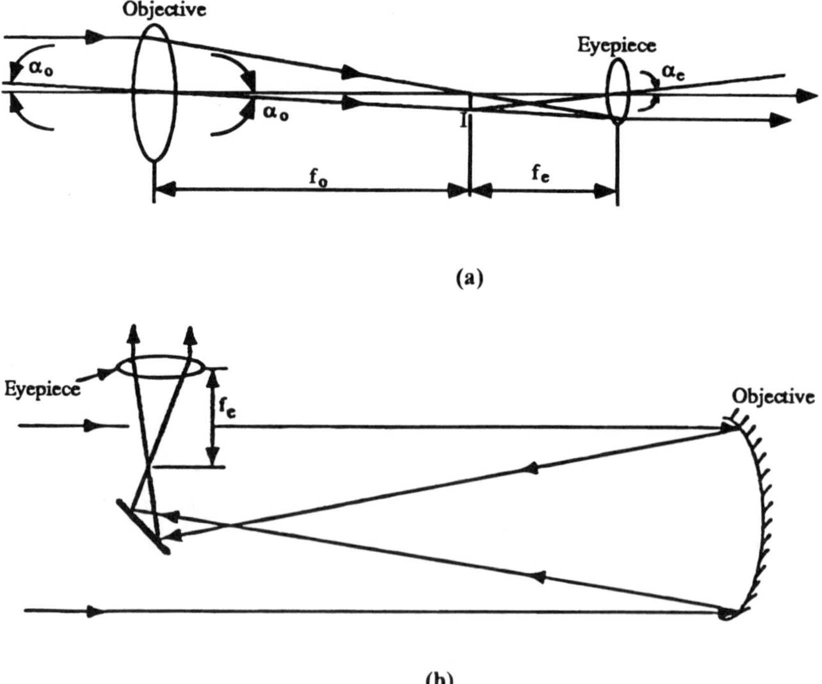

Fig. 2.23 Simple (a) refractive and (b) reflective telescopes.

$$h = -\alpha_o f_o \tag{2.9.8}$$

This is examined using the eyepiece. The angle subtended by this image from the first principal point of the eyelens is

$$\alpha_e = h/f_e \tag{2.9.9}$$

The angular magnification is

$$m_a = \frac{\alpha_e}{\alpha_o} = -f_o/f_e \tag{2.9.10}$$

The image is real and inverted.

High angular magnification involves increasing the objective focal length while decreasing that of the eyepiece. Figure 2.14 indicates that when only one lens is used, angular magnification for real images is unity since angle $Q'AM'$ subtended by the image is equal to angle QAM subtended by the object. Limits to the worthwhileness of too-high magnification are discussed in Chapter 10 and in examples in Parts 4, 5, and 6.

The image may be erected by use of an auxiliary lens or prism for terrestrial applications. However, this does not imply improved resolution of small detail.

2.10 SELFOC LENSES

Graded-index (GRIN) optical fibers are described in Section 1.7 in Chapter 1. Our intent here is to show that such a fiber is actually a lens as well.

The imaging properties of lenses depend on their ability to focus. Parallel rays at the input, for example, independent of the radial height at which they enter the system, must be focused to a single point, called the *focal point*, at the output. If this requirement is satisfied, then such an element can be used for image formation in accordance with the thin lens equations, since there would be one image plane only.

In Section 1.7, we showed that for paraxial rays at a parabolic index fiber input at radial coordinate r_0, their distance away from the axis varies cosinusoidally according to

$$r(z) = r_o \cos \sqrt{\frac{n_1}{n_0}} z \tag{2.10.1}$$

while the slope of the ray varies as

$$r'(z) = -\sqrt{\frac{n_1}{n_0}}\, r_0 \sin \sqrt{\frac{n_1}{n_0}}\, z \tag{2.10.2}$$

In a parabolic index or selfoc fiber medium, refractive index varies as

$$n(r) = n_0 - \frac{1}{2} n_1 r^2 \tag{2.10.3}$$

and it is assumed that, for a fiber radius a,

$$n_0 >> n_1 a^2$$

The last condition, also shown in Chapter 1, implies that although input paraxial rays propagate cosinusoidally through the fiber, their zero crossings across the optical axis are very far apart because $\sqrt{n_1/n_0}$ in Eq. (2.10.1) is so very small. As such, even in the fiber the rays are still essentially paraxial. Therefore, when they exit the fiber, their angles of incidence to the fiber/air interface at the fiber exit are very small. For this reason, refraction effects from the fiber to air are neglected and the exit rays are assumed to propagate in air in the same direction in which they were propagating upon exiting the fiber. For a fiber of length L, this direction is

$$r'(L) = -\sqrt{\frac{n_1}{n_0}}\, r_0 \sin \sqrt{\frac{n_1}{n_0}}\, L \tag{2.10.4}$$

This slope must be equal to $r(L)/f$, as shown in Fig. 2.24. Therefore,

$$f = \frac{r(L)}{r'(L)} = -\sqrt{\frac{n_0}{n_1}} \cot \sqrt{\frac{n_1}{n_0}} L \tag{2.10.5}$$

Note that focal length is independent of r_0. Therefore, all paraxial rays entering the fiber are focused at F, regardless of their initial elevation r_0. Thus, selfoc fibers are also lenses. Note too that such lenses can be either positive or negative, depending on the phase of the cosine cycle at fiber exit. In Fig. 2.24, the cosine is negative, causing the focal length to be positive.

Ordinary glass convex lenses require parabolic surfaces for ideal geometric optics imaging, as pointed out in Eq. (2.3.7). Here it is seen that selfoc fibers of *planar* surfaces can also provide geometric optics perfect imaging because of the parabolic nature of their refractive index. This causes the phase differences between input and output resulting from different initial heights r_0 to be the same.

Perhaps the most successful commercial application of GRIN lenses is in photocopiers. Single lenses in photocopiers are replaced by two row arrays of 1-cm-long, 1-mm-diameter cylindrical selfoc lenses. An assembly of 600 to 2000 rods can obtain a field of view as wide as a 36-inch sheet of paper while reducing overall paper-to-image optical system length to 70 mm. Alignment is critical so that the image from

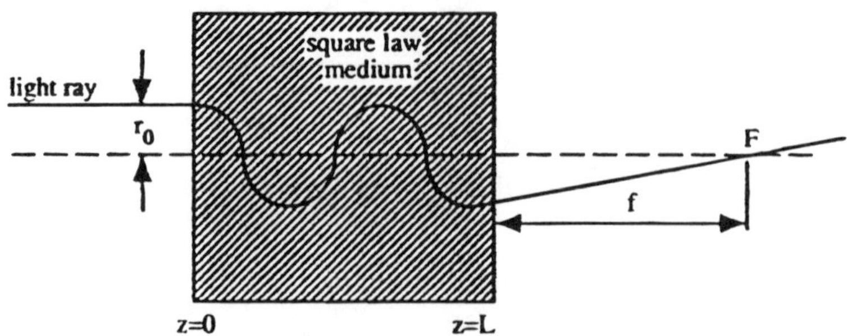

Fig. 2.24 Graded-index (GRIN) lens.

one rod should not overlap the image from another. Even small angles between the rods can cause image mismatch.

2.11 MIRRORS

Application of Snell's law to convert the thin lens equation for refractive optics to that for reflective optics has been considered in Example 2.3. Equation (2.4.15) is identical to the refractive optics version relating s, s', and f. For a *convex* mirror of radius or curvature r, it has been shown there that $f = r/2$. It is important to clarify the rules for signs when imaging with mirrors [2.1]:

1. As for refractive optics, distances measured from left to right are positive.
2. Incident rays travel from left to right and reflected rays from right to left.
3. Focal length is measured from focal point to vertex.
4. Radius is measured from vertex to center of curvature. Thus, r is negative for concave mirrors and positive for convex mirrors and f is positive for concave mirrors and negative for convex mirrors.
5. Object (s) and image (s') distances are measured from object and image, respectively, to the vertex. Therefore, s and s' are both positive and the object and image are real when they lie to the left of the vertex, while they are negative and virtual when they lie to the right.

As is the case for refractive optics, an object distance of less than f gives rise to a virtual image since the reflected rays never converge and Eq. (2.4.15) can be used for reflective optics to describe imaging.

2.12 TRANSMITTING AND RECEIVING OPTICS

In imaging systems, receiver field of view is often a critical design parameter. In active optics systems, transmitter beamwidth involves the same geometrical optics principles, but in reverse direction. Conceptually, it is often simpler to visualize the transmitter optics since the transmitter can be visualized as an object source. Receiver optics are identical, but with ray directions or arrows reversed. Here, transmitter optics with a single lens are considered from the standpoint of geometrical optics. The same treatment is applicable to a mirror as well, as demonstrated in Fig. 2.6. With regard to a transmitter, α is transmission beamwidth, while for a receiver α is field of view. In either case, location of source or receiver with respect to focal length plays a key role in determining limits of transmitted or received beam divergence. Here, sources or imagers very small compared to optics diameter are considered. The more general case is given in Example 3.5 in Section 3.4.

Consider, for example, Fig. 2.25(a) where the light source is closer to the lens than to the primary focal plane. As a result, the lens causes incoming rays to diverge. Transmitter beamwidth, α, is determined by the angle of maximum divergence. Divergences considered include those for chief rays as well as those emanating from maximum vertical coordinates of the source. In Fig. 2.25(a) α is determined by the latter.

For receiver field of view, Fig. 2.25(a) implies, if arrows are reversed, that the closer the detector is to the receiver optics relative to focal plane, the wider the receiver field of view.

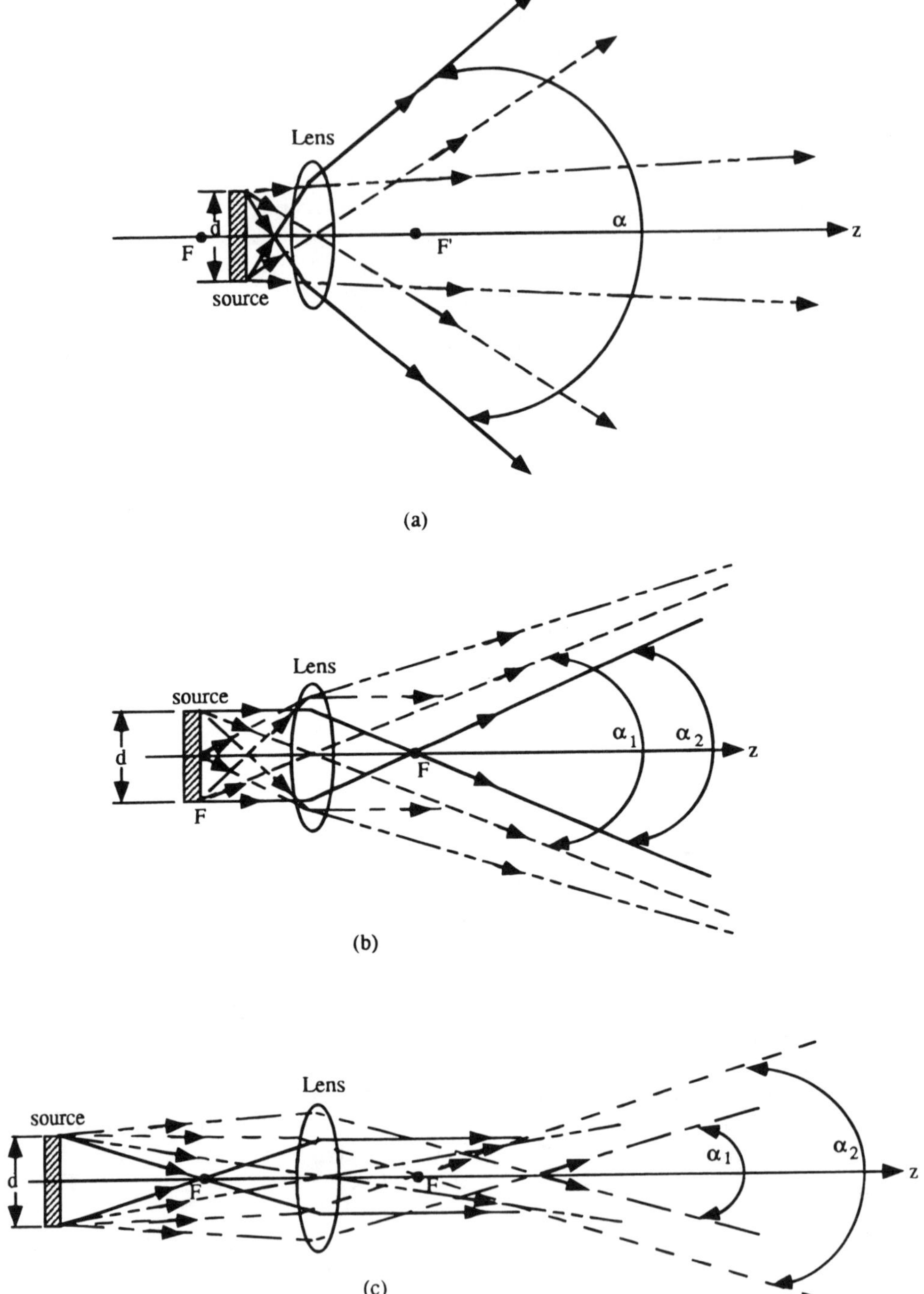

Fig. 2.25 Transmitter beamwidth (or receiver field of view if arrows are reversed) for devices of diameter *d* located at distances from the lens that are (a) less than its focal length, (b) equal to its focal length, and (c) greater than its focal length.

In Fig. 2.25(b) the source is in the primary focal plane. Beamwidth determined by chief rays is α_1; beamwidth determined by parallel ray tracing is α_2. From the geometry, both are identical and are equal to

$$\alpha_1 = \frac{d}{f} = \alpha_2 \tag{2.12.1}$$

where d is source diameter. For a receiver arrows are reversed and d is detector diameter. When a source or detector is in the optics focal plane, transmitter beamwidth and receiver field of view are less than when the source or detector is closer to the optics than to the focal plane.

Figure 2.25(c) illustrates the case for a source further away from the lens than its focal plane. Because of the distance, incoming light is caused to converge by the lens. Rays emanating from extreme vertical coordinates of the source determine α_2, while α_1 is determined according to the parallel ray tracing method. In Figs. 2.25(b) and (c), α_2 is identical and equal. However in Fig. 2.25(c), $\alpha_1 > \alpha_2$. Therefore, transmitter beamwidth in Fig. 2.25(c) is equal to α_1. For a receiver instead of a source, arrow directions are simply reversed again.

The treatment here indicates that for small sources imagers, transmitter beamwidth or receiver field of view are determined essentially by device size and optics focal length. Optics diameter, however, is also important, but rather from the standpoint of image brightness (Section 3.4) and signal-to-noise ratio (Chapter 11). For larger sources or imagers, as in Example 3.5, optics diameter also affects beamwidth and field of view.

In general, minimum transmitter beamwidth or receiver field of view is given by Eq. (2.12.1), when a small transmitter or receiver is in the optics focal plane. Note that as the device approaches a point source, or detector, output beam or field of view approaches perfect collimation ($\alpha = 0$) according to the definition of focal point. As the detector or source is moved out of the focal plane, field of view or beamwidth for such cases increases (Exercise 2.19). One method of calculating field of view or beamwidth for such cases is by using optical matrices, which are described next.

Field of view can be varied if zoom lenses are used. Such devices consist of a number of optical elements together. As separations between them are varied, effective overall focal length changes. In this way, field of view can be narrowed and magnification increased so as to "zoom in" on a particular object or target. Although such effects were considered for a pair of lenses in Section 2.7, more sophisticated tools are required in order to expedite treatment for larger numbers of optical elements. One method is matrix optics.

2.13 MATRIX OPTICS

In the beginning of Section 2.9, we pointed out that the *change* in ray direction deriving from passage through a lens is the same regardless of initial direction. If r is a radial coordinate as in Fig. 2.26, and if ray slope dr/dz is designated by r', then for an incoming ray parallel to the optical axis by definition $r'_{in} = 0$ and, for a positive thin lens, $r'_{out} = -r_{in}/f$. The change in ray slope produced by the lens is therefore equal to $-r_{in}/f$, and the ray is refracted toward the optical axis.

Consider as an example the propagation of a ray through a straight section of a homogeneous medium of length L followed by a thin lens of focal length f. This corresponds to propagation between planes 1 and 3 in Fig. 2.26. Since the effect of the straight section is merely that of increasing the coordinate r by Lr', for a thin lens the output ray height r_3 at plane 3 and input height r_1 at plane one can be related by

$$r_3 = r_1 + Lr_1' \tag{2.13.1}$$

The output slope of the ray is affected by the lens so that

$$r_3' = r_1' - \frac{r_3}{f} = r_1' - \frac{1}{f}(r_1 + Lr_1') = -\frac{1}{f}r_1 + \left(1 - \frac{L}{f}\right)r_1' \tag{2.13.2}$$

where Eq. (2.13.1) is substituted into Eq. (2.13.2). In matrix form,

$$\begin{bmatrix} r_3 \\ r_3' \end{bmatrix} = \begin{bmatrix} 1 & L \\ -\frac{1}{f} & 1 - \frac{L}{f} \end{bmatrix} \begin{bmatrix} r_1 \\ r_1' \end{bmatrix} = [M] \begin{bmatrix} r_1 \\ r_1' \end{bmatrix} \tag{2.13.3}$$

Matrix $[M]$ can be decomposed into two matrices. For a thin lens by itself, $L = 0$ and the ray relationship between planes 3 and 2 is

$$\begin{bmatrix} r_3 \\ r_3' \end{bmatrix} = \begin{bmatrix} 1 & 0 \\ -\frac{1}{f} & 1 \end{bmatrix} \begin{bmatrix} r_2 \\ r_2' \end{bmatrix} = [M_2] \begin{bmatrix} r_2 \\ r_2' \end{bmatrix} \tag{2.13.4}$$

The first line of this matrix indicates that radial coordinates r_2 at lens input and r_3 at lens output are identical, as expected for a thin lens. For propagation through homogeneous space a distance L but without a lens, $f \to \infty$ and, from Eq. (2.13.3),

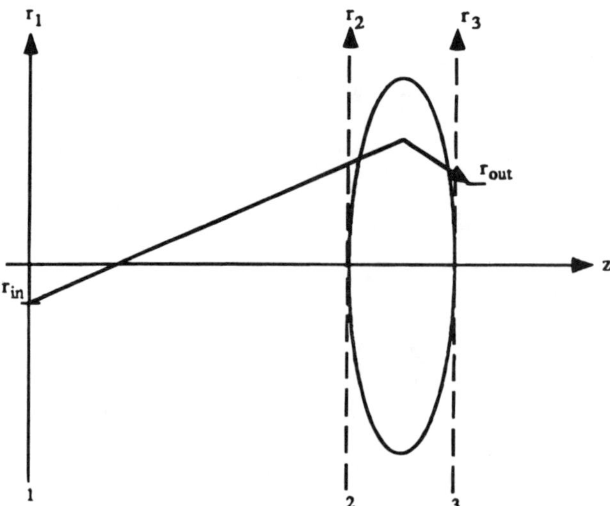

Fig. 2.26 Geometry for the optical matrix method of ray tracing.

$$\begin{bmatrix} r_2 \\ r_2' \end{bmatrix} = \begin{bmatrix} 1 & L \\ 0 & 1 \end{bmatrix} \begin{bmatrix} r_1 \\ r_1' \end{bmatrix} = [M_1] \begin{bmatrix} r_1 \\ r_1' \end{bmatrix} \tag{2.13.5}$$

The second line of this last matrix indicates there is no change in ray direction or slope, as expected in homogeneous space and as illustrated between planes 2 and 1 in Fig. 2.26. Matrices (2.13.4) and (2.13.5) can be combined to yield

$$\begin{bmatrix} r_3 \\ r_3' \end{bmatrix} = [M_2] \begin{bmatrix} r_2 \\ r_2' \end{bmatrix} = [M_2]\,[M_1] \begin{bmatrix} r_1 \\ r_1' \end{bmatrix} \tag{2.13.6}$$

where

$$[M_2]\,[M_1] = [\mathrm{M}] \tag{2.13.7}$$

in Eq. (2.13.3). This development indicates that the matrices appear in *reverse* order to that encountered left to right through the system. The concept is similar to matrix representation of output and input voltages and currents in electronics, where each network is represented by a matrix, and matrix multiplication in reverse order allows one to obtain a relationship between output voltage and current of the last network and input voltage and current of the first network. The same concept holds for ray propagation as shown here in Eq. (2.13.6).

Example 2.7

Prove Eq. (2.13.7) through matrix multiplication.

Solution

Matrix multiplication rules are such that

$$\begin{bmatrix} a_{11} & \cdots & a_{1p} \\ a_{i1} & \cdots & a_{ip} \\ a_{m1} & \cdots & a_{mp} \end{bmatrix} \begin{bmatrix} b_{11} & \cdots & b_{ij} & \cdots & b_{in} \\ & & \cdots & & \\ b_{p1} & \cdots & b_{pj} & \cdots & b_{pn} \end{bmatrix} = \begin{bmatrix} c_{11} & \cdots & c_{in} \\ \cdots & c_{ij} & \cdots \\ c_{m1} & \cdots & c_{mn} \end{bmatrix}$$

where

$$c_{ij} = a_{i1}b_{ij} + a_{i2}b_{2j} + \cdots + a_{ip}b_{pj} = \sum_{k=1}^{p} a_{ik}b_k$$

For the 2 × 2 matrices involved here, and assuming the matrices above from left to right are M_2, M_1, and M, one obtains

$$c_{11} = a_{11}b_{11} + a_{12}b_{21} = 1\cdot 1 + 0\cdot 0 = 1$$

$$c_{21} = a_{21}b_{11} + a_{22}b_{21} = \left(-\frac{1}{f}\right)\cdot 1 + 1\cdot 0 = -1/f$$

$$c_{12} = a_{11}b_{12} + a_{12}b_{22} = 1 \cdot L + 0 \cdot 1 = L \tag{2.13.8}$$

$$c_{22} = a_{21}b_{12} + a_{22}b_{22} = \left(-\frac{1}{f}\right) L + 1 \cdot 1 = 1 - L/f$$

Consequently,

$$[M] = [M_2][M_1] = \begin{bmatrix} c_{11} & c_{12} \\ c_{21} & c_{22} \end{bmatrix} = \begin{bmatrix} 1 & L \\ -\frac{1}{f} & 1 - L/f \end{bmatrix}$$

as in Eq. (2.13.3).

Example 2.8

For an object at a distance s from a lens of focal length f, find image distance by application of the optical matrix technique.

Solution

From Eq. (2.13.3), substituting s for L, the matrix for object plane to lens output is

$$M_a = \begin{bmatrix} 1 & s \\ -\frac{1}{f} & 1 - \frac{s}{f} \end{bmatrix} \tag{2.13.9}$$

From Eq. (2.13.5), substituting s' for L, the matrix appropriate for propagation from lens output plane to image plane is

$$M_b = \begin{bmatrix} 1 & s' \\ 0 & 1 \end{bmatrix} \tag{2.13.10}$$

If the object and image planes are designated by subscripts "o" and "i" respectively, then

$$\begin{bmatrix} r_1 \\ r_1' \end{bmatrix} = [M_b][M_a] \begin{bmatrix} r_o \\ r_o' \end{bmatrix}$$

$$= \begin{bmatrix} 1 - \frac{s'}{f} & s + s'\left(1 - \frac{s}{f}\right) \\ -\frac{1}{f} & 1 - \frac{s}{f} \end{bmatrix} \begin{bmatrix} r_o \\ r_o' \end{bmatrix} \tag{2.13.11}$$

The middle matrix in Eq. (2.13.11) is that of ray propagation from object to image plane. To understand this result, consider an object plane point $r_o = b$. By the parallel ray method of ray tracing in Fig. 2.14, rays starting out at this point with slopes r_o' $= 0$ and $r_o' = -b/s$ must both intersect at the same point in the image plane. Therefore, for the upper ray in Fig. 2.14 designated here in the image plane as r_{i1}, using the upper equation in matrix Eq. (2.13.11) with $r'_o = 0$, yields the following image plane coordinate:

$$r_{i1} = \left(1 - \frac{s'}{f}\right)b + \left(s + s' - \frac{s's}{f}\right)\cdot 0 = \left(1 - \frac{s'}{f}\right)b \tag{2.13.12}$$

For the second or chief ray in Fig. 2.14, designated in the image plane as r_{i2}, using again the upper equation in matrix Eq. (2.13.11) but with $r'_o = -b/s$, the image plane coordinate is

$$r_{i2} = \left(1 - \frac{s}{f}\right)b + \left(s + s' - \frac{s's}{f}\right)\left(-\frac{b}{s}\right) \tag{2.13.13}$$

Since both rays must intersect in the image plane, $r_{i1} = r_{i2}$. Consequently,

$$\left(1 - \frac{s'}{f}\right)b = \left(1 - \frac{s'}{f}\right)b + \left(s + s' - \frac{s's}{f}\right)\left(-\frac{s}{f}\right) \tag{2.13.14}$$

or

$$s + s' - \frac{s's}{f} = 0$$

$$s'\left(1 - \frac{s}{f}\right) = -s$$

$$s' = \frac{s}{\dfrac{s}{f} - 1} = \frac{sf}{s - f}$$

$$\frac{1}{s'} = \frac{1}{f} - \frac{1}{s} \tag{2.13.15}$$

Example 2.9

Consider an infinite series of lens pairs of focal lengths f_1 and f_2 as shown in Fig. 2.27, each lens being equidistant by a distance L. This is equivalent to the problem of propagation inside an optical resonator such as a laser (Chapter 4) with mirrors of radii of curvature $r_1 = 2f_1$ and $r_2 = 2f_2$, where L is the length of the resonator and focal lengths f_1 and f_2 are in accordance with Eq. (2.4.15). The radiation is reflected back and forth an infinite number of times. This is equivalent to propagation *through* an infinite series of lens pairs as in Fig. 2.27.

Using matrix analysis, find conditions involving L, f_1, and f_2, which are required so that radiation does not escape from the resonator. In other words, the radiation inside the resonator must not diverge so that r_s or r_{s+1} becomes greater than the radius of the mirror in plane s or plane $s+1$.

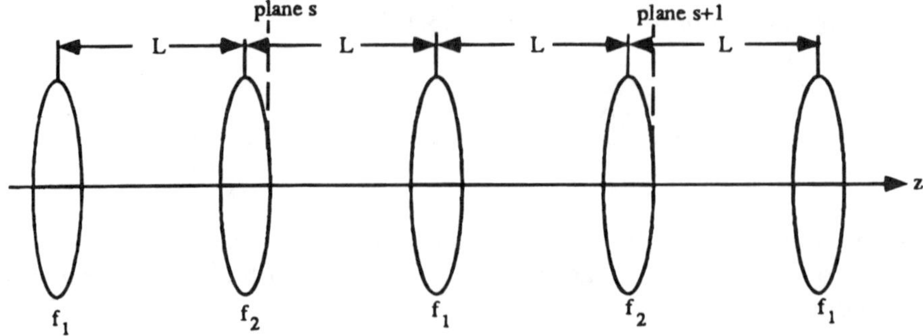

Fig. 2.27 Infinite series of lens pairs analogous to ray propagation inside an optical resonator.

Solution

The transmission matrix, using Eq. (2.13.3), is related to input and output by

$$\begin{bmatrix} r_{s+1} \\ r'_{s+1} \end{bmatrix} = \begin{bmatrix} 1 & L \\ -1/f_1 & 1 - L/f_1 \end{bmatrix} \begin{bmatrix} 1 & L \\ -1/f_2 & 1 - L/f_2 \end{bmatrix} \begin{bmatrix} r_s \\ r'_s \end{bmatrix}$$

$$= \begin{bmatrix} A & B \\ C & D \end{bmatrix} \begin{bmatrix} r_s \\ r'_s \end{bmatrix} \tag{2.13.16}$$

where

$$\begin{aligned} A &= 1 - L/f_2 \\ B &= L(2 - L/f_2) \\ C &= -\left(\frac{1}{f_1} + \frac{1}{f_2} - \frac{d}{f_1 f_2}\right) = -\frac{1}{f'} \\ D &= -\left[\frac{L}{f_1} - \left(1 - \frac{L}{f_1}\right)\left(1 - \frac{L}{f_2}\right)\right] \end{aligned} \tag{2.13.17}$$

From the first part of Eq. (2.13.16)

$$r'_s = \frac{1}{B}(r_{s+1} - Ar_s) \tag{2.13.18}$$

and thus

$$r'_{s+1} = \frac{1}{B}(r_{s+2} - Ar_{s+1}) \tag{2.13.19}$$

Using the second part of Eq. (2.13.16) in Eq. (2.13.19) and substituting r'_s from Eq. (2.13.18) yields

$$r_{s+2} - (A + D)r_{s+1} + (AD - BC)r_s = 0 \quad (2.13.20)$$

which is a difference equation governing the propagation through the infinite series of lens pairs. Using Eq. (2.13.17) we can show that $AD - BC = 1$. Consequently, Eq. (2.13.20) can be rewritten as

$$r_{s+2} - 2br_{s+1} + r_s = 0 \quad (2.13.21)$$

where

$$b = \frac{1}{2}(A + D) = \left(1 - \frac{d}{f_2} - \frac{d}{f_1} + \frac{d^2}{2f_1f_2}\right) \quad (2.13.22)$$

Equation (2.13.21) is the equivalent, in terms of difference equations, of the differential equation $r'' + Gr = 0$, whose solution is $r(z) = r(0)\exp[\pm j\sqrt{G}z]$. It is thus prudent to try a solution of the form [2.3]

$$r_s = r_o e^{jsq}$$

which, when substituted in Eq. (2.13.21), leads to

$$e^{2jq} - 2be^{jq} + 1 = 0 \quad (2.13.23)$$

Therefore, solving this quadratic equation yields

$$e^{jq} = b \pm j\sqrt{1 - b^2} = e^{\pm j\theta} \quad (2.13.24)$$

where $\cos\theta = b$.

The general solution of Eq. (2.13.21) can be taken as a linear superposition of $\exp(js\theta)$ and $\exp(-js\theta)$ solutions or, equivalently, as

$$r_s = r_{max}\sin(s\theta + \alpha) \quad (2.13.25)$$

where $r_{max} = r_0/\sin\alpha$, and α can be expressed using Eq. (2.13.19) in terms of r_0 and r'_0.

The condition for beam confinement or stable oscillation is that θ be a real number, between r_{max} and $-r_{max}$. According to Eq. (2.13.25) the necessary and sufficient condition for θ to be real is that [2.4]

$$|b| \leq 1 \quad (2.13.26)$$

In terms of the system parameters, Eq. (2.13.22) can be used to reexpress Eq. (2.13.26) as

$$-1 \leq 1 - \frac{L}{f_2} - \frac{L}{f_1} + \frac{L^2}{2f_1f_2}$$

or

$$0 \leq \left(1 - \frac{L}{2f_1}\right)\left(1 - \frac{L}{2f_2}\right) \leq 1 \tag{2.13.27}$$

Under such conditions a laser resonator is stable. If, however, $|b| > 1$, the solution is in the form of [2.5]

$$r_s = c_1 e^{\alpha_1 s} + c_2 e^{\alpha_2 s} \tag{2.13.28}$$

where $e^{\alpha_{1,2}} = b \pm \sqrt{b^2 - 1}$. If $|b| > 1$, the magnitude of either exponent in Eq. (2.13.18) exceeds unity and beam radius increases as a function of radial distance s, i.e., the beam escapes sideways from the resonator because at some point r_s is greater than the radius of either mirror in the resonator.

In summary, matrix optics is a technique to trace rays analytically through large numbers of elements and spacings. Most of this is done today digitally with commercial software.

2.14 ABERRATIONS

Ray aberrations are deviations from the paths described by the Gaussian formulas. These take several forms.

Chromatic aberrations derive from the fact that dielectric constants and therefore refractive indices, including the real parts, vary somewhat with wavelength. Since focal length is a function of refractive index, this means that focal length varies with wavelength; that is, there are many focal lengths, each at a different wavelength. As a result, since image distance s' depends on focal length, there are many image planes, each for a different wavelength and each having a different magnification. Such lenses are called chromatic, meaning that their properties vary with color or wavelength.

A usual procedure to prevent such image distortion is to use one or more additional lenses, often cemented together, where the material of each lens is such that overall variations of refractive index are essentially nullified. If for one lens material n' increases with wavelength, then for another lens a material is used whose refractive index decreases correspondingly over the same wavelength region, thereby leading to an overall refractive index independent of wavelength. Such lens combinations are called *achromatic*. They require planning and workmanship and are more expensive than simple individual chromatic lenses.

The aberrations discussed next are called *monochromatic* aberrations because they exist for any specified color and refractive index.

If spherically shaped instead of parabolically shaped lenses are used, *spherical* aberrations usually result since only parabolic surfaces yield focal lengths independent of radial heights of incoming rays. Such aberrations can be manifested in both longitudinal and lateral directions.

Coma is a phenomenon whereby magnification varies with the radial coordinate of incoming rays. This again derives from the use of spherical rather than parabolic surfaces. Thus, when oblique rays are incident on a lens with coma, rays passing through edge portions of the lens are imaged at a different height than those passing through the center. If magnification for outer rays through a lens is greater than that for central rays, the coma is positive. If the reverse is true the coma is negative.

Except for a few special cases, no lens combination with spherical surfaces is completely free of both these spherical aberrations and coma. An optical system free of both is *aplanatic*, that is, image quality is identical throughout the field of view.

Astigmatism results when rays in the horizontal or sagittal plane and those in the vertical or tangential plane do not cross in the same spot in the image plane. Effects of astigmatism on the image can be limited by employing a stop in front of the lens as in Fig. 2.3 so as to make the optics more paraxial.

Distortion implies nonuniform lateral magnification. The image of an off-axis point is formed further from the axis or closer to the axis than would be expected from the paraxial expressions derived previously. Decrease of magnification toward the edge of the field results in *barrel distortion*. On the other hand, *pincushion distortion* corresponds to greater magnification at the borders.

No optical system can be free of all monochromatic aberrations. Therefore, it is customary to treat each separately. Certain ones can be made to vanish, but not all simultaneously. The business of minimizing total aberrations in lens design is a specialty in itself, as described in Ref. 2.2, for example. It is beyond the scope and purpose of this text. Many commercial software programs are available for this purpose. In general the greater the number of non-parabolic optical elements the more pronounced are the aberrations. Zoom lenses, for example, are generally of poorer quality than individual achromatic lenses.

REFERENCES

2.1. F. A. Jenkins and H. E. White, *Fundamentals of Optics* (3rd ed.), McGraw-Hill, New York, 1957.

2.2. W. J. Smith, *Modern Optical Engineering* (2nd ed.), McGraw-Hill, New York, 1990.

2.3. S. Ramo, J. R. Whinnery, and T. Van Duzer, *Field and Waves in Communication Electronics*, Wiley, New York, 1965.

2.4. M. Kline and I. W. Kay, *Electromagnetic Theory and Geometrical Optics*, Wiley, New York, 1965.

2.5. A. Yariv, *Optical Electronics* (3rd ed.), Holt, Rinehart and Winston, New York, 1985.

EXERCISES

2.1 From the geometry shown below, use Fermat's principle to show $\theta_1 = \theta_2$ for a smooth n-n' boundary.

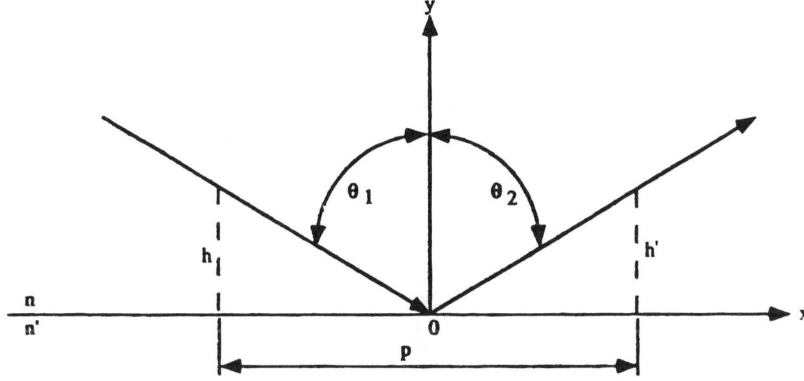

2.2 Consider a lens of thickness d, focal length f, and parabolic surface $z(r) = r^2/\{4f[(n_2/n_1) - 1]\}$ as shown in the figure below. Use Fermat's principle to calculate image distance s' for a ray traversing through the lens from the object point M at a distance s from the lens. Assume the actual optical path (solid line) and equivalent optical path (broken line) are identical.

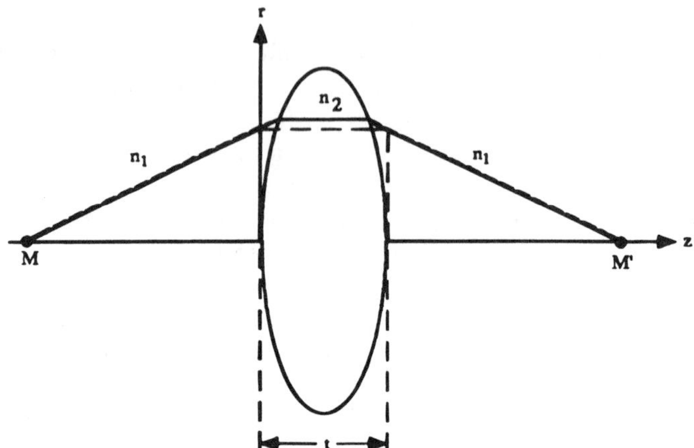

2.3 A lens in air is characterized by $r_1 = 15$ cm and $r_2 = 20$ cm. Is it a positive (convex) lens or a negative (concave lens)?

2.4 A convex lens of index 1.5 is immersed in a medium of index 1.6. Does the lens behave as a positive or negative focal length lens?

2.5 Using the fact that all paraxial rays are in phase at the focal point of a thin lens, obtain an expression for the lens surface.

2.6 Two lenses in series of focal lengths f_1 and f_2, respectively, are to form a beam expander, that is, a collimated (parallel) beam of diameter d_1 at the input to the first lens is to be converted to a collimated beam of diameter d_2 at the output of the second lens. What should the lens separation be and how is the expansion ratio d_2/d_1 related to the lens focal lengths?

2.7 A positive lens of focal length 10 cm is used to image an object 5 cm high located 40 cm to the left of the primary focal point. Find the position and size of the image using the Newtonian form of the thin lens equation.

2.8 Prove Helmholtz's law, Eq. (2.2.5). Use Fig. 2.15.

2.9 Solve Example 2.4 using the equivalent lens method.

2.10 Using Fig. 2.19, derive the Newtonian form of the thin lens equation for a thick lens by way of similar triangles. From it derive the Gaussian form. Assume $n = n''$.

2.11 An airborne reconnaissance system features a TV camera tube imaging system facing downward but protected by a circular perspex ($n = 1.49$) canopy. The camera is at the center of the canopy, which is characterized by inner and outer radii of 163 and 165 mm, respectively. The camera features a zoom lens, where focal length can be varied from 20 to 150 mm. The zoom lens is at a distance of 133 mm above the canopy. Assume a height of 1 km for the camera. If the camera is facing straight down, what is the effect of the canopy on image magnification in percent for extreme values of focal length? What should be done to minimize the effects of the canopy?

2.12 A ray parallel to the z axis ($r' = 0$) is incident to a *short* length L of selfoc fiber. After traversing a distance f after exiting the fiber, the output ray intersects the z axis. What should be the value of f so that the phase difference from the fiber input to f is independent of the radial coordinate r_0 at the fiber input? Assume $L << 0.1\sqrt{\frac{n_o}{n_1}}$.

2.13 An object 4 cm high is at a distance s from a concave mirror with a radius of curvature of 30 cm. What is image distance for $s = 60$ cm? $s = 10$ cm? What are the magnifications, and what types of images are obtained?

2.14 If one looks through a microscope with a relaxed eye, the image is viewed as if it were at infinity. What is the magnification for a 16-mm focal length objective?

2.15 The objective in Exercise 2.14 is placed in a 160-mm tube followed by an eyepiece so as to form a compound microscope, where tube length is essentially from the objective to the primary focal plane of the eyepiece where the image of the objective is formed. What is the overall magnification of the microscope if eyepiece focal length is 12.7 mm?

2.16 (a) Determine magnification of a telescope consisting of a 1.4-m focal length objective and a 1.4-cm focal length eyepiece. (b) A Barlow lens is inserted immediately in front of the eyepiece of part (a) so as to double magnification. What should be the focal length of the Barlow lens?

2.17 What is the acceptance angle at the center of a selfoc lens? Assume the rays are never actually incident on the core/cladding interface.

2.18 A 0.2-mm-diameter light-emitting diode (LED) with 20° beamwidth is to be used as a source in a 1-mrad beamwidth transmitter. What focal length and minimum diameter are required for the lens?

2.19 (a) What is minimum field of view when a 1/2-in. TV tube is used for star-gazing with a large-diameter 1.4-m focal length objective? (b) What is field of view when observing an object 5 km distant? (c) Repeat part (b) for an object 100 m distant. (d) What conclusion is reached by comparing the results of parts (a), (b), and (c)?

2.20 What is the matrix for a selfoc lens of length L?

2.21 Prove Eqs. (2.7.4), (2.7.6), and (2.7.7) using the matrix method.

Part Two

IMAGE BRIGHTNESS AND SIGNAL-TO-NOISE RATIO

CHAPTER

3

Radiometry and Photometry

Radiometry and photometry are similar terms in that both deal with the laws governing the transfer of optical radiation from one location to another—with or without the aid of optical systems. Traditionally, these subjects have included the description of the geometrical laws of radiation transfer, the behavior of radiant energy when transferred through and detected by optical systems, and the nature of radiation sources. Discussion of sources is deferred to Chapter 4.

Specifically, radiometry treats electromagnetic radiant energy of any wavelength, whereas photometry restricts itself to describing the laws of radiation transfer in the visible region of the spectrum, taking into account the physiological sensation that such radiation produces in an average human eye. If the human eye is regarded as merely a detector of electromagnetic radiation with a given spectral response, then photometry can be considered as just a special case of radiometry in which the characteristics of the detector are taken into account in computing the effect of the incident radiation. It is only natural, because of the spectral response of the human eye and its ready availability as a photodetector, that the development of the field of photometry preceded that of radiometry. As a result, however, two different sets of units are in use today, one describing the more general radiometric quantities and the other describing photometric quantities, which are applicable to visible optics only.

Because of its generality, the emphasis here is on radiometry [3.1–3.4]. The fundamental radiometric quantities are defined first followed by a presentation of the laws governing the transfer of radiant energy in the geometrical optics approximation. Photometry is presented last as a special case of the discussion of optical radiation given a specific detector (the eye) with a specific spectral response lying in the visible region of the spectrum. In this sense, our study of photometry is not only important as being the yardstick of visible optics but it also provides us with an example of the interaction of electromagnetic radiation energy with a detector having a specified spectral response. In this case the spectral response is that of the human visual system, or a detector-filter combination with similar spectral response. The reader is cautioned that radiometry is a discipline generally based on the ray theory of incoherent radiation. For coherent radiation, distribution of radiant power depends on phase relations, which can produce interference, as shown in Part 3. We concentrate here on radiometry for incoherent radiation.

3.1 FUNDAMENTAL RADIOMETRIC QUANTITIES

Electromagnetic fields can be characterized by *radiant energy* U measured in joules, or, more meaningfully in many instances, by *radiant energy density* u measured in joules per cubic meter.

Symbolically, u and U are related as

$$u \triangleq \frac{\partial U}{\partial V} \qquad (\mathrm{J \cdot m^{-3}}) \tag{3.1.1}$$

where V stands for volume.

Partial rather than total derivatives are used for the description of radiometric quantities since radiant energy may vary simultaneously with wavelength, position, direction, polarization, and time, as well as with volume.

The term *radiant power* (or flux) describes the rate at which radiant energy is transferred from one location to another. Radiant power, denoted P, is assumed to flow along noninterfering rays, and is defined as

$$P \triangleq \frac{\partial U}{\partial t} \qquad (\mathrm{W}) \tag{3.1.2}$$

3.1.1 Geometrical Relations

We generally are concerned with radiant power as it emanates in a given direction from an extended surface, as it passes through various components of an optical system, or as it impinges on a detection device. To handle these situations some basic concepts of radiant power in relation to the geometry involved must be developed. One of the most fundamental geometrical concepts involved in radiant power flow calculations is that of a *solid angle.*

Since the remaining radiometric quantities involve the concept of flux per unit solid angle, it is worthwhile to recall first the definition of a solid angle. By definition, a cone is the volume swept out when a straight line passing through a fixed vertex is moved through every point on a closed, nonintersecting curve, as illustrated in Fig. 3.1. The solid angle Ω in units of steradians (sr) subtended by a cone is given by the expression

$$\Omega = \frac{a}{r^2} \tag{3.1.3}$$

where a is the area intercepted by the cone on the surface of a sphere of radius r

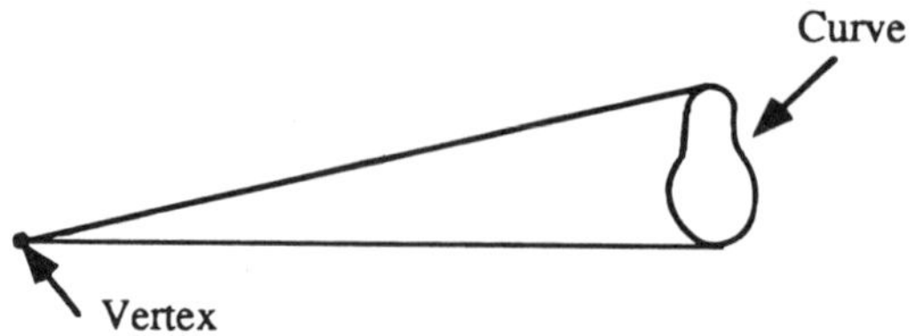

Fig. 3.1 Cone.

centered on the cone vertex, as shown in Fig. 3.2. The solid angle subtended by an entire sphere is thus equal to 4π sr.

As an example of a radiometric parameter utilizing the concept of solid angle consider *radiant intensity* J, defined as the radiant power P radiated from a point source in a given direction per element of solid angle $d\Omega$:

$$J = \frac{\partial P}{\partial \Omega} \qquad (\mathrm{W \cdot sr^{-1}}) \tag{3.1.4}$$

Many optical sources can be regarded, for all practical purposes, as point sources. A source is uniformly *isotropic* if it radiates equally in all directions. For this special source, Eq. (3.1.4) yields

$$P = \int_{\substack{\text{all}\\\text{directions}}} J\, d\Omega = J \int_{\substack{\text{all}\\\text{directions}}} = 4\pi J \tag{3.1.5}$$

or

$$J_{\text{uniform isotropic point source}} = \frac{P}{4\pi} \qquad (\mathrm{W \cdot sr^{-1}}) \tag{3.1.6}$$

In general, however, a major concern is nonisotropic point sources whose radiant intensities are functions of direction, in which case J must remain within the integral sign in Eq. (3.1.5).

When the dimensions are such that a radiation source cannot be considered as a point source, its radiating surface area must be considered. To aid in the description of these "extended" sources the concept of *radiance* is defined. Radiance is a measure of the radiant power per unit projected area dA_p of the source per unit solid angle $d\Omega$ in a particular direction (the direction of the rays defining $d\Omega$). Mathematically,

$$N = \frac{\partial^2 P}{\partial A_p \partial \Omega} \qquad (\mathrm{W \cdot m^{-2} \cdot sr^{-1}}) \tag{3.1.7}$$

Frequently, it is of interest to consider the rate at which radiant energy is emitted

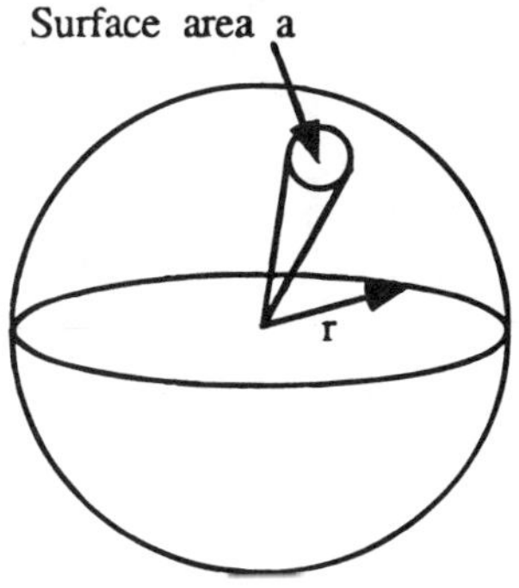

Fig. 3.2 Solid angle.

from a unit projected area of an extended source into a hemisphere without further regard to specific directivity. This geometry is described as *radiant emittance* W:

$$W = \frac{\partial P}{\partial A_p} \qquad (\mathrm{W \cdot m^{-2}}) \tag{3.1.8}$$

where the total radiation dP from the elementary projected area dA_p is to include all that which is radiated into a solid angle of 2π sr about dA. From Eqs. (3.1.6), (3.1.7), and (3.1.8), we obtain

$$J = \int_{\substack{\text{source}\\ \text{area}}} N \, dA_p \qquad (\mathrm{W \cdot sr^{-1}}) \tag{3.1.9}$$

$$W = \int_{2\pi \mathrm{sr}} N \, d\Omega \qquad (\mathrm{W \cdot m^{-2}}) \tag{3.1.10}$$

These relationships establish radiance as the fundamental quantity for geometrical computations.

Determination of radiance by Eq. (3.1.7) and emittance by Eq. (3.1.8) assumes a projected area A_p that is the component of the area normal to the direction of propagation as shown in Fig. 3.3, that is,

$$dA_p = dA \cos\theta \tag{3.1.11}$$

where θ is the angle between the outward surface normal of elemental area dA and the direction of propagation.

Thus Eqs. (3.1.7) and (3.1.8) are more properly written

$$N = \frac{\partial^2 P}{\cos\theta \; \partial A \; \partial \Omega} \qquad (\mathrm{W \cdot m^{-2} \cdot sr^{-1}}) \tag{3.1.12}$$

$$W = \frac{\partial P}{\cos\theta \; \partial A} \qquad (\mathrm{W \cdot m^{2}}) \tag{3.1.13}$$

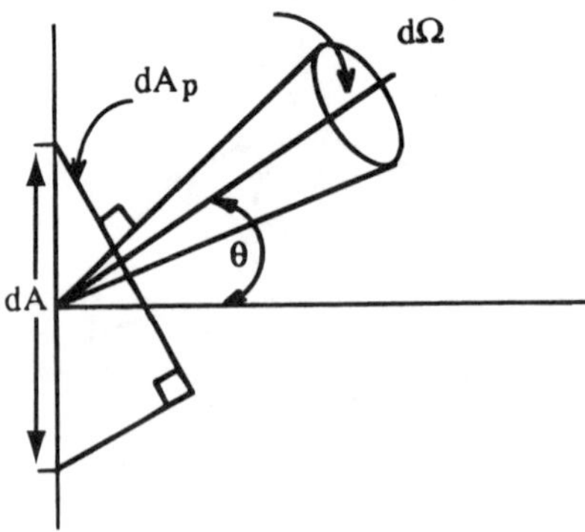

Fig. 3.3 Radiating and projected radiating areas.

Because radiance decomposes radiation propagation to basic components involving both area of emittance as well as directivity of propagation, radiance is generally the most fundamental radiometric parameter and is the basic parameter used in calculations of power transmittance to a receiver or images, as shown in Section 3.2 and Eq. (3.2.2).

To complete the group of terms needed to adequately describe the flow of radiant power, a measure of the radiant power incident on or received by an elemental surface area is needed. This quantity is called *irradiance* and it is denoted by the symbol H. Specifically, irradiance is a measure of the radiant power per unit area *received* by an elemental surface area A_r. Mathematically,

$$H = \frac{\partial P}{\partial A_r} \qquad (\mathrm{W \cdot m^{-2}}) \tag{3.1.14}$$

Note that while H has the same units as W it refers to radiation incident on a receiving surface while W refers to radiation leaving an emitting surface. To avoid confusion, the elemental receiving area in Eq. (3.1.14) is denoted dA_r. Note that the radiometric parameter irradiance is identical to the *time-averaged Poynting vector* in Eq. (1.3.11b).

For detection of total received power as in optical communication, there is no particular importance to variation of received power over the receiver and small "point" detectors are used to receive total radiation focused optically to a "point." If larger detectors are used, the output signal is proportional to irradiance averaged over the detector surface area, or to total received power. However, image information is the *variation* of radiated power over the image plane. "Point" detectors of total radiation cannot be used to see variation (unless they sample the image spatially). Therefore, received power per unit area is of paramount importance in the image signal-to-noise ratio (SNR) rather than total received power. For this reason *irradiance* is critical in imaging system analysis.

When recording the irradiance at some location, the term *exposure* is used to refer to the accumulation of energy incident on the recording material area over a period of time. Although not considered a fundamental radiometric quantity, exposure is related to irradiance as

$$E = \int_{t_1}^{t_2} H(t)\, dt \qquad (\mathrm{J \cdot m^{-2}}) \tag{3.1.15}$$

3.1.2 Spectral Radiometric Quantities

The radiometric quantities described in the preceding section have been defined as differential with respect to volume, time, area, solid angle, and/or direction. Frequently, these quantities are also differential with respect to wavelength (or frequency). When such is the case the subscript λ (or ν) is used to indicate differentiation with respect to wavelength (or frequency). For example, if radiant power is a function of wavelength, P_λ and P_ν are defined as

$$P_\lambda = \frac{\partial P}{\partial \lambda} \qquad (\mathrm{W \cdot \mu m^{-1}}) \tag{3.1.16}$$

$$P_\nu = \frac{\partial P}{\partial \nu} \qquad (\mathrm{W \cdot Hz^{-1}}) \tag{3.1.17}$$

(It may be considered preferable to refer all quantities to frequency only and not wavelength because if energy is conserved ν does not vary with the medium in which the energy is propagating, whereas λ does. However, λ is popular by tradition.)

In this context, the total radiant power between wavelengths λ_1 and λ_2 is given by

$$P = \int_{\lambda_1}^{\lambda_2} P_\lambda \, d\lambda \tag{3.1.18}$$

Other examples of spectral radiometric quantities are *spectral radiant intensity*,

$$J_\lambda = \frac{\partial J}{\partial \lambda} = \frac{\partial^2 P}{\partial \Omega \, \partial \lambda} \qquad (\mathrm{W \cdot \mu m^{-1}}) \tag{3.1.19}$$

spectral radiance,

$$N_\lambda = \frac{\partial N}{\partial \lambda} = \frac{\partial^3 P}{\cos\theta \, \partial A \, \partial \Omega \, \partial \lambda} \tag{3.1.20}$$

and so on.

A complete tabulation of the fundamental radiometric quantities is given in Table 3.1.

3.2 THE FUNDAMENTAL RADIOMETRIC EQUATION

On the basis of the foregoing definitions, incoherent radiation from a source, or at a receiver, in an optical system can be described completely by specifying the radiance

TABLE 3.1 *The Fundamental Radiometric Quantities*

Quantity	Symbol	Definition	Units
Radiant energy	U		J
Radiant energy density	u	$\partial U/\partial v$	$\mathrm{J \cdot m^{-3}}$
Radiant power	P	$\partial U/\partial t$	W
Radiant intensity	J	$\partial P/\partial \Omega$	$\mathrm{W \cdot sr^{-1}}$
Radiant emittance	W	$\partial P/\partial A \cos\theta$	$\mathrm{W \cdot m^{-2}}$
Radiance	N	$\partial^2 P/\cos\theta \, \partial A \, \partial \Omega$	$\mathrm{W^{-2} \cdot m^{-2} \cdot sr^{-1}}$
Irradiance	H	$\partial P/\partial A_r$	$\mathrm{W \cdot m^{-2}}$
Spectral radiant power	P_λ P_ν	$\partial P/\partial \lambda_r$ $\partial P/\partial \nu$	$\mathrm{W \cdot \mu m^{-1}}$ $\mathrm{W \cdot Hz^{-1}}$
Spectral radiant intensity	J_λ J_ν	$\partial J/\partial \lambda$ $\partial J/\partial \nu$	$\mathrm{W \cdot sr^{-1} \cdot \mu m^{-1}}$ $\mathrm{W \cdot sr^{-1} \cdot Hz^{-1}}$
Spectral radiant emittance	W_λ W_ν	$\partial W/\partial \lambda$ $\partial W/\partial \nu$	$\mathrm{W \cdot m^{-2} \cdot \mu m^{-1}}$ $\mathrm{W \cdot m^{-2} \cdot Hz^{-1}}$
Spectral radiant emittance	N_λ N_ν	$\partial N/\partial \lambda$ $\partial N/\partial \nu$	$\mathrm{W \cdot m^{-2} \cdot sr^{-1} \cdot \mu m^{-1}}$ $\mathrm{W \cdot m^{-2} \cdot sr^{-1} \cdot Hz^{-1}}$

as a function of wavelength (or frequency), position, direction, time, and polarization. (Since most optical radiation beams are steady, time is usually not a significant parameter; also, for incoherent radiation, polarization can be neglected). Thus, if the spectral radiance N_λ is known at each point on the surface of a source or receiver and for each ray between source and receiver, the total radiant power P in the beam itself at any point in the optical system is, from Eq. (3.1.20)

$$P = \iiint N_\lambda \cos\theta \, d\lambda \, d\Omega \, dA \tag{3.2.1}$$

where the integration is carried out over any convenient reference surface that intersects the beam and N_λ [$= N_\lambda(\lambda)$] is the spectral radiance at the source for a ray through the surface element dA within the element of solid angle $d\Omega$, as illustrated in Fig. 3.3. If the medium between the source and the region where P is calculated causes energy loss by reflection, scattering, or absorption this effect can be included in Eq. (3.2.1) by defining a *generalized spectral transmittance* $\tau(\lambda)$. Thus,

$$P = \iiint N_\lambda \tau \cos\theta \, d\lambda \, d\Omega \, dA \tag{3.2.2}$$

where τ[$= \tau(\lambda)$] is wavelength dependent with magnitude between 0 (no energy reaches the region at which P is calculated) and 1 (no loss).

Until now, total transmitted power has been considered, where $d\Omega$ refers to transmitted beamwidth and dA refers to emitting area.

To determine total *received* power, Eq. (3.2.2) is also used. Area dA now refers to emitting area *seen* by the imager. The emitting area seen by the receiver is either the total emitting area if all of it is within the field of view, or only part of the emitting area if the emitting area extends beyond the field of view. In the latter case, the emitting area seen by the receiver is approximately the receiver field of view in steradians multiplied by the square of the range z. The solid angle of integration in Eq. (3.2.2) for received power must also be clarified. If the receiving optics are large enough to intercept the whole transmitted beam, then $d\Omega$ still refers to transmitted beamwidth. However, often the receiver intercepts only a portion of the transmitted beam. In this case, $d\Omega$ refers to the solid angle subtended by the imager, which is approximately A_r/z^2. This is illustrated in Examples 3.1 and 3.4.

Example 3.1

A point source emits uniformly 40 W · sr^{-1} toward a 3-cm-diameter lens. How much power is collected when the lens is 50 m distant from the source?

Solution

Since a point rather than an extended source is under consideration, the appropriate radiometric parameter is intensity rather than radiance.

$$P = \int J \, d\Omega = J\Omega$$

The solid angle Ω_L subtended by the lens is approximately the area of the lens (πr_L^2) divided by the distance (z) squared. Therefore, the power collected by the lens is

$$P = 40 \text{ W}\cdot\text{sr}^{-1} \cdot \frac{\pi(0.015 \text{ m})^2}{(50 \text{ m})^2} \text{ sr} = 1.13 \times 10^{-5} \text{ W}$$

3.2.1 Specular and Diffuse Surfaces

It is difficult to describe the properties of radiating and reflecting (i.e., receiving) surfaces in a general way. Therefore, it is useful to develop relationships describing surfaces in certain idealized configurations. The two most important of these are *specular* and *diffuse* surfaces. A *specular* surface is one that closely obeys the requirements of Snell's law. Examples of such surfaces are mirrors, clear lenses, and highly glossed objects. These have smooth surfaces. The radiance of a specular radiating surface or the irradiance of a specular reflecting surface is a sensitive function of the angle from which it is viewed (in the case of a radiator) or irradiated (in the case of a reflector).

When a plane element of surface area has a radiance (or irradiance) independent of the viewing (or irradiating) angle, it is said to be *uniformly diffuse.* Physically speaking, radiation striking a diffuse surface is scattered in all directions in such a way that the surface appears to have the same radiance regardless of the angle at which it is irradiated or the angle from which it is viewed. These are usually "rough" surfaces as described in Section 1.4.1 with regard to Lambert's law. A white blotter is a typical example of a diffuse reflecting surface while ground glass diffuses optical radiation passing through it.

To describe mathematically a uniformly diffuse surface it is noted from Eqs. (3.1.4) and (3.1.12) that

$$N = \frac{\partial J}{\partial A \cos \theta} \tag{3.2.3}$$

Thus, in order for radiance N to be constant (i.e., independent of angle θ in Fig. 3.3) the radiant intensity J must vary as $\cos\theta$:

$$J(\theta) = J(0) \cos \theta = \int N(0) \cos \theta \, dA \tag{3.2.4}$$

where θ is the angle between the normal to the surface area dA and $J(0)$ and $N(0)$ are values of radiant intensity and radiance, respectively, for $\theta = 0$ (i.e., normal to dA).

Equation (3.2.4) is known as *Lambert's law,* as described in Section 1.4.1, and a uniformly diffuse surface is often referred to as a Lambertian surface or source. Some translucent materials such as opal glass provide almost uniformly diffuse transmission. If a reflecting surface reflects *all* incident energy uniformly it is said to be a *perfect diffuser.* Equations (3.2.3) and (3.2.4) imply that as a Lambertian source is viewed from different angles, the projected source area changes accordingly. However, these changes are cancelled by identical changes in direction intensity of the radiation, so that radiance or brightness of Lambertian sources or surfaces is independent of viewing angle.

An additional property of Lambertian sources is illustrated by the experimental setup shown in Fig. 3.4. An opening of area dA_0 in screen S is irradiated by a Lambertian source that can be rotated so as to vary the angle θ between the source normal and the screen normal. The flux hitting an area element dA_1 that is parallel to screen S and centered on the opening is measured as a function of the angle θ. From the definition of radiance, the power incident on the element dA_1 is given by

$$d^2P = N(0)\, dA_p\, d\Omega \tag{3.2.5}$$

Here the projected area element is given by $dA_p = dA_0$ for *any* angle θ, and the solid angle subtended by elemental area dA_1 is given by $d\Omega = dA_1/z^2$. Thus Eq. (3.2.5) becomes

$$d^2P = \frac{N(0)\, dA_0\, dA_1}{z^2} \tag{3.2.6}$$

which is independent of the orientation angle θ.

3.2.2 Radiant Power from a Lambertian Source

Consider a uniformly diffuse or rough plane surface source of area dA radiating into a circular cone as illustrated in Fig. 3.5. The cone has its apex at dA and is generated by rotating the line r about the z axis at a planar angle θ. The power $dP(\theta)$ radiated within the differential solid angle $d\Omega$ (defined between the cone of generating angle θ and the cone of angle $\theta + d\theta$) is, from Eq. (3.2.2), and ensuing discussion,

$$dP(\theta) = N \cos\theta\, dA\, d\Omega_r \tag{3.2.7}$$

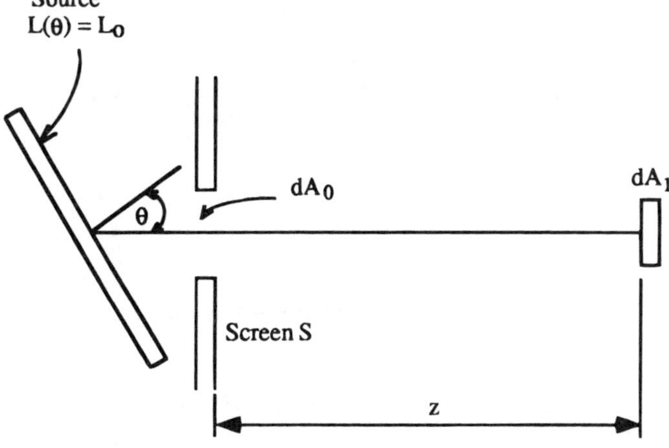

Fig. 3.4 Experiment involving a Lambertian source.

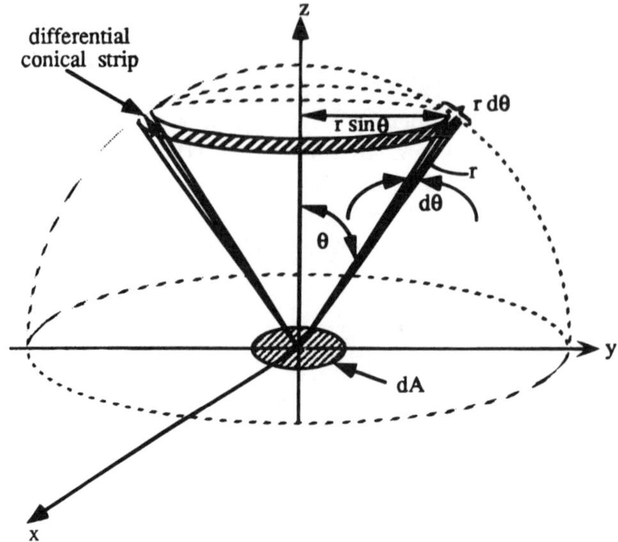

Fig. 3.5 Power radiation from a uniformly diffuse (Lambertian) source.

where, by definition,

$$d\Omega_r = \frac{\text{area of sphere of radius } r \text{ enclosed between the two zones}}{r^2} \tag{3.2.8}$$

$$= \frac{2\pi(r \sin\theta)(r\, d\theta)}{r^2} = 2\pi \sin\theta\, d\theta$$

Substituting Eq. (3.2.8) into Eq. (3.2.7) yields

$$dP(\theta) = 2\pi N\, dA \cos\theta \sin\theta\, d\theta \tag{3.2.9}$$

and the power radiated into a cone of generating angle θ is

$$P(\theta) = \int_0^\theta dP(\theta) = \int_0^\theta 2\pi N\, dA \cos\theta \sin\theta\, d\theta$$

Since the source is Lambertian, $N = N(0)$ is constant and is not a function of θ. Therefore it and dA can be removed from the integral sign giving

$$P(\theta) = 2\pi N\, dA \int_0^\theta \cos\theta \sin\theta\, d\theta = \pi N\, dA \sin^2\theta \tag{3.2.10}$$

The total radiant power into a hemisphere about dA is obtained from Eq. (3.2.10) by setting $\theta = \pi/2$ giving

$$P = \pi N\, dA \tag{3.2.11}$$

The corresponding radiant emittance is

$$W = \frac{P}{dA} = \pi N \tag{3.2.12}$$

By comparing Eq. (3.2.10) with Eq. (3.2.2), it follows that the effective solid angle of radiation into a cone of half-planar angle θ from a Lambertian source is

$$\Omega_L = \pi \sin^2\theta \tag{3.2.13}$$

Note that for a specular source the cosθ term is dropped from Eqs. (3.2.7) and (3.2.9) and the effective solid angle is (see Exercise 3.3)

$$\Omega_s = 2\pi(1 - \cos\theta) \tag{3.2.14}$$

The radiant power and emittance radiated by a specular source into a hemisphere are twice those of a Lambertian source (Exercise 3.3).

Inspection of Eq. (3.2.13) indicates that while for specular sources solid angles by definition [Eq. (3.1.3) and Fig. 3.2] refer to areas intercepted on the *surface* of a sphere by cones, for Lambertian sources solid angles essentially refer to cross-sectional areas intercepted by such cones. For cone beamwidths of small planar angles there is little difference. This is relevant to laser beams, for example, which may often be characterized by milliradian-order beamwidths.

Example 3.2

Consider a source such as a laser radiating into a small planar angle beamwidth of 2θ. Determine equivalent solid angles if the source is considered Lambertian or specular.

Solution

From Fig. 3.6 the arc of a spherical surface of radius r intercepted by a cone of half-angle θ is $2r\theta$.

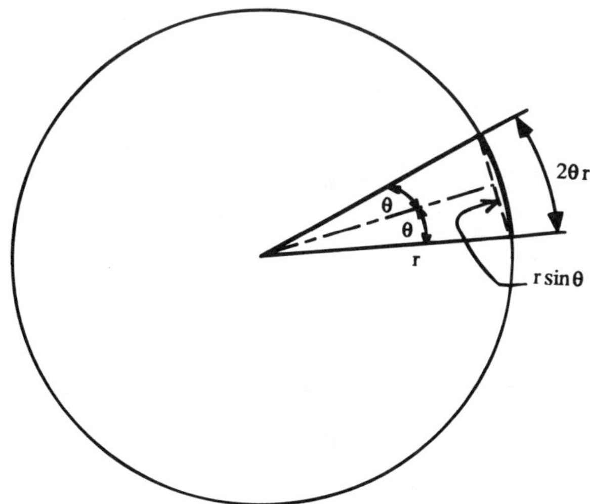

Fig. 3.6 Comparison between cross-sectional area of cone and surface area of sphere intercepted by cone.

If the source is specular, the solid angle of interest involves the area on the surface of the sphere intersected by the cone. In view of the arc of this circle, its area is approximately $(\pi/4)(2\theta r)^2 = \pi\theta^2 r^2$. The solid angle is therefore approximately $\Omega_s \approx \pi\theta^2$ for a specular source. The value $\pi\theta^2$ is slightly larger than the exact value $2\pi(1 - \cos\theta)$ in Eq. (3.2.14). This approximation is quite valid, however, for small angles θ, and only when θ approaches a radian or so does the discrepancy become increasingly noticeable.

If the source is Lambertian, then the solid angle involves the cross-sectional area whose diameter is the dashed line in Fig. 3.6. The radius of this circle is $r \sin\theta$, yielding the solid angle $\Omega_L = \pi \sin^2\theta$ given in Eq. (3.2.13). For small values of θ, $\Omega_L \approx \pi\theta^2 \approx \Omega_s$. The approximation $\pi \sin^2\theta \approx \pi\theta^2$ is quite valid for small values of θ, and only when θ approaches about 0.256 radians does the discrepancy become increasingly noticeable. In general, $\pi\theta^2 > 2\pi(1 - \cos\theta) > \pi \sin^2\theta$. However, for small values of θ all three expressions are essentially identical so that $\Omega_L \approx \Omega_s$.

Example 3.3

In Fig. 3.7 elemental area dA_r receives light from a point source p. The distance from p to dA_r is r. Find irradiance at dA_r.

Solution

As shown in Fig. 3.7, the solid angle subtended at p by dA_r is

$$d\Omega \approx \frac{dA_r \cos\phi}{r^2} \tag{3.2.15}$$

If the radiantly intensity of p in the direction of dA_r is J, then the power radiated into the solid angle $d\Omega$ is

$$dP = J\, d\Omega \approx \frac{J\, dA_r \cos\phi}{r^2} \tag{3.2.16}$$

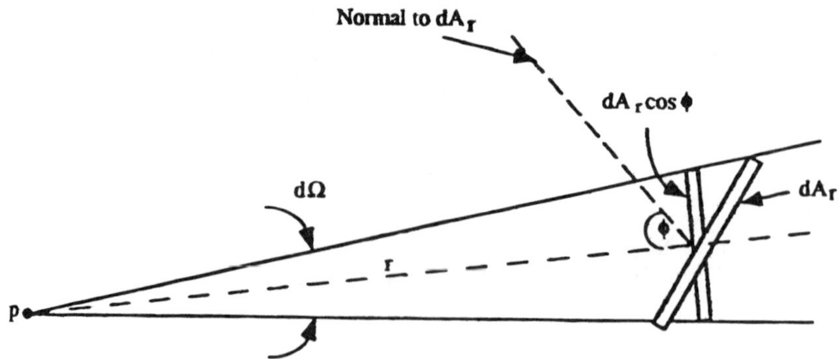

Fig. 3.7 Irradiance of surface element by point source.

and via Eq. (3.1.13) the irradiance is

$$H = \frac{\partial P}{\partial A_r} \cong \frac{J \cos \phi}{r^2} \tag{3.2.17}$$

which is known as the *inverse-square law* and holds when the distance between p and dA_r is large enough compared to dA_r so that the irradiated area subtends only a small angle $d\Omega$ when viewed from the source.

If a real source is too large to be treated as a point source then it becomes uncertain what r and ϕ are in Eq. (3.2.17) since they vary for different parts of a finite source. Also, for a finite source, the irradiance at dA_r is the sum of radiant power emitted from many points so that it would tend to fall off more slowly with increasing distance than is indicated by Eq. (3.2.17). This is illustrated later in Example 3.5.

Example 3.4

An outdoor scene in Fig. 3.8 is illuminated by solar irradiance H_s. Assuming an average power reflectance ρ (defined in Section 1.4.3) for objects in the scene, and assuming that at optical wavelengths their surface roughness is such that they are uniform diffuse reflectors, what is the average irradiance reaching the receiving optics situated at a distance z from the scene and at an angle θ with respect to the normal to the outdoor scene? Assume a receiver field of view of Ω_r sr and that the scene fills the entire field of view. The area of receiving optics is A_r.

Solution

This is a two-step problem. First object radiance must be determined, and then from it image irradiance is determined. For solar irradiance H_s, emittance of the scene consists of reflected solar light and is equal to ρH_s. Since the objects viewed are diffuse, their average radiance is $N = W/\pi = \rho H_s/\pi$ for zero degree direction of reflection.

Power received by the imager is, from Eq. (3.2.2),

$$P_r = N\Omega A \cos \theta = \frac{\rho H_s}{\pi} \Omega A \cos \theta$$

The object source subtends the entire receiver field of view Ω_r. Therefore, the object scene area seen by the imager is $A = \Omega_r z^2$. The angle subtended by the receiver is $\Omega = A_r/z^2$. Note that Ω is from the scene to the receiver while Ω_r is from the receiver to the scene, as depicted in Fig. 3.8. By substitution,

$$P_r = \frac{\rho H_s}{\pi} \frac{A_r}{z^2} \cdot \Omega_r z^2 \cos \theta = \frac{\rho H_s A_r \Omega_r \cos \theta}{\pi} \tag{3.2.18}$$

Average irradiance incident to the *receiver optics* is

$$H_r = \frac{P_r}{A_r} = \frac{\rho H_s \Omega_r \cos \theta}{\pi} \tag{3.2.19}$$

This example illustrates application of radiometry to typical imaging situations. H_r

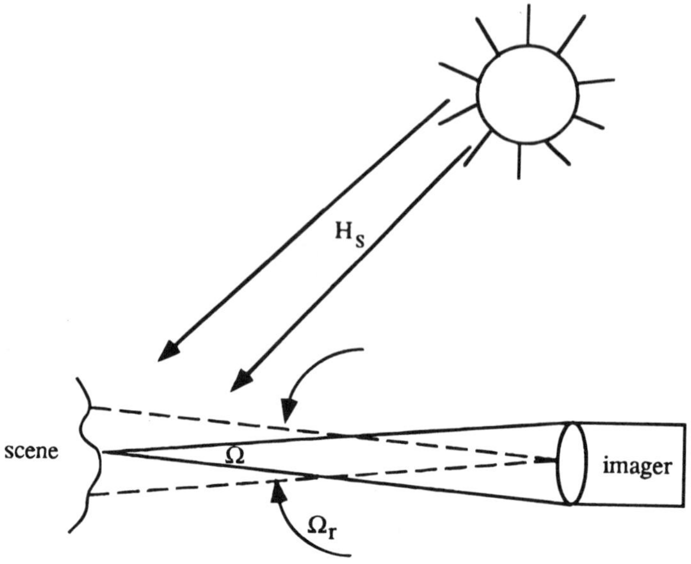

Fig. 3.8 Outdoor imaged scene.

is not the irradiance in the image plane behind the receiver optics. The irradiace there is $H_r A_r / dA'$, where dA' is image area. However, a more direct expression is considered in Section 3.4 following the radiance theorem.

3.3 THE RADIANCE THEOREM

An important question that arises frequently when radiant energy is transferred through an optical system is the determination of radiance N' and irradiance H' at the output (i.e., in the image) of the system given the radiance N of the object at the input.

To answer this question consider the typical case of an optical system forming an image of area dA' of a uniformly diffuse object surface of area dA and radiance N, the normal to dA being the optical axis of the system. The system is characterized by an entrance and exit pupil as shown in Fig. 3.9. Entrance pupil is the image of the aperture stop as seen from the object side of the lens in object space. Exit pupil is the image of the clear aperture of the lens limiting the field of view. Since all light transmitted by an optical system must pass through the aperture stop, all light in the image space appears to pass through the exit pupil. Between entrance and exit pupils in Fig. 3.9 lies any optical system such as lenses and/or mirrors.

The radiant power dP entering the optical system within the annular ring of the entrance pupil between θ and $\theta + d\theta$ is given by Eq. (3.2.9) as

$$dP = 2\pi N \, dA \cos\theta \sin\theta \, d\theta \qquad (3.3.1)$$

The radiant power dP emerging from the exit pupil of the optical system within θ' and $\theta' + d\theta'$ is

$$dP_i = \tau' dP = \tau' 2\pi N \, dA \cos\theta \sin\theta \, d\theta \qquad (3.3.2)$$

where $\tau'(<1)$ is transmittance through the optical system.

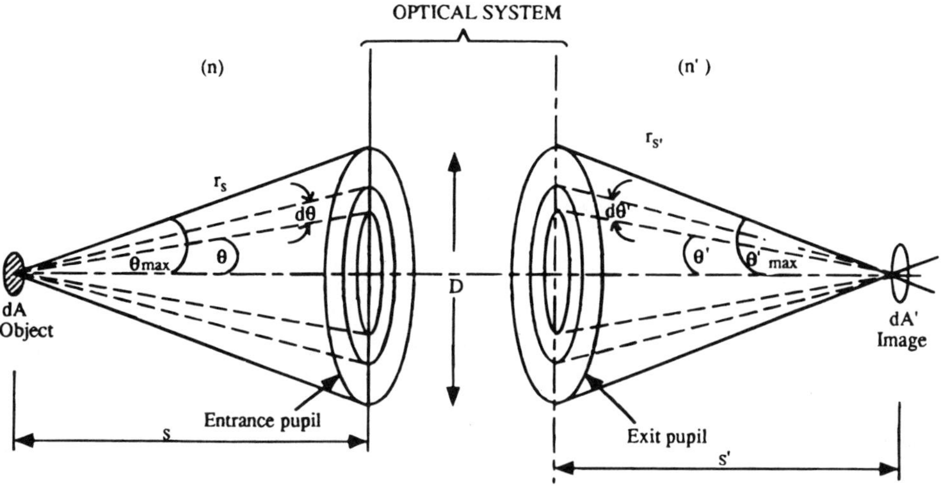

Fig. 3.9 Geometry for derivation of radiance theorem.

To accurately image linear source elements of dA that lie perpendicular to the optical axis, Abbe's law Eq. (2.2.1) must be satisfied; hence,

$$nh \sin \theta = n'h' \sin \theta' \tag{3.3.3}$$

where n and n' are, respectively, the indices of refraction at the input (object space) and output (image space) of the optical system and h and h' are, respectively, the source and image element heights about the optical axis. Squaring Eq. (3.3.3) yields

$$n^2h^2 \sin^2\theta = n'^2(h')^2 \sin^2\theta'$$

or

$$n^2(\pi h^2) \sin^2\theta = n'^2(\pi h'^2) \sin^2\theta'$$

If dA is considered circular, then

$$n^2 \, dA \sin^2\theta = n'^2 \, dA' \sin^2\theta' \tag{3.3.4}$$

This is differentiated with respect to the variables θ and θ'

$$n^2 \, dA^2 \sin \theta \cos \theta \, d\theta = n'^2 \, dA'^2 \sin \theta' \cos \theta' \, d\theta'$$

from which

$$dA \sin \theta \cos \theta \, d\theta = \left(\frac{n'}{n}\right)^2 dA' \sin \theta' \cos \theta' \, d\theta'$$

Substituting this expression in Eq. (3.3.2) yields for the radiant power in the image

$$dP_i = \tau'\left(\frac{n'}{n}\right)^2 2\pi N\, dA' \sin\theta' \cos\theta'\, d\theta' \tag{3.3.5}$$

or

$$dP_i = 2\pi N'\, dA' \sin\theta' \cos\theta'\, d\theta' \tag{3.3.6}$$

where image radiance is

$$N_i = \tau\left(\frac{n'}{n}\right)^2 N \tag{3.3.7}$$

This indicates that the image area dA' (which can be projected on a screen or observed directly by a detector or the eye) has radiance given by Eq. (3.3.7). The ratio of image radiance to object radiance is therefore $\tau(n'/n)^2$ where τ is the transmittance of the optical system and is a measure of its losses. When $n = n'$ and the losses are negligible, Eq. (3.3.4) tells us that the radiance throughout an optical system is invariant. Equation (3.3.4) is known as the *radiance theorem*. When dealing with visible light only (i.e., in photometry) it is referred to as the *luminance* or *brightness theorem*.

3.4 IMAGE IRRADIANCE

From Eq. (3.2.10) total radiant power P at the entrance pupil of the optical system in Fig. 3.9 is

$$P = \pi N\, dA \sin^2\theta_{max} \tag{3.4.1}$$

The radiant power P' emerging from the exit pupil is therefore

$$P' = \tau' P = \tau' \pi N\, dA \sin^2\theta_{max} \tag{3.4.2}$$

Using Eq. (3.3.4) this becomes

$$P' = \tau' \pi N \left(\frac{n'}{n}\right)^2 dA' \sin^2\theta'_{max} \tag{3.4.3}$$

Hence, the mean irradiance at the image plane is, in terms of object radiance,

$$H_i = \left(\frac{P'}{dA'}\right) = \tau' \pi N \left(\frac{n'}{n}\right)^2 \sin^2\theta'_{max} \tag{3.4.4}$$

In photographic and detection systems when maximum irradiance (illumination) of the photographic plate or detector surface is desired, large θ'_{max} in conjunction with low system transmission loss is required. It is recognized from Fig. 3.9, however, that an increase in θ'_{max} implies a larger aperture and therefore increased aberration. Alternatively, θ'_{max} can be increased by decreasing image distance s'. Since lateral magnification is $|m| = s'/s$, this implies decreased magnification; that is, the same radiant power P' is distributed over decreased image area, thus increasing irradiance of the image at the expense of image size. Radiant power is conserved.

For practical reasons $n = n'$ in most cases. However, lens immersion in situations where the image space is given as $n' > n$ has been used as a means of increasing the irradiance in the image plane in some infrared systems by decreasing image area. The effect on dA' can explain physically why the ratio (n'/n) is to the second power. Note that $N' \sin \theta'_{max}$ is the image plane numerical aperture. For a lossless system, this product remains constant for a given object plane numerical aperture in keeping with Abbé's law.

Although Eq. (3.4.4) is a *basic* equation for image irradiance and is always valid, it is often much more convenient to use a form of this equation that deals with image magnification rather than θ'_{max}. Such forms involve approximations. From Fig. 3.9, $\sin \theta'_{max} = D/2r'_s$. In terms of optical path, by Fermat's principle $r'_s = s'$ for a thin lens. Therefore,

$$H_i = \tau' W \left(\frac{n'}{n}\right)^2 \frac{D^2}{4(s')^2} \tag{3.4.5}$$

where it is assumed that at optical wavelengths most objects are diffuse reflectors. For a real image ($s' \geq f$), the equality holds only when the object is at infinity. Let $b = s'/f$ express the extent to which the object is in from infinity. Then

$$H_i = \tau' W \left(\frac{n'}{n}\right)^2 \frac{D^2}{4b^2 f^2} = \frac{\tau' W}{4b^2 F^2} \left(\frac{n'}{n}\right)^2 \tag{3.4.6}$$

where $F = f/D$ is commonly known as the *f-number*. Since

$$b = \frac{s'}{f} = s'\left(\frac{1}{s'} + \frac{1}{s}\right) = 1 + m$$

where m is lateral magnification,

$$H_i = \frac{\tau' W}{4F^2(1 + m)^2} \left(\frac{n'}{n}\right)^2 \tag{3.4.7}$$

Equations (3.4.5) and (3.4.7) are approximations for Eq. (3.4.4) and pertain to objects at any distance. However, if the object is very distant, that is, $s \rightarrow \infty$, then $s' \rightarrow f$ and $b = 1$. In this case Eq. (3.4.6) reduces to

$$H_i = \frac{\tau' W}{4F^2} \left(\frac{n'}{n}\right)^2 \tag{3.4.8}$$

Another approximation approach used by some workers is to consider the difference in physical path between r'_s and s' in Fig. 3.9, that is, $r'^2_s = (s') + D^2/4 = f^2 + D^2/4$ for a very distant object. According to this approach, Eq. (3.4.6) is amended to

$$H_i = \tau' W \left(\frac{n'}{n}\right)^2 \frac{D^2}{4(f^2 + D^2/4)} = \frac{\tau' W}{4F^2 + 1} \left(\frac{n'}{n}\right)^2 \tag{3.4.9}$$

for very distant objects.

Equations (3.4.8) and (3.4.9) are both used but by different workers in the field. Each one is based on a different understanding of the relationship between r_s' and s'. For large *f*-numbers, there is little difference between Eqs. (3.4.8) and (3.4.9), corresponding to small values of θ'_{max} for which $\tan\theta'_{max} \cong \sin\theta'_{max}$. Both equations indicate that large *f*-numbers result in decreased irradiance of the image. In photography the *f*-number can be varied by the use of irises as in Fig. 2.3 to change effective lens diameter. In this case magnification or image size is not varied, but radiative power collected by the lens is altered. For given required exposures as defined in Eq. (3.1.14), larger *f*-numbers give rise to a requirement to increase exposure time as a result of the decreased image irradiance. This requires high-speed films, where speed is reciprocal to exposure.

Exercises 3.12 and 3.13 involve numerical calculations and comparisons of image irradiances according to the basic and various approximation forms of Eq. (3.4.4) derived here. It is seen that the various approximation forms [Eqs. (3.4.5), (3.4.7)] that can be used to describe image irradiance for a near object all yield rather close results. The various forms to describe image irradiance for a very distant object [Eqs. (3.4.8), (3.4.9)] yield results similar to each other and are different from those that pertain to near objects.

We have already established in Example 3.4 that objects with diffuse surfaces illuminated with irradiance H have an average radiance $\rho H/\pi$. This can be substituted into Eq. (3.4.4). Next, we consider how object scene emittance W in the above equations is affected by artificial illumination (for solar irradiance illumination see Example 3.4).

Example 3.5

Consider the irradiance produced by a source placed at the front focus of an optical system. A common example is the searchlight, which is composed of a source, often a carbon arc, placed at the focus of a parabolic reflector.

A schematic searchlight, using lossless, transmitting optics, is shown in Fig. 3.10. A planar Lambertian source of diameter $2r_s$ and radiance N is placed at the focus of a lens of diameter D and focal length f. Light from each point on the surface of the source is collimated by this lens. Assume $r_s << D$. The beam from the searchlight is seen to diverge at a half-angle α_1, given by Eq. (2.12.1):

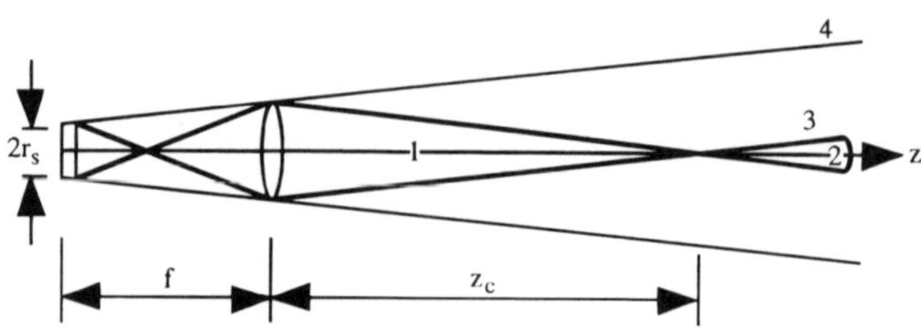

Fig. 3.10 Irradiance produced in various regions by searchlight in front focal plane.

$$\tan \alpha_1 = \frac{r_s}{f} = \frac{d}{2f} \tag{3.4.10}$$

where α and d are defined in Fig. 2.25(b). Calculate irradiance at any point to the right of the lens. Note that this question does not deal with an image, and therefore is more general than the treatment above. (Note also that here the image plane is at $s' \to \infty$.)

Solution

For any point z on the optical axis to the right of the lens, the half-angle subtended by the lens is

$$\tan \theta'_{\max} = \frac{D}{2z} \tag{3.4.11}$$

Since magnification is proportional to distance z, if one looks back at the lens from a given distance z in region 1 in Fig. 3.10, magnification is small and the source appears smaller than the lens, while in region 2 the source appears to be larger than the lens because of the larger magnification. The distance at which the source appears to just fill the lens aperture is called z_c. At this distance $\theta'_{\max} = \alpha_1$ and, by equating Eqs. (3.4.10) and (3.4.11), we obtain

$$z_c = \frac{fD}{2r_s} = \frac{fD}{d} \tag{3.4.12}$$

For $z < z_c$, since the source appears to be smaller than the lens, the planar angle limiting irradiance is that produced by the *source*, that is, α_1, rather than that produced by the lens ($\theta'_{\max}$). Consequently, in accordance with Eq. (3.2.10) and Fig. 3.5, emittance from the object reaching the entrance pupil of the lens is

$$W_s = \pi N_s \sin^2\alpha_1 \tag{3.4.13}$$

where α_1 is a constant independent of z, as defined in Eq. (3.4.10) and the subscript "s" refers to the searchlight source. Corresponding irradiance in image space is, for $\tau = 1$,

$$H_1 = W_s\left(\frac{n'}{n}\right)^2 \sin^2\alpha_1 = W_s\left(\frac{n'}{n}\right)^2 \frac{r_s^2}{r_s^2 + f^2} \tag{3.4.14}$$

Since r_s is a property of source dimension and f that of the lens, in region 1 in Fig. 3.10 irradiance is constant.

In region 2, where the source appears larger than the lens because of the higher magnification, the radiation entering this image space is limited spatially not by the source but by the lens, or the angle $\theta'_{\max}$ in Eq. (3.4.11). Hence,

$$H_2 = W_s\left(\frac{n'}{n}\right)^2 \sin^2\theta'_{\max} = W_s\left(\frac{n'}{n}\right)^2 \frac{D^2}{D^2 + 4z^2} \tag{3.4.15}$$

These equations, rather than Eq. (3.4.5) or (3.4.9), are appropriate here because according to Fermat's principle optical paths are equal for all rays traversing from a point in the object to the corresponding point in the image plane for an image only. Here, irradiance is being considered for any z. In region 2, $z > z_c$ and, for typically small dimension lenses or reflectors, Eq. (3.4.15) reduces to

$$H_i \cong W_s\left(\frac{n'}{n}\right)^2 \frac{D^2}{4z^2} \tag{3.4.16}$$

In region 2 irradiance falls off approximately according to the inverse-square law.

The differences between irradiance in the various regions derive only from the solid angle subtended by the "apparent" source as seen from the various regions in the lens output.

The normal irradiance at off-axis points can be determined with the help of Fig. 3.10. In regions 1 and 2 the irradiance does not change with the transverse distance. Thus the irradiance anywhere in region 1 is given by Eq. (3.4.14), and the irradiance anywhere in region 2 is given by Eq. (3.4.15). In region 4, which lies exterior to the volume bounded by the rays most transversely transmitted by the system, the irradiance is zero. As the point of observation is moved transversely outward through region 3, the irradiance drops monotonically from its on-axis value to zero.

Object scene emittance is simply the product of reflection coefficient ρ and irradiance appropriate to the region illuminated.

3.5 RADIOMETRY IN AN ATTENUATING MEDIUM

Attenuating constant α has been defined in Section 1.3.2 for conducting and for dielectric media. When imaging through the atmosphere, for example, one has to be concerned with absorption at various wavelengths but not with conduction losses unless the atmosphere has been broken down or ionized by high-intensity radiation. When imaging through seawater, conduction must also be considered. An additional form of attenuation is light scattering by particulates present in the propagation channel. Examples are aerosols in the atmosphere and salt particulates, amoeba, etc., in the sea. Light propagating in a given direction is scattered by such particulates to other directions. Consequently, irradiance in the initial direction is diminished or attenuated.

It is advantageous to start by considering the spectral irradiance H_λ, the amount of radiant power per unit wavelength of radiation incident or passing through a unit area of a surface held normal to the direction of power flow. For simplicity consider a semi-infinite scattering medium as shown in Fig. 3.11, where the power flow is along the z axis. The medium is assumed to be of negligible conductivity. As a result of the scattering losses, H_λ will decrease by an amount of dH_λ over a distance dz' into the scattering medium. The scattering coefficient of the medium for radiation of wavelength λ, designated $S_A(\lambda, z)$ is defined then by

$$\frac{dH_\lambda}{dz'} \triangleq -S_A(\lambda, z')H_\lambda(z') \tag{3.5.1}$$

where $H_\lambda(\lambda, z')$ is spectral irradiance observed at position z' in the scattering medium ($\mathrm{W \cdot m^{-2} \cdot \mu m^{-1}}$), $dH_\lambda(\lambda, z')$ is decrease in spectral irradiance in the z' direction

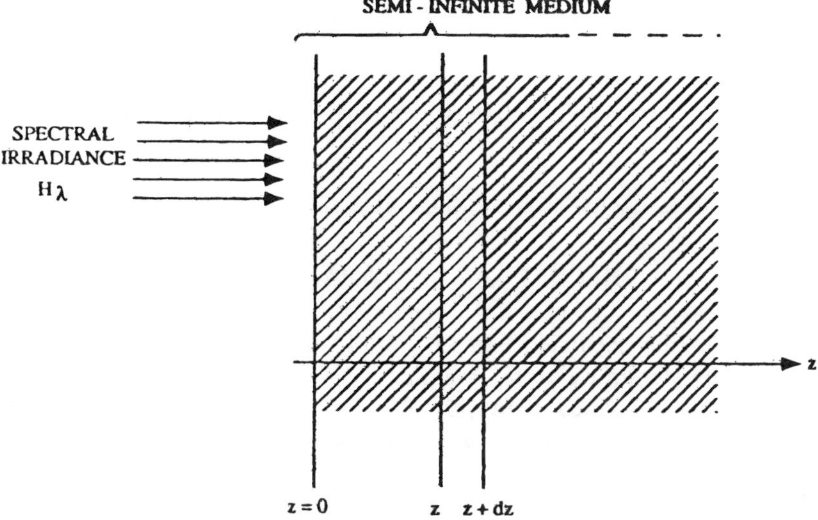

Fig. 3.11 Geometry for analyzing scattering attenuation in a semi-infinite scattering medium.

caused by scattering over distance dz', and $S_A(\lambda, z')$ is the scattering coefficient (m^{-1}) which serves as a proportionality constant.

From Eq. (3.5.1)

$$\frac{dH_\lambda(\lambda, z')}{H_\lambda(\lambda, z')} = -S_A(\lambda, z')\,dz' \tag{3.5.2}$$

or

$$\ln H_\lambda(\lambda, z')|_0^z = -\int_0^z S_A(\lambda, z')\,dz' \tag{3.5.3}$$

Therefore,

$$H_\lambda(\lambda, z) = H_\lambda(\lambda, 0)\exp\left[-\int_0^z S_A(\lambda, z')\,dz'\right] \tag{3.5.4}$$

When the scattering medium is uniform, $S_A(\lambda, z)$ is not a function of z. In this case Eq. (3.5.4) can be written as

$$H_\lambda(\lambda, z) = H_\lambda(\lambda, 0)\exp[-S_A(\lambda)z] \tag{3.5.5}$$

The quantity $S_A(\lambda)z$ is called the *optical scattering depth*. The distance z at which the optical scattering depth is unity is called the *depth of penetration*. At this distance the observed irradiance is e^{-1} (= 36.8%) of its initial value at $z = 0$.

If all attenuation losses are included, then in general

$$H_\lambda(\lambda, z) = H_\lambda(\lambda, 0)\exp\left[-\int_0^z \alpha_\lambda(\lambda, z')\,dz'\right] \tag{3.5.6}$$

where Eq. (3.5.6) is known as the *Beer-Lambert law* and

$$\alpha_\lambda(\lambda, z) = S_A(\lambda, z) + A_A(\lambda, z) + \text{others} \tag{3.5.7}$$

is overall attenuation coefficient. In Eq. (3.5.7), $A_A(\lambda, z)$ is absorption coefficient describing losses stemming from imaginary components of dielectric constant or refractive index as described already in Eq. (1.3.30) or (1.3.33b). Other forms of attenuation can include conduction losses, for example, and other types of loss mechanisms, if any, characteristic of the propagation channel. These various coefficients generally vary with position and local conditions such as weather for the atmospheric propagation channel. Scatter coefficients vary with particulate size distributions, concentrations, and chemical contents, particularly as regards outer coatings. These vary with humidity, wind speed and direction, air temperature, etc. Absorption coefficients vary with concentration of absorbing atoms or molecules, and also with weather parameters such as temperature. Consequently, attenuation is often not uniform.

Radiant transmittance through the propagation channel is obtained from Eq. (3.5.6)

$$\tau(\lambda, z) = \frac{H_\lambda(\lambda, z)}{H_\lambda(\lambda, 0)} = \exp\left[-\int_0^z \alpha_\lambda(\lambda, z')\,dz'\right] \tag{3.5.8}$$

A relation involving $\alpha(\lambda, z)$ and the radiance can be obtained by dividing Eq. (3.5.6) by the solid angle subtended by the source of radiation; thus

$$N_\lambda(\lambda, z) = \frac{H_\lambda(\lambda, z)}{\Omega_{\text{source}}} = N_\lambda(\lambda, 0)\exp\left[-\int_0^z \alpha_\lambda(\lambda, z')\,dz'\right] \tag{3.5.9}$$

where $N_\lambda(\lambda, z)$ is the observed or detected radiance at position z in the attenuating medium. The validity of Eq. (3.5.9), however, is rather questionable. Exercise 3.6 considers those conditions in which Eq. (3.5.9) can be used.

3.6 REFLECTION COEFFICIENTS

Reflection coefficients vary with wavelength. In system design and analysis it is important to consider spectral considerations and determine at which wavelength image quality or contrast is likely to be best. This often depends on wavelength dependencies of object and background reflectivities. For example, vegetation usually exhibits low reflectance in the visible because of sunlight absorption by chlorophyll. In the near-infrared region, this is reversed; absorption decreases while reflectance increases very strongly. The difference can indeed be striking even over short wavelength intervals.

Tables of reflection coefficients for various materials are available in the literature [3.5]–[3.7]. More detail is presented in Chapter 10.

3.7 FUNDAMENTAL PHOTOMETRIC QUANTITIES

If the properties of the detector are involved in the analysis of electromagnetic radiant power transfer, the treatment usually becomes more complicated. At the very least, involving the detector requires that irradiance at the detecting surface be weighted with spectral characteristics of the detector. In the case of a fabricated detector, such as an antenna or photodiode, the task is not as difficult because the detector characteristics can be specified and maintained within fairly rigid bounds by appropriate manufacturing techniques. Such devices were not available in early times, however, and the only available detector was the human eye.

The use of the eye as a detector of course limited observation of radiant power to the visible spectrum only. However, it caused other difficulties as well. For example, there is no "standard" eye. The spectral characteristics of the human eye depend on the level of retinal irradiance, the field of view, the state of adaptation of the eye (i.e., whether it has prepared itself for high or low irradiance), and the direction in which it is oriented. In addition, all these parameters vary from observer to observer. Thus, there are many ramifications involved in using the characteristics of the human visual system to describe visual radiometry. For convenience a set of "average" characteristics has been standardized for computational purposes. These are usually given in the form of two *luminosity functions* as shown in Fig. 3.12, one for normal light levels (daytime or *photopic vision*) and the other for very low light levels (moonlight or scotopic vision). For photopic vision the various wavelengths appear in their corresponding colors, whereas for scotopic vision the irradiance is of such a low level that the retina cannot distinguish colors and all wavelengths appear gray. These two

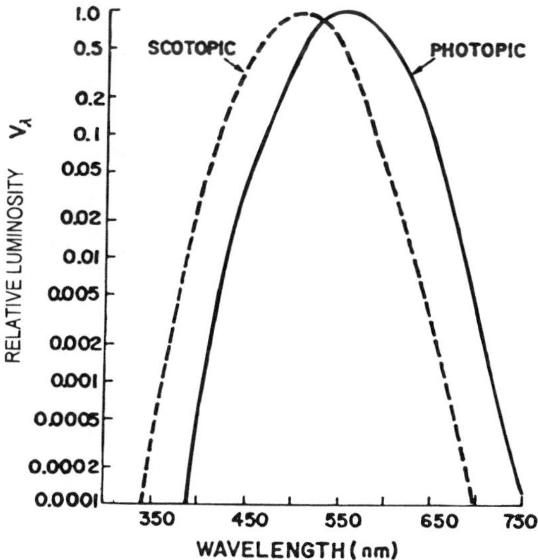

Fig. 3.12 Relative luminosity functions of the standard observer for photopic (daytime) vision and scotopic (nighttime) vision. To convert the ordinate to absolute luminosity values, multiply the curve for photopic vision by 685 $\text{lm}\cdot\text{W}^{-1}$ and for scotopic vision by 1750 $\text{lm}\cdot\text{W}^{-1}$.

different spectral responses derive from the rod (scotopic vision) and cone (photopic vision) structure of the human eye, described in more physiological detail in Chapter 12.

Visual radiometry is referred to as *photometry*. To emphasize the fact that the eye is involved in photometry, different terms are defined. For example, radiant power is referred to as *luminous power* or *luminous flux* and is denoted by the symbol F. In mks units, a unit of luminous flux is called a *lumen* to distinguish it from the watt, which is the radiometric unit for radiant power. The lumen is similar to the watt except that it is modified by the characteristics of the eye in the following way. Since

$$P = \int_0^\infty P_\lambda(\lambda)\, d\lambda \tag{3.7.1}$$

is the description of radiant power then

$$F = \int_0^\infty K(\lambda) P_\lambda(\lambda)\, d\lambda \tag{3.7.2}$$

where

$$K(\lambda) = 685 V(\lambda) \qquad (\mathrm{lm \cdot W^{-1}}) \tag{3.7.3}$$

is the *absolute luminosity function* and $V(\lambda)$, the *relative luminosity function*, describes the shape of the photopic (daytime) vision curve in Fig. 3.12. The absolute luminosity is a function that expresses the ratio of luminous flux $F_\lambda(\lambda)$ to radiant power $P_\lambda(\lambda)$ over the spectral bandwidth of the standard human visual system. For scotopic (nighttime) vision, $K(\lambda) = 1764V(\lambda)$, where $V(\lambda)$ describes the shape of the scotopic curve in Fig. 3.12. The peak absolute luminosity of 685 $\mathrm{lm \cdot W^{-1}}$ for photopic vision occurs at $\lambda \cong 556$ nm (yellow-green), whereas the peak of 1746 $\mathrm{lm \cdot W^{-1}}$ for scotopic vision occurs at $\lambda \cong 511$ nm (blue-green). Note that scotopic vision responsivity is much greater than that of photopic vision, as explained further in Chapter 12. This greater responsivity (1746 to 685 $\mathrm{lm \cdot W^{-1}}$) enables us to have some vision under dark conditions, after adaptation.

Other radiometric quantities are modified to photometric quantities in a fashion identical to that in Eq. (3.7.2). For example, *luminous intensity* I is derived from radiant intensity J by

$$I = \int_0^\infty K(\lambda) J_\lambda(\lambda)\, d\lambda \tag{3.7.4}$$

In terms of photometric quantities, then, Eq. (3.1.4) becomes

$$I = \frac{\partial F}{\partial \Omega} \; (\mathrm{lm \cdot sr^{-1}} \text{ or cd}) \tag{3.7.5}$$

Weighting all the radiometric quantities by $K(\lambda)$ in this way produces the set of photometric quantities defined in Table 3.2. Along with the mks photometric units listed in this table, the cgs and English units are also indicated. English units are

TABLE 3.2 *Definitions of Fundamental Photometric Quantities and Their Relationships to the Fundamental Radiometric Quantities*

Radiometric Quantity		Definition	Photometric Quantity		Definition	mks	cgs	English
Radiant energy	U	—	Luminous energy	Q	—	lm·s	lm·s	lm·s
Radiant energy density	u	$\frac{\partial U}{\partial V}$	Luminous energy density	q	$\frac{\partial O}{\partial V}$	$\mathrm{lm \cdot s \cdot m^{-3}}$	$\mathrm{lm \cdot s \cdot m^{-2}}$	$\mathrm{lm \cdot s \cdot ft^{-3}}$
Radiant power	P	$\frac{\partial U}{\partial t}$	Luminous flux	F	$\frac{\partial O}{\partial t}$	lm	$\mathrm{lm \cdot sr^{-1} = cd}$	lm
Radiant intensity	J	$\frac{\partial P}{\partial \Omega}$	Luminous intensity	I	$\frac{\partial E}{\partial t \Omega}$	$\mathrm{lm \cdot sr^{-1} = cd}$	$\mathrm{lm \cdot sr^{-1} = cd}$	$\mathrm{lm \cdot sr^{-1} = cd}$
Radiant emittance	W	$\frac{\partial P}{\partial A \cos\theta}$	Luminous emittance	L	$\frac{\partial F}{\partial A \cos\theta}$	$\mathrm{lm \cdot m^{-2}}$	$\mathrm{lm \cdot cm^{-2}}$	$\mathrm{lm \cdot ft^{-2}}$
Radiance	N	$\frac{\partial^2 P}{\partial \cos\theta \, \partial A \, \partial \Omega}$	Luminance	B	$\frac{\partial^2 F}{\cos\theta \, \partial A \, \partial \Omega}$	$\mathrm{lm \cdot cm^{-2} \cdot sr^{-1} = cd \cdot m^{-2} = nit}$	$\mathrm{lm \cdot cm^{-2} \cdot sr^{-1} = cd \cdot cm^{-2} = stilb}$	$\mathrm{lm \cdot ft^{-2} \cdot sr^{-1} = fL}$
Irradiance	H	$\frac{\partial P}{\partial A_r}$	Illuminance	E	$\frac{\partial F}{\partial A_r}$	$\mathrm{lm \cdot m^{-2} = lx = mc}$	$\mathrm{lm \cdot cm^{-2} = phot}$	$\mathrm{lm \cdot ft^{-2} = fc}$

used to a great extent by illumination and television engineers. All photometric units are defined in terms of the candela (cd), the luminous intensity I of a standard source: One candela is equal to 1/60 of the luminous intensity of a 1-cm^2 section of a blackbody at the temperature of solidification of platinum (2042 K). (Blackbody sources are described in Chapter 4.) Thus, in terms of this standard source the lumen is defined as the flux emitted into a unit solid angle by a uniform isotropic point source whose luminous intensity in all directions is one candela. Since there are 4π (=12.57) sr subtended by a spherical surface, the total flux emitted by a uniform isotropic point source of one candela intensity is 4π lumens.

Because of its generality, we are concerned in this text with radiometry rather than photometry. The latter has been presented here briefly only as background reference material.

More detailed data on the luminosity curves in Fig. 3.12 are tabulated in the literature [3.4].

REFERENCES

3.1. R. W. Boyd, *Radiometry and the Detection of Optical Radiation*, Wiley, New York, 1983.
3.2. F. Grum and R. J. Becherer, *Optical Radiation Measurements*, Academic Press, New York, 1979.
3.3. W. L. Wolfe, "Radiometry," in *Applied Optics and Optical Engineering*, Vol. VIII, R. Kingslake, Ed., Academic Press, New York, 1980.
3.4. *RCA Electro-Optics Handbook*, RCA, Harrison, NJ, 1974.
3.5. D. E. Bowker, R. E. Davis, D. L. Myrick, K. Stacy, and W. T. Jones, *Spectral Reflectances of Natural Targets for Use in Remote Sensing Studies*, NASA Ref. Publication 1134, NASA Langley Research Center, Hampton, VA, June 1985.
3.6. L. Levi, *Applied Optics Vol. 2,* Wiley, New York, 1980, pp. 117–118.
3.7. R. A. McClatchey, R. W. Fenn, J. E. A. Selby, F. E. Voltz, and J. S. Garing, "Optical properties of the atmosphere," in *Handbook of Optics*, W. Driscoll, Ed., McGraw-Hill, New York, 1978, pp. 14–54.

EXERCISES

3.1 Determine the radiant power radiated from a light source into a cone of semi-angle 45°. The source is of mean diameter 0.2 cm and radiance 20 $W \cdot cm^{-2} \cdot sr^{-1}$. Assume a diffuse source of surface roughness equivalent to that of a Lambertian source.

3.2 If the source from Exercise 3.1 is used in conjunction with a focusing lens of diameter 3 cm, what is the radiant power collected by the lens when the source is placed at a distance of 10 cm from the lens along its optical axis?

3.3 How much power from a specular source of radiance N and radiating area dA is radiated into a hemisphere? What is the radiant emittance? For radiation into a cone of semi-angle θ, what is equivalent solid angle for a specular source?

3.4 Repeat Exercises 3.1 and 3.2 for a specular source. What can be concluded regarding Lambertian and specular source radiation of small beamwidth?

3.5 Radiation from a Lambertian source of area dA and radiance N is incident on a surface of area dA_r as in the figure given here. Find the irradiance on dA. This is known as the cos to the fourth power law.

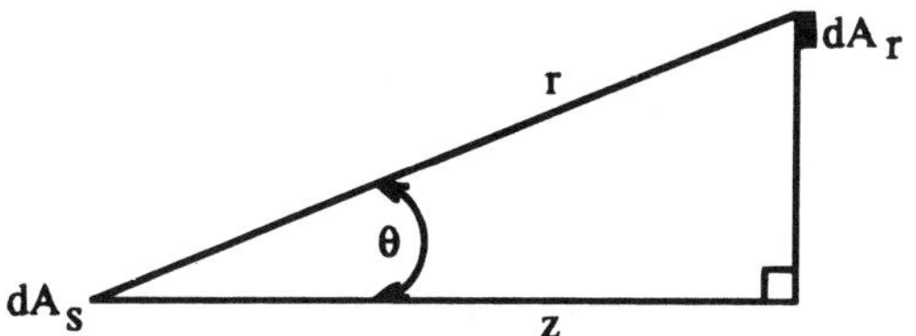

3.6 The main surface of a sphere is coated with a rough (diffuse) surface that reflects according to Lambert's law. Through a very small aperture of area dA passes radiation of power P from a Lambertian source that enters the sphere. The radiation is reflected many times inside the sphere until it becomes uniformly distributed over the interior surface. A detector with which to measure the irradiance is located at another aperture of area dA_r as shown in the figure below. Assume a reflection coefficient (power) ρ for the interior surface. What is the irradiance incident on the detector? Assume no direct path from source to detector. Such a sphere is called an *integrating sphere* and is used to facilitate measurement for nonuniform irradiance by converting it to uniform irradiance.

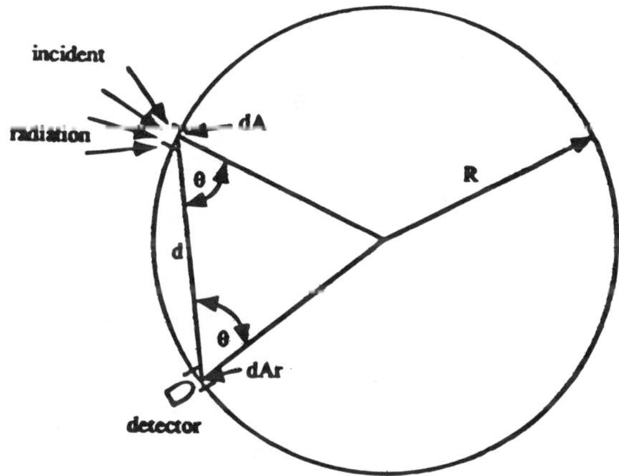

3.7 A scene at a mean distance z is illuminated by a uniform Lambertian light source of power P at mean angle of incidence θ. The scene is imaged by an imager of area A_r adjacent and parallel to the light source. If the beamwidth of the light source is Ω, what is the average irradiance incident on the imager, assuming a mean reflectivity (power) ρ and assuming surface roughness at optical wavelengths corresponding to Lambert's law to characterize objects in the scene? Consider two cases: (1) $A_s \cos\theta/z^2 < \Omega$; and (2) $A_s \cos\theta/z^2 > \Omega$, where A_s is scene area. Compare answers to the inverse-square law. (*Hint*: The problem is to be done in two steps. First consider radiance of source and power transfer to scene, then consider radiance of scene and power transfer to image.)

3.8 What happens in Exercise 3.7 if the light source is inclined at an angle ϕ_s with respect to the direction to the scene?

3.9 In an experiment to measure variations in angle of arrival a laser beam of narrow beamwidth 2α is transmitted a distance z to a retroreflector of projected area A_r. The retroreflector ideally returns each ray in the direction from which it is incident. Since the retroreflector is not completely ideal, a tolerance of ϕ is assumed. The retroreflected beam is imaged by an imager adjacent to the laser and the variations in angle of arrival caused by the atmosphere are observed. If the laser power is P, what is the irradiance at the imagers? Consider two cases: (1) $2\alpha z > A_r^{1/2}$; and (2) $2\alpha z < A_r^{1/2}$. Assume reflectivity (power) of ρ and uniform power density in transmitted laser beams.

3.10 Use the information given in Exercise 3.9 for a specular (or mirror-like) target instead of a retroreflector.

3.11 Pulses from a Nd:YAG laser (wavelength $\lambda = 1.06\ \mu\text{m}$) are transmitted to a satellite of 1-m^2 cross section at about 500 km altitude in order to ascertain altitude more precisely according to round-trip propagation time. Assume satellite diffuse reflectivity of 70% and a 1-m radius for the receiver aperture. Furthermore, assume a narrow-diffraction-limited beamwidth exiting a 1-mm radius laser partially transmitting mirror. Ignoring atmospheric attenuation, what is the minimum pulse energy required from the laser if at least 1000 photons per pulse must be received at the receiver in order for the return pulse to be detected? Assume 95% transmission each for the transmitting and receiving optics. (Diffraction-limited beamwidth is equal to 1.22 λ/D for a circular aperture as shown in Section 7.4. Here, D is aperture diameter.)

3.12 If the lens in Exercise 3.2 is of transmissivity 0.9 and focal length 9 cm, what is image irradiance? Compare calculations involving Eqs. (3.4.4), (3.4.5), (3.4.7), (3.4.8), and (3.4.9). Discuss discrepancies.

3.13 Repeat Exercise 3.12 for a light source that is 90 cm distant from the lens.

3.14 Do Eqs. (3.4.14) and (3.4.15) yield identical results for $z = z_c$?

3.15 Assuming the scene in Exercise 3.8 is in region 2 of Fig. 3.10, what is the image irradiance?

3.16 If the propagation medium exhibits attenuation, such as that defined by attenuation constant α, under which conditions if any does the following relation hold true:

$$N_\lambda\ (\lambda, z) = N(\lambda, 0)\ \exp[-\int_0^Z \alpha_\lambda(\lambda z)\ dz']$$

where $N_\lambda(\lambda, 0)$ is radiance at the beginning of the absorbing medium and $N_\lambda(\lambda, z)$ is radiance a distance z into the absorbing medium. Is the above a general relation?

CHAPTER

4

Sources of Optical Radiation

Sources in the optical region of the spectrum can be classified under two broad categories: (1) thermal and (2) nonthermal. The first category includes a wide class of natural and artificial sources whose radiation obeys blackbody radiation laws. The second class includes electronic sources such as various arc lamps, and quantum sources such as lasers. Laser radiators are available at various frequencies from millimeter waves up and into the ultraviolet and even x-ray region. We will begin with thermal sources.

4.1 BLACKBODY RADIATION

A blackbody is defined as one that completely absorbs all radiation incident on it. Since no radiation is reflected or transmitted, its spectral radiation is ideally zero at all wavelengths. Therefore it appears black. A whitebody is one that reflects completely radiation at all wavelengths. When illuminated with relatively high irradiance, its spectral radiation is ideally high at all wavelengths and it appears white at all wavelengths. A graybody is one that absorbs only a portion α', reflects a portion ρ, and transmits a portion τ' of the radiation incident on it. From conservation of energy,

$$\rho + \alpha' + \tau' = 1 \tag{4.1.1}$$

From a practical standpoint, essentially all media are graybodies. It is however convenient to model highly absorbing and highly reflecting bodies as blackbodies and whitebodies, respectively. The radiant power reflection coefficient ρ is identical to that defined in Section 1.4.3 for a single interface. When the radiant power transmission coefficient τ defined there for an interface is applied to a whole body or medium, then $\tau = \alpha' + \tau'$, i.e., the radiant energy entering a body consists of two components. One is absorbed by it while the other is transmitted through it. Using the concept of attenuation coefficient $\alpha(\lambda, z')$ defined in Eq. (3.5.8), note that $\tau' = \exp[-\int_0^z \alpha(\lambda, z')\, dz']$. The parameters of Eq. (4.1.1) are summarized in Table 4.1.

The "absorptivity" $\alpha'(\lambda)$ measures the fraction of incident radiation absorbed by a system and should not be confused with the "absorption coefficient" $A_A(\lambda, z)$, which was defined in Section 3.5, or the attenuation constant $\alpha(\lambda, z')$ defined in Eq. (3.5.8) and in Eq. (1.3.27a). The units of $A_A(\lambda, z)$ and $\alpha(\lambda, z)$ are *reciprocal meters*, whereas absorptivity in Eq. (4.1.1) is *dimensionless*. From Eq. (4.1.1),

TABLE 4.1 *Definitions of α', ρ, τ', and ϵ*

Symbol	Name	Definition
α'	Radiant absorptivity	Ratio of radiant energy absorbed by a body to that incident on it
ρ	Radiant reflectivity	Ratio of the radiant energy reflected by a body to that incident on it
τ'	Radiant transmitivity	Ratio of radiant energy transmitted *through* a body to that incident on it
ϵ	Radiant emissivity	Ratio of the rate of radiant energy emission from a body, as a consequence of its *temperature* only, to the corresponding emission from a *blackbody* at the same temperature

$$\alpha'(\lambda) = 1 - \rho - \exp \int_0^t [-\alpha(\lambda, z')\, dz']$$

where t is thickness of the body considered.

Note that by definition $\alpha'(\lambda) = 1$ for a blackbody and $\rho(\lambda) = 1$ for a whitebody.

If the material on which the radiant energy is incident is opaque, then $\tau' = 0$ and $\alpha' + \rho = 1$. In the definitions of Table 4.1 no specification was made as to whether specular or diffuse quantities are indicated; hence, the definitions hold as long as the same specification is made for each of the quantities. Even so, care must be exercised in this regard. For example, the meaning of a value of ρ for a specularly reflecting surface is easily specified: Radiant energy impinges on the surface and a certain fraction ρ of it is reflected in a direction determined by Snell's law. On the other hand, the meaning of a value of ρ for a diffusely reflecting surface or for a partially transparent body with internal scattering is not at all clear. This difficulty in creating an accurate description arises from the complicated way in which the energy is "reflected" from a diffuse surface and from the physical difficulty to distinguish clearly the transmitted, scattered, and reflected portions of the radiation after incidence. Thus, care must be taken when discussing "reflectivity" in any given situation involving a nonspecular surface. Similar arguments hold when considering the "transmissivity"; the meaning of a value of τ' for a nonscattering homogeneous medium is clear but its meaning for any other medium is suspect.

If a given body is heated to a temperature above that of its surroundings, then after the source of heat has been removed the body cools until it has reached thermodynamic equilibrium with its surroundings. In other words, the energy absorbed is now radiated. This process is characterized by the radiant *emissance* or *emissivity* ϵ of the medium under consideration. Since a perfect absorber is also a perfect radiator, a perfectly black surface has the highest possible radiant emittance W_{bb}, where the subscript stands for blackbody. Thus any real surface is characterized by an emittance $W < W_{bb}$ for a given temperature. As defined in Table 4.1, the ratio of W to W_{bb} is the radiant emissivity; i.e.,

$$\frac{W}{W_{bb}} = \epsilon \tag{4.1.2}$$

which ideally is wavelength independent.

If a body has reached an equilibrium temperature, it is then emitting the same amount of radiation as it absorbs, otherwise its temperature would change. Under such conditions

$$W = \alpha' W_{bb} \tag{4.1.3}$$

which implies from Eq. (4.1.2) that

$$\alpha' = \epsilon \tag{4.1.4}$$

under conditions of thermal equilibrium. Equation (4.1.4) is referred to as *Kirchhoff's radiation law.* It is consistent with conservation of energy. It is also actually the basis for an accepted religious law in the Babylonian Talmud (Pesachim 74a).

A graybody is a surface for which the emissivity is less than unity and does not vary rapidly with wavelength. Such a body will have a spectral energy distribution very similar to that of a blackbody, as given by Planck's radiation law presented next, except that its radiant emittance will always be less than that of a perfect blackbody. A blackbody and a graybody at the same temperature, therefore, will appear to be the same color by virtue of their thermal radiation, the blackbody being somewhat brighter.

With this in mind, the radiation and spectral properties of ideal blackbodies are presented next, with their applicability to graybodies being simply a matter of multiplying by the radiant emissivity of the given material. These parameters have been tabulated for various materials. Essentially, because of its dependence upon reflectivity in Eq. (4.1.1), ϵ is determined by the outer surface of the given material, regardless of what lies underneath. Most metals, for example, are good reflectors. Hence, by virtue of Eq. (4.1.1) α or ϵ is very small. Yet, if the metal is coated with even a very thin layer of nonmetallic paint, or water, emissivity becomes that of the paint or water because reflectivity is determined by the paint or water. Hence, radiant emissivity can change significantly with humidity, dew, etc. Since reflectivity can vary with wavelengths, so too can emissivity. Table 4.2 lists typical emissivities of various materials.

4.2 PLANCK'S LAW

The spectral radiant density of a blackbody is described by Planck's law [4.2]–[4.4]:

$$U_\nu = \frac{8\pi h\nu^3}{c^3}\frac{1}{e^{h\nu/kT} - 1} \qquad (\mathrm{J{\cdot}m^{-3}{\cdot}Hz^{-1}}) \tag{4.2.1}$$

where

ν = electromagnetic wave frequency (Hz)
h = Planck's constant ($6.63 \cdot 10^{-34}$ J·s)
k = Boltzmann's constant ($1.38 \cdot 10^{-23}$ J·K)
T = temperature (K)
c = velocity of electromagnetic radiation in vacuum ($3 \cdot 10^8$ m·s^{-1}).

The radiation described by Eq. (4.2.1) exists within an isothermal enclosure of the blackbody material, which is in thermodynamic equilibrium.

TABLE 4.2 *Typical Emissivities of Various Materials (after [4.1])*

Material		Temp. (°C)	ϵ
Aluminum:	Polished sheet	100	0.05
	Sheet as received	100	0.09
	Anodized sheet, chromic acid process	100	0.55
	Vacuum deposited	20	0.04
Brass:	Highly polished	100	0.03
	Rubbed with 80-grit emery	20	0.20
	Oxidized	100	0.61
Copper:	Polished	100	0.05
	Heavily oxidized	20	0.78
Gold:	Highly polished	100	0.02
Iron:	Cast, polished	40	0.21
	Cast, oxidized	100	0.64
	Sheet, heavily rusted	20	0.69
Magnesium:	Polished	20	0.07
Nickel:	Electroplated, polished	20	0.05
	Electroplated, no polish	20	0.11
	Oxidized	200	0.37
Silver:	Polished	100	0.03
Stainless steel:	Type 18-8, buffed	20	0.16
	Type 18-8, oxidized at 800°C	60	0.85
Steel:	Polished	100	0.07
	Oxidized	200	0.79
Tin:	Commercial tin-plated sheet iron	100	0.07
Brick:	Red common	20	0.93
Carbon:	Candle soot	20	0.95
	Graphite, filed surface	20	0.98
Concrete		20	0.92
Glass:	Polished plate	20	0.94
Lacquer:	White	100	0.92
	Matte black	100	0.97
Oil lubricating (thin film on nickel base):			
	Nickel base alone	20	0.05
	Film thickness of 0.001, 0.002, 0.005 in.	20	0.27,0.46,0.72
	Thick coating	20	0.82
Paint oil:	Average of 16 colors	100	0.94
Paper:	White bond	20	0.93
Plaster:	Rough coat	20	0.91
Sand		20	0.90
Skin, human		32	0.98
Soil:	Dry	20	0.92
	Saturated with water	20	0.95
Water:	Distilled	20	0.96
	Ice, smooth	−10	0.96
	Frost crystals	−10	0.98
	Snow	−10	0.85
Wood:	Planed oak	20	0.90

Blackbody spectral radiant energy density can be described in terms of wavelength rather than frequency by requiring that

$$U_\nu d\nu = U_\lambda d\lambda \qquad (4.2.2)$$

which is an equation saying that the radiant energy in a certain frequency interval $d\nu$ must be the same as that within a corresponding wavelength interval $d\lambda$. Since $\nu = c/\lambda$, then

$$d\nu = \frac{-c}{\lambda^2}\,d\lambda \tag{4.2.3}$$

where a minus sign indicates that an increase in wavelength corresponds to a decrease in frequency and vice versa. With the significance of the minus sign understood, combination of Eqs. (4.2.1) and (4.2.3) yields

$$U_\nu d\nu = \frac{8\pi hc^3}{c^3\lambda^3}\,\frac{1}{e^{hc/\lambda kT}-1}\,\frac{c}{\lambda^2}\,d\lambda \tag{4.2.4}$$

and in view of Eq. (4.2.2),

$$U_\lambda = \frac{8\pi hc}{\lambda^5}\,\frac{1}{e^{hc/\lambda kT}-1} \qquad (\mathrm{J\cdot m^{-3}\cdot \mu m^{-1}}) \tag{4.2.5}$$

where wavelength is measured in micrometers.

Planck's law is the basis for almost all considerations involving thermal radiation. However, when considering the flow of energy or power transfer from place to place it is more convenient to use the value of spectral radiance of the source at hand. Since Planck's law describes the radiant energy density within an isothermal blackbody enclosure at thermodynamic equilibrium, then for each point of the interior of this enclosure and for each spectral component the spectral energy density U_ν is associated with a flow of radiant power that is the same in all directions; i.e., the radiant power flows through 4π sr. Therefore, the power flow through a unit solid angle is obtained by multiplying the spectral energy density by the velocity of flow c and dividing by 4π sr giving the spectral radiance

$$N_\nu = \frac{U_\nu}{4\pi}\cdot c = \frac{2h\nu^3}{c^2}\,\frac{1}{e^{h\nu/kT}-1} \qquad (\mathrm{W\cdot m^{-2}\cdot sr^{-1}\cdot Hz^{-1}}) \tag{4.2.6}$$

A small opening in the side of the enclosure will allow a small portion of the contained blackbody radiation to escape. The state of thermodynamic equilibrium is negligibly perturbed if the size of the opening is small compared to the dimensions of the enclosure. The opening acts now as a blackbody radiator since all incident radiation entering the enclosure will eventually be absorbed through multiple reflections at the opaque enclosure walls and very little of it escapes again from the small opening. The spectral radiance of such a source approaches that given by Eq. (4.2.6). In terms of wavelength, Eq. (4.2.6) becomes, using Eq. (4.2.3) again,

$$N_\lambda = \frac{2hc^2}{\lambda^5}\,\frac{1}{e^{hc/\lambda kT}-1} \qquad (\mathrm{W\cdot m^{-2}\cdot sr^{-1}\cdot \mu m^{-1}}) \tag{4.2.7}$$

where the wavelength is measured in micrometers. If N_λ (N_ν) is integrated over all wavelengths (frequencies) the total radiance N_{bb} of a blackbody source over all wavelengths (frequencies) is obtained. Thus, from Eq. (4.2.6)

$$N_{bb} = \frac{2h}{c^2} \int_0^\infty \frac{\nu^3}{e^{h\nu/kT} - 1} \, d\nu \tag{4.2.8}$$

Letting

$$x = \frac{h\nu}{kT}$$

$$dx = \frac{h}{kT} \, d\nu$$

Eq. (4.2.8) becomes

$$N_{bb} = \left(\frac{2h}{c^2}\right)\left(\frac{kT}{h}\right)^4 \int_0^\infty \frac{x^3}{e^x - 1} \, dx \tag{4.2.9}$$

This integral can be evaluated using the general relationship

$$\int_0^\infty \frac{x^p \, dx}{e^x - 1} = \int_0^\infty \left(x^p e^{-x} \sum_{t=0}^{\infty} e^{-tx}\right) dx$$

$$= p! \sum_{t=1}^{\infty} \frac{1}{t^{p+1}} \tag{4.2.10}$$

For $p = 3$,

$$\int_0^\infty \frac{x^3 \, dx}{e^x - 1} = 6 \sum_{t=1}^{\infty} \frac{1}{t^4} = \frac{\pi^4}{15} \tag{4.2.11}$$

Consequently, Eq. (4.2.9) becomes

$$N_{bb} = \left(\frac{2\pi^4 k^4}{15c^2h^3}\right) T^4 = \sigma' T^4 \tag{4.2.12}$$

where

$$\sigma' = \frac{2\pi^4 k^4}{15c^2h^3} = 1.8 \times 10^{-8} \qquad (\mathrm{W \cdot m^{-2} \cdot K^{-4} \cdot sr^{-1}}) \tag{4.2.13}$$

The same result as Eq. (4.2.12) is obtained by integrating N_λ in Eq. (4.2.7) over all wavelengths.

Since N_{bb} in Eq. (4.2.12) is independent of direction, then corresponding to Lambert's law radiant emittance W_{bb} is

$$W_{bb} = \pi N_{bb} = \pi\sigma' T^4 \qquad (\mathrm{W \cdot m^{-2}}) \tag{4.2.14}$$

where

$$\sigma = \pi\sigma' = \frac{2\pi^5 k^4}{15c^2h^3} = 5.67 \times 10^{-8} \qquad (\text{W}\cdot\text{m}^{-2}\cdot\text{K}^{-4}) \tag{4.2.15}$$

Equation (4.2.14) is known as the *Stefan-Boltzmann law* and the factor σ in Eq. (4.2.15) is called the *Stefan-Boltzmann constant.*

Because at optical wavelengths the surfaces of most materials or bodies are sufficiently rough to be diffuse, blackbody radiant emittance is $N_\lambda \pi$. If we differentiate either N_λ or W_λ with respect to wavelength (Exercise 4.1), we find that peak spectral energy density occurs at a wavelength λ_m given by

$$\lambda_m T = 2.897 \times 10^3 \ (\mu\text{m}\cdot\text{K}) \tag{4.2.16}$$

This relationship is known as Wien's displacement law because it shows (Fig. 4.1) that the peak of the plot of W_λ versus λ is displaced toward a shorter wavelength as T increases. This is the reason objects begin to glow if they are sufficiently heated. Thus, by measuring λ_m it is possible to determine the temperature of a given object. Graybodies exhibit the same spectral dependencies as blackbodies; they are simply reduced in magnitude according to ϵ of the outer layer.

Many relationships can be developed. For example, peak radiation from two objects at different temperatures, say, 600 and 300 K, will vary as the ratio of absolute temperatures raised to the fifth power, that is, $(600/300)^5 = 32$ (see Exercise 4.5). Furthermore, full width-half maximum points λ_{short} and λ_{long} can be defined for N_λ and W_λ where

$$\lambda_{\text{short}} T = 1.8 \times 10^3 \ (\mu\text{m}\cdot\text{K}) \tag{4.2.17}$$

$$\lambda_{\text{long}} T = 5.1 \times 10^3 \ (\mu\text{m}\cdot\text{K}) \tag{4.2.18}$$

such that 60% of the total area under the spectral radiance or emittance curves lies between λ_{short} and λ_{long}. The remaining 40% of the area under the curve is split such

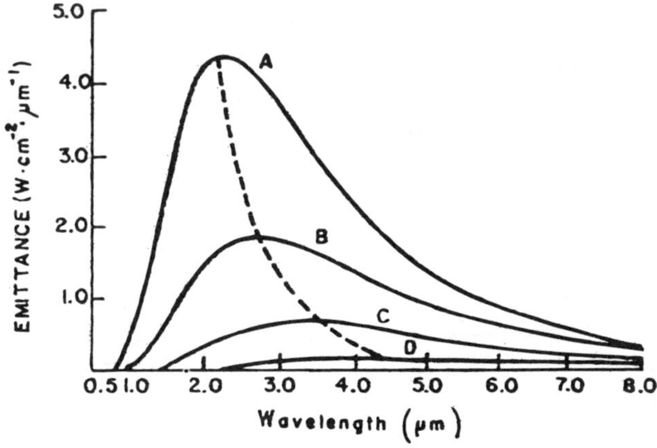

Fig. 4.1 Blackbody radiation curves for temperatures: A = 1273 K, B = 1073 K, C = 873 K, D = 673 K.

that 37% lies at wavelengths greater than λ_{long} and 3% lies at wavelengths shorter than λ_{short}. From $\lambda = 0$ to $\lambda = \lambda_m$ lies 25% of the total area under the curve.

From a historical standpoint, the two earliest attempts at explaining curves such as Fig. 4.1 were the empirical radiation laws of Wien and Rayleigh-Jeans which lacked quantum considerations. They are special cases of the more general relation developed by Max Planck. For example, at low frequencies (below the peak value of U_ν), Planck's law [Eq. (4.2.1) or Eq. (4.2.5)] can be approximated to

$$U_\nu \cong \frac{8\pi\nu^2 kT}{c^3} \qquad \text{W}\cdot\text{m}^{-3}\cdot\text{Hz}^{-1} \tag{4.2.19}$$

$$U_\lambda \cong \frac{8\pi kT}{\lambda^4} \qquad \text{W}\cdot\text{m}^{-3}\cdot\mu\text{m}^{-1} \tag{4.2.20}$$

These approximations are referred to as the *Rayleigh-Jeans law.*

These attempts were not accepted and were even ridiculed to some extent because they predict that at very high frequencies or very short wavelengths thermal emission blows up. In that sense they were referred to sometimes as "the ultraviolet catastrophe."

It is useful to consider effects of thermal emission from the standpoint of everyday experience. For example, if a typical room temperature on the order of 300 K is assumed, Wien's law predicts N_λ and W_λ peak at the wavelength $\lambda_m = 9.66$ μm, which is far beyond the visible spectrum. Despite the broad thermal emission spectrum, $\lambda_{short} = 6$ μm. Only 3% of the total area under the curve is at wavelengths below 6 μm. Therefore, since essentially the whole thermal emission spectrum is at rather long infrared wavelengths, as seen in Fig. 4.2, such radiation is simply not seen by the human visual system. Consequently, our typical surroundings exhibit essentially two different spectra of radiation. At shorter wavelengths there is visible light emanating from the sun, moon, stars, etc., or suitable man-made sources which is reflected by the objects around us, thus enabling us to see such objects. In addition, there is thermal emission originating in every object at temperatures above 0 K, which is *self-emission.* Since the latter is essentially entirely in the thermal infrared (see the Introduction for a description of spectral regions) it is not seen by us but is detected and measured with infrared detectors (Chapter 6). Only for temperatures on the order of thousands of degrees kelvin does thermal radiation spread to the visible.

Another interesting property that can be observed in Figs. 4.1 and 4.2 is that blackbody curves for different temperatures do not intersect. In other words, a blackbody at temperature $T_1 > T_2$ exhibits a higher emittance or radiance at all wavelengths and frequencies than does a blackbody at temperatures T_2. Further, as can be seen in Fig. 4.1, for a blackbody of temperature T not only does the spectrum shift to the left as T increases, it also increases greatly. The increase at wavelength λ_m is according to T^5 (see Exercise 4.5).

Another typical phenomenon with which most are familiar is the greenhouse effect. For example, if one closes a car entirely on a hot summer day for several hours under a hot sun and then opens it, one encounters a recognizable heat wave. The heat derives from the continued incidence of solar energy through the transparent windows of the car, thus heating up the inside of the car. This generates an increase in thermal emission inside the car, which takes the form of infrared radiation. Since the glass windows are not transparent to radiation at wavelengths longer than 2 to 3 μm, such radiation is then trapped inside the car and cannot escape. It keeps increasing with prolonged heating by visible solar radiation until the car door is

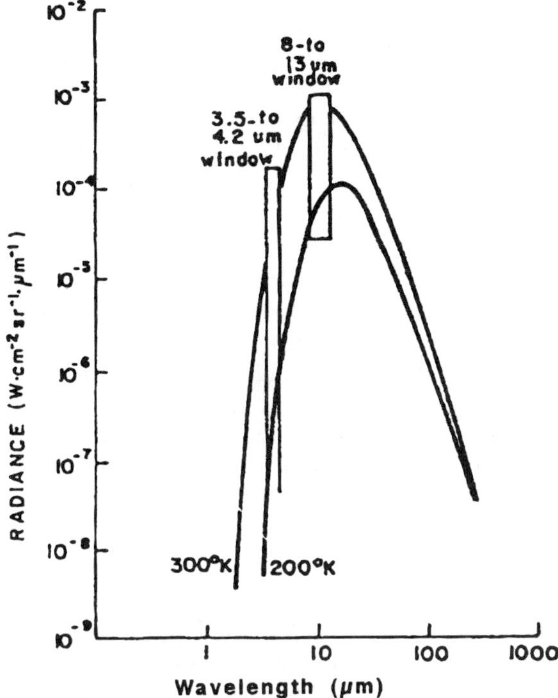

Fig. 4.2 Radiances as functions of wavelength for 200 and 300 K blackbodies showing two primary infrared absorption windows.

opened. At this point, in an effort to reach thermodynamic equilibrium with surroundings outside the car, much of the heat is encountered as an outgoing heat wave by the person opening the car. This same greenhouse effect is utilized to maintain elevated temperatures in greenhouses or hothouses for agricultural purposes.

The greenhouse effect is also manifested in the earth's atmosphere. As described in the Introduction, there is little atmospheric absorption at visible and near-infrared wavelengths. Thus, except for scattering losses, the atmosphere is quite transparent to solar radiation, which heats up the surface of the earth during daytime, causing thermal infrared emission by the surface. However, much of this thermal emission is centered around 10-μm wavelength. Although there is relatively little absorption in the 8- to 14-μm atmospheric window, there is some absorption by CO_2 which, along with very strong absorption by water vapor in the 5- to 7-μm absorption band and at wavelengths beyond 13 μm as well as in the 8- to 13-μm window itself, heats up the atmosphere. Since the concentration of CO_2 in the atmosphere is steadily increasing because of car exhaust, industrial waste, etc., so too is the temperature of the atmosphere, thus giving rise to fears of melting of solar ice caps. Because the concentration of CO_2 is much less than that of water vapor, the rate of this atmospheric heating or greenhouse effect is very slow.

Other everyday experiences involving Planck's law include household heating. When electric heaters are used the heat is usually felt only in given directions. However, if radiators are used they heat the air, which is in continual motion. The resulting heat diffusion gives rise to much more uniform or "central" heat. The reason is that radiator temperatures are such that the thermal emission is in the infrared, thus causing absorption by the atmospheric constituent gases such as H_2O and CO_2 and heating of the air, which diffuses throughout the surroundings.

Applications of thermal radiation in imaging are many and varied. Unlike reflected radiation, thermal emission requires no source of radiation to illuminate the object scene. Thermal imaging is therefore entirely passive. Only the "self-radiation" of the object scene is received by the imager. Thermography is a well-accepted medical diagnostic technique involving no risk to the patient from exposure to radiation such as x-rays. "Hot points" are often an indication of locations that should be examined further by other techniques such as biopsies.

Night vision is a prime application of thermal vision. Since there is little difference in absolute temperature between day and night (say, from 300 to 272 K), thermal vision makes it possible to almost turn night into day. Because thermal emission depends only on temperature and not on illumination, thermal imaging is possible in absolute darkness. This is in contradiction to visible and near-infrared nighttime imaging with SLS (starlight systems), which depend on scene illumination by light from the stars. The small light irradiance reflected by the object scene is then amplified so as to produce a rather bright image. However, in absolute darkness SLSs do not work since there is no light to amplify. They cannot image with thermal radiation unless object temperatures reach thousands of degrees because SLS devices are limited by photocathodes to wavelengths less than about 1 μm. Thermal imaging is applied strongly in military and remote sensing applications. Image irradiance is obtained by substituting thermal radiance into Eq. (3.4.4) or an appropriate approximation of it.

4.3 NATURAL THERMAL SOURCES

Many of nature's sources can be modeled as thermal as summarized in [4.5] and the references cited there.

The spectral irradiance H_λ of the sun at the outer reaches of the earth's atmosphere is somewhat similar to that produced within the 0.2- to 3.0-μm region by 5900 K blackbody radiation. This is evident from Fig. 4.3 where experimentally measured data are presented for comparison. Note from Fig 4.3 that within earth's atmosphere, absorption effects produce irregular deviations from the blackbody curve. Also, solar radiation peaks at visible wavelengths and falls off quite rapidly outside this spectral region. For wavelengths beyond 3 μm the blackbody approximation still applies, but at different temperatures; for example, at 4, 5, and 11.1 μm the corresponding blackbody temperatures are 5600, 5300, and 5040 K [4.6].

Variation of solar irradiance with zenith angle of the sun can be found in the literature [4.6]. Even though the sun appears to us to be an orange-red color, that is not its main color (see Exercise 4.2). Scattering in the atmosphere is, in clear weather, much stronger at shorter wavelengths. This causes the sky to appear blue. Transmission is much better at longer wavelengths, so the sun appears to be orange-red. In strong haze and fog the particulate scatter is much more wavelength neutral. Under such conditions both the sky (scattered light) and the sun (transmitted light) appear to be much more "whitish." This is explained in more detail in Section 15.3. On a clear day about 20% of the total irradiance at earth's surface derives from skylight, or sunlight scattered and diffused by the atmosphere.

Radiation from the brighter stars can be important at night should they come into the imager field of view. As an example, the radiation from Sirius produces a peak irradiance at 3 μm of $5 \cdot 10^{-7}$ $W \cdot m^{-2} \cdot \mu m^{-1}$ before entering earth's atmosphere. Detailed data on stellar irradiance are available in the literature [4.7–4.9].

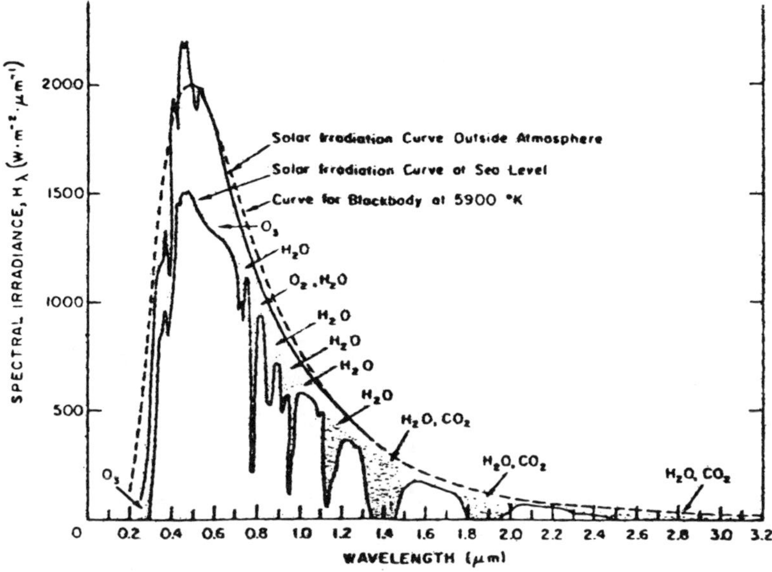

Fig. 4.3 Solar spectral irradiance with sun at zenith. Absorption bands shaded (after [4.6]). Scattering is neglected.

The spectral distribution of radiation reflected from the moon and the planets is similar to that of the sun, differing primarily in magnitude. This magnitude is determined by the distance traversed by the radiation, the *albedo* (ratio of total reflected radiation to total incident radiation) properties of the reflecting body, and its orientation with respect to the sun and the receiver. The dominant source of such radiation in earth's sky is that reflected from the moon. Its peak irradiance just outside earth's atmosphere is about $5 \cdot 10^{-3}$ $W \cdot m^{-2} \cdot \mu m^{-1}$ at about 0.5 μm wavelength [4.7]. Venus, the brightest planet as seen from earth, has a peak spectral irradiance on the order of $5 \cdot 10^{-6}$ $W \cdot m^{-2} \cdot \mu m^{-1}$ [4.7]. Further data of this kind are available in the literature [4.7–4.9]. Also, data on earth-reflected solar radiation as a function of earth–sun orientation have been tabulated [4.10].

The effects of source radiance or emittance on irradiance of the images of such sources can be determined via the equations derived in Section 3.4 by substituting appropriately for object radiance or emittance. The effects of sources on scene illumination and irradiance of the images of object plane scenes can be determined using the results obtained in Example 3.4. If the radiation is focused on the scene, Example 3.5 is appropriate.

More detail on atmospheric scattering and path luminance is presented in Part 5.

Another source of radiation, often undesired because it reduces contrast, is path luminance or radiance of the atmosphere between object and receiver planes. Figure 4.4 illustrates measurements of the radiance of the atmosphere for various solar elevation and observation angles. It is important to note that atmospheric radiance consists essentially of two sources: (1) scattered sunlight or moonlight at short wavelengths and (2) thermal emission by the atmosphere itself at longer wavelengths. Between 2- to 4-μm wavelength skylight and path luminance are minimum in both day and nighttime. At night the normal emission component at long infrared wavelengths is hardly varied, but the sky radiance spectrum is about 5 to 6 orders

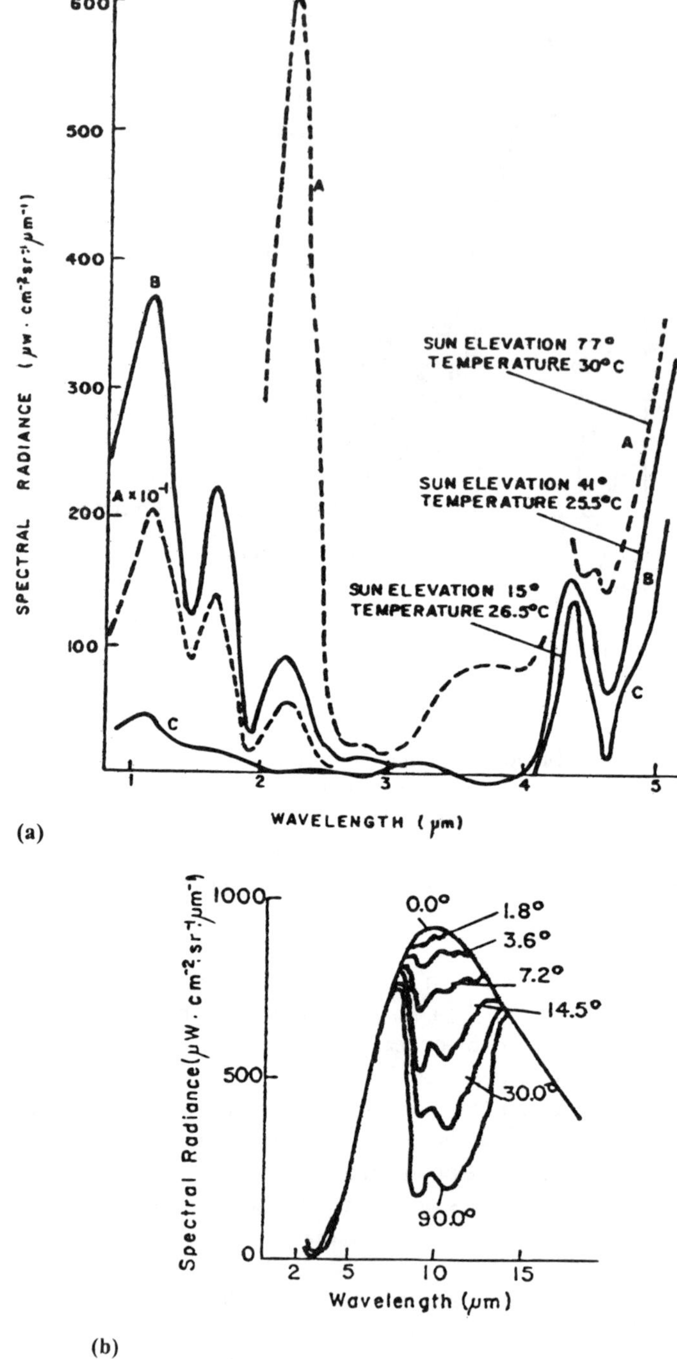

Fig. 4.4 Spectral radiances for clear skies measured at Cocoa Beach, Florida (after [4.11]), show solar radiation scattered at shorter wavelengths and thermal emission at longer wavelengths. (a) Daytime; (b) nighttime for several angles of elevation above horizon after [4.11].

of magnitude less intense even for a full moon than in daytime. This means that while solar radiation is at primarily visible wavelengths, night sky irradiance is at primarily infrared wavelengths. In both cases the path luminance minimum between 2- to 4-μm wavelengths suggests that often the best contrast can be obtained in that spectral region provided there is sufficient object or target illumination.

Example 4.1

Assume an airplane with an engine of 1-m^2 cross section at 800 K is to be imaged by an infrared imager with spectral range centered at 4-μm wavelength by an optical filter. If the irradiance in the image of the plane must be at least twice that of the atmospheric background (300 K) in order for the plane to be detected on the display, what is the maximum range at which the image of the plane can be detected? Assume a planar field of view of 60 mrad for the imager. Assume $\epsilon = 1$ for both the plane and atmospheric background. Assume $\tau' = e^{-\alpha z}$ through the atmosphere.

Solution

Let subscripts a and A refer to airplane and atmosphere, respectively. The geometry is similar to that shown in Fig. 3.8 except that there the target fills up the entire field of view. If z is the distance from the imager to the airplane, then a transverse distance of 60 mrad·z fills up the entire field of view, while the airplane engine occupies only 1 m of this. The ratio of airplane engine area A_a to entire field of view area is

$$\frac{A_a}{\Omega_r z^2} \approx 1/[z^2 \pi \sin^2(30\text{ mrad})] = 1/(2.83 \times 10^{-3} z^2) \qquad (4.3.1)$$

where Ω_r is receiver field of view (sr) as in Fig. 3.8. Now, airplane power received is $P_a = \tau' N_a A_a \Omega$, where N_a is airplane engine radiance, A_a is 1 m^2, and $\Omega = A_r/z^2$, where A_r is receiver area. Therefore, airplane engine irradiance at the entrance pupil is

$$H_a = \frac{P_a}{A_r} = \frac{\tau' N_a A_a}{z^2} \qquad (4.3.2)$$

Power received there from nearby atmospheric thermal background is

$$P_A = \frac{N_A(\Omega_r z^2) A_r}{z^2} \qquad (4.3.3)$$

where $\Omega_r z^2$ is area seen by imager at distance z. Therefore, irradiance of atmosphere on entrance pupil is $H_A = N_A \Omega_r \approx N_A \pi \sin^2(30\text{ mrad}) = 2.83 \times 10^{-3} N_A$. Note that H_a and H_A refer to irradiance on the optics entrance pupil of area A_r and not to the image. In the image plane, where the camera is located, irradiance in the image of the airplane includes both that from the airplane (H_a) as well as that of the intervening atmosphere or path luminance (H_A) between the object and entrance pupil planes. Although irradiance from both the airplane and the atmospheric background fall on the entire entrance pupil plane, the optics focuses the irradiance from the airplane into an image area that is only $A_a(z^2\Omega_r)^{-1} \approx (z^2\pi \sin^2 30 \text{ mrad})^{-1} = (2.83 \times 10^{-3}z^2)^{-1}$ of the area of that of the atmospheric background filling up the entire field of view. Consequently, for irradiances in the image plane

$$\frac{H'_a}{H'_A} = \frac{H_a}{H_A}\left(\frac{z^2\Omega_r}{A_a}\right) \tag{4.3.4}$$

where primes refer to image plane. It is assumed here that $A_a \leq z^2\Omega_r$, i.e., the aircraft image fills up only a portion of the scene. In the image plane it is required that

$$\frac{H'_a + H'_A}{H'_A} \geq 2 \tag{4.3.5}$$

This implies $H'_a \geq H'_A$ or $H'_a/H'_A \geq 1$ or

$$\frac{H_a}{H_A}\left(\frac{z^2\Omega_r}{A_a}\right) \geq 1$$

$$\frac{\tau' N_a A_a}{z^2 N_A \Omega_r}\frac{(z^2\Omega_r)}{A_a} \geq 1$$

$$\frac{\tau' N_a}{N_A} \geq 1 \tag{4.3.6}$$

The ratio of image plane irradiances reduces to apparent object plane radiances in accordance with Eq. (3.4.4). Therefore, since receiver optical passband is identical for both airplane image and atmospheric background, then at a given wavelength $\lambda = 4$ μm

$$e^{-\alpha z}\frac{e^{hc/\lambda k T_A} - 1}{e^{hc/\lambda k T_a} - 1} \geq 1$$

where $T_A = 300$ K and $T_a = 800$ K. Therefore,

$$e^{-\alpha z}\frac{1.63 \times 10^5}{89} \geq 1 \rightarrow e^{-\alpha z} \geq 5.46 \times 10^{-4} \tag{4.3.7}$$

or

$$z \leq \frac{\ln(5.46 \times 10^{-4})}{-\alpha} = \frac{7.51}{\alpha} \tag{4.3.8}$$

This result is imaging range and varies with wavelength according to $\alpha(\lambda)$. Note that even though airplane radiance is N_a, the radiance in Eq. (4.3.6) seen a distance z away is $\tau' N_a$. The latter is called *apparent radiance*.

In real life, the imaging range is likely to be shorter because emissivity of the engine is generally considerably less than unity since metallic parts exhibit high reflectivity unless painted with nonmetallic paint. Also, the irradiance of the image of the airplane engine can be affected by quantization of the image into pixels or picture elements. For example, if the size of the airplane engine image is less than that of a pixel, then the whole pixel is illuminated but with a decrease in irradiance since the radiant power of the airplane image is averaged across the whole pixel area, which is greater than that of the optical image of the engine. In addition there are atmospheric blur effects to be considered in Part 5.

Example 4.2

Consider Example 4.1 again but from the standpoint of a detector rather than imager. A detector detects total received power. As shown in Chapter 2, detector field of view is minimum when the detector is situated at the back focal point of the optics. This decreases the received atmospheric background power which fills the field of view. Develop an equation from which *detection range* can be obtained.

Solution

Both the atmospheric background power as well as the radiant power received from the aircraft fill up the entrance pupil. The difference from imaging, however, is that both sources of radiation are focused onto the detector at the focal point and what is detected is total radiant power rather than irradiance spatial distribution. The fact that the airplane fills only a small portion of the field of view and all radiant power from it is focused into a small image area is irrelevant, for no image is obtained. For the airplane $P_a = \tau' N_a A_a A_r / z^2$, whereas for the atmospheric background $P_A = N_A \Omega_r z^2 A_r / z^2 = N_A \Omega_r A_r$. Therefore, the ratio of received radiant powers is equal to the ratio of irradiances incident on the optics, since receiver area A_r is identical for both:

$$\frac{P_a}{P_A} = \frac{H_a}{H_A} = \frac{\tau' N_a A_a}{z^2 N_A \Omega_r} \tag{4.3.9}$$

or

$$\frac{\exp(-\alpha z)}{z^2}\left(\frac{N_a}{N_A}\frac{A_a}{\Omega_r}\right) \geq 2 \tag{4.3.10}$$

This has to be solved numerically to obtain detection range z. Note that the detection range is different from that for imaging because the criterion in imaging is not total radiant power received but rather image irradiance distribution, which depends on the size of the aircraft image relative to total scene area. If the airplane area is smaller than the total scene area ($A_a < \Omega_r z^2$), then detection range is less than imaging range. For imaging, the focusing action of the optics causes the image of the aircraft to appear brighter than it would be if airplane image irradiance were to be averaged over the entire field of view. Hence, the ratio of image plane irradiance is multiplied by the ratio of target area to total imaged area in order to obtain the ratio of incident radiant powers for detection purposes.

The condition that the aircraft fill only a portion of the field of view requires $z^2 > A_a/\Omega_r = 1/(\pi \sin^2 0.03)$ or $z > 18.81$ m in this example. Otherwise, the aircraft area "seen" is equal to $\Omega_r z^2$ since only a portion of the aircraft fills up the entire field of view. In this latter case, A_a is therefore effectively reduced to $\Omega_r z^2$.

In summary, for longer distances such that the object fills only a portion of the field of view, imaging range depends on ratio of target to background radiances, while detection range depends on ratio of target to background received radiant powers. In both cases, the radiant target and background powers in the optics and sensor planes each are identical, except for any optics transmission losses. However, for imaging, irradiances in the optics and image planes are different because essentially the same target and background radiant powers are spread over the same optics area but over different image plane areas.

4.4 ARTIFICIAL SOURCES

Artificial sources include primarily various types of gas discharge sources such as arc lamps and fluorescent lamps as well as incandescent lamps, light-emitting diodes, lasers, etc.

Many of the devices such as gas discharges and incandescent lamps involve heating which, in itself, gives rise to thermal emission. Figure 4.5 compares spectral radiant emittance of a tungsten-filament incandescent lamp at 3000 K to those of a graybody (wavelength-independent emissivity) and blackbody at the same temperature. The filament is heated by current to incandescence. Most of the spectrum is in the infrared rather than visible. Thus, incandescent lamps are rather inefficient radiators of visible radiation.

Fluorescent lamps are much better simulators of daylight, with most of their spectrum being in the visible. They also exhibit high luminous efficiency and long life. The large size of the radiating area contributes significantly to the total illumina-

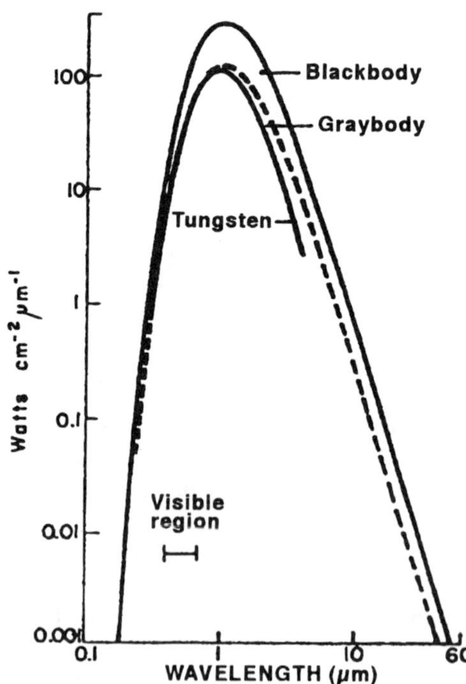

Fig. 4.5 Spectral emittance of blackbody, graybody, and tungsten ribbon operating at 3000 K (after [4.12]). © John Wiley & Sons, Inc.

tion. The gas mixture gives rise to radiation at spectral lines corresponding to radiative deexcitation transitions and recombination, in addition to the thermal emittance. A typical spectrum is shown in Fig. 4.6.

Other discharge sources also exhibit combinations of spectra deriving from both quantum radiative transitions as well as thermal radiation, as shown in Fig. 4.7.

Thermal infrared sources include Nernst glowers (Fig. 4.8) and globars (Fig. 4.9), as well as commercial blackbodies. As a rough estimate, radiance of Nernst glowers is nearly that of a blackbody at the operating temperature with emissivity in excess of 75%.

4.5 THE LASER

The laser can be an extremely broad subject, most of which is beyond the scope of this book. Here we will concentrate basically on two subjects that can have relevance to imaging, particularly *active imaging*, in which an object or scene is illuminated with a laser beam as discussed at the end of Chapter 17. The two subjects of importance are spectral and spatial properties. These two subjects determine the irradiance distribution and coherence of a laser beam. In previous chapters for the sake of simplicity a laser beam has been modeled as if it exhibited uniform irradiance. This is not so even at the laser output, and certainly not so after propagation through the turbulent or scattering channel. As mentioned at the beginning of Chapter 3 with regard to radiometry, the effects of diffraction in radiometry are not considered here. Attention will be focused, however, on the Gaussian spatial properties of laser intensity distribution. This follows a very brief introduction to the mechanism of lasing and the spectral properties of laser radiation. The latter depends strongly on the laser resonator.

The term *laser* stands for *l*ight *a*mplification by *s*timulated *e*mission *r*adiation. When an atom is excited so that energy of an electron is in an elevated state such as E_2 in Fig. 4.10, deexcitation usually takes place shortly after excitation but typically on a spontaneous basis whereby the electron experiences a completely spontaneous decrease in energy from E_2 to E_1. The drop in energy results in emission of a photon of energy $E_2 - E_1 = h\nu_{21}$, exactly equal to the energy decrease of the electron, if the transition from E_2 to E_1 is radiative. If a photon of energy $h\nu_{21}$ is incident on an atom with an electron already excited from E_1 to E_2, then emission of a second

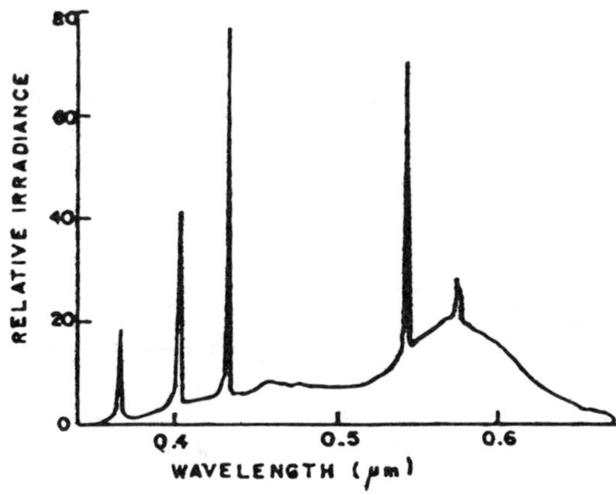

Fig. 4.6 Typical spectrum of fluorescent lamp (after [4.12]). © John Wiley & Sons, Inc.

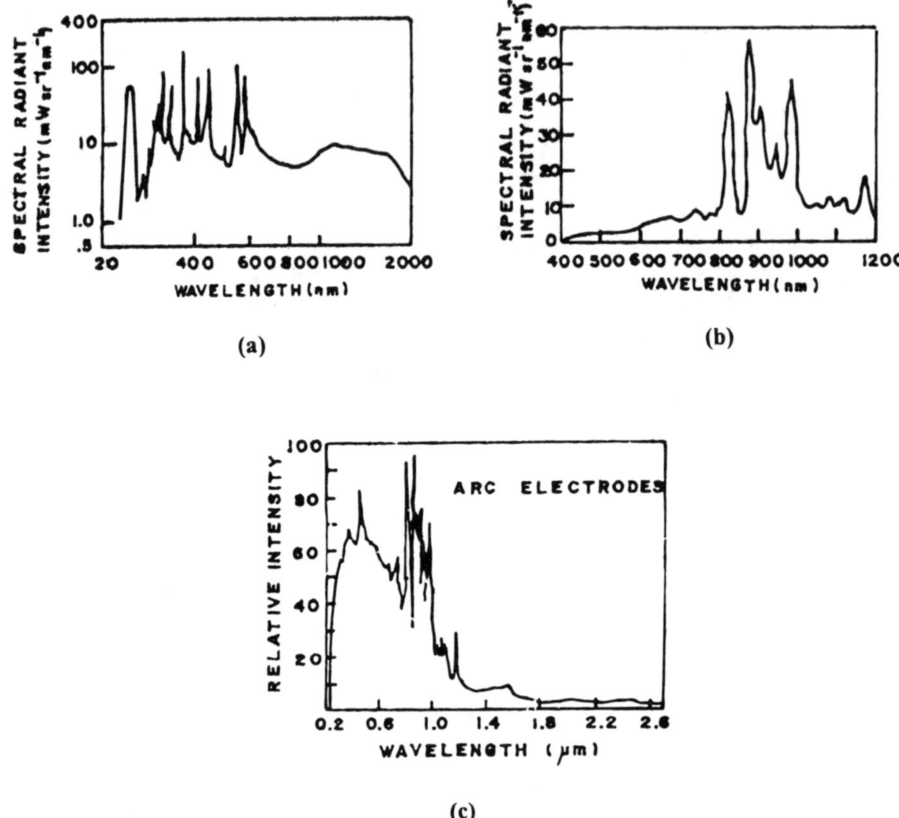

Fig. 4.7 Typical intensities of various arc lamps: (a) 200-W mercury short arc; (b) 150-W xenon short arc; (c) 20-kW xenon short arc (after [4.13]).

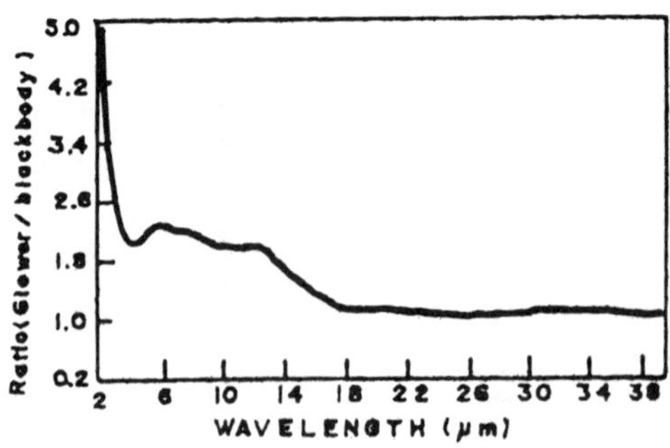

Fig. 4.8 Ratio of Nernst glower to 1173 K blackbody irradiances as a function of wavelength.

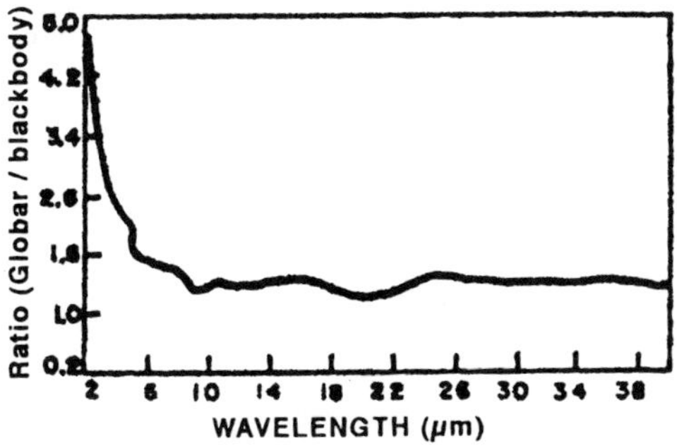

Fig. 4.9 Ratio of Globar to 1173 K blackbody irradiances as a function of wavelength.

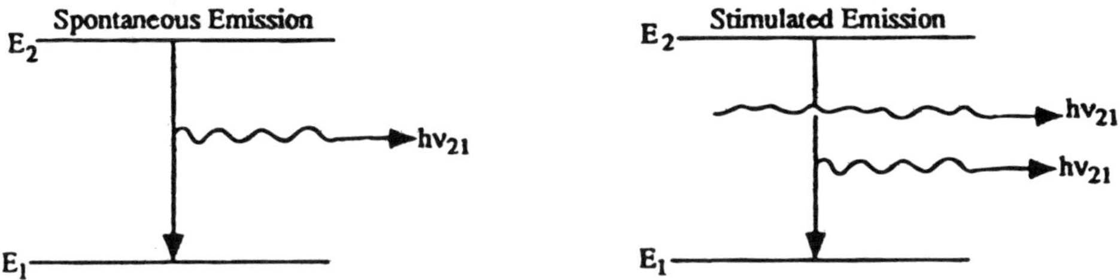

Fig. 4.10 Spontaneous and stimulated emission processes.

photon of energy $h\nu_{21}$ is thereby stimulated, and is accompanied by deexcitation of the electron from E_2 to E_1.

The photon resulting from the stimulated emission is of the same phase as that of the photon stimulating the emission. Since one stimulating photon gives rise to an additional photon, light amplification by stimulated emission is taking place. To improve efficiency of such phenomena it is necessary to (1) cause excitation from E_2 to E_1 and (2) supply photons of the energy required for the energy transition of interest so as to give rise to stimulated emission. The first process is called *pumping*, and can be done by supplying energy to the lasing medium by external means such as optical excitation via some sort of lamp, or via electrical excitation by applying voltage to a gas discharge, which gives rise to excitation collisions between electrons and neutral atoms. In semiconductor devices application of voltage can permit excitation of electrons from the valence to the conduction band or of holes from the conduction to the valence band. Photons that are absorbed by atoms with electrons at E_1 rather than E_2 are lost to the stimulated emission processes. Therefore, enough pumping is required so that the population of E_2 is greater than that of E_1 in order for more light to be amplified rather than absorbed. Such pumping gives rise to *population inversion*. In gas discharge lasers the energy levels E_1 and E_2 usually do not involve ground or low-level metastable states. In semiconductor lasers E_2 and E_1 often involve conduction and valence bands, although they can involve donor and/or acceptor levels.

4.5.1 Laser Spectra

Photons to stimulate the emission often arise initially from spontaneous emission. To improve efficiency of stimulated emission, a resonator is formed as shown in Fig.

4.11 via a mirror on each side of the lasing medium so that light is reflected back and forth through the laser medium an infinite number of times. If the optical path length of the resonator is l, then resonance can take place only at wavelengths where l is equal to an integral number of half-wavelengths, or

$$\lambda_m = 2ml = \frac{c}{\nu_m} \tag{4.5.1}$$

where m is an integer. Each frequency or wavelength represents a longitudinal mode of the laser.

Neglecting phase components of mirror reflection coefficients, at resonance frequencies a round-trip of a photon from mirror 1 to mirror 2 and back again to mirror 1 must involve a phase shift equal to an integral number of times 2π, or

$$2\beta l = 2m\pi \tag{4.5.2}$$

leading to

$$2\left(\frac{2\pi}{\lambda_m}\right)l = 2m\pi$$

or

$$\frac{2l\nu_m}{c} = m$$

so that

$$\nu_m = \frac{mc}{2l} \tag{4.5.3}$$

This leads to a frequency band between laser modes

$$\Delta f_m = \nu_m - \nu_{m-1} = \frac{c}{2l} \tag{4.5.4}$$

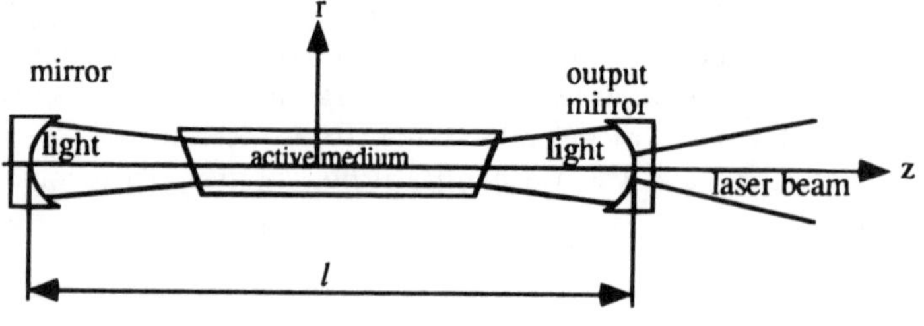

Fig. 4.11 Typical laser resonator. (Output laser beam here exits the right-side mirror.)

Each longitudinal mode is of bandwidth approximately equal to

$$df_m = \nu_m/Q \tag{4.5.5}$$

where Q is the quality factor of each mode and is equal to

$$Q = \frac{2\pi l}{\lambda\delta} \tag{4.5.6}$$

where δ represents the fractional amount of radiant power loss per transit within the resonator. One mirror is of maximum reflectivity. The mirror at the other end is typically of (power) reflectivity on the order of $\rho \cong 80$ to 90%. The radiant power not reflected back into the resonator by this mirror is usually the major source of resonator power loss δ and *actually composes the output beam* of the laser as shown on the right side of Fig. 4.11. Other losses may stem from diffraction at the mirror edges or reflectivity losses at the other mirror.

As a result

$$df_m = \frac{c\delta}{2\pi l} \tag{4.5.7}$$

A typical laser spectrum is shown in Fig. 4.12.

Example 4.3

Consider a HeNe laser ($\lambda = 0.6328\ \mu\text{m}$) with $\delta = 0.1$ resulting from 10% power transmission through one mirror, which is the output aperture. Assume the refractive index of the gas discharge is unity. If the resonator is 1 m long, what is the output spectrum of the laser?

Solution

$$\Delta f_m = \frac{c}{2l} = \frac{3 \times 10^8}{2(1)} = 150\ \text{MHz}$$

$$df_m = \nu_m/Q = \frac{c\delta}{2\pi l} = \frac{3 \times 10^8(0.1)}{2(1)} = \frac{3 \times 10^7}{2\pi} = 4.77\ \text{MHz} << \Delta f_m$$

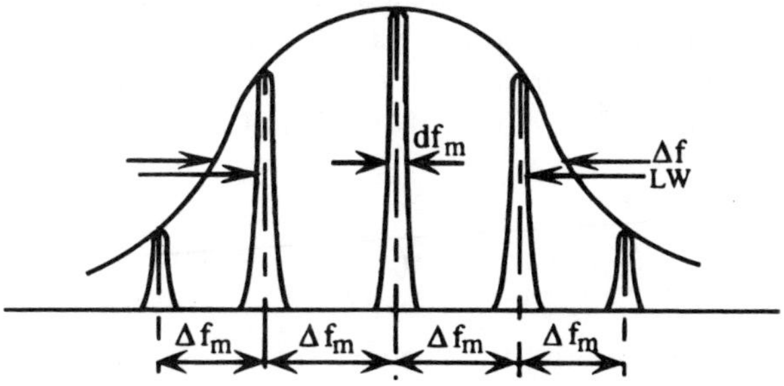

Fig. 4.12 Typical laser spectrum.

Note that compared to the frequency difference Δf_m between longitudinal modes in Fig. 4.12, each longitudinal mode itself is of negligible bandwidth.

The envelope to the spectrum in Fig. 4.12 derives from the radiation which can stimulate the desired photon emission. This radiation may derive from photon emission as a result of recombination and deexcitation in gas discharges, for example, or from an arc lamp or light-emitting diode for solid-state lasers. The half-power or half-maximum bandwidth is used to define the frequency linewidth (LW) of lasers. Since the laser spectrum is not continuous but consists essentially of discrete longitudinal modes, the laser linewidth is actually limited only to those modes falling within the half-power linewidth of the laser. Lasing may occur at other modes not included in LW, but radiant power in them is below the half-maximum threshold usually used to define laser linewidth. For example, in Fig. 4.12 there are three modes within the half-maximum bandwidth of the laser. Accordingly, the laser linewidth is essentially only $2\Delta f_m + 3df_m$, or 314 MHz. If nine modes were to fall within the half-power bandwidth of the laser, then actual laser linewidth would be only $8\Delta f_m + 9df_m$. To increase Δf_m and thereby decrease the number of longitudinal modes, it is usually necessary to decrease the resonator length l. This is what is usually done when single longitudinal mode operation is desired. In single-mode lasers, linewidth reduces to df_m, the linewidth of one individual mode. Since there are fewer longitudinal modes, laser output power is reduced as well.

Radiative transitions of frequency ν_0, which give rise to the photons that stimulate emission, are very much broadened in frequency. For example, in gas discharges such frequency broadening is caused generally by Doppler shifts arising from thermal velocity of the neutral excited molecules. In such cases the Doppler-broadened transition, which forms the envelope in Fig. 4.12, is Gaussian in form and is given by [4.15]:

$$G(\nu - \nu_0) = \frac{2}{\Delta f}\left(\frac{\ln 2}{\pi}\right)^{1/2} \exp[-4 \ln(\nu - \nu_0)^2/(\Delta f)^2] \tag{4.5.8}$$

where Δf is the half-power linewidth of the Doppler-broadened radiation in Fig. 4.12, which supplies the photons stimulating the emissions in the lasing medium and is equal to

$$\Delta f = 2\nu_0\left(\frac{2kT \ln 2}{Mc^2}\right)^{1/2} \tag{4.5.9}$$

The resonance frequencies are not broadened because they are limited only to those permitted according to resonator optical path length l, as described in Eq. (4.5.3). It is only these frequencies rather than all of $G(\nu - \nu_0)$ in which light is amplified by stimulated emission. In Eq. (4.5.9) M is mass of the atoms involved. From Fig. 4.12 it is clear that

$$\Delta f > (p - 1)\Delta f_m + pdf_m = \text{LW} \tag{4.5.10}$$

where p is the number of longitudinal modes that fit within Δf and LW is actual laser linewidth.

Although the above simplified treatment involves atomic transitions, it can be applied similarly to molecular or vibrational transitions as well.

4.5.2 Resonator Stability

Resonator stability requires that the radiation reflected back and forth between the mirrors in Fig. 4.11 stay in the resonator rather than be reflected outside the resonator. Stability depends on mirror separation l as well as radii of curvature. It is customary to model the back and forth reflections of light between the mirrors as if the light were propagating through an infinite sequence of lens pairs, as shown in Example 2.9, where the focusing action of each mirror is represented by the focusing action of the corresponding lens in Fig. 2.27. Using the treatment of Example 2.9, relation (2.13.27) there, which describes requirements for light not to be reflected out of the resonator, can be rewritten as follows:

$$0 \leq \left(1 - \frac{l}{R_1}\right)\left(1 - \frac{l}{R_2}\right) \leq 1 \tag{4.5.11}$$

since mirror focal length is one-half radius of curvature [Eq. (2.4.15). In relation (4.5.11), R_1 and R_2 are radii of curvature for mirrors 1 and 2, respectively. Relation (4.5.11) is described graphically in Fig. 4.13. If this relationship is satisfied, the resonator is stable, that is, light is not directed out of the resonator.

Example 4.4

Consider stabilities of confocal ($l = R_1 = R_2$), parallel-plane ($R_1 = R_2 = \infty$), and concentric ($R_1 = R_2 = l/2$) resonators.

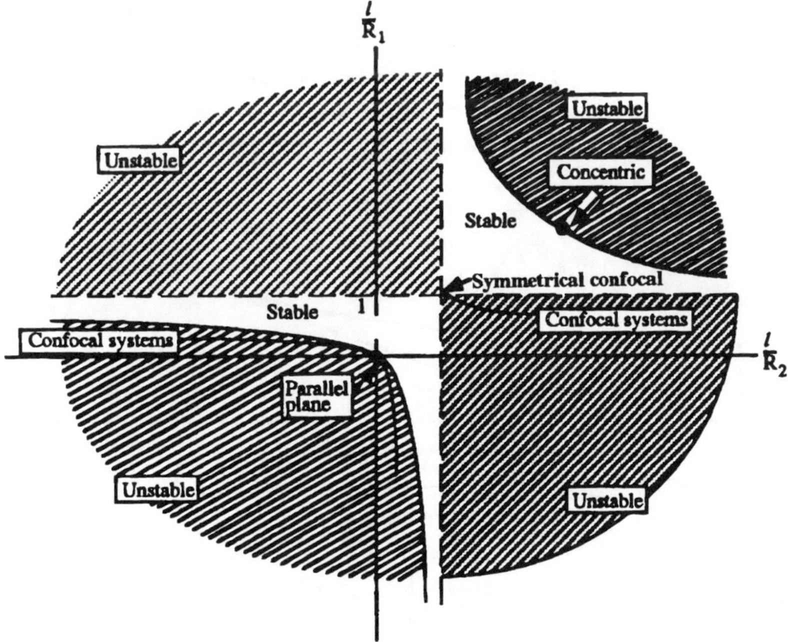

Fig. 4.13 Stability diagram for laser resonators [from Eq. (4.5.11)].

Solution

For a confocal resonator, relation (4.5.11) yields only the solution $l/R_1 = l/R_2 = 1$. This is the point (1,1) in Fig. 4.13. It is on the border between stable and unstable regions, so that even slight inaccuracies in alignment or aberrations can easily cause the resonator to be unstable.

The parallel-plane resonator is also known as a *Fabry-Perot resonator.* The mirrors exhibit no curvature. Since $R_1 = R_2 = \infty$, then $l/R_1 = l/R_2 = 0$. The origin is also on the border between stable and unstable regions and is therefore not a good location to operate.

The concentric resonator is stable only at the point $l/R_1 = l/R_2 = 2$. The point (2,2) is also on the border between stable and unstable regions.

In principle, resonators should be designed to operate within regions of stability rather than on the borders of stable regions. Examples might be points such as $l/R \cong 0.6, 1.4$, etc.

4.5.3 Gaussian Irradiance

As a result of the focusing action of the resonator mirrors, irradiance between the mirrors is generally greater than at the mirrors depicted in Fig. 4.11. The spatial distribution of the irradiance is Gaussian. This also means that in images of laser beams irradiance is Gaussian rather than uniform. Laser irradiance is greater at the beam center than at the beam edges. This can affect the quality of active imaging.

If r is a radial spatial coordinate, the electric field varies as

$$E(r) = \frac{Aw_0}{w(z)} \exp[-r^2/w^2(z)] \tag{4.5.12}$$

where A is a constant, $w(z)$ describes beam radius as a function of axial coordinate z, and w_0 is the minimum value of $w(z)$ or beam radius. In Eq. (4.5.12) minimum beam radius occurs at $z = 0$, that is, $w(0) = w_0$, as shown in Fig. 4.14.

Beam radius $w(z)$ extends by definition in the transverse direction from the z axis along the center of the beam until the field descends to e^{-1} of its value at

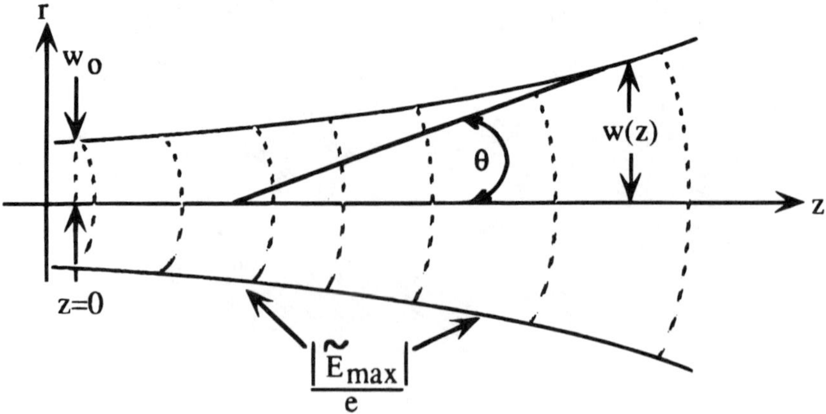

Fig. 4.14 Gaussian irradiance distribution of laser beam.

beam center ($r = 0$). In terms of irradiance, then, beamwidth is defined according to e^{-2} of its value at $r = 0$. In Eq. (4.5.12),

$$w(z) = w_0 \left[1 + \left(\frac{\lambda}{\pi w_0^2} \right)^2 z^2 \right]^{1/2} \tag{4.5.13}$$

In the beam far field where $z >> (\pi/\lambda)w_0^2$,

$$w(z) \cong \frac{\lambda}{\pi w_0} z \tag{4.5.14}$$

and laser half-beamwidth is

$$\theta \cong w(z)/z = \frac{\lambda}{\pi w_0} \tag{4.5.15}$$

If R_2 refers to output mirror in Fig. 4.11, the distance of beam waist from it is

$$\frac{l(R_1 - l)}{R_1 + R_2 - 2l}$$

Example 4.5

An argon laser beam of wavelength 0.5145 μm is transmitted 3.85×10^5 km to the moon. If minimum beam waist inside the laser is of 0.25-cm radius, what is the beam radius of the light beam incident on the moon?

Solution

Using Eq. (4.5.14)

$$w(z) \cong \frac{\lambda}{\pi w_0} z = \frac{0.5145 \times 10^{-6} \text{ m } (3.85 \times 10^8 \text{ m})}{\pi(2.5 \times 10^{-3} \text{ m})} = 25.2 \text{ km}$$

Example 4.6

What should w_0 be so as to limit laser beam radius at the moon to 100 m? What conclusion is reached by comparing this value for w_0 to that in the previous example?

Solution

From Eq. (4.5.14),

$$w_0 \cong \frac{\lambda z}{\pi w(z)} = \frac{0.5145 \times 10^{-6}(3.85 \times 10^8)}{\pi(100)} = 63 \text{ cm}$$

To obtain a smaller beam radius in the far field, it is necessary to increase w_0.

Focusing Gaussian beams is depicted in Fig. 4.15. The ratio of beam waist at lens input to that at output is [4.15]

$$\frac{w_{01}^2}{w_{02}^2} = \left(1 - \frac{z_1}{f}\right)^2 + \left(\frac{\pi w_{01}^2}{\lambda f}\right)^2 \tag{4.5.16}$$

Note that because of the Gaussian irradiance distribution, the output beam waist does not necessarily occur in the output focal plane, but rather at [4.15]

$$z_2 = f + \frac{(z_1 - f)f^2}{(z_1 - f)^2 + \left(\dfrac{\pi w_{01}^2}{\lambda f}\right)^2} \tag{4.5.17}$$

as shown in Fig. 4.15. If $z_1 \cong f$, then from Eq. (4.5.16)

$$\frac{w_{01}}{w_{02}} \cong \frac{\pi w_{01}^2}{\lambda f} \tag{4.5.18}$$

and

$$z_2 = f + \frac{(z_1 - f)f^2}{(z_1 - f)^2 + \left(\dfrac{w_{01}}{w_{02}}\right)^2} \tag{4.5.19}$$

Therefore, for a focused laser beam where $w_{01} >> w_{02}$, then $z_2 \cong f$. This result differs from geometrical optics where if a source is located at the front focal point of a lens, its radiation will pass through the lens and be imaged (or focused) at infinity. However, in laser optics, if the preceding conditions are satisfied, laser radiation will focus at the back focal point. On the other hand, if $w_{01} << w_{02}$, then

$$z_2 = f + \frac{f^2}{z_1 - f} \tag{4.5.20}$$

so that if z_1 equals f, then z_2 approaches the geometrical optics limit of infinity. The

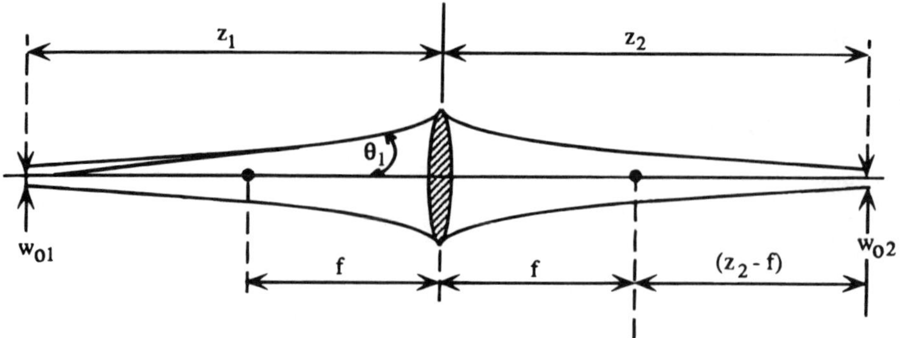

Fig. 4.15 Beam waist transformation through thin lens.

question as to whether outgoing beam waist is larger or smaller than the incoming one depends largely on focal length. The smaller the focal length, the smaller the outgoing beam waist. It is focused to a smaller point.

Example 4.7

(a) Suppose a lens is situated so that w_{01} is in the front focal plane. What should lens focal length be in order to decrease the beam waist by a factor of 10?

(b) What should focal length be if $w_{01} = 1$ mm and $\lambda = 10.6$ μm? What focal length should be used if the laser is HeNe ($\lambda = 0.6328$ μm) instead of CO_2 ($\lambda = 10.6$ μm)?

(c) In both cases what should be minimum lens diameter D? Explain the result.

Solution

(a) From Eq. (4.5.16)

$$\left(\frac{w_{01}}{w_{02}}\right)^2 = 10^2 = (1 - 1)^2 + \left(\frac{\pi w_{01}^2}{\lambda f}\right)^2$$

(b) From Eq. (4.5.18) $f = \pi w_{01}^2/10\lambda = \pi(10^{-3})^2/10(10.6 \times 10^{-6}) = 2.96$ cm for the CO_2 laser. If $\lambda = 0.6328$ μm, then $f = 49.6$ cm.

(c) Using Eq. (4.5.13) $D = 2w(z) = 2w_{01}\,[1 + (\lambda z/\pi w_{01}^2)^2]^{1/2}$. For the CO_2 laser,

$$D = 2(10^{-3})\left[1 + \left(\frac{10.6 \times 10^{-6}}{\pi(10^{-3})^2}\right)(2.96 \times 10^{-2})^2\right]^{1/2}$$

$$= 2 \times 10^{-3}(1 + 0.01)^{1/2} = 2.008 \text{ mm}$$

For the HeNe laser

$$D = 2(10^{-3})\left[1 + \frac{0.633 \times 10^{-6}\,(0.496)^2}{\pi(10^{-3})^2}\right]^{1/2}$$

$$= 2 \times 10^{-3}(1 + 0.01)^{1/2}\text{m} = 2.008 \text{ mm}$$

Minimum lens diameters are identical. The reason is that from part (b) $f = \pi w_{01}^2/10\lambda$ so that for $z = f$

$$w(z) = w_{01}\left[1 + \left(\frac{\lambda z}{\pi w_{01}^2}\right)^2\right]^{1/2} = w_{01}\left[1 + \left(\frac{\lambda}{\pi w_{01}^2}\right)\left(\frac{\pi w_{01}^2}{10\lambda}\right)\right]^{1/2}$$

$$= w_{01}\left[1 + \left(\frac{1}{10}\right)^2\right]^{1/2} = w_{01}\left[1 + \left(\frac{w_{01}}{w_{02}}\right)^{-2}\right]^{1/2}$$

which is independent of wavelength.

If the incoming laser beam waist is in the lens front focal plane, Eq. (4.5.18) indicates

$$w_{02} \cong \frac{\lambda f}{\pi w_{01}}$$

In other words, if we want to focus with the lens to a minimum beam waist (w_{02}), which is small so as to increase irradiance, then in order to do so it is necessary to widen the incoming beam waist.

In imaging applications involving laser beams, very often the laser has propagated a long distance prior to being incident on the receiver lens, so that $z_1 >> f$. In such cases Eq. (4.5.16) reduces to

$$w_{02} = w_{01}\frac{f}{\sqrt{z_1^2 + \left(\frac{\pi w_{01}^2}{\lambda}\right)}} \cong \frac{w_{01}f}{z_1} \tag{4.5.21}$$

so that $w_{02} << w_{01}$. Again, the wider the incoming beam the narrower the outgoing beam. Furthermore, from Eq. (4.5.17) the output beam from the lens is focused at

$$z_2 \cong f + \frac{z_1 f^2}{z_1^2 + \left(\frac{\pi w_{01}^2}{\lambda f}\right)^2} \approx f + \frac{f^2}{z_1} \tag{4.5.22}$$

which is very close to the back focal plane, as expected from geometrical optics since z_1 is very large.

In the image of the laser beam r and $w(z)$ are scaled according to magnification. Only in the case of $z_1 = f$ and $w_{01} >> w_{02}$ does the location of output beam focus differ significantly from that predicted on the basis of geometrical optics.

When dealing with Gaussian laser beams it is of interest to consider what lens diameter is required in order to collect most of the irradiance incident on the lens. Following Eq. (4.5.12), the irradiance varies as

$$H(r, z) = H_0(0, z)\exp[-2r^2/w^2(z)] \tag{4.5.23}$$

where H_0 is irradiance along the beam axis. Radiant power collected by a lens of diameter D is, therefore,

$$P = \int_{-D/2}^{D/2} H(r, z) 2\pi r \, dr = 2\pi \, H_0 \int_{-D/2}^{D/2} \exp(-2r^2/w^2) r \, dr \qquad (4.5.24)$$

Let $x = r^2$. Then $r = x^{1/2}$ and $dr = (1/2)x^{-1/2}$. Substituting into Eq. (4.5.24)

$$P = 2\pi H_0 \int_0^{(D/2)^2} \exp[-2x/w^2(z)] \, dx \qquad (4.5.25)$$

$$= \pi w^2(z) \, H_0[1 - \exp(-D^2/2w^2(z))]$$

Thus, only for an infinite-diameter lens does collected radiant power equal its maximum value of $\pi w^2(z) H_0$ or

$$H_0 = \frac{P}{\pi w^2(z)} \qquad (4.5.26)$$

Theoretically, as radial coordinate r increases, irradiance never reaches zero. When lens diameter exactly equals beamwidth as defined by e^{-2} fall-off points, then in Eq. (4.5.23) $r = w(z)$ and $H = H_0 e^{-2}$. The fractional amount of radiant power collected from Eq. (4.5.25) is, for $D = 2w(z)$,

$$P|_{D=2w(z)} = 1 - \exp(-2) = 86.5\%$$

If, however, $D = 3w(z)$, then

$$P|_{D=3w(z)} = 1 - \exp(-4.5) = 99\%$$

Therefore it is generally desirable to choose a lens diameter that is at least 1.5 times the nominal laser beam diameter of $2w(z)$. This also minimizes the amount of light diffracted at the edges of the lens.

4.6 COHERENCE LENGTH

One of the basic parameters limiting use of sources for various applications is source coherence. Most imaging is carried out with incoherent light, and in many applications coherence is undesirable because it gives rise to diffraction effects added to the image which degrade the image and limit the information to less than that which would be available were the image to be obtained with incoherent light. However, for many of the applications connected with Fourier optics and radar imaging, coherence is desired. For these reasons it is important to define coherence properties.

A coherent light source is one that produces radiation with waves vibrating in phase. This means that such radiation exhibits a clear, ordered wavefront since the radiation is propagating in phase. There are two kinds of coherence—temporal and spatial. The first is considered here; the second in Chapter 7. Coherence effects on imaging are considered in Chapter 9.

Coherence is generally described in terms of correlation functions involving complex field amplitude (CFA). For temporal coherence, the autocorrelation at a given spatial point is considered as a function of time:

$$\Gamma_{11}(\tau) \triangleq \langle U_1(\bar{r}_1, t)U_1^*(\bar{r}_1, t + \tau)\rangle / 2\eta_0 \tag{4.6.1}$$

where $\bar{r}_1$ is a position vector of a space point taken with respect to some arbitrary origin, $U_1(\bar{r}_1, t)$ is CFA at space point $\bar{r}_1$ at time t, and $U_1(\bar{r}_1, t + \tau)$ is CFA at space point $\bar{r}_1$ at time $t + \tau$. The angular brackets indicate time average, so that

$$\Gamma_{11}(\tau) = \lim_{T\to\infty} \frac{1}{2T} \int_{-T}^{T} U_1(\bar{r}_1,t)U_1^*(\bar{r}_1,t + \tau)\, dt \tag{4.6.2}$$

When $\tau = 0$, $\Gamma_{11}(0)$ is the time-averaged field irradiance H_1 at point $\bar{r}_1$ and is also the maximum value of $\Gamma_{11}(\tau)$. Complex degree of temporal coherence is defined as the normalized autocorrelation function:

$$\gamma_{11}(\tau) \triangleq \frac{\Gamma_{11}(\tau)}{\Gamma_{11}(0)} = \frac{\Gamma_{11}(\tau)}{H_1(\bar{r}_1)} \tag{4.6.3}$$

It follows from this definition that $\gamma_{11}(0) = 1$ and $\gamma_{11}(t > t_c) = 0$, where t_c is coherence time. Thus, 0 and t_c determine extremum values of complex degree of coherence. In general,

$$t_c = (\mathrm{LW})^{-1} \tag{4.6.4}$$

where LW is linewidth in hertz of the optical source. The physical concept behind Eq. (4.6.4) is the following. Consider a radiation source emitting at two wavelengths, λ_1 and λ_2. Radiation at both wavelengths propagates coincidentally, that is, radiation at both wavelengths exists at a given point in space (such as $\bar{r}_1$). As can be seen in Fig. 4.16, after propagating a short distance of only a few wavelengths, the phase difference between the two waves is small and some phase correlation exists. However, as the distance from the source increases, so too does the phase difference between the two wavelengths, and the phase correlation decreases. The larger the difference in wavelength or frequency of the waves, the less the phase correlation between them, and the shorter the distance at which the correlation approaches zero. Correlation length or coherence length is defined as

$$l_c = ct_c = c/\mathrm{LW} \tag{4.6.5}$$

where c is speed of light in the usual propagation channel of air or vacuum. Although

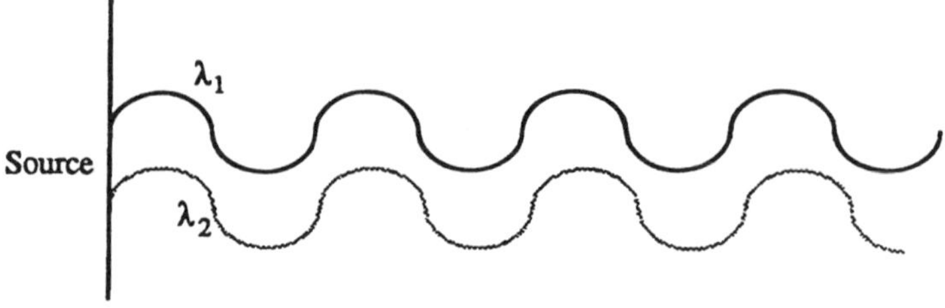

Fig. 4.16 Propagation at two slightly unequal wavelengths.

in Fig. 4.16 radiation at the two wavelengths is shown separately, in reality radiation at both wavelengths is coincident. Thus, for propagation distances much larger than l_c, or time durations of propagation much larger than t_c, wavefront is essentially obliterated because of the differences in phase at the different wavelengths. This does not mean that the complex degree of temporal coherence is unity for $t < t_c$ and zero for $t > t_c$. Clearly, $\gamma_{11}(\tau)$ decreases gradually as τ increases. Coherence time and coherence length are simply convenient criteria with which to give some idea as to the distance radiation can propagate and still exhibit some semblance of wavefront definition. The larger the linewidth of the source, the shorter the coherence time and length.

Lasers exhibit much narrower linewidths than other optical sources and thus yield much longer coherence lengths. However, as seen in the exercises at the end of this chapter, these coherence lengths are usually on the order of several meters or less, except for single-mode lasers that have been specially designed for high coherence. Such devices can be designed for coherence lengths on the order of several and even tens of kilometers.

Although the preceding explanation is presented in terms of propagation distance from the source, the same explanation also pertains to path difference from a source. For example, interferometry deals with interference arising from light propagating over two different optical paths with an optical path difference. The same explanation regarding Fig. 4.16 holds true for an optical path difference beginning at some point away from the source. As long as the optical path difference is less than l_c, correlation of phase at one wavelength relative to that of the other still exists and mixing of the two waves so as to permit interference between them is possible. Consequently, in interferometry it is important for the optical path length to be less than the coherence length of the source.

Another application is in heterodyne detection, which is discussed in Chapter 6. Here, signal and local oscillator beams are mixed so as to yield a frequency difference or an i-f frequency. It is important to use a source of coherence length greater than optical path difference between the signal and local oscillator beams.

Note that l_c and t_c depend only on source linewidth and not source frequency. Compared to sources at lower frequencies, laser coherence lengths are very short. For example, while laser linewidths are typically on the order of several megahertz to even hundreds of megahertz, klystron linewidths at microwave or millimeter wave frequencies are typically at the order of only several or tens of kilohertz, thus easily yielding much longer coherence lengths and facilitating heterodyne detection over long distances.

Another parameter connected with source linewidth is *monochromaticity*, defined as LW/f_0, where f_0 is carrier frequency. For a laser at optical wavelengths, ν_0 is on the order of 10^{13} to 10^{15} Hz, so that laser monochromaticity approaches zero and lasers are therefore much more monochromatic than microwave or millimeter devices. However, no source can be completely monochromatic because any modulation generates sidebands, which increase linewidth. Simply turning a laser on and off is pulse modulation. For example, suppose a laser were completely monochromatic (single frequency). By Fourier integral

$$\delta(f - f_0) = \int_{-\infty}^{\infty} f(t)e^{-j2\pi(f-f_0)t}\,dt = \int_{-\infty}^{\infty} (1)e^{-j2\pi(f-f_0)t}\,dt \tag{4.6.6}$$

which requires that $f(t)$ exist over all time, because any change in the limits of the integral will produce a result other than a delta function. Thus, only if a completely monochromatic source were emitting from $t = -\infty$ to $t = \infty$ would the source stay completely monochromatic without sidebands.

As seen in Example 4.3 and Exercises 4.13 and 4.14, lasers are not really monochromatic. However, because of their degree of monochromaticity they are much more monochromatic than other sources and are therefore usually considered to be quasimonochromatic.

REFERENCES

4.1. R. D. Hudson, *Infrared System Engineering*, Wiley, New York, 1969, Chap. 2. Reprinted by permission of John Wiley & Sons, Inc.

4.2. T. P. Merrit and F. F. Hall, Jr., "Blackbody radiation," *Proc. IRE*, Vol. 47, September 1959, pp. 1435–1441.

4.3. B. M. Oliver, "Thermal and quantum noise," *Proc. IEEE*, Vol. 53, May 1965, pp. 436–454.

4.4. F. E. Nicodemus, "Directional reflectance and emissivity of an opaque surface," *Appl. Opt.,* Vol. 4, July 1965, pp. 767–773.

4.5. N. S. Kopeika and J. Bordogna, "Background noise in optical communication systems," *Proc. IEEE*, Vol. 58, October 1970, pp. 1571–1577.

4.6. P. R. Gast, "Solar irradiance," in *Handbook of Geophysics and Space Environments,* L. Valley, Ed., Air Force Geophysics Laboratory, Cambridge, MA, 1965, Sec. 16.1.

4.7. R. C. Ramsey, "Spectral irradiance from stars and planets above the atmosphere from 0.1 to 100 microns," *Appl. Opt.,* Vol. 1, July 1962, pp. 465–471.

4.8. M. Ross, *Laser Receivers,* Wiley, New York, 1965, pp. 281–293.

4.9. W. K. Pratt, *Laser Communication Systems*, Wiley, New York, 1969, Chap. 8.

4.10. W. L. Wolfe and G. J. Zissis, Eds., *Infrared Handbook*, Environmental Research Institute of Michigan, Ann Arbor, 1989.

4.11. E. E. Bell, L. Eisner, J. Young, and R. A. Oetjen, "Spectral radiance of sky and terrain at wavelengths between 1 and 20 microns. II. Sky measurements," *J. Opt. Soc. Am.,* Vol. 50, 1960, pp. 1313–1320.

4.12. R. W. Boyd, *Radiometry and the Detector of Optical Radiation,* Wiley, New York, 1983. Reprinted by permission of John Wiley & Sons, Inc.

4.13. *Electro-Optics Handbook*, RCA, Harrison, NJ, 1974.

4.14. F. Grum and R. J. Becherer, *Optical Radiation Measurements,* Vol. 1, *Radiometry,* Academic Press, New York, 1979.

4.15. A. Yariv, *Optical Electronics* (4th ed.), Holt, Rinehart, and Winston, New York, 1988.

EXERCISES

4.1 Derive Wien's law and justify an approximate solution.

4.2 Assuming the sun radiates very much like a blackbody at 5900 K, determine the wavelength at which solar irradiance is maximum. What is the true color of the sun? The sun appears to be of different color at the earth's surface because of selective scattering of shorter wavelengths to produce a blue sky (scattered light) and less scattering (better transmission) at longer wavelengths to make the sun appear to be orange or red, which it is *not*.

4.3 Compute total emittance of the sun.

4.4 Equations (4.2.16) and (4.2.17) cannot be proved analytically. Using analytical means, however, develop a proof up to the point that only graphical or numerical techniques can be used to finish the proof. To check the validity of your analytical work, substitute Eqs. (4.2.16) and (4.2.17) into your last step and see to what extent the last equation holds.

4.5 Prove

$$\frac{N_\lambda(\lambda_{m2},T_2)}{N_\lambda(\lambda_{m1},T_1)} = \left(\frac{T_2}{T_1}\right)^5 = \left(\frac{\lambda_{m1}}{\lambda_{m2}}\right)^5$$

4.6 Derive the Rayleigh-Jeans law from Planck's law.

4.7 Compare solar irradiance at the moon's surface with that produced there by a 1-GW Nd:YAG laser ($\lambda = 1.06$ μm) located on earth. Assume a diffraction-limited laser beam whose aperture D is 1 cm in diameter. Neglect losses in the earth's atmosphere. Assume the sun to be a blackbody. Is it possible to see the laser irradiance at the moon with a sunlight background? (*Note:* Half-angle of diffraction-limited circular beam is $1.22\lambda/D$, as proven in Section 7.4).

4.8 Suppose a 1-nm-wide optical filter is available to decrease the solar irradiance background in Exercise 4.7. At which wavelength would you want it to be centered to improve your chances of seeing the laser irradiance against the sunlight background? Using such a filter, can the laser irradiance at the moon be seen despite the sunlight background? Assume uniform laser irradiance.

4.9 Within the field of view of a thermal imaging system are graybodies with emissivities of $\epsilon = 0.95$ and $\epsilon = 0.05$. Temperatures vary from 20°C to 26°C. Contrast derives from differences in irradiance in the image plane. Which provides greater contrast: the differences in emissivity or in temperature?

4.10 Thermal images of the human anatomy such as in mammography involve no risk to the patient because the patient is not exposed to radiation. For a 5° field of view infrared lens, what is the irradiance in the image plane at the lens output in such applications? Assume the image plane is relatively close to the focal plane for simplicity and that human anatomy at 37°C fills up the field of view.

4.11 A planar slab of material is illuminated by 10 W. The material exhibits a reflectivity (power) of 0.1 and an absorption coefficient of 2 cm^{-1}. If 7 W of radiant power exits the slab of material from the back side, determine the width of the slab.

4.12 A graybody of aperture diameter d_0 is controlled by a radiometer that collects all radiant power emitted by the graybody all across the spectrum. If the graybody is at temperature T (K) and emits radiant power P, what is the variation of its reflectivity with temperature?

4.13 In Example 4.3, what is the maximum permissible value of Δf (bandwidth of radiation stimulation emission) so that the laser is single mode? What would be the coherence length in this case?

4.14 What is the spectrum of an argon (gas discharge) laser at 0.488-μm wavelength if there are 2.9% losses per transit within a 0.7-m-long resonator? (Atomic weight of argon is 40). Assume 300 K temperature. What is overall laser linewidth? What is coherence length?

4.15 If a resonator does not satisfy $|b| \leq 1$ [Eq. (2.13.26)], what is the solution for r_s [Eqs. (2.13.21) and (2.13.23)] and what is the implication?

4.16 Determine whether the following resonators are stable:
(a) $L = 2m$, $R_1 = (4/3)m$, $R_2 = 1m$
(b) $L = R_2/2$, $R_1 = \infty$ (half-confocal)
(c) $L = R_2$, $R_1 = \infty$ (half-concentric)

4.17 For what resonator lengths is the resonator shown below stable? Assume unity refractive index.

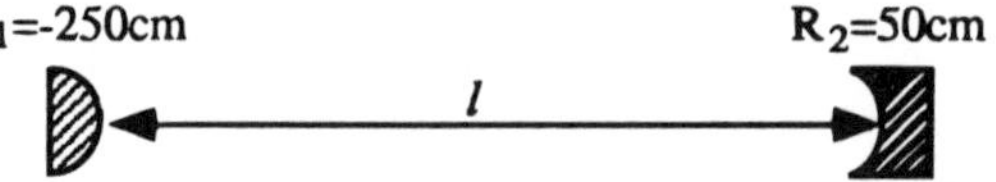

4.18 If mirror radii of curvature are identical, what is the requirement for resonator stability?

4.19 To achieve output HeNe laser beamwidths of 1 mrad and 1 μrad, what should beam width be inside the laser? Assume no radiation is incident at the edges of the output aperture, so that diffraction is neglected.

4.20 A HeNe laser exhibits an incoming beam waist $w_{01} = 0.436$ mm to a 26-cm focal length lens situated 70 cm away from the location of w_{01}. What is the narrowest laser beamwidth at the lens output and where is it located?

4.21 What is the error regarding w_{02} if a mistake is made and it is assumed $z_1 = f$?

4.22 We want to image an object "designated" by a Nd-YAG laser ($\lambda = 1.06$ μm) transmitted a distance of 4 km. Object dimensions are on the order of 2 m. Therefore, it is important that the radius of the beam incident on the object be less than 1 m, so as to avoid illuminating additional objects. If beam waist inside the laser resonator is 0.2 mm, what can be done to meet the task requirements? Assume laser beam is not incident on laser aperture edges or any lens edges, so that diffraction can be neglected.

4.23 If the designated object in Exercise 4.22 is viewed through an 89-cm aperture with a 1400-mm focal length, how does the image of the laser irradiance incident on the object vary spatially? Assume the imager is located a distance of 2 km in a direction of reflectivity θ from the object, which at optical wavelengths is assumed to be of uniformly diffuse surface. Unlike the previous question, consider what happens if the atmosphere exhibits an average extinction coefficient α.

4.24 If at $t = 0$, a signal of CFA $f(t) = \exp(j2\pi f_0 t - t/\tau)$ is emitted, what is the coherence length of the source?

CHAPTER

5

Noise

Image quality is generally either photon limited or noise limited. Both are concerned with fluctuations in image irradiance, which may derive from photon statistics and/or random motion of charge carriers. The basic concept of noise is considered here. Extension to actual imaging systems is delayed to Chapter 11 where spatial frequency effects are considered too.

Electronic noise in photoelectronic imaging systems derives primarily from two sources: quantum noise and Johnson noise. The former results from fluctuations in the arrival rates of photons into the imaging system, which gives rise to fluctuations in the generation rates of optically generated current carriers. Fluctuations in generation rate of current carriers exist also with regard to bias current or dark current and are known as *shot noise*. All such fluctuations exhibit similar statistics and are lumped together here under the general heading of *quantum noise*. In addition, there is a noise current component resulting from random motion of current carriers based on their thermal energy. This noise current is known as Johnson noise. Note that both cases are concerned with variations in current, the first deriving from fluctuations in generation rate and the second from random motion. Generally, nonsignal direct currents themselves are not considered "noise" because they can be removed via a capacitor. However, variations in direct currents are time varying and therefore not removable with a capacitor. For this reason, *noise current is defined as standard deviation of current from its average value*. Both types of noise are considered in turn here for a single detector. This is extended in Chapter 11 to imaging systems.

The material in this chapter and the next is basic to optical communication systems as well as both passive and active imaging.

5.1 QUANTUM NOISE

When discrete particles arrive at random times there are fluctuations in rate of arrival. Such processes are characterized by a Poisson distribution [5.1]. For photons of energy $h\nu$ and received radiation power P, the average rate of arrival is

$$\bar{r} = P/h\nu \tag{5.1.1}$$

where ν is electromagnetic wave frequency.

Since photons are themselves generated randomly as a result of deexcitation or recombination processes at their source (both natural and artificial as in Chapter 4), there are fluctuations in the rate at which they arrive at the receiver. At relatively low frequencies such as microwave or radio waves, $\bar{r}$ is invariably so large that it is difficult to sense the discreteness of the radiation. Therefore, there essentially are little or no recognizable fluctuations in arrival rate since the wave appears to be continuous. However, at optical frequencies, the arrival rate is much less because the energy of each photon is so much greater than at radio or microwave frequencies. At low radiant powers in particular, the discreteness or photon nature of the radiant power is much more apparent. For example, for 100-pW radiant power at 0.5 μm wavelength

$$\bar{r} = \frac{P\lambda}{hc} = \frac{10^{-10}(5 \times 10^{-7})}{6.6 \times 10^{-34}(3 \times 10^{8})} = 2.5 \times 10^{8}\ \text{s}^{-1}$$

Consider time intervals $\Delta t << (\bar{r})^{-1}$. On time scales such as nanoseconds here the discreteness of photon arrival rate and fluctuations in it would be very evident. Assuming statistical stationarity, the probability of receiving a photon in the time Δt is $\bar{r}\Delta t$. The probability of not receiving a photon is $1 - \bar{r}\Delta t$. The probability of no arrivals during a time $t > \Delta t$ is (see Exercise 5.1)

$$P_0(t) = \lim_{\Delta t \to 0} (1 - \bar{r}\Delta t)^{t/\Delta t} = e^{-\bar{r}t} \tag{5.1.2}$$

where $t/\Delta t$ is the number of independent intervals over which no arrival must occur. The probability of exactly one arrival during the period t is the probability of one photon arrival in the interval from time τ to time $\tau + d\tau$ and none before or after. This is summed over all τ where τ varies over the interval t:

$$\begin{aligned} P_1(t) &= \int_0^t p_0(\tau)(\bar{r}\, d\tau)P_0(t - \tau) \\ &= \int_0^t e^{-\bar{r}\tau}(\bar{r}\, d\tau)e^{-\bar{r}(t-\tau)} \\ &= \bar{r}\, e^{-\bar{r}t} \int_0^t d\tau = \bar{r}t\, e^{-\bar{r}t} \end{aligned} \tag{5.1.3}$$

The probability of two photon arrivals is the probability of receiving one up to time τ, one during τ and $\tau + d\tau$, and none thereafter. This too is summed over all τ with τ varying over the interval t so that

$$P_2(t) = \int_0^t P_1(\tau)(\bar{r}\, d\tau)P_0(t - \tau) = \frac{(\bar{r}t)^2}{2} e^{-\bar{r}t} \tag{5.1.4}$$

Continuing in such a fashion, the probability of exactly n photons arriving in the time t is

$$P_n(t) = \frac{(\bar{r}t)^n}{n!} e^{-\bar{r}t} \tag{5.1.5}$$

as expected for a Poisson distribution. A Poisson distribution is characterized by a variance equal to its expectation:

$$\overline{(\delta n)^2} = \bar{n} \tag{5.1.6}$$

where $\delta n = n - \bar{n}$. When $n >> 1$, the Poisson distribution approaches a Gaussian distribution according to the law of large numbers.

These statistics carry over too to the rate of current carrier generation in the detector arising from the photon absorption. If there are m current carriers generated during the time T, where T is the effective duration of detector impulse response, then the resulting instantaneous current is

$$i = qm/T \tag{5.1.7}$$

where q is electron charge. Average current is

$$I = q\bar{m}/T \tag{5.1.8}$$

The variance in current is, using this last result,

$$\overline{(\delta i)^2} = \frac{q^2}{T^2}\overline{(\delta m)^2} = \frac{q^2}{T^2}\bar{m} = \frac{q}{T}I \tag{5.1.9}$$

This constitutes noise current power.

A detector is a device that converts radiant power at its input to current at its output. Various types of detectors are considered in Chapter 6. If $F(f)$ is the temporal frequency response of the detector, where f is a baseband electronic (rather than optical wave) frequency of output current, then noise bandwidth is defined

$$B_n = \frac{\int_0^\infty |F(f)|^2 \, df}{|F(0)|^2} \tag{5.1.10}$$

where it is assumed that detector response is maximum at dc or zero frequency. Equation (5.1.10) tells us that if the detector is modeled as having an ideal electronic frequency response in the form of a rectangle of amplitude equal to dc response out to frequency B_n, then the area of this rectangle is equal to $\int_0^\infty |F(f)|^2 \, df$. This is illustrated in Fig. 5.1. Now, effective time duration of impulse response is inversely proportional to bandwidth and, because of negative frequencies, is defined such that

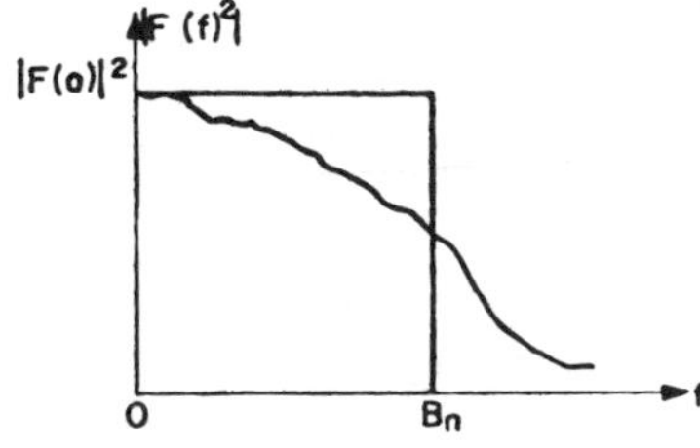

Fig. 5.1 Definition of noise bandwidth. Area of rectangle is $B_n|F(0)|^2 = \int_0^\infty |F(f)|^2 \, df$.

$$\frac{1}{T} = \frac{\int_{-\infty}^{\infty} |F(f)|^2 \, df}{|F(0)|^2} = 2B_n \tag{5.1.11}$$

Fast detector response implies large bandwidth.

By substituting this result into Eq. (5.1.9),

$$\overline{(\delta i)^2} = 2qIB_n \tag{5.1.12}$$

Because of the Poisson properties, noise current power or current variance is seen to be directly proportional to average current.

The number of current carriers actually generated by the incident radiant power is not equal to the number of photons incident on the detector. Some photons are reflected by the detector and so are not absorbed. Others may be absorbed but may fail to produce current carriers for reasons to be discussed in Chapter 6. Quantum efficiency describes the probability of a current carrier being generated by an incident photon and, for n incident photons giving rise to m current carriers, is equal to

$$\eta = m/n \tag{5.1.13}$$

Consequently, Eq. (5.1.7) becomes

$$i = \eta qn/T \tag{5.1.14}$$

and, in view too of Eq. (5.1.1), Eq. (5.1.8) becomes

$$I = \eta q \frac{\bar{n}}{T} = \eta q\bar{r} = \frac{\eta qP}{h\nu} \tag{5.1.15}$$

The variance of the detected current described in Eq. (5.1.12), which is quantum noise current power, becomes

$$i_{nq}^2 = \overline{(\delta i)^2} = 2qB \frac{\eta q}{h\nu} P \tag{5.1.16}$$

where the subscript n for noise bandwidth has been dropped. Both average current and variance of it are proportional to incident radiant power P. For the former this can be written as

$$I = [\eta q/(h\nu)]P = \Re P \tag{5.1.17}$$

where $\Re$ is detector responsivity. This parameter describes how large a current response is generated by the detector per unit incident radiant power. Some detectors, such as photomultipliers and avalanche photodiodes, are characterized by internal gain so that each current carrier generated by incident light gives rise to G current carriers at detector output. In general

$$I = \frac{\eta q}{h\nu} GP \tag{5.1.18}$$

$$i_{nq}^2 = 2qB \frac{\eta q}{h\nu} G^2 P \tag{5.1.19}$$

where G is unity if internal current carrier multiplication is unity.

Incident radiant power can include not only signal power P_s but also background radiant power P_b such as skylight, thermal radiation from the atmosphere or receiver itself, and so on. Therefore,

$$P = P_s + P_b \tag{5.1.20}$$

Quantum noise resulting from incident radiant power is

$$\overline{i_{nq}^2} = 2qB \frac{\eta q}{h\nu} G^2(P_s + P_b) \tag{5.1.21}$$

If dark current shot noise is included

$$\overline{i_{nq}^2} = 2qBG^2(I_s + I_b + i_d) = 2qBG^2\left[\frac{\eta q}{h\nu}(P_s + P_b) + i_d\right] \tag{5.1.22}$$

where i_d is dark current. This current exists even with no radiation incident on the detector. An example of dark current is bias current. However, even with no bias a dark current exists, however small it may be, as a direct current associated with thermal energy of charge carriers.

Equation (5.1.22) is a general equation for quantum or shot noise power. Note that since ν is input optical wave frequency and not output electronic frequency, quantum noise is considered "white"; that is, it exhibits no electronic frequency dependence. The reason is that optical wave frequencies are too large for electrical circuits to pass currents that vary fast enough so as to follow the input carrier wave frequency. Optical wave frequencies are on the order of 30 THz (10-μm wavelength) to 1000 THz (0.3-μm wavelength) while commercially available detectors exhibit picosecond rise time at best. (Although femtosecond (10^{-15} s) detectors are under development, they are not commercially available as yet.) Consequently, the white noise nature of quantum noise may be more easily apparent if Eq. (5.1.22) is rewritten as

$$\overline{I_{nq}^2} = 2qBG^2[\Re(P_s + P_b) + i_d] \tag{5.1.23}$$

Typical photodetector circuit and equivalent are shown in Fig. 5.2.

5.1.1 Generation-Recombination Noise

In photoconductor detectors, bias current consists of both electrons and holes. Recombination causes random fluctuations in total current. Such fluctuations are in addition to those arising from random generation of current carriers as expressed in Eq. (5.1.23).

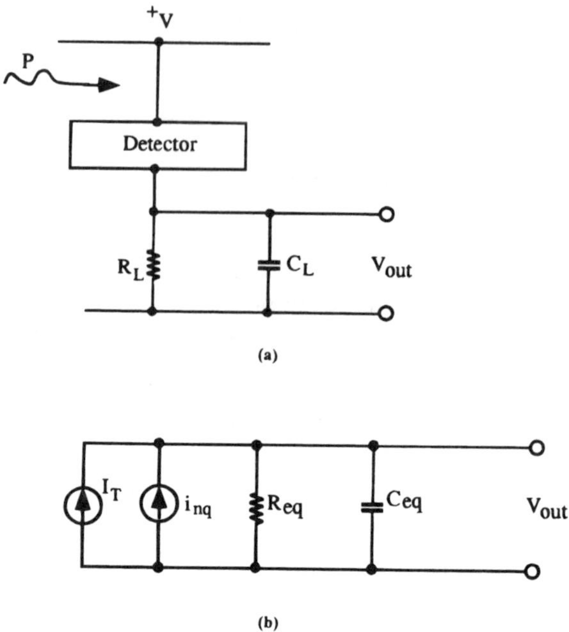

Fig. 5.2 Photodetector circuit including quantum noise source: (a) schematic circuit design; (b) equivalent circuit showing total average current I_T and quantum noise. R_{eq} includes R_L and equivalent resistance of detector in parallel. C_{eq} includes C_L, capacitance of detector, and other parasitic capacitances.

The total generation-recombination variance of such current from its average value under time spans of interest is then approximately twice the value of Eq. (5.1.23). A more exact expression for generation-recombination noise power is [5.2]

$$I^2_{ng\text{-}r} = \frac{4qI_T(\tau_0/\tau_d)B}{1 + 4\pi^2 f^2 \tau_0^2} \tag{5.1.24}$$

where τ_0 is average current carrier lifetime before recombination takes place and τ_d is drift time for current carriers over the distance between electrodes of the semiconductor crystal; I_T is total average current. In such devices internal gain is

$$G = \tau_0/\tau_d \tag{5.1.25}$$

At low electronic frequency ($f << \tau_0^{-1}$) generation-recombination (g-r) noise is simply twice the value expressed in Eqs. (5.1.22) and (5.1.23) and is "white." As baseband frequencies increase, g-r noise is no longer "white" but decreases with increasing frequency according to Eq. (5.1.24).

5.1.2 Photon-Limited Detection

Photon-limited detection refers to the ideal situation where the dominant noise is fluctuations in the signal itself. All other noise sources—background quantum noise, dark current shot noise, Johnson noise, etc.—are negligible. The signal may be constant or time varying; if it is constant then it should, of course, not be put through a capacitor. Signal current is, from Eqs. (5.1.14) through (5.1.18),

$$i_s = \frac{\eta q}{h\nu} G P_s \tag{5.1.26}$$

If the detector is a photoconductor, signal fluctuations are manifested as generation-recombination noise. Then for photon-limited detection

$$\overline{i_{ng\text{-}r}^2} = 4 q i_s B G^2 = 4q\left(\frac{\eta q}{h\nu}\right) P_s B G^2 \tag{5.1.27}$$

For other detectors, the noise current power for photon-limited detection is

$$i_{nq}^2 = 2q\left(\frac{\eta q}{h\nu}\right) P_s B G^2 \tag{5.1.28}$$

At the detector *output* the signal-to-noise ratio (SNR) is

$$\text{SNR} = \frac{i_s^2}{\overline{i_{ng\text{-}r}^2}} = \frac{\left(\frac{\eta q}{h\nu}\right)^2 G^2 P_s^2}{4q\left(\frac{\eta q}{h\nu}\right) P_s B G^2} = \frac{\eta P_s}{4 h\nu B} \tag{5.1.29}$$

for the first case and

$$\text{SNR} = \frac{i_s^2}{\overline{i_{nq}^2}} = \frac{\eta P_s}{2 h\nu B} \tag{5.1.30}$$

for the second case. These results represent best possible SNRs using simple direct detection and refer to SNR at detector *output*. However, they can also be interpreted as indicating an equivalent noise power

$$P_{nq} = 2h\nu B/\eta \tag{5.1.31a}$$

or

$$P_{ng\text{-}r} = 4h\nu B/\eta \tag{5.1.31b}$$

at the detector input where P_s, P_{nq}, and $P_{ng\text{-}r}$ represent equivalent radiant electromagnetic wave powers rather than current power. Eqs. (5.1.31) indicate that for constant quantum efficiency, effects of quantum noise (or g-r noise) increase with electromagnetic frequency and thus are not of significance at radio wave, microwave, or millimeter wave frequencies.

In photon-limited detection, the SNR is limited by the discrete nature of the photons themselves, which give rise to the signal fluctuations. Note that in Eqs. (5.1.29) and (5.1.30) the ratio $P_s/h\nu$ is the average photon arrival rate. The greater the arrival rate, the less noticeable the discrete nature of the photons, and the better the resulting SNRs.

Example 5.1

Compare the magnitude of background radiation current fluctuations to background radiation current itself. Assume unity internal gain.

Solution

$$\overline{(\delta i_b)^2} = 2qI_bB \tag{5.1.32}$$

where I_b is average background current. Therefore,

$$\frac{\overline{(\delta i_b^2)}^{1/2}}{I_b} = (2qB/I_b)^{1/2} \tag{5.1.33}$$

Since $qB = 1.6 \times 10^{-19}B << 1$, the fluctuations in background radiation current are orders of magnitude less than the background radiation current itself.

Example 5.2

Consider detection of an optical signal in the presence of strong natural background radiation $P_b >> P_s$. Can a signal still be detected? Consider both constant and time-varying examples of P_s.

Solution

Natural background radiation such as thermal emission, daylight, moonlight, etc., are essentially non–time varying. (Artificial sources such as incandescent lamps, fluorescent lamps, etc., usually exhibit a small degree of amplitude modulation at the frequency of the power system: 60 Hz in North America, 50 Hz in other continents.) If $P_b >> P_s$, then the output SNR is

$$\mathrm{SNR} = \frac{i_s^2}{I_b^2 + \overline{i_n^2}} = \frac{[(\eta q/h\nu)GP_s]^2}{\left(\dfrac{\eta q}{h\nu}GP_b\right)^2 + 2qBG^2\left(\dfrac{\eta q}{h\nu}P_b\right)} \tag{5.1.34}$$

where the total noise from background radiation includes both the time-varying $(\overline{i_n^2})$ and the non–time-varying (I_b^2) components. Now, if P_s is constant, detector output currents cannot be put through a capacitor because the signal current, which is dc, would be wiped out. In that case the SNR is as described in Eq. (5.1.34) because the constant background radiation current is noise relative to the constant signal current. On the basis of the previous example, background radiation current fluctuations are much less than the background radiation current itself. In this case, both signal and noise currents are dc and

$$\text{SNR} \cong \frac{i_s^2}{I_b^2} = \frac{(\eta q/h\nu)^2 G P_s^2}{(\eta q/h\nu)^2 G^2 P_b^2} = \left(\frac{P_s}{P_b}\right)^2 \tag{5.1.35}$$

However, if the signal is modulated, then detector current can be passed through a capacitor, thus wiping out I_b only. The SNR now is the ratio of time-varying signal current to background radiation current fluctuations only, that is,

$$\text{SNR} = \frac{i_s^2}{\overline{i_b^2}} = \frac{(\eta q/h\nu)^2 G^2 \tilde{P}_s^2}{2qBG^2(\eta q/h\nu)P_b} = \frac{\eta \tilde{P}_s^2}{2h\nu BP_b} \tag{5.1.36}$$

where $\tilde{P}_s$ is the modulation signal power only, which consists of sideband signal power and not carrier frequency signal power. Minimum detectable signal power, assuming unity SNR, is

$$\tilde{P}_{s\,\min} = (2h\nu BP_b/\eta)^{1/2} \tag{5.1.37}$$

Since at optical wavelengths $h\nu$ is on the order of 10^{-19} J, then P_b in Eq. (5.1.37) is multiplied by a number very much less than P_b itself. In this case it is possible to detect modulated signals of radiant powers $\tilde{P}_s$ that are much less than the constant power P_b (see Exercise 5.5). Detection of weak optical signals in the presence of strong background radiation is called BLIP (background-limited photon) detection and is possible because the optical signal is time varying while the background is constant. The noise of interest, therefore, is not the constant current I_b but rather its fluctuations because these are not removed by a capacitor.

5.2 JOHNSON NOISE

Johnson, or Nyquist, noise describes fluctuations in voltage or current across a dissipative circuit element caused by thermal motion of charge carriers. Such dissipative elements include typical circuit elements such as resistors, as well as detectors. Consider a cube of dimensions L whose walls are at temperature T. Electromagnetic fields inside this enclosure are periodic in L. A typical field satisfies Helmholtz's equation:

$$\nabla^2 E_x(x,y,z,t) - \frac{1}{v_p^2}\frac{\partial^2 E_x(x,y,z,t)}{\partial t^2} = 0 \tag{5.2.1}$$

The periodicity in L is exemplified in a Fourier series

$$E(x,y,z,t) = \sum_{l,m,n=-\infty}^{\infty} E_{l,m,n}(t)\exp j[k_x x + k_y y + k_z z] \tag{5.2.2}$$

where

$$k_x = 2\pi l/L$$
$$k_y = 2\pi m/L \qquad (5.2.3)$$
$$k_z = 2\pi n/L$$

and the wave propagation constant β in Eq. (1.3.16) is designated as k where

$$k^2 = \frac{\omega^2}{v_p^2} = k_x^2 + k_y^2 + k_z^2 \qquad (5.2.4)$$

and a time dependence $\exp(j\omega t)$ is assumed for each field component. In Eq. (5.2.4), k can represent radius of a sphere in k space, as shown in Fig. 5.3. According to Eq. (5.2.3) two adjacent modes differ in their value of k by

$$\Delta k = 2\pi/L \qquad (5.2.5)$$

Modes describe the spatial variation of an electric field. At a given frequency, ω, there are many possible spatial variations or modes of the electromagnetic (EM) fields. This is true of *waves* as in metallic or dielectric waveguides. The cube of this example, as well as other blackbodies and graybodies, exhibits such wave propagation characteristics since energy is reflected back and forth internally between dielectric interfaces or electrodes. However, we must distinguish between waves on the one hand and electrical current or voltage on the other hand. The electrical current or voltage resulting from such waves are all of one mode only. Therefore, to determine average noise power or variance deriving from thermal energy of charge carriers in an electronic circuit, it is necessary to determine the average thermal energy per mode because current in a wire or voltage exhibits one mode only.

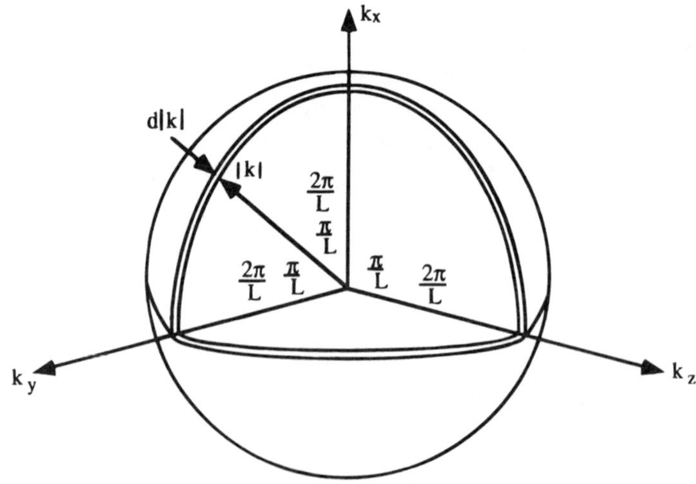

Fig. 5.3 Wave vector radius in k space according to Eq. (5.2.4).

From Eq. (5.2.5) it follows that a volume $dk_x dk_y dk_z = (2\pi/L)^3$ is associated with each mode in k space. For each wave vector k there are two possible directions of polarization. Therefore, as shown in Fig. 5.3, the number of modes whose wave vectors have magnitudes between zero and k is obtained by dividing the total volume in k space $[(4/3)\pi k^3]$ by the volume per mode $[(2\pi/L)^3]$ and doubling the result to account for polarization. Accordingly

$$N_k = \frac{2\left(\frac{4}{3}\pi k^3\right)}{(2\pi/L)^3} = \frac{k^3L^3}{3\pi^2} \tag{5.2.6}$$

Since $k = 2\pi\nu/c$ and L^3 is cube volume,

$$\frac{N_\nu}{V} = \frac{8\pi\nu^3}{3c^2} \qquad (\mathrm{m}^{-3}) \tag{5.2.7}$$

Result (5.2.7) is then independent of circuit element geometry and describes the number of modes per unit volume between the frequencies $\nu = 0$ and $\nu = \nu$. The mode density per unit bandwidth is

$$m(\nu) = \frac{1}{V}\frac{dN_\nu}{d\nu} = \frac{8\pi\nu^2}{c^3} \qquad (\mathrm{m}^{-3}\cdot\mathrm{Hz}^{-1}) \tag{5.2.8}$$

Now, Planck's law [Eq. (4.2.1)] describes for good absorbers ($\epsilon = \alpha \approx 1$), such as detectors or resistors, average thermal energy per unit volume per unit bandwidth. If Planck's law is divided by Eq. (5.2.8), one obtains the average thermal energy per mode. This is

$$\overline{E} = \frac{8\pi h\nu^3}{c^3}\frac{1}{\exp(h\nu/kT)-1} \div \frac{8\pi\nu^2}{c^3} \tag{5.2.9a}$$

$$= \frac{h\nu}{\exp(h\nu/kT) - 1} \qquad \text{J per mode}$$

Now, 1 J is equal to 1 W·s or $\mathrm{W}\cdot\mathrm{Hz}^{-1}$. Therefore,

$$\overline{E} = \frac{h\nu}{\exp(h\nu/kT) - 1} \qquad \mathrm{W}\cdot\mathrm{Hz}^{-1} \text{ per mode} \tag{5.2.9b}$$

is the average thermal power spectral density per mode as derived from Planck's law. This is the thermal power spectral density that can be associated with average Johnson noise current deriving from thermal energy.

Electronic frequencies that are compatible with detector rise times are much smaller than those at which thermal radiation peaks in accordance with Wien's law. Therefore, Johnson noise occurs at frequencies $f << kT/h$, where electronic or baseband frequency f has been substituted for EM wave frequency ν. By applying a Taylor series expansion for the exponent in Eqs. (5.2.9) one obtains

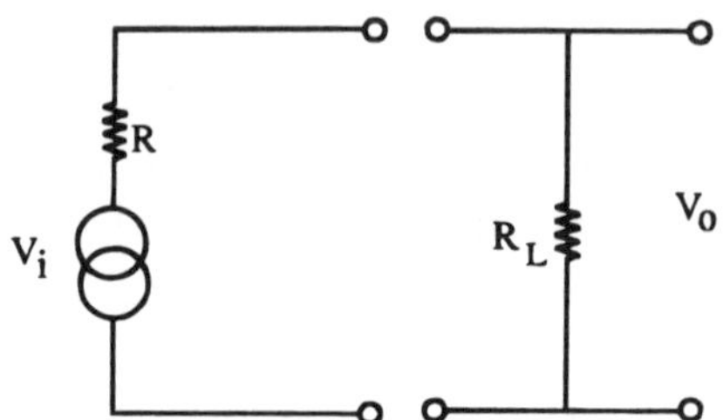

Fig. 5.4 Johnson noise representation.

$$\overline{E} \cong \frac{hf}{1 + \dfrac{hf}{kT} - 1} = \frac{hf}{hf/kT} = kT \qquad \mathrm{W \cdot Hz^{-1}} \tag{5.2.10}$$

since $hf/kT << 1$ at electronic frequencies. This average energy gives rise to time-varying Johnson noise current, which is "white" as long as $hf << kT$. To describe the Johnson noise voltage, let the detector be represented as a noise source with internal resistance R, as in Fig. 5.4. Output voltage is

$$V_0 = \frac{R_L}{R + R_L} V_i \tag{5.2.11}$$

and over a bandwidth B power consumption of the load resistor R_L is

$$P_0 = kTB = \frac{V_0^2}{R_L} = \left(\frac{R_L}{R + R_L}\right)^2 \frac{V_i^2}{R_L} = \frac{R_L}{(R + R_L)^2} V_i^2 \tag{5.2.12}$$

Consequently, if detector Johnson noise power is delivered to a matched load,

$$P_0 = kTB = V_i^2/4R_L$$

Detector Johnson noise voltage is

$$V_i^2 = \overline{V}_{nj}^2 = 4R_L kTB \tag{5.2.13}$$

and is maximum noise power available from the detector for a matched load. Thevenin and Norton equivalents are shown in Fig. 5.5.

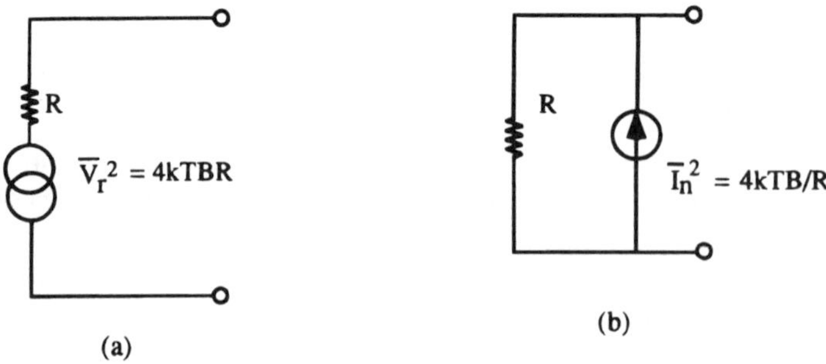

Fig. 5.5 (a) Thevenin and (b) Norton equivalents for Johnson noise.

Although Johnson noise is white for $f << h/kT$, Eq. (5.2.9a) indicates that as frequency increases, Johnson noise power spectral density approaches zero according to L'Hopital's rule. The spectral dependence of Johnson noise is therefore as shown in Fig. 5.6, where photon-limited quantum and g-r noise [Eqs. (5.1.31)] are shown as well.

When considering a receiver, bandwidth may well be limited to values much smaller than that for the detector by itself, particularly if electronic filtering is used so as to decrease noise at electronic frequencies where there is no signal. Johnson noise is not limited to the detector alone, but applies to amplifiers and any other circuit elements. For example, if an amplifier exhibits effective input noise temperature T_A, and the detector noise temperature is T_d, then the combination yields a Johnson noise current power

$$\overline{i_{nj}^2} = 4k(T_d + T_A)B/R_L \tag{5.2.14}$$

where T_A is related to noise figure F and noise temperature 290°K by

$$T_A = 290(F - 1) \tag{5.2.15}$$

Johnson noise is governed by Gaussian statistics, whereas quantum noise is governed by Poisson statistics. The latter approaches Gaussian statistics at larger noise current levels, according to the law of large numbers.

5.3 1/*f* NOISE

In addition to Johnson noise, there is an electronically generated noise found in essentially all electronic devices. It is known as flicker noise, contact noise, current noise, and $1/f$ noise. At very low baseband frequencies, it is dominant. Although many theories have been presented, it has not been completely clarified. It is characterized by a power spectrum which decreases with decreasing electronic frequency, that is,

$$\langle \overline{i_{nf}^2} \rangle \propto (I_T^\alpha BG/f^\beta)^2 \tag{5.3.1}$$

where I_T is total dc current, $\alpha \cong 2$, and $\beta \cong 1$ in electron tube detectors and

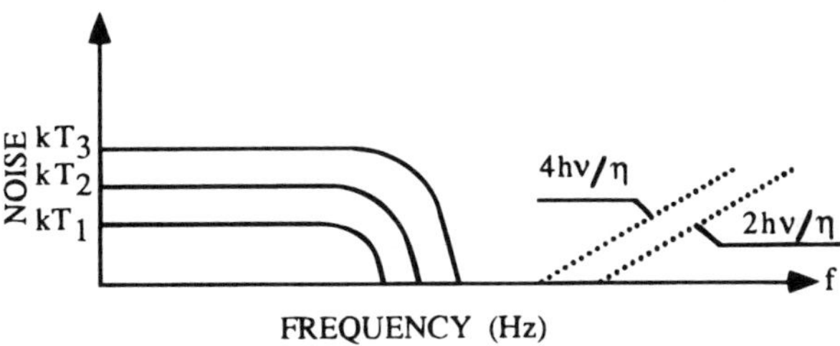

Fig. 5.6 Johnson noise spectral density dependence with temperature as a parameter. Here, $T_3 > T_2 > T_1$. Dotted lines represent quantum and generation-recombination noise equivalent power spectral densities in photon-limited detection.

$1.25 \leq \alpha \leq 4$ and $1 \leq \beta \leq 3$ for semiconductor devices. At baseband frequencies on the order of several kilohertz, $1/f$ noise is usually already negligible.

A typical baseband noise spectrum is shown in Fig. 5.7.

5.4 ELECTRONIC SIGNAL-TO-NOISE RATIO: DETECTION VERSUS IMAGING

5.4.1 Detection

For large baseband bandwidth total electronic noise power is essentially the sum of the quantum and Johnson noise, and constant currents other than the signal:

$$\overline{i_n^2} = \overline{i_{nq}^2} + \overline{i_{nj}^2} + I_b^2 + i_d^2 = \{2qG^2[\Re(P_s + P_b) + i_d]$$
$$+ 4kT/R_L\}B + (\Re P_b)^2 + i_d^2 \quad (5.4.1)$$

For photoconductive detectors, generation-recombination noise is used instead of quantum noise, that is,

$$\overline{i_n^2} = \overline{i_{ng\text{-}r}^2} + \overline{i_{nj}^2} + I_b^2 + i_d^2 \quad (5.4.2)$$

where $i_{ng\text{-}r}$ is defined in Eq. (5.1.24) and is about twice the quantum noise. If the signal is time varying then I_b^2 and i_d^2 can be removed with a capacitor.

Detector internal gain on the order of a hundred or more generally tends to render Johnson noise insignificant when compared to quantum noise or the signal itself. Signal current power is

$$i_s^2 = \left(\frac{\eta q}{h\nu}\right) P_s^2 \quad (5.4.3)$$

Output SNR for large-bandwidth detection is

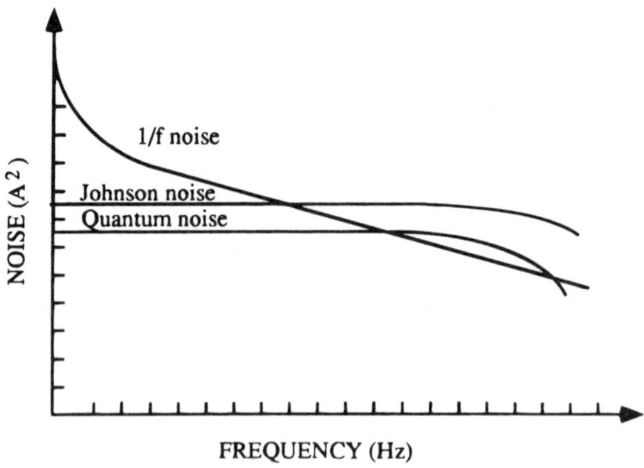

Fig. 5.7 Typical baseband detector output noise spectrum for unity internal gain.

$$\text{SNR} = \frac{i_s^2}{i_n^2} \qquad (5.1.4)$$

The above refers to radiation *detection* and to imaging involving image scanning with one or a small number of detectors.

5.4.2 Imaging

For imaging, as opposed to detection, various factors must be considered. In detection, radiation from the entire field of view is focused onto a detector which generates currents proportional to the total received signal and background radiation powers. In imaging, however, irradiance distinctions are made for various parts of the scene. The information, in fact, is the spatial variation of irradiance over the scene. Such distinctions are nonexistent in simple detection because the irradiances from the entire scene are added to form one output current. Consequently, in imaging there is no distinction between signal and background radiations and currents because they all are part of the scene. Background radiation degrades the contrast in the scene, as discussed in Chapter 10, but not the video or electronic SNR of the scene itself.

In addition, because in electro-optical imaging the scene is normally scanned (either electronically, mechanically, or optically), the signal current (including background) is time varying as long as the scene irradiance is spatially varying. Thus, constant noise sources such as dark current itself are removed with a capacitor, and only time-varying noise sources need be considered. This does not apply to photographic systems because they are not electronic and there is no scanning.

Here, electronic noise ratios concerned with detectors and imaging systems have been considered. This is expanded in Chapter 11 to include the spatial frequency dependencies of the SNR. However, detection schemes must be considered first as in the next chapter.

REFERENCES

5.1. B. M. Oliver, "Thermal and quantum noise," *Proc. IEEE*, Vol. 53, May 1965, pp. 436–454.

5.2. A. Yariv, *Optical Electronics,* (3rd ed.), Holt, Rinehart and Winston, New York, 1985.

EXERCISES

5.1 Prove Eq. (5.1.2).

5.2 Six-microwatt radiant power at 1.3-μm wavelength incident on a detector gives rise to 5 μA. (a) What is the detector responsivity? (b) What is the quantum efficiency?

5.3 Assume responsivity and wavelength, as in Exercise 5.2. If $P_b = 1$ μW, find I_b and $(\delta \bar{I}_b^2)^{1/2}$ for a 1-MHz receiver bandwidth. What is the ratio between background radiation current and fluctuations of it?

5.4 A light pulse of energy E and wavelength λ is incident on an optical detector of quantum efficiency η at that wavelength. (a) What is the average number of current carriers generated? For a noiseless detector: (b) What is the probability of error (probability that during the pulse time no charge carrier is generated by the incident light)? (c) For a bit error probability of 10^{-9} or less, what is the required optical pulse energy E? (d) For an ideal detector, how many photons are required during the light pulse for an error probability of 10^{-9} or less?

5.5 Consider received laser pulses of 10-nW radiant power at 1-μm wavelength, returned from a target surface together with 1 μW of natural background at the same wavelength. Can the signal be detected, despite the large background? Neglect other noise sources compared to background radiation fluctuations current, and assume 10% quantum efficiency and 1-MHz electronic bandwidth. What is the SNR if the laser radiation is constant rather than pulsed?

5.6 Compare the magnitude of dark current to average time-varying fluctuations of it.

5.7 Typical photodiode capacitance is on the order of 10 pF. For a series resistance of 50 Ω, what is the detector's time constant? For a 1-μm wavelength, compare the time constant to the reciprocal of the optical frequency. What does the comparison imply concerning the capability of an electronic circuit including such a detector to react to the carrier frequency at this wavelength?

5.8 (a) Determine Johnson noise current for a receiver of noise temperature 300 K and dynamic resistance 1 MΩ. Assume 1-MHz bandwidth as in Exercise 5.3. Compare Johnson noise current with that of background radiation current fluctuations in Exercise 5.3. (b) Repeat part (a) for 1-kΩ dynamic resistance. (c) Compare the result of part (b) with $(\overline{\delta I_b^2})^{1/2}$ if internal gain = 100; how does internal gain affect the relationship between quantum and Johnson noise?

CHAPTER

6

Detector Concepts and Fundamentals

The previous chapter considered noise regarding both electro-optical radiation detection and imaging. In this chapter, detection concepts are extended to include comparisons of both direct and heterodyne detection schemes, both of which can be used in passive and active imaging systems. Criteria are presented with which to compare detector devices. This is followed by a brief comparison of detector types.

As in the previous chapter, this material is relevant to imaging as well as optical communication. Its relevance to imaging pertains particularly to video signal-to-noise ratios (SNRs) and noise-limited imaging in Chapter 11.

6.1 DETECTION SCHEMES

Radiation is usually detected with either a direct detection or heterodyne receiver.

6.1.1 Direct Detection

A schematic for a direct detection receiver is shown in Fig. 6.1. The purpose of the optical filter is to remove background radiation at wavelengths other than those of the desired signal. The lens or alternatively a mirror is used to collect radiation over a large area and focus it onto the detector. A typical photodetector circuit is shown in Fig. 5.2. SNRs developed in the previous chapter are relevant actually to direct detection receivers.

From the Poynting vector in Eq. (1.3.11b), the time-averaged radiant signal power incident on the detector is

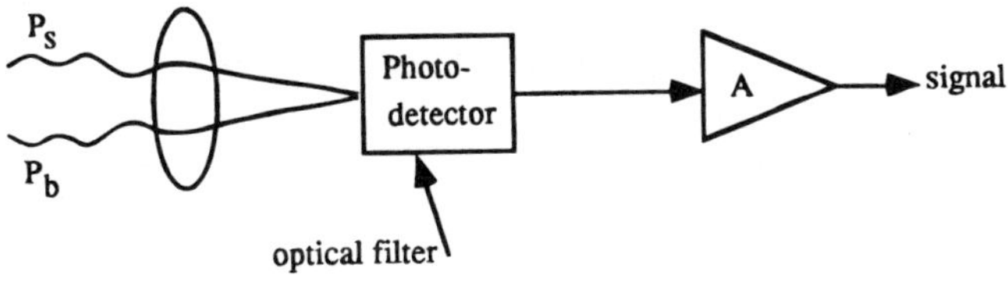

Fig. 6.1 Schematic of direct detection receiver, where A is amplifier gain. Other parameters are as defined in Chapter 5.

$$\overline{P}_s = \frac{1}{2}\,\mathrm{Re}(\overline{E}_s \times \overline{H}_s^*)A_r = \frac{|E_s|^2}{2\eta_0}\,A_r \tag{6.1.1}$$

where $\overline{E}_s$ and $\overline{H}_s$ are the electric and magnetic fields of the signal beam and A_r is the optics area over which radiation incident on the receiver is collected. Assume, for example, that radiation is propagating in the z direction and that polarization is such that

$$\overline{E}_s = a(t)\cos(\omega_s t + \phi_s - \beta_z z)\hat{a}_x \tag{6.1.2}$$

$$\overline{\mathrm{H}}_s = \frac{\mathrm{a}(t)}{\eta_0}\cos(\omega_s t + \phi_s - \beta_z z)\hat{a}_y \tag{6.1.3}$$

where ω_s is signal carrier frequency, ϕ_s is an initial phase term, and $a(t)$ represents electric field amplitude, which can be time varying. Instantaneous irradiance through the atmosphere or free space is

$$\overline{P}_{is} = \overline{E} \times \overline{H}^* = \frac{|\overline{E}|^2}{\eta_0} \tag{6.1.4}$$

and, therefore, instantaneous received signal radiant power is

$$\overline{P}_{is}(t) = [a(t)\cos(\omega_s t + \phi_s - \beta_z z)]^2\,\frac{A_r\hat{a}_z}{\eta_0} \tag{6.1.5}$$

Instantaneous signal current is

$$\begin{aligned} i_s(t) &= \Re G P_{is}(t) = \frac{\eta q}{\eta \nu}\,G a^2(t)\cos^2(\omega_s t + \phi_s - \beta_z z)\,\frac{A_r}{\eta_0} \\ &= \frac{\eta q}{\eta \nu}\,G\,\frac{A_r}{\eta_0}\,a^2(t)\left[\frac{1 + \cos 2(\omega_s t + \phi_s - \beta_z z)}{2}\right] \end{aligned} \tag{6.1.6}$$

where G is detector internal signal gain and $\Re$ is responsivity, as defined in Eq. (5.1.17). Since ω_s is in the optical range, being typically on the order of 10^{14} to 10^{15} rad·s^{-1}, it is too large to be followed by electrical signal currents and voltages at the detector output [6.1], which are limited typically by picosecond order rise times in commercial detectors. Although femtosecond (10^{-15} s) response detectors have been developed, practical application also requires femtosecond response time amplifiers and circuit elements, which are still far away. In view of these considerations, as exemplified in Exercise 5.7, the cosine term in Eq. (6.1.6) is neglected since it is at a frequency well beyond receiver electronic bandwidth.

Consequently, for all practical purposes

$$i_s(t) = \frac{\eta q}{2h\nu}\,\frac{GA_r}{\eta_0}\,a^2(t) = K\Re G a^2(t) \tag{6.1.7}$$

where K is a constant (equal to $A_r/2\eta_0$) and it is assumed that the electronic frequency spectrum of $a^2(t)$ is within the receiver electronic bandwidth. Since $a(t)$ modulates field amplitude it represents *amplitude modulation* (AM). If it were to modulate directly the radiant power, it would be called *intensity modulation*. The latter occurs when a mechanical chopper, for example, is placed at the receiver input. The radiant power is then incident on the detector when the rotating chopper blade is not blocking the detector, but is not incident when the blade is blocking the chopper, thus giving rise to intensity modulation.

An important implication of Eq. (6.1.7) is that since there is no term in $i_s(t)$ that includes the carrier frequency, direct detection receivers are capable of detecting neither *frequency modulation* (FM) nor *phase modulation* (PM), both of which would require the output current signal to show changes according to the argument of the cosine term. This is important in active imaging and in communication.

The SNR in direct detection, using Eqs. (6.1.7) and (5.4.1) for nonphotoconducting detectors, is

$$\mathrm{SNR} = \frac{\overline{i_s^2}}{\overline{i_n^2}} = \frac{[K\Re G\overline{a^2(t)}]^2}{\{2q^2[\Re(P_{is} + P_{ib})G^2 + i_d G^2] + 4kT/R_L\}B + \Re^2 P_b^2 G^2 + i_d^2} \tag{6.1.8}$$

where $1/f$ noise is neglected for large electronic bandwidths. If the signal is time varying it is useful to remove the last two noise terms, which are direct current powers, with a capacitor. For photoconductors, the generation-recombination (g-r) noise replaces the quantum noise. Note that if internal detector gain G is sufficiently high, Johnson noise can be neglected and SNR improved.

Assuming signal amplitude is time varying and therefore that the constant current terms i_b and i_d are removable with a capacitor, if the dominant noise stems from *fluctuations* in background current (rather than i_b itself) then, for an unbiased detector,

$$\mathrm{SNR} = \frac{\eta \tilde{P}_s^2}{2h\nu B P_b} \tag{5.1.36}$$

as shown in Chapter 5 for BLIP detection. The term $\tilde{P}_s$ represents the time-varying component of signal radiant power. For a signal-to-noise threshold of unity, the minimum detectable signal is

$$\tilde{P}_{s\,\min} = \left(2h\nu B \frac{P_b}{\eta}\right)^{1/2} \tag{5.1.37}$$

which may be orders of magnitude weaker than P_b itself, as shown in Exercise 5.3.

A convenient parameter with which to compare detectors is *noise equivalent power* (NEP). Equation (5.1.37) indicates that $\tilde{P}_s$ min varies according to $B^{1/2}$. Therefore, in comparing detectors it is useful to normalize according to unit bandwidth to the one half power. NEP is defined as the value of rms input radiant signal power required to produce unity SNR at detector output. It is the input *wave* signal power equivalent to noise *current* output power. To see this, simply reverse the term NEP to (input) *power equivalent* to *noise* (at output). For BLIP detection

$$\text{NEP} = (2h\nu P_b/\eta)^{1/2} \qquad \text{W}\cdot\text{Hz}^{-1/2} \tag{6.1.9}$$

on the basis of Eq. (5.1.37). The lower the NEP, the more sensitive the detector.

In general, NEP is a function of both noise and responsivity, that is,

$$\text{NEP} = \frac{i_n}{\Re B^{1/2}} \tag{6.1.10}$$

For visible and near-infrared wave detectors, manufacturers prefer to measure NEP in darkness so as to characterize the detector according to its own internal noise with $P_b = 0$. Such internal noise may derive from dark current shot noise or from Johnson noise (see Exercises 6.2 and 6.3).

Equation (6.1.9) indicates that for BLIP detection NEP varies as $P_b^{1/2}$. However, from radiometric considerations the value of P_b collected varies directly with the area of the receiver optics. Thus,

$$P_b \propto A_r \tag{6.1.11}$$

and from Eq. (6.1.9)

$$\text{NEP} \propto A_r^{1/2} \tag{6.1.12}$$

This result indicates that minimum detectable signal under background limited detection (BLIP) conditions varies according to $A_r^{1/2}$. The smaller the receiver optics, the weaker the background radiant powers that can be detected under such conditions (i.e., the more sensitive the detection).

This dependence on A_r is particularly relevant to thermal imaging in the medium and far infrared where large background radiation levels stemming from thermal radiation are the primary source of detector noise even in absolute darkness. Under such conditions it is useful to introduce a different parameter for detector sensitivity which normalizes according to the square root of detector area. This parameter is D^* and is discussed in Section 6.2. NEP is governed by Eq. (6.1.9).

It is important to note that in direct detection there is essentially no connection between optical wave bandwidth at receiver input and electronic bandwidth at the output. The former is limited by the spectral transmission of an optical filter or a window material enclosing the detector to protect it. The latter is limited by time response of receiver circuit elements, including the detector itself and output amplifier. The output amplifier amplifies all signal *and* noise currents and thus amplification is not included in SNR, although amplifier noise temperature should be included in Johnson noise (Eq. 5.2.14).

Example 6.1

Consider Examples 4.2 and 4.1 in Chapter 4 but with a spectral filter of linewidth $\Delta\lambda$ at the receiver input. At which wavelength should the filter be centered so as to maximize SNR? For simplicity, consider only wavelengths at which airplane and atmospheric backgrounds peak, as well as 3, 4, 8 and 12 μm, and neglect atmospheric transmission losses.

Solution

Thermal emission of the engine is essentially constant for constant temperature. Example 4.2 considers photon-limited or BLIP detection, but for a non-time-varying signal. Therefore, the signal cannot pass through a capacitor, and it is the background current itself and not fluctuations of it that limits detection. From Eq. (5.1.35), at the detector output

$$\mathrm{SNR} \cong \frac{i_s^2}{i_b^2} = \left(\frac{P_s}{P_b}\right)^2$$

From Example 4.2,

$$P_s = P_a = \tau' N_a A_a \frac{A_r}{z^2}$$

and

$$P_b = P_A = N_A \Omega_r A_r$$

Therefore,

$$\mathrm{SNR} = \left(\frac{\tau' N_a A_a}{N_A \Omega_r z^2}\right)^2 \tag{6.1.13}$$

Here it is assumed that $\tau' = 1$. Therefore,

$$\mathrm{SNR} = \left(\frac{N_a A_a}{N_A \Omega_r z^2}\right)^2 \tag{6.1.14}$$

and peaks at a wavelength where N_a/N_A is maximized. The same conclusion holds for imaging SNR in Example 4.1. Substituting,

$$\frac{N_a}{N_A} = \frac{\exp(hc/\lambda k T_A) - 1}{\exp(hc/\lambda k T_a) - 1} \tag{6.1.15}$$

Results are listed in Table 6.1. Plane and atmospheric background radiations peak at 3.62- and 9.66-μm wavelengths, respectively, as can be seen from Wien's law [Eq. (4.2.16)]. It is clear from the data of Table 6.1 that although signal power peaks at the 3.62-μm wavelength, SNR for both BLIP detection and imaging increases with decreasing wavelength. The reason is that although at wavelengths less than 3.62 μm the signal falls off for colder backgrounds, the background noise falls off even

TABLE 6.1 *Equation (6.1.15) as a Function of Wavelength*

λ(μm)	3	3.62	4	5	8	9.66	12
N_a/N_A	22081	4001	1828	4115	47.4	26.2	15.4

more. This can be seen conceptually in Fig. 6.2. The practical limit occurs at around the 3-μm wavelength, for at shorter wavelengths atmospheric background radiation actually increases as a result of scattered sunlight, moonlight, etc., as shown in Fig. 4.4.

This is also true with regard to target contrast with respect to atmospheric background in thermal imaging [6.2]. From Table 6.1 it should be clear that proper wavelength filtering can have a crucial effect on SNR. The latter conclusion does not change if wavelength dependence of atmospheric transmission is also included. However, the atmospheric transmission can alter significantly the wavelength dependence shown in Table 6.1.

Most imaging systems involve direct detection by the imager of the light received from the scene. Usually, such systems are passive. However, often the scene is illuminated, as in Exercises 3.7 and 3.8 for example. In such active or *laser radar* systems, heterodyne detection can also be used to increase sensitivity to low light levels [6.3] and to image through scattering media [6.4] such as biological tissue [6.5, 6.6] and the atmosphere.

For the sake of completeness, heterodyne detection is presented next. Although it is used for active imaging, it is used primarily in communication. Certain concepts such as effective receiver area are essential to active imaging when affected by distortions introduced by the atmosphere. Concepts of atmospheric coherence diameter are treated more fully in Part 5. As discussed in Section 17.8.1, heterodyne detection can be an effective means to reduce or eliminate blur resulting from received small-angle forward-scattered light caused by particulates in imaging through scattering channels such as the atmosphere or biological media.

6.1.2 Heterodyne Detection

Following the above introduction, a schematic for *heterodyne detection* is shown in Fig. 6.3. The collecting optics such as lenses or mirrors are omitted so as not to complicate the drawing. In addition to the signal and background radiant powers P_s and P_b collected after propagating through the atmosphere, radiant power P_o from a local oscillator such as a laser is added to a heterodyne receiver and incident on each detector. Figure 6.3 depicts two detectors, so as to maximize the received signal.

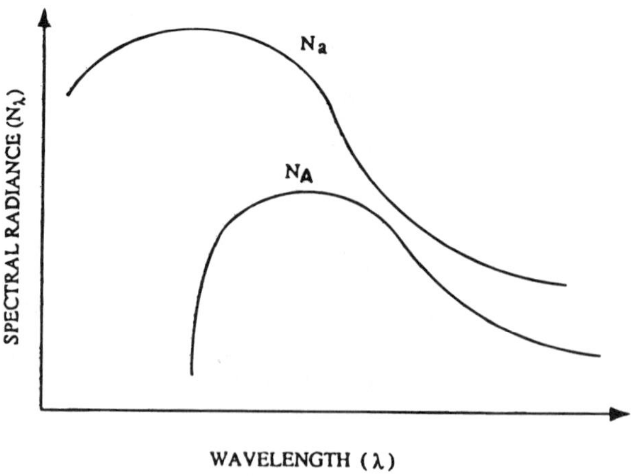

Fig. 6.2 Simplified comparison of airplane (800 K) and atmospheric background (300 K) wavelength spectra.

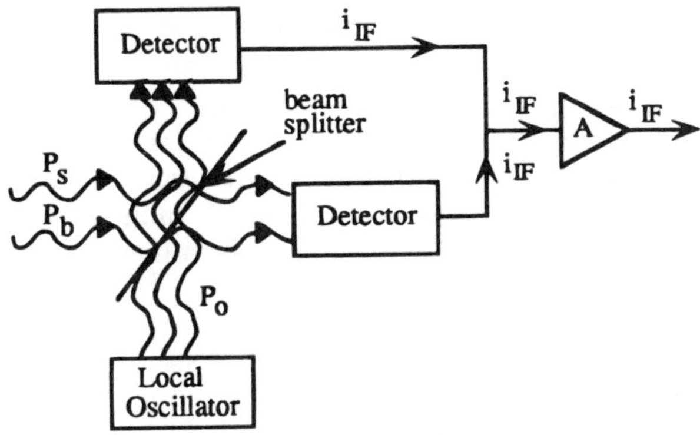

Fig. 6.3 Schematic of heterodyne receiver.

However, if one can afford a few decibel loss, one detector is often sufficient. A beamsplitter is needed so as to align P_s and P_0 so they can be coincident on the detectors. Outputs of both detectors are summed.

EM fields of the signal received from afar have already been described in Eqs. (6.1.2) and (6.1.3). Local oscillator electric field is

$$\overline{E}_0 = A_0 \cos(\omega_0 t + \phi_0 - \beta_z z)\hat{a}_x \tag{6.1.16}$$

where ω_0 and ϕ_0 are local oscillator EM frequency and initial phase, respectively, and the z direction is that of incidence on a given photodetector. The $z = 0$ plane is placed at the input to each detector. Most detectors are linearly responsive to input radiant power, with detector responsivity $\Re$ being the proportionality constant. Consequently, such devices are square-law detectors, because they respond to the square of the electric field of the light wave at the input as seen in Eq. (6.1.5). Therefore, the output current of a heterodyne receiver is

$$i = \Re G \frac{|E|^2}{\eta_0} Ar = \frac{\Re G A_r}{\eta_0} |\overline{E}_s + \overline{E}_0 + \overline{E}_b|^2 \tag{6.1.17a}$$

or

$$\begin{aligned}
i(t) &= \frac{\Re G A_r}{\eta_0} [a(t)\cos(\omega_s t + \phi_s) + A_0 \cos(\omega_0 t + \phi_0) \\
&\quad + \sum_b A_b \cos(\omega_b t + \phi_b)]^2 \\
&= \frac{\Re G A_r}{\eta_0} \left\{ \frac{a^2(t)}{2} (1 + \cos 2(\omega_s t + \phi_s)] + \frac{A_0^2}{2} [1 + \cos 2(\omega_0 t + \phi_0)] \right. \\
&\quad + \left(\sum_b \frac{A_b^2}{2} \right)^2 [1 + \cos(2\omega_b t + \phi_b)]^2 + 2a(t)A_0 \cos(\omega_s t + \phi_0)\cos(\omega_0 t + \phi_0) \\
&\quad + 2 \sum_b a(t) A_b \cos(\omega_s t + \phi_s)\cos(\omega_b t + \phi_b) \\
&\quad \left. + 2 \sum_b A_0 A_b \cos(\omega_0 t + \phi_0)\cos(\omega_b t + \phi_b) \right\}
\end{aligned} \tag{6.1.17b}$$

In the preceding equations, A_b, ω_b, and ϕ_b are radiant background electric field amplitude, frequency, and initial phases at each frequency.

As explained regarding Eq. (6.1.6) for direct detection, optical frequencies are too fast for electronic circuits to follow. Therefore, the terms involving the frequencies $2\omega_s$, $2\omega_0$, and $2\omega_b$ can be neglected. The last three terms in Eq. (6.1.17b) are cross products between the various terms. These involve sums and differences of their frequencies. The sum frequencies are too large to be followed by detector electronic circuits. Equation (6.1.17b) simplifies to

$$i(t) = \frac{\Re G A_r}{\eta_0}\left\{\frac{a^2(t)}{2} + \frac{A_0^2}{2} + \left(\sum_b \frac{A_b^2}{2}\right)^2 + a(t)A_0 \cos[(\omega_s - \omega_0)t + \phi_s - \phi_0]\right.$$
$$+ a(t)\sum_b A_b \cos[(\omega_s - \omega_b)t + \phi_s - \phi_b]$$
$$\left. + A_0 \sum_b A_b \cos[(\omega_0 - \omega_b)t + \phi_0 - \phi_b]\right\} \tag{6.1.18}$$

For reasons concerned with spatial coherence to be described in Section 6.1.5, heterodyne receivers are essentially immune to background radiation effects since such radiation is of very weak coherence. Therefore,

$$i(t) = \frac{\Re G A_r}{\eta_0}\left[\frac{a^2(t)}{2} + \frac{A_0^2}{2} + \sum \frac{A_b^2}{2} + a(t)A_0 \cos(\omega_{IF}t + \phi_{IF})\right] \tag{6.1.19}$$

where *intermediate frequency* $\omega_{IF} = \omega_s - \omega_0$ and *intermediate phase* $\phi_{IF} = \phi_s - \phi_0$. If ω_s and ϕ_s vary with time, so too do ω_{IF} and ϕ_{IF}. Consequently, unlike direct detection, heterodyne receivers can detect frequency and phase modulation as well as amplitude modulation $[a(t)]$ provided the frequency difference $\omega_s - \omega_0 = \omega_{IF}$ is sufficiently small so as to pass through the detector electronic and amplifier circuits. If ω_{IF} is too large, the cosine term in Eq. (6.1.19) disappears and so too does the ability to follow frequency or phase modulation. In heterodyne detection, the frequency and phase information at the signal frequency ω_s is simply translated down the frequency axis to the frequency ω_{IF}, which is made appropriate to receiver electronics time response.

If $\omega_s = \omega_0$, then ω_{IF} is zero. Such detection is called *homodyne*, and is used chiefly in phase modulation schemes.

In Eq. (6.1.18), the IF current is

$$i_{IF}(t) = \frac{\Re G A_r}{\eta_0} a A_0 \cos(\omega_{IF}t + \phi_{IF}) \tag{6.1.20}$$

where it is assumed that both the signal and local oscillator fields are of constant amplitude. If $\omega_s \neq \omega_0$, the IF term is time varying. Consequently,

$$\overline{i_{IF}^2} = (\Re G A_r/\eta_0)^2(a^2 A_0^2/2) \tag{6.1.21}$$

From Eq. (6.1.1), $a^2 = 2\eta_0 P_s/A_r$ and $A_0^2 = 2\eta_0 P_0/A_r$. In view of this and Eq. (6.1.21),

$$\bar{I}_{IF}^2 = 2\Re^2 G^2 P_s P_0 \tag{6.1.22}$$

The IF signal-to-noise is

$$\text{SNR} = \frac{2\Re^2 G^2 P_s P_0}{2qB[\Re(P_s + P_0 + P_b) + i_d]G^2 + 4FkT_0 B/R_L} \tag{6.1.23}$$

For a photoconductive detector, the first term in the denominator of Eq. (6.1.23) is doubled, since it then represents g-r noise instead of shot noise. In Eq. (6.1.23) the dc noises such as $\Re(P_s + P_0 + P_b)$ and i_d do not appear since they are removable with a capacitor. Noises of interest are time varying, including fluctuations in current deriving from fluctuations in local oscillator radiant power. Hence, $\tilde{P}_s$ again refers to the time-varying or sideband component of signal radiant power.

In practical receivers, P_0 can easily be made large enough to make the quantum noise deriving from it the dominant noise source [see, however, Exercise 6.6(b)]. Under these conditions

$$\text{SNR} = \frac{\Re P_s}{qB} = \frac{\eta P_s}{h\nu B} \tag{6.1.24}$$

For a photoconducting detector, SNR $= \eta P_s/2h\nu B$.

Homodyne receivers can provide SNRs 3 dB better (see Exercise 6.8) than heterodyne receivers.

The minimum detectable IF signal is, from Eq. (6.1.24),

$$P_{s\,\min} = \frac{h\nu B}{\eta} \tag{6.1.25}$$

yielding an NEP for heterodyne detection

$$\text{NEP} = \frac{P_{s\,\min}}{B} = \frac{h\nu}{\eta} \tag{6.1.26}$$

Note that as a consequence of Eq. (6.1.25), for heterodyne detection, units of NEP are $\text{W}\cdot\text{Hz}^{-1}$ instead of $\text{W}\cdot\text{Hz}^{-1/2}$ as in direct detection. For an ideal detector $\eta = 1$ and ideal heterodyne NEP then is $h\nu$, that is, minimum detectable energy is that of a single photon, assuming unity SNR.

6.1.3 Advantages of Heterodyne Detection

As compared to direct detection, heterodyne detection exhibits the following advantages:

1. It can detect FM and PM provided ω_{IF} is compatible with receiver electronic bandwidth.

2. Since the local oscillator beam is part of the receiver and does not arrive from afar, the dominant noise arises from fluctuations in P_0 rather than from background power, thus providing discrimination against background and other unwanted light. The discrimination is seen to be even greater after spatial coherence effects are considered in Section 6.1.5.
3. The IF conversion process provides gain so that the IF signal output of the detector may be made large enough to override both thermal and quantum (or g-r) noise. This can be visualized in the following way. In direct detection, signal current power $\overline{i_s^2}$ is proportional to $\overline{P_s^2}$, whereas for heterodyne detection, signal current power $\overline{i_{IF}^2}$ is proportional to the product P_0P_s. Therefore, a weak P_s received from a very distant source, for example, can be compensated for in heterodyne detection by increasing P_0 correspondingly. The conversion gain is P_0/P_s. Consider a situation where P_s is 1 pW. Then $i_s^2 \propto P_s^2 = 10^{-24}$ W^2. If a milliwatt local oscillator is used in heterodyne detection, $i_{\mathrm{IF}}^2 \propto P_0P_s = 10^{-15}$ W^2. The conversion gain P_0/P_s is 10^9. The improvement in SNR is by many orders of magnitude, as shown in Exercise 6.6. Thus, heterodyne detection is a technique capable of permitting detection of radiant signal powers P_s much weaker than those possible in direct detection. However, this powerful potential is limited by the practical efficiency of the IF conversion process. These practical problems are addressed next. It is also limited by the maximum radiant power that can be received by the detector without destroying it. For this reason, P_0 is generally limited to about a milliwatt.

6.1.4 Disadvantages of Heterodyne Detection

Heterodyne detection is a distributed phenomenon in that it takes place over detector dimensions (~1 to 100 mm) extremely large in comparison with light wavelength. This is opposite to the situation at millimeter or microwave wavelengths. The aA_0 product dependence of $i_{\mathrm{IF}}(t)$ in Eq. (6.1.20) involves a scalar dot product of the signal and local oscillator radiant electric field over the front surface of the photodetector. The detector surface may be decomposed into incremental areas, each one yielding its own incremental current di_{IF}. The total IF current is then the sum of the individual IF current increments over the detecting surface. Each incremental detector surface area illuminated by both the signal and local oscillator light beams produces its own differential quantity of IF current. If these differential currents are generated with the same phase angle, their vector addition is simply their sum. However, in reality each differential IF current is characterized by its own phase angle. Two differential detector areas giving rise to opposing phase angles yield an overall IF current of magnitude equal to their IF current difference rather than sum. Therefore, it is important in heterodyne and homodyne detection for the phase difference between both light beams to be as uniform as possible over the detector surface so that the differential IF currents add to one another rather than cancel one another. Such a uniform mutual phase relation in space is called *mutual spatial coherence*.

Heterodyne and homodyne detection are referred to as *coherent detection*. At microwave frequencies where wavelength is so much greater than detector dimensions such spatial coherence over the detector area usually takes place naturally. However, at optical wavelengths, which are so much smaller than detector dimensions (micrometers or less as compared to millimeters), special provisions must be made so as to minimize IF conversion loss by maximizing mutual spatial coherence between P_s and P_0. These provisions are [6.1]:

1. Both beams should have the same transverse spatial mode structure. Ideally both should be in the fundamental (TEM_{00}) mode. For example, if one beam mode is TEM_{00} and the other is the TEM_{01} or TEM_{10} structure, then over half of the detector surface where fields of both waves are in the same direction a positive IF current is generated, while over the other half where fields of both waves are in opposite directions a negative IF current is generated. This negative current cancels part or all of the positive IF current component. If both beams are of the TEM_{01} or TEM_{10} mode, then zero intensity is incident along a line crossing the middle of the detector surface, thus decreasing total IF current.
2. Both beams should be coincident and of equal diameter, since only detector surface area where both beams overlap can yield an IF current. Areas covered by one beam alone yield direct direction currents rather than IF signals. These are noise relative to the IF signal.
3. Pointing vectors of both beams should be coincident. Otherwise, wavefronts are not coincident and phase difference between the two waves varies as a function of position across the detector surface (see Exercise 6.10). The greater the angle between the two beams, the smaller the area over which phase difference between the two beams is constant.
4. Both wavefronts should have the same radius of curvature. Otherwise, wavefronts cannot coincide and phase difference varies here too as a function of position across the detector surface.
5. Both beams should be identically polarized. The IF current derives from the dot product of the two electric fields. Therefore, it is maximum only when both fields are aligned, falling off as the cosine of the angle between them.

The conditions listed are required in order to maximize the mixing process producing IF current. Eqs. (6.1.20), (6.1.22), (6.1.23) and (6.1.24) refer only to idealized conditions where IF current is indeed maximum. Misalignments, even if on the order of milliradians or less can reduce IF current most significantly and can even cause it to disappear.

The disadvantage of coherent detection (heterodyne or homodyne) is the difficulty of fulfilling so exactly the five conditions just listed. Hence, it is generally desirable not to undertake coherent detection except for detection of extremely weak $\tilde{P}_s$ signals or for detection of frequency or phase shifts or modulation which cannot be done with direct detection.

The advantage of direct detection is its practical simplicity. From a practical standpoint, heterodyne detection is quite feasible in the laboratory using a stable table and positioning equipment permitting delicate alignment. In the field, however, the wavefront of the signal beam may be altered by atmospheric effects, and this will affect the ability to match the wavefronts with the local oscillator beam. The local oscillator is situated within the receiver itself, so that wavefronts in its beam are not affected like those of the distant signal beam. A beam (plane wave or spherical) propagating in the z direction exhibits a spatial phase dependence of $\psi = \pm\beta z = \pm 2\pi z/\lambda$. A spatial phase distortion of $\Delta\psi = \pm 2\pi\Delta z/\lambda$ caused by the atmosphere, for example, is more significant at shorter wavelengths than at longer wavelengths for a given Δz. Indeed, Fig. 6.4 compares a CO_2 laser beam (10.6 μm wavelength) laser beam and a HeNe (0.633 μm wavelength) beam propagated with the same

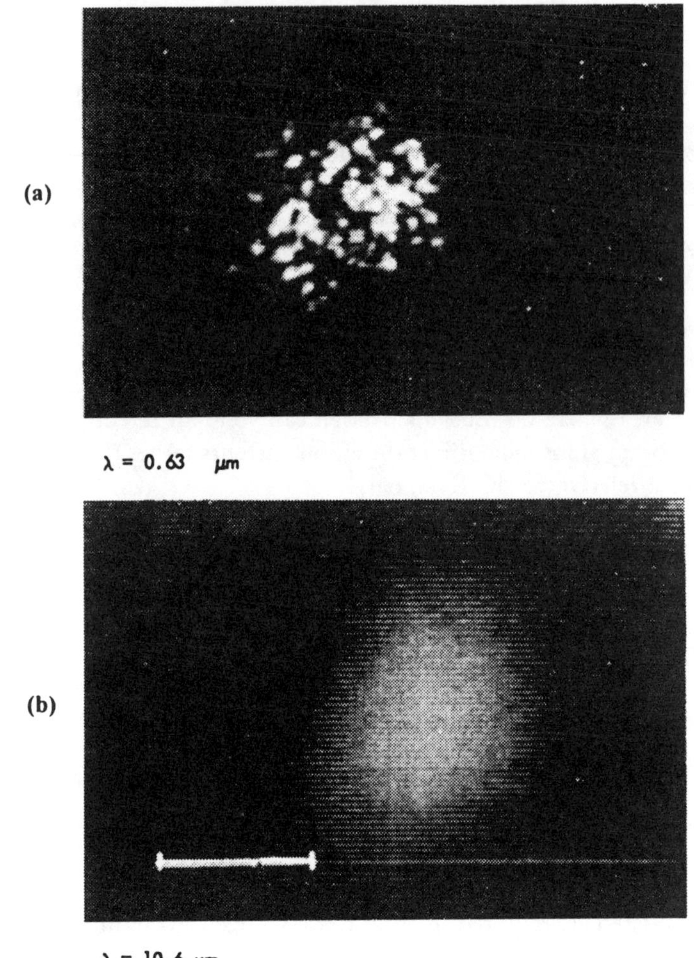

Fig. 6.4 Comparison of (a) He-Ne and (b) CO_2 laser beams traversing the same atmospheric path (5.05 km) with the same optics (after [6.7]).

optics over the same atmospheric path (5.05 km). The wavefront at the shorter wavelength is essentially destroyed. Consequently, any mutual coherence with a local oscillator beam would be essentially zero. The longer wavelength, however, is essentially unaffected and can be used in coherent detection. Consequently, outside the laboratory coherent detection is in general much more feasible at longer rather than at shorter wavelengths.

Another limitation to be considered regarding implementation of coherent detection is temporal coherence. The coherence length (Section 4.6) of the signal source must be greater than the path length. Otherwise, wavefront at one wavelength is uncorrelated with wavefront at another wavelength, and overall signal beam wavefront is destroyed. In laser Doppler velocimetry, for example, the same laser is used for both $\tilde{P}_s$ and P_0, with a beamsplitter used to separate the beams as in Fig. 6.5. A frequency shift such as a Doppler shift can be detected via heterodyne detection only. The frequency shift is proportional to the relative velocity between the source and

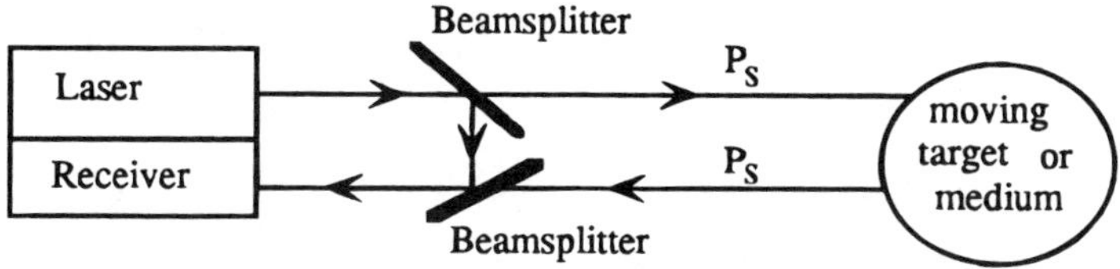

Fig. 6.5 Laser velocimetry setup.

target in the direction of the laser beam. Here, the path difference between P_s and P_0 must be less than the coherence length of the laser beam in order to maintain mutual coherence between the different wavelengths composing the laser beam.

6.1.5 Antenna Properties of a Coherent Receiver [6.8]

An optical heterodyne or homodyne receiver is also an antenna. It can be characterized by an effective aperture or capture cross section as a function of signal wave direction.

In common with antennas, it can be shown [6.8] that the effective aperture integrated over all possible arrival directions is essentially λ^2. More specifically,

$$A_e\Omega_e \sim \lambda^2 \tag{6.1.27}$$

where A_e is the effective area over which both signal and local oscillator wavefronts exhibit constant phase difference, and Ω_e is the effective solid angle field of view formed by the two beams. This means there is an inverse trade-off between directional tolerance or field of view of a coherent receiver and its effective aperture or capture area. Outside the laboratory, A_e is generally limited by atmospheric effects that distort the wavefront of P_s [6.3, 6.9].

As a consequence of the small value of λ^2 at optical wavelengths, coherent detection is virtually useless to detect thermal radiation, which is effectively incoherent, without addition of an integrator (considered next). This can be shown in the following way. Consider a blackbody radiator as the signal beam. In view of Eq. (3.2.1), the radiant power collected by a coherent receiver at a distance z is

$$P_s = N_\nu A_{bb} A_e \Delta\nu / z^2 \tag{6.1.28}$$

where A_e/z^2 is the solid angle subtended by the coherent receiver and A_{bb} is the emitting area "seen" by the receiver, the value of which is $\Omega_e z^2$. In view of Eqs. (4.2.6), (6.1.27), and (6.1.28).

$$\begin{aligned} P_s &= \frac{1}{2}\left[\frac{2h\nu}{\lambda^2[(\exp h\nu/kT) - 1]}\right]\Omega_e z^2 \frac{A_e}{z^2}\Delta\nu \\ &= \frac{h\nu\Delta\nu\Omega_e A_e}{\lambda^2[\exp(h\nu/kT) - 1]} = \frac{h\nu\Delta\nu}{\exp(h\nu/kT) - 1} \end{aligned} \tag{6.1.29}$$

where the 1/2 in the first step of this equation is the portion of the randomly polarized blackbody radiant power that is in the direction of polarization of the local oscillator. If Eq. (6.1.29) is substituted into Eq. (6.1.24), we obtain

$$\text{SNR} = \frac{\eta P_s}{h\nu B} = \frac{\eta}{\exp(h\nu/kT) - 1}\frac{\Delta\nu}{B} \tag{6.1.30}$$

We now come to the interesting question as to whether optical ($\Delta\nu$) and electronic (B) bandwidths are related in coherent detection. It has been shown following Eq. (6.1.12) that no such relationship exists for direct detection. For heterodyne detection of a *wide* spectral source such as thermal radiation, however, the IF detected radiation is that which exists on either side of the local oscillator frequency, up to the limits in the frequency domain deriving from the electronic bandwidth of the receiver. Larger intermediate frequencies cannot be detected. The electronic bandwidth of the receiver is filled up by a *continuous distribution* of intermediate frequencies deriving from frequency differences between those of the local oscillator and the source (P_s). A blackbody source exhibits no single frequency but rather a continuous spectrum. Therefore, $\Delta\nu = B$ for detection of a thermal source and

$$\text{SNR} = \frac{\eta}{\exp(h\nu/kT) - 1} \tag{6.1.31}$$

For detection of a nonthermal source, IF spectral purity depends on the monochromaticity of both signal and local oscillator wave. For a SNR of unity or more in detection of thermal radiation, Eq. (6.1.31) indicates

$$T \geq \frac{h\nu}{k\ln(1 + \eta)} \cong \frac{h\nu}{k\eta} \tag{6.1.32}$$

where the last approximation holds for $\eta << 1$. The quantity $h\nu/k$ corresponds to temperatures on the order of 25,000 K for the middle of the visible. Because solar radiation is approximated by a blackbody radiator of temperature 5900 K, a heterodyne receiver with visible local oscillator radiation pointed directly at the sun will not produce a detectable IF signal deriving from the sun without use of an integrator. This also means that if P_s derives from some other source such as a visible wave laser, for example, but the coherent receiver is pointed directly at the sun so that background radiation P_b is sunlight, the efficiency of mixing involving such background radiation with the local oscillator is so poor that there are no real beat frequencies from the background radiation. Consequently, such background radiation also has virtually no effect on IF SNR involving received laser radiation. The reason is that signal and local oscillator waves from lasers exhibit much greater spatial coherence than does background radiation from the sun (as described in Chapter 7). Therefore, A_b in Eq. (6.1.18) can be neglected and coherent receivers are virtually immune to interference by natural background radiation.

We should point out, however, that this conclusion should be reexamined in the infrared, where at about a 10-μm wavelength $h\nu/k$ is about 1250 K instead of 25,000 K.

6.1.6 Effects of Integration on Signal-to-Noise Ratio

If detector output is put through an integrating circuit, noise fluctuations are smoothed and reduced. As a result, for sine wave signals as in heterodyne detection, SNR improves by the factor $(B\tau_i)^{1/2}$, where τ_i is integration time. As a result, thermal radiation can be detected via heterodyne detection, but not in real time [6.10].

Such integration techniques with heterodyne detection can be used to measure spectra of various sources simply by varying local oscillator wavelength. At each local oscillator wavelength, a narrow portion limited to the IF bandwidth B around ω_0 is detected. By frequency scanning via changes in ω_0 the whole source spectrum can be measured with high optical frequency resolution on the order of the electronic bandwidth B. The smaller the value of B, the finer the optical frequency resolution. Decreasing B may necessitate increasing integration time so as to maintain SNR.

Example 6.2

Assume atmospheric conditions such that the attenuation coefficient is 0.2 km^{-1} and the diameter of mutual spatial coherence is limited to 1 mm. A laser beam at 10.6-μm wavelength is transmitted 5 km to a homodyne receiver with a 0.01 m^2 receiving area and 10-mrad full field of view, with a detector quantum efficiency of 50%. The transmitted laser beam has 1 mW power and a 1-mrad beamwidth. Laser aperture is 1 cm. For optimum IF conversion, what is the SNR? Assume a 1-MHz electronic bandwidth.

Solution

This example encompasses many concepts discussed in past chapters. The signal power transmission τ_a' through 5 km of atmosphere is $\exp(-0.2\ \text{km}^{-1} \cdot 5\ \text{km}) = 0.368$. The laser beamwidth α of 1 mrad gives rise to a beam diameter of 5 m after 5 km. This is much larger than the receiver area. However, atmospheric effects (spatial coherence) limit A_e to $(\pi/4)(10^{-3})^2 = 7.85 \times 10^{-7}\ \text{m}^2$. The coherent receiver field of view is limited to $\Omega_e \cong \lambda^2/A_e = (10.6 \times 10^{-6})^2/7.85 \times 10^{-7}\ \text{m}^2 = 0.143$ mrad $<< 10$ mrad. With this field of view the coherent receiver can see a source diameter of 72 cm at 5-km distance. Since the laser transmitter aperture is smaller, then all of the transmitter area can be seen by the receiver.

Laser transmitter radiance is $N_L = \dfrac{P_t}{A\Omega}$ where $\Omega \approx \dfrac{\pi}{4}\alpha^2$. Here, P_t is transmitted power, A is area of laser aperture, and α is beamwidth. Received laser radiant power is

$$P_s = N_L \tau_a' A \frac{A_e}{z^2} = \frac{P_d \tau_a'(\pi/4)}{(\pi/4)\alpha^2} \frac{(10^{-3})^2}{(5 \times 10^3)^2} = \frac{10^{-3}(0.368)10^{-6}}{(10^3)^2(5 \times 10^3)^2}$$
$$= 1.47 \times 10^{-11}\ \text{W}$$

where for coherent detection A_e is used instead of the much larger A_r. For heterodyne detection, maximum SNR is

$$\text{SNR} = \frac{2\eta P_s}{h\nu B} = \frac{2(0.5)1.47 \times 10^{-11}}{6.62 \times 10^{-34}\left(\dfrac{3 \times 10^8}{10.6 \times 10^{-6}}\right)10^6} = 784$$

6.2 DETECTORS

Before discussing detector types and devices, it is important to establish criteria with which to compare them. Detector figures of merit include responsivity, NEP, speed of response, etc., which have already been introduced. Other criteria include detectivity

(D), normalized detectivity (D^*), and noise equivalent temperature difference for thermal imaging such as at infrared wavelengths.

Detectivity is simply the reciprocal of NEP. Since NEP represents for 1-Hz electronic bandwidth the equivalent noise radiant power, D or 1/NEP represents for 1-Hz electronic bandwidth SNR for 1 W of signal power. Units of D are $\mathrm{Hz}^{1/2}\cdot\mathrm{W}^{-1}$ for direct detection.

At ultraviolet (UV), visible, and near IR wavelengths, limiting noise in direct detection derives usually from dark current shot noise or from Johnson noise, both of which are internal, unlike background noise. To obtain smaller values of NEP, it is advantageous to manufacturers to measure NEP under conditions where there is no background radiation. The detector is normally in darkness. Such NEP values are the lowest possible and represent sensitivity limitations resulting from inherent noise. (It is to the manufacturer's advantage to characterize his detector with the lowest possible NEP.)

At thermal imaging wavelengths, even in the dark, the major source of noise is background radiation. This background radiation is thermal emission.

As can be seen from Planck's law and the various blackbody radiation formulas, such radiation depends only on temperature and emissivity and exists even in absolute darkness. BLIP detection is most pertinent to this situation where major noise derives from thermal background radiation.

As shown in Eq. (6.1.12), NEP varies according to $A_r^{1/2}$ in BLIP detection, regardless of the nature of the background radiation. Generally, detection area is determined by the optics area, rather than by the detector itself. Therefore, since D varies according to $A_r^{-1/2}$, a large receiving optics area yields a smaller value of detectivity than does a smaller receiving optics area for the same detector. In such situations it is useful to use a detector figure of merit that is independent of the optics receiving area. Normalized or specific detectivity is used for this purpose and it is the product of D and $A_r^{1/2}$

$$D^* = \frac{A_r^{1/2}}{\mathrm{NEP}} \tag{6.2.1}$$

in units of $\mathrm{cm}\cdot\mathrm{Hz}^{1/2}\cdot\mathrm{W}^{-1}$. Its significance is that it is independent of receiving area [6.11] and permits comparison of detectors under identical operating conditions.

In view of Eq. (6.1.9) for BLIP conditions as well as definition (6.2.1)

$$D^* = [\eta(\nu)A_r/(2h\nu P_b(\nu))]^{1/2} = \left[\int_{-\infty}^{\infty}\left(2N_\nu(\nu)h\nu\,\frac{\Omega_r}{\eta(\nu)}\right)d\nu\right]^{-1/2} \tag{6.2.2a}$$

with A_r cancelling out. This result assumes that background radiation of radiance $N_\nu(\nu)$ fills the entire receiver field of view Ω_r and that the receiver detects radiation all across the EM spectrum. When that background radiation is blackbody, then Eq. (4.2.6) is substituted for $N_\nu(\nu)$. For an ideal detector $\eta(\nu)$ equals unity. Equation (6.2.2) can be used to determine D^* for an ideal detector sensitive to all EM frequencies or wavelengths and which is in thermal equilibrium with the radiation background. If such thermal equilibrium does not exist, as is usually the case, then NEP and D^* must be calculated in terms of fluctuations in emitted and absorbed power. The average power spectral density of such fluctuations is represented by $\bar{E}(\lambda, T)$ in Eqs. (5.2.9).

Quantum detectors are sensitive only to photon energies above $h\nu_c$, where ν_c is a cutoff frequency (equal to c/λ_c) determined by the specific detector and is discussed below for various detector types. In that case, the limits of the integral in Eq. (6.2.2a) are from ν_c to infinity. Figure 6.6 compares D^* for ideal and actual photovoltaic and photoconductive quantum detectors. In the ideal BLIP case, D^* is least (or worst) when $\lambda \approx 1.52\lambda_m$ instead of the value of λ_m characterizing background thermal emission [Wien's law, Eq. (4.2.16)] because of the wavelength dependencies of $\overline{E}$ and ideal responsivity $\Re_{\text{ideal}}$. Figure 6.6 indicates how close actual quantum detectors in the thermal infrared approach ideal D^*. For an ideal photodetector D^* can be written as

$$D^*_{\text{ideal}}(\lambda) = \frac{\Re_{\text{ideal}}(\lambda)}{\sqrt{2q\int_0^{\lambda_c} W(\lambda)\Re_{\text{ideal}}(\lambda)\overline{E}(\lambda, T)\, d\lambda}} \tag{6.2.2b}$$

where $\Re_{\text{ideal}}(\lambda) = q/(h\nu) = (\lambda/1.24)\quad \text{A}\cdot\text{W}^{-1}$. If the detector has generation-recombination noise the denominators in Eqs. (6.2.2) are $\sqrt{2}$ larger.

Infrared radiometry often is concerned with measuring temperatures of emitters (usually passive) rather than actual radiant power. Thermal imaging contrast, in particular, is usually dependent on temperature differences between objects in a scene, since emissivities of many materials may often be similar. To this end, it is useful in thermal imaging to consider noise equivalent temperature difference (NETD) rather than NEP. NETD is the temperature difference which produces a difference in radiant power at the receiver input that is just equal to electronic noise at the detector output. NETD is related to NEP such that

$$\text{NETD} = \frac{(\text{NEP})_{\text{BLIP}}}{dP_b/dT} \tag{6.2.3}$$

and is the minimum temperature difference measurable as limited by noise. Now

$$P_b = N_\nu A_r \Omega_r \Delta\nu$$

which, for blackbody radiation, equals

$$\begin{aligned} P_b &= A_r\Omega_r \frac{2h\nu^3\Delta\nu}{c^2[\exp(h\nu/kT) - 1]} \\ &\cong A_r\Omega_r 2h\nu^3\Delta\nu[\exp(-h\nu/kT)]/c^2 \end{aligned} \tag{6.2.4}$$

so that

$$\frac{dP_b}{dT} \cong \frac{h\nu}{kT^2} P_b \tag{6.2.5}$$

When this result is substituted into Eq. (6.2.3)

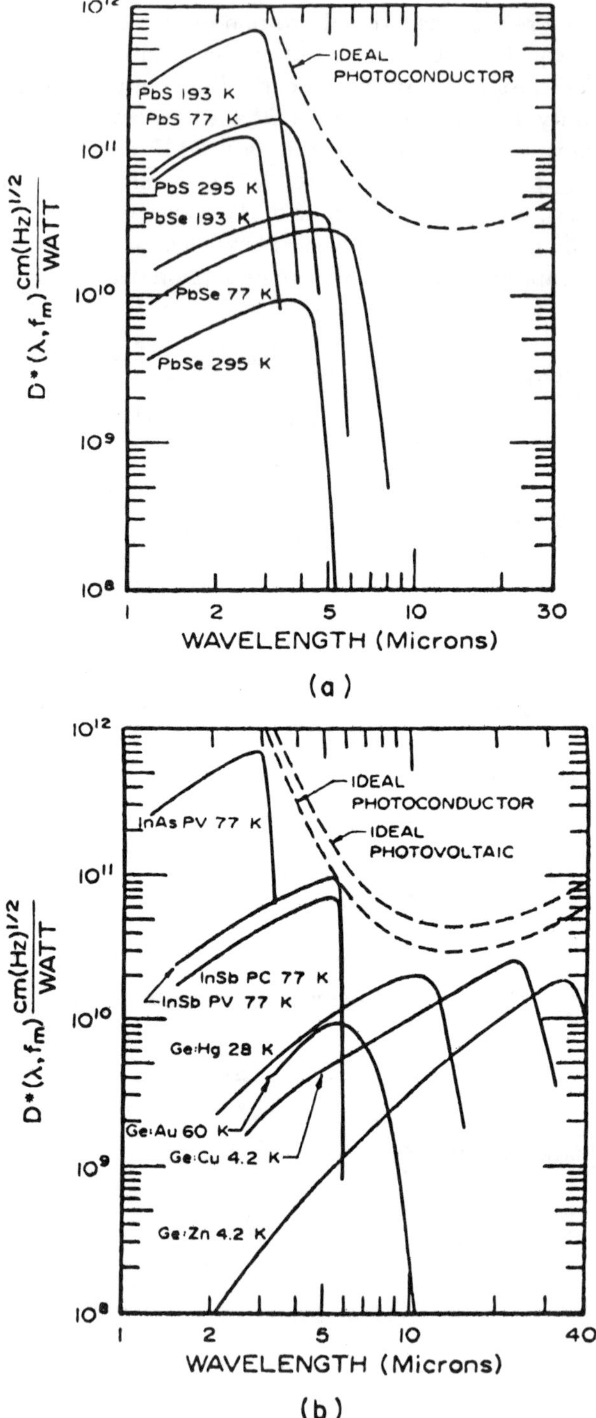

Fig. 6.6 *D** for ideal and actual quantum detectors for 295°C background and π sr field of view. (Courtesy of Santa Barbara Research Center.)

$$\mathrm{NETD} \cong \frac{kT^2}{h\nu P_b}(\mathrm{NEP})_{\mathrm{BLIP}} = kT^2\left[\frac{2}{(\eta h\nu P_b)}\right]^{1/2} \tag{6.2.6}$$

where Eq. (6.1.9) has been used for $(\mathrm{NEP})_{\mathrm{BLIP}}$ and NETD is in units of $\mathrm{K \cdot Hz^{-1/2}}$. Note that NETD varies with temperature. For refrigerated or cryogenic receivers with quantum detectors, NETD is typically below 0.07°C.

Having defined parameters with which to compare detectors, we now consider actual detector types. Detectors can be divided into two categories: quantum and thermal detectors. Quantum detectors are those which respond to photon absorption, in contrast to thermal detectors which respond to heating of the detector by incident radiation. The responsivity of quantum detectors is very wavelength dependent corresponding to processes of photon absorption. Thermal detectors are quite insensitive to wavelength, as is device heating. Thermal detectors themselves respond to radiation all across the EM spectrum. However, detectors are housed with a window material in front of the detecting surface in order to protect the device. Transparency of each window material is limited to a given wavelength range, so that thermal detectors are limited practically to specific wavelength ranges according to window material selected by the consumer.

In general, quantum detectors are more sensitive than thermal detectors and thus detect lower radiant power levels. Also, they are faster, and thus exhibit larger electronic bandwidths. Thermal detectors are of most interest generally in the longwave thermal infrared (3- to 5-μm and 8- to 14-μm wavelength) where quantum detectors are quite expensive.

Quantum detectors are discussed first. This is followed by brief discussions regarding thermal detectors.

6.2.1 Quantum Detectors

Quantum detectors consist essentially of two types: electron tube and semiconductor devices.

6.2.1.1 Electron Tubes

Most electron tube detectors are photoemissive, that is, they detect radiation via the photoelectric effect. Photons absorbed by the cathode give rise to electron emission by the cathode, thereby increasing the current in the detector circuit. Photon-generated electrons are called *photoelectrons*, and their initial kinetic energy is equal to

$$E_i = h\nu - \phi \tag{6.2.7}$$

where ϕ is the work function of the cathode surface material. Equation (6.2.7) implies a cutoff minimum frequency (for $E_i = 0$) of

$$\nu_c = \phi/h \tag{6.2.8}$$

or cutoff (maximum) wavelength

$$\lambda_c = ch/\phi \tag{6.2.9a}$$

Work function ϕ is in electron volts and is equal to qW, where W is work function in volts. Substituting into Eq. (6.2.9a)

$$\lambda_c = \frac{ch}{qW} = 1.24/W \tag{6.2.9b}$$

where mks units are used for the constants and λ_c is in micrometers. Usually λ_c is between 0.6 and 1.2 μm, depending on cathode material. At longer wavelengths each photon possesses insufficient energy to liberate an electron from the cathode. Consequently, electron tube detectors are limited to wavelengths up to and including this portion of the near infrared, but not the thermal infrared. Equation (6.2.7) implies that the photoelectric effect depends not on total incident energy but on energy of individual photons. Therefore, such devices are quantum rather than thermal detectors.

Metals were the first materials from which photoelectron emission was detected. Reflectivity for visible and near-infrared photons is high, thus limiting quantum efficiency since only a small number of incident photons are actually absorbed. Furthermore, of those that are absorbed, only a small percentage actually give rise to photoelectron emission. The reason is that the energy loss of excited current carriers due to electron scattering within the cathode material is rapid. Photoelectrons generated only within a few atomic layers of the surface have enough energy remaining when they reach the cathode surface to be emitted. Photons absorbed further down are simply lost since photoelectrons generated by them are not emitted. Typical quantum efficiencies for metal photocathodes are therefore a fraction of a percent to a few percent.

Work function of the cathode surface determines wavelength cutoff λ_c in Eqs. (6.2.9). Since for metals ϕ exceeds 2 eV, metal photocathodes sensitive beyond 0.6 μm cannot be fabricated. Consequently, semiconductor photocathodes were developed in which photon absorption is more efficient and relaxation time for energy losses longer. The much smaller free-electron concentration in semiconductors as compared to metals gives rise to a rate of photoelectron energy loss due to electron scattering which is much less than that in metals. As a consequence of this, as well as the lower reflectivity of semiconductors (less conductivity than in metals), η for semiconductor photocathodes is much higher, as shown in Table 6.2 and Fig. 6.7.

To simplify identification of photocathode materials, the *S number* has been introduced. A list of some of the more important photocathodes with their *S* numbers, λ_c, and peak quantum efficiencies is given in Table 6.2. Typical spectral quantum efficiency curves are shown in Fig. 6.7.

Photoemissive detectors can be divided into three categories, all of which contain photocathodes: vacuum tubes, gas-filled tubes, and photomultiplier tubes.

TABLE 6.2 *Photocathodes*

Photocathode	*S* Number	Long Wavelength Threshold (μm)	Peak Quantum Efficiency (%)
Ag–O–Cs	*S*-1	1.2	0.5
Cs3Sb	*S*-11	0.67	20
Bi–Ag–O–Cs	*S*-10	0.78	5
$Na_2KSb(Cs)$	*S*-20	0.87	30
$K_{22}CsSb$		0.66	30
$K_2CsSb(O)$		0.78	35
Negative electron affinity			
GaAs(CsO)		0.95	50
$InAs_{1-x}P_x$		0.95	12
$InAs_{1-x}P_x$		1.06	3

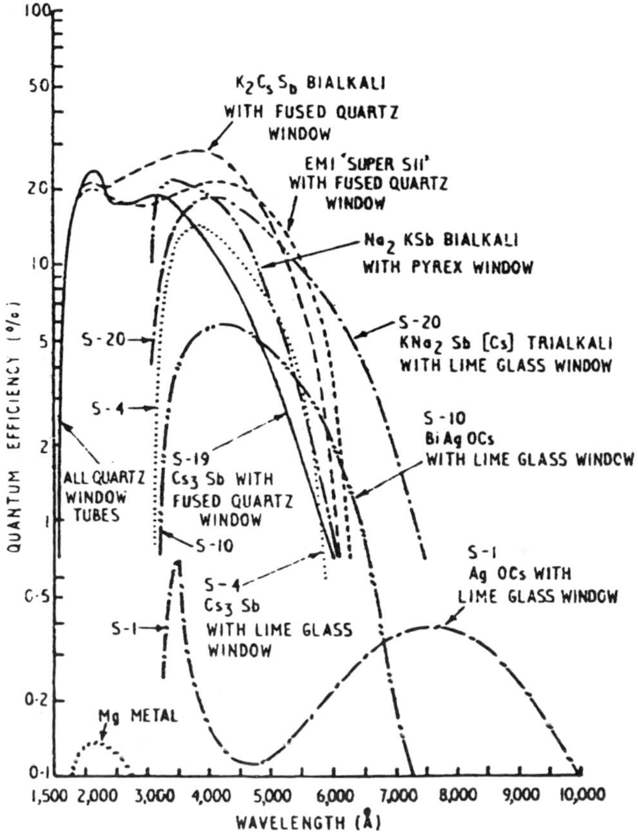

Fig. 6.7 Typical spectral sensitivity curves of commonly used photocathodes (after [6.12]).

Vacuum tubes are the simplest. They are diodes, consisting simply of a photocathode and anode in an evacuated tube. Electronic bias, typically 50 to 150 V, is generated between the two electrodes. Photons absorbed by the photocathode gives rise, via the photoelectric effect, to photoelectrons, which are propelled via the bias field to the anode. The change in current brought about by the incident photons is the detected signal. Thus, as with any detector, the input is radiant power and the output is current. The photocathode may be opaque, in which case the light must be incident from the side or the direction of the anode. Often, the photocathode is a large curved surface forming a third to a half of a cylinder, with a pole along the cylinder axis serving as the anode. In this way, radiation is collected over a large area and the resulting current is "focused" onto the anode. Alternatively, the photocathode may be transparent, in which case radiation may be incident from behind the photocathode.

For simplicity, consider a vacuum photodiode with parallel-plate electrodes of separation d. If at time $t = 0$ an electron is ejected from the photocathode at $z = 0$, it is accelerated toward the anode with a force

$$m\ddot{z} = qV/d \tag{6.2.10}$$

where m is electron mass and V is bias voltage. Via integration of this equation, electron velocity is

$$\dot{z}(t) = qVt/(md) + v_i \tag{6.2.11}$$

where v_i is initial electron velocity and, from Eq. (6.2.7), is equal to

$$v_i = (2E_i/m)^{1/2} = [2(h\nu - \phi)/m]^{1/2} \tag{6.2.12}$$

Electron position is given by

$$z(t) = qVt^2/(2md) + v_i t \tag{6.2.13}$$

When the electron reaches the anode

$$z(t) = d = qVT^2/(2md) + v_i T \tag{6.2.14}$$

where, from Eq. (6.2.14), electron transit time T is given by

$$T = \frac{md}{qV}[(2qV/m + v_i^2)^{1/2} - v_i] \tag{6.2.15}$$

Each current pulse generated by a photoelectron is

$$i(t) = q\dot{z}(t)/d = \begin{cases} q^2Vt/(md^2) + I_0 & \text{for } 0 \leq t \leq T \\ 0 & \text{otherwise} \end{cases} \tag{6.2.16}$$

where I_o is qv_i/d and is the current at $t = 0$.

From Eq. (6.2.15) it is clear that the smaller the d/V ratio, the faster the speed of response. Hence, electrode separation and bias are critical parameters [6.13]. Vacuum photodiode rise times on the order of nanoseconds are quite typical.

In addition, from Eq. (6.2.16) it is clear that good responsivity requires the bias field (ratio V/d) to be large. Hence, generally, both responsivity and speed of response improve with bias field. Also, from both equations it is clear that the greater the initial electron energy $E_i = h\nu - \phi$ and initial electron velocity, the better the detector performance in terms of both speed of response as well as responsivity. This occurs at short wavelengths in particular. However, this wavelength dependence is small compared to effects of the electric field V/d [6.14] (see Exercise 6.13).

Lifetime is limited by photocathode lifetime. For semiconductor photocathodes, the photocathode is generally a thin film on the cathode surface. It can be damaged if voltage and/or illumination levels are set too high. The reason is that in vacuum phototubes vacuum levels are generally only on the order of 10^{-6} torr, which are very far from absolute vacuums. Use of Avogadro's number indicates that at standard temperature there are about 10^{18} residual gas molecules per liter under such vacuum pressures. If voltage is set too high, kinetic energies of more and more electrons increase beyond ionization energies of the gas atoms and collisions between electrons and residual gas atoms give rise to increased ionization of gas atoms. Electrons produced as a result of such ionization proceed to the anode and positive ions to the cathode. The relatively large mass of the positive ions can cause damage to the thin film photocathode coating under relatively high voltage conditions. High illumination levels give rise to large numbers of photoelectrons and these, in turn, under relatively high voltage conditions, to large numbers of positive ions and resultant cathode destruction as a consequence of positive ion bombardment.

Another problem with high voltage and illumination levels is that they give rise to space charge, which causes detector response to be nonlinear [6.15].

Vacuum photodiodes exhibit NEP typically on the order of 10^{-12} to 10^{-14} $W \cdot Hz^{-1/2}$, and time responses of nanosecond order.

Gas-filled photodiodes are similar to vacuum phototubes except that they contain a low-pressure ($\cong$1 to 50 torr) gas, usually inert (neon or argon) so as to prevent interaction with electrodes. The purpose of the gas is to increase the number of ionizing collisions between photoelectrons and gas atoms, thus providing internal signal amplification. Typically, for such devices $G \cong 10$ at rated voltages. Especially in such devices, with much higher gas atom concentration, it is imperative not to use voltages that are too high so as to prolong photocathode life. Electronic bandwidth is usually limited to about 15 kHz as a result of the long transit times associated with ionizing collision processes followed by secondary emission of electrons from the cathode deriving from bombardment by the slow positive ions. At increased voltages and/or small load resistors (short RC values) time response can be shortened by orders of magnitude [6.13, 6.14]. These devices are often used to detect audio signals in motion pictures, thus synchronizing the audio with the visual signal.

Photomultipliers are perhaps the most sensitive detectors known. They exhibit internal signal amplification on the order of $G \cong 10^5$ to 10^9, thereby making Johnson noise negligible as compared to quantum noise [see Eq. (6.1.8)]. Such high gain is accomplished by including additional electrodes, called dynodes, between anode and cathode. The dynodes are coated so that they have a high secondary emission coefficient. Consequently for each incident photoelectron several secondary electrons are released. A high-voltage power supply maintains the photocathode at a negative high voltage, usually in the range of 1 to 3 kV, and a voltage divider chain is used to ensure that successive dynodes lie at successively higher potentials. A single photoelectron leaving the photocathode is thus accelerated to the first dynode where it causes the ejection of several secondary electrons, each of which is accelerated by the bias field toward the second dynode where additional electrons are ejected by the same process. Additional amplification takes place similarly at subsequent stages. The gain G of a photomultiplier is the average number of electrons leaving the anode for each photoelectron leaving the photocathode. If δ is the average number of secondary electrons leaving dynode surface per incident primary electron, and if there are N dynodes, then internal gain is

$$G = \delta^N \tag{6.2.17}$$

where δ increases with kinetic energy of primary electrons and is typically on the order of 3 to 5. It generally behaves according to a Poisson distribution. The higher the potential difference between dynodes, the higher the value of δ. Typically, a photomultiplier contains 10 to 15 dynodes, and about 100 to 200 V is maintained between successive electrodes. For example, if $\delta = 4$ and $N = 10$, then $G = 1.05 \times 10^6$. However, gain for the noise is somewhat higher, so that SNR is reduced by the factor $(\delta - 1)/\delta$ from its quantum noise limit (see Exercise 6.14). The increased gain for noise is a result of shot noise fluctuations in secondary-emission rates from the dynodes in addition to those of the cathode. However, without the high gain, in many cases the signal would be swamped by Johnson noise, so that the SNR reduction is generally worthwhile.

NEP for photomultipliers is typically on the order of10^{-16} W·Hz$^{1/2}$, and rise time is of nanosecond order.

Because of the high photomultiplier gain, it is very important to limit illumination levels so as to protect the photocathode from positive ion bombardment resulting from ionizing collisions of photoelectrons with residual gas atoms. Consequently, photomultipliers are used with special housings which limit received light to one direction only.

Photoemissive surfaces are applied frequently in imaging. Examples are image intensifiers or converters. Simple devices consist of a semitransparent photoemissive cathode located in the image plane formed by the optics. The optical image then gives rise to a photoelectron image at the output surface of the photocathode. The photoelectrons are propelled by a strong bias field (tens of kilovolts) and focused with electron optics onto a phosphor screen. The kinetic energy of the absorbed electrons is such that for each electron the phosphor produces many photons, the *photon gain* often reaching orders of magnitude. Spectral response at the input is that of the photocathode. Often near-infrared photoemissive materials are used, such as S-1. Phosphor type is selected so that spectral response at output matches closely that of the human visual system (scotopic in particular for night use). In this way a near-infrared image, for example, is converted via the photocathode with high efficiency to a visible one (usually green).

Image tubes use an insulating target electrode on which the photoelectrons are accelerated. This produces a charge image, which is read out with an electron beam.

Another type of electron tube detector that seems promising, particularly for near UV wavelengths, involves a different mechanism of detection instead of the photoelectric effect. In these devices the UV radiation is detected by an enclosed gas, rather than by a photocathode. In such devices"ordinary"cathodes are used instead of photocathodes, that is, cathodes with relatively high photoelectric work functions are used. The purpose of such cathodes is to limit electron emission to primary electrons, rather than photoelectrons or secondary emission electrons which are undesired. The primary electrons are emitted as a result of bias and not photons. The tubes are filled with a low-pressure gas, and voltage is set to maximize the rate of excitation rather than ionization collisions of electrons with gas atoms.

The electric field is affected strongly by electrode separation, electrode shape, and gas pressure, all of which affect the rate of excitation collisions. Most such collisions result in excitation to the low-lying metastable levels, which have long lifetimes. Exposure of the electron tube gas cell to light, particularly UV radiation, results in *photoionization of excited states*, thereby increasing current in the bias circuit as a consequence of photon absorption. The quantum structure of the gas permits spectral response selection according to gas selection, with responsivity improving with wavelength decrease. This leads to near solar blind UV response [6.13]. Very high responsivities have been reported, with very high gain resulting from subsequent ionization collisions of photoionization electrons with neutral gas atoms. Because the gas is not broken down, internal noise is very low, leading to very low values of NEP limited by measurement instrumentation rather than the detector itself [6.13–6.18]. UV sensitivities are between those of photomultipliers and photodiodes.

6.2.1.2 Semiconductor Detectors

Semiconductor devices may be simple photoconducting slabs of material, or photodiodes. In both cases, a photon of energy $h\nu$ greater than band-gap energy E_g can be absorbed, thus exciting an electron from the valance to conduction band or a hole from conduction to valence band. This implies a long wavelength or high frequency cutoff given by

$$\nu_c = E_g/h \tag{6.2.18a}$$

$$\lambda_c = hc/E_g \tag{6.2.18b}$$

where ν_c determines the lower limit to the integral in Eq. (6.2.2a) for quantum detectors, particularly at infrared wavelengths. In many ways, the unoccupied state remaining in the valence band can be considered a particle of charge $+q$ (where electron charge is $-q$). This unoccupied electron state is known as a *hole*. Electrons and holes drift under influence of an applied electric field, giving rise to a current. Since they drift in opposite directions, they both produce a current in the same direction.

Semiconductor devices can be intrinsic, extrinsic, or free carriers. The first two depend on the energy gap E_g under consideration. *Intrinsic* materials are of relatively low impurity concentrations. Consequently, E_g is essentially the entire forbidden energy gap between the valence and conducting bands. This energy gap is relatively large, leading to short cutoff wavelengths λ_c. For example, at room temperature λ_c is about 1.1 and 1.8 μm for silicon and germanium, respectively. These materials, in an intrinsic state, are incapable of detecting thermal infrared radiation. However, materials such as PbTe, PbSe, and InSb exhibit cutoff wavelengths of 4.1, 4.6, and 5.4 μm, respectively, so they are useful in the shortwave 3- to 5-μm wavelength thermal imaging band.

Extrinsic materials have high impurity levels. This permits photoexcitation of charge carriers from a bound impurity state *within* the forbidden energy gap (such as donor or acceptor levels E_v and E_a, respectively) to a free conduction state (such as the conduction band E_c for electrons or the valence band E_v for holes, respectively). In this case, the energy gap E_c-E_d or E_a-E_v involved in the transition is much smaller than the forbidden energy gap E_g, thus increasing λ_c considerably. Typical transition energies are in the range of 0.01 to 0.5 eV. Consequently, at room temperatures such impurities are largely ionized as a result of thermal energy, often leading to significant increases in electrical conductivity and dark current without light being incident on the detector. Therefore, such devices are operated at low temperatures, typically in the 4 to 30°K range, so that ionization from bound to free states is a result of photon absorption rather than ambient thermal energy. Examples of extrinsic devices include Ge doped with Au, Hg, Cd, Cu, Zn, or Ga, which exhibit wavelength cutoffs of approximately, 5, 10, 19, 27, 37, and 120 μm, respectively. D^* is typically on the order of 2×10^{10} cm·Hz$^{1/2}$·W^{-1}.

In general, for light of photon energy greater than appropriate E_g, semiconductor devices are opaque because of photon absorption. However, for incident photons of energy less than E_g, semiconductor materials are transparent, since the photons are not absorbed. For example, intrinsic Ge with a cutoff wavelength of 1.8 μm is opaque to visible light, but is transparent to thermal infrared photons. Hence, Ge is often used to fabricate lenses for use at thermal imaging wavelengths where ordinary glass is opaque.

Comparisons of actual extrinsic semiconductor detectors of normalized detectivities with ideal D^* are shown in Fig. 6.6.

Free carrier devices are fabricated from materials with high carrier mobility, such as InSb [6.12]. Incident photons give rise to intraband transitions within the conduction band itself, changing electron mobility and, consequently, resistance of the material, thereby altering current in the circuit. Such phenomena are most apparent at long wavelengths and at very low temperatures.

Figure 6.8 shows a schematic of a *photoconductor* of dc resistance R_d. The dc output voltage is

$$V_0 = \frac{R_L}{R_d + R_L} V_s \tag{6.2.19}$$

Incident light of radiant power P increases photoconductor conductivity and therefore decreases R_d. The resulting voltage change is

$$\Delta V_0 = -\frac{R_L V_s}{(R_d + R_L)^2} \Delta R_d \tag{6.2.20a}$$

Since R_d is decreased, output voltage and current are increased by incident photon absorption, the current increase being

$$\Delta I = -\frac{V_s}{(R_d + R_L)^2} \Delta R_d \tag{6.2.20b}$$

Time response and internal gain are determined by transit time τ_d through the device and by mean lifetime τ_0 of carriers in their free conduction state [see Eq. (5.1.24)]. For simplicity, consider an *n*-type photoconductor with electron mobility much greater than that of holes. Assume monochromatic radiation of radiant power P is absorbed. Let N be average number of free electrons, and assume ΔN is the contribution to N stemming from photoexcited electrons whose number is much greater than that of thermally excited electrons. Time evolution of ΔN is determined by

$$\Delta N = \frac{\eta P}{h\nu} - \frac{\Delta N}{\tau_0} \tag{6.2.21}$$

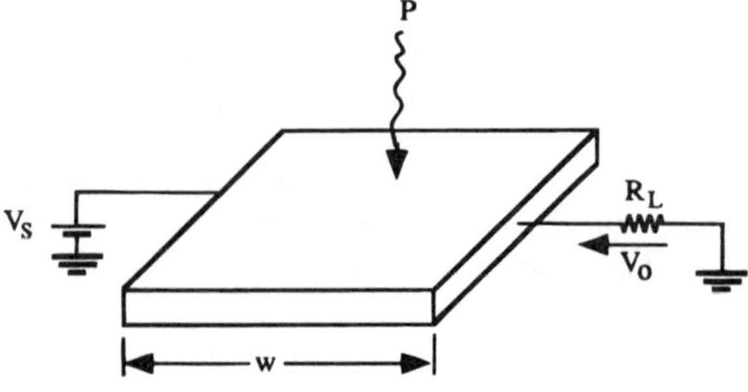

Fig. 6.8 Photoconducting detector circuit.

In the steady state

$$\Delta N = \frac{\eta P \tau_0}{h\nu} \tag{6.2.22}$$

Electron drift velocity magnitude is

$$v_n = \mu_n E = \mu_n V_D / w \tag{6.2.23}$$

where V_D is potential difference over the detector itself and dimension w shown in Fig. 6.8, and μ_n is electron mobility. Signal current is, from the previous equations,

$$\Delta I = \Delta N q \frac{v_n}{w} = \frac{\eta q}{h\nu} P\left(\frac{\tau_0}{\tau_d}\right) \tag{6.2.24}$$

where transit time $\tau_d = w/v_n$. The factor (τ/T) is therefore the fraction of photoconductor length drifted by average photoexcited carrier before recombining. It represents internal signal gain G. Since detector bandwidth B_D is $1/\tau_0$,

$$GB_D = (\tau_0/\tau_d)\tau_0^{-1} = T^{-1} = \frac{v_n}{w} = \frac{\mu_n V_D}{w^2} \tag{6.2.25}$$

At a given bias voltage, gain-bandwidth product is constant. It can be increased by increasing electric field, but for a given field it is constant.

Values of G greater than unity require a flow of more than one electron through the device for each photon absorbed. This is possible if conduction band electrons leave the device at the positive electrode and are replaced by additional electrons entering from the negative electrode. Indeed, gains as large as 1000 have been measured [6.11].

Photoconductive detectors are biased externally and produce changes in conductivity and therefore current under light illumination. *Photovoltaic* detectors, on the other hand, are not necessarily biased, and they produce signal voltages in devices such as *semiconductor photodiodes* as a result of light illumination, which increases free charge for $h\nu > E_g$ and therefore alters built-in potential.

A photovoltaic homojunction consists of a p-type and an n-type region in an intrinsic semiconductor as shown in Fig. 6.9. Light absorbed by the device produces

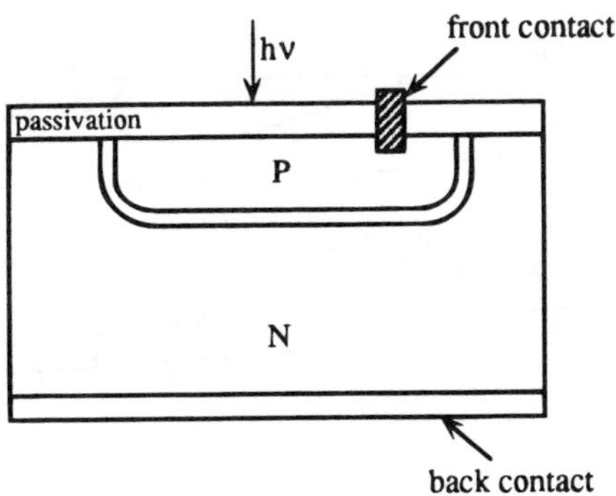

Fig. 6.9 Typical photodiode construction (p and n layers can be reversed). p and n layers separated by depletion layer.

electron-hole pairs. The minority carriers then diffuse to the junction, where they are swept across the depletion region by the junction field, forward biasing the device to produce either an open-circuit voltage or a short circuit current, as shown in Fig. 6.10.

The current through the *p–n* junction can be calculated by summing the four contributions shown in Fig. 6.10. Two contributions result from negatively charged carriers (electrons) and are labeled i_{nd} for diffusion current and i_{ng} for generation current. The diffusion current i_{nd} is made up of those electrons in the conduction band of the *n*-type material that diffuse into the junction with sufficient energy to surmount the potential barrier separating the *n*-type and *p*-type regions. Since the height of this potential barrier decreases with increasing applied voltage V_a, this current is given by

$$i_{nd} = i_{nd,0}e^{qV_a/kT} \tag{6.2.26}$$

where $i_{nd,0}$ denotes the electron diffusion current in the absence of an applied voltage.

The other contribution due to electrons is the generation current i_{ng}. It results from those few electrons within the *p*-type material that are thermally excited from the valence band to the conduction band. If these thermally generated electrons encounter the junction, they are pulled into the *n*-type region, independent of the existence of an applied voltage V_a. There are analogous contributions to the current as a result of the motion of holes. The hole diffusion current is given by

$$i_{pd} = i_{pd,0}e^{qV_a/kT} \tag{6.2.27}$$

and the hole generation current i_{pg} is independent of V_a. The total current through the junction is given by the sum of these contributions

$$i = i_{pd} + i_{nd} - i_{pg} - i_{ng} = (i_{pd,0} + i_{nd,0})e^{qV_a/kT} - (i_{pg} + i_{ng}) \tag{6.2.28}$$

No current can exist at the junction when applied voltage is zero, and thus the two terms in parentheses must be equal. Defiing the saturation current by

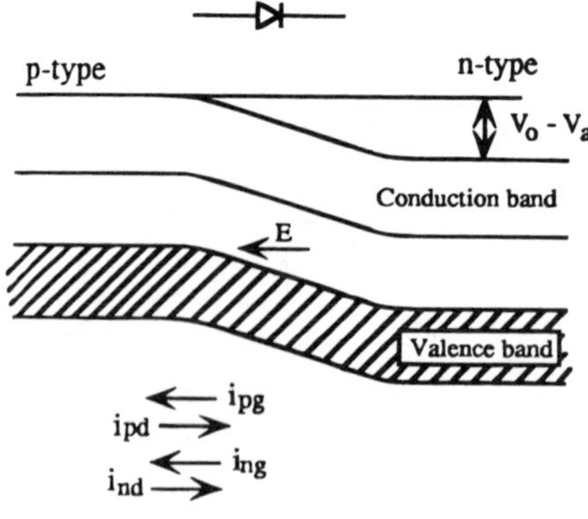

Fig. 6.10 Four contributions to current through a *p-n* junction (after [6.11]). © John Wiley & Sons, Inc.

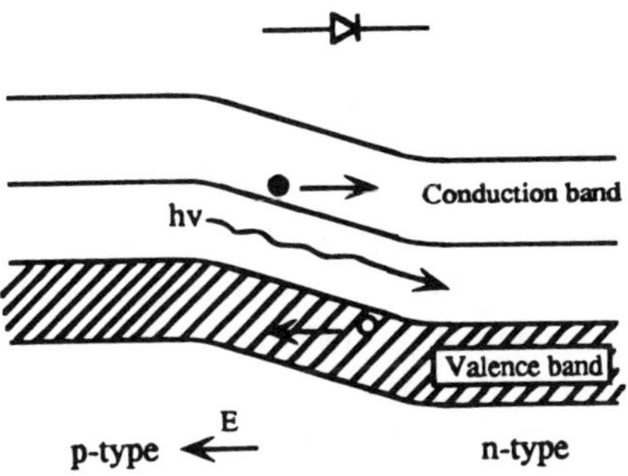

Fig. 6.11 A photon is absorbed within the depletion layer, creating an electron–hole pair.

$$i_{sat} = i_{pg} + i_{ng} = i_{pd,0} + i_{nd,0} \tag{6.2.29}$$

Eq. (6.2.28) can be reexpressed as

$$i = i_{sat}(e^{qV_a/kT} - 1) \tag{6.2.30}$$

The value of i_{sat} depends on the area of the junction and on the carrier mobilities and recombination rates. It is typically 10^{-7} to 10^{-9} A for silicon photodiodes. It is higher for germanium because E_g and V_0 are smaller.

The current-voltage relation given by Eq. (6.2.30) is modified if photoexcited carriers are present in the space-charge region; this change in the electrical characteristics induced by the presence of radiation forms the basis of the use of the *p-n* junction as a radiation detector. If the photon energy $h\nu$ of the incident light is greater than the band-gap energy E_g, the photon can be absorbed within the depletion layer, creating an electron-hole pair. The electron and hole thus created are accelerated in opposite directions by the built-in electric field, and give rise to a current to the left in Fig. 6.11. Since by convention this constitutes a current flow in a negative sense, the current-voltage relation becomes, for incident radiant power P,

$$i = \frac{-\eta q P}{h\nu} + i_{sat}(e^{qV_a/kT} - 1) \tag{6.2.31}$$

Equation (6.2.31) is displayed graphically in Fig. 6.12 for several values of applied radiant powers.

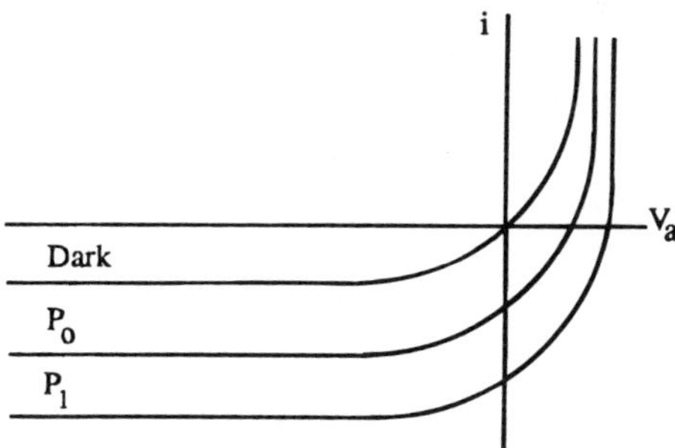

Fig. 6.12 Current-voltage relation for a *p-n* junction photodiode, where radiant power $P_1 > P_0$.

The modification of the current-voltage relation of a p–n junction diode by the presence of light falling onto the diode can be used in several distinct ways to detect radiation.

Reverse bias can be used to increase the volume of the depletion layer and thus the volume in which detection can take place. This increases quantum efficiency. The greater field magnitude, as compared to zero bias, causes detection charge carriers to move faster between device contacts, thus improving speed of response responsivity. The drawback is that bias increases bias current, which acts as dark current.

Photodiodes can also be used with no bias. The short-circuit current or the open-circuit voltage can then be measured; or, more generally, the output voltage or current into some specified load can be measured.

The photocurrent measured at either zero voltage or at a fixed bias voltage varies linearly with optical power, according to responsivity $\eta q/(h\nu)$. However, due to the exponential voltage dependence in Eq. (6.2.31) the open-circuit voltage depends nonlinearly on the absorbed radiant power and is given by [6.11]:

$$V = \frac{kT}{q} \ln\left(1 + \frac{\eta q P}{h\nu i_{sat}}\right) \tag{6.2.32}$$

Since it is usually desirable for a radiation detector to display a linear response, it is conventional to use a semiconductor photodiode as a current source.

Quantum response and responsivity vary strongly with wavelength, as shown typically in Fig. 6.13. Detection takes place within the depletion layer or within a diffusion length away from it shown in Fig. 6.9. If the photon is absorbed further away, although electron–hole pairs may be formed, the minority signal carrier will likely undergo recombination before it reaches the depletion layer where it can be swept across to the other side. If such recombination takes place, there is no addition to current in the circuit or voltage across the device, so such photons are lost to the detection process. Therefore it is important to widen the depletion layer, thereby increasing the volume within the device in which photons can actually be detected and affect the external circuit. One way to widen the depletion layer is to add an i-layer or intrinsic layer between the p and n regions, thereby widening the depletion layer considerably. Such devices are called *p-i-n diodes*. It is also important to minimize photon absorption above and below the depletion layer. In general, more energetic photons are absorbed faster and thus produce electron–hole pairs in the front surface region. This occurs principally at shorter wavelengths, and results in quantum efficiency and responsivity decreases because of surface recombination losses. This significantly limits semiconductor response to ultraviolet radiation, although zinc phosphide photoconductors have exhibited fine ultraviolet response [6.19, 6.20]. For the above reason it is important to limit the front layer (p-layer in Fig. 6.19) depth. Longer wavelength photons are absorbed less rapidly and often may not be absorbed until after they pass through the depletion layer and penetrate well into the other side, thereby decreasing responsivity at longer wavelengths in Fig. 6.13. Finally, the long wavelength limit is reached at λ_c according to Eq. (6.2.18b) where $h\upsilon = E_g$. For wavelengths longer than λ_c, the semiconductor material becomes semitransparent because essentially no photon absorption is taking place. As explained earlier, the value λ_c depends on whether the material is intrinsic or extrinsic.

If correction is made for reflection loss at the front surface of a silicon photodiode, quantum efficiencies very close to 100% can be achieved [6.21]. Peak responsivities

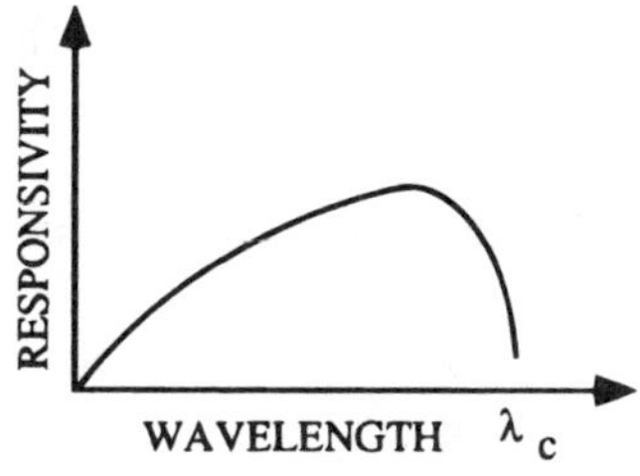

Fig. 6.13 Typical spectral responsivity of semiconductor photodiode.

are on the order of a few hundred mA·W^{-1}. However, one limitation to possible use as an absolute radiometric standard is the effect of environment on photodiode performance as a result of surface effects [6.22, 6.23]. Semiconductor atoms such as Ge or Si form covalent bonds in four directions. Atoms in the bulk can generally form such bonds with identical atoms of the same material. Surface atoms, on the other hand, can form such bonds only on three sides with identical atoms since in the direction external to the surface there are no semiconductor atoms with which to form such bonds. As a result, ambient molecules such as H_2O, O_2, N_2, etc., give rise to weaker covalent bonds with the semiconductor surface atoms in the direction external to the surface. When the ambient changes, so too do the elements chemisorbed or adsorbed in this way to the surface. Such processes change surface charge distributions and give rise to free charge redistribution through diffusion processes. For example, if silicon photodiodes are kept in even modest vacuums on the order of 0.01 torr for a day or two, because some ambient impurities are desorbed the devices exhibit significant quantum efficiency improvement, much better uniformity of response over the detection area, and a broader detection area [6.22, 6.23]. Spectral response is altered, yielding a shift of λ_c to shorter wavelengths because populations in impurity bonds within E_g are diminished. Changes in electrical characteristics involve increases in short-circuit current and open-circuit voltage; the diode I-V curve becomes more ideal (nonlinear). These changes are attributed to desorption in vacuum of previously adsorbed or chemisorbed surface impurities, thus decreasing surface recombination losses and making the front surface more uniform. Loss of charge stemming from the surface impurities gives rise to redistribution of the remaining charge through diffusion, thus affecting depletion layer width. Passivation layers are supposed to prevent such interaction with the ambient environment, but even diodes that have been "hermetically sealed" exhibit such behavior. The larger the photodiode surface area, the more pronounced its sensitivity to ambient environmental effects.

Semiconductor imaging devices such as charge-coupled device (CCD) cameras can also be affected by operation in a vacuum [6.24] such as space.

Ionizing radiation can give rise to photodesorption of such surface impurities, thus resulting in optical and electronic property changes similar to those induced by a vacuum [6.25]. Both vacuum desorption and ionizing radiation should be considered in space applications.

Internal gain in photodiodes is normally unity. *Avalanche photodiodes*, however, exhibit gains of up to several hundred.

When operated at a sufficiently large negative bias voltage (such as kilovolts), photoexcited carriers within the space-charge region of a junction between fairly heavily doped *p*- and *n*-type materials are accelerated to kinetic energies sufficiently large that collisions with neutral atoms can lead to the creation of electron–hole pairs.

These free carriers are also accelerated by high bias to high energies, leading to a large current amplification by this avalanching effect, typically on the order of a few hundred.

The avalanche photodiode (APD) is similar in construction to the ordinary photodiode except that special care is taken to obtain very uniform junctions. This is required because of the strong dependence of G on bias field in the avalanche region. While internal gain causes the signal current power to increase according to $2G^2$, quantum noise current power increases according to G^{2+x}, where excess noise factor x generally varies between 1 and 2. The excess noise stems from randomness of the inelastic scattering collisions which produce the additional electron-hole pairs. The value of G in avalanche photodiodes varies strongly with temperature and with bias. Generally, for small values of G, Johnson noise in Eq. (6.1.8) is larger than quantum noise. As G increases, both signal and quantum noise increase, thus diminishing the effects of Johnson noise and improving SNR. This improvement continues until quantum noise becomes comparable with Johnson noise. Further increases in G cause SNR to actually *decrease* because of the excess noise. (See Exercise 6.15.)

A more general model of the excess APD noise, which includes the gain dependence of the excess noise, is [6.26]

$$G^* = k_{\text{eff}}G + (1 - k_{\text{eff}})(2 - 1/G) \tag{6.2.33}$$

where k_{eff} is the ratio of hole to electron ionization coefficients and is equal to about 0.02 at about 800-nm wavelengths, although values of 0.002 have been achieved.

A new type of vacuum tube detector has been developed that combines both a vacuum photodiode and a large-area avalanche silicon photodiode. The photodiode acts as a detector and the APD provides the gain. These devices produce internal gains as high as 10^6 via a two-stage process in which incoming photons pass through a glass window and generate photoelectrons at the photocathode via the photoelectric effect. A short distance away is the avalanche photodiode. The photocathode voltage is many kilovolts (typically 8 kV) negative relative to the avalanche photodiode, which acts as an anode. The strong field between the electrodes accelerates the photoelectrons emitted from the photocathode and boosts their energy by a factor of about 2000. When these photoelectrons impact the avalanche photodiode anode, secondary electrons are released in the silicon. Because of the high kinetic energy of the photoelectrons impacting the avalanche photodiode, each incoming photoelectron produces about 2000 such secondary electrons. In addition, the large reverse bias of the APD also produces a current amplification via avalanching on the order of 500, thus yielding an overall internal gain on the order of 10^6. These devices are called *vacuum avalanche photodiodes* (VAPDs). They exhibit gain similar to that of photomultipliers. Advantages over photomultipliers include lower noise, a greater dynamic range of linear response, and higher quantum efficiency.

The excess noise factors of VAPDs are typically on the order of 1.1, while those of photomultipliers (Exercise 6.14) are on the order of 1.4 or higher.

Schottky barrier photodiodes composed of platinum silicide have become significant near-infrared detectors in the 1- to 5-μm wavelength range and have led to extensive improvements in the 3- to 5-μm thermal imaging window. Rectification derives from contact potential between metal and insulator. Capacitance is often very small (<0.1 pF) leading to fast time response. Platinum silicide focal plane arrays at liquid nitrogen temperatures exhibit NETD [Eq. (6.2.6)] on the order of 0.02°C, similar to that of HgCdTe photodiodes and extrinsic photoconducting semiconductors used

TABLE 6.3 *Performance and Operation Parameters of Some Intrinsic Semiconductor Detectors (from [6.27])*

Detector Material	Operation Mode	Operation Temp.	Time Constant	Peak Wavelength	Usable Wavelength	D°(500 K) (cm·Hz$^{1/2}$·W^{-1})
Si	PV	300	10^{-2}	0.9	0.7–1.1	~10^{7}
Si	Avalanche	300	~10^{-5}	0.9	0.2–1.1	<2·10^{7}
Ge	PV	300	~10^{-5}	1.5	0.4–1.9	5·10^{9}
PbS	PC	300	>100	2.4	1.1–3.5	1.5·10^{9}
InAs	PV	77	0.5	3.1	1.8–3.8	2·10^{10}
PtSi	Schottky barrier	<90	~10^{3}	0.9	0.8–5.0	2·10^{9}
PbSe	PC	300	1–5	3.9	1–5.0	2·10^{8}
InSb	PV	77	0.1–2	4.9	2–5.5	5·10^{10}
CdHgTe	PV	195	~2	4.5	1–5.5	4·10^{10}
CdHgTe	PC	77	0.1–2	10.5	8–14	2.5·10^{10}
CdHgTe	PV	77	0.5–2	10.5	8–11.5	2·10^{10}
PbSnTe	PV	77	~0.5	11	8–11.5	2·10^{10}

also in thermal imaging. Infrared photons energize electrons from the PtSi layer that then have a probability of tunneling through the PtSi-Si Schottky barrier. Because tunneling probability is an exponential function of photon energy, quantum efficiency decays with wavelength from about 5% at 1.3-μm wavelength to about 0.1% at 5-μm wavelength. The focal plane arrays of these devices are characterized by high detector uniformity and therefore very low spatial noise. This compensates somewhat for the low infrared quantum efficiency.

Because of the low quantum efficiencies, PtSi focal plane arrays are sometimes replaced with InSb photodiode focal plane arrays because the much higher quantum efficiency of the latter can permit greater imaging ranges.

Various intrinsic and extrinsic semiconductor detector properties are summarized in Tables 6.3 and 6.4, respectively [6.27].

6.2.2 Thermal Detectors

Radiant energy absorption by an object gives rise to a temperature increase. Monitoring of this temperature increase yields information about the radiant power absorbed.

TABLE 6.4 *Performance and Operation Parameters of Extrinsic Semiconductor Detectors (from [6.27])*

Material Dopant	Operation Time (K)	Time Constant (ns)	Peak Wavelength (μm)	Usable Wavelength (μm)	D°(500 K) (cm·Hz$^{1/2}$·W^{-1})
Si:Zn	70–110	10–50	2.4	0.9–2.5	8·10^{9}
Si:Te	<120	50–100	3.5	0.9–4.2	2·10^{10}
Si:In	≤50	~100	4.6	0.9–7	1·10^{10}
Si:Mg	<50	15	11.5	0.9–12	5·10^{10}
Si:Ga	<30	~50	15	0.9–17	2.5·10^{10}
Si:Bi	<30	1–10	17.5	0.9–22	2.5·10^{10}
Si:As	<30	~100	24	0.9–28	1·10^{10}
Ge:Cu	4.2	1–10	21	1.5–27	1·10^{10}
Ge:Zn	4.2	0.5–5	37	1.5–45	2.5·10^{10}
Ge:Be	4.2	0.1–10	50	1.5–65	BLIP
Ge:Cd	4.2	1–50	90	1.5–100	BLIP
Ge:In (stressed)	4.2	1–10	110	1.5–240	BLIP

Thermal detectors achieve this by utilizing a material with a strongly temperature-dependent property such as electrical conductivity or thermal expansion. Such devices generally exhibit flat spectral response, being affected by radiant power heating and not by wavelength. However, disadvantages compared to quantum detectors are lower responsivity and much slower time of response. Both factors are critical for imaging systems. The first affects the range at which an object can be seen. The second affects the time exposure and frame rate permissible so that the image appears continuous to the human visual system rather than a set of discrete images.

Some of the thermal detectors are described briefly below. They are used mostly in the 8- to 14-μm wavelength infrared atmospheric window because of the expense of infrared quantum detectors.

A simple *bolometer* consists of a resistor with a very small thermal capacity, whose surface is blackened.

Metal bolometers have been constructed either from thin foils or evaporated layers, such as nickel, bismuth, and antimony. When operated at room temperature, D^* values of $\sim 5 \times 10^9$ cm·Hz$^{1/2}$·W^{-1} have been obtained, with response times of approximately 10 ms [6.12]. Responsivity (voltage) is on the order of 30 to 80 V·W^{-1}.

Thermistor bolometers are fabricated from sintered mixtures of various semiconducting oxides (e.g., cobalt, nickel, and manganese) mounted on a sapphire substrate. Their performance is very similar to that of the metal bolometers but generally they are more robust.

Improved performances can be obtained from cooled bolometers (often to liquid helium temperature) constructed from germanium, silicon and gallium arsenide. Responsivities reach 10^4 V·W^{-1}.

The *Golay cell* is an important thermal detector often used as a standard for other infrared devices. A volume of low-pressure gas such as xenon is heated by incident radiation falling on a thin metal membrane. Part of the wall of the cavity is silvered to form a mirror that deflects light as it expands or contracts. The light beam is directed onto a photocell and the output signal monitored.

The wavelength response is broad, from the visible to the microwave region, but the response time is slow, ~15 ms. However, the main disadvantage is the extreme sensitivity of the device to vibrations which give rise to deflection of the light beam away from the photocell.

On the other hand, the main advantage of the *thermopile* is its ruggedness and high reproducibility. A thermopile consists of more than one *thermocouple*. A thermocouple consists of two dissimilar materials that, when heated, produce voltage across the two open leads (thermovoltaic effect). Thermopiles are fabricated with thin film techniques so as to produce high-density thermocouple junctions and detector arrays. No bias is required, so that Johnson noise predominates.

Original devices were fabricated from two thin metal wires or stripes, such as copper-constantan or manganin-constantan. Typical responsivities are about 0.5 V·W^{-1} with time constants on the order of 10 ms. NEP is about 2×10^{-10} W·Hz$^{-1/2}$ [6.28]. Sensitivity can be improved if two semiconductors form the junction, yielding responsivities of 10 to 50 V·W^{-1} with $D^* \cong 10^9$ cm·Hz$^{1/2}$ [6.12].

One method of fabricating a semiconductor thermopile involves fusing the two materials onto the tips of a pair of gold pins and then completing the circuit by melting a gold foil between the pins. The foil is then blackened to form the receiving area. Typical alloys suitable for such a construction are 33%Te, 32%Ag, ~27%Cu, 7%Se, and 1%S for the positive side and 50%Ag_2Se, 50%Ag_2S for the negative. The devices may be mounted in air or vacuum. The response will be faster but the

TABLE 6.5 *Performance and Operation Parameters of Some Thermal Detectors (from [6.27])*

Detector Type	Operation Temperature (K)	Time Constant (ms)	Usable Wavelength (μm)	D°(500 K) (cm·Hz$^{1/2}$ ·W^{-1})
Thermocouples	300	10–100	0.8–20	$\leq 1.5 \cdot 10^9$
Thermopiles	300	0.5–400	0.8–35	$< 5 \cdot 10^9$
Bolometers	300	1–100	0.8–40	$< 5 \cdot 10^9$
Superconductor-B	16	~0.5	1–1000	$\sim 10^{10}$
Pyroelectric	300	0.1–100	0.6–35	$\leq 8 \cdot 10^8$

sensitivity lower if the detector is used in air. Detectivities of 13×10^9 cm·Hz$^{1/2}$·W^{-1} can be achieved with these devices [6.12].

The thermal detector most extensively used in imaging is the *pyroelectric* detector. Arrays of miniaturized devices can be fabricated into *pyroelectric vidicons*. The pyroelectric effect is observed in insulating materials such as triglycine sulfate (TGS), lithium tantalate, strontium barium niobate, and polyvinylidene fluoride. At room temperature, the magnitude of the polarization is often found to be strongly temperature dependent. Thus, as the temperature T of the detector is varied, the electric dipole moment of the crystal must change, leading to the motion of bound charge. If electrodes are placed on the surfaces of the crystal, the motion of bound charge within the crystal can induce current through an external circuit. The magnitude of this current varies according to dT/dt. Thus, a pyroelectric detector does not respond to constant input power, since only the time derivative of the detector's temperature leads to a signal current.

This rate of change dependence permits much faster time response from that of bolometers or thermopiles. Furthermore, the dT/dt dependence means that pyroelectric detectors respond only to chopped or pulsed radiation; constant background radiation has no effect.

The performance of these devices at low frequencies is within an order of magnitude of that of an ideal thermal detector, with a D^* of about 2×10^9 cm·Hz$^{1/2}$·W^{-1} at 10 Hz for a 10^9 W load resistor. One other important property of the pyroelectric sensor compared to most other thermal detectors is that with suitable circuitry the response time can be less than 1 μs. When a very low load resistor such as 50Ω is shunted across the detector, nanosecond response time can be obtained, but at the expense of responsivity, which drops then to the order of 10^{-4} V·W^{-1}.

The ability to operate these devices at room temperature with a good detectivity, and their relative cheapness, has meant that they are gradually replacing other thermal detectors.

Various thermal detector properties are summarized in Table 6.5.

REFERENCES

6.1. D. E. De Lange, "Optical heterodyne detection," *IEEE Spectrum*, Vol. 5, October 1968, pp. 77–85.

6.2. N. S. Kopeika, "General wavelength dependence of imaging through the atmosphere," *Appl. Opt.*, Vol. 20, May 1, 1981, pp. 1532–1536.

6.3. L. G. Kazovsky and N. S. Kopeika, "Heterodyne detection through rain, snow, and turbid media: effective receiver size at optical through millimeter wavelengths," *Appl. Opt.*, Vol. 22(5), 1983, pp. 706–710.

6.4. D. Sadot and N. S. Kopeika,"Imaging through the atmosphere: practical instrumentation-based theory and verification of aerosol MTF," *J. Opt. Soc. Am. A*, Vol. 10(1), 1993, pp. 172–179.

6.5. M. Toida, M. Kondo, T. Ichimura, and H. Inaba, "Two-dimensional coherent detection imaging in multiple scattering media based on the directional resolution capability of the optical heterodyne method," *Appl. Phys.*, Vol. B52, 1991, pp. 391–394.

6.6. M. Toida, I. Ichimura, and H. Inaba, "The first demonstration of laser computed tomography achieved by coherent detection imaging for biomedical applications," *IEICE Trans.*, Vol. E74, June 1991, pp. 1692–1694.

6.7. H. Raidt and D. H. Höhn, "Instantaneous intensity distribution in a focused laser beam at 0.63 μm and 10.6 μm propagating through the atmosphere," *Appl. Opt.*, Vol. 14, November 1975, pp. 2747–2749.

6.8. A. E. Siegman, "The antenna properties of optical heterodyne receivers," *Proc. IEEE*, Vol. 54, October 1966, pp. 1350–1356.

6.9. D. L. Fried,"Optical heterodyne detection of an atmospherically distorted wavelength," *Proc. IEEE*, Vol. 55, January 1967, pp. 57–67.

6.10. Th. de Graauw and H. van de Stadt, "Infrared heterodyne detection of the moon, planets, and stars at 10 μm," *Nature*, Vol. 246, December 3, 1973, pp. 73–75.

6.11. R. W. Boyd, *Radiometry and the Detection of Optical Radiation*, Wiley, New York, 1983. Reprinted by permission of John Wiley & Sons, Inc.

6.12. P. N. J. Dennis, "Photodetectors," *Sci. Prog. Oxf.*, Vol. 66, 1979, pp. 267–294. Reprinted with permission from Science Reviews Ltd.

6.13. M. Cohen and N. S. Kopeika, "A near UV envelope detector in the prebreakdown regime based on photoionization of excited gas atoms," *Meas. Sci. Technol.*, Vol. 5, May 1994, pp. 540–547.

6.14. N. S. Kopeika and G. Eytan, "Photoionization of excited atoms in gas-filled photodiodes: improved detection with microsecond-order risetimes," *IEEE Trans. Plasma Sci.*, Vol. PS-6, 1978, pp. 139–157.

6.15. N. S. Kopeika, T. Karcher, and C. S. Ih, "Gas discharge response to light: dependence of linearity on space charge for optogalvanic and excited-atom photoionization signals," *Appl. Opt.*, Vol. 18, 1979, pp. 3513–3516.

6.16. N. S. Kopeika, R. Gellman, and A. P. Kushelevsky, "Improved detection of ultraviolet radiation with gas-filled phototubes through photoionization of excited atoms," *Appl. Opt.*, Vol. 16, 1977, pp. 2470–2476.

6.17. N. S. Kopeika, R. Shuker, Y. Yerachmiel, Y. Gabai, and C. S. Ih, "Observation of Cooper minima in excited S-state photoionization cross sections in neon and argon," *Phys. Rev. A.*, Vol. 28, 1983, pp. 1517–1527.

6.18. N. Yackerson and N. S. Kopeika, "Dynamic nonoptogalvanic signal polarity and magnitude in prebreakdown gas discharges," *IEEE J. Quantum Electron.*, Vol. AE-21, 1985, pp. 1728–1735.

6.19. K. Irwin, N. S. Kopeika, and R. G. Hunsperger, "Ultraviolet photoconductive detectors in Zn_3P_2," *Electron. Lett.* Vol. 15, 1979, p. 718.

6.20. S. Hava, "Surface effects and grain-boundary domination in thin-film Zn_3P_2 photoconductivity, *J. Appl. Phys.*, Vol. 54, 1986, pp. 4097–4102.

6.21. R. Korde and J. Geist, "Quantum efficiency stability of silicon photodiodes," *Appl. Opt.*, Vol. 26, 1987, pp. 5284–5290.

6.22. N. S. Kopeika, S. Hava, I. Hirsh, and E. Hazout, "Significant photodiode quantum efficiency improvement and spectral response alteration through surface effects in vacuum," *IEEE Trans. Electron. Dev.*, Vol. ED-31, 1984, pp. 1198–1206.

6.23. S. Hava, "Comparison of vacuum surface effects on commercial photodiodes spatial response uniformity via a computerized laser scanning system," *Appl. Opt.*, Vol. 26, 1987, pp. 121–126.

6.24. S. Hava,"Improvement of dark signal and near IR response of imaging CCD under vacuum surface operation," *Solid State Electron.*, Vol. 32(11), 1989, pp. 1048–1050.

6.25. S. Hava and N. S. Kopeika, "Short wavelength responsivity improvement and long wavelength responsivity degradation in photodiodes as a result of gamma irradiation," *Opt. Eng.*, Vol. 26, 1987, pp. 959–962.

6.26. P. P. Webbs, R. J. McIntyre, and J. Conradi, "Properties of avalanche photodiodes," *RCA Rev.*, Vol. 35(6), 1974, pp. 234–278.

6.27. K. J. Stahl, "Infrared detectors" in 1991 *Photonics Design and Applications Handbook*, 119, pp. 17–171. Reprinted with permission from Laurin Publishing Co, Inc.

6.28. C. L. Wyatt, *Electro-Optical System Design for Information Processing*, McGraw-Hill, New York, 1991.

EXERCISES

6.1 From Eq. (6.1.6) determine the time-averaged signal current for direct detection.

6.2 What is the NEP for a detector limited by noise fluctuations deriving from dark current? Assume direct detection of a time-varying signal so that the direct current i_d is removed with a capacitor.

6.3 What is the NEP for a detector limited by Johnson noise? Assume direct detection.

6.4 Assume $P_s^2(t) = A^2(1 + m \cos \omega_s t)^2$. What is the resulting mean square signal current and what portion is to be used for SNR in envelope detection?

6.5 (a) A Ge photodetector is reverse-biased to 0.1 μA with a 1-kΩ load resistor. Assume negligible background radiation. For nanowatt pulses at 1 μm, what is the SNR in direct detection? Which is the dominant noise? Assume 50% quantum efficiency at that wavelength, a 1-kHz electronic bandwidth, unity internal signal amplification, and 300 K noise temperature. (b) If $G = 100$, what are the SNR and the dominant noise?

6.6 (a) Consider a weak HeNe laser ($\lambda = 633$ nm) signal of 10^{-10} W modulation power and 10-MHz modulation bandwidth to be detected by a detector with 50% quantum efficiency at that wavelength. Assume a 1-kΩ load resistor and that $G = 1$. If 300 K Johnson noise is dominant, find SNR for direct detection. (b) Assuming a 1 mW local oscillation power, determine the SNR for heterodyne detection.

6.7 A radiant power signal $P_s(t) = 10^{-8}\,(1 + m \sin 2\pi \times 10^8 t)$ at 1 μm wavelength is incident on a 100 cm^2 detector with 0.1 A·W^{-1} responsivity and a dark current-limited NEP of 10^{-14} W · Hz$^{-1/2}$ at this wavelength. If $m = 0.5$, what is SNR? Assume a 100-MHz baseband bandwidth. Repeat for $m = 1$ and $m = 0.1$.

6.8 Show that homodyne receivers can provide SNR 3 dB better than those obtainable in heterodyne detection.

6.9 For 50% detector quantum efficiency, what NEP is achievable in heterodyne detection at 0.8-μm wavelength? What is the minimum value of P_s required for a SNR of 10 dB, assuming a 10 kHz baseband bandwidth?

6.10 Consider two plane waves incident on a coherent receiver but from different directions. For simplicity, assume one wave is incident normal to the receiver, and the other at an angle θ. Assume that the center of each beam is incident to the same point on the receiver. Let that point be the origin. Develop an expression for the value of radial coordinate in the plane of the receiver at which a phase difference of π radians occurs relative to that at the origin. For a detector radius of 1 mm, how must θ be limited so as to prevent IF current cancellation? Assume $\lambda = 0.633$-μm wavelength (HeNe laser).

6.11 We want to detect an 800 K blackbody source via heterodyne detection. Compare SNR with and without an integrator assuming an ideal detector and optimum conversion. Assume $\lambda = 10$ μm, an electronics bandwidth of 1 MHz, and an integration time of 1 s in the comparison.

6.12 Beamsplitters in an optical heterodyne receiver have identical reflectances ρ and transmissions τ'. Assume no absorption losses. For given signal and local oscillator radiant powers, what should be the values of ρ and τ' so as to maximize detected IF current?

6.13 Consider a vacuum photodiode with parallel-plane electrodes of separation 5 mm under 250-V bias. Assume an S-20 photocathode. Compare electron transit time and response per absorbed photon for 0.80- and 0.40-μm wavelengths.

6.14 Show that in photomultipliers quantum-limited SNR is reduced by $(\delta - 1)/\delta$ from its unity internal gain value.

6.15 What is NEP for an avalanche photodiode with optimum gain?

Part Three

THE OPTICAL TRANSFER FUNCTION

CHAPTER

7

Diffraction

Laws governing image formation were presented in Chapter 2. They derive from geometrical optics in which, according to the Eikonal equation, diffraction is neglected. However, diffraction at the *edges* of apertures such as lenses, mirrors, and other optical elements causes deflections of light rays in directions different from those in accordance with geometrical optics, thereby giving rise to image blur. Diffraction effects are an example of *physical optics*, in which light is treated as waves rather than rays.

In this chapter diffraction effects are considered in general, particularly Fraunhofer diffraction, which can be expressed in the form of spatially varying Fourier transforms. This is fundamental to optical data processing, the basic elements of which are introduced here. Chapter 8 then considers the effects of diffraction on image quality, expressed in the form of the optical transfer functions deriving from the Fourier optics described here.

According to the ray theory of geometrical optics discussed in Chapter 2, light rays in a homogeneous, isotropic medium travel in straight lines. Opaque objects and screens are therefore expected to cast sharp shadows when placed in the path of a bundle of rays. In reality, however, this is not so. A careful inspection of shadow regions reveals systems of fringes, that is, patterns of light and dark lines. The phenomenon responsible for this behavior, not in accordance with Snell's law, is known as *diffraction*.

Diffraction lies at the basis of physical optics and plays a major role in determining resolution. Since considerable difficulties confront a general vector formulation and treatment of diffraction when the electromagnetic field is characterized fully by the electric and magnetic field vectors, rigorous solutions have been obtained for only a few cases of obstacles possessing simple geometries. Fortunately, however, at optical wavelengths a simple scalar diffraction theory can be developed which produces results that agree well with observations in regions removed a few wavelengths or more from the diffracting boundaries. In this chapter the fundamentals of scalar diffraction theory are developed and results are used to demonstrate some important behavior of electro-optical systems caused by diffraction effects.

7.1 THE FUNDAMENTAL EQUATION OF SCALAR DIFFRACTION

As a consequence of ray theory the field vectors $\overline{E}$ and $\overline{H}$ at optical frequencies are transverse to the diffraction of power flow. When field vectors are linearly polarized in one direction, their amplitudes can be represented by

$$V(x, y, z, t) = \text{Re}\{U(x, y, z)\}e^{j\omega t} \tag{7.1.1}$$

where U is a complex scalar function of position (i.e., a phasor) which satisfies the time-independent Helmholtz or wave equation

$$(\nabla^2 + k^2)U = 0$$

where $k = \omega/v$ is again the magnitude of the wave vector. It must be obeyed by the complex amplitude of any monochromatic TEM optical disturbance traveling through free space. The term V represents either E or H, and U is the corresponding complex field amplitude (CFA).

Given a number of sources generating fields as depicted in Fig. 7.1, the total field or intensity at an observation point P can be obtained by summing the individual contributions of each source. To show this, we assume that U possesses continuous first and second partial derivatives on the surface S and within it, and that the volume v enclosed within S is source free. Next another auxiliary complex scalar function G satisfying the wave equation and also possessing single-valued continuous first and second partial derivatives on S and within is considered.

G represents spatial impulse response and is referred to as *Green's function.* Accordingly,

$$(\nabla^2 + k^2)G = -4\pi\delta(r - r_0) \tag{7.1.2}$$

where $\delta(r - r_0)$ is an impulse source located at r_0 outside S. Next, the wave equation is multiplied by G:

$$G\nabla^2 U + Gk^2 U = 0 \tag{7.1.3}$$

and Eq. (7.1.2) is multiplied by U:

$$U\nabla^2 G + Gk^2 U = -4\pi U\delta(r - r_0) \tag{7.1.4}$$

The last equation is subtracted from Eq. (7.1.3), yielding

$$G\nabla^2 U - U\nabla^2 G = 4\pi U\delta(r - r_0) \tag{7.1.5}$$

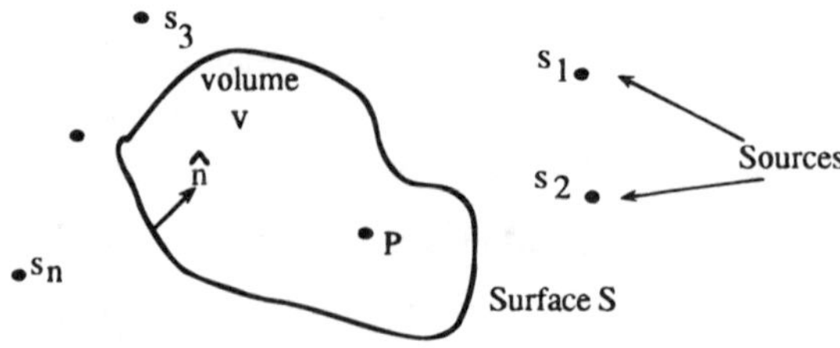

Fig. 7.1 Geometry for calculating total optical disturbances at a point P.

This result is integrated over volume v yielding

$$\int_v (G\nabla^2 U - U\nabla^2 G)\, dv = 0 \tag{7.1.6}$$

since r_0 is zero outside v. Now, by vector identity

$$G\nabla^2 U - U\nabla^2 G = \nabla(G\nabla U - U\Delta G) \tag{7.1.7}$$

When the right-hand side of Eq. (7.1.7) is substituted into Eq. (7.1.6), the divergence equation can be used to convert Eq. (7.1.6) into an integral over the closed surface S:

$$\oint_s \left(U \frac{\partial G}{\partial n} - G \frac{\partial U}{\partial n} \right) dS = 0 \tag{7.1.8}$$

where $\hat{n}$ is inward normal and $\partial/\partial n$ denotes differentiation along the inward normal.

For three-dimensional propagation, the solution to Eq. (7.1.2) is

$$G(r) = \exp(\pm jkr)/r \tag{7.1.9}$$

which is recognized as describing propagation of a spherical wave from Eq. (1.3.40) in Chapter 1, assuming zero attenuation. A condition necessary for the Green's function, Eq. (7.1.9), is the Sommerfeld radiation condition:

$$\lim_{r\to\infty} r \left(\frac{\partial U}{\partial n} - jkU \right) = 0$$

the validity of which is considered later. Equation (7.1.9) must now be substituted into Eq. (7.1.8). However, it is clear from this choice that G has a singularity at $r = 0$ that violates the requirement that G be continuous within S. To avoid this problem, P is surrounded with a sphere of radius ρ and surface S' as shown in Fig. 7.2. Consider now $S + S'$ as the surface enclosing v, allowing $\rho \to 0$ later on in

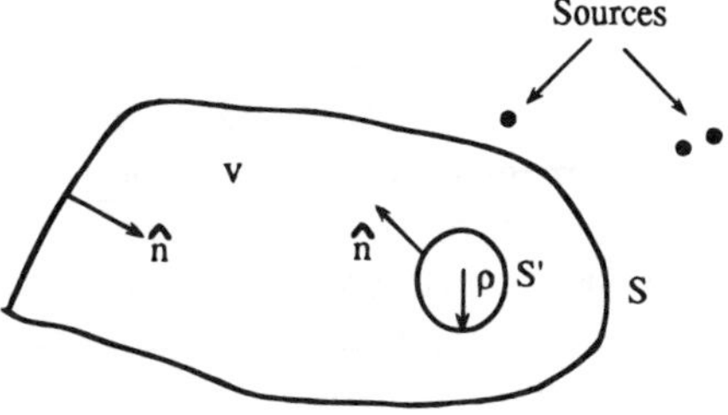

Fig. 7.2 Geometry for treating the singularity of G' at P.

order to obtain the required results, that is, the integration of Eq. (7.1.8), which may now be written in the following form:

$$\oint_s \left[U \frac{\partial}{\partial n}\left(\frac{e^{-jkr}}{r}\right) - \frac{e^{-jkr}}{r}\frac{\partial U}{\partial n} \right] dS + \oint_{s'} \left[\frac{\partial}{\partial n}\left(\frac{e^{-jkr}}{r}\frac{\partial U}{\partial n}\right) \right] dS' = 0 \quad (7.1.10)$$

where the forward rather than backward propagation sign in Eq. (7.1.9) is used. Since S' is a sphere with center at P the distance ρ from this center to any point on S' coincides with the direction of the unit normal to S'. Therefore, $\partial/\partial n = \partial/\partial r$ on S', and Eq. (7.1.10) becomes

$$\oint_s \left[U \frac{\partial}{\partial n}\left(\frac{e^{-jkr}}{r}\right) - \frac{e^{-jkr}}{r}\frac{\partial U}{\partial n} \right] dS = - \oint_{\substack{\text{sphere}\\ \text{of radius}\\ r=\rho}} \left[U\left(\frac{-jk}{\rho} - \frac{1}{\rho^2}\right)e^{-jk\rho} - \frac{e^{-jk\rho}}{\rho} - \frac{e^{-jk\rho}}{\rho}\frac{\partial U}{\partial n} \right] \rho^2\, d\Omega \quad (7.1.11)$$

where $d\Omega = \dfrac{dS}{\rho^2}$ is the element of solid angle subtended by dS at P.

As $\rho \to 0$, Eq. (7.1.11) becomes

$$\oint_s \left[U \frac{\partial}{\partial n}\left(\frac{e^{-jkr}}{r}\right) - \frac{e^{-jkr}}{r}\frac{\partial U}{\partial n} \right] dS = \lim_{\rho\to 0} \oint_{\substack{\text{sphere}\\ \text{of radius } r=\rho}} U\, d\Omega = 4\pi U(P)$$

Hence,

$$U(P) = \frac{1}{4\pi} \oint_s \left[U \frac{\partial}{\partial n}\left(\frac{e^{-jkr}}{r}\right) - \frac{e^{-jkr}}{r}\frac{\partial U}{\partial n} \right] dS \quad (7.1.12a)$$

which can also be written in an alternative form:

$$U(P) = \frac{1}{4\pi} \oint_s \left[U\nabla\left(\frac{e^{-jkr}}{r}\right) - \frac{e^{-jkr}}{r}\nabla U \right] \cdot \hat{n}\, dS \quad (7.1.12b)$$

since $\partial U/\partial n = \Delta U \cdot \hat{n}$. Eqs. (7.1.12) are forms of the *inhomogeneous reduced wave equation*. They are also two forms of the *Helmholtz-Kirchhoff integral theorem*, which is the basic tool used in the treatment of scalar diffraction. The Helmholtz-Kirchhoff integral theorem yields the field magnitude U. With this parameter, the light irradiance (which, on the basis of the Poynting vector, is proportional to $U \cdot U^*$) can be evaluated at any point inside an arbitrary source-free closed surface S from a knowledge of U and $\partial U/\partial n = \nabla U \cdot \hat{n}$ on the surface S.

In the study of diffraction phenomena we are concerned mostly with the passage of optical radiation through apertures of various shapes in opaque screens as demonstrated in Fig. 7.3, where the source of radiation is to the left of the screen and outside the closed surface comprised of S_1, S_2 and S_3.

It is advantageous then to choose the surface of integration in a fashion that will render evaluation of the integral (7.1.12) as easy as possible. Thus, a portion of S is to be a surface S_1 bounded by the aperture. For a plane screen, S_1 is a plane surface; obviously, if the screen is a part of a spherical surface the choice of S_1 to coincide with the spherical surface facilitates the integration in that case. A second portion of S is S_2, which coincides with a portion of an infinitely long screen. Finally, the surface $S_1 + S_2$ is closed with S_3, a portion of a sphere of radius R centered at the observation point P as shown in Fig. 7.3.

To proceed further, it is necessary to postulate some boundary conditions. In this regard we follow the assumptions of Kirchhoff:

1. The perturbation of the incident field by the aperture is negligible. This is equivalent to saying that, on S_1, both U and $\partial U/\partial n$ have values equal to those of the incident wave in the absence of the screen. This assumption is reasonable for all regions of an aperture with dimensions much greater than the wavelength except near the boundary of the aperture. The reason is that diffraction takes place primarily only within a wavelength or so of the aperture edge. (Apertures of dimensions much less than wavelength do not pass such radiation.)
2. Since the screen is opaque U and $\partial U/\partial n$ vanish on S_2 except near the boundary of the aperture.
3. Allowing $R \to \infty$ relaxing the strict monochromaticity of the radiation U and $\partial U/\partial n$ can be made arbitrarily small. By allowing $R \to \infty$, U and $\partial U/\partial n$ are made smaller on S_3; however, S_3 is increased so that the integral over S_3 does not necessarily vanish. The argument here is that by allowing the radiation to start at some time $t = 0$ (i.e., we are not dealing with strictly monochromatic radiation), we can assume that R is large enough so that the

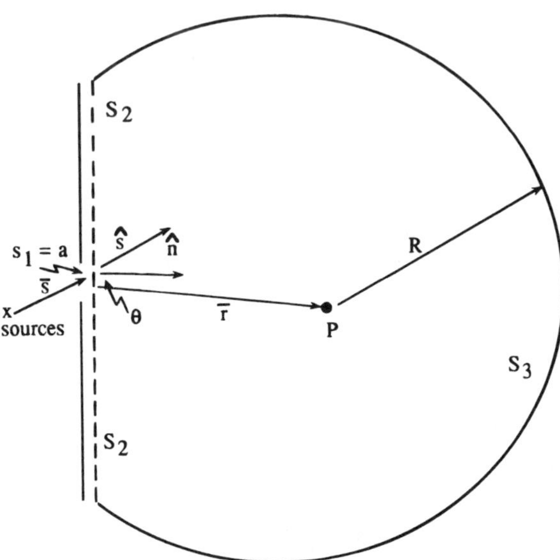

Fig. 7.3 Geometry of a typical situation in which diffraction is evident.

field cannot propagate over the distance R and reach S_3 in the time we are concerned with calculating $U(P)$.

A more rigorous argument showing the vanishing contribution of S_3 to Eq. (7.1.12) and one that eliminates the inconsistency of relaxing monochromaticity required by assumption 3 is the following. On S_3

$$G = \exp(-jkR)/R$$

Therefore,

$$\frac{\partial G}{\partial n} = -\frac{\partial G}{\partial R} = -\frac{\partial}{\partial R}\left(\frac{e^{-jkR}}{R}\right) = \frac{jke^{-jkR}}{R} + \frac{e^{-jkR}}{R^2}$$
$$= \left(jk + \frac{1}{R}\right)\frac{e^{-jkR}}{R} = \left(jk + \frac{1}{R}\right)G$$

and, as a result,

$$\lim_{R\to\infty} \frac{\partial G}{\partial n} = jkG$$

Thus in Eq. (7.1.12) the contribution of S_3 becomes,

$$I_3 = \int_{s3}\left(U\frac{\partial G}{\partial n} - \frac{G\partial U}{\partial n}\right) dS = \int_{s3}\left(UjkG - G\frac{\partial U}{\partial n}\right) dS$$
$$= \int_{s3} GR\left(jkU - \frac{\partial U}{\partial n}\right) r\, d\Omega$$

since on S_3, $dS = R^2\, d\Omega$. Now on S_3, $|RG| = |e^{-jkR}| = 1$. Thus, integral I_3 will vanish if

$$\lim_{R\to\infty}\left(jkU - \frac{\partial U}{\partial n}\right) = 0$$

thus satisfying the *Sommerfeld radiation condition* after Eq. (7.1.9). To satisfy this condition U must decay with R at least as a diverging spherical wave. In practice, the radiation illuminating the aperture S_1 in Fig. 7.3 consists of spherical wave or a linear combination of such waves. Thus, the Sommerfeld radiation condition is satisfied in practice and the contribution I_3 to the integral is zero.

In view of these assumptions, the integration of Eqs. (7.1.12) can be performed over $S_1 \triangleq a$ (where a stands for aperture) instead of S:

$$U(P) = \frac{1}{4\pi}\int_a\left(U\frac{\partial}{\partial n}\frac{e^{-jkr}}{r} - \frac{e^{-jkr}}{r}\frac{\partial U}{\partial n}\right) dS \tag{7.1.13}$$

The Helmholtz-Kirchhoff theory can be simplified further by applying the principle of optical path learned in Chapter 2 [Eq. (2.1.1)]. Thus, according to Eq. (2.1.1), the field distribution over an area in space such as an aperture can be described by

$$U = A(x, y, z) \exp[-jk_0 S(x, y, z)] \tag{7.1.14}$$

where $S(x, y, z)$ is the "optical path." Using Eq. (7.1.14) in Eq. (7.1.12a), we obtain

$$\begin{aligned}\frac{\partial U}{\partial n}\, \hat{n} \cdot \nabla U &= \hat{n} \cdot [-jk_0\, Ae^{-jk_0 S}\, \nabla S + \nabla A e^{-jk_0 S}] \\ &= -jk_0\, U\hat{n} \cdot \nabla S + \hat{n} \cdot \nabla A\, \frac{U}{A} \\ &= -jk_0\, U\hat{n} \cdot \nabla S + \frac{1}{A}\frac{\partial A}{\hat{\partial} n}\, U \end{aligned} \tag{7.1.15}$$

where $\hat{n}$ is the normal to the aperture shown in Fig. 7.3.

Now for short wavelengths k_0 is large and the second term in Eq. (7.1.15) can be neglected compared to the first; therefore,

$$\frac{\partial U}{\partial n} \approx -\, jk_0 U\, \hat{n} \cdot \nabla S \tag{7.1.16}$$

But from the Eikonal equation, Eq. (1.6.8)

$$\nabla S = n\hat{s} = \frac{k}{k_0}\, \hat{s}$$

so that

$$k_0\, \nabla S = k\hat{s} \tag{7.1.17}$$

where $\hat{s}$ is a unit vector in the direction of the Poynting vector tangential to the ray and n is the refractive index.

Substituting Eq. (7.1.17) into Eq. (7.1.16) yields

$$\frac{\partial U}{\partial n} \approx -jkU\, \hat{n} \cdot \hat{s} \tag{7.1.18}$$

Consider next the factor $\dfrac{\partial G}{\partial n} = \dfrac{\partial}{\partial n} \exp(-jkr)/r$ in Eq. (7.1.12a); it may be written as

$$\frac{\partial G}{\partial n} = \hat{n} \cdot \nabla G = \hat{n} \cdot \frac{r}{dr}\left(\frac{e^{-jkr}}{r}\right) \hat{r} \tag{7.1.19}$$

or

$$\frac{\partial G}{\partial n} = \left(\frac{-jk}{r} - \frac{1}{r^2}\right) e^{-jkr}\, \hat{n} \cdot \hat{r} \tag{7.1.20}$$

where $\hat{r}$ is a unit vector in the direction of $\bar{r}$, the vector from the observation point to the aperture.

Equation (7.1.20) may be rewritten as follows:

$$\frac{\partial G}{\partial n} = \left(jk + \frac{1}{r}\right) \frac{e^{-jkr}}{r}\, \hat{n} \cdot \hat{r} \tag{7.1.21}$$

where now $\hat{r}$ is a unit vector from the point on the aperture to the observation point, as shown in Fig. 7.3.

Substituting Eqs. (7.1.18) and (7.1.21) in Eq. (7.1.13),

$$U(P) = \frac{1}{4\pi} \int_a U_a \frac{e^{-jkr}}{r} \left[\left(jk + \frac{1}{r}\right) \hat{n} \cdot \hat{r} + jk\, \hat{n} \cdot \hat{s}\right] dS$$

where U_a denotes the complex amplitude distribution exiting the aperture. Consistent with the short-wavelength approximation, k is very large and

$$U(P) = \frac{jk}{4\pi} \int_a U_a \frac{e^{-jkr}}{r} (\hat{n} \cdot \hat{r} + \hat{n} \cdot \hat{s})\, dS \tag{7.1.22}$$

Equation (7.1.22) is known as the *Fresnel-Kirchhoff diffraction integral.*

When the distance of the field point P to the aperture is much greater than the dimensions of the aperture $\hat{n} \cdot \hat{r} \cong \cos\theta$ all points of the aperture, and when the phase across the aperture is constant (i.e., when the aperture is parallel to the wavefront) $\hat{n} \cdot \hat{s} = 1$ over the entire aperture. In this case Eq. (7.1.22) becomes

$$U(P) = \frac{jk}{4\pi} \int_a U_a \frac{e^{-jkr}}{r} (1 + \cos\theta)\, dS \tag{7.1.23}$$

Note that $\hat{n} \cdot \hat{s} = 1$ holds for plane waves normally incident on the aperture or for distant point sources when their normally incident spherical wavefronts can be regarded as nearly plane over the aperture.

Example 7.1

A point source is located a distance R behind an opaque circular disk of radius a. What is the irradiance reaching a point P on the optical axis a distance R behind the disk?

Solution

For this geometry, Eq. (7.1.22) is appropriate rather than Eq. (7.1.23). From Eq. (7.1.22),

$$U(P) = \frac{jk}{4\pi} \int_a U_a(r) \frac{e^{-jkr}}{r} (\hat{n} \cdot \hat{r} + \hat{n} \cdot \hat{s})\, dS$$

where $U_a(r) = A \exp(-jkr)/r$, A being a constant. In Fig. 7.4, $\hat{n} \cdot \hat{r} = \hat{n} \cdot \hat{s} = \cos\theta = R/r$, and $dS = \rho\, d\rho\, d\phi$. Since $r^2 = \rho^2 + R^2$, then $2r\, dr = 2\rho\, d\rho$. Consequently,

$$U(P) = \frac{JA}{\lambda} \int_0^{2\pi} \int_{(a^2+R^2)^{1/2}}^{\infty} \frac{e^{-2jkr}}{r^2} \frac{R}{r} r\, dr\, d\theta = \frac{j2\pi Ar}{\lambda} \int_{(a^2+R^2)^{1/2}}^{\infty} \frac{e^{-j2kr}}{r^2} dr$$

This can be integrated by parts (using $u = r^{-2}$, $dv = [\exp(-j2kr)]\, dr$) to yield

$$U(P) = -\frac{kAR}{j2k} \left\{ \frac{e^{-j2kr}}{r^2} \Big|_{(a^2+R^2)^{1/2}}^{\infty} - \int_{(a^2+R^2)^{1/2}}^{\infty} \frac{2e^{-j2kr}}{r^3} + \cdots \right\}$$

After several iterations this converges to

$$U(P) = \frac{AR}{2} \frac{e^{-jk(a^2+R^2)^{1/2}}}{a^2 + R^2} \left[1 - \frac{j}{k(a^2 + R^2)^{1/2}} + \cdots \right]$$

Note that at optical wavelengths $R >> \lambda$ and that $k >> (a^2 + R^2)^{-1/2}$. Therefore,

$$U(P) \approx \frac{A\ Re^{-j2k(a^2+R^2)^{1/2}}}{2(a^2 + R^2)}$$

and the time-averaged irradiance at P is, from Eq. (1.3.11b),

$$H(P) = \frac{U(P)\, U^*(P)}{2\eta_0} = \frac{A^2R^2}{8\eta_0(a^2 + R^2)}$$

Irradiance at P reaches its maximum when $R = a$. An important conclusion here is that even though there is no direct line of sight from the point source to P, nevertheless there is light at P. Light reaches P through diffraction, primarily near the edges of the disk.

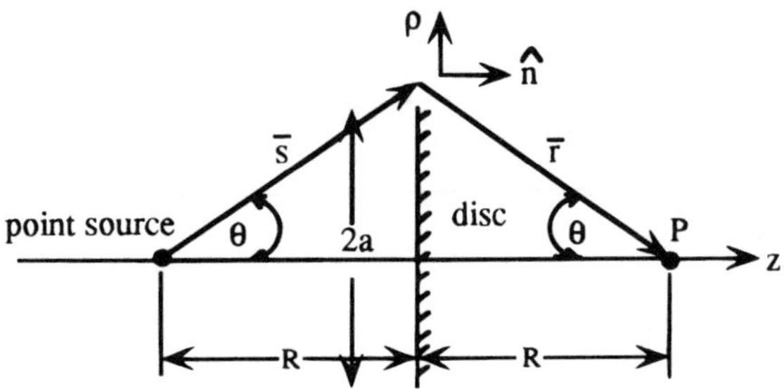

Fig. 7.4 Geometry for Example 7.1.

7.2 FRAUNHOFER AND FRESNEL DIFFRACTION

Often in optics the need arises to evaluate the diffraction of waves incident normally on a diffracting aperture. In that case $\hat{n} \cdot \hat{s} = 1$ in Eq. (7.1.22). If attention is confined to the central region of the resulting diffraction pattern where most of the energy is concentrated, then $\hat{n} \cdot \hat{r} \cong 1$ also for all points within this region. The Fresnel-Kirchhoff integral [Eq. (7.1.22)] becomes, since $\cos\theta$ then $\rightarrow 1$,

$$U(P) = \frac{j}{\lambda} \int_a U_a \frac{e^{-jkr}}{r} dS \tag{7.2.1}$$

Consider the geometry shown in Fig. 7.5, which depicts a plane wave normally incident on an aperture a in an infinite plane screen. Let dS be an incremental area in that aperture, and let P be an observation point in a plane a distance z_0 away from the plane of the aperture. From the geometry

$$r^2 = z_0^2 + (x_d - x_s)^2 + (y_d - y_s)^2 = z_0\left[1 + \left(\frac{x_d - x_s}{z_0}\right)^2 + \left(\frac{y_d - y_s}{z_0}\right)^2\right]^{1/2} \tag{7.2.2}$$

Assuming z_0 is very large compared to the numerators in Eq. (7.2.2), then through the binomial approximation r can be approximated by

$$r \approx z_0 + [(x_d - x_s)^2 + (y_d - y_s)^2/2z_0 \tag{7.2.3}$$

The value of r must be substituted into Eq. (7.2.1). For the denominator, it is sufficient to use $r \cong z_0$. However, for the numerator the more exact approximation of Eq. (7.2.3) is more appropriate since even small errors in r can change phase significantly at optical wavelengths. Consequently, after such substitution

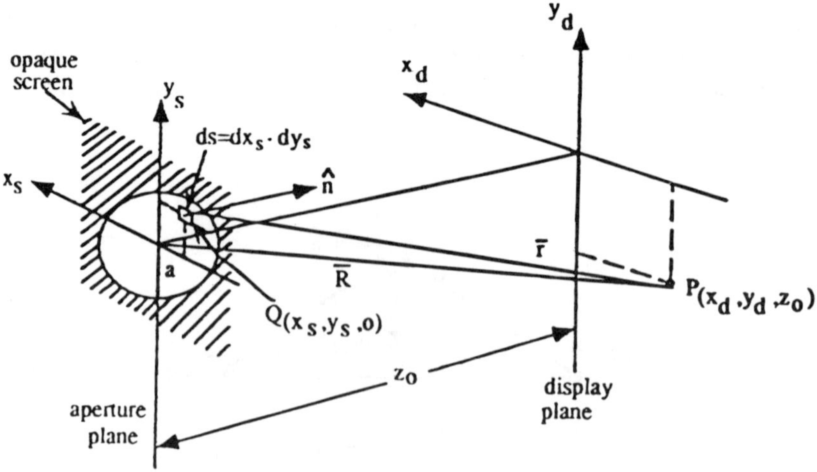

Fig. 7.5 Diffraction Geometry.

$$U(P) = C_1 \iint_a U(x_s, y_s) \exp\left\{\frac{jk_0}{2z_0}[(x_d - x_s)^2 + (y_d - y_s)^2]\right\} dx_s\, dy_s \quad (7.2.4)$$

where

$$C_1 = \frac{j \exp(-jkz_0)}{\lambda z_0} \quad (7.2.5)$$

and $k = k_0$ for propagation through air. Equation (7.2.4) can also be written

$$U(P) = C_1 \int_{-\infty}^{\infty}\int_{-\infty}^{\infty} U_t(x_s, y_s) \exp\left\{-\frac{jk_0}{2z_0}[(x_d - x_s)^2 + (y_d - y_s)^2]\right\} dx_s\, dy_s \quad (7.2.6)$$

where

$$U_t = U_i \cdot t \quad (7.2.7)$$

and

$$t(x_s, y_s) = \begin{cases} 1, & x_s \cdot y_s \in a \\ 0, & \text{otherwise} \end{cases} \quad (7.2.8)$$

In this case, U_i is CFA incident to the aperture and t is spatial transmittance through the plane of the aperture. CFA U_t is then CFA transmitted through the aperture plane. The square brackets in the exponent in Eq. (7.2.6) can be opened up, leaving

$$U(P) = C_2 \int_{-\infty}^{\infty}\int_{-\infty}^{\infty} (x_s, y_s) \exp\left\{-\frac{jk_0}{2z_0}(x_s^2 + y_s^2)\right] \exp\left[\frac{jk_0}{z_0}(x_s x_d + y_s y_d)\right] dx_s\, dy_s \quad (7.2.9)$$

where

$$C_2 = \frac{je^{-jk_0 z_0}}{\lambda z_0} \exp\left[-\frac{jk_0}{2z_0}(x_d^2 + y_d^2)\right] \quad (7.2.10)$$

By defining spatial Fourier frequencies as

$$\omega_x = 2\pi f_x = -kx_d/z_0 = -2\pi x_d/(\lambda z_0) \quad (7.2.11a)$$

$$\omega_y = 2\pi f_y = -ky_d/z_0 = -2\pi y_d/(\lambda z_0) \quad (7.2.11b)$$

Eq. (7.2.9) can be written as

$$U(f_x, f_y, z_0) = C_2 \int_{-\infty}^{\infty} \int_{-\infty}^{\infty} \left[u_t(x_s, y_s) \exp\left[-\frac{jk_0(x_s^2 + y_s^2)}{2z_0} \right] \right] \times \exp[-j2\pi(f_x x_s + f_y y_s)]\, dx_s\, dy_s \quad (7.2.12)$$

where units of f_x and f_y are cycles per unit width. We can see from Eq. (7.2.12) that CFA at observation point P behind the aperture is equal to the Fourier transform of the product of the CFA exiting the aperture with the quadratic exponent shown in the square brackets.

The concept of spatial (Fourier) frequencies requires some elaboration. Electrical signals, for example, are amplitudes of voltage or current as a function of time. Therefore, Fourier units are reciprocal to time, and are cycles$\cdot$s^{-1} or Hz. With regard to diffraction, we are interested in determining the diffraction pattern, that is, the variation of CFA $U(P)$ as a function of position, where $U(P)$ is a spatially rather than time-varying signal. Spatial coordinates in the plane in which the diffraction pattern is observed are (x_d, y_d). Equations (7.2.11) tell us that these spatial coordinates can also be related to spatial Fourier frequencies whose units are reciprocal to position in much the same way as electrical Fourier frequencies are reciprocal to time. Therefore, in both cases the Fourier units are reciprocal to the domain of the information being sought. There are, however, several differences between Fourier concepts in optics and electronics:

1. Electrical signals are one dimensional, whereas spatial signals are two dimensional (or even three dimensional on occasion).
2. Fourier transforms in electronics are purely mathematical tools that do not exist physically.

Two-dimensional spatial Fourier transforms, on the other hand, actually do exist physically in the spatial plane (x_d, y_d, z_0) or (f_x, f_y, z_0). The human visual system cannot see spatial variation of complex field amplitude because it responds to radiant power rather than field. Consequently, the irradiance of the Fourier transform in the plane (x_d, y_d) can actually be seen, with the irradiance at each point (x_d, y_d) being equal to $U(2\pi x_d/\lambda z_0,\ 2\pi y_d/\lambda z_0) \cdot U^*(2\pi x_d/\lambda z_0,\ 2\pi y_d/\lambda z_0)/2\eta_0$ according to the definition of a time-averaged Poynting vector.

Equation (7.2.12) describes *Fresnel* diffraction. However, when z_0 is very large, that is, $z_0 >> (\pi/\lambda)(x_{s\,\max}^2 + y_{s\,\max}^2)$, where $x_{s\,\max}$ and $y_{s\,\max}$ characterize the edges of the diffracting aperture in plane (x_s, y_s), then the quadratic exponent in Eq. (7.2.12) approaches zero much faster than does $f_x x_s$ or $f_y y_s$ because, as z_0 increases, the diffraction pattern expands. This means that $f_x(\propto x_d)$ and $f_y(\propto yd)$ increase with increasing z_0 while x_s and y_s are limited by aperture size to $x_{s\,\max}$ and $y_{s\,\max}$. In this situation, Eq. (7.2.12) can be approximated as

$$U(f_x, f_y, z_0) = C_2 \int_{-\infty}^{\infty} \int_{-\infty}^{\infty} U_t(x_s, y_s) \exp[-j2\pi(f_x x_s + f_y y_s)]\, dx_s\, dy_s \quad (7.2.13)$$

This very far-field situation is referred to as *Fraunhofer diffraction*. Equation (7.2.13) indicates that Fraunhofer diffraction CFA is the *spatial Fourier transform* of the CFA exiting the aperture.

If U_t is simply a uniform CFA with no spatial variation, then $U(f_x, f_y)$ in the (x_d, y_d) plane is simply the spatial Fourier transform of the aperture itself.

Example 7.2

A rectangular aperture of dimensions 10 mm in the x_s direction and 4 mm in the y_s direction is illuminated by a uniform plane wave of wavelength 0.63 μm. At what distance behind the aperture is the diffraction pattern equal to the Fourier transform of the aperture? What does this Fourier transform look like?

Solution

$$z_0 >> \frac{\pi}{\lambda}(x_{s\,max}^2 + y_{s\,max}^2) = \frac{\pi}{0.63 \times 10^{-6}}[(5 \times 10^{-3})^2 + (2 \times 10^{-3})^2] >> 145 \text{ m}$$

Clearly, this football-field distance is too large for convenient laboratory application, and a way of reducing it considerably is considered next. The CFA diffraction pattern itself is, from Eqs. (7.2.13) and (7.2.11),

$$U\left(\frac{x_d}{\lambda z_0}, \frac{y_d}{\lambda z_0}, z_0\right) = C_2 \int_{-a}^{a}\int_{-b}^{b} A \exp\left[-j2\pi\left(\frac{x_d x_s}{\lambda z_0} + \frac{y_d y_s}{\lambda z_0}\right)\right] dx_s dy_s$$

where A is CFA amplitude in the aperture plane and Eq. (7.2.8) has been used to define integral units in which a and b are 5 and 2 mm, respectively. The intergrand is separable into two independent integrals and integration yields:

$$U\left(\frac{x_d}{\lambda z_0}, \frac{y_d}{\lambda z_0}, z_0\right) = C_2 A \left.\frac{e^{-j2\pi f_x x_s}}{j2\pi f_x}\right|_{-a}^{a} \left.\frac{e^{-j2\pi f_y y_s}}{j2\pi f_y}\right|_{-b}^{b}$$
$$= C_2 Aab \,\text{sinc}(2\pi f_x a)\,\text{sinc}(2\pi f_y b)$$

where f_x and f_y have been substituted for $x_d/\lambda z_0$ and $y_d/\lambda z_0$. The result is sinc functions along both the x_d and y_d axes, similar to the example shown in Fig. 7.6(c). Since $a > b$, zeros of the sinc function in the y_d direction are closer together than those in the x_d direction. Spatial transmission of the rectangular aperture is analogous to pulses in the x and y directions, the Fourier transforms of which are sinc functions. Irradiance of the diffraction pattern varies as

$$H\left(\frac{x_d}{\lambda z_0}, \frac{y_d}{\lambda z_0}, z_0\right) = \frac{C_2^2 A_0^2 a^2 b^2}{2\eta_0}\,\text{sinc}^2\left(\frac{2\pi x_d a}{\lambda z_0}\right)\text{sinc}^2\left(\frac{2\pi y_d b}{\lambda z_0}\right)$$

Zeros along the x_d and y_d axes occur at $x_{d0} = \pm\frac{\lambda z_0 m}{2a}$ and $y_{d0} = \pm\frac{\lambda z_0 n}{2b}$, where m and n are integers. The narrower the aperture in a given direction, the more slowly varying is the resultant sinc function. It is spread over a larger area. Here, since $b < a$, $\Delta x_{d0} > \Delta y_{d0}$, where Δx_{d0} and Δy_{d0} refer to the distances between zero crossings.

Note that if, for example, x_s is the horizontal axis, it is diffraction by the horizontal edges of the aperture that gives rise to the sinc function in the vertical direction of argument $2\pi f_y b$, where b is the (vertical) separation between the horizontal boundaries. The horizontal edges diffract light upward and downward. Similarly, it is the vertical edges that give rise to the sinc function of argument $2\pi f_x a$ in the horizontal direction, where a is (horizontal) separation between the vertical edges.

Irradiance examples for various aperture shapes are shown in Fig. 7.6. Note that CFA can be negative, but irradiance cannot. An example for a parallelogram aperture is shown in the solution for Exercise 7.2.

7.3 DIFFRACTION THROUGH A LENS

As shown in the previous example, the Fraunhofer condition for far-field distance z_0 requires very large distances. It is, very conveniently, possible to circumvent this distance if a lens is used. As shown in Chapter 2 (Fig. 2.5a) a positive lens images

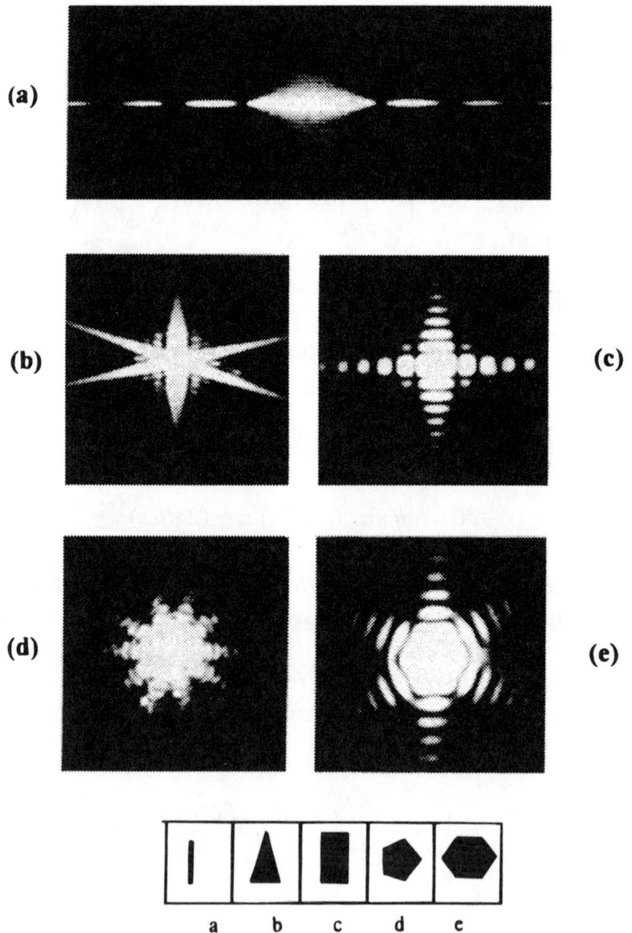

Fig. 7.6 Photographs of the Fraunhofer diffraction patterns (Fourier transforms) of various apertures: (a) slit aperture; (b) triangular aperture; (c) rectangular aperture; (d) pentagonal aperture; and (e) hexagonal aperture (from [7.1] with permission).

in the back focal plane a plane wave from infinity, that is, it converts a plane wave to a spherical wave. Therefore, the image observed in its back focal plane corresponds to that of an object an infinite distance in front of the lens. Consequently, if an aperture is illuminated by a uniform plane wave, the CFA exiting the aperture is identical to that which would exist if the aperture were at $s \rightarrow \infty$. As a result, the Fraunhofer or far-field condition is obtained in the back focal plane of the lens. To see this, one must consider first the phase change introduced by a positive lens, as shown in Fig. 7.7.

Phase change from front vertex plane to back vertex plane introduced by the lens is $-2n_1k_0z$ for the path in air on each side of the lens and $-k_0n_2(d_0 - 2z)$ for the path through the lens itself, that is,

$$\begin{aligned}\Delta\phi(\rho) &= -2n_1k_0z - k_0n_2(d_0 - 2z)\\ &= -k_0n_2d_0 + 2(n_2 - n_1)k_0z \end{aligned} \tag{7.3.1}$$

Substituting for z from Eq. (2.3.7) for parabolic surfaces or spherical surfaces with large radii of curvature,

$$\begin{aligned}\Delta\phi(\rho) &= -k_0n_2d_0 + \frac{2k_0(n_2 - n_1)\rho^2n_1}{4f_l(n_2 - n_1)}\\ &= -k_0n_2d_0 + \frac{k_0\rho^2}{2f_l} \end{aligned} \tag{7.3.2}$$

where f_l designates focal length and is used instead of f so as to avoid confusion with spatial frequencies. Phase transmission through the lens is therefore

$$t(\rho) = \exp(-jk_0nd_0)\exp(jk_0\rho^2/2f_l)$$

$$t(\xi, \eta) = \exp(-jk_0nd_0)\exp\left[\frac{jk_0}{2f_l}(\xi^2 + \eta^2)\right] \tag{7.3.3}$$

where ξ and η are Cartesian coordinates in the plane of the lens and, for convenience,

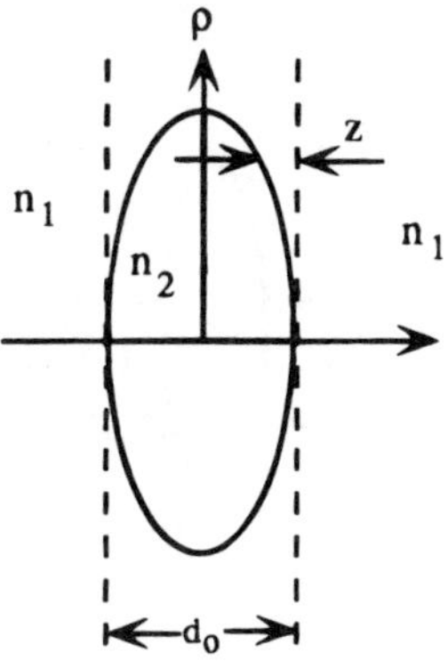

Fig. 7.7 Phase transmission through a lens.

$n_2 = n$ since the medium outside the lens is assumed to be air ($k = k_0$, $n_1 = 1$ in Fig. 7.7).

Next, propagation over the entire path from the aperture plane (x_s, y_s) through the lens plane (ξ, η) to the display plane (x_d, y_d), as shown in Fig. 7.8, must be evaluated.

It will be shown below that under certain conditions the positive quadrative exponent in the lens phase transmission described in Eq. (7.3.3) leads to cancelling of the negative quadratic exponent in the square brackets in Eq. (7.2.12), thus leading to Fraunhofer diffraction or optical Fourier transform similar to Eq. (7.2.13).

Propagation from the aperture plane to the front vertex plane of the lens a distance z_0 from the aperture away is given by Eq. (7.2.6) if the coordinates (ξ, η) for the plane of the lens are used instead of (x_d, y_d). Therefore if $U_i(\xi, \eta)$ represents CFA at the lens input plane (front vertex plane) and U_t as previously is CFA at the output of the aperture, then

$$U_i(\zeta, \eta) = C_1 \int_{-\infty}^{\infty} \int_{-\infty}^{\infty} U_t(x_s, y_s) \exp\left\{ -\frac{jk_0}{2z_0} [(\xi - x_s)^2 + (\eta - y_s)^2] \right\} dx_s dy_s \qquad (7.3.4)$$

which can be recognized as a convolution integral:

$$U_i(\zeta, \eta) = C_1 U_t(\zeta, \eta) * [-jk_0(\xi^2 + \eta^2)/2z_0] \qquad (7.3.5)$$

where * represents convolution.

The CFA at the lens output (back vertex plane), $U_0(\xi, \eta)$, is the product of $U_i(\xi, \eta)$ at the lens input and the lens phase transmission $t(\xi, \eta)$ in Eq. (7.3.3). Therefore, putting lens transmission first,

$$U_0(\xi, \eta) = C_3 \exp\left[\frac{jkl_0}{2f_l} (\xi^2 + \eta^2) \right]$$
$$\times\ U_t(x_s, y_s) * \exp\left[-\frac{jk_0}{2z_0} (\xi^2 + \eta^2) \right] \qquad (7.3.6)$$

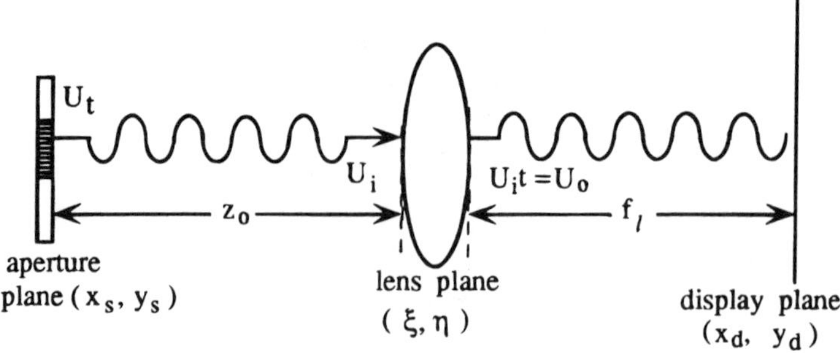

Fig. 7.8 Geometry for Fourier transform via a positive lens.

or

$$U_0(\xi, \eta) = C_3 U_t(x_s, y_s) * \exp\left[-\frac{jk_0}{2}(\xi^2 + \eta^2)\left(\frac{1}{f_l} - \frac{1}{z_0}\right)\right] \quad (7.3.7)$$

where

$$C_3 = C_1 \exp(-jk_0 n d_0) \quad (7.3.8)$$

We can see immediately from the preceding that when $z_0 = f_l$, that is, the aperture is placed in the *front* focal plane of the lens, the negative quadratic exponent at the right-hand end of Eq. (7.3.6), which gives rise to Fresnel diffraction in the plane of the lens, is cancelled by the lens phase transmission positive quadratic exponent. The lens thus removes the Fresnel diffraction entirely. In this case the CFA exiting the lens is simply

$$U_0(\xi, \eta) = C_3 U_t(x_s, y_s) \quad (7.3.9)$$

U_0 represents effective CFA which, within a constant, is Fourier transformed in the back focal plane of the lens according to Eq. (7.2.13) as if it originated an infinite distance in front of the lens. This removal of the quadratic phase indicates the result is Fraunhofer rather than Fresnel diffraction. This is now shown mathematically. We want to relate CFA in the lens back focal plane (x_d, y_d) to that in the aperture plane (x_s, y_s) a general distance z_0 in front of the lens.

Returning to the integral rather than the convolution form, so as to obtain a more general result, at the lens output Eqs. (7.3.3) and (7.3.4) yield

$$\begin{aligned} U_0(\xi, \eta) &= U_t(\xi, \eta) t(\xi, \eta) \\ &= C_3 \int_{-\infty}^{\infty}\int_{-\infty}^{\infty} U_t(x_s, y_s) \exp\left\{-\frac{jk_0}{2z_0}[(\xi - x_s)^2 + (\eta - y_s)^2]\right\} \\ &\quad \times \exp\left[-\frac{jk_2}{2f_l}(\xi^2 + \eta^2)\right] dx_s dy_s \end{aligned} \quad (7.3.10)$$

where propagation from the aperture plane a general distance z_0 in front of the lens to the lens rear vertex has been considered.

Next, propagation from the lens output to the display or observation plane which is the back focal plane of the lens is described.

Complex field amplitude in the back focal plane can also be determined via Eq. (7.2.6), which relates CFA in the display or observation plane to that at the exit of an aperture. In this case, the lens itself also serves as an aperture since in the absence of background radiation, it is only radiation transmitted through the lens that is transmitted to the back focal region. To use Eq. (7.2.12) for this purpose, $U_0(\xi, \eta)$ at the lens aperture output replaces $U_t(x_s, y_s)$, f_l replaces z_0, and (ξ, η) replaces (x_s, y_s). Now designating $U_{fl}(x_d, y_d)$ as the CFA in the lens *back* focal plane, and considering that distance from the aperture to the observation or display plane is $\approx z_0 + f_l$,

$$U_{fl}(x_z, y_d, z_0 + f_l) = C_4 \int_{-\infty}^{\infty}\int_{-\infty}^{\infty} U_0(\xi, \eta) \times \exp\left\{-\frac{jk_0}{2f_l}[(x_d - \xi)^2 + (y_d - \eta)^2]\right\} d\xi\, d\eta \quad (7.3.11)$$

where

$$C_4 = \frac{je^{-jk_0 f_l}}{\lambda f_l} \quad (7.3.12)$$

Substituting Eq. (7.3.10) into Eq. (7.3.11), we get

$$\begin{aligned} U_{fl}(x_d, y_d, z_0 + f_l) = {} & C_5 \int_{-\infty}^{\infty}\int_{-\infty}^{\infty}\left\{\int_{-\infty}^{\infty} U_t(x_s, y_s) \exp\left\{-\frac{jk_0}{2z_0}[(\xi - x_s)^2 + (\eta - y_s)^2]\right\}\right. \\ & \left.\times \exp\left[-\frac{jk_0}{2f_l}(\xi^2 + \eta^2)\right]\right\} dx_s dy_s \\ & \times \exp\left\{-\frac{jk_0}{2f_l}[(x_d - \xi)^2 + (y_d - \eta)^2]\right\} d\xi\, d\eta \end{aligned} \quad (7.3.13)$$

where

$$C_5 = C_3 C_4 = -\exp[-jk_0(nd_0 + z_0 + f_l)]/\lambda^2 z_0 f_l \quad (7.3.14)$$

The exponents in Eq. (7.3.13) can be rewritten and rearranged to yield (see Exercise 7.4)

$$\begin{aligned} & -\frac{jk_0}{2z_0}[(\xi - x_s)^2 + (\eta - y_s)^2] - \frac{jk_0}{2f_l}[(x_d - \xi)^2 + (y_d - \eta)^2 - (\xi^2 + \eta^2)] \\ & = \frac{-jk_0}{2z_0}\left\{\left[\xi - \left(x_s + \frac{z_0}{f_l}x_d\right)\right]^2 + \left[\eta - \left(y_s + \frac{z_0}{f_l}y_d\right)\right]^2\right\} \\ & \quad -\frac{jk_0}{2f_l}(x_d^2 + y_d^2)\left(1 - \frac{z_0}{f_l}\right) + \frac{jk_0}{f_l}(x_s x_d + y_s y_d) \end{aligned} \quad (7.3.15)$$

The third term on the right-hand side can be taken outside the integral in Eq. (7.3.13). Changing the order of integration in Eq. (7.3.13), that equation may now be rewritten as

$$U_{fl}(x_d, y_d, z_0 + f_l) = C_5 \exp\left[-jk_0 \frac{(x_d^2 + y_d^2)}{2f_l}\left(1 - \frac{z_0}{f_l}\right)\right]$$

$$\times \int_{-\infty}^{\infty}\int_{-\infty}^{\infty} \exp\left\{\frac{-jk_0}{2z_0}\left[\left(\xi - \left(x_s + \frac{z_0}{f_l}x_d\right)\right]^2\right.\right.$$

$$\left.+\left[\eta - \left(y_s + \frac{z_0}{f_l}y_d\right)\right]^2\right\} d\xi\, d\eta$$

$$\times \int_{-\infty}^{\infty}\int_{-\infty}^{\infty} U_t(x_s, y_s) \exp\left[\frac{jk_0}{f_l}(x_s x_d + y_s y_d)\right] dx_s dy_s \qquad (7.3.16)$$

By using the definite integral

$$\int_{-\infty}^{\infty} e^{-a^2(x - x_0)^2} dx = \pi^{1/2}/a$$

the second integral (over $d\xi$, $d\eta$) is equal to $-j2\pi z_0/k_0 = -j\lambda z_0$. When this is multiplied by C_5 in Eq. (7.3.14), their product with the first exponent in Eq. (7.3.16) is

$$C_6(x_d, y_d) = -j\lambda z_0 C_5 = \frac{j}{\lambda f_l} \exp[-jk_0(nd_0 + z_0 + f_l)]$$

$$\times \exp\left[\frac{jk_0}{2f_l}\left(\frac{z_0}{f_l} - 1\right)(x_d^2 + y_d^2)\right] \qquad (7.3.17)$$

which involves a quadratic phase function of display position unless, as shown in Eq. (7.3.7), the aperture is placed in the front focal plant of the lens, that is, $z_0 = f_l$. Accordingly, CFA in the observation (back focal) plane becomes

$$U_{fl}(x_d, y_d, z_0 + f_l) = C_6(x_d, y_d) \int_{-\infty}^{\infty}\int_{-\infty}^{\infty} U_t(x_s, y_s) \exp\left[\frac{\mathrm{j}k_0}{\mathrm{f}_l}(x_s x_d + y_s y_d)\right] dx_s dy_s$$

$$(7.3.18)$$

Put in a more familiar form,

$$U_{fl}\left(\frac{x_d}{\lambda f_l}, \frac{y_d}{\lambda f_l}, z_0 + f_l\right) = C_6(x_d, y_d)\mathbf{F}[U_t(x_s, y_s)] \qquad (7.3.19)$$

where Fourier spatial frequencies f_x and f_y are defined

$$f_x = -\frac{x_d}{\lambda f_l}, \qquad f_y = -\frac{y_d}{\lambda f_l} \qquad (7.3.20)$$

and the symbol **F** stands for *Fourier transform*. In general $C_6(x_d, y_d)$ as defined in Eq. (7.3.17) is a function of position in the lens back focal plane, hereby also known as the Fourier plane. However, if the aperture plane (x_s, y_s) is positioned in the front focal plane of the lens so that $z_0 = f_l$, then C_6 reduces to a constant independent of display position:

$$C_6|_{z0=fl} = \frac{j}{\lambda f_l} \exp[-jk_0(nd_0 + 2f_l)] \tag{7.3.21}$$

We can see from Eq. (7.3.19) that *spatial diffraction pattern in the back focal plane of a lens is, within a constant, equal to the spatial Fourier transform of the CFA in the front focal plane*. The exponent in Eq. (7.3.21) is related to the optical path traversed from front to back focal point. If the aperture (x_s, y_s) is located in the lens front focal plane, and illuminated with spatially uniform irradiance, then CFA in the front focal plane is proportional to the aperture itself and CFA in the back focal plane is proportional to the spatial Fourier transform of the aperture. The examples shown in Fig. 7.6 for far-field Fraunhofer diffraction of an aperture without a lens are therefore most relevant to optical Fourier transforms with a lens. The main differences are that lensless far-field Fraunhofer diffraction requires very large distances (z_0 approximately equal to hundreds of meters) and yields large diffraction patterns, while optical Fourier transforms with a lens require small distances on the order of twice the lens focal length and yield extremely small Fourier transforms since they are located in the back focal plane of the lens. To be viewed without a microscope, the irradiance distribution of the Fourier transform is generally enlarged optically.

A significant difference between geometrical and physical (diffraction) optics should be noticeable here. Geometrical optics as in Fig. 2.5a predicts a point image of zero radius in the back focal plane of a lens illuminated with a plane wave. However, diffraction causes a microscopic Fourier transform of the aperture times C_6, instead of a point, to exist in the back focal plane. (Diffraction is thus seen to spread an otherwise "point" image. This causes *blur*, and its effects on image quality are described in the next chapter.)

If there is a uniformly illuminated aperture in the front focal plane, then the Fourier transform in the back focal plane is of that aperture. If there is an aperture preceding the back focal plane (even behind the lens) but the aperture is not in the front focal plane, that is, $z_0 \neq f_l$, then Eq. (7.3.19) indicates CFA in the back focal plane is still not a point as predicted by geometrical optics but rather a Fourier transform of the aperture distorted by $C_6(x_d, y_d)$, which varies with position. Nevertheless, time-averaged *irradiance* in the back focal plane, even in this case, is an undistorted version of $|\mathbf{F}[U_t(x_s, y_s)]|^2$ since time-averaged Poynting vector calculations yield $|C_6|^2 = (\lambda f_l)^{-2}$, which is independent of position. Finally, even if there is no aperture in front of or behind the lens, the lens itself acts as an aperture since radiation not passing through it is not focused to the back focal region. In this case, Eq. (7.3.19) is still relevant if zero is substituted for z_0, in which case $U_t(x_s, y_s) = U_t(\xi, \eta)$ which is the shape of the lens itself (usually circular) for uniform plane wave illumination. (Fourier transforms for circular apertures are treated in the next section.) Depending on geometry, Fourier transforms of the lens may be superimposed on Fourier transforms of apertures in the plane (x_s, y_s).

From Eq. (7.3.18) we can see that the only difference between the Fourier and the *inverse* Fourier transform is the sign of the exponent. This means that a lens can also be used to convert a Fourier transform to an inverse Fourier transform, where coordinate axes in the inverse Fourier transform plane are in directions opposite to those in the Fourier transform plane. This is further developed later.

7.4 FOURIER TRANSFORMS IN POLAR COORDINATES

Perhaps the simplest class of functions separable in polar coordinates is comprised of those possessing circular symmetry. The function U_t is circularly symmetric if it can be written as a function of radius r alone, that is,

$$U_t(r, \theta) = U_R(r) \tag{7.4.1}$$

Such functions are of particular interest here since most optical systems have precisely this type of symmetry. A transformation to polar coordinates in both the x_s, y_s and f_x, f_y planes is made as follows:

$$\begin{aligned} r_s^2 &= x_s^2 + y_s^2 & x_s &= r_s\cos\theta \\ \theta &= \tan^{-1}(y_s/x_s) & y_s &= r_s\sin\theta \\ \rho^2 &= f_x^2 + f_y^2 & f_X &= \rho\cos\theta \\ \phi &= \tan^{-1}\frac{f_Y}{f_X} & f_Y &= \rho\sin\phi \end{aligned} \tag{7.4.2}$$

For the present the transform is written as a function of both radius and angle:

$$\mathbf{F}(U_t) = U_t(\rho, \phi)$$

Applying the coordinate transformations of Eqs. (7.4.2), the Fourier transform of U_t can be written

$$\mathbf{F}(U_t) = \int_0^{2\pi} d\theta \int_0^{\infty} dr_s \cdot r_s U_R(r_s) \exp[-j2\pi r_s\rho(\cos\theta\cos\phi + \sin\theta\sin\phi)] \tag{7.4.3}$$

or

$$\mathbf{F}(U_t) = \int_0^{\infty} dr_s \cdot r_s U_R(r_s) \int_0^{2\pi} d\theta \exp[-j2\pi r_s\rho\cos(\theta - \phi)] \tag{7.4.4}$$

Now to simplify Eq. (7.4.4) use is made of the Bessel function identity

$$J_0(a) = \frac{1}{2\pi}\int_0^{2\pi} \exp[-ja\cos(\theta - \phi)]\, d\theta \tag{7.4.5}$$

where J_0 is a Bessel function of the first kind and zero order. Substituting Eq. (7.4.5) into Eq. (7.4.4), the dependence of the transform on angle θ is seen to disappear, leaving $U_t(\rho, \phi) = U_t(\rho) = \mathbf{F}[U_t]$ as the following function of radius ρ:

$$U_t(\rho) = \mathbf{F}[U_t] = 2\pi \int_0^\infty r_s U_R(r_s) J_0(2\pi r_s \rho)\, dr_s \tag{7.4.6}$$

Thus the Fourier transform of a circularly symmetric function is itself circularly symmetric (no dependence on θ) and can be found by performing the one-dimensional manipulation of Eq. (7.4.6). This particular form of the Fourier transform occurs frequently enough to warrant a special designation; the expression (7.4.6) is accordingly referred to as the *Fourier-Bessel transform*, or alternatively, as the *Hankel transform of zero order*. For brevity the former terminology is adopted.

By means of arguments identical with those used earlier, the *inverse* Fourier transform of a circularly symmetric function $U_t(\rho)$ can be expressed as

$$U_R(r_s) = 2\pi \int_0^\infty \rho U_t(\rho) J_0(2\pi r_s \rho)\, d\rho \tag{7.4.7}$$

Since the last two equations are similar in form, for circularly symmetric functions, there is no difference between the transform and inverse transform operations, unlike the sign difference for Fourier and inverse Fourier transforms in Cartesian coordinates.

When using Eq. (7.4.6) for the Fourier-Bessel transform, the reader should remember that it is no more than a special case of the two-dimensional Fourier transform, and therefore any familiar property of the Fourier transform has an entirely equivalent counterpart in the terminology of Fourier-Bessel transforms.

Example 7.3

Consider the Fourier transform of a circular aperture of radius a, as in Fig. 7.9. The term $U_R(r)$ then equals unity, with r varying from 0 to a.

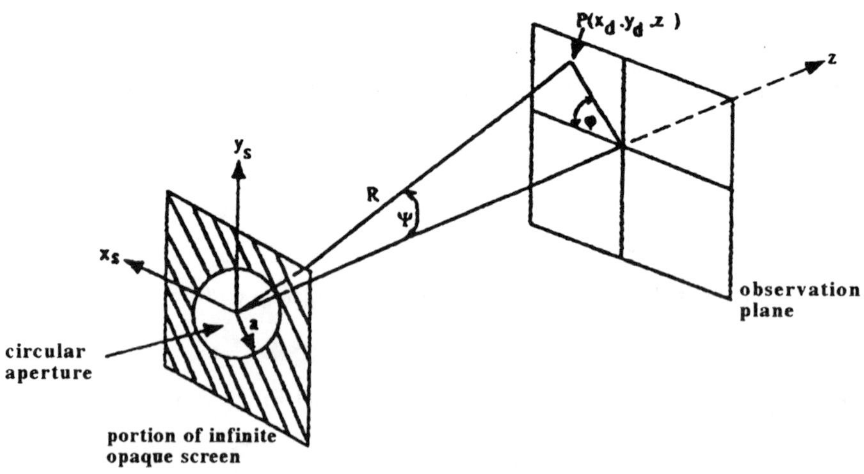

Fig. 7.9 Geometry for Fourier transform of circular aperture.

Solution

From the Fourier-Bessel transform equation, Eq. (7.4.6),

$$U_t(\rho) = \mathbf{F}\{U_t(r_s)\} = 2\pi K \int_0^{\infty} r_s(1) J_0(2\pi r_s \rho)\, dr_s$$

$$= \frac{2\pi K}{(2\pi\rho)^2} \int_0^{a} (2\pi r_s \rho) J_0(2\pi r_s \rho)\, dr_s \tag{7.4.8}$$

where $K = C_2$ or C_6, depending on whether the diffraction is lensless (C_2) or through a lens (C_6). The $U_R(r_s)$ term is assumed here to be unity. It represents CFA in the aperture plane. Since

$$\int X J_0(X)\, dX = X J_1(X) \tag{7.4.9}$$

where $J_1(X)$ is Bessel function of the first order, Eq. (7.4.8) becomes

$$\mathbf{F}\{U_t(r_s)\} = \frac{2\pi K}{(2\pi\rho)^2} [(2\pi r_s \rho) J_1(2\pi r_s \rho)|_0^a = K\pi a^2 \left[\frac{2J_1(2\pi\rho a)}{2\pi\rho_a}\right] \tag{7.4.10}$$

The corresponding irradiance is

$$H(\rho) = a^4 \pi^2 |K|^2 \left\{\frac{2J_1(2\pi\rho a)}{2\pi\rho a}\right\}^2 \tag{7.4.11}$$

At $\rho = 0$, $2J_1(2\pi\rho a)/(2\pi\rho a)$, is indeterminate. Application of L'Hopital's rule shows, however, that

$$\lim_{\rho\to 0} \left\{\frac{2J_1(2\pi\rho a)}{2\pi\rho a}\right\} = \lim_{\rho\to 0} \left\{\frac{2\pi a J_0(2\pi\rho a) - 2\pi a J_2(2\pi\rho a)}{2\pi a}\right\} = 1$$

where the formula, $2J_n'(X) = J_{n-1}(X) - J_{n+1}(X)$ has been used. Note also that $J_2(0) = 0$ while $J_0(0) = 1$. Thus, at $\rho = 0$,

$$H_0(\rho) = \pi^2 |K|^2 a^4 \tag{7.4.12}$$

The normalized irradiance

$$\frac{H(\rho)}{H(0)} = \left\{\frac{2J_1(2\pi a\rho)}{2\pi a\rho}\right\}^2 \tag{7.4.13}$$

is shown in Fig. 7.10. It appears similar to an isotropic sinc^2 function, except that the latter is periodic, while Eq. (7.4.13) is aperiodic.

The first zero of $H(\rho)/H(0)$ occurs when

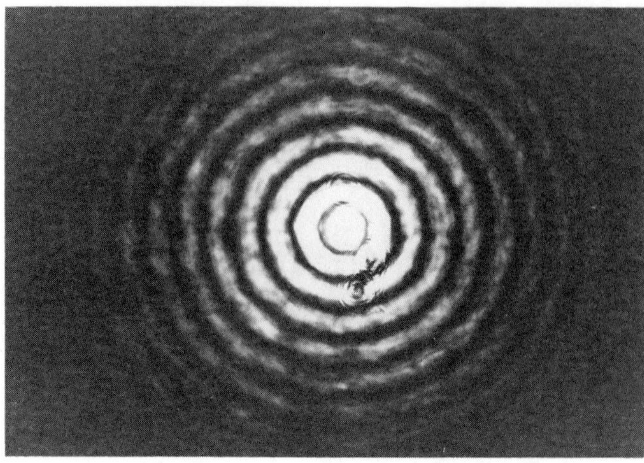

Fig. 7.10 Irradiance distribution for diffraction (Fourier transform) by a circular aperture.

$$2\pi a\rho = 3.83$$

$$= a\left(\frac{2\pi}{\lambda}\right)\frac{(x_d^2 + y_d^2)^{1/2}}{z} = \frac{2\pi a}{\lambda}\tan\psi \approx \frac{2\pi a}{\lambda}\psi = 3.83$$

where z equals z_0 for lensless Fraunhofer diffraction or f_l when a lens is used, and ψ is defined in Fig. 7.9. Therefore, at the first zero,

$$\psi \approx \frac{3.83\lambda}{2\pi a} = 0.61\frac{\lambda}{a} = 1.22\frac{\lambda}{D} \tag{7.4.14}$$

where $D = 2a$ is the diameter of the circular aperture.

The radiation from a distant point source such as a star incident at the circular aperture of an optical system such as a telescope will produce on a screen in the image plane a diffraction pattern irradiance with a shape like that shown in Fig. 7.10. Although geometrical optics predicts a point image, light diffracted by aperture edges gives rise to a spreading of the "point" image and to the accompanying rings. This is developed further in Chapter 8. If an adjacent point source is viewed simultaneously, two such diffraction patterns will be produced as shown in Fig. 7.11. The question is then how much overlap in the two patterns is allowed before an observer is unable to differentiate between the principal maxima in the patterns. The

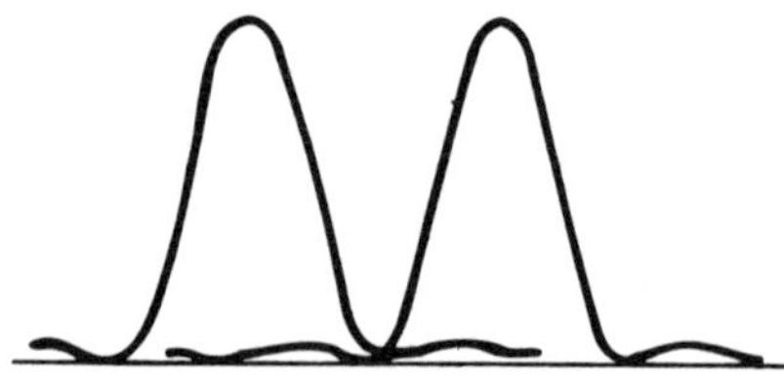

Fig. 7.11 Irradiance distribution for diffraction by a circular lens of light from two adjacent point sources.

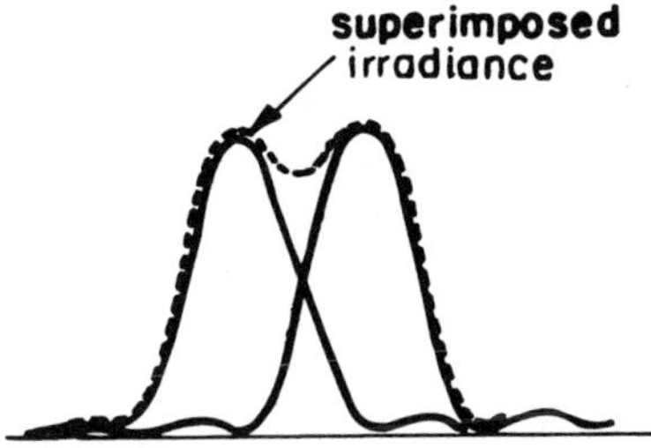

Fig. 7.12 Superimposed irradiance distribution for imaging of two point sources near Rayleigh criterion limit.

Rayleigh criterion for resolution answers this question. According to this criterion two quasimonochromatic noncoherent point sources of the same wavelength are said to be just resolved if the maximum intensity of one source occurs at the position of the first intensity minimum of the diffraction pattern of the other source as depicted in Fig. 7.12. This criterion leads to the conclusion that in order to resolve the principal maxima, the angular separation between them should be $\psi \geq 1.22\lambda/D$.

This can be shown with the aid of Fig. 7.13, which shows two identical quasimonochromatic point sources M and Q located a distance S_1 in front of an optical system represented by a simple thin lens. Due to the diffraction by the circular aperture of the lens the image produced by the two point sources will be a diffraction pattern in which the intensity is a superposition of the patterns of the individual sources centered at the points M' and Q'. Thus in order to be able to resolve the two sources their angular separation should not be less than

$$\psi_{\min} = \frac{1.22\lambda}{D}$$

or their linear separation be less than

$$d = \frac{1.22\ \lambda s_1}{D} \tag{7.4.15}$$

For this reason, for example, good quality images cannot be obtained with microwave

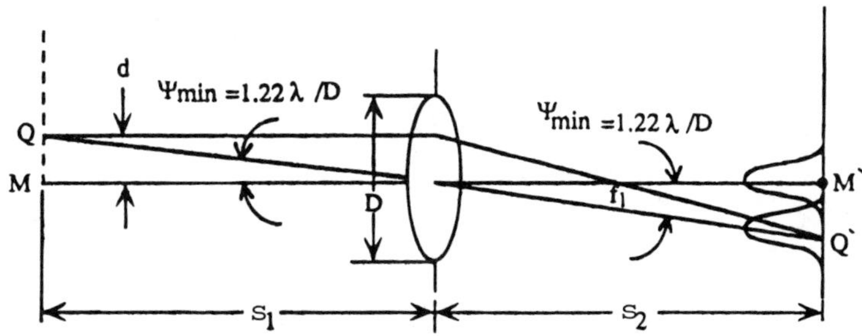

Fig. 7.13 Geometry for resolving two point sources.

radiation unless effective antenna apertures are on the order of kilometers so that an increase in D corresponds to increase in λ from optical wavelengths.

Diffraction by a circular aperture is of great significance because most lenses are circular. Since the lens itself is an aperture, if the edge of the lens is illuminated by a plane wave, the focused illumination does not give rise to a "point" at the back focal point, but rather to the "rings" of Fig. 7.10. About 84% of the total irradiance is enclosed within the first zero defined by Eq. (7.4.14). The bright central disk defined in this manner is known as *Airy's disk.* The smaller the ratio of λ/D, the more shrunken the size of Airy's disk, and the better the ability of the instrument to resolve one point of an image from another. This is considered further in the next chapter with regard to diffraction-limited imaging, and this limitation is formulated there in the form of an optical transfer function.

To determine diffraction pattern resulting from plane wave illumination of a circular lens including its edges, Eq. (7.3.19) can be used with $z_0 = 0$, resulting in

$$\begin{aligned} U_{fl}\left(\frac{x_d}{\lambda f_l}, \frac{y_d}{\lambda f_l}, f_l\right) &= C_6(x_d, y_d)|_{z_0=0}\mathbf{F}[U_t(\xi, \eta)] \\ &= \frac{j}{\lambda f_l}\exp[-jk_0(nd_0 + f_l)]\exp\left[-\frac{jk_0}{2f_l}(x_d^2 + y_d^2)\right] \\ &\times \frac{\pi D^2}{4}\left[\frac{2J_1(\pi\rho D)}{\pi\rho D}\right] \end{aligned} \tag{7.4.16}$$

where $\rho = (f_x^2 + f_y^2)^{1/2} = r_d/(\lambda f_l)$, r_d being the radial coordinate in the back focal plane equal to $(x_d^2 + y_d^2)^{1/2}$. The argument of the first-order Bessel function is $(\pi r_d/f_l)(D/\lambda)$. Therefore, as D/λ increases, the rings shrink and the image of the plane wave resembles more and more a point image, as is expected from geometrical optics. *Note that this conclusion is consistent with Eq. (1.6.10),* illustrating the relevance of the Eikonal equation, which defines conditions under which diffraction can be ignored.

Diffraction by a circular lens is important not only for imaging, but also for laser beam propagation. The output aperture of a laser is a partially reflecting and partially transmitting mirror usually of circular shape, as in Fig. 4.11. The output laser beamwidth is then limited by diffraction to the semi-angle 1.22 λ/D, where D is diameter of the laser aperture. Such beamwidths are referred to as *diffraction-limited.*

It is important to keep in mind that when the two sources as in Figs. 7.11, 7.12, and 7.13 are *coherent*, the Fraunhofer diffraction pattern *cannot be* obtained by adding the *individual* irradiance distribution corresponding to each individual source. This is so because the net electric field amplitude at any point P of the diffraction pattern is

$$U(P) = U_1(P) + U_2(P)$$

The irradiance at P will be

$$\begin{aligned} H(\rho) &= \frac{UU^*}{2n_0} = (U_1 + U_2)(U_1^* + U_2^*)/2\eta_0 \\ &= [U_1U_1^* + U_2U_2^* + U_1U_2^* + U_2U_1^*]/2\eta_0 \end{aligned}$$

$$= [|U_1|^2 + |U_2|^2 + (1/2)\mathrm{Re}(U_1U_2^*)]/2\eta_0$$

$$= H_1 + H_2 + (1/2)\mathrm{Re}(U_1U_2^*)/2\eta_0 \qquad (7.4.17)$$

where $(1/2)\ \mathrm{Re}(U_1U_2^*) = \langle U_1U_2\rangle$, the cornered brackets indicating time average, and H_1 and H_2 are irradiances of each source separately. When the two sources are incoherent U_1 and U_2 are uncorrelated and $\langle U_1U_2\rangle = 0$, in which case net irradiance is the sum of irradiances arising from each source separately. However, when the two sources are coherent, U_1 and U_2 are correlated and $\langle U_1U_2\rangle \neq 0$, in which case net irradiance is not simply the sum of the two irradiances. Coherence is treated at the end of this chapter.

7.5 ARRAY THEOREM [7.1]

An aperture may often consist of a one-dimensional or two-dimensional array of identical apertures, all of them in the plane (x_s, y_s) in Fig. 7.8. Examples can include diffraction gratings, which are composed of identical slits, often in large densities. They give rise to combined diffraction involving mixing of similar diffraction patterns. Such interference effects can be handled with a single equation, the *array theorem*. This unifying theorem is developed as follows. It is developed here for a one-dimensional array for simplicity. The Fourier transform of a two-dimensional array in which x_s and y_s dependencies are separable simply involves multiplication of the individual one-dimensional arrays.

Let $U_t(x_s)$ be CFA transmitted through a single aperture of the array, as previously. Let $U_{tn}(x_s)$ be overall CFA for a one-dimensional array of n identical apertures in the x_s direction. The $U_t(x_s - x_{sn})$ term is the CFA transmitted through aperture number n, where x_{sn} is the coordinate in the x_s direction located at the center of the nth aperture. Now

$$U_t(x_s - x_{sn}) = \int U_t(x_s - \alpha)\delta(\alpha - x_{sn})\, d\alpha \qquad (7.5.1)$$

according to the combing property of delta functions.

This integral exists only for $\alpha = x_{sn}$. Equation (7.5.1) is also a convolution integral. A large number N of such apertures can be represented as

$$U_{tn}(x_s) = \sum_{n=1}^{N} U_t(x_s - x_{sn}) = \sum_{n=1}^{N} \int U_t(x_s - \alpha)\delta(\alpha - x_{sn})\, d\alpha \qquad (7.5.2)$$

where $A(\alpha)$ is now defined as a function that describes the array itself, and is equal to

$$A(\alpha) = \sum_{n=1}^{N} \delta(\alpha - x_{sn}) \qquad (7.5.3)$$

It exists only at the center ($\alpha = x_{sn}$) of each individual aperture, and thus characterizes the array itself. Substituting this expression into Eq. (7.5.2) results in a compact convolution integral

$$U_{tn}(x_s) = \int U_t(x_s - \alpha)A(\alpha)\, d\alpha \tag{7.5.4}$$

Convolution in the spatial domain implies multiplication of Fourier transforms in the Fourier domain. Accordingly,

$$\mathbf{F}[(U_{tn})(x_s)] = \mathbf{F}\{U_t(x_s)\}\mathbf{F}\{A(x_s)\} \tag{7.5.5}$$

Therefore, to determine the Fourier transform for an array of identical apertures, one simply multiplies the Fourier transform for one aperture by the Fourier transform of the array function, the array function being simply a train of impulse functions defined in Eq. (7.5.3). In this connection the following relations may be useful to remember:

$$\mathbf{F}\{\delta(x)\} = 1 \tag{7.5.6}$$

$$\mathbf{F}\{U(x - a)\} = e^{-j2\pi f_x a}\mathbf{F}\{U(x)\} \tag{7.5.7}$$

Example 7.4

Consider a one-dimensional array of identical slits as shown in Fig. 7.14. The distance from the center of one slit to that of an adjacent slit is $2(a + b)$.

Solution

From Eq. (7.5.3), and using the above two Fourier relations

$$A(x_s) = \sum_{n=0}^{N-1} \delta[x_s - 2n(a + b)] \tag{7.5.8}$$

and

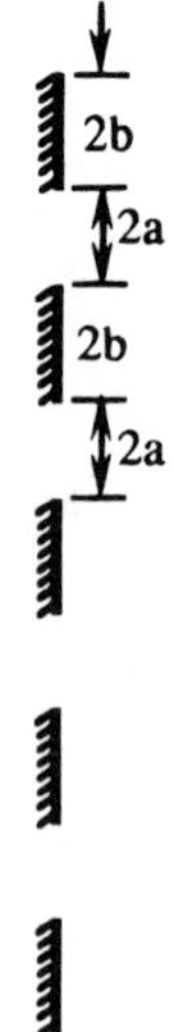

Fig. 7.14 Diffraction geometry of identical slits of width $2a$ and separation $2b$.

$$\mathbf{F}\{A(x_s)\} = \sum_{n=0}^{N-1} \exp[-j2\pi(b + a)x_d n/\lambda f_l] \tag{7.5.9}$$

where definition (7.3.20) is used for f_x. Equation (7.5.9) is a geometric series whose sum equals

$$\mathbf{F}\{A(x_s)\} = \frac{1 - \exp[-j4\pi N(a + b)x_d/\lambda f_l]}{1 - \exp[-j4\pi(a + b)x_d/\lambda f_l]} \tag{7.5.10}$$

For a slit aperture (see Exercise 7.1), assuming Fourier transform via a lens,

$$\mathbf{F}\{u_t(x_s)\} = C_6 A_a a \operatorname{sinc}(2\pi x_d a/\lambda f_l) \tag{7.5.11}$$

where A_a is the uniform incident CFA. Therefore, the diffraction grating of Fig. 7.14 yields a Fourier transform or Fraunhofer diffraction pattern in the back focal plane described by the product of the last two equations:

$$\mathbf{F}\{U_{tn}(x_s)\} = C_6 A_a a \frac{1 - \exp[-j4\pi N(a + b)x_d/\lambda f_l]}{1 - \exp[-j4\pi(a + b)x_d/\lambda f_l]} \cdot \operatorname{sinc}(2\pi x_d a/\lambda f_l) \tag{7.5.12}$$

The human visual system sees irradiance rather than field. To determine irradiance distribution for this CFA, note that

$$(1 - e^{-j\theta})(1 - e^{j\theta}) = 1 - e^{j\theta} - e^{-j\theta} + 1 = 2(1 - \cos\theta)$$

Using this result, irradiance distribution in the back focal plane is

$$\begin{aligned} H(x_d/\lambda f_l) &= |\mathbf{F}\{U_{tn}(x_s)\}|^2/2\eta_0 \\ &= 2\frac{A_a^2 a^2}{\eta_0 \lambda^2 f_l^2}\left\{\frac{1 - \cos[2\pi N2(a + b)x_d/\lambda f_l]}{1 - \cos[2\pi 2(a + b)x_d/\lambda f_l]}\right\} \cdot \operatorname{sinc}^2(2\pi a x_d/\lambda f_l) \end{aligned} \tag{7.5.13}$$

where irradiance incident to the diffraction grating itself is only $A_a^2/2\eta_0$. From trigonometry, Eq. (7.5.13) can be simplified further to

$$H(x_d/\lambda f_l) = \frac{2A^2 a^2}{\eta_0 \lambda^2 f_l}\left\{\frac{\sin[2\pi N(a + b)x_d/\lambda f_l]}{\sin[2\pi(a + b)x_d/\lambda f_l]}\right\}^2 \cdot \operatorname{sinc}^2(2\pi a x_d/\lambda f_l) \tag{7.5.14}$$

To analyze this result, notation is simplified to

$$H(x_d/\lambda f_l) = KH_1(x_d)[(\sin Nqx_d)/(\sin qx_d)]^2 \tag{7.5.15}$$

where $H_1(x_d)$ is irradiance for Fraunhofer diffraction by a single slit, and K is a constant. The expression in square brackets derives from the array function only and thus is independent of aperture shape. It applies to any array of identical apertures, regardless of the aperture shape. This term is referred to here as $B(x_d)$.

The aperture shape gives rise to $H_1(x_d)$, which is manifested as an envelope function in the Fourier domain that encloses $B(x_d)$. In this example, $H_1(x_d)$ is a sinc function, corresponding to the Fourier transform of a slit.

When slit width $2a$ is very small, $H_1(x_d)$ approaches unity over a broad range of x_d and irradiance distribution is determined largely by $B(x_d) = [(\sin Nqx_d)/(\sin qx_d)]^2$. A plot of this function for $N = 3$, 4, and 5 is shown in Fig. 7.15. Primary maxima occur when the denominator is zero, or $qx_d = \pm m\pi$, where m is called the *order of interference* and equals 0, 1, 2, 3, ... The value of $B(x_d)$ at the primary maxima is N^2 since, by L'Hopital's rule,

$$\lim_{qx_d \to m\pi} \frac{\sin(Nqx_d)}{\sin(qx_d)} = \lim_{qx_d \to m\pi} \frac{Nq \cos(Nqx_d)}{q \cos(qx_d)} = N \tag{7.5.16}$$

Since locations of primary maxima are located where

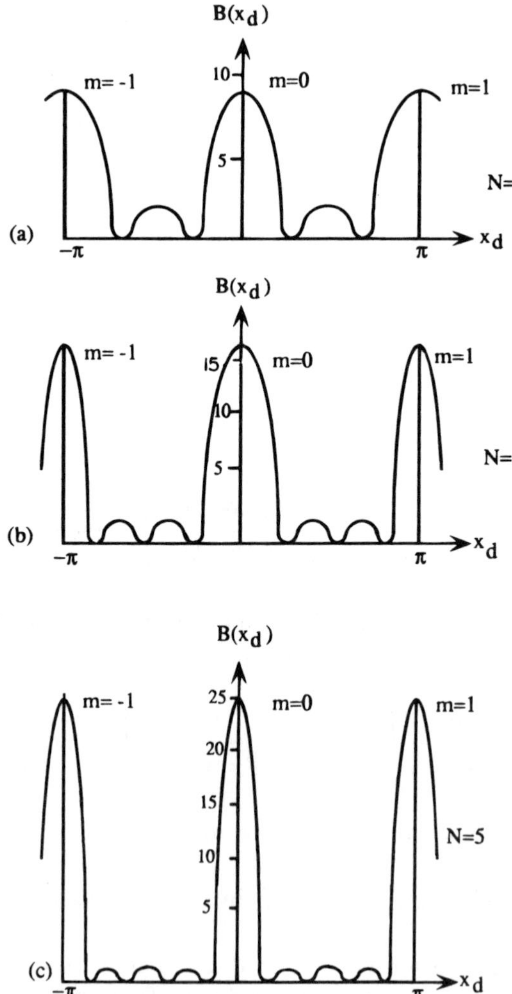

Fig. 7.15 Plot of $B(x_d)$ for various values of N.

$$x_d = \pm m\pi/q = \pm\frac{m\pi\lambda f_l}{2\pi(a + b)} = \pm\frac{mf_l}{2(a + b)}\lambda \tag{7.5.17}$$

these locations vary directly with wavelength.

Secondary maxima occur when the numerator of $B(x_d)$ is maximum. This occurs when $Nqx_d = \pm m\pi/2$. Minima of zero irradiance occur when the numerator is zero, or $Nqx_d = \pm m\pi$. The numerator will pass through zero N times more frequently than the denominator so that between consecutive primary maxima there are $N - 1$ zero values of irradiance and $N - 2$ secondary maxima. Examples are shown in Fig. 7.16. As N increases, primary maxima narrow.

Based on Eq. (7.5.17), diffraction gratings with very high slit densities or large values of N are used to determine wavelength spectra of light sources, or to vary transmitted wavelength of white light sources. Such instruments are called *monochromators*. Linewidth narrows with increased values of N. However, the order of interference m also plays a critical role in wavelength measurement and accuracy (see Exercises 7.11 and 7.12).

7.6 OPTICAL COMPUTING

The Fourier transform properties of Fraunhofer diffraction permit data processing operations carried out optically that can, in many situations, be more advantageous than digital signal processing. The primary advantages of optical techniques over electronic analog and digital signal processing are best realized when the information to be processed has two degrees of freedom. A picture, for example, has two independent variables (the x and y position coordinates) and one dependent variable (the irradiance

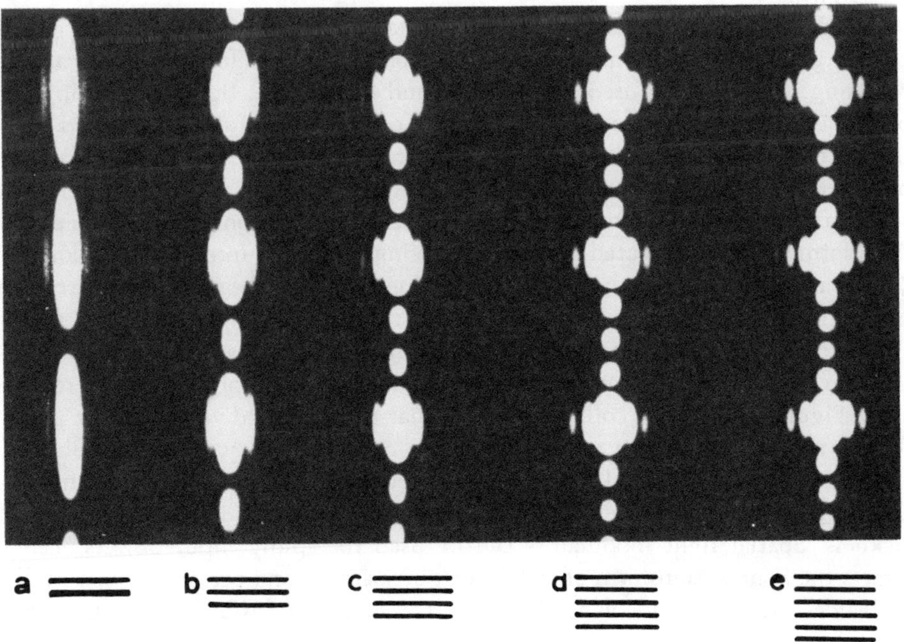

Fig. 7.16 Fraunhofer diffraction patterns of various numbers of slits from two to six (from [7.1] with permission).

at the position specified). Optical systems correspondingly can have two degrees of freedom. Thus, for handling pictorial data, they can be intrinsically superior to electric or electronic systems, which have only time as an independent variable. This two-dimensional processing property can be particularly valuable in radar applications where previously velocity and position data had to be processed separately in order for sufficient accuracy in the estimation of each to be obtained.

Many of the basic operations that can be performed by optical computers can also be done with electronic systems. However, if large quantities of wideband data are to be operated on, the electronic techniques can often be prohibitively expensive and time consuming. The optical computer can be advantageous here for certain types of operations.

The optical computer is faster than a digital computer because it operates on signals that are functions of position, not time. The input to an optical data processor is a pattern of light irradiance that varies over some area, not light irradiance that varies as a function of time. This has two important consequences:

1. The optical computer is able to handle directly signals that are functions of two independent variables, because it is just as easy to make the density of an optical mask vary over two dimensions as one.
2. Normally time-consuming operations, such as integration, can be performed virtually instantaneously because all the input data are available simultaneously, the integration being over space, not time.

Thus, optical data processing can be very appropriate for such lengthy operations as the calculation of complicated Fourier transforms and the processing of side-looking radar signals.

In addition to mathematical operations such as differentiation, integration, and producing Fourier transforms, the two classes of applications for which optical signal processing is most often used are filtering and correlation. Before discussing these operations, it is prudent to first discuss the optical computer itself, based on the Fourier transform properties of Fraunhofer diffraction by a converging lens. A typical optical computer is depicted in Fig. 7.17. Light coming from a cw laser (a few milliwatts is enough for most applications) passes through a convex lens and a pinhole, which forms a low-pass spatial filter for removing unwanted intensity variations, that is, the spatial filter passes only the central portion of the laser beam Gaussian irradiance distribution. The light transmitted through the pinhole is essentially of uniform irradiance over a broad central portion, with rings further out as a result of diffraction by the pinhole. The uniform irradiance over the broad central portion is of interest here.

In Figure 7.17 L_1 is a collimating lens that renders the diverging light parallel. Next in the plane $P_1(x_s, y_s)$ is the input film or transparency, often contained in a liquid gate assembly. A fluid whose index of refraction matches that of the film base and/or emulsion is used to eliminate phase noise such as random variations in film thickness. Spatial light modulators can be used to rapidly input objects. Lens L_2 extracts the Fourier transform of the input transparency and presents it as a complex amplitude distribution in plane $P_2(x_d, y_d)$, the Fourier transform (FT) plane in the back focal plane of L_2.

It is in plane P_2 that various spatial filters can be placed to operate on the Fourier transformed signal. Spatial light modulators can rapidly change spatial filters. Lens L_3 extracts the inverse Fourier transform and presents the processed image in plane

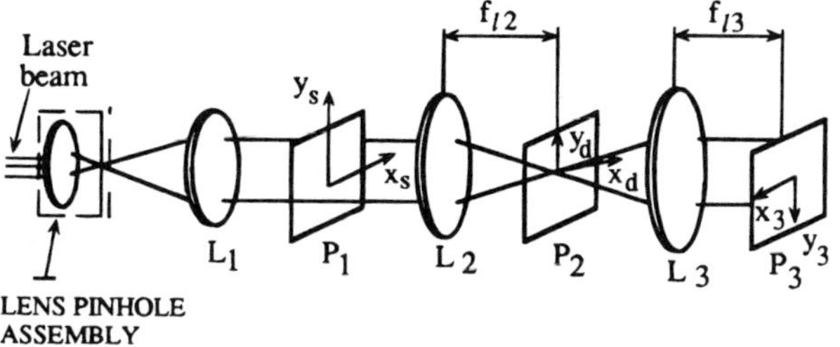

Fig. 7.17 Typical optical data processor.

$P_3(x_3, y_3)$. Note that plane P_2 is located a distance behind L_2 equal to the focal length of L_2; Similarly P_3 is one focal length behind L_3.

To understand how lens L_3 extracts the inverse Fourier transform, let us revisit Section 7.3. Recall that in plane P_2 CFA is

$$U_{fl_2}\left(\frac{x_d}{\lambda f_{l_2}}, \frac{y_d}{\lambda f_{l_2}}\right) = C_6 \mathbf{F}\{U_t(x_s, y_s)\} \tag{7.3.19}$$

$$= C_{6L_2} \int_{-\infty}^{\infty} \int_{-\infty}^{\infty} U_t(x_s, y_s) \exp\left[\frac{jk_0}{f_{l_2}}(x_s x_d + y_s y_d)\right] dx_s\, dy_s \tag{7.3.18}$$

where C_6 is defined in Eq. (7.3.21) and the notation C_{6L_2} refers to lens L_2, and

$$f_x = -\frac{x_d}{\lambda f_{l_2}}, \qquad f_y = -\frac{y_d}{\lambda f_{l_2}} \tag{7.3.20}$$

Therefore CFA in plane P_2 is proportional to the Fourier transform of CFA in plane P_1 provided that the axes (f_x, f_y) are *opposite* to axes (x_d, y_d). Similarly in plane $P_3(x_3, y_3)$ the CFA is

$$U_{fl_3}(x_3, y_3) = C_{6L_2} C_{6L_3} \int_{-\infty}^{\infty} \int_{-\infty}^{\infty} \left[U_{fl_2}\left(\frac{x_d}{\lambda f_{l_2}}, \frac{y_d}{\lambda f_{l_2}}\right) t(x_d, y_d)\right]$$

$$\times \exp\left[\frac{jk_0}{f_{l_3}}(x_d x_3 + y_d y_3)\right] dx_d\, dy_d \tag{7.6.1}$$

where $t(x_d, y_d)$ is the spatial transmission of the filter plane in plane P_2.

The first square-bracketed expression in Eq. (7.6.1) is the CFA transmitted by the filter in plane P_2. The integral itself is the *inverse* Fourier transform of the square-bracketed expression in the integrands provided that now

$$f_{x_3} = +\frac{x_d}{\lambda f_{l_3}}, \qquad f_{y_3} = +\frac{y_3}{\lambda f_{l_3}} \tag{7.6.2}$$

that is, the spatial frequency axes f_{x_3} and f_{y_3} are in the *same* directions as x_3 and y_3, respectively. However, we will demonstrate shortly that directions of axes x_3 and y_3 should be opposite to those of x_2 and y_2 in order to visualize CFA in plane P_3 as inverse Fourier transformation of the CFA exiting plane P_2.

Now, since in Eq. (7.6.1)

$$dx_d = \lambda f_{l_3}\, df_{x3}, \qquad dy_d = \lambda f_{l_3}\, df_{y_3} \tag{7.6.3}$$

Eq. (7.6.1) can be written in the form of an inverse Fourier transform

$$U_{fl_3}(x_3, y_3) = C_{6L_2}C_{6L_3}\lambda^2 f_{l_3}^2 \int_{-\infty}^{\infty}\int_{-\infty}^{\infty}\left[U_{fl_2}\left(\frac{x_d}{\lambda f_{l_2}}, \frac{y_d}{\lambda f_{l_2}}\right)t(x_d, y_d)\right] \times \exp[j2\pi(f_{y_3}x_3 + f_{y_3}y_3)]\, dfx_3\, dfy_3 \tag{7.6.4}$$

where, from Eq. (7.3.21), the constants are equal to

$$C_{6L_2}C_{6L_3} = -\frac{1}{\lambda^2 f_{l_2} f_{l_3}} \exp[-jk_0(n_2d_{02} + n_3d_{03} + 2f_{l_2} + 2f_{l_3})] \tag{7.6.5}$$

where subscripts 2 and 3 refer to lenses L_2 and L_3, respectively. Consequently, constants outside the integral sign in Eq. (7.6.4) are lumped together as

$$C_7 = C_{6L_2}C_{6L_3}\lambda^2 f_{l_3}^2 = -\frac{f_{l_3}}{f_{l_2}} \exp[-jk_0(n_2d_{02} + n_3d_{03} + 2f_{l_2} + 2f_{l_3})] \tag{7.6.6}$$

The ratio f_{l_3}/f_{l_2} simply is lateral magnification of the inverse transform accomplished by the $L_2 - L_3$ collimator optics. However, the negative value of C_6 implies the inverse Fourier transform is reversed; hence, in Fig. 7.17 axial directions in plane P_3 have been reversed from their analogs in plane P_2. The exponent in Eq. (7.6.6) stems from phase change in the z direction for propagation from plane P_1 to P_3. In summary, if no spatial filter is placed in plane P_2, the CFA obtained in plane P_3 is the image of that in plane P_1 but reversed in direction and magnified optically by the ratio f_{l_3}/f_{l_2}, as is to be expected (see Exercise 2.6) on the basis of geometrical optics without considering Fraunhofer diffraction or optical Fourier transforms. It is, however, the spatial filtering possibilities in the Fourier transform plane P_2 that are so intriguing and are considered next.

Some applications of the optical computer do not require all of the components shown here. Integration, for example, does not require L_3. The detector is placed directly in the Fourier transform plane P_2.

Now let us see how some of the mathematical operations discussed earlier can be performed with the optical computer.

Differentiating and integration are easy and virtually instantaneous when performed optically. If $U_t(x_s, y_s)$ is differentiated with respect to one of its variables, say, x_s, then, from Fourier transform theory, we obtain using Eq. (7.3.19)

$$\mathbf{F}\left\{\frac{\partial}{\partial x_s} U_t(x_s, y_s)\right\} = C_6 j2\pi f_x \mathbf{F}\{U_t(x_s, y_s)\}$$

$$= C_6 \frac{j2\pi x_d}{\lambda f_{l_2}} \mathbf{F}\{U_t(x_s, y_s)\} \quad (7.6.7)$$

Thus, to take the derivative of a function with respect to the x_s direction, one multiplies the spectrum of $U_t(x_d/\lambda f_{l_2}, y_d/\lambda f_{l_2})$ by a filter whose transmission is proportional to x_d, and then takes the inverse Fourier transform. Figure 7.18 shows a plot of the filter and its physical realization. Negative values of the filter are achieved by shifting the relative phase θ of the light by $\lambda/2$. This can be done by depositing a layer of clear dielectric on glass substrate—similar to the process for putting antireflection coatings on lenses and optical plates.

Consider integration. Inspection of the Fourier transform theory shows that the definite integral of a function is equal to the Fourier transform evaluated at the origin $x_d = y_d = 0$. Thus, to instantaneously integrate a two-dimensional function, one merely measures the light level on the optical axis in the Fourier transform plane. This point is often called the *dc spot*.

Consider frequency filtering. Since the radial distance from the origin in the transform plane P_2 is proportional to the spatial frequency [Eq. (7.3.20)], the irradiance of light at any point is proportional to the irradiance of that spatial frequency component in the Fourier transform. Note that both positive and negative frequencies are present. By blocking or passing the optical signal at different radial distances from the origin in the transform plane P_2, *spatial frequency filtering* is achieved. For example, a sharp cutoff, low-pass filter is constructed by simply punching a hole in an opaque card. When inserted in the Fourier transform plane, only spatial frequencies within the clear aperture can get through. This filter has nearly infinitely steep sides, a fact of optics that is extremely difficult and expensive to do electronically. Figure 7.19 shows several other filter configurations and their electronic equivalents. Modulation techniques can be used to make spatial and alternating current frequencies correspond.

Such binary (because they are everywhere either clear or opaque) amplitude filters have been used by geophysicists extensively in removing unwanted and confusing noise structure in seismic data [7.2].

The earliest commercial application of optical data processing was to the analysis and filtering of seismic records obtained in the geophysical exploration for oil. This constitutes one of the most widespread single uses of optical processing techniques [7.3].

Another application of spatial filtering is image restoration. This could be useful in such diverse applications as sharpening TV pictures or photographs from space or

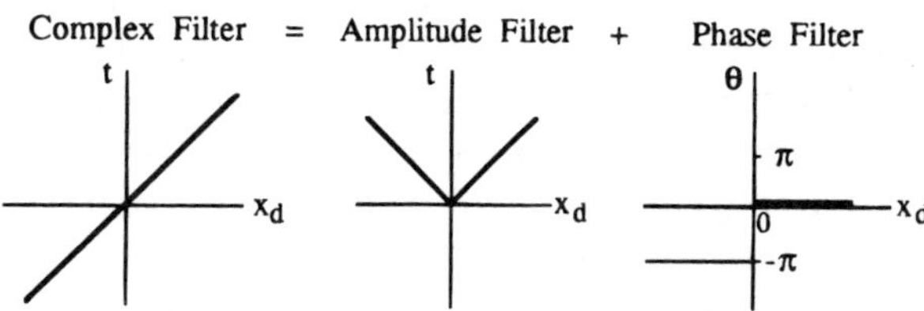

Fig. 7.18 Fourier transform plane filter for differentiation.

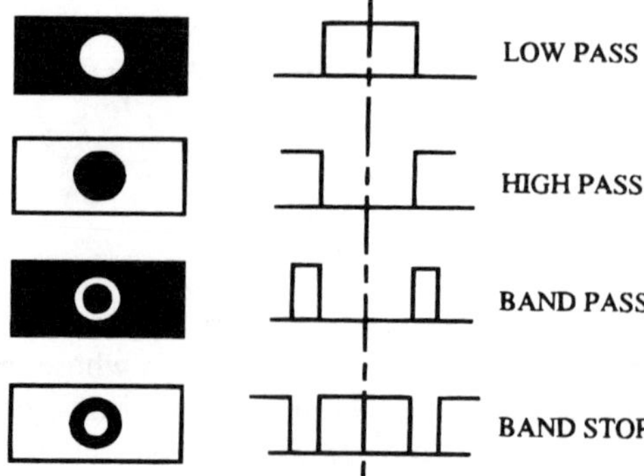

Fig. 7.19 Spatial frequency filters in Fourier transform plane and electronic equivalents.

for increasing the resolution of medical images. Spatial filtering techniques can correct images suffering from such things as out-of-focus adjustment, blur caused by camera or object motion during the exposure, and loss of resolution caused by atmospheric turbulence and aerosol forward scattering during exposure. Such blurring phenomena are characterized by a reduction of the higher spatial frequency spectrum in the pictures. If a spatial filter can be prepared that shades the lower frequencies and passes the higher frequencies with just the right proportions, then the fuzzy edges become sharp. If the spatial filter transfer functions of the optical system causing the blur (including atmosphere and image motion) were $M(x_d, y_d)$, then the desired mask should be $1/M(x_d, y_d)$. Thus, if $o(x_s, y_s)$ represents the irradiance of the original scene, the irradiance of the unfiltered image in plane P_1 is

$$i'(x_s, y_s) = m(x_s, y_s)*o(x_s, y_s) \tag{7.6.8}$$

where * again represents convolution and $m(x_s, y_s)$ represents the inverse Fourier transform of $M(x_d, y_d)$. In essence, $m(x_s, y_s)$ represents the *image system spatial impulse response*, the meaning of which is developed in the next chapter. Here, we simply point out that a spatial impulse is a point source. Image system response $m(x_s, y_s)$ is simply the image of a point source. Because the image system is not ideal, $M(x_s, y_s)$ is not a point image but is *spread*. Now, if a spatial filter of transmission $1/m(x_d, y_d)$, called an *inverse filter*, can be synthesized and inserted in the Fourier transform plane P_2 in Fig. 7.17, the light passing through the filter will form in P_2 the image spectrum

$$M(x_d, y_d) \cdot I(x_{d_a}, y_d) \cdot 1/M(x_d, y_d) = I(x_d, y_d) \tag{7.6.9}$$

where I represents the Fourier transform or spectra of (i). In the output plane P_3 the inverse Fourier transform $i(x_3, y_3)$ is produced rather than the previously distorted and blurred $i'(x_s, y_s)$. In other words, the degraded image $i'(x_s, y_s)$ has been deconvolved from the *point spread function* $m(x_s, y_s)$ in the optical computer to give a reproduction of the original undistorted scene. Although we show in Chapter 18 that the restored scene will not be perfect because of the noise present in any real system, considerable enhancement or restoration of the picture can still be accomplished for low-noise situations. The noise can be minimized through use of Wiener filters to reduce noise bandwidth, as described in Chapter 18 for digital image restoration.

Other data processing operations that can be carried out with an appropriate spatial filter in plane P_2 of Fig. 7.17 are convolution and cross-correlation. These operations can be simplified by preparing a special filter unique to the data to be processed. These filters are called Vander Lugt filters [7.4], after their inventor. They represent a real breakthrough and the technique is described in the next section. One of the primary limitations to widespread use of optical data processing is the difficulty of fabricating the appropriate filter for many of the objects or operations described here. This limitation causes much of the data processing to be done digitally rather than optically, since filter fabrication is a relatively rather lengthy process. Vander Lugt filters, however, are unique to the data to be processed and contain the required phase as well as amplitude spatial filtering.

7.7 VANDER LUGT FILTER [7.4]

Vander Lugt filters are placed in plane P_2 of the optical data processor in Fig. 7.17. They are used to do correlation or convolution involving the spatial signal in plane P_1 of Fig. 7.17. Vander Lugt filters are fabricated as shown in Fig. 7.20.

Convolution and correlation by definition involve two signals. One is that in the input plane P_1 in Fig. 7.17. It is designated $g(x_s, y_s)$. The spatial signal to which the Vander Lugt filter is unique is placed in plane P_4 in Fig. 7.20. It is designated $h(x_4, y_4)$. This latter signal is to be correlated or convolved with $g(x_s, y_s)$ by means of the Vander Lugt filter formed by interference of the two beams in plane P_2 in Fig. 7.20. To record this interference pattern, film or some other material capable of recording the irradiance pattern of the interference is placed in plane P_2 in Fig. 7.20. One interference beam is the Fourier transform of $h(x_4, y_4)$ produced by lens L_4. This CFA is equal to $C_{6L_4}H(x_d/\lambda f_{l4}, y_d/\lambda f_{l4})$, where H signifies the Fourier transform of h and C_{6L4} is described by Eq. (7.3.21) referenced to lens L_4. The beam interfering with this is an off-axis uniform plane wave incident on the film at an angle θ as shown in Fig. 7.20.

This latter beam serves as a reference beam and the angle θ will be seen to be critical with regard to the capability of spatially resolving the convolution or correlation result from unwanted irradiance. The planes of constant phase in the reference beam are

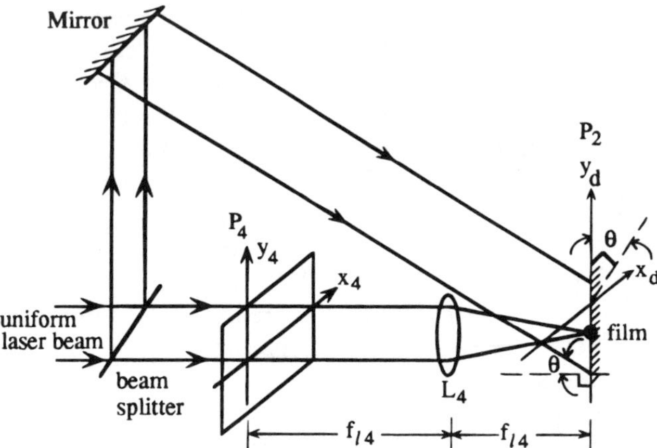

Fig. 7.20 Geometry for fabricating Vander Lugt filters. Dotted line is reference beam wavefront perpendicular to direction of propagation.

$y_d\cos\theta$ and parallel to it since the direction of propagation is $y_d\sin\theta$ which is perpendicular to them. The reference beam in the plane of the film is then

$$U_r(x_d, y_d) = r_0\exp(-j2\pi y_d\sin\theta/\lambda) \tag{7.7.1}$$

This expression describes a plane wave whose phase in plane P_2 is changing according to $y_d\sin\theta$ because of the diagonal direction of the reference beam. Equation (7.7.1) can also be written as

$$U_r(x_d, y_d) = r_0\exp(-j2\pi\alpha y_d) \tag{7.7.2}$$

where α represents a spatial frequency equal to $\sin\theta/\lambda$.

The total CFA incident on the film is then the sum of those representing the two beams

$$U_{p2}(x_d, y_d) = C_{6L_4}H\left(\frac{x_d}{\lambda f_{l4}}, \frac{y_d}{\lambda f_{l4}}\right) + r_0e^{-j2\pi\alpha y_d} \tag{7.7.3}$$

This yields an irradiance equal to

$$\begin{aligned} H_{p2}(x_d, y_d) &= U_{p2}(x_d, y_d)\cdot U_{p2}^*(x_d, y_d)/2\eta_0 \\ &= \frac{1}{2\eta_0}\left[\frac{1}{(\lambda f_{l4})^2}\left|H\left(\frac{x_d}{\lambda f_{l4}}, \frac{y_d}{\lambda f_{l4}}\right)\right|^2 + r_0^2\right. \\ &\quad + r_0C_{6L_4}H\left(\frac{x_d}{\lambda f_{l4}}, \frac{y_d}{\lambda f_{l4}}\right)e^{j2\pi\alpha y_d} \\ &\quad \left. + r_0C_{6L_4}^*H^*\left(\frac{x_d}{\lambda f_{l4}}, \frac{y_d}{\lambda f_{l4}}\right)e^{-j2\pi\alpha y_d}\right] \end{aligned} \tag{7.7.4}$$

where in the first term $(\lambda f_{l4})^{-2}$ is equal to $|C_{6L_4}|^2$. The third and fourth terms on the right-hand side involve mixing of the two CFAs incident on the film and yield irradiance proportional to the CFA H or H^*. The mixing transfers the spatial frequency spectrum of H from that around the origin in plane P_2 to the higher carrier spatial frequency α, which increases with θ. (Note that in Fig. 7.20, $\theta < \pi/2$.) This remarkable ability to record a *complex* function such as H or H^* including phase on an *irradiance*-sensitive material such as photographic film forms the basis of holography and optical pattern recognition, correlation, and convolution.

The third and fourth terms in Eq. (7.7.4) are the mixing terms and are produced by the interference of the two input light patterns. This requires that both input waves be coherent, otherwise H_{P_2} would derive from the first two terms only which comprise the sum of the individual irradiances as in Eq. (7.4.17). Coherent systems are thus linear in amplitude [Eq. (7.7.3)] but not irradiance [Eq. (7.7.4)], whereas noncoherent systems are linear in both amplitude (total CFA = sum of CFAs) and in irradiance (total irradiance equals sum of individual irradiances). Vander Lugt filters and holography require coherence. In the present case, the desired information is in the third and fourth terms of Eq. (7.7.4), which result from mixing of the two beams, and their existence depends on the coherence of the two beams. (The subject of coherence is addressed more fully in Section 7.9.)

The film in plane P_2 in Fig. 7.20, on which the interference pattern is recorded, is developed and a positive transparency is made. This is the Vander Lugt filter. It is placed in plane P_2 in the optical computer described in Fig. 7.17. Critical to the application of the filter to optical data processing is its transmission. Typical transmission as a function of exposure [defined in Eq. (3.1.14)] is as shown in Fig. 7.21. The region between the dotted lines is that in which transmission is linear with exposure. Irradiance and time exposures must be set so as to operate within this region, that is, the film is used as a square-law detector. Accordingly, transmission of the Vander Lugt filter is, for a given appropriate time exposure,

$$t(x_d, y_d) \propto H_{p_2}(x_d, y_d) \tag{7.7.5}$$

where $H_{P_2}(x_d, y_d)$ consists of the four terms described in Eq. (7.7.4). Referring to Fig. 7.17, an input signal $g(x_s, y_s)$ is placed in plane P_1. Incident on plane P_2 is the CFA $C_{6L_2}G(x_d/\lambda f_{l_2}, y_d/\lambda f_{l_2})$, where G designates the Fourier transform of g and C_{6L_2} is $C_6(f_{l_2})$. The CFA exiting the Vander Lugt filter is then the product of the CFA incident on it and the transmission of the filter. In view of Eqs. (7.7.4) and (7.7.5), the CFA exiting the Vander Lugt filter is

$$\begin{aligned} U_{fl_2}(x_d, y_d) \propto {} & \frac{C_{6L_2}G}{2\eta_0\lambda^2 f_{l_4}^2}|H|^2 + \frac{C_{6L_2}r_0}{2\eta_0} \\ & + C_{6L_2}C_{6L_4}r_0GHe^{j2\pi\alpha y_d}/2\eta_0 \\ & + C_{6L_2}C_{6L_4}^{*}r_0GH^*e^{-j2\pi\alpha y_d}/2\eta_0 \end{aligned} \tag{7.7.6}$$

where C_{6L_4} is $C_6(f_{l_4})$. Lens L_3 in Fig. 7.17 then gives rise to the inverse Fourier transform of Eq. (7.7.6) in plane $P_3(x_3, y_3)$. Products in Eq. (7.7.6) therefore give rise to convolutions of inverse transforms in plane P_3. As seen from Eqs. (7.7.4), each term in plane P_3 must also be multiplied by $C_{6L_3}\lambda^2 f_{l_3}^2$. Each term in Eq. (7.7.6) contains the coefficient C_{6L_2}, which, when multiplied by $C_{6L_3}\lambda^2 f_{l_3}^2$, yields C_7 as defined in Eq. (7.6.6). The magnitude of C_7 is f_{l_3}/f_{l_2}. Accordingly,

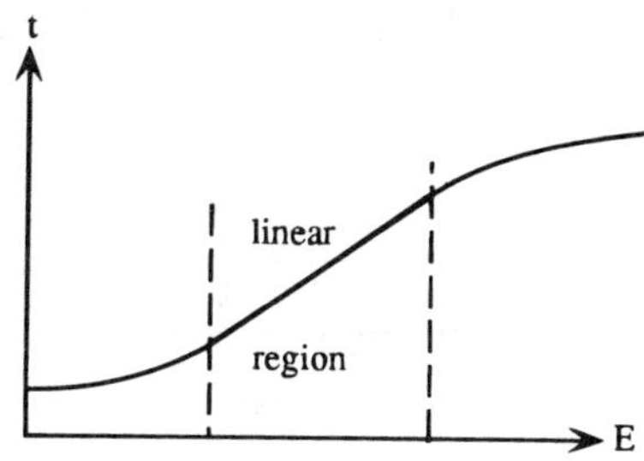

Fig. 7.21 Typical transmission of positive transparency as a function of exposure.

$$U_{fl_3}(x_3, y_3) \propto \frac{C_7}{2\eta_0\lambda^2 f_{l4}} [h(x_3, y_3)*h(-x_3, -y_3)*g(x_3, y_3)] + \frac{C_7 r_0^2}{2\eta_0} g(x_3, y_3)$$
$$+ C_7 C_{6L_4} r_0 [g(x_3, y_3)*h(x_3, y_3)*\delta(x_3, y_3 + \alpha\lambda f_{l_3})]$$
$$+ C_7 C^*_{6L4} r_0 [g(x_3, y_3)*h(-x_3, -y_3)*\delta(x_3, y_3 - \alpha\lambda f_{l_3})] \quad (7.7.7)$$

The first two terms on the right-hand side of Eq. (7.7.7) are centered at the origin of plane P_3. The third term, because of delta function convolution, is centered at $(0, -\alpha\lambda f_{l_3})$ in plane P_3 and is therefore displaced from the origin upward in the $-y_3$ direction as shown in Fig. 7.22. Since the displacement depends on $\alpha = \sin\theta/\lambda$, it is clear that if $\theta = 0$ in the Vander Lugt filter fabrication there is no such displacement. Hence, the reference beam in Fig. 7.20 should be incident at an angle $\theta \neq 0$.

Similarly the fourth term is displaced a distance $\alpha\lambda f_{l_3}$ downward in the $+y_3$ direction, and here too, were it not for the off-axis direction θ in Fig. 7.20 used in Vander Lugt filter fabrication, the fourth term would be centered at the origin in the output plane P_3 where it would be essentially indistinguishable from the first two terms.

The effect of the angle θ can be viewed in the following way. If the complex function H has an amplitude distribution $|\mathbf{H}(x_d, y_d)|$ and a phase distribution $\psi(x_d, y_d)$ so that $\mathbf{H} = |\mathbf{H}| \exp(-j\psi)$, then Eq. (7.7.6) can be rewritten in the following form:

$$U_{fl_4}(x_d, y_d) \propto \frac{C_{6L_2} G}{2\eta_0\lambda^2 f_{l4}^2} |H|^2 + \frac{C_{6L_2} r_0^2 G}{2\eta_0}$$
$$+ \frac{2C_{6L_2} r_0 G}{2\eta_0} |C_{6L_4} H| \cos(2\pi\alpha y_d - \psi) \quad (7.7.8)$$

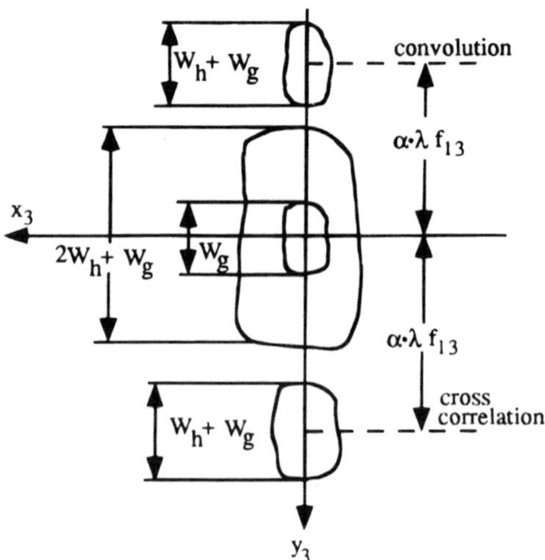

Fig. 7.22 Output plane P_3 of optical computer using Vander Lugt filter after [7.5].

This form illustrates the method by which the interference process in which the Vander Lugt filter is fabricated permits recording of both amplitude and phase of H. They are recorded as amplitude and phase modulations of a *high* spatial frequency carrier introduced by the tilted reference beam. In the previous chapter heterodyne detection was seen as a mixing of two temporal frequencies, which produced a difference or intermediate frequency (IF). Here, low spatial frequency information (H) near the origin is mixed with a high spatial frequency carrier producing a spatial frequency sum (instead of difference) so that the low spatial frequency information is carried by the high spatial frequency carrier. Consequently, the information terms in P_3 are displaced from the origin.

With regard to output plane P_3, in order to understand the significance of each term in Eq. (7.7.7), consider the following definitions:

$$\text{Convolution: } g*h = \iint g(\xi, \eta)h(x_3 - \xi, y_3 - \eta)\, d\xi\, d\eta$$

$$\text{Cross correlation: } g \oplus h = \iint g(\xi, \eta)h^*(x_3 + \xi, y_3 + \eta)\, d\xi\, d\eta$$

Therefore, the first term in Eq. (7.7.7) involves correlation of h with h and convolution with g, all of which is centered at the origin. The second term is simply proportional to g, also centered at the origin.

The third term involves convolution of g with h, and the fourth term involves correlation of g with h, both terms being displaced from the origin. These two terms are the desired convolution and correlation terms. By virtue of the diagonal angle of incidence θ in Fig. 7.20, each can be displaced sufficiently so as to not overlap with those at the origin. In Fig. 7.22, W_h and W_g represent spatial frequency bandwidths of h and g, respectively. Thus, in order to prevent overlap it is required that [7.5]

$$(W_h + W_g)/2 + W_h + W_g/2 < \alpha\lambda f_{l_3}$$

or

$$\sin\theta > \frac{1}{f_{l_3}}\left(W_g + \frac{3}{2}W_h\right) \tag{7.7.9}$$

Applications of Vander Lugt filtering are many and varied. Examples can be identification of fingerprints, missile batteries, etc. In the case of fingerprints, $h(x_4, y_4)$ can be a transparency of a specific fingerprint. The input $g(x_s, y_s)$ to the optical computer can be any fingerprint transparency in a data bank. When the input $g(x_s, y_s)$ is equal to $h(x_4, y_4)$, then correlation is maximum and the owner of the fingerprint at hand has been identified. Similarly, for missiles, $h(x_4, y_4)$ can be a transparency of a missile battery, while $g(x_s, y_s)$ can be a set of reconnaissance transparencies. When the input transparencies $g(x_s, y_s)$ include a missile battery, then correlation with $h(x_4, y_4)$ is maximum and the presence of a missile battery at a specific location has been identified.

An example of optical pattern recognition is shown in Fig. 7.23 for the word "radar". A corresponding dot appears in the output plane P_3 wherever "radar" appears in the input signal $g(x_s, y_s)$. To understand the mechanism of the correlation, consider that the fourth term in Eq. (7.7.4) contains H^* $(x_d/\lambda f_{l4}, y_d/\lambda f_{l4})$. In plane P_3 (x_s, y_s) wherever $g(x_s, y_s) = h(x_s, y_s)$ (in this case "radar"), then in plane P_2 for the fourth term in Eq. (7.7.6) $\mathbf{GH}^* = \mathbf{HH}^* = |\mathbf{H}|^2$. The phase of $\mathbf{H}^*$ is equal and opposite to that of $\mathbf{H}$. The resultant wavefront $\mathbf{HH}^*$ exiting P_2 is therefore a *plane* wave. Lens L_3 then focuses it in plane P_3 to a point rather than extended image. Essentially what is taking place is that when $g(x_s, y_s)$ in plane P_1 equals $h(x_s, y_s)$, the Vander Lugt filter in plane P_2 with transmission proportional to $\mathbf{H}^*$ is then a *matched filter* to this particular signal only. Wherever in plane P_1 $g(x_s, y_s) \neq h(x_s, y_s)$, then the corresponding transmission of the filter in plane P_2 does not cancel the spatial modulation of G and the CFA exiting the filter is therefore not a plane wave to be focused by L_2 to a point in plane P_3. Correlation in plane P_3 is essentially zero.

Parallel processing and real-time operation are two of the principal advantages of optical data processing over other technologies. However, in optical data processing the input and filter plane data must be available as transparencies. The data must be introduced into the optical computer as a transparency that can be changed in real time. Materials and devices capable of such performance are referred to as *spatial light modulators* or *light valves*. More widespread use of optical data processing is largely dependent on advances in light valve technology. Small compact optical computers have passed military field tests in recent years.

7.8 HOLOGRAPHY

Holography is a method of recording two mutually coherent beams in which one beam is the object beam and the other a reference beam. The purpose of recording a hologram is not optical data processing but rather a lensless method of forming an image [7.7–7.9]. The uniqueness of a hologram as compared to an ordinary photographic image, for example, is that the latter contains amplitude information only. If $a(x, y)$ represents the CFA of a given scene, then a photograph records only $|a(x, y)|^2$. All phase information is lost. A hologram, on the other hand, because it involves interference of $a(x, y)$ with a reference beam records both amplitude and phase of $a(x, y)$ as seen within its field of view by a square-law photosensitive device such as film.

The film is exposed within its region of linearity (Fig. 7.21) whereby transmission is proportional to irradiance for a given time exposure. The transparency resulting

THE DEVELOPMENT OF RADAR DURING WORLD WAR II BROUGHT RADAR FROM A LABORATORY CONCEPT TO A MATURE DISCIPLINE IN JUST A FEW SHORT YEARS. SINCE 1945 RADAR TECHNOLOGY HAS BECOME SO SOPHISTICATED THAT THE BASIC RECTANGULAR PULSE RADAR SIGNAL IS NO LONGER SUFFICIENT IN THE DESIGN OF MANY NEW RADAR SYSTEMS. MORE COMPLEX RADAR SIGNALS MUST BE TAILORED TO SPECIFIC REQUIREMENTS.

Fig. 7.23 Example of optical pattern recognition. (a) Input plane pattern and (b) output correlation plane pattern. The position of each of the output correlation peaks of light corresponds to one of the locations of the word "radar" in the input paragraph (after [7.6]).

from the film is the hologram. To view the holographic image, the hologram is illuminated with a reference wave, which causes $a(x, y)$ (both amplitude and phase) to be reconstructed. Since hologram reconstruction involves $a(x, y)$ in its entirety, depth information is included as well. Furthermore, it is even possible to view the holographic scene from different physical points of view and to see around objects in the scene. This ability, however, is limited by the solid angle subtended by the object when the hologram is recorded. Hologram reconstruction cannot supply more information than that recorded in the hologram itself.

Typically examples of types of holograms are shown in Fig. 7.24. In Fig. 7.24(a) the object beam is irradiance reflected by the scene and in part (b) the object beam is irradiance transmitted through the scene. In Fig. 7.24(c) a lens is inserted one focal length away from the film so that the object beam is the Fourier transform of light reflected from the scene. In each case, a reference beam is incident on the film as well. In Fourier transform holography, the image upon reconstruction is not that of the scene itself but rather the spatial Fourier transform of the scene. An additional lens must be inserted in the reconstruction process so as to inverse transform the resulting CFA into that of the scene itself.

The mathematical treatment is similar to that of the Vander Lugt filter since both involve recording of interference patterns. Here, it will be considered from a simple, basic point of view with a plane wave split by the beamsplitter.

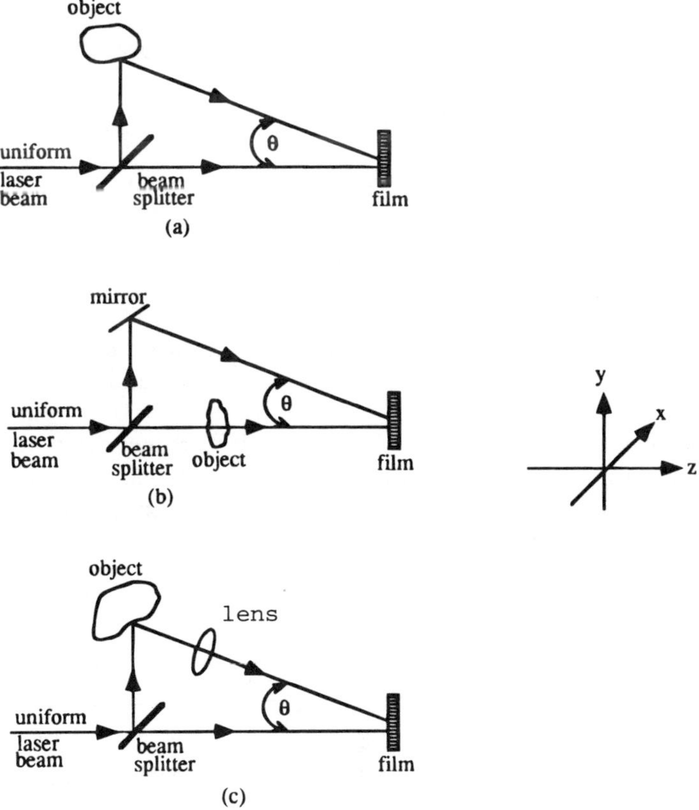

Fig. 7.24 Setups for recording various types of holograms: (a) reflection hologram; (b) transmission hologram; (c) Fourier transform reflection hologram. Assume object distance is z_s.

In Fig. 7.24 all examples are off-axis holograms, that is, there exists an angle θ between the object and reference beams incident on the film. Accordingly, the CFA incident on the film is

$$U(x, y) = Ae^{-j2\pi\alpha y} + a(x, y) \tag{7.8.1}$$

where, as in Eq. (7.7.2), $\alpha = \sin\theta/\lambda$. Since the film is used in its linear region of exposure and acts as a square-law device with regard to field amplitude, transmission of the resulting transparency is proportional to exposure or irradiance:

$$t(x, y) \propto [A^2 + |a(x, y)|^2 + Aa(x, y)e^{j2\pi\alpha y} + Aa^*(x, y)e^{-j2\pi\alpha y}]/2\eta_0 \tag{7.8.2}$$

The first term is the reference beam irradiance, which contains no information regarding $a(x, y)$. The second term is the magnitude of the scene irradiance and is what would be observed in an ordinary photographic image. The third and fourth terms contain the required information that can be seen after hologram reconstruction. The developed film is illuminated with a uniform reference plane wave CFA whose value is B as in Fig. 7.25. The CFA exiting the hologram transparency is then proportional to the product of B and $t(x, y)$ in Eq. (7.8.2), or

$$U_r(x, y) \propto \frac{B}{2\eta_0}[A^2 + |a(x, y)|^2 + Aa(x, y)e^{j2\pi\alpha y} + Aa^*(x, y)e^{-j2\pi\alpha y} \tag{7.8.3}$$

By writing $a(x, y) = |a(x, y)\,[\exp(-\phi(x, y))]$, the preceding equation can be written

$$U_r(x, y) \propto \frac{B}{2\eta_0}[A^2 + |a(x, y)|^2 + 2A|a(x, y)|\cos[2\pi\alpha y - \phi(x, y)] \tag{7.8.4}$$

which indicates that the amplitude *and* phase of light arising from the object have been recorded, respectively, as amplitude and phase modulations at the high spatial carrier frequency, α. In this way they can be separated spatially from the first two terms whose spatial frequency components are at low frequencies near the origin.

The third term in Eq. (7.8.3) involves spatial modulation $a(x, y)$ of a plane wave CFA, AB. The linear exponential factor $\exp(j2\pi\alpha y) = \exp(jky\sin\theta)$ indicates that this image is diffracted off the transparency axis at an angle θ as shown in Fig. 7.25

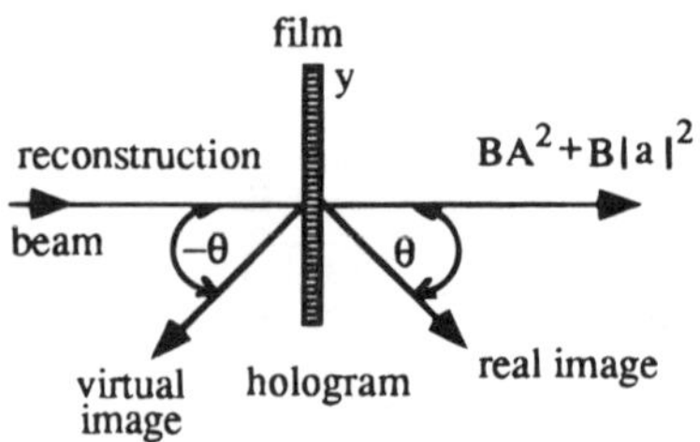

Fig. 7.25 Hologram reconstruction.

and is propagating in the $-y \sin \theta$ direction. The fourth term in Eq. (7.8.3) involving spatial modulation $a^*(x, y)$ of the plane wave AB is deflected off axis by an angular distance θ. The third and fourth terms thus exhibit spatial modulation of plane waves by the original CFA, including both amplitude and phase which provide depth information. These images are diffracted off axis from the reconstruction beam. Thus, the image can be observed without looking directly into the reconstruction beam. The first and second terms in Eqs. (7.8.3) and (7.8.4) are low spatial frequency components and are carried directly on the reconstruction beam rather than off axis.

A real image can be projected onto a viewing system and therefore involves a direction component in the direction of B. A virtual image cannot be seen on an observation screen, but rather involves a direction component opposite to B. It can be seen without a viewing screen.

The holograms described here are off axis, that is, the reference and object beams in Fig. 7.24 are not coincident but are inclined to each other at an angle θ, thus permitting diffraction at the same angle θ of real and virtual images away from the reconstruction or zero-order beam. This is a variation attributed to Leith and Upatnieks [7.10, 7.11] at the University of Michigan from the original scheme suggested by Nobel prize winner Dennis Gabor, which was on axis [7.7–7.9].

When viewing the holographic image it is important to focus one's eyes in the appropriate direction but not to focus on the hologram itself. The image is projected in space at a distance equal to z_s, and not on the hologram itself. The hologram is simply a recording of the interference pattern and contains only constructive and destructive interference fringes separated from one another by distances on the order of wavelength. Only a small portion of the interference pattern comprising the hologram is required in order to specify $a(x, y)$ uniquely. Therefore, the hologram can actually be cut and only a small portion illuminated by the reconstruction beam in order to reconstruct $a(x, y)$ or $a^*(x, y)$. This procedure would result in loss of signal-to-noise ratio in the reconstructed image since all the illuminated fringes contribute to the holographic image, but not in loss of information. This makes it possible to produce holographic memories. For example, one can produce a hologram of each page in a telephone directory, and attach a microscopic portion of each hologram on a card in which holograms are arranged according to page number in the phone book. A laser can be used to reconstruct the image of each individual page at a time by using an optical deflector (usually acousto-optic) to deflect the laser beam at different angles so as to illuminate each hologram separately.

The field of view in the hologram image depends on the solid angle subtended by the illuminated scene with reference to the film. Field of view is often narrow under ordinary hologram fabrication circumstances. Special techniques are required in order for the object beam in Fig. 7.24 to carry 360° information. These usually involve placing film as a cylinder 360° around an enclosed object scene, and then illuminating the whole scene from above. Partially reflecting cones around the object scene between it and the film cylinder provide reflection of radiation from above to the film [7.12–7.14].

Because holography is essentially an interferometric process, it is not limited to optical waves but can be carried out at millimeter wave and microwave frequencies as long as a square-law device for recording the interferogram can be used over large areas in place of film. One method is to use a single detector and to move it extensively over the interference plane so as to scan over many interference fringes, which increase in size because of the wavelength increase. This has been studied extensively by N. H. Farhat et al. at the University of Pennsylvania [7.15–7.17]. In such situations a visible reconstruction wavelength is different from that used in the recording process. This gives rise to some distortions in reconstructed image quality as a result of different magnifications in the lateral and longitudinal directions.

Central to optical computing, Vander Lugt filters, and holography is the requirement of coherence in order to permit mixing of spatial signals to take place. The optical path difference must be limited by coherence length; spatial spread must be limited by coherence diameter. Coherence is treated next.

7.9 COHERENCE

Temporal coherence was described in Section 4.6. It is necessary in heterodyne detection in order to permit mixing of signal and local oscillator temporal or electromagnetic frequency components so as to yield an IF component. Mixing of spatial frequency components is described in this chapter regarding Vander Lugt filters and holography. The present treatment is on spatial coherence, which is necessary to carry out the latter operations. Spatial coherence implies a fixed phase relationship across the full diameter of a cross section of a light beam.

The mathematical treatment [7.18, 7.19] is similar to that for temporal coherence given in Section 4.6. There, the complex degree of temporal coherence, $\gamma_{11}(\tau)$, was defined in Eq. (4.6.3) as the normalized autocorrelation function $\Gamma_{11}(\tau)/\Gamma_{11}(0)$ or $\Gamma_{11}(\tau)/H_1$ where H_1 is the time-averaged irradiance at the point $\bar{r}_1$ and is also the maximum value of the autocorrelation function $\Gamma_{11}(\tau) \triangleq \langle U_1(\bar{r}_1, t)U_1^*(\bar{r}_1, t + \tau)\rangle$, where angular brackets indicate long time average.

Now, generalizing this concept of correlation leads to a convenient technique for quantitatively describing the partial coherence properties associated with *two* complex field amplitudes at *two* different points in space. This type of coherence is referred to as *spatial coherence*. Thus, the *cross-correlational function* between the complex field amplitudes at two separate points in space at different instances of time is defined as the *mutual coherence function*

$$\Gamma_{12}(\tau) = \langle U_1(\bar{r}_1, t)U_2^*(\bar{r}_2, t + \tau)\rangle /2\eta_0 \qquad (7.9.1)$$

and the corresponding *complex degree of coherence* as

$$\gamma_{12}(\tau) = \frac{\Gamma_{12}(\tau)}{\sqrt{\Gamma_{11}(0)\Gamma_{22}(0)}} = \frac{\Gamma_{12}(\tau)}{\sqrt{H_1H_2}} \qquad (7.9.2)$$

where as before, $\Gamma_{11}(0)$ is the irradiance at $\bar{r}_1$ and equals H_1. Similarly, $\Gamma_{22}(0)$ is the irradiance H_2 at $\bar{r}_2$. When $\bar{r}_1 = \bar{r}_2$, that is, when the two space points are coincident, Eq. (7.9.2) reduces to Eq. (4.6.3). The complex degree of coherence as defined by Eq. (7.9.2) is therefore more general than that defined in Eq. (4.6.3) in the sense that it is also a measure of the coherence between the complex field amplitudes at *two different* space points; that is, it is a measure of the spatial as well as temporal coherence properties of the radiation field. Note that Eq. (7.9.1) corresponds to the definition of correlation preceding Eq. (7.7.9). When cross-correlation at two separate points is considered simultaneously, Eq. (7.9.1) reduces to $\Gamma_{12}(0)$, which is the *mutual irradiance function* (often called *mutual intensity function*). It is used to characterize spatial coherence.

The Young double-pinhole interference experiment serves to illustrate spatial coherence properties of radiation. Consider an extended quasimonochromatic source of mean wavelength $\overline{\lambda}$. The term *quasimonochromatic* means that the linewidth of the radiation $\Delta\nu$ is very small compared to mean frequency $\overline{\nu}$, and that all path differences in the experiment satisfy the condition $\Delta l << c/\Delta\nu$. In other words, path differences are less than coherence length l_c defined in Eq. (4.6.5). Suppose light from such a source reaches the two pinholes P_1 and P_2 as shown in Fig. 7.26. The irradiance at point P on a viewing screen can intuitively be expected to be dependent on the coherence between the radiation diffracted at P_1 and the radiation diffracted at P_2.

Let the complex field amplitudes at P_1 and P_2 caused by an infinitesimal emission element ds_m of the source be V_{m1} and V_{m2}, respectively. Since they are caused by the same source element, V_{m1} and V_{m2} will be coherent with each other as long as the difference between the distance of P_1 and P_2 from ds_m does not exceed the coherence length associated with emission of radiation from the m'th source element.

Next let f_1 and f_2 be the complex field amplitudes produced at a point P of the viewing screen in Fig. 7.26 by a CFA of unit magnitude and zero phase at P_1 and P_2, respectively. The CFA produced at P via P_1 and P_2 is then $(V_{m1}f_1 + V_{m2}f_2)$. The irradiance at P caused by the element ds_m is therefore

$$\begin{aligned} dH_p &= (V_{m1}f_1 + V_{m2}f_2)(V_{m1}f_1 + V_{m2}f_2)^*/2\eta_0 \\ &= [|V_{m1}|^2|f_1|^2 + |V_{m2}|^2|f_2|^2 + V_{m1}V_{m2}^*f_1f_2^* + V_{m1}^*V_{m2}f_1^*f_2]/2\eta_0 \quad (7.9.3) \\ &= |V_{m1}|^2|f_1|^2 + |V_{m2}|^2|f2_2|^2 + 2\,\mathrm{Re}(V_{m1}V_{m2}^*f_{m1}f_2^*) \end{aligned}$$

Since the different infinitesimal elements ds_m of the source all emit incoherently with each other, the total irradiance at P due to the entire source is the sum of the irradiance caused by the individual element:

$$\begin{aligned} H_p = \sum_m dH_p &= \left(\sum_m |V_{m1}|^2\right)|f_1|^2 + \left(\sum_m |V_{m2}|^2\right)|f_2|^2 + 2\,\mathrm{Re}\left\{\left(\sum_m V_{m1}V_{m2}^*\right)_1 f_2^*\right\} \\ &= H_1|f_1|^2 + H_2|f_2|^2 + 2\,\mathrm{Re}\left\{\left(\sum_m V_{m1}V_{m2}^*\right)f_1f_2^*\right\} \quad (7.9.4) \end{aligned}$$

where

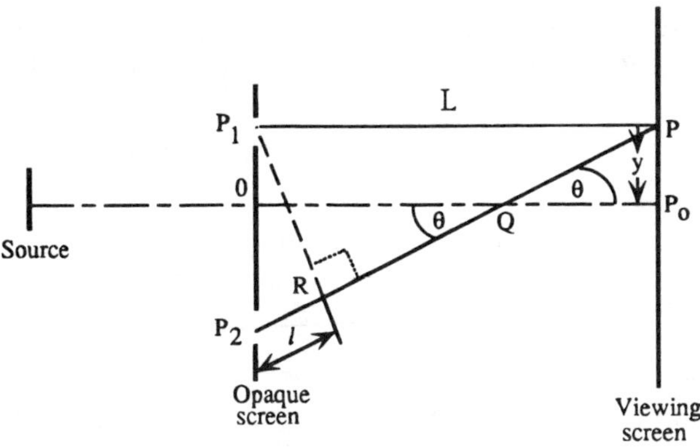

Fig. 7.26 Young's double-pinhole experiment (Assume P_1 and P_2 are almost symmetrical around the OP_0 axis.)

$$H_1 \triangleq \sum_m |V_{m1}|^2,\ H_2 \triangleq \sum_m |V_{m2}|2 \tag{7.9.5}$$

are the total irradiances at P_1 and P_2 due to the entire extended source S. Note that the form of Eq. (7.9.4) is the same regardless of the shape of the primary extended source S.

As shown earlier, the complex quantity

$$\Gamma_{12} \triangleq \sum_m V_{m1} V_{m2} \tag{7.9.6}$$

in Eq. (7.9.4) is called the *mutual irradiance* (or *intensity*) of P_1 and P_2. When P_1 and P_2 coincide, $\Gamma_{11} = \Gamma_{22} = H_1 = H_2$. Using this terminology a complex degree of coherence between P_1 and P_2 in analogy with Eq. (7.9.2) is defined as follows:

$$\gamma_{12} = \gamma_{12}(0) \triangleq \frac{\Gamma_{12}}{\sqrt{\Gamma_{11}\Gamma_{22}}} = \frac{\sum_m V_{m1} V^*_{m2}}{\sqrt{H_1 H_2}} \tag{7.9.7}$$

From the Schwarz inequality,

$$|\sum_m V_{m1} V^*_{m2}|^2 \leq \left(\sum_m |V_{m1}|^2 \sum_m |V_{m2}|^2 = H_1 H_2\right) \tag{7.9.8}$$

we can see that

$$\gamma_{12} \leq 1 \tag{7.9.9}$$

Using Eq. (7.9.7), Eq. (7.9.4) can be rewritten as

$$H_p = H_1|f_1|^2 + H_2|f_2|^2 + 2\sqrt{H_1 H_2}\ \mathrm{Re}\ \{\gamma_{12} f_1 f_2^*\} \tag{7.9.10}$$

When P coincides with the point P_0 on the viewing screen in Fig. 7.26, where P_0 is equidistant from P_1 and P_2, then $f_1 = f_2$ and Eq. (7.9.10) becomes

$$H_p = |f_1|^2[H_1 + H_2 + 2\sqrt{H_1 H_2}\ \mathrm{Re}\{\gamma_{12}\}] \tag{7.9.11}$$

Now, if P is moved away from P_0 by a small displacement y, then

$$f_2 = f_1 e^{-j\bar{k}_0 l} \tag{7.9.12}$$

where, from Fig. 7.26, $l = P_1P_2 \sin\theta$. Noting that $\Delta P_1P_2R \sim \Delta QP_2O \sim \Delta QPP_0$, it is clear that

$$0Q = P_1P_2/(2\tan\theta),\ QP_0 = y/\tan\theta$$

Adding, $0Q + QP_0 = L = [(P_1P_2/2) + y]/\tan\theta$. This implies

$$P_1P_2 = 2\,(L\tan\theta - y)$$

leading to the result

$$l = P_1P_2\sin\theta = (2\sin\theta)(L\tan\theta - y) \qquad (7.9.13)$$

Consequently, the irradiance at P can be expressed as

$$\begin{aligned}H_p &= |f_1|^2[H_1 + H_2 + 2\sqrt{H_1H_2}\mathrm{Re}\{\gamma_{12}\exp[j\bar{k}_0(2\sin\theta)(L\,tan\theta - y)]\}] \\ &= |f_1|^2[H_1 + H_2 + 2\sqrt{H_1H_2}\mathrm{Re}\{|\gamma_{12}|\exp(j\phi)\exp[j\bar{k}_0(2\sin\theta)(L\,tan\theta - y)]\}] \\ &= |f_1|^2[H_1 + H_2 + 2\sqrt{H_1\,H_2}|\gamma_{12}|\cos[\phi + \bar{k}_0 2\sin\theta\cdot(L\,tan\theta - y)] \quad (7.9.14)\end{aligned}$$

where $|\gamma_{12}| > 0$ for $l < l_c$. From Eq. (7.9.14) it is clear that when $\gamma_{12} = 0$, the irradiance does not vary with y, that is, $H_p \propto H_1 + H_2$. This is the case when no interference fringes are produced; in other words, the complex field amplitudes at P_1 and P_2 are mutually incoherent. However, when $\gamma_{12} \neq 0$, and θ is small, H_p varies as a function of y between

$$H_{p_{max}} = |f_1|^2(H_1 + H_2 + 2\sqrt{H_1H_2}|\gamma_{12}|)$$

and

$$H_{p_{min}} = |f_1|^2(H_1 + H_2 - 2\,\sqrt{H_1H_2}\,|\gamma_{12}|) \qquad (7.9.15)$$

An interference pattern is formed on the viewing screen. The visibility of fringes in the interference pattern is defined in a manner analogous to the standing wave ratio in electric networks:

$$V = \frac{H_{p_{max}} - H_{p_{min}}}{H_{p_{max}} + H_{p_{min}}} = \frac{2\sqrt{H_1H_2}|\gamma_{12}|}{H_1 + H_2} \qquad (7.9.16)$$

It is equal to the ratio of irradiance of the cosine term to that of the uniform background. In particularly, when $H_1 = H_2$ (i.e., for equal irradiance at P_1 and P_2) the fringe visibility V and the modulus $|\gamma_{12}|$ of the complex degree of coherent are equal. This provides a convenient method for measuring $|\gamma_{12}|$. $|\gamma_{12}|$ is often referred to as the *degree of coherence*.

Equation (7.9.14) is a very important result. It describes interference in the viewing screen plane. The interference is characterized by the cosine term. Interference contrast depends on irradiance of the cosine term relative to the constant background $H_1 + H_2$. The interference varies too with distance P_1P_2, through dependence of the cosine argument on θ. For small P_1P_2, $\tan\theta$ is small and the cosine argument varies rapidly with y. As P_1P_2 increases, $\tan\theta$ increases much more than $\sin\theta$, and dependence of the interference with y decreases. For very large P_1P_2, H_p tends toward a constant almost independent of position in the viewing screen plane because the value of the cosine term changes very slowly with y. Thus, over a given dimension y, aperture separation P_1P_2 also plays a role in determining interference visibility or contrast. However, while the cosine argument depends on P_1P_2, the cosine coefficient is $|\gamma_{12}(0)|$.

7.9.1 The Van Cittert-Zernike Theorem

The Van Cittert-Zernike theorem provides a method for calculation of degree of coherence in the radiation field of an extended quasimonochromatic source. Thus, to derive the theorem, we need to determine the complex degree of coherence γ_{12} for two points P_1 and P_2 on a screen or surface irradiated by an extended quasimonochromatic source s as shown in Fig. 7.27.

Assume for simplicity that s is planar with its plane parallel to an observation screen and that the intervening medium is homogeneous and isotropic. Assume also that the maximum linear dimension of s is small compared to the distance R between the source and the screen and that the angles between the lines joining P_1 and P_2 to any point of the source are small. The source is now divided into a set of N elementary sources $\{ds_1, ds_2, \ldots, ds_N\}$, centered at points $\{p_1, p_2, \ldots, p_N\}$, which we assume to be statistically independent in their emissions, that is, incoherent with respect to each other. The size of the element ds_m is assumed to be less than the mean wavelength of the quasimonochromatic radiation so that interference effects of radiation arriving at the screen from different parts of an elementary source cannot take place.

Let $U_{m1}(t)$ and $U_{m2}(t)$ be, respectively, the complex field amplitudes at P_1 and P_2, generated by sources ds_m. If we assume the polarizations of all elemental sources

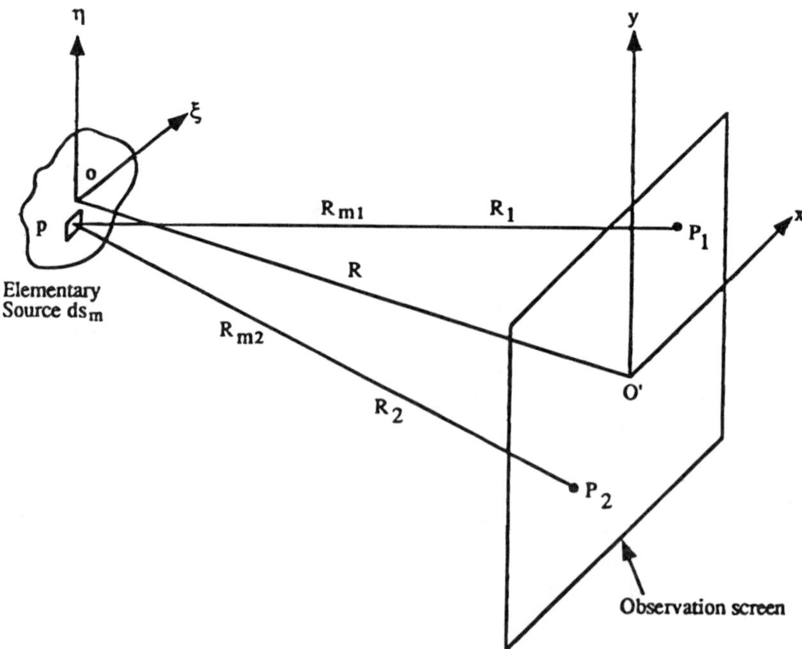

Fig. 7.27 Geometry for derivation of the Van Cittert-Zernike theorem.

are identical, then the total complex field amplitude at P_1 and P_2 is a simple summation; namely,

$$U_1(t) = \sum_m U_{m1}(t) \qquad \text{and} \qquad U_2(t) = \sum_m U_{m2}(t) \tag{7.9.17}$$

respectively. The mutual irradiance for P_1 and P_2 (i.e., the cross-correlation of the total complex field amplitudes at P_1 and P_2) is therefore

$$\begin{aligned} H_{12} &= \langle U_1(t)U_2^*(t)\rangle/2\eta_0 = \langle \sum_m U_{m1}(t) \sum_n U_{n2}^*(t)\rangle/2\eta_0 \\ &= [\sum_m \langle U_{m1}(t)U_{m2}^*(t)\rangle + \sum_{m\neq n}\sum \langle U_{m1}(t)U_{n2}^*(t)\rangle]/2\eta_0 \end{aligned} \tag{7.9.18}$$

Since the complex field amplitudes produced at a point due to different elements of the source are incoherent (i.e., statistically independent) the double sum in Eq. (7.9.18) vanishes.

Denoting the distances of P_1 and P_2 from the infinitesimal element d_{sm} by R_{m1} and R_{m2}, and assuming ds_m behaves like a point source of mean frequency $\overline{\nu}$ $(= c/\lambda)$, the following specific expressions for $U_{m1}(t)$ and $U_{m2}(t)$ result from spherical wave propagation [Eq. (1.3.40)] where propagation time is included:

$$\begin{aligned} U_{m1}(t) &= A_m\left(t - \frac{R_{m1}}{v}\right)\frac{\exp\left[j2\pi\overline{\nu}\left(t - \frac{R_{m1}}{v}\right)\right]}{R_{m1}} \\ U_{m2}(t) &= A_m\left(t - \frac{R_{m2}}{v}\right)\frac{\exp\left[j2\pi\overline{\nu}\left(t - \frac{R_{m2}}{v}\right)\right]}{R_{m2}} \end{aligned} \tag{7.9.19}$$

where v is the phase velocity in the medium between the source and the screen and $A_m(t)$ is a complex quantity characterizing the strength and the phase of radiation emitted from the source element ds_m. According to Eq. (7.9.19),

$$\langle U_{m1}(t)U_{m2}^*(t)\rangle = \left\langle A_m\left(t - \frac{R_{m1}}{v}\right)A_m^*\left(t - \frac{R_{m2}}{v}\right)\right\rangle \times \frac{\exp[j2\pi\overline{\nu}(R_{m2} - R_{m1})]}{R_{m1}R_{m2}} \tag{7.9.20}$$

If the path difference $(R_{m2} - R_{m1})$ in Eq. (7.9.20) is small compared to the coherence length of the source element ds_m, then Eq. (7.9.20) can be written as

$$\langle U_{m1}(t)U_{m2}^*(t)\rangle = \langle A_m(t)A_m^*(t)\rangle \frac{\exp[j2\pi\overline{\nu}(R_{m2} - R_{m1})]}{R_{m1}R_{m2}} \tag{7.9.21}$$

Combining Eqs. (7.9.21) and (7.9.18),

$$\Gamma_{12} = \sum_m \langle A_m(t)A_m^*(t)\rangle \frac{\exp[j2\pi\bar{\nu}(R_{m2} - R_{m1})]}{R_{m1}R_{m2}} \tag{7.9.22}$$

The quantity $\langle A_m(t)A_m^*(t)\rangle/2\eta_0$ characterizes irradiance from the element ds_m. If it is assumed that $ds_m \to 0$ so that the source can be regarded as effectively continuous, then denoting irradiance per unit area at point p_m of the source by $H(p_m)$,

$$\langle A_m(t)A_m^*(t)\rangle \triangleq H(p_m)ds_m(2\eta_0) \tag{7.9.23}$$

Using Eq. (7.9.23), the cross-correlation function Eq. (7.9.22) becomes

$$\Gamma_{12} = 2\eta_0 \int_s H(p) \frac{\exp[-j\bar{k}_0(R_1 - R_2)]}{R_1R_2} ds \tag{7.9.24}$$

where k_0 is the wave number or propagation factor and R_1, R_2 are, respectively, the distances of the observation points P_1 and P_2 from the source point p. Using Eq. (7.9.24) in Eq. (7.9.2), the complex degree of coherence is

$$\gamma_{12} = \frac{1}{\sqrt{H_1H_2}} \int_s H(p) \frac{\exp[-j\bar{k}_0(R_1 - R_2)]}{R_1R_2} ds \tag{7.9.25}$$

where H_2 and H_2 can be obtained from Eq. (7.9.24) by letting $R_1 = R_2$:

$$H_1 = 2\eta_0 \int_s \frac{H(p)}{R_1^2} ds, \qquad H_2 = 2\eta_0 \int_s \frac{H(p)}{R_2^2} ds \tag{7.9.26}$$

7.9.2 Coherence Diameter

Note that the integral in Eq. (7.9.25) is similar to the diffraction theory Fraunhofer diffraction integral of Eq. (7.2.13). To see this relationship more clearly return to Fig. 7.27 and note $R = OO'$. In addition, let the coordinates of a point p on the source be designated by ζ, η and the coordinates P_1 and P_2 by X_1, Y_1 and X_2, Y_2, respectively. Then

$$\begin{aligned} R_1^2 &= (X_1 - \zeta)^2 + (Y_1 - \eta)^2 + R^2 \\ &= R^2\left[1 + \frac{(X_1 - \zeta)^2}{R} + \frac{(Y_1 - \eta)^2}{R}\right] \end{aligned}$$

or

$$R_1 \approx R + \frac{(X_1 - \zeta)^2}{2R} + \frac{(Y_1 - \eta)^2}{2R} \tag{7.9.27}$$

where the binomial in the last step is consistent with the condition postulated earlier that the angles between the lines joining P_1 and P_2 to the point p of the source are small. Similarly,

$$R_2 \approx R + \frac{(X_2 - \xi)^2}{2R} + \frac{(Y_2 - \eta)^2}{2R} \tag{7.9.28}$$

The phase factor in Eq. (7.9.25) can be expressed in the familiar form of Eqs. (7.3.2) and (7.3.3). Thus, calculating

$$(R_1 - R_2) \approx \frac{(X_1^2 + Y_1^2) - (X_2^2 + Y_2^2)}{2R} - \frac{(X_1 - X_2)\xi + (Y_1 - Y_2)\eta}{R} \tag{7.9.29}$$

and letting

$$-\frac{X_1 - X_2}{R} \triangleq \alpha, \qquad -\frac{Y_1 - Y_2}{R} \triangleq \beta \tag{7.9.30}$$

$$\psi \triangleq [\bar{k}(X_1^2 + Y_1^2) - (X_2^2 + Y_2^2)]/2R \tag{7.9.31}$$

Eq. (7.9.25) becomes a Fourier transform of the source irradiance distribution normalized by total irradiance of the source (or by the value of the Fourier transform at zero spatial frequency):

$$\gamma_{12} = \frac{e^{-j\psi} \iint_s H(\zeta,\eta)\exp[-jk_0(\alpha\xi + \beta\eta)]\, d\xi\, d\eta}{\iint_s H(\xi,\eta)\, d\xi\, d\eta} \tag{7.9.32}$$

where γ_{12} represents the complex degree of coherence of the source. It can be used to define the *coherence diameter* of the emitted radiation reaching the observation plane.

For example, consider a flat *uniform* source of radius a with center at o in Fig. 7.27. Equation (7.9.32) gives

$$\gamma_{12} = \left[\frac{2J_1(u)}{u}\right] e^{-j\psi} \tag{7.9.33}$$

where ψ is given by Eq. (7.9.31) and J_1 is Bessel's function of the first kind and first order. From Eq. (7.4.10),

$$u = \frac{2\pi}{\lambda}(d/R)a \tag{7.9.34}$$

where, from Eqs. (7.9.29) and (7.9.30), d is separation between points P_1 and P_2 in the observation plane and a is source radius.

From the properties of the Bessel function of the first order, $|2J_1(u)/u|$ has a maximum value of unity at $u = 0$ and decreases to zero at $u = 3.83$. Thus, as u increases (i.e., as the separation between P_1 and P_2 on the screen increases), the complex degree of coherence $|\gamma_{12}|$ will behave as shown in Fig. 7.28. Because it is dependent on the distance between P_1 and P_2, $|\gamma_{12}|$ is a measure of spatial coherence of source radiation as it reaches the observation plane.

Using Eq. (7.9.34) and Fig. 7.27, the separation $d_0(= 2r_0)$ between points P_1 and P_2 for which the first zero of coherence occurs is found in the following manner:

$$u = 3.83 = \frac{2\pi}{\bar{\lambda}} \frac{d_0 a}{R} \rightarrow$$

$$d_0 = 1.22\, \bar{\lambda}(R/2a) = 1.22\bar{\lambda}/\theta_s \tag{7.9.35}$$

where $2a/R \cong \theta_s$ is the angular spread of the source as seen from center point O' in the observation screen in Fig. 7.27.

Often the *diameter of coherence* $d_c(= 2r_c)$ associated with a uniform source of radius a at a screen a distance R from the source is a useful source parameter. It is defined as the observation point separation when $|\gamma_{12}| = 0.88$, which occurs for $u = 1$ in Fig. 7.28. Thus, from Eq. (7.9.34),

$$u = 1 = \frac{2\pi}{\bar{\lambda}} \frac{2}{R} d_c \tag{7.9.36}$$

so that

$$d_c = \bar{\lambda}R/(2\pi a) = \bar{\lambda}/(\pi\theta_s) \tag{7.9.37}$$

The dependence of coherence diameter on small values of θ_s implies d_c increases as the radiation incident on the observation plane becomes more and more paraxial.

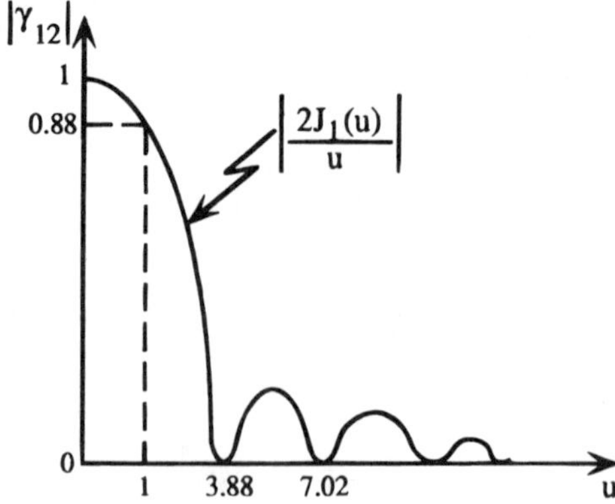

Fig. 7.28 Variation of complex degree of coherence on an observation plane for a flat uniform source.

The phase differences from the source to various points on the observation plane thereby decrease, thus improving spatial coherence and its diameter.

In Young's pinhole experiment, pinhole separation *d* clearly affects γ_{12} (Fig. 7.28) and thus the intensity of the cosine term in Eq. (7.9.14) and the visibility of the interference pattern in Eq. (7.9.16).

Coherence diameter is an essential parameter in Fourier optics, for it is a measure of the spatial extent over which Fourier transforms, inverse Fourier transforms, Vander Lugt filters, cross correlations, convolutions, etc., can be obtained. This is so because in Fourier or inverse Fourier transform planes, at radial distances from the origin

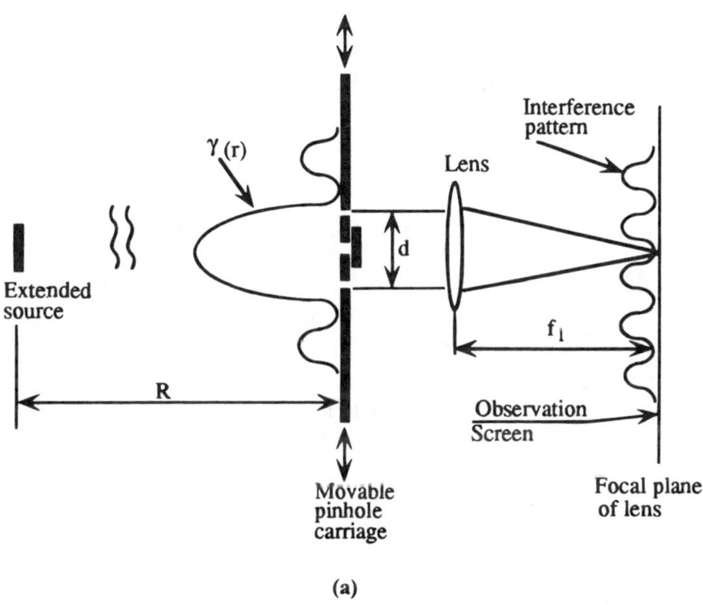

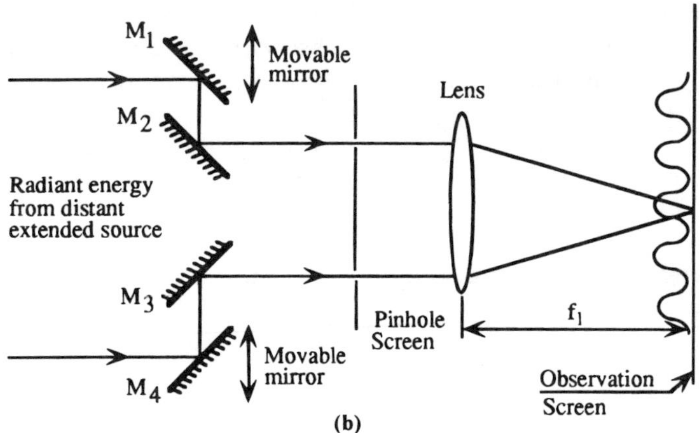

Fig. 7.29 Measurement of diameters of distant extended sources: (a) basic arrangement; (b) Michelson stellar interferometer (M_1, M_4 movable; M_2, M_3, fixed).

that are less than coherence radius $d_c/2$, at each point in space essentially one value of phase is obtained for a given CFA. At points further away from the origin, a collection of phases is obtained at each point, so that Fourier and inverse Fourier transforms are not defined at such locations. Thus, relatively narrow linewidth nonlaser sources can also be used for Fourier optics provided small pinholes are used to effectively decrease a in Eq. (7.9.36) so as to increase coherence diameter by making the beamwidth wider and the wavefront more planar. It is also necessary for interferometer path *differences* in Vander Lugt filter fabrication to be shorter than the coherence length of the source.

The analysis here is based on spherical wave propagation as specified in Eq. (7.9.19). This is *inappropriate to lasers*, which exhibit $|\gamma_{12}(0)| \cong 1$ over the spatial extent of the beam if spatial coherence is not distorted by the propagation channel. If the laser beam is expanded, the spatial coherence essentially remains unchanged and coherence diameter also expands. Single-mode lasers have a higher degree of spatial coherence than do multimode ones, and in a two-pinhole experiment they exhibit higher fringe visibility all across the beam.

An important implication here is that even "incoherent" sources such as the sun or blackbody radiators still have some degree of partial coherence (Exercise 7.22).

7.9.3 Measurement of Diameters of Extended Sources

The analysis leading to Eq. (7.9.35) can be put to practical use if the geometry of Fig. 7.27 is modified to include a second screen in front of and parallel to the observation screen. If the new screen contains two pinholes whose separation can be gradually altered, then interference fringes would appear on the observation screen provided that the pinhole separation satisfies $d < d_0$. Obviously, for $d = d_0$ the complex field amplitudes at the two pinholes are incoherent, thereby causing the fringe visibility of the interference pattern on the observation screen to vanish. This phenomenon, illustrated in Fig. 7.29(a), was applied successfully by Michelson to determine the diameter of distant extended sources. Michelson used a refinement of the basic arrangement shown in Fig. 7.29(a). His device is shown in Fig. 7.29(b) and is known as the *Michelson stellar interferometer* because it was used by Michelson principally to measure the diameters of stars.

By rotating M_1 and M_4 as indicated in Fig. 7.29(b), the effective distance between the two pinholes is altered. In operation, the observer merely increases the effective pinhole separation until the fringes on the observation screen disappear; according to Eq. (7.9.35) the separation at this point then corresponds to

$$d_0 \approx 1.22 \frac{\bar{\lambda} R}{2a} \tag{7.9.38}$$

from which we find the extended source diameter to be

$$2a = 1.22\bar{\lambda}R/d_0 \tag{7.9.39}$$

REFERENCES

7.1. G. O. Reynolds, J. B. DeVelis, G. B. Parrent, Jr., and B. J. Thompson, *The New Physical Optics Notebook*, SPIE, Bellingham, WA, 1989.

7.2. M. B. Dobrin, "Optical processing in the earth sciences," *IEEE Spectrum*, Vol. 5, September 1968, pp. 59–66.

7.3. P. L. Jackson, "Analysis of variable density seismograms by means of optical diffraction," *Geophysics*, Vol. 30, 1965, pp. 1144–1178.

7.4. A. Vander Lugt, "Signal detection by complex spatial filtering," *IEEE Trans. Info. Theory*, Vol. IT-10, February 1984, pp. 1300–1319.

7.5. J. W. Goodman, *Introduction to Fourier Optics*, McGraw-Hill, New York, 1968.

7.6. D. Casasent, "Optical data processing for engineers, Part I: Fundamentals, techniques, and system architectures," *Electro-Optical Systems Design*, pp. 26–36, Feb. 1978.

7.7. D. Gabor, "A new microscope principle," *Nature*, Vol. 161, 1948, p. 777.

7.8. D. Gabor, "Microscopy by reconstructed wavefronts," *Proc. Roy. Soc.*, Vol. A197, 1949, p. 454.

7.9. D. Gabor, "Microscopy by reconstructed wavefronts: II," *Proc. Phys. Soc.*, Vol. B64, 1951, p. 449.

7.10. E. N. Leith and J. Upatnieks, "Reconstructed wavefronts and communication theory," *J. Opt. Soc. Am.*, Vol. 52, 1962, p. 1123.

7.11. E. N. Leith and J. Upatnieks, "Wavefront reconstruction with diffused illumination and three-dimensional objects," *J. Opt. Soc. Am.*, Vol. 54, 1964, p. 1295.

7.12. R. J. Annulli and J. T. Ziewacz, "Single-beams 360° holograms," *Am. J. Phys.*, Vol. 45, May 1977, pp. 493–494.

7.13. K. Murata and K. Kunugi, "Cone-shaped cover for 360° holography," *Appl. Opt.*, Vol. 16, July 1977, pp. 1798–1800.

7.14. O. D. D. Soares and J. C. A., Fermandes, "Cylindrical hologram of 360° field of view," *Appl. Opt.*, Vol. 21, September 1982, pp. 3194–3196.

7.15. N. H. Farhat and Y. Shen, "Microwave imaging of objects in the presence of severe clutter," *J. Opt. Soc. Am.*, Vol. A2, 1985, p. 20.

7.16. F. Mandelhorn and N. H. Farhat, "Multiplex methods for mapping amplitude and phase of microwave wavefronts," *J. Opt. Soc. Am.*, Vol. 64, 1974, p. 559.

7.17. N. H. Farhat, "Principles of broad-band coherent imaging," *J. Opt. Soc. Am.*, Vol. 67, 1977, p. 1015.

7.18. M. J. Beran, *The Theory of Partial Coherence*, Prentice Hall, Englewood Cliffs, NJ, 1968.

7.19. L. Mandel and E. Wolf, "Coherence properties of optical fields," *Rev. Mod. Phys.*, Vol. 37, 1965, p. 231.

7.20. M. Born and E. Wolf, *Principles of Optics* (4th ed.), Pergamon Press, New York, 1970.

EXERCISES

7.1 Consider a rectangular aperture illuminated by a uniform circular beam of light as shown below. What is the irradiance of the resulting Fraunhofer diffraction pattern? Assume incident light is of complex field amplitude (CFA) A.

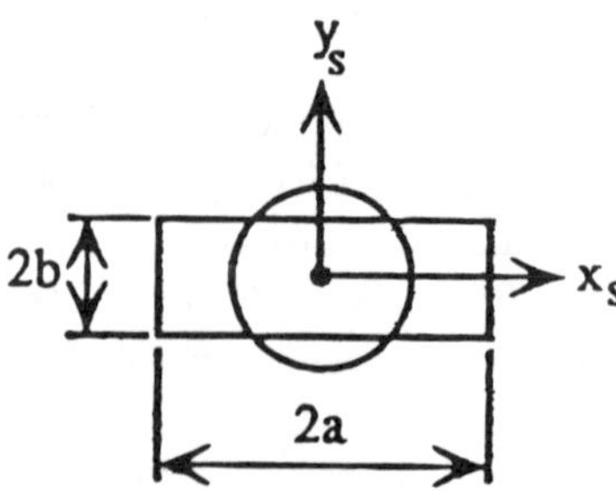

7.2 Consider the parallelogram aperture shown below. If all of it is illuminated uniformly, draw the irradiance pattern of the resulting Fraunhofer diffraction.

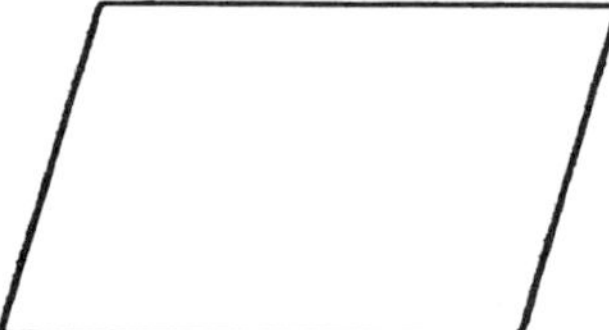

7.3 Compare phase transmission of a selfoc lens of short length $L << (\pi/2)(n_0/n_1)^{1/2}$ to that of an ordinary convex lens. What is common to both?

7.4 Prove Eq. (7.3.15).

7.5 What is the irradiance of optical Fourier transforms?

7.6 If a rectangular aperture of dimensions $2a \times 2b$ is placed adjacent to a lens and illuminated with a uniform plane wave of amplitude A, what are the CFA and irradiance in the back focal plane of the lens?

7.7 The aperture of Exercise 7.1 is illuminated by a uniform plane wave of wavelength λ_1, and later by a uniform plane wave of wavelength $2\lambda_1$. What difference is there in the Fourier transforms?

7.8 Two stars are a distance 1.5×10^8 km apart. At what distance can they be resolved by the unaided human eye? Compare this result to that obtainable using the 200-in. telescope on Mt. Palomar with a TV system at its output. Use the model for the human eye that is shown below and comment on the effect of the refractive index of the medium preceding the retina ("detectors"). If that refractive index increased, would resolution improve? Assume a 5-mm diameter for the "lens" of the eye.

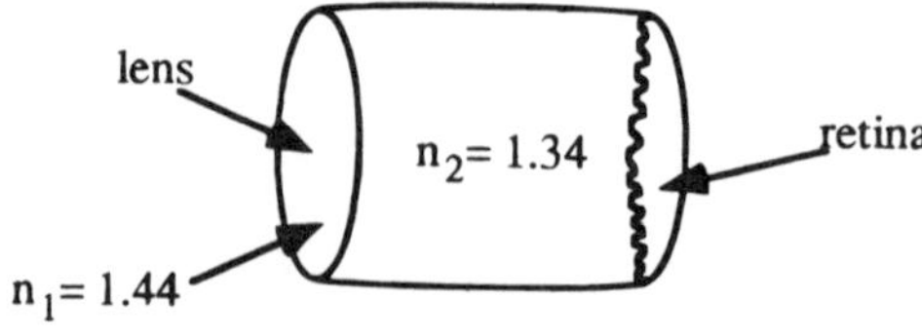

7.9 Consider the apertures shown below illuminated by a uniform plane wave of CFA A located in the front focal plane of a lens. Determine CFA and irradiance distribution in the back focal plane of a lens. Assume the edges of the lens are not illuminated. In parts (b), (c), and (d) circles are of radius c.

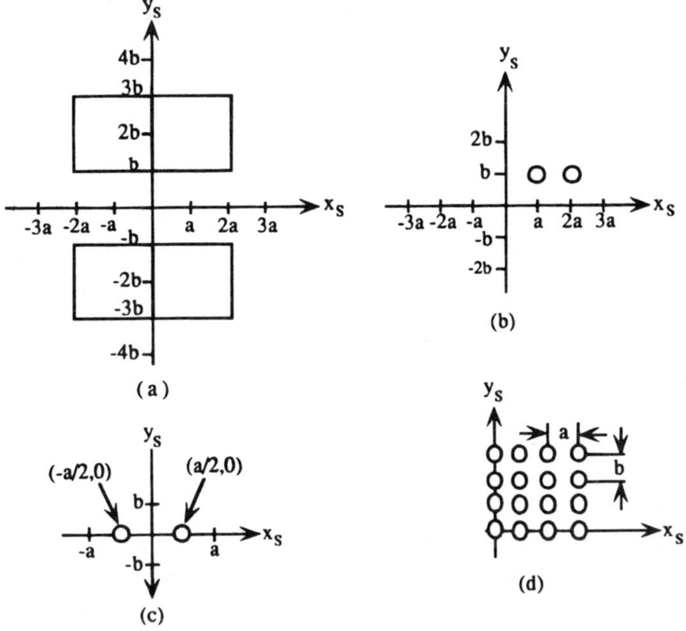

7.10 Develop an equation with which to determine the spatial half-maximum width of the primary maxima in Fig. 7.15.

7.11 Assume using Rayleigh's criterion that, in a monochromator diffraction pattern, principal maxima for wavelengths $\lambda \pm \Delta\lambda/2$ are just resolved when the principal maxima for that wavelength coincide with adjacent minima for λ. Develop an expression for wavelength resolution for the m'th interference order.

7.12 A broad spectrum optical source is measured with a monochromator to reach maximum irradiance at a wavelength of 900 nm. Is it valid to assume that conclusion is necessarily correct? Which questions arise and how can they be answered? What if the source is of narrow linewidth?

7.13 To measure the wavelength of a laser, a "square wave" diffraction grating is used. The grating contains 13,400 slits per inch. Laser wavelength is estimated to be on the order of 20 μm. In a plane parallel to the grating but located 50 cm behind it, where would the first primary maximum be located if the grating is illuminated by the laser?

7.14 A transparency contains an image plus constant background irradiance b that reduces contrast. What spatial filter can be located in plate P_2 in Fig. 7.17 to yield improved contrast if the original transparency is located in plane P_1?

7.15 How can the transparency of a slit grating (as in Fig. 7.14) be converted through Fourier filtering into a transparency of a sine grating? Assume N to be very large.

$$\textit{Hint:} \sum_{n=-\infty}^{\infty} e^{j2\pi n f_r r} = \sum_{n=-\infty}^{\infty} \delta\left(f_r - \frac{n}{r}\right)$$

7.16 Vertical film strips are joined to form an image with a large horizontal field of view. Each vertical film strip is of a different portion of the overall image. However, the overall image contains periodic vertical lines where the film strips have been joined. These lines are bothersome to observers and detract from the ability to quickly identify various detail in the scene. How should the vertical lines be removed? The distance between consecutive lines is d.

7.17 A transparency is characterized by periodic noise of the form $\cos(ay_s)$ when placed in plane P_1 of an optical computer and illuminated with a plane wave. How can this noise be removed?

7.18 Since optical detectors and imagers respond to irradiance rather than field amplitude, it is not an easy task to preserve phase information. However, by means of spatial filtering it is possible to overcome this problem in certain circumstances. Show that for the signal $U_t(y_s) = \exp[j\alpha\phi_1(y_s)] \cong 1 + j\alpha\phi_1(y_s)$ for $\alpha << \phi_1^{-1}$, it is possible to preserve phase information through the Fourier plane filter $t(y_d) = j$ for $|y_2| < \epsilon$ and unity for $|y_2| > \epsilon$ if ϵ is sufficiently small. The imaginary value of ϵ can be implemented via a dielectric coating causing proper $(\pi/2)$ phase change in the area of the origin. Note that $j = \exp(j\pi/2)$.

7.19 A spatial signal $A(1 + \cos\pi x_s/L)$ is placed in plane P_1 of an optical computer as shown in Fig. 7.17. If in plane P_2 an opaque disk with a radius of $\pi f_{l_2}/(Lk_o)$ is placed, what is the output CFA in plane P_3?

7.20 A diffraction grating is placed in front of, parallel to, and adjacent to film in a photographic camera. The slits are in the horizontal direction. A picture is taken. However, instead of rolling the film so that the next picture can be taken on previously unexposed film, the photographer rotates instead the diffraction grating so that the slits are at an angle θ to their previous direction. The grating is still in front of, parallel to, and adjacent to the exposed film. A new picture is taken on the same film as the old one. This is *image multiplexing*. How can the images on the same film be separated from one another? In other words, suggest a technique for demultiplexing images. Assume the openings in the diffraction grating ($2a$ in Fig. 7.14) are very small, on the order of wavelengths, and thus the number of slits is very large. If L designates the vertical dimension of the film, find an expression for the minimum number of slits required.

7.21 In fabricating a Vander Lugt filter a student made a mistake. Instead of putting plane P_4 in Fig. 7.20 in the front focal plane of lens L_4, he put the transparency $h(x_4, y_4)$ immediately adjacent to lens L_4. To salvage this filter for correlation purposes, the input $g(x_s, y_s)$ in Fig. 7.17 is placed a distance z_0 in front of L_2 rather than necessarily in the front focal plane of L_2. What should be the value of z_0?

7.22 Determine coherence diameter surrounding an arbitrary point on a screen illuminated by sunlight. Assume *mean* wavelength to be 0.55 μm (Fig. 4.3, Exercise 4.2) and neglect variation of brightness across the sun's disk.

7.23 (a) Determine coherence area deriving from an Hg arc lamp at 0.633-μm wavelength at a distance 1 m from the source. Assume the output aperture is 3 mm and that the beam is diffraction limited. (b) If a 30-μm pinhole is placed in front of the lamp, how does this affect the results for part (a)? Compare in both cases coherence diameter with diffraction-limited beamwidth diameter.

7.24 A 0.5-mm-diameter pinhole is used as a source for a double-slit interference experiment using a sodium lamp ($\lambda = 0.589\ \mu m$). If the distance from the source to the plane of the slits is 0.5 m, what is the minimum value of maximum slit spacing such that interference fringes are just observable in the observation plane?

CHAPTER

8

Diffraction-Limited Imaging

After a brief introduction to Fourier optics in the previous chapter, we now return to imaging with incoherent light. In Chapter 2 geometrical optics was used to describe image formation and location. Diffraction effects were neglected, and distortions from ideal image quality were attributed to chromatic and spherical aberrations. At this point, another limitation to image quality can be considered, namely, diffraction. Diffraction causes parallel light rays incident on the edges of optical elements such as lenses or mirrors to be deflected away from the focal point, thus degrading image quality. This causes spreading of point images and loss of resolution as described on p. 244 and in Section 7.4 for coherent light. Here we consider primarily incoherent light which is normally used in imaging.

Diffraction-limited imaging refers to a situation in which image quality is limited by such diffraction. It is very common, particularly when geometrical optics aberrations are minimized through use of achromatic elements and parabolic rather than spherical surfaces. Moreover, diffraction-limited imaging is a convenient method in which to introduce the use of system concepts such as impulse response and transfer function to characterize image quality. These system concepts are introduced here and emphasized throughout the remainder of this book.

8.1 APERTURE SIZE: IMPULSE RESPONSE

Although incident plane waves or parallel rays are useful to determine focal point, most imaging involves objects and images at finite distances. Consequently, the waves incident and refracted by lenses are *spherical* rather than plane waves, as illustrated in Chapter 2. When objects are at finite distances, the treatment of plane wave illumination of apertures in the previous chapter is inapplicable. Nevertheless, many of the concepts and results obtained with regard to Fourier optics will be seen to be relevant to diffraction-limited imaging.

Consider a typical imaging system similar to the one shown in Fig. 8.1. For simplicity the treatment here is one dimensional. The aperture in the figure is the lens itself. Of course, field stops or iris diaphragms when used can be the limiting aperture in place of the lens, depending on which limits field of view. We wish to consider the effect of the aperture on image quality for a point object of zero width. Such a point object represents a spatial impulse function input to an imaging system.

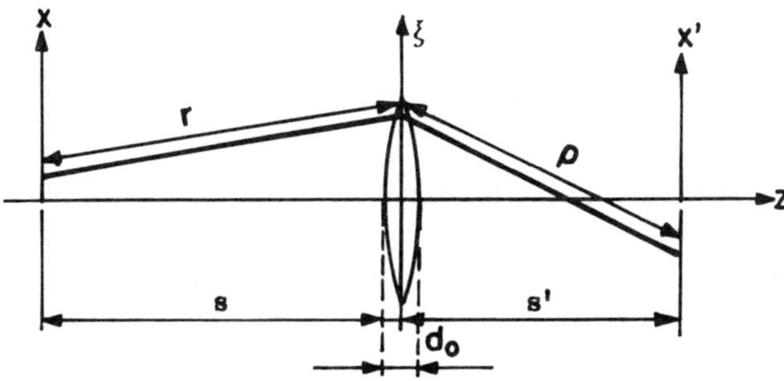

Fig. 8.1 Geometry for impulse response determination.

The image of the point object is the impulse response of the imaging system. Spatial impulse response can then be used to define the spatial frequency transfer function, analogous to methods used to define system concepts in electronics. Transfer functions can then be used to determine the size of spatial detail resolved by an imaging system in much the same way as transfer functions in electronics are used to determine the shape of output signal for a given input signal.

Ideally, a point image of zero width for a point object is desirable because it enables detail on the size order of zero width to be resolved by the imaging system. We will see shortly that diffraction by aperture edges makes this impossible, even if the lens is ideally parabolic and achromatic. This limitation to spatial narrowness of impulse response imposed by diffraction is the subject of the present treatment.

Let ψ_o and ψ_i represent object and image plane CFA, respectively. In accordance with the Fresnel-Kirchhoff diffraction integral [Eq. (7.2.1)]

$$\psi_i(x') = K \int_{-\infty}^{\infty} \int_{-\infty}^{\infty} \psi_o(x) \frac{e^{-jk_0 r}}{r} A(\xi) e^{jk_0 \xi^2/2f_l} \frac{e^{-jk_0 p}}{p} \, dx \, d\xi \tag{8.1.1}$$

where

$$K = j \exp(-jk_0 n d_0)/\lambda \tag{8.1.2}$$

and $A(\xi)$ is the aperture function in the plane of the lens. As in the previous chapter the aperture function is unity for space within the aperture and zero outside the aperture. Note that the constant K also includes the constant part of the lens transmission function given by Eq. (7.3.3), where n is the refractive index of the lens material. Outside the lens, unity refractive index is assumed. The first exponent component in Eq. (8.1.1) represents spherical wave propagation from an object point to the lens aperture plane. The second exponential is the coordinate-dependent component of the lens transmission function. The last exponential component represents spherical wave propagation from the wavefront exiting the lens to the image plane. We integrate over object and lens aperture spaces. Defining $R^2 = x^2 + s^2$, and $(R')^2 = (x')^2 + (s')^2$, from Fig. 8.1

$$r^2 = (x - \xi)^2 + s^2 = R^2 - 2x\xi + \xi^2 \tag{8.1.3}$$

$$p^2 = (x' - \xi)^2 + (s')^2 = (R')^2 - 2x'\xi^2 \tag{8.1.4}$$

For paraxial rays such that $R, R' << \xi$, the binomial approximation can be used to yield

$$r \cong R - x\xi/R + \xi^2/2R \tag{8.1.5}$$

$$p \cong R' - x'\xi/R' + \xi^2/2R' \tag{8.1.6}$$

The last two results can now be substituted into (8.1.1). As in the previous chapter, the full expressions for r and p must appear in the exponents in Eq. (8.1.1) since even small errors in distances can significantly affect phase at optical wavelengths. However in the denominator the approximations $r \cong R$ and $p \approx R'$ are quite sufficient. Accordingly, Eq. (8.1.1) is rewritten

$$\psi_i(x') = \frac{K}{RR'} \int_{-\infty}^{\infty} \int_{-\infty}^{\infty} \psi_o(x) \exp[-jk_0(R - x\xi/R + \xi^2/2R)] \exp(j\xi^2/2f_l)$$
$$\times \exp[-jk_0(R' - x'\xi/R' + \xi^2/2R')]\, d\xi\, dx \tag{8.1.7}$$

Although R is a function of x by definition, $\exp(jkR)$ can essentially be removed from this integral since, for paraxial rays, $x << s$ and therefore R varies very little with x. Furthermore, since the previous treatment is to be applied toward the impulse response, light amplitude at a coordinate x' in the image plane must consist of contributions from only a tiny region of object space, centered on the ideal geometrical object point x. Combining constant factors outside the integral, we obtain

$$\psi_i(x') = K_1 \int_{-\infty}^{\infty} \int_{-\infty}^{\infty} \psi_o(x) A(\xi) \left[\frac{-jk_0\xi^2}{2}\left(\frac{1}{s} + \frac{1}{s'} - \frac{1}{f_l}\right)\right]$$
$$\times \exp\left[jk_0\xi\left(\frac{x}{s} + \frac{x'}{s'}\right)\right] d\xi\, dx \tag{8.1.8}$$

where in the denominators of the exponent the approximations $R \cong s$ and $R' \approx s'$ have again been used. As a final simplification, attention is restricted to the image plane where, from Eq. (2.4.14),

$$\frac{1}{s} + \frac{1}{s'} - \frac{1}{f_l} = 0$$

In Eq. (8.1.8),

$$K_1 = K \exp[-jk_0(R + R')]/RR' \tag{8.1.9}$$

Under this condition, and rearranging,

$$\psi_i(x') = K_1 \int_{-\infty}^{\infty} \int_{-\infty}^{\infty} \psi_o(x) \exp\left(\frac{jk_0\xi x}{s}\right) A(\xi) \exp\left(\frac{jk_0\xi x'}{s'}\right) d\xi\, dx \tag{8.1.10}$$

This equation involves two Fourier transform components. Integrating over the object plane first yields

$$\psi_i(x') = K_1 \int \tilde{\psi}(\xi/\lambda z) A(\xi) \exp\left(\frac{jk\xi x'}{s'}\right) d\xi \tag{8.1.11}$$

where $\tilde{\Psi}_o(\xi/\lambda z)$ is the Fourier transform of $\Psi_o(x)$. To obtain impulse response, let $\Psi_o(x) = \delta(x)$, yielding

$$\tilde{\psi}_o(\xi/\lambda s) = \int \delta(x) \exp(-jk_0 \xi x/s)\, dx = 1 \tag{8.1.12}$$

Using this relation, Eq. (8.1.11) reduces to impulse response $u(x')$, which is equal to

$$u(x') = \psi_i(x')\,|_{\psi_o(x)=\delta(x)} = K_1 \int_{-\infty}^{\infty} A(\xi) \exp\left(\frac{jk_0 \xi x'}{s'}\right) d\xi \tag{8.1.13}$$

Equation (8.1.13) is an important result. It states that for *diffraction-limited imaging* amplitude distribution in the image of a point, or impulse response, is given by the *Fourier transform* of the aperture distribution function $A(\xi)$. Here, Fourier spatial frequency is radian frequency $-k_0\xi/s'$. Note that the constant K_1 contains a spherical phase factor R'. After passage through the lens, any deviation of phase front from a reference sphere is a spherical aberration as discussed briefly at the end of Chapter 2 and as illustrated in Fig. 8.2. Image degradations as a result of aberrations can also be characterized by transfer functions. The reader is referred to Ref. [8.1]. For the present, we will assume that the lens is ideally parabolic and achromatic.

The two-dimensional impulse response, assuming separability, yields CFA distribution in the image plane

$$u(x', y') = K_1 \int_{-\infty}^{\infty} \int_{-\infty}^{\infty} A(\xi, \eta) \exp\left[\frac{-jk_0}{s'} (\xi x' + \eta y')\right] d\xi\, d\eta \tag{8.1.14}$$

which suggests that the image plane is a Fourier transform plane with inverted axes. This is conceptually misleading since, in accordance with system theory, the image plane is the output plane and should involve the image, which is the *inverse* transform of the transfer function. Consequently, to cast Eq. (8.1.14) into more convenient form, "reduced" coordinates [8.2] are introduced, having dimensions of radian spatial frequency

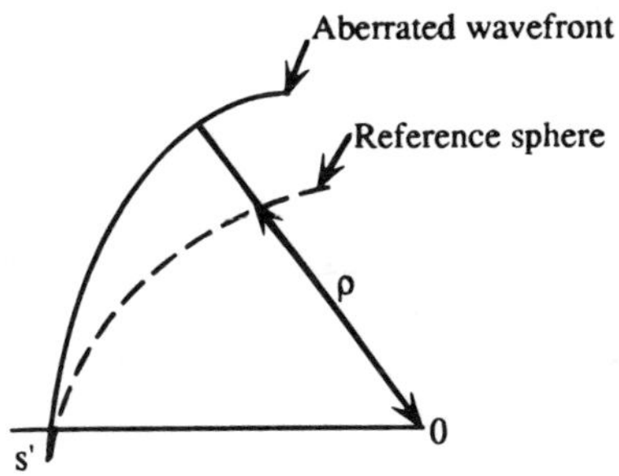

Fig. 0.2 Aberration function.

$$\beta = k_0\xi/s', \qquad \gamma = k_0\eta/s' \tag{8.1.15}$$

and, for apertures exhibiting spherical symmetry,

$$\alpha^2 = \beta^2 + \gamma^2 = \left(\frac{k_0\rho}{s'}\right)^2 \tag{8.1.16}$$

where

$$\rho^2 = \xi^2 + \eta^2 \tag{8.1.17}$$

and ξ and η are spatial coordinates in the aperture plane. Further, $A(\xi, \eta)$ is redefined as $A(\beta, \gamma)$. In other words, $A(\beta, \gamma)$ represents a mapping of the aperture function into a spatial frequency plane. The shape of the aperture function is unchanged. It is not a Fourier transform of the aperture. This mapping is done so as to permit conceptualization of the aperture plane as an imaginary Fourier transform plane with the output plane being the image plane. This is analogous to electronics where the Fourier transform plane is entirely imaginary and is not located at the signal output. Here, in image system theory, the Fourier transform plane does not physically exist as it did in the previous chapter for Fraunhofer diffraction, but is strictly imaginary, as in electronics. Therefore, it is more convenient to "move" it to the aperture plane so that the image plane can be the output plane. That is the meaning of this mapping. Equation (8.1.15) indicates the mapping is strictly linear. For example, if maximum aperture coordinate in the ξ direction is ξ_{max}, then $\beta_{max} = k_0\xi_{max}/s'$ is an aperture limit in the (β, γ) aperture spatial frequency plane. In view of these considerations, Eq. (8.1.14) can be rewritten as

$$u(x', y') = K_2 \int_{-\infty}^{\infty} \int_{-\infty}^{\infty} A(\beta, \gamma) \exp[j(\beta x' + \gamma y')]\, d\beta\, d\gamma \tag{8.1.18}$$

where K_2 combines constant factors outside the integral. In other words, impulse response (output) $u(x', y') \propto \mathbf{F}\{A(\xi, \eta)\} \propto \mathbf{F}\{A(\beta, \gamma)\}$. Equations (8.1.14) and (8.1.18) describe amplitude distributions of imaging systems when viewing a point object under aberration-free situations. However, visual systems both human and artificial react to irradiance rather than CFA. For ordinary imaging of objects illuminated essentially incoherently, $|u(x', y')|^2$ adds linearly and is related to the irradiance impulse response or spread function of the system in the form

$$s(x', y') = u(x', y')u^*(x', y')/2\eta_0 \tag{8.1.19}$$

This point spread function (PSF) describes spreading of the image irradiance distribution from that of a point image and represents irradiance impulse response for incoherent imaging. Amplitude impulse response does not depend on coherence of illumination. Hence, Eq. (8.1.18) is relevant for all diffraction-limited imaging and does not depend on the coherence of the light. Equation (8.1.19) refers only to incoherent imaging. Coherent imaging is described later in this chapter. For the present we continue with ordinary imaging via essentially incoherent light.

8.2 OPTICAL TRANSFER FUNCTION FOR INCOHERENT IMAGING

In electronics, the transfer function is redefined as the Fourier transform of the impulse response. The *optical transfer function* (OTF) is defined similarly, but is normalized to its own maximum value which normally occurs at zero spatial frequency, that is,

$$\text{OTF} = \tau(\omega) = \tau(\omega_x, \omega_y) = \frac{\int_{-\infty}^{\infty}\int_{-\infty}^{\infty} s(x', y') \exp[-j(\omega_x x' + \omega_y y')]\, dx'\, dy'}{\int_{-\infty}^{\infty}\int_{-\infty}^{\infty} s(x', y')\, dx'\, dy'}$$

$$= \frac{S(\omega_x, \omega_y)}{S(0, 0)} = \text{MTF} \exp(j\text{PTF}) \tag{8.2.1}$$

where $S(\omega_x, \omega_y)$ is the Fourier transform of the spread function or impulse response $s(x', y')$. Ordinarily, for artificial imaging systems, the OTF is maximum at zero frequency. (The human visual system, described in Chapter 12, exhibits maximum response at spatial frequencies that are low but are not zero.) Equation (8.2.1) is applied broadly in this text. According to Eq. (8.2.1), the magnitude of the OTF is typically as shown in Fig. 8.3. This magnitude is called the *modulation transfer function* (MTF). The name refers to *spatial modulation* of light irradiance. If there is no such modulation, irradiance is uniform and there is no image. Thus, spatial modulation is related to image quality and MTF is a measure of the ability of an imaging component or system to transfer such spatial modulation from input (object plane) to output (image plane). This relationship is formulated shortly.

There is also a *phase transfer function* (PTF) whose importance to resolution is not nearly as widespread as that of MTF but which, nevertheless, often cannot be neglected. Spatial phase determines image position and orientation, rather than size of detail. If a target is displaced bodily in the image plane such that each part of the target image is displaced by the same amount, the target image is not distorted. However, if portions of the target image are displaced more or less than other portions, then the target image is distorted. This information is contained in the PTF.

The physical implications of the OTF are very analogous to those of electronics transfer functions. Both permit determination of output for any given input. In both cases, since the Fourier transform of a delta function is a constant (unity) that contains all frequencies (temporal in electronics and spatial in optics) at equal (unity) amplitude, a delta function input permits indication of the frequency response of the system, that is, which frequencies pass through the system unattenuated and which are

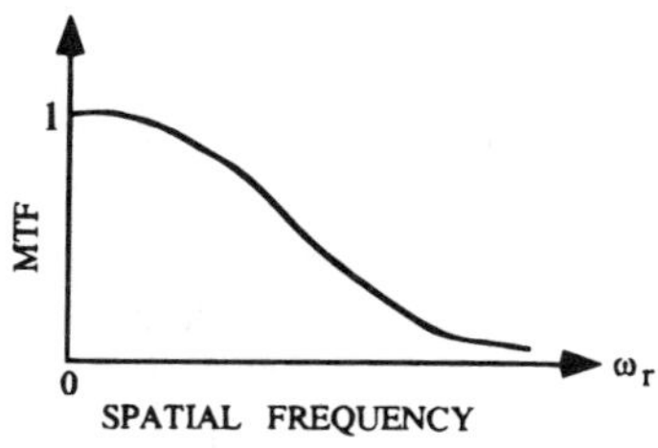

Fig. 8.3 Typical MTF curve.

attenuated and by how much. Perfect fidelity or resemblance between input and output requires infinite bandwidth so that an impulse function is obtained at output for impulse function input. In imaging, an impulse formation at output is a point image. This means that, to obtain a point image for a point object, infinite spatial frequency bandwidth is required of the imaging system. Physically, this means that because of diffraction effects optical elements such as lenses, field stops, and mirrors must be of infinite diameter so as to eliminate diffraction at edges. This, of course, is not realizable.

As a consequence of Eqs. (8.1.18) and (8.1.19) the smaller the aperture (or β_{max}, γ_{max}) the greater the relative amount of light diffracted and the poorer the image quality. Diffraction, for example, causes deflections of rays by aperture edges so that parallel rays incident on the edges are imaged elsewhere than in the lens back focal plane. The multitude of focal points arising from diffraction by aperture edges, in addition to those deriving from refraction through lens surfaces and reflection by mirrors, causes image blur. In the next section and in the examples to follow, mathematical formulations of spatial frequency cutoffs imposed by diffraction at aperture edges are presented as functions of aperture dimensions. These yield upper limits to image quality. Image quality can be degraded further by aberration and by external effects such as image motion and vibration and atmospheric effects described in the latter part of this book.

The preceding statement pertains to quasimonochromatic radiation of wavelength λ. The heterochromatic form of the optical transfer function has the form

$$\tau_b(\omega) = \int_{\lambda_1}^{\lambda_2} \tau(\omega)\, d\lambda \tag{8.2.2}$$

where $\tau(\omega)$ is OTF at wavelength λ and ω is radian spatial frequency.

8.3 OPTICAL TRANSFER FUNCTION FOR DIFFRACTION-LIMITED INCOHERENT IMAGING

For the specific case of diffraction-limited imaging, in view of Eq. (8.1.19) involving the product of $u(x, y)$ and $u^*(x, y)$, OTF is also described via convolution in the spatial frequency domain by

$$\tau(\omega_x, \omega_y) = \frac{\displaystyle\int_{-\infty}^{\infty}\int_{-\infty}^{\infty} A(\beta, \gamma)A^*(\beta - \omega_x, \gamma - \omega_y)\, d\beta\, d\gamma}{\displaystyle\int_{-\infty}^{\infty}\int_{-\infty}^{\infty} |A(\beta, \gamma)|^2\, d\beta\, d\gamma} \tag{8.3.1}$$

There are, therefore, two ways in which OTF can be obtained for diffraction-limited incoherent imaging. First, the CFA impulse response $u(x', y')$ must be obtained from the aperture function $A(\beta, \gamma)$. Then, one can either multiply $u(x', y')$ by $u^*(x', y')$ and thus obtain the spread function or irradiance impulse response and Fourier transform it as indicated by Eq. (8.2.1), or one can convolve the Fourier transforms of $u(x', y')$ and $u^*(x', y')$, which are $A(\beta, \gamma)$ and $A^*(\beta, \gamma)$, respectively, with each other, as indicated in Eq. (8.3.1). In both cases, the final result must be normalized by its maximum value, which is assumed to occur at zero frequency. These are summarized in Table 8.1. In practice $2\eta_0$ is usually absorbed in the constant in the irradiance PSF.

Fourier transforms of various apertures have been considered in the examples and exercises of Chapter 7. Thus, in these cases $u(x', y')$ is readily obtainable and, from it, $s(x', y')$. However, often Fourier or inverse Fourier transforms of $s(x', y')$ are much more complicated than those of $u(x', y')$. Hence, the convolution method of determining OTF, described by Eq. (8.3.1), is often the much simpler and faster of the two methods. Now, in the next two sections two examples are considered: one-dimensional and two-dimensional apertures.

8.4 SLIT APERTURE

The first example involves a slit very long in the x direction, so that diffraction by the edges at $\pm x_{max}$ is neglected, and only that at $\pm y_{max} = \pm a$ is considered. This example is representative of a slit of width $2a$ being placed adjacent to a lens, or of a cylindrical lens of width $2a$, shown in Fig. 8.4, being used. In the latter case, the cylindrical lens itself is the aperture since radiation that does not propagate through it does not appear in the image. For such an aperture

$$A(\xi) = \begin{cases} 1, & |\xi| \leq a \\ 0, & \text{elsewhere} \end{cases} \tag{8.4.1}$$

or

TABLE 8.1 *Impulse Responses and OTF for Diffraction-Limited Incoherent Imaging*

CFA in point source image

$$u(x', y') = K_1 \int_{-\infty}^{\infty} \int_{-\infty}^{\infty} A(\beta, \gamma) e^{i(\beta x' + \gamma y')} d\beta \, d\gamma$$

$$\beta = k_0 \xi / s', \qquad \gamma = k_0 \eta / s'$$

$$\alpha^2 = \beta^2 + \gamma^2 = (k_0 \rho / s')^2$$

Irradiance of point spread function

$$s(x', y') = K_1^2 |u(x', y')|^2 / 2\eta_0$$

$$= K_2^2 \left| \int_{-\infty}^{\infty} A(\beta, \gamma) \, e^{i(\beta x' + \gamma y')} \, d\beta \, d\gamma \right|^2 / 2\eta_0$$

Optical transfer function

$$\tau(\omega_x, \omega_y) = \frac{\int_{-\infty}^{\infty} \int_{-\infty}^{\infty} s(x', y') e^{-j\overline{w} \cdot \overline{r}'} dx \, dy}{\int_{-\infty}^{\infty} \int_{-\infty}^{\infty} s(x', y') \, dx \, dy}$$

$$= \frac{\int_{-\infty}^{\infty} \int_{-\infty}^{\infty} A(\beta, \gamma) A^*(\beta - \omega_x, \gamma - \omega_y) \, d\beta \, d\gamma}{\int_{-\infty}^{\infty} \int_{-\infty}^{\infty} |A(\beta, \gamma)|^2 \, d\beta \, d\gamma}$$

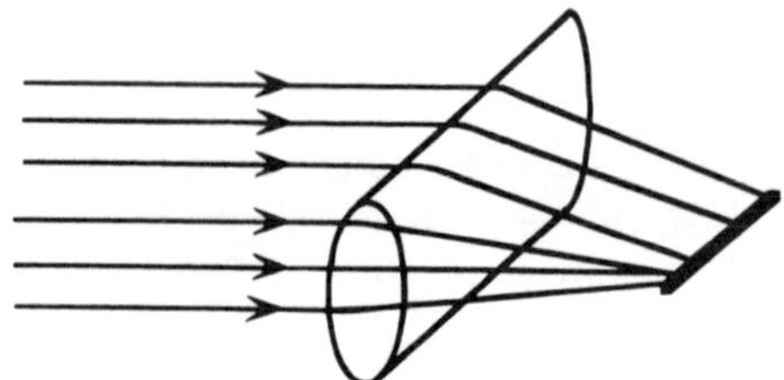

Fig. 8.4 Typical cylindrical lens focusing light in only one-direction instead of two. For incident parallel rays, a line image instead of point image is obtained according to geometrical optics.

$$A(\beta) = \begin{cases} 1, & |\beta| < \beta_0 = k_0 a/s' \\ 0, & \text{elsewhere} \end{cases} \tag{8.4.2}$$

Accordingly, from Eqs. (8.1.18) and (8.1.19), respectively

$$u(x') = K_2 \int_{-\beta_0}^{\beta_0} (1) e^{j\beta x'} d\beta = 2K_2\beta_0 \operatorname{sinc}(\beta_0 x') \tag{8.4.3}$$

$$s(x') = u(x')u^*(x') = 4K_2^2\beta_0^2 (\operatorname{sinc}^2 \beta_0 x')/2\eta_0 \tag{8.4.4}$$

To determine OTF, one can calculate the Fourier transform of Eq. (8.4.4) or utilize convolution of $A(\beta)$ with $A^*(\beta) = A(\beta)$ here. The latter method is much easier. The denominator in Eq. (8.3.1) is

$$\int_{-\infty}^{\infty} |A(\beta)|^2 \, d\beta = \int_{-\beta_0}^{\beta_0} (1) \, d\beta = 2\beta_0 \tag{8.4.5}$$

The numerator is

$$\int_{-\infty}^{\infty} A(\beta)A^*(\beta - \omega_x) \, d\beta = \int_{\omega_x - \beta_0}^{\beta_0} (1)(1) \, d\beta = 2\beta_0 - \omega_x \tag{8.4.6}$$

as seen in Fig. 8.5 where the hatched area represents overlap between $A(\beta)$ and $A(\beta - \omega_x)$. Dividing result (8.4.6) by result (8.4.5), we obtain as shown in Fig. 8.5.

$$\tau(\omega_x) = \begin{cases} \dfrac{S(\omega_x)}{S(0)} = \dfrac{2\beta_0 - \omega_x}{2\beta_0} = 1 - |\omega_x|/2\beta_0, & |\omega_x| \leq 2\beta_0 \\ 0, & |\omega_x| > 2\beta_0 \end{cases} \tag{8.4.7}$$

The physical significance is the following. If an "ideal" cylindrical lens is used to image a line, then the image obtained is not a line of zero thickness as predicted by geometrical optics but rather a line whose thickness (in the vertical direction in Fig. 8.4) is the sinc^2 function described in Eq. (8.4.4). This is called a *line spread function*. The wider the aperture (i.e., the greater the value of $2a$), then the greater the value of β_0 and the narrower the central lobe of $s(x')$. Only if a and β_0 approach infinity

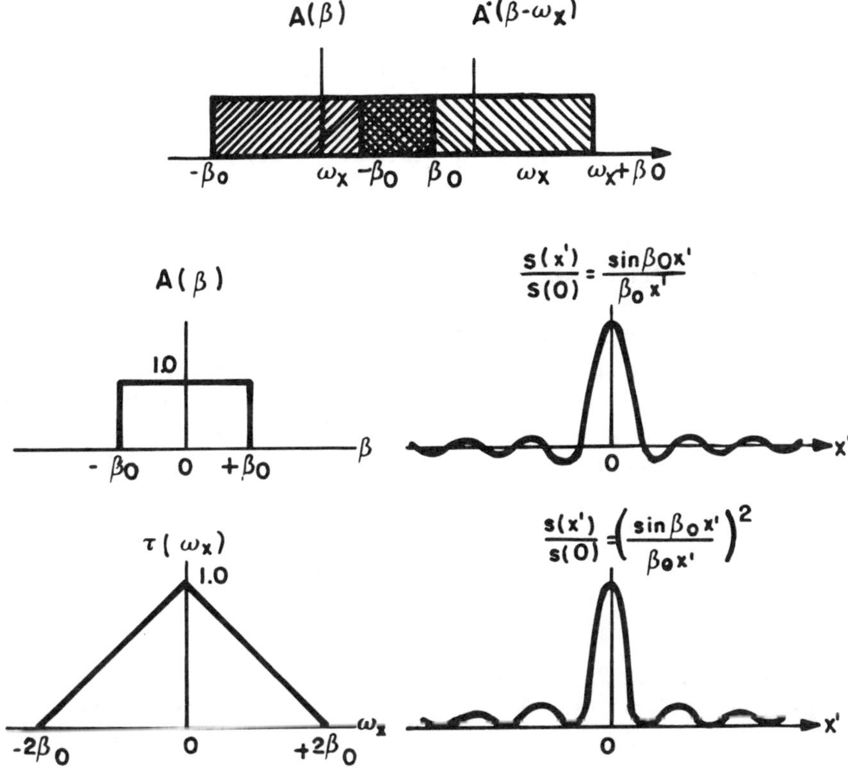

Fig. 8.5 Pictorial representations of impulse response and OTF for slit aperture diffraction-limited incoherent imaging.

does the line spread function approach zero thickness. In the latter situation according to implications of the Eikonal equation the relative amount of light diffracted at edges would approach zero, thus giving rise to the geometrical optics approximation of a line image of zero thickness.

The OTF indicates that in the frequency domain resolution is limited to a spatial frequency bandwidth of $2\beta_0 = 2(2\rho a)/\lambda s'$ in either the positive or negative x direction (vertical here). The greater the aperture width, the greater the spatial frequency bandwidth of the imaging system and the greater the fidelity or resemblance between object and image. Only if β_0 and a approach infinity would spatial frequency bandwidth also approach infinity. Another important implication from β_0 is that for *diffraction-limited imaging, higher spatial frequency bandwidth or better resolution is obtainable at shorter wavelengths.*

The previous example indicates how much simpler the convolution technique often is for determining OTF rather than Fourier transforming spread functions [such as the sinc squared function in Eq. (8.4.4)]. The convolution integral is simply integration over the overlapping area between $A(\beta, \gamma)$ and $A^*(\beta - \omega_x, \gamma - \omega_y)$. Thus, the numerator of Eq. (8.3.1) is simply the overlap of two aperture functions, one centered at the origin and the other centered at (ω_x, ω_y). The denominator simply normalizes this area of overlap by the total area of the aperture. This is a general result for diffraction-limited imaging, i.e.,

$$\mathrm{OTF} = \frac{\text{area of aperture overlap}}{\text{total aperture area}} \tag{8.4.8}$$

This approach can easily be shown to yield result (8.4.7) in the previous example. It is used in the next example.

8.5 CIRCULAR APERTURE

This example pertains to a circular lens or mirror which, if no field stops or irises are used, is the limiting aperture. Consider a lens of diameter D. In this example because of circular symmetry it is more convenient to use polar rather than Cartesian coordinates. Therefore, in accordance with Eq. (8.1.16) with ρ as the radial coordinate in the spatial aperture domain,

$$\alpha = \frac{k_0\rho}{s'}, \qquad \alpha_0 = \frac{k_0 D/2}{s'} \tag{8.5.1}$$

where $|\alpha_0|$ is the radial limit of the aperture in spatial-frequency space. This formulation reduces the problem to that of a single coordinate. The physical implication of this is that in the spatial-frequency domain here Fourier bandwidth is independent of direction and is determined simply by distance from the origin.

In the spatial domain, using Eq. (7.4.6),

$$\begin{aligned} u(a) &= 2\pi K_1 \int_0^\infty A(\alpha) J_0(\alpha a)\alpha\, d\alpha \\ &= 2\pi K_1 \int_0^{\alpha_0} (1) J_0(\alpha a)\alpha\, d\alpha \end{aligned}$$

yielding, using Eq. (7.4.10),

$$u(a) = K_1 \pi \alpha_0^2 \left[\frac{2J_1(\alpha_0\rho)}{\alpha_0\rho}\right] \tag{8.5.2}$$

and, using Eq. (8.1.19),

$$\frac{s(a)}{s(0)} = \left[\frac{2J_1(\alpha_0\rho)}{\alpha_0\rho}\right]^2 \tag{8.5.3}$$

To determine OTF, it is much more convenient to use convolutions rather than the Fourier transform result of Eq. (8.2.1).

To determine aperture overlap, the aperture is displaced a spatial-frequency distance ω on the α axis, as shown in Fig. 8.6. (In the figure the α axis is horizontal, but it can be in any direction because of the circular symmetry.) Aperture radius is α_0, and semi-angle of sector overlap is ϕ. Area of overlap is

$$A = 2[\phi\alpha_0^2 - (\alpha_0 \sin \phi)(\alpha_0 \sin \phi)] \tag{8.5.4}$$

The first term in the square brackets is the area of sector OAB. (Note that if ϕ approaches π, the sector becomes a full circle of area $\pi\alpha_0^2$). The second term in the square brackets is the area of the triangle OAB. The difference yields half the overlap. Since the area of the aperture is $\pi\alpha_0^2$, then from Eq. (8.4.8)

$$\begin{aligned} \text{OTF} = \tau(\omega) &= \frac{2[\phi\alpha_0^2 - \alpha_0^2 \sin \phi \cos \phi]}{\pi\alpha_0^2} \\ &= 2(\phi - \sin \phi \cos \phi)/\pi \end{aligned} \tag{8.5.5}$$

Now, from Fig. 8.6, $\alpha_0 \cos \phi = \omega/2$, implying that

$$\cos \phi = \omega/2\alpha_0 \tag{8.5.6}$$

$$\sin \phi = (1 - \cos^2\phi)^{1/2} = [1 - (\omega/2\alpha_0)^2]^{1/2} \tag{8.5.7}$$

Substituting the last two into Eq. (8.5.5) yields

$$\begin{aligned} \tau(\omega) &= \frac{2}{\pi}\left[\cos^{-1}\left(\frac{\omega}{2\alpha_0}\right) - \frac{\omega}{2\alpha_0}\sqrt{1 - \left(\frac{\omega}{2\alpha_0}\right)^2}\right], \ \omega \leq 2\alpha_0 \\ &= 0, \ \omega \geq 2\alpha_0 \end{aligned} \tag{8.5.8}$$

which is the OTF of an ideal circular diffraction-limited lens. Results are shown

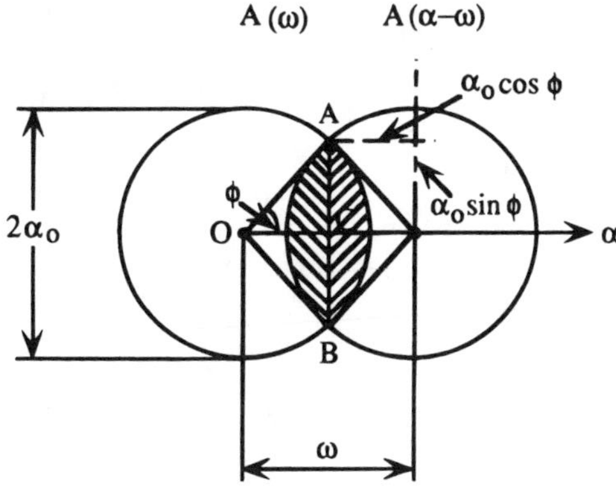

Fig. 8.6 Geometry to determine OTF of circular aperture.

pictorially in Fig. 8.7. Equation (8.5.3) implies that when looking through an ideal circular lens at a star, for example, instead of seeing a point of zero radius as expected from geometrical optics because the radiation is also incident to the lens edges one instead sees a spreading of the point (Airy disk) with irradiance of the rings being much less than that of the Airy disc. The greater the value of α_0 or aperture radius, the narrower is the point spread function and the wider the spatial-frequency bandwidth in a given direction, which is equal to $2\alpha_0 = 2\pi D/\lambda s'$. Again, as with cylindrical lenses, *better resolution is obtained* at *shorter* wavelengths since α_0 increases. This is a *general property* of diffraction-limited imaging.

If an obstruction is placed in the center of the aperture, as for example in the reflecting telescope in Fig. 2.23, the effect of the obstruction is to lower the OTF at low spatial frequencies and to raise it at high spatial frequencies. An example is considered in Exercise 8.7 for a rectangular aperture. Computer calculations show similar tendencies for circular aperture and circular obstruction. Use of variable transmission filters at the aperture center to bring about alteration in OTF is called *apodization*.

Point spread functions for square and circular apertures are shown in Fig. 8.8 for no apodization.

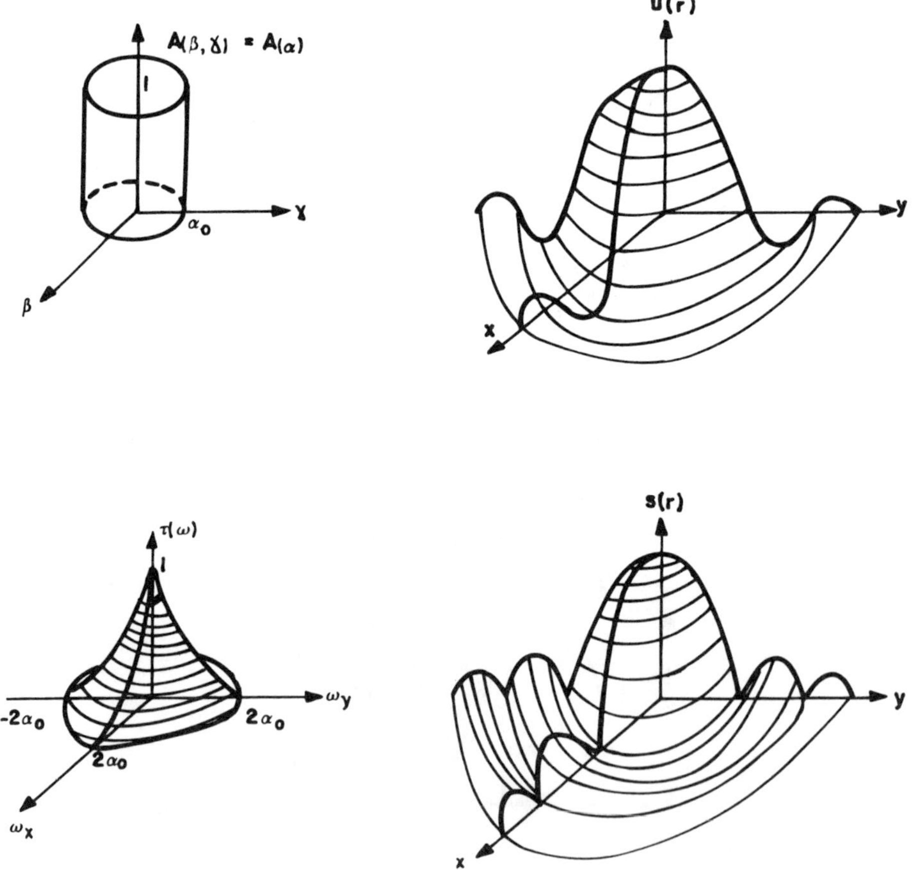

Fig. 8.7 Pictorial representation of impulse responses and OTF for diffraction-limited imaging through a circular aperture.

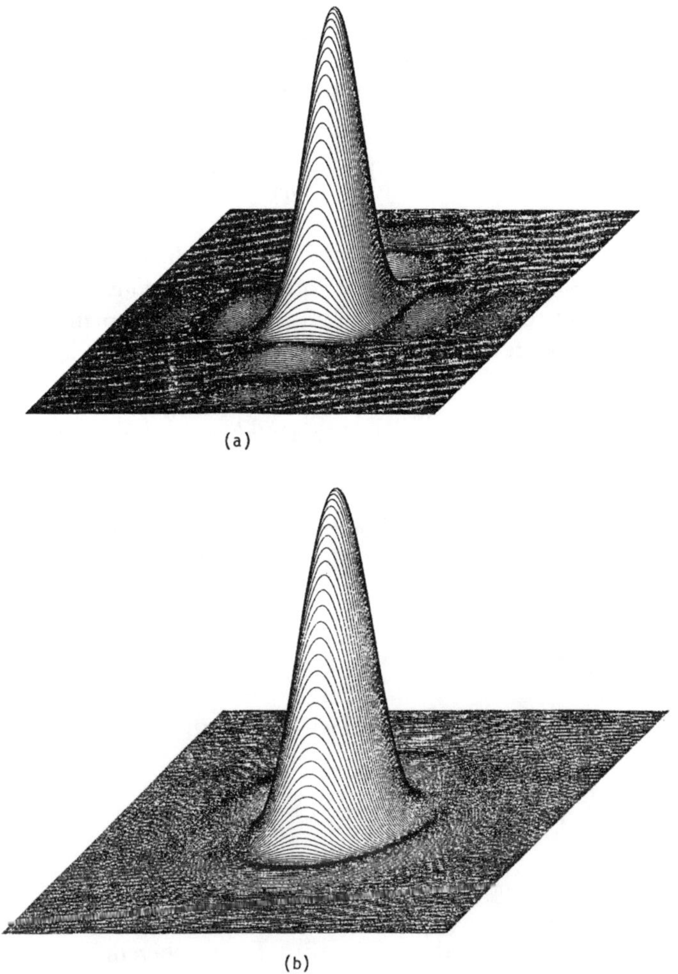

Fig. 8.8 Point spread function for (a) square aperture and (b) circular aperture.

8.6 COHERENCE AND LINEARITY

The applicability of Eq. (8.1.19) is limited to incoherent illumination. Only under such conditions does the irradiance of several point objects add linearly. In Section 7.9 the theory of partial coherence was introduced. Now we want to consider the role of coherence on irradiance spread function. To do so, we should recall the mutual coherence function defined in Eq. (7.9.1). For complex field amplitudes ψ_1 at (x_1, y_1) and ψ_2 at (x_2, y_2), the *mutual coherence function* is their mathematical cross-correlation

$$\Gamma_{12}(\tau) = \langle \psi_1(x_1, y_1, t)\psi_2(x_2 y_2, t + \tau) \rangle \tag{8.6.1}$$

where the angular brackets denote a long time average. If the same instant of time is involved then Eq. (8.6.1) is called a *mutual irradiance function*. (Often in the literature the term *mutual intensity* is used. However, in terms of the radiometric units defined in Chapter 3 and the Poynting vector in Eqs. (1.3.11), *mutual irradiance* is more correct.) Using the definition of $\Gamma_{12}(\tau)$ and the fact that the CFAs ψ_1 and ψ_2 must satisfy the Helmholtz wave equation, it can be shown that the mutual coherence function in a vacuum also satisfies the wave equation [8.3, 8.4]:

$$\nabla^2\Gamma(x_1, y_1, x_2, y_2, \tau) = \frac{1}{c^2}\frac{\partial^2\Gamma}{\partial\tau^2}(x_1, y_1, x_2, y_2, \tau) \tag{8.6.2}$$

where the Laplacian operates separately on the coordinates (x_1, y_1) and (x_2, y_2), thereby resulting in a pair of coupled wave equations. The mutual coherence function propagates from object to image plane. In *general*, the image plane CFA distribution is, apart from a constant factor,

$$\psi_i(x', y', \tau) = \int_{-\infty}^{\infty}\int_{-\infty}^{\infty} \psi_o(x, y, \tau)u(x' - x, y' - y)\, dx\, dy \tag{8.6.3}$$

where $\psi_o(x, y, \tau)$ is object plane CFA, u is again impulse amplitude response, and quasimonochromatic radiation is assumed. Time-averaged irradiance in the image plane is given by

$$H_i(x', y') = \langle\psi_i(x', y', \tau)\psi_i^*(x', y', \tau)\rangle/2\eta_0 \tag{8.6.4}$$

Substituting for ψ_i, apart from constant factors including η_0,

$$H_i(x', y') \propto \int_{-\infty}^{\infty}\int_{-\infty}^{\infty}\int_{-\infty}^{\infty}\int_{-\infty}^{\infty} \langle\psi_o(x, y, \tau)\psi_o^*(x_1, y_1, \tau)\rangle$$

$$\times\, u(x' - x, y' - y)u^*(x' - x_1, y - y_1)\, dx\, dx_1\, dy\, dy_1 \tag{8.6.5}$$

This general result is considered now in turn for incoherent and coherent object illumination.

8.6.1 Incoherent Illumination

Incoherent illumination refers to ordinary sunlight, thermal emission, or radiation deriving from such sources as incandescent or fluorescent lamps or gas discharge sources described in Chapter 4 in contrast to sources such as lasers, which exhibit much greater degrees of temporal and spatial coherence. As shown at the end of Chapter 7, even "incoherent" sources such as sunlight (Exercise 7.22) exhibit partial coherence. Nevertheless, for the sake of simplicity, incoherent sources are considered in their limit of zero mutual coherence, an approximation that is valid for very weak coherence even though such sources give rise to interference patterns as in Young's experiment in Section 7.9 (see Exercise 8.9).

If $\psi_o(x, y, \tau)$ is incoherent, each point in the object is assumed to be statistically independent timewise of every other point, yielding an object plane mutual irradiance function [8.5]

$$\langle\psi_o(x, y, \tau)\psi_o^*(x, y, \tau)\rangle = H_o(x, y)\delta(x - x_1, y - y_1) \tag{8.6.6}$$

for an object located at (x_1, y_1). This implies a *lack of correlation* between all pairs of points. Here, $H_o(x, y)$ is object irradiance distribution. Equation (8.6.5) then becomes, after substitution of Eq. (8.6.6),

$$H_i(x', y') = \int_{-\infty}^{\infty}\int_{-\infty}^{\infty} H_o(x, y)\delta(x - x_1, y - y_1)u(x' - x, y' - y) \times u^*(x' - x_1, y' - y_1)\, dx\, dx_1\, dy\, dy_1 \tag{8.6.7}$$

Performing the x_1 and y_1 integration results in

$$H_i(x', y') = \int_{-\infty}^{\infty}\int_{-\infty}^{\infty} H_o(x, y)u(x' - x, y' - y)u^*(x' - x)(y' - y)\, dx\, dy \propto \int_{-\infty}^{\infty}\int_{-\infty}^{\infty} H_o(x, y)|u(x' - x, y' - y)|^2\, dx\, dy \tag{8.6.8}$$

or

$$H_i(x', y') = \int_{-\infty}^{\infty}\int_{-\infty}^{\infty} H_o(x, y)s(x' - x, y' - y)\, dx\, dy \tag{8.6.9}$$

Image *irradiance* distribution is *linear* with object *irradiance* distribution. This result is consistent with Eq. (8.1.19) for incoherent illumination.

8.6.2 Coherent Illumination

When object plane CFA $\psi_o(x, y, \tau)$ is coherent, then for stationary quasimonochromatic fields the object plane mutual irradiance function becomes separable into the product

$$\Gamma_o(x_1, y_1, x_2, y_2, 0) = \psi_o(x_1, y_1)\psi_o^*(x_2, y_2)/2\eta_0 \tag{8.6.10}$$

As a result, Eq. (8.6.5) becomes for coherent illumination, neglecting constant factors,

$$H_i(x', y') \propto \int_{-\infty}^{\infty}\int_{-\infty}^{\infty}\int_{-\infty}^{\infty}\int_{-\infty}^{\infty} \psi_o(x,y)\psi_o^*(x_1,y_1)u(x' - x,y' - y) \times u^*(x' - x_1,y' - y_1)\, dx\, dy\, dx_1\, dy_1 \propto \left|\int_{-\infty}^{\infty}\int_{-\infty}^{\infty} \psi_o(x,y)u(x' - x,y' - y)\, dx\, dy\right|^2 \tag{8.6.11}$$

The object CFA distribution is convolved with the amplitude impulse response and the result squared to determine image irradiance distribution within constant factors. For both incoherent and coherent radiation, image CFA distribution is linear with object CFA distribution [Eq. (8.6.3)], but for coherent imaging the irradiances are nonlinear [Eq. (8.6.11)]. This result is consistent with Eq. (7.4.17) and the ensuring discussion.

The treatments here neglect possible nonlinearities in detectors and detection electronics.

8.7 RESOLUTION OF TWO POINTS: COHERENT VERSUS INCOHERENT IMAGING

One of the criteria by which resolution is addressed is the ability of an imaging system to resolve two closely spaced points. A good quality imaging system is one in which fine detail can be resolved. On the basis of the preceding considerations concerning incoherent and coherent imaging, the ability to resolve two closely spaced points is considered now for both types of illumination.

We pointed out in Eq. (8.6.2) that mutual coherence also propagates through an imaging system. The general Green's function solution of the coupled wave equations involving both points (x_1, y_1) and (x_2, y_2) under the quasimonochromatic approximation yields the mutual irradiance function [8.5]:

$$\Gamma(x_1, y_1, x_2, y_2, 0) = \int_{-\infty}^{\infty}\int_{-\infty}^{\infty} \Gamma(\bar{S}_1, \bar{S}_2, 0)\, \frac{\partial G_2(\bar{S}_1, x_1, y_1)}{\partial n} \frac{\partial G_2^*(\bar{S}_2, x_2, y_2)}{\partial n}\, d\bar{S}_1\, d\bar{S}_2 \quad (8.7.1)$$

where $\bar{S}_1$ and $\bar{S}_2$ are the variables on the object surface. Now, from Section 7.1,

$$\frac{\partial G}{\partial n} \cong jkG = \frac{j2\pi}{\lambda} G = \frac{j2\pi}{\lambda} \frac{e^{-jkr}}{r} \quad (8.7.2)$$

Using the paraxial approximation of Eq. (7.2.3) in the geometry of Fig. 8.1, we obtain

$$r \cong z + [(\xi - x)^2 + (\eta - y)^2]/2s \quad (8.7.3)$$

and substituting it into Eq. (8.7.2) results in

$$\frac{\partial G}{\partial n} \cong \frac{j2\pi}{\lambda z} e^{-jkz} \exp[-jk(\xi - x)^2 + (\eta - y)^2]/2s \quad (8.7.4)$$

By utilizing the Gaussian focusing condition $(1/s) + (1/s') = 1/f_l$ as was done in Section 8.1, the resulting mutual irradiance function in the image plane is, using the analysis of Section 8.1,

$$\Gamma_i(x_1', y_1', x_2', y_2', 0) = K_2 \int_{-\infty}^{\infty}\int_{-\infty}^{\infty}\int_{-\infty}^{\infty}\int_{-\infty}^{\infty} \Gamma_o(x_1, y_1, x_2, y_2, 0) \exp[jk(x_1^2 - x_2^2) + (y_1^2 - y_2^2)]/2s$$
$$\times\, u\left(\frac{x_1'}{s'} + \frac{y_1'}{s'} + \frac{x_1}{s} + \frac{y_1}{s}\right) u^*\left(\frac{x_2'}{s'} + \frac{y_2'}{s'} + \frac{x_2}{s} + \frac{y_2}{s}\right) dx_1\, dy_1\, dx_2\, dy_2 \quad (8.7.5)$$

where K_2 is a complex constant and CFA impulse response is

$$u\left(\frac{x_1'}{s'} + \frac{y_1'}{s'} + \frac{x_1}{s} + \frac{y_1}{s}\right) = \int A(\xi, \eta) \exp\left[-jk\xi\left(\frac{x_1'}{s'} + \frac{x_2'}{s'} + \frac{x_1}{s} + \frac{x_2}{s}\right)\right] d\xi\, d\eta$$
$$(8.7.6)$$

Equation (8.7.5) is similar to Eq. (8.1.10), and Eq. (8.7.6) is similar to Eq. (8.1.13) for one-dimensional relationships between image and object CFA distributions. Equation (8.7.5) relates mutual irradiance in the image to mutual irradiance in the object, utilizing the amplitude imaging response of the aperture lens, which is spatially stationary or *isoplanatic.*

Consider now two equally bright point objects, along the x axis as in Fig. 8.9, separated by a distance $2b$. The problem in this example is reduced now to one dimension. The mutual irradiance in the object plane, $\Gamma_o(x_1, x_2)$, is determined from Eq. (8.6.10) and evaluated under conditions of stationary phase. Within a constant factor, and considering possible mutual illumination coherence between the point sources,

$$\Gamma_o(x_1, x_2, 0) = H_o\gamma(x_1, x_2)\{[\delta(x_1 - b) + \delta(x_1 + b)] \times [\delta(x_2 - b) + \delta(x_2 + b)]\} \tag{8.7.7}$$

where $\gamma(x_1, x_2)$ is complex degree of coherence and H_0 is normalized object irradiance. Mutual irradiance in the image plane derives from Eq. (8.7.5)

$$\Gamma_i(x_1', x_2', 0) = \int_{-\infty}^{\infty}\int_{-\infty}^{\infty} \Gamma_o(x_1, x_2, 0)\exp[-jk(x_1^2 - x_2^2)/s] \times u\left(\frac{x_1'}{s'} + \frac{x_1}{s}\right) \cdot u^*\left(\frac{x_2'}{s'} + \frac{x_2}{s}\right) dx_1\, dx_2 \tag{8.7.8}$$

Irradiance in the image is determined by setting $x_1' = x_2' = x'$. Hence,

$$H_i(x') = \int\int \Gamma_o(x_1, x_2, 0)u\left(\frac{x'}{s'} + \frac{x_1}{s}\right)u^*\left(\frac{x'}{s'} + \frac{x_2}{s}\right) dx_1\, dx_2 \tag{8.7.9}$$

Equation (8.7.7) is substituted into the above, yielding

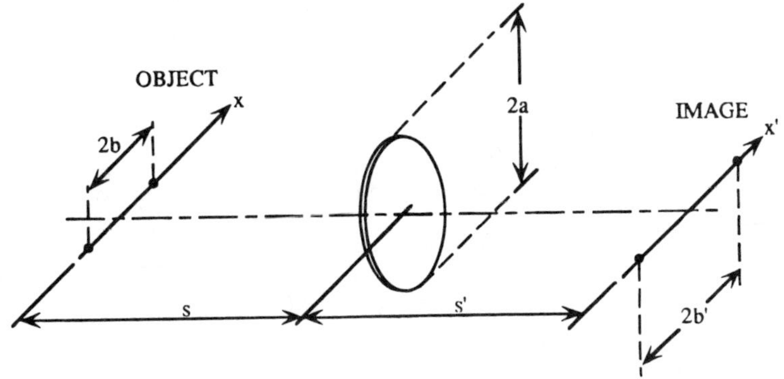

Fig. 8.9 Schematic of basic imaging system with two point objects.

$$H_i(x') = H_o \int\int \gamma(x_1, x_2)\{[\delta(x_1 - b) + \delta(x_1 + b)][\delta(x_2 - b) + \delta(x_2 + b)]\}$$

$$\times\, u\left(\frac{x_1'}{s'} + \frac{x_1}{s}\right)u^*\left(\frac{x_2'}{s'} + \frac{x_2}{s}\right) dx_1\, dx_2 \tag{8.7.10}$$

Since by definition, in the object plane $\gamma(b, b) = \gamma(-b, -b)$, then

$$H_i(x') = \{|u(x' + b')|^2 + |u(x' - b')|^2 + 2\,\mathrm{Re}[\gamma(b', -b')u(x' + b')u^*(x' - b')]\} \tag{8.7.11}$$

where $b' \propto (s'/s)b$, and (s'/s) is magnitude of lateral magnification. Now, an ideal circular lens of diameter $2a$ yields a diffraction-limited amplitude impulse response

$$\frac{u(x')}{u(0)} = \pi a^2\left[\frac{2J_1(ka|x'|s')}{kax'/s'}\right] \tag{8.7.12}$$

as can be seen from Eq. (8.5.2). Substituting this result into Eq. (8.7.11) yields, within a constant,

$$H_i(x') = \pi^2 a^4 H_o \left\{\left\{\frac{2J_1[ka|x'| - b')/s']}{ka(|x'| - b')/s'}\right\} + \left\{\frac{2J_1[ka(|x'| + b')/s']}{ka(|x'| + b')/s'}\right\}^2 + 2\,\mathrm{Re}\gamma(b', -b')\left[\frac{2J_1[ka(|x'| - b')/s']}{ka(|x'| - b')/s'}\right] \cdot \left[\frac{2J_1[ka(|x'| + b')/s']}{ka(|x'| + b')/s'}\right]\right\} \tag{8.7.13}$$

This general result is now applied to the two extreme cases of completely incoherent and coherent radiation.

In the *incoherent* limit, $\gamma(b', -b') = 0$ in the image plane, thereby yielding there simply the separate irradiances of the two point sources:

$$H_i(x')|_{\mathrm{incoh}} = \pi^2 a^4 H_o \left\{\left\{\frac{2J_1[(ka(|x'| - b')/s']}{ka(|x'| - b')/s'}\right\}^2 + \left\{\frac{2J_1[ka(|x|' + b')/s']}{ka(|x'| + b'/s)}\right\}^2\right\} \tag{8.7.14}$$

The two-point resolution limit imposed by the Rayleigh criterion requires a minimum image plane separation

$$2b'_{\min R} = 3.83s'/ka = 0.61\lambda s'/a \tag{8.7.15}$$

as developed in Eq. (7.4.14).

An additional criterion for resolution is the *sparrow criterion*. This criterion requires minimum separation between image points such that the second derivative of image irradiance midway between the two image points is zero. The reasoning can be understood from examination of Figs. 7.11, 7.12, and 7.13 from Chapter 7.

The Sparrow criterion thus requires

$$\left.\frac{\partial^2 H_i(x')}{\partial x^2}\right|_{x'=0} = 0 \qquad (8.7.16)$$

This requires taking the derivative of Eq. (8.6.14) twice and involves Bessel functions of orders 2 and 3. The final result implies minimum separation [8.5]:

$$2b'_{\min S} = 2.98s'/ka = 0.473\lambda s'/a \qquad (8.7.17)$$

This is less than that of the Rayleigh criterion in Eq. (8.7.15).

A numerically calculated comparison of the Rayleigh and Sparrow criterion is shown in Fig. 8.10.

In the *coherent* limit, $\gamma(b', -b') = 1$, and Eq. (8.7.13) becomes

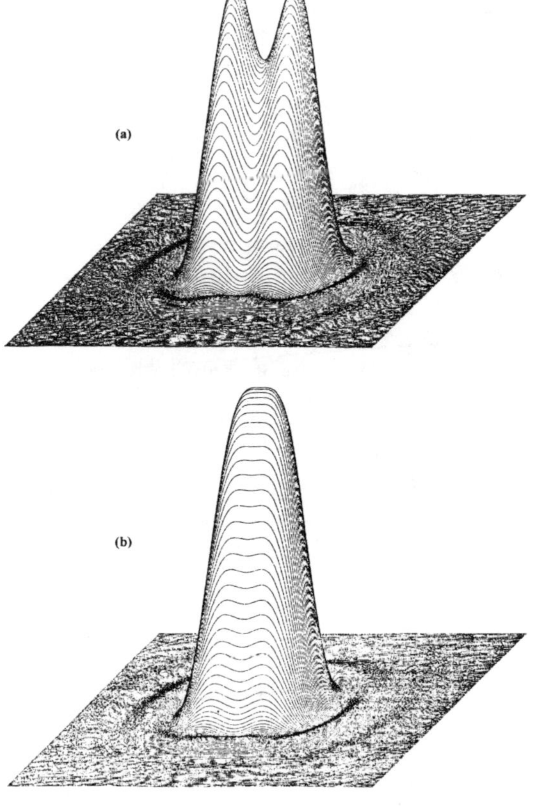

Fig. 8.10 Numerical calculation of resolution of two point images for circular aperture according to (a) Rayleigh criterion and (b) Sparrow criterion.

$$H_i(x')_{\text{coh}} = \pi^2 a^4 H_o \left\{ \frac{2J_1[ka(|x'| - b')/s']}{ka(|x'| - b)/s'} + \frac{2J_1[ka(|x'| + b')/s']}{ka(|x'| + b')/s'} \right\}^2 \tag{8.7.18}$$

The Sparrow criterion yields [8.5]

$$2b'_{\min S} = 4.630s'/ka = 0.732\lambda s'/a \tag{8.7.19}$$

Through comparison of Eqs. (8.7.17) and (8.7.19), it is clear that resolution with *incoherent* imaging is significantly better than that with *coherent* imaging. Incoherent imaging permits distinction between much finer detail. The reason is that *diffraction* by edges in the object plane gives rise to ringing or multiple images of the same edge in the image plane, thereby widening line and point spread functions through addition of sinc function or Bessel function type ringing. Consequently, incoherent imaging is commonly used not only because of convenience but also because it yields much better resolution of fine detail. Additional degradation of coherent image quality stems from speckle, which is a dynamic granular effect that arises from diffraction of light by aerosols and fine particulates that are in motion in the air. Thus, diffraction, which is made possible by coherence, tends to degrade image quality.

This book concentrates on incoherent imaging because of its generally finer resolution as well as simplicity since ordinary incoherent illumination is used. Unless otherwise stated here, imaging henceforth is incoherent. This does not mean incoherent imaging is always superior in every respect [see Exercise 8.9(d)]. In applications where diffraction and interference are important, so too is coherence.

Example 8.1

Consider an image of two similar laser beams. A convenient resolution criterion when dealing with Gaussian distributions is to assume they are resolved when the peaks of the two distributions are separated by an angle equal to the full-width-at-half-maximum irradiance of one of the distributions. What should be the angular separation of the two laser beams in order for both of them to be resolved?

Solution

From Eq. (4.5.23) laser irradiance varies transverse to the beam axis as

$$H(r, z) = H_o(o, z) \exp[-2r^2/w^2(z)]$$

where H_o is irradiance along the beam axis and r is the radial coordinate. Assuming a long focal length f_l, the above simplifies in image space to

$$H(r, z) = H_o(o, z) \exp[-2\theta_2^2/\theta_{02}^2] \tag{8.7.20}$$

where from Eq. (4.5.15)

$$\theta_2 \approx \lambda/(\pi w_{02}) \tag{8.7.21a}$$

is the angle defined by beam width in the optics plane and vertex in the image plane and

$$\theta_{02} = r/z_2 \tag{8.7.21b}$$

Here, z_2 approaches zero in the image plane and is maximum in the optics plane. The above refers to the region of large z_2, that is, near the optics plane. In object space, in the far-field approximation $z_1 >> \pi w_{01}^2/\lambda$ and, using subscript 1 for object space parameters,

$$\theta_1 = \lambda(\pi w_{01})^{-1} \tag{8.7.22a}$$

$$\theta_{01} = r/z_1 \tag{8.7.22b}$$

where z_1 is zero in the laser output plane and is directed toward the optics plane, so that both z_1 and z_2 are maximum there. The beam divergence semi-angle θ_1 is defined in Fig. 4.15. The two beam waists are related by $w_{02}/w_{01} \approx f_l/z_1$, as shown in Eq. (4.5.21). We assume that angular separation between the two laser beams is unchanged by the lens, in which case it is sufficient to consider far field object space only.

The field in Eq. (8.7.20) is reduced to half its maximum value when, using subscript 1 for object space parameters,

$$\exp(-2\theta_1^2/\theta_{01}^2) = 1/2$$

or

$$-2\theta_1^2/\theta_{01}^2 = -\ln 2$$

This implies a half-angle for each laser beam of $\theta_1 = (\theta_{01}/2)(2 \ln 2)^{1/2}$ or a full angular separation of

$$2\theta_1 \leq \theta_{01}\,(2 \ln 2)^{1/2} \tag{8.7.23}$$

is required in order to distinguish between both laser beams. Consequently, the beam waist in the object plane is limited to

$$w_{01} \geq [\lambda/(\pi\theta_{01})](2/\ln 2)^{1/2} \tag{8.7.24}$$

using Eq. (8.7.22a).

8.8 IMAGE QUALITY IN THE SPATIAL-FREQUENCY DOMAIN

Image quality can be determined in the spatial domain via the overall spread function of the imaging system, that is, for incoherent imaging, assuming linearity and stationarity (and revisiting Section 8.6),

$$H_i(x', y') = \int_{-\infty}^{\infty}\int_{-\infty}^{\infty} H_o(x, y)s(x' - x, y' - y)\,dx\,dy \tag{8.6.9}$$

This is analogous to the relationship in electronics whereby output is determined by convolution of input with network impulse response. Differences from electronics are that imaging is two dimensional rather than one and that while subscripts i and o are input and output, respectively, in electronics, in imaging i signifies image, which is output, and o signifies object, which is input.

In the spatial-frequency domain, if the following are Fourier transform pairs:

$$H_o(f_x, f_y) = \int_{-\infty}^{\infty}\int_{-\infty}^{\infty} H_o(x, y) \exp[-j2\pi(f_x x + f_y y)]\, dx\, dy \tag{8.8.1}$$

$$\tau(f_x, f_y) = \frac{\int_{-\infty}^{\infty}\int_{-\infty}^{-\infty} s(x', y') \exp[-j2\pi(f_x x' + f_y y')]\, dx'\, dy}{\int_{-\infty}^{\infty}\int_{-\infty}^{\infty} s(x', y')\, dx'\, dy'} \tag{8.8.2}$$

$$H_i(f_x, f_y) = \int_{-\infty}^{\infty}\int_{-\infty}^{\infty} H_i(x', y') \exp[-j2\pi(f_x x' + f_y y')]\, dx'\, dy' \tag{8.8.3}$$

then

$$H_i(f_x, f_y) = \tau(f_x, f_y) H_o(f_x, f_y) \tag{8.8.4}$$

In other words, the optical transfer function describes relative transmittance of various spatial frequencies. For a point object, a point image can be obtained only if τ equals unity all across the spatial-frequency spectrum. As seen earlier, this is impossible because of diffraction. The spatial-frequency filtering described by the OTF is a fundamental tool in characterization of image quality. Its use is demonstrated in the following example.

Example 8.2

A square wave object shown in Fig. 8.11 is of irradiance distribution

$$H_o(x) = \begin{cases} 1 & x < X/4 \\ 0 & X/4 < x < 3X/4 \\ 1 & 3X/4 < x < X \end{cases} \tag{8.8.5}$$

$$H_o(x \pm X) = H_o(x)$$

Such an object is often called a *Ronchi ruling*. We wish to obtain a square wave image

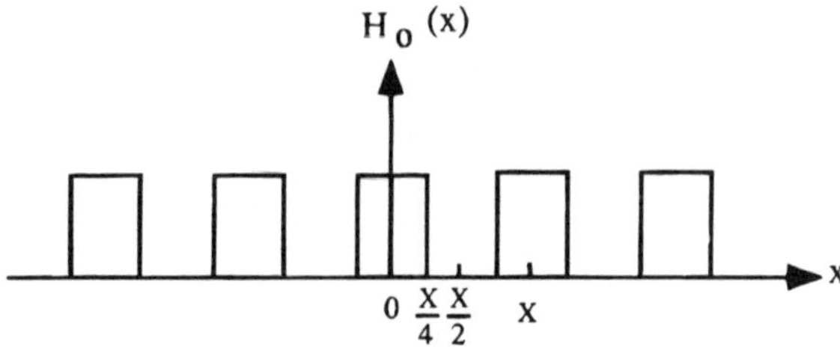

Fig. 8.11 Irradiance transmittance of a square wave.

and are interested in the effect of aperture size on image quality. Rather than consider image quality of this object through spatial domain techniques such as Eq. (8.6.9), image quality will be considered in the spatial-frequency domain as described by Eq. (8.8.4). The question addressed is what is the minimum lens diameter required so that image irradiance distribution is not uniform?

Solution

The object spectrum is obtained via Eq. (8.8.1):

$$\begin{aligned} H_o(f_x) &= \int_{-\infty}^{\infty} H_o(x) \exp(-j2\pi f_x x)\, dx \\ &= \sum_{n=-\infty}^{\infty} \int_{nX-X/4}^{nX+X/4} \exp(-j2\pi f_x x)\, dx \\ &= (X/2)\, \text{sinc}\left(\frac{2\pi f_x X}{4}\right) \sum_{-\infty}^{\infty} \exp(-j2\pi n f_x X) \end{aligned} \tag{8.8.6}$$

By using the Fourier identity [8.5]

$$\sum_{n=-\infty}^{\infty} \exp(-j2\pi n f_x X) = \sum_{n-\infty}^{\infty} \delta(f_x - n/X) \tag{8.8.7}$$

Eq. (8.8.6) becomes

$$H_o(f_x) - (X/2)\, \text{sinc}(2\pi f_x X/4) \sum_{n=-\infty}^{\infty} \delta(f_x - n/X) \tag{8.8.8}$$

This object spectrum is discrete and consists only of spatial frequencies that are multiples of X^{-1}. The sinc function envelope is zero for even harmonics and changes sign on every other odd harmonic, as shown in Fig. 8.12.

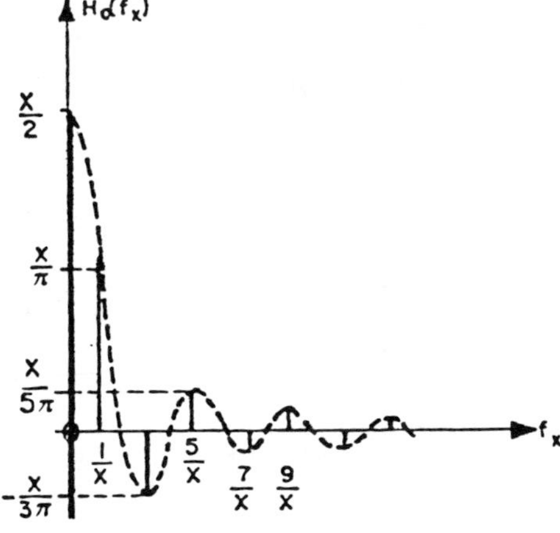

Fig. 8.12 Spectral plot of a Ronchi ruling.

Now, if the spatial-frequency bandwidth in the f_x direction of the imaging system, $2\alpha_0 = kD/s'$, is less than X^{-1}, then only the impulse at zero spatial frequency is passed by the imaging system, in which case image irradiance distribution according to Eq. (8.8.4) is constant at $X\tau(0)/2$. If, however, the first harmonics only, at $\pm X^{-1}$, are also passed by the imaging system, then the image is a cosine wave of irradiance $(2X/\pi)\tau(X^{-1})$ against the constant background $X\tau(0)/2$. This requires a lens diameter D of size $\lambda s'/X < D < 2\lambda s'/X$. The periodic variation of the square wave object can be seen in the image, but the image is a cosine rather than a square wave. As lens diameter increases, so too does Fourier bandwidth and the square wave character of the image becomes increasingly evident since more and more harmonics are passed by the lens as its diameter increases.

This example illustrates how image quality depends on OTF spatial-frequency bandwidth. It is analogous to passing a square wave in time through an electrical filter. If the filter bandwidth is small and passes only the first harmonic, a cosine wave instead of square wave appears at the filter output. As bandwidth increases, the filter passes more harmonics and the output resembles more and more a square wave.

Example 8.3

Using the Rayleigh criterion, consider resolution angles for diffraction-limited imaging with a circular lens of diameter D with and without a square mask over the lens as in Fig. 8.13. Compare separately in vertical and horizontal directions.

Solution

Using Fig. 7.9, the resolution angle for a circular aperture using the Rayleigh criterion is

$$\psi_{\min} = \frac{3.83\lambda}{\pi D} = 1.22\lambda/D \tag{8.8.9}$$

where 3.83 is the argument of the first-order Bessel function which yields the first zero. This expression is derived from Eq. (7.4.15) and leads to the point spread function [Eq. (8.5.3)]. Now, if the square mask is put over the lens, each side of the square is of length $D/\sqrt{2}$ defining $\beta_0 = kD/2\sqrt{2}\,s' = \gamma_0$ so that from Eqs. (8.1.18) and (8.1.19)

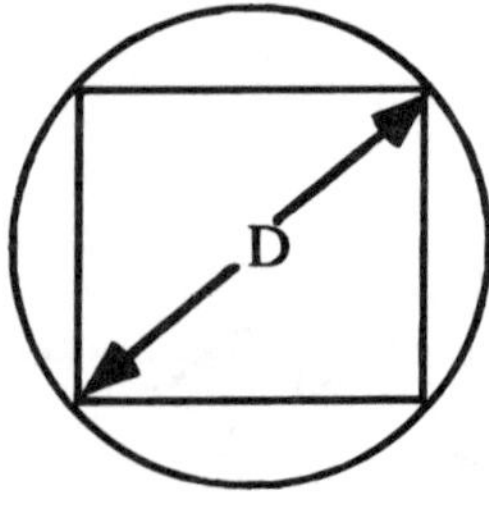

Fig. 8.13 Geometry for Example 8.3.

$$\frac{s(x')}{s(0)} = \text{sinc}^2\beta_0 x' \tag{8.8.10}$$

$$\frac{s(y')}{s(0)} = \text{sinc}^2\gamma_0 y' \tag{8.8.11}$$

The first zero of the sinc function occurs when its argument is π. Using the same analysis as in Section 7.4, this occurs when

$$\beta_o x' = \pi = kDx'/2\sqrt{2}\,z' = 2\pi D\,\frac{\psi_x'}{2\sqrt{2}\lambda} \tag{8.8.12}$$

where $\psi' = \arctan x'/s'$. This value of ψ' is determined from the above equation as

$$\psi_x' = \sqrt{2}\lambda/D \approx 1.41\lambda/D \tag{8.8.13}$$

A similar analysis in the y' direction leads to $\psi' = 1.41\lambda/D$. Both ψ_x' and ψ_y' are slightly larger than ψ_{min}' in Eq. (8.8.9) for the situation in which no mask is used, thus implying resolution is degraded by the factor $1.41/1.22 = 1.16$ in the x and y directions when the mask is used. This is not surprising since the square aperture is smaller than the circular one.

In both cases, the great majority of the image irradiance is contained between the first zeros, thus justifying the use of Eqs. (8.8.9) and (8.8.13) as resolution angles. Note from this example and from Section 8.7 that the resolution angle in the spatial domain derives from the point spread function. The narrower the point spread function, the smaller the resolution angle.

8.9 SYSTEM PROPERTIES OF SPREAD FUNCTIONS AND OPTICAL TRANSFER FUNCTIONS

Spread functions describe the widening of incoherent images of point objects. Such spreading limits the fineness of detail that can be resolved in the image. Because the spread function represents system response to a point object, the spread function represents the smallest detail that can be resolved. OTFs describe such resolution limitations in the spatial-frequency domain. Although this chapter has dealt with such image degradation stemming from diffraction-limited imaging, impulse response (spread function) and transfer function are *general* system engineering concepts applicable to any type of blur. Optical components other than lenses or apertures can also be characterized according to spread function or OTF for point object inputs. Such components include television cameras, monitors, TV electronics, image intensifiers, photographic systems, etc. Any of these can be characterized by spread function and OTF for a point object input.

The following chapters deal with such image system characterization.

8.10 CASCADE PROPERTIES OF OPTICAL TRANSFER FUNCTIONS

As in electronics, an important advantage of utilization of transfer system concepts is that for incoherent imaging overall image system OTF can be determined by cascading the various component OTFs such as those characterizing optics, imaging sensor, atmospheric channel, image motion and vibration, electronics, the human eye, etc., provided that the OTFs are independent of one another. Cascading of OTFs is often *approximated* by cascading of MTFs. An example of a situation where cascading is inappropriate is when the optics consists of a combination of apertures or lenses. In this case it is appropriate to multiply the OTFs of each aperture only after each is limited to the size of the *limiting* aperture alone. The reason is that propagation of an image through the optics involves interaction between the apertures, that is, their effects are not independent of one another unless they are separated by a diffuser. Similarly, a lens combination where one lens balances aberrations of another is one in which the effects of each lens are not independent of the effects of another. In such cases image quality for the combination may be superior to that of either by itself. It is prudent to characterize the optics with an overall OTF rather than to cascade the individual component lens OTFs.

By permitting OTF cascading for different components such as optics, electronics, image motion, optical channel, human visual system, etc., with which to characterize an imaging system overall, transfer function concepts are a powerful tool with which to quantitatively characterize image resolution. The relationship between system OTF and size detail that can be resolved is the subject of the next few chapters. It is often called target acquisition.

REFERENCES

8.1. C. S. Williams and O. A. Becklund, *Introduction to the Optical Transfer Function*, Wiley, New York, 1989.
8.2. E. L. O'Neill, *Introduction to Statistical Optics*, Addison-Wesley, Reading, MA, 1963.
8.3. M. Born and E. Wolf, *Principles of Optics*, Pergamon Press, New York, 1964.
8.4. M. J. Beran and G. B. Parrent, Jr., *Theory of Partial Coherence*, Prentice Hall, Englewood Cliffs, NJ, 1964.
8.5. G. O. Reynolds, J. B. DeVelis, G. B. Parrent, Jr., and B. J. Thompson, *The New Physical Optics Notebook*, SPIE, Bellingham, WA, 1989.

EXERCISES

8.1 Apply Eq. (8.1.10) to the geometrical optics approximation in which diffraction is neglected as D/λ approaches infinity.

8.2 In many telescopes the central portion of a circular aperture is obstructed by the presence of a secondary mirror. Consider a telescope mirror objective of radius $D/2$, and a secondary obstructing mirror of radius $\epsilon D/2$, where $0 < \epsilon < 1$. Find the normalized point spread function of the telescope, assuming ideal optics.

8.3 A rectangular aperture of dimensions $2a \times 2b$ is placed adjacent to a lens. What is the resulting normalized point spread function if the lens is of diameter $D > 2a, 2b$ and is "ideal" (parabolic and achromatic)? Assume the entire aperture is covered by the lens.

8.4 The aperture of Exercises 7.9(b) and 7.9(c) is placed adjacent to an "ideal" lens of diameter $D > 2(c + a)$. Find the normalized point spread function.

8.5 Find the OTF analytically for the aperture of Exercise 8.3.

8.6 Find the OTF graphically for the aperture of Exercise 8.4, assuming $a > c/2$.

8.7 Find normalized $u(x', y')$, $s(x', y')$ for the aperture shown below for incoherent imaging and sketch the OTF. What happens to the OTF sketch as the stop increases in size?

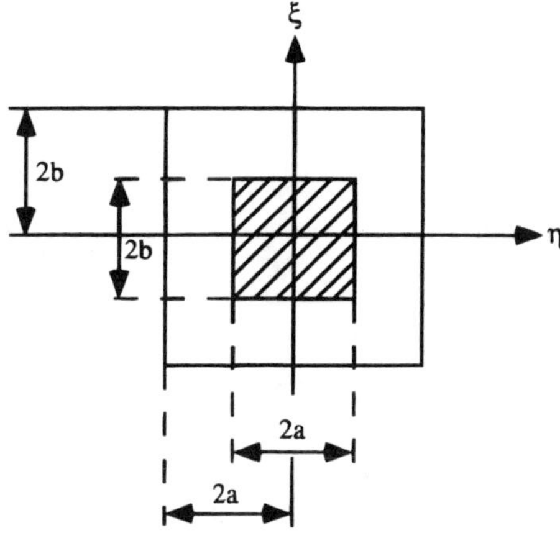

8.8 A stop is placed along the bottom half of a circular ideal lens as demonstrated below. Find the OTF for incoherent imaging.

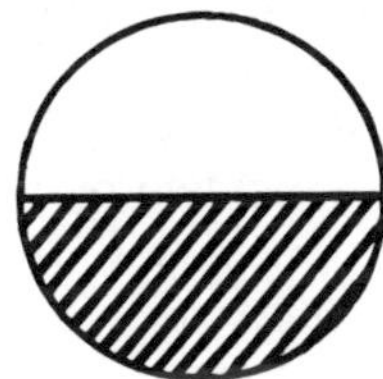

8.9 An aperture consists of two narrow slits spaced a distance $2d$ apart. Slit width is on the order of wavelength.

(a) What is the CFA impulse response of the aperture?

(b) What is its spread function for incoherent imaging?

(c) A uniform plane wave is incident on the aperture. The source of the plane wave is of diameter $D' >> 2d$. If the aperture is placed adjacent to an ideal lens of diameter $D > 2d$ and of focal length f_l, what is the resulting image for (1) coherent and (2) incoherent imaging?

(d) What is the visibility of the interference pattern in the image plane in both cases? Which illumination yields better contrast (visibility)?

(e) For incoherent imaging what must slit separation $2d$ be to change the sign of the visibility? What can be learned from this about source diameter?

(f) For incoherent imaging, what can be determined concerning $|\gamma_{12}|$?

CHAPTER

9

Modulation Contrast Function

9.1 INTRODUCTION

In Chapter 8 the optical transfer function was introduced. It was used first to describe limitations on image quality stemming from diffraction by various shape and size apertures. Such diffraction at aperture edges gives rise to deflection of light incident on such aperture edges. This deflection causes rays incident on aperture edges to take exit paths different from those of rays incident on the middle portions of the aperture. The latter rays behave according to Snell's law and form the image. The diffracted rays give rise to image blur. In Section 8.9, it was suggested that the transfer function approach can be used to describe image quality in *general* for an imaging system and for its various components and not only for diffraction-limited imaging. In general, the optical transfer function (OTF) is equal to [Eq. 8.2.1)]

$$\tau(\omega_x, \omega_y) = \text{MTF} \exp(j\text{PTF}) = \frac{\int_{-\infty}^{\infty}\int_{-\infty}^{\infty} s(x', y') \exp[-j(\omega_x x' + \omega_y y')]\, dx'\, dy'}{\int_{-\infty}^{\infty}\int_{-\infty}^{\infty} s(x', y')\, dx'\, dy'} \tag{8.2.1}$$

where $s(x', y')$ and $\tau(\omega_x, \omega_y)$ are irradiance impulse response (or spread function) and OTF for the imaging system or any of its components, respectively (assuming component OTFs are independent of one another), and MTF and PTF are modulation and phase transfer functions, respectively, as described in Section 8.2. Spread function is a system concept. It refers to widening of the image of a point object. It is not limited to aperture effects or the diffraction-limited imaging considered in the previous chapter but includes all blurring regardless of the source. In general, image irradiance distribution with incoherent light in the spatial domain is described by:

$$H_i(x', y') = \int_{-\infty}^{\infty} H_o(x, y) s(x' - x, y' - y)\, dx\, dy \tag{8.6.9}$$

In the spatial frequency domain [9.1]

$$H_i(f_x, f_y) = \tau(f_x, f_y)H_o(f_x, f_y) \tag{8.8.4}$$

where $H_i(f_x, f_y)$ and $H_o(f_x, f_y)$ are spatial Fourier transforms of image and object irradiance distributions $h_i(x', y')$ and $h_o(x, y)$, respectively.

In this chapter we show that the mathematical concept of the transfer function is accompanied by a real physical phenomenon called *contrast*. This physical concept must first be defined, and then applied to OTFs. Since it is a real physical phenomenon it also provides a method for measuring transfer functions of complicated instruments such as TV systems, photographic cameras, image intensifiers, etc. In this way, where no analytical solutions for transfer functions are possible, transfer functions can be determined by direct measurement in order to characterize imaging systems or instrumentation. While analytical solutions are often possible for diffraction-limited transfer functions as shown in the previous chapter, they are not appropriate to the image recording systems mentioned above.

9.2 MODULATION CONTRAST

Contrast deals with relative differences in irradiance levels. If such differences are relatively large, an image is considered to be of good contrast. Various definitions of contrast have been used in the past. Here, the specific definition of *modulation contrast* (MC) is used. It is defined as

$$\text{MC} = \frac{H_{\max} - H_{\min}}{H_{\max} + H_{\min}} \tag{9.2.1}$$

and is similar to the definition of visibility introduced in Chapter 7 [Eq. (7.9.16)] to describe contrast of an interference pattern. Modulation contrast varies from zero to unity. The term *modulation* is considered in the spatial sense. If there is no spatial modulation of irradiance, an image consists of uniform irradiance, which implies zero information. Good image contrast requires high spatial frequency modulation.

Consider a one-dimensional sinusoidally varying object wave

$$H_o(x) = b + a \cos 2\pi f_x x \tag{9.2.2}$$

as depicted in Fig. 9.1. Physically, irradiance minima appear relatively "darkish" and

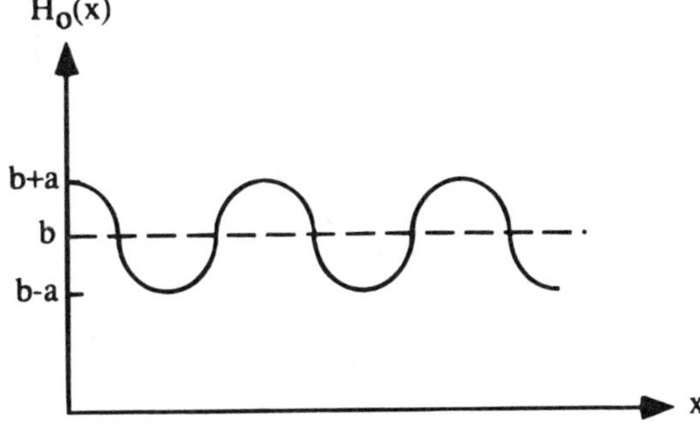

Fig. 9.1 One-dimensional object irradiance.

irradiance maxima appear relatively "whitish." That is a physical implication of the irradiance differences. A spatial square wave would consist of black and white stripes with sharp edges, as considered in Example 8.2. Here, instead of sharp edges the whitish and blackish (actually shades of gray rather than absolute white and black) portions of the object blend into each other in a sinusoidal fashion, rather than discretely as in a square wave.

Using Eq. (9.2.1) modulation contrast of the object in Fig. 9.1 is

$$\mathrm{MCO} = \frac{(b + a) - (b - a)}{(b + a) + (b - a)} = \frac{a}{b} \tag{9.2.3}$$

which is the ratio of cosine irradiance a to background irradiance b. The greater the irradiance of the cosine and/or the weaker the irradiance of the background, the better the contrast of the cosine maxima (light gray levels)relative to minima (dark gray levels) or of the cosine relative to background. This result is identical to Eq. (7.9.16).

To determine modulation contrast in the image, the image itself must be determined first. A convenient method is to use Eq. (8.6.9). Substituting for object irradiance distribution of Eq. (9.2.2), Eq. (8.6.9) becomes

$$H_i(x') = \int_{-\infty}^{\infty} (b + a \cos 2\pi f_x x) s(x' - x)\, dx \tag{9.2.4}$$

In the spatial frequency plane, from Eqs. (8.8.4) and (9.2.4) above

$$H_i(f_x) = \left\{ b\delta(f_x) + \frac{a}{2} [\delta(f_x - f_o) + \delta(f_x + f_o)] \right\} \cdot \mathbf{S}(f_x) \tag{9.2.5}$$

Eq. (9.2.4) can be evaluated by taking the inverse Fourier transform of Eq. (9.2.5). [Because of the delta functions this operation is relatively simpler than direct evaluation of Eq. (9.2.5)]. This inverse Fourier transform yields

$$H_i(x') = \int_{-\infty}^{\infty} H_i(f_x) e^{j2\pi f_x x'} dx' =$$

$$b\mathbf{S}(0) + \frac{a}{2} [\mathbf{S}(f_o) e^{j2\pi f_x x'} + \mathbf{S}(-f_o) e^{-j2\pi f_o x'}] \tag{9.2.6}$$

where $\mathbf{S}(f_x)$ represents the Fourier transform of $s(x')$.

Now, since the complex conjugate of OTF is

$$\mathbf{S}^*(f) = \int s(x') e^{-j2\pi f x'} dx' = \mathbf{S}(-f) \tag{9.2.7}$$

when this result is substituted into Eq. (9.2.6), the latter includes the sum of two conjugates and reduces to

$$H_i(x') = b\mathbf{S}(0) + \left(\frac{a}{2}\right) 2\ \mathrm{Re}[\mathbf{S}(f_o)e^{j2\pi f_o x'}] \tag{9.2.8}$$

By designating

$$\mathbf{S}(f_x) = |\mathbf{S}(f_x)| \exp[j\phi(f_x)] \tag{9.2.9}$$

image irradiance distribution is described as

$$\begin{aligned} H_i(x') &= b\mathbf{S}(0) + a|\mathbf{S}(f_o)| \cos[2\pi f_o x' + \phi(f_o)] \\ &= b\mathbf{S}(0)[1 + a|\mathbf{S}(f_o)| \cos[2\pi f_o x' + \phi(f_o)]/b\mathbf{S}(0)] \end{aligned} \tag{9.2.10}$$

Thus, for a linear imaging system (incoherent light), the image of a cosine is a cosine of the same spatial frequency, with possible changes in phase of the cosine, which imply changes in position.

The different irradiance of cosine and background in the image gives rise to a change in contrast in the image plane. Following definition (9.2.1), modulation contrast in the image plane is

$$\begin{aligned} \mathrm{MCI} &= \frac{[b\mathbf{S}(0) + a|\mathbf{S}(f_o)|] - [b\mathbf{S}(0) - a|\mathbf{S}(f_o)|]}{b\mathbf{S}(0) + a|\mathbf{S}(f_o)| + b\mathbf{S}(0) - a|\mathbf{S}(f_o)|} \\ &= a|\mathbf{S}(f_o)|/b\mathbf{S}(0) = (a/b)|\mathbf{S}(f_o)|/\mathbf{S}(0) \end{aligned} \tag{9.2.11}$$

Once again, modulation contrast is equal to the ratio of cosine to background irradiance. In the image, however, modulation contrast has been altered from that of the object by two values of the imaging system OTF, namely, $|\mathbf{S}(f_o)|$ and $\mathbf{S}(0)$.

9.3 MODULATION CONTRAST FUNCTION: SINE WAVE RESPONSE

Modulation contrast function (MCF) is defined [9.1] as the ratio of image (output) modulation contrast to that of the object (input). For a cosine object,

$$\mathrm{MCF} = \frac{\mathrm{MCI}}{\mathrm{MCO}} = \frac{|\mathbf{S}(f_o)|}{\mathbf{S}(0)} \tag{9.3.1}$$

where Eqs. (9.2.11) and (9.2.3) have been used. In view of Eq. (8.2.1) or (8.8.3), the MCF is identical to the modulation transfer function (MTF) at the cosine spatial frequency. In Section 8.2 MTF was defined as magnitude of the optical transfer function. Equation (8.2.1) or (8.8.2) indicates that $\tau(f) = \mathbf{S}(f)/\mathbf{S}(0)$. Therefore, Eq. (9.3.1) indicates that for a cosinusoidal spatially varying object of spatial frequency f_o, MCF $= |\tau(f_o)|$. For such a sinusoidal object, MCF is called *sine wave response* [9.2]. Equation (9.3.1) indicates that the sine wave response is equal to the imaging system MTF at that discrete spatial frequency. Therefore, one method of measuring the MTF of an imaging system is to expose it to many sine wave objects, each of different spatial frequency. By measuring image and object modulation contrasts, system MTF can be determined at each discrete spatial frequency, thus leading to a graph of system MTF or sine wave response as a function of spatial frequency. Image and object modulation contrasts can be measured using a radiometer to measure maximum and minimum image and object irradiance at each spatial frequency.

The importance of the MCF is twofold: (1) It indicates the mathematical concept of the OTF and its magnitude, the modulation transfer function, is associated with the very real physical phenomenon called contrast transfer from object to image plane; (2) it also provides a method of measuring MTF, which is important where analytical solutions are not feasible. For example, one can prepare a single resolution chart, containing sine wave objects of different spatial frequencies, and image it with a given photoelectronic (TV) or photographic imaging system. The ratio of image modulation contrast to object modulation contrast is then the MTF at each spatial frequency.

Sine wave resolution charts, however, are difficult to fabricate. Blackish and whitish areas must blend into each other not sharply but in a spatially sinusoidal fashion. Nowadays, such resolution charts are commercially available, although they are orders of magnitude more expensive than square wave resolution charts. Examples of the latter are shown in Fig. 9.2. While sine wave frequencies are cycles per unit width, square wave spatial frequencies are denoted as lines or line pairs per unit width, where two lines (one black and one white) equal one line pair.

A square wave can be decomposed into a Fourier series of *sine* waves, where the first harmonic is the square wave frequency. This has already been done in Example 8.1 and leads to the following question. Because the sine wave response is identical to MTF, is the same true of the square wave response? This would be a great simplification since resolution charts as in Fig. 9.2 are not only so inexpensive, but they can actually be easily made nowadays with word processors. This question is addressed next.

9.4 MODULATION TRANSFER FUNCTION: SQUARE WAVE RESPONSE

Consider a periodic real object, such as but not limited to a square wave. It can be represented by a Fourier series of the form

$$H_o(x) = b + \sum_{n=-\infty}^{\infty} a_n \cos(2\pi n f_o x + \phi_n) \tag{9.4.1}$$

As in the previous example, the image can be obtained by convolution of the object irradiance distribution with the image system spread function, which may also be expressed as an inverse transform of the product of their Fourier transforms, that is,

$$H_i(x') = \int_{-\infty}^{\infty} H_o(f_x)\mathbf{S}(f_x)\exp(-j2\pi f_x x')\, df_x \tag{9.4.2}$$

By taking the Fourier transform of Eq. (9.4.1) and substituting it into Eq. (9.4.2), we obtain

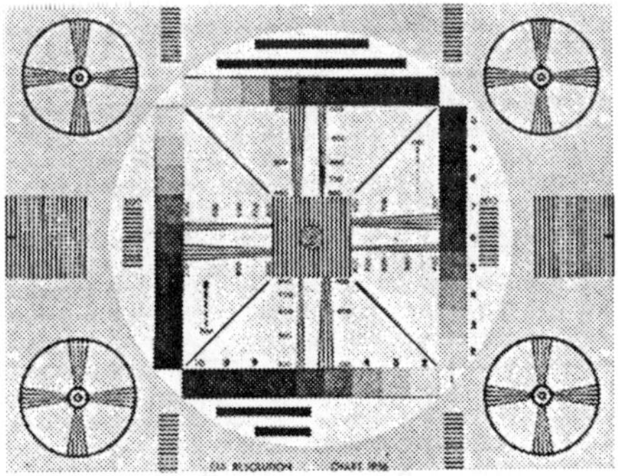

Fig. 9.2 USAF, NBS, Igor Limansky, and EIA resolution charts. (The usual Limansky resolution chart consists of nine individual charts arranged in a 3 × 3 array.)

$$
\begin{aligned}
H_i(x') &= b\mathbf{S}(0) + \sum_{n=-\infty}^{\infty} (a_n/2) \int \left[\delta(f_x - nf_o)e^{j\phi_n} + \delta(f_x + nf_o)e^{-j\phi_n} \right] \\
&\quad \times \mathbf{S}(f_x) \exp(-j2\pi f_x x')\, df_x \\
&= b\mathbf{S}(0) + \sum_{n=-\infty}^{\infty} (a_n/2) \left[\mathbf{S}(nf_o)e^{j\phi_n}e^{j2\pi nf_o x'} + \mathbf{S}(-nf_o)e^{-j\phi_n}e^{-j2\pi nf_o x'} \right] \quad (9.4.3)
\end{aligned}
$$

where symmetry of impulse response has been considered. In view of Eq. (9.2.9), the preceding equation reduces to

$$H_i(x') = b\mathbf{S}(0) + \sum_{n=-\infty}^{\infty} a_n|\mathbf{S}(nf_o)| \cos 2\pi[nf_o x' + \phi_n + \phi(f_x)] \tag{9.4.4}$$

The image described above is in precisely the same form as the object described in Eq. (9.4.1). Each cosine term is preserved, with possible changes in relative amplitude $|\mathbf{S}(nf_o)|$ and phase $[\phi(f)]$ imposed by the imaging system. Modulation contrast of the image is, by definition (9.2.1),

$$\text{MCI} = \frac{b\mathbf{S}(0) + \sum_{n=\mp\infty}^{\infty} a_n|\mathbf{S}(nf_o)| - \left[b\mathbf{S}(0) - \sum_{n=-\infty}^{\infty} a_n|\mathbf{S}(nf_o)|\right]}{b\mathbf{S}(0) + \sum_{n=-\infty}^{\infty} a_n|\mathbf{S}(nf_o)| + b\mathbf{S}(0) - \sum_{n=-\infty}^{\infty} a_n|\mathbf{S}(nf_o)|)}$$

$$= \sum_{n=-\infty}^{\infty} a_n|\mathbf{S}(nf_o)|/b\mathbf{S}(0) \tag{9.4.5}$$

Modulation contrast of the object [Eq. (9.4.1)] is

$$\text{MCO} = \frac{b + \sum_{n=-\infty}^{\infty} a_n - \left(b - \sum_{n=-\infty}^{\infty} a_n\right)}{b + \sum_{n=-\infty}^{\infty} a_n + \left(b - \sum_{n=-\infty}^{\infty} a_n\right)} = \frac{\sum_{n=-\infty}^{\infty} a_n}{b} \tag{9.4.6}$$

and the MCF is equal to the ratio MCI/MCO, or

$$\text{MCF} = \frac{\sum_{n=-\infty}^{\infty} a_n|\mathbf{S}(nf_o)|}{\left[\mathbf{S}(0) \sum_{n=-\infty}^{\infty} a_n\right]} \tag{9.4.7}$$

Since the a_n terms in the numerator and denominator do not cancel, for a periodic real object such as a square wave MCF $\neq$ MTF $= |\mathbf{S}(nf_o)|\mathbf{S}(0)$. In other words, while sine wave response equals MTF, square wave response does not [9.2]. Nevertheless, since resolution charts as in Fig. 9.2 are easy to fabricate and readily available, they are widely used to characterize the resolution capability of imaging components and systems. They are manufactured such that MCO equals unity out to very high spatial frequencies where the black-white line pairs become extremely narrow. One then simply measures MCI at the output of the instrumentation, and this is essentially equal to MCF or, in this case, square wave response. Square wave response is commonly featured on the data sheets of most imaging instrumentation instead of MTF or OTF. It behooves us, therefore, to understand the meaning of square wave response. It is used as an approximation to MTF, but is *not* MTF. Since a square wave contains in principle an infinite number of harmonics, while a sine wave contains only one, the energy input of a square wave object is greater than that of a sine wave object. Consequently, square wave response is greater than sine wave response. In other words, square wave response is an approximation to MTF or sine wave response, but the approximation is upward from MTF. This approximation favors the manufacturer rather than the consumer, because square wave response is higher than MTF is. The increase of square wave over sine wave response is several percent, depending on spatial frequency. Often, the square wave response is a sufficient approximation to MTF. From Fourier analysis, if square wave response $S_q(f)$ has been measured out to very high spatial frequencies, MTF for spatially continuous (nonsampled) images such as those obtained on film can be calculated from [9.2]

$$\mathrm{MTF}(f_r) = (\pi/4)[S_q(f_r) + S_q(3f_r)/3 - S_q(5f_r)/5 + S_q(7f_4)/7 \cdots\cdot + S_q(nf_r)/n \pm \cdots] \qquad (9.4.8)$$

where f_r is spatial frequency. Alternatively, the square wave response can be decomposed into a Fourier series. The first harmonic at each spatial frequency is then MTF.

A comparison of MTF and square wave response is shown in Fig. 9.3.

9.5 CONTRAST TRANSFER

The equivalence between sine wave response and MTF has many implications for system design. By definition (9.3.1), for a sine wave object of spatial frequency f_o, the image plane modulation contrast is

$$\begin{aligned}\mathrm{MCI}(f_o) &= \mathrm{MCF}(f_o)\mathrm{MCO}(f_o)\\ &= |\tau(f_o)|\mathrm{MCO}(f_o)\end{aligned} \qquad (9.5.1)$$

where $|\tau(f_o)| \approx \mathrm{MTF}\,(f_o)$, which determines contrast transfer at each spatial frequency.

Note that the relationship between $H_i(f_x, f_y)$ and $H_o(f_x, f_y)$ in Eq. (8.8.4) is in the spatial-frequency domain. Both are direct Fourier transforms of $H_i(x', y')$ and $H_o(x', y')$ respectively. However, MCI and MCO in Eq. (9.5.1) are both in the *spatial* domain and, as in sine wave and square wave response, contrast calculation is in the

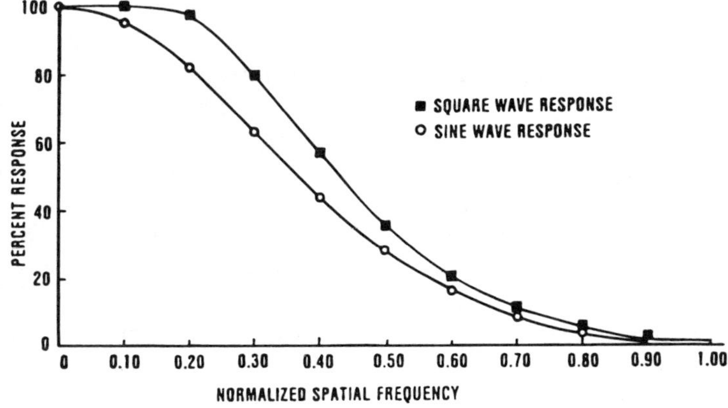

Fig. 9.3 Comparison of sine wave response and square wave response.

spatial domain only. Spatial frequency dependence, such as on f_o here, derives from object dimensions proportional approximately to $(2f_o)^{-1}$, as discussed further in the next chapter. In this manner, if object scene contrast and system MTF are known, image plane contrast is determined from Eq. (9.5.1). This relationship is correct for sine wave objects and approximately correct for others. This contrast transfer is critical in determining the size detail that is resolved in the image of the object scene, as discussed in the next chapter for resolution that is limited by contrast. Basic to implementation of Eq. (9.5.1) is knowledge of contrast in the object scene itself.

We showed in Section 3.4, following the radiance theorem, that image irradiance is directly proportional to object emittance or radiance, with proportionality factors involving optics geometry that is the same for each object in the scene. Therefore, in both object and image spatial domains modulation contrast between two objects of radiance N_1 and N_2 is

$$\mathrm{MCO} = \mathrm{MCI} = (N_2 - N_1)/(N_2 + N_1) \tag{9.5.2}$$

This assumes no attenuation between object and image space. Since $N = \int N_\lambda(\lambda)\, d\lambda$, Eq. (9.5.2) implies that contrast can vary strongly with wavelength. Good system design requires imaging at wavelengths of maximum image contrast. The wavelength dependence of contrast can vary strongly between object and image scenes, according to the wavelength dependencies of instrumentation and image channel (particularly the atmosphere) optical transfer functions. Therefore, system analysis also includes overall wavelength dependence of the image plane modulation contrast, which depends on both wavelength dependencies of MCO as well as of overall system MTF. The first is considered next; the latter in the rest of this book.

9.6 OBJECT PLANE MODULATION CONTRAST

In the visible and near-infrared regions, where thermal radiation is negligible, modulation contrast depends on contrast in the reflection coefficients of the objects being viewed. Assuming uniform illumination of the object scene, modulation contrast between objects of reflectivities $\rho_1(\lambda)$ and $\rho_2(\lambda)$ is given by (see Exercise 9.1)

$$\mathrm{MCO} = \frac{[\rho_1(\lambda) - \rho_2(\lambda)]}{[\rho_1(\lambda) + \rho_2(\lambda)]} \tag{9.6.1}$$

The absolute value notation is in keeping with Eq. (9.2.1). Typical reflectivities for various objects is given in Fig. 9.4 and more detailed data are available in the literature [9.3]–[9.6]. It is interesting to note that vegetation exhibits low reflectivity in the visible, where chlorophyll strongly absorbs sunlight. In the near infrared (0.8–1.7 μm wavelength), where solar irradiance is much diminished, so too is absorptance by vegetation. Since the sum of reflectance, absorptance, and transmittance must be unity [see Eq. (4.1.1)], the near-infrared decrease in absorptance by vegetation is accompanied by a strong increase in reflectance. Therefore, imaging of objects with a vegetative background often yields poor object scene contrast in the near infrared. Rare metals (unlike oxidized ones) generally exhibit low emissivity and absorptance (see Table 4.2). and therefore generally exhibit high reflectivities. Many objects are painted. Metallic paints often behave like pure metals and exhibit high reflectivities. Other types of paint generally exhibit high emissivity and absorptance, and low reflectivity.

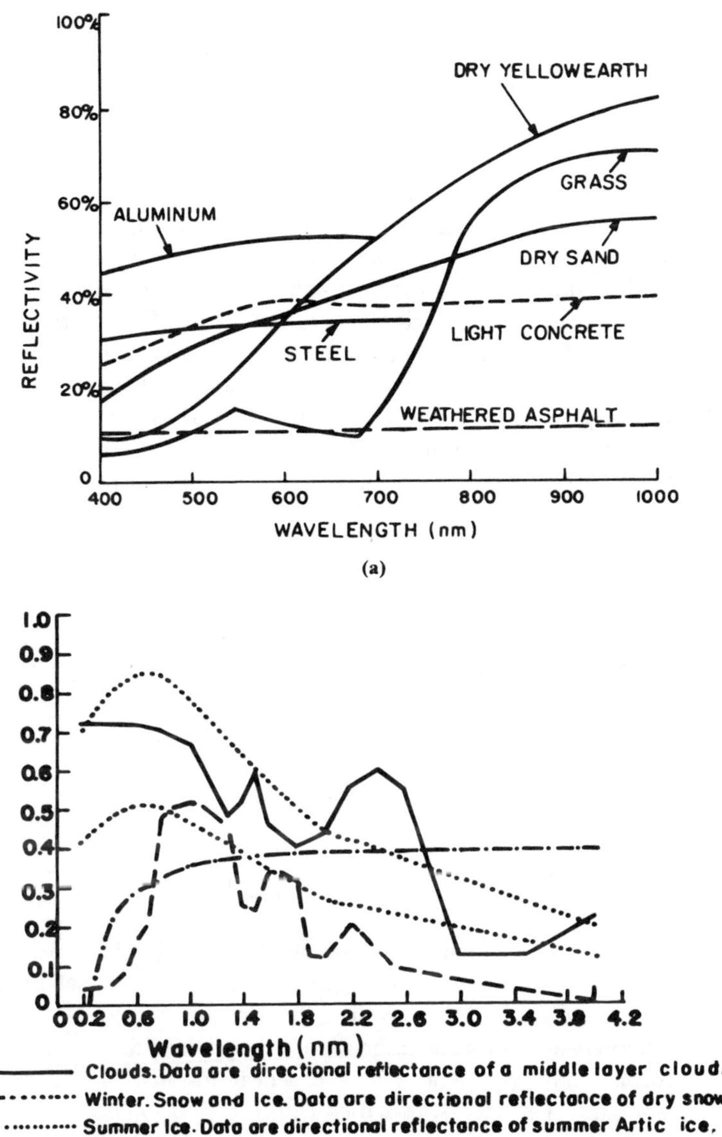

(a)

(b)

Fig. 9.4 Spectral reflectives for various materials [(a) after {9.3}, (b) after {9.4}].

Thermal imaging contrast behaves spectrally as

$$\mathrm{MCO}(\lambda) = \frac{|[1 - a(\lambda)]|}{[1 + a(\lambda)]} \tag{9.6.2}$$

where

$$a(\lambda) = \frac{\epsilon_2[\exp(hc/\lambda kT_1) - 1]}{\epsilon_1[\exp(hc/\lambda kT_2) - 1]} \tag{9.6.3}$$

as shown in the solution to Exercise 9.2. In the above ϵ_1, ϵ_2 and T_1, T_2 are the emissivities and absolute temperatures, respectively, of two objects whose modulation contrast is being considered. For thermal imaging of passive scenes, the emissivity differences often determine thermal contrast primarily [9.7] as demonstrated in Exercise 9.4. For similar weather conditions there is little difference between daytime and nighttime thermal imaging [9.7] since, on an *absolute* temperature scale, the temperature differences of objects between daytime and nighttime is small.

At long wavelengths such as millimeter and submillimeter wavelengths where the exponents in Eq. (9.6.3) are much less than unity, a Taylor series approximation yields, for imaging of thermal emissions, an object plane contrast on the order of

$$\mathrm{MCO}(\lambda) \approx \frac{|1 - \epsilon_2(hc/\lambda kT_1)|}{1 + \epsilon_1(hc/\lambda kT_2)} \tag{9.6.4}$$

which approaches unity at very long and very short wavelengths. At infrared wavelengths where the exponents in Eq. (9.6.3) are quite large

$$a(\lambda) \approx (\epsilon_2/\epsilon_1)\exp[hc\Delta T/(\lambda kT_1T_2)] \tag{9.6.5}$$

where $\Delta T = |T_2 - T_1|$. Substitution into Eq. (9.6.2) then yields

$$\mathrm{MCO}(\lambda) \approx \frac{|1 - (\epsilon_2/\epsilon_1)\exp[hc\Delta T/(\lambda kT_1T_2)]|}{1 + (\epsilon_2/\epsilon_1)\exp[hc\Delta T/(\lambda kT_1T_2)]} \tag{9.6.6}$$

If ϵ_2/ϵ_1 is wavelength independent, this last result favors imaging at shorter wavelengths [9.8]. Eq. (9.6.6) approaches a magnitude of unity as wavelength decreases. The limit, however, occurs where wavelengths have decreased to such an extent that reflective irradiance becomes greater than thermal irradiance [9.8]. This occurs typically for normal room temperatures in the near infrared at around a 2–3 μm wavelength as shown in Chapter 4, in which case Eq. (9.6.1) becomes more relevant than Eq. (9.6.2).

We should emphasize here that although Wien's law determines wavelength λ_m at which the irradiance of each object is its maximum, contrast depends on irradiance *difference*. For a constant emissivity ratio ϵ_2/ϵ_1, this difference increases as wavelength *decreases more and more* from λ_m where thermal irradiance peaks according to Wien's law. Consequently, if there is enough signal, it can be worthwhile to thermal image at shorter and shorter wavelengths, until reflectance becomes too significant [9.8].

The magnitude of Eq. (9.6.6) also approaches unity for large temperatures. This is most appropriate for imaging of *active* targets heated internally, against passive backgrounds.

If temperatures of object and background are very similar, then at such wavelengths $a(\lambda) \approx \epsilon_2(\lambda)/\epsilon_1(\lambda)$ and

$$\mathrm{MCO}(\lambda) \approx \frac{|\epsilon_1(\lambda) - \epsilon_2(\lambda)|}{\epsilon_1(\lambda) + \epsilon_2(\lambda)} \tag{9.6.7}$$

and contrast at those wavelengths is of emissivities alone. If reflectance varies with wavelength (Fig. 9.4), then so too must absorptance and emissivity according to the laws of conservation of energy. However, only limited data are available yet concerning the wavelength dependence of emissivity for various materials.

Note that Eqs. (9.6.3) and (9.6.6) imply thermal contrast maxima at relatively very long and short wavelengths and thermal contrast minima in the mid- to far-infrared regions (see Exercise 9.6).

The preceding treatment for thermal imaging object scene contrast assumes that irradiance reaching the imager is thermal only and that reflected irradiance is negligible. This is often not true in the short wavelength range of 3 to 5 μm, as shown in Chapter 4. The irradiance of each object scene object is actually

$$H = \epsilon(\lambda)H_T(\lambda, T) + \rho(\lambda)H_s(\lambda) \tag{9.6.8}$$

where $H_s(\lambda)$ is the irradiance (usually sunlight), illuminating the object, and $H_T(\lambda, T)$ is thermal irradiance. A more general expression for modulation contrast in the object scene is therefore

$$\mathrm{MCO}(\lambda) = \frac{|\epsilon_1(\lambda)H_{T_1}(\lambda, T_1) + \rho_1(\lambda)H_s(\lambda) - \epsilon_2(\lambda)H_{T_2}(\lambda, T_2) + \rho_2(\lambda)H_s(\lambda)|}{\epsilon_1(\lambda)H_{T_1}(\lambda, T_1) + \rho_1(\lambda)H_s(\lambda) + \epsilon_2(\lambda)H_{T_2}(\lambda, T_2) + \rho_2(\lambda)H_s(\lambda)} \tag{9.6.9}$$

This is particularly relevant in the short wavelength region of the 3- to 5-μm atmospheric absorption window.

In this section, the concept of modulation contrast has been applied to the object plane. Image plane modulation contrast, however, also depends on the propagation channel, the MTF of the instrumentation, and the MTF of the apertures involved. The effect of the latter is considered in the following example.

Example 9.1

A square wave object, shown in Fig. 8.11 in Example 8.2, is imaged from a distance through a slit aperture of width $2a$. The lens is aberration free. Both the square wave and slit length are in the vertical direction. What is the image irradiance spectrum if

$$\frac{2a}{sf_l} = \frac{2\lambda}{X(s - f_l)} \tag{9.6.10}$$

where λ is wavelength? Sketch the image spectrum. Determine the image modulation contrast and system modulation contrast function.

Solution

From Eq. (8.8.6), the object spectrum is

$$H_o(f_x) = (X/2)\ \mathrm{sinc}\ (2\pi f_x X/4) \sum_{n=-\infty}^{\infty} \delta(f_x - n/X)$$

where x and f_x are in the horizontal direction. The slit aperture yields an optical transfer function [from Eq. (8.4.7)]

$$\tau(f_x) = \begin{cases} 1 - |2\pi f_x|/(2\beta_0) & |f_x| \le \beta_0/\pi \\ 0 & |f_x| > \beta_0/\pi \end{cases}$$

where $\beta_0 = ka/s'$. Now from the Gaussian focusing relationship

$$\frac{1}{s'} = \frac{1}{f_l} - \frac{1}{s} = \frac{s - f_l}{sf_l} \rightarrow \beta_0 = \frac{2\pi a}{\lambda}\left(\frac{s - f_l}{sf_l}\right) = \frac{2a}{sf_l} \cdot \frac{\pi(s - f_l)}{\lambda}$$
$$= 2\pi/X \tag{9.6.11}$$

from Eq. (9.6.10).

The object spectrum and $\tau(f_x)$ are sketched in Fig. 9.5. Only three harmonics are involved in the image spectrum.

The image spectrum is, using Eqs. (8.8.8) and (8.4.7),

$$\begin{aligned} H_i(f_x) = H_o(f_x) \cdot \tau(f_x) &= \tau(f_x) \cdot \{(X/2)\delta(f_x) + \text{sinc}(\pi/2)\delta(f_x - 1/X) \\ &\quad + \text{sinc}(-\pi/2)\delta(f_x + 1/X)\} \\ &= (X/2)[\delta(f_x) + (2/\pi)(1 - \pi/X)(2\pi/X)\delta(f_x - 1/X) \\ &\quad + (2/\pi)(1 + \pi/X)(2\pi/X)\delta(f_x + 1/X)] \\ &= (X/2)\{\delta(f_x) + (1/\pi)[\delta(f_x - 1/X) + \delta(f_x + 1/X)]\} \end{aligned} \tag{9.6.12}$$

The image irradiance distribution therefore is the inverse Fourier transform of the above,

$$H_i(x') = X/2 + (2/\pi)\cos(2\pi x'/X) \tag{9.6.13}$$

Because of the limited spatial-frequency bandwidth of the slit aperture, the image of the square wave is limited to a cosine plus background. There is no square wave image corresponding to the square wave object.

Image plane modulation contrast is

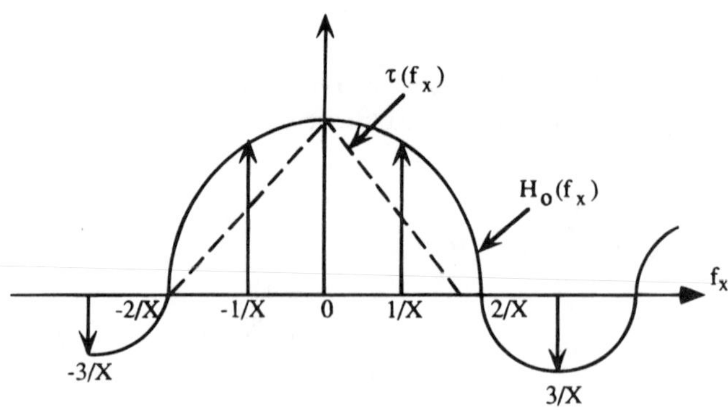

Fig. 9.5 Object spectrum and OTF for Example 9.1.

$$\text{MCI} = \frac{H_{i\max} - H_{i\min}}{H_{i\max} + H_{i\min}} = \frac{[(X/2) + 2/\pi] - [(X/2) - 2/\pi]}{X/2 + 2/\pi + X/2 - 2/\pi}$$

$$= \frac{2/\pi}{X/2} = 4/(\pi X) \tag{9.6.14}$$

Object plane modulation contrast is

$$\text{MCO} = \frac{H_{o\max} - H_{o\min}}{H_{o\max} + H_{o\min}} = \frac{1 - 0}{1 + 0} = 1 \tag{9.6.15}$$

Therefore, the modulation contrast function is

$$\text{MCF} = \text{MCI/MCO} = 4/(\pi X) \tag{9.6.16}$$

This example does not involve wavelength dependencies. It does, however, illustrate how the modulation contrast function is determined. Most imaging systems use photoelectronic or photographic instrumentation to record the image in addition to optics to form the image. As a result, MTFs often must be measured rather than calculated analytically in order to determine OTF and contrast transfer. This is often carried out with bar or sine wave resolution charts or by measuring point spread functions directly with a point object and then Fourier transforming $s(x', y')$. Considerations for such measurements are discussed next.

9.7 OTHER METHODS TO MEASURE OTF AND MTF

The use of resolution charts to measure MTF was described in Sections 9.3 and 9.4. Here, two other methods are considered: point spread function (PSF) measurement and edge response.

9.7.1 Point Spread Function: Dependence on Pixel Size

To measure PSF, it is important to determine which size object can be considered a point object. This is affected primarily by instrumentation pixel size. The term *pixel* stands for "picture element." It is the smallest element recordable in image space. The brightness value represents average irradiance over that small portion of the image scene. Pixel size is often related to detector size. If only a portion of the detector is illuminated, the output current is equivalent to that obtained for the same total radiant power absorbed by the detector but averaged over the entire detector area. No detail smaller than a pixel can be resolved in the image. In practice, dead space between pixels violates isoplanaticism (spatial stationarity) which is a requirement for linear systems. Nevertheless, dead space is small and such effects are often neglected in image system characterization. However, it is possible to overcome distortions in MTF measurement arising from this lack of isoplanaticism by using a "white" spatial noise random object [9.9]. This gives rise to a white noise random image. The ratio of the image power spectral density to object power spectral density is equal to the square of the system MTF. In this way the lack of isoplanaticism is overcome by the spatial randomness of the object and image [9.9].

A point image is actually a whole pixel. Consequently, to measure PSF directly rather than with the random noise technique, a point object when projected to image plane dimensions according to lateral magnification should be no larger than an image pixel. The image of the point object should be real. For example, if pixel size is 10×10 μm as may be typical in a charge-coupled device (CCD) TV camera, and if lateral magnification s'/s is 0.2, then the "point" object should be of dimensions no larger than 50×50 μm. This can be accomplished by using a pinhole having a size smaller than 50 μm in front of a larger source and thereby imaging only a very small portion of the source (see Exercise 9.8). The illuminated 50 μm pinhole then represents a "point" source. Because of aberrations, imager imperfections, mechanical vibrations, or atmospherically related blur, the actual spread function obtained is usually larger than one pixel and represents overall system impulse response or spread function.

Pixel size strongly affects system MTF. If a pixel represents a point image, then pixel size and shape are minimum spread function. A best case optical transfer function for such an imaging system is then a normalized Fourier transform of the pixel shape. In the preceding CCD example, the impulse response or PSF is a rectangle of size $2a \times 2b$ to represent one pixel, where $a = b = 5$ μm. OTF as determined by pixel size is then $\text{sinc}(2\pi f_x a)\,\text{sinc}(2\pi f_y b)$. This represents an upper limit to CCD OTF. The smaller the pixel dimensions, the larger the spatial-frequency bandwidths. In this example, maximum values of f_x and f_y occur when each sinc reaches its first zero, that is, $(2a)^{-1} = (2b)^{-1}$, respectively. As a and b decrease, pixel size MTF broadens. This is to be expected since a decrease in pixel size means smaller detail can be resolved, corresponding to larger spatial-frequency bandwidth. Because of device, optics, and propagation channel imperfections as stated, it is difficult for pixel size MTF to be realized as system MTF. Nevertheless, a point source corresponding to image plane size less than that of a pixel is a good way to measure system MTF.

Although mathematically the sinc functions go out to spatial frequencies higher than the first zero, this is *false resolution* and is physically not possible because such high spatial frequencies would imply pixel dimensions smaller than the actual size $2a \times 2b$. Therefore, actual OTF extends from zero spatial frequency out to only the first zeros of the sinc functions, which represent cutoff frequencies $f_{cx} = (2a)^{-1}$ and $f_{cy} = (2b)^{-1}$.

9.7.2 Edge Response

Since a delta function is the one-dimensional derivative of a step function, line spread function can be determined from edge response of a discrete transition from black to white or vice versa. Mathematically, for an edge $E(x)$ in object space

$$\delta(x) = \frac{dE(x)}{dx} \tag{9.7.1}$$

Therefore, from Fourier identities, in the spatial-frequency domain

$$\mathbf{F}\{\delta(x)\} = f_x \mathbf{F}\{E(x)\} \tag{9.7.2}$$

Consequently, by taking the inverse Fourier transform of Eq. (9.7.2), the line spread function (LSF) in the image plane is

$$\text{LSF}(x') = \frac{d}{dx} E(x') \tag{9.7.3}$$

where $E(x')$ is edge response. As a result, line spread function (LSF) for an imaging system can be obtained by imaging a bar (as in a bar chart) and then taking the spatial derivative of the transition from white to black. Other edges used can be of buildings, knives, etc. Alternatively, one can first Fourier transform the edge response directly and then multiply the transform by $(j2\pi f_r)$, where f_r equals f_x in the x direction and f_y in the y direction. The advantages over determining MCF from square wave response are as follows:

1. Square wave response is obtained from a bar chart only at discrete spatial frequencies corresponding to square wave widths, while edge response yields LSF from which, through Fourier transform, the OTF can be obtained on a continual basis for all spatial frequencies up to system bandwidth limitations.
2. Square wave response yields MTF only; PTF is lost. On the other hand, Fourier transform of LSF yields OTF. However, edge response is disadvantageous to point spread function (PSF) since it yields OTF in one direction only, whereas the PSF yields two-dimensional OTF.

REFERENCES

9.1. G. O. Reynolds, J. B. DeVelis, G. B. Parrent, Jr., and B. J. Thompson, *The New Physical Optics Notebook: Tutorials in Fourier Optics*, SPIE, Bellingham, WA, 1989.

9.2. J. W. Coltman, "The specification of imaging properties by response to a sine wave input," *J. Opt. Soc. Am.*, Vol. 44, June 1954, pp. 468–471; and J. Bhushnan and A. K. Jaiswal, "Limitations in the determination of the sine wave response from the square-wave response," *Opt. Comm.*, Vol. 12(2), October 1974, pp. 181–182.

9.3. R. B. Toobin, "Atmospheric optics-reflectance," in *Handbook of Geophysics and Space Environments*, S. L. Valley, Ed., Air Force Cambridge Research Laboratory, Cambridge, MA, 1965, Sec. 7.2.

9.4. W. L. Wolfe and G. J. Zissis, Eds., *The Infrared Handbook Revised*, Environmental Research Institute of Michigan, Ann Arbor, 1989.

9.5. L. Levi, *Applied Optics Vol. 2*, Wiley, New York, 1980.

9.6. D. E. Bowker, R. E. Davis, D. L. Myrick, K. Stacy, and W. T. Jones, "Spectral reflectance of natural targets for use in remote sensing studies," Reference Publication 1139, NASA, June 1985.

9.7. A. Shushan, Y. Meningberg, I. Levy, and N. S. Kopeika, "Prediction of thermal image quality as a function of weather forecasts," *Opt. Eng.*, Vol. 30, November 1991, pp. 1709–1715.

9.8. N. S. Kopeika, "The general dependence of imaging through the atmosphere," *Appl. Opt.*, Vol. 20, May 1, 1981, pp. 1532–1536.

9.9. A. Daniels, G. D. Boreman, and A. D. Ducharme, "Random transparency targets for modulation transfer function measurement in the visible and infrared," *Opt. Eng.*, Vol. 34(3), March 1995, pp. 860–868.

EXERCISES

9.1 Obtain an expression for modulation contrast for a scene in the visible and near infrared, illuminated by sunlight, and consisting of objects of reflectivities ρ_{max} and ρ_{min}, respectively.

9.2 Determine modulation contrast for thermal imaging of a scene consisting of objects of emissivities ϵ_1 and ϵ_2 and temperatures T_1 and T_2, respectively. (a) Assume a very wideband thermal imager. (b) Assume a narrowband thermal imager.

9.3 Resolution of many imaging instruments, including TV systems, is characterized by a square wave response. In such cases, the line spread function may be more appropriate than the point spread function. Given point spread function $s(x',y')$, how can line spread function be determined from it?

9.4 A very wideband thermal imager views passive objects of emissivities ϵ_1 and ϵ_2, respectively. During the day solar irradiance generates a temperature difference ΔT in degrees kelvin between object temperatures. Assuming a mean emissivity of 50% and mean temperature 290 K, does thermal contrast derive primarily from emissivity or temperature difference if $\Delta\epsilon > 0.3$ and $\Delta T < 10$ K? What implication [9.7] is there concerning comparison of night-vision and day-vision thermal imaging?

9.5 From the standpoint of object plane contrast, is it preferable to image light concrete structure against dry sand backgrounds at 900-nm or at 500-nm wavelength?

9.6 Assume $T_1 = 290$ K, $T_2 = 300$ K, $\epsilon_1 = 0.1$, and $\epsilon_2 = 0.95$. Plot the wavelength dependence of MCO over the 2- to 20-μm wavelength region. What is the short wavelength limit to MCO? What is the long optical wavelength limit? Calculate MCO at 200 μm and 2 mm wavelengths and compare to theory.

9.7 Show that for similar emissivities at infrared wavelengths, MCO $\approx \Delta T/(\Delta T + 2\lambda k T_1 T_2/hc)$, where ΔT is temperature difference between objects at temperatures T_1 and T_2.

9.8 A 20-μm-diameter pinhole is available to measure the spread function of a CCD imaging system containing pixels of size 10 × 10 μm. The pinhole is placed in front of a halogen lamp and the light exiting the pinhole is imaged with a 10-cm focal length lens placed 12 cm in front of the CCD. What should be the distance between the source–pinhole combination and the TV system in order to measure the spread function of the latter?

CHAPTER

10

Contrast-Limited Resolution and Target Acquisition

In the previous two chapters the optical transfer function was introduced and its physical implications concerning contrast derived. In the present chapter, these results are applied toward determining image resolution in a quantitative manner. By *resolution* we mean the smallest size detail that can be resolved. This can be related to system optical transfer function and spatial frequency bandwidth. However, in reality image quality and resolution are very subjective parameters, varying considerably with each person's visual system characteristics and its dependence on background irradiance—system signal-to-noise ratio. Therefore, the quantitative formulations presented in this and the next few chapters should be viewed as representative rather than as absolute quantities. A more appropriate approach is stochastic rather than deterministic and is presented as well. *System design integrating the material from the previous eight chapters is emphasized strongly in the exercises and solutions.*

10.1 SPATIAL FREQUENCY BANDWIDTH

In electronics the significance of temporal frequency bandwidth with regard to fidelity or resemblance of output to input is well accepted. The greater the bandwidth, the better the fidelity. A similar situation exists in terms of spatial-frequency bandwidth and image quality. The greater the spatial-frequency bandwidth, the better the image quality. In electronics, bandwidth is generally defined rather arbitrarily for the sake of convenience according to half-maximum power values of a system's temporal frequency response. For imaging systems, spatial-frequency bandwidth is also defined arbitrarily. However, for the sake of convenience, spatial-frequency bandwidth limits differ from the half-power limitations in electronics but are determined by required system threshold contrast values for photon-limited imaging or by threshold signal-to-noise ratio (SNR) for noise-limited imaging.

10.2 LIMITING RESOLUTION REQUIRED THRESHOLD CONTRAST

Each imaging system has a threshold contrast required at its output according to the needs of the observer or instrumentation for whom the image is intended. We can see from the resolution bar charts of Fig. 9.2 that as spatial frequency increases (or bar widths narrow), the contrast between black and white line pairs decreases. Actually, at higher spatial frequencies the lines (or bars) appear to be neither purely white nor black but rather whitish and darkish shades of gray because they are sufficiently narrow so as to appear to blend into one another. Finally, as line pairs narrow a high spatial-frequency limit is reached at which the "white" line is no longer distinguishable from the "black" line. Both have become identical shades of gray. Higher spatial frequencies, where line pairs are even more narrow, exhibit an identical situation where "black" and "white" lines are indistinguishable.

The highest spatial frequency at which "black" and "white" are distinguishable is called here $f_{r\ \max}$. It is the maximum *usable* spatial frequency and represents limiting resolution. Spatial-frequency bandwidth is from zero spatial frequency to $f_{r\ \max}$. This is shown schematically in Fig. 10.1(a) for a typical imaging system. The lower curve represents required threshold contrast and is qualitatively similar to that of the human visual system discussed in Chapter 12 and shown in Fig. 12.17 for photopic vision. (The human visual system threshold reaches a minimum at an angular spatial frequency of about 2 cycles per degree.) Intersection of the *overall* system MTF and required threshold contrast plots determines the value of $f_{r\ \max}$. Overall system MTF must also include contrast reduction factors such as modulation contrast of the object (MCO) itself versus background (Section 9.6), atmospheric MTF, vibration MTF, path luminance MTF (at end of this chapter), as well as all hardware MTFs, for example.

Although MTF is used in the literature [10.1–10.3] to determine $f_{r\ \max}$ as in Fig. 10.1(a), *image plane modulation contrast* is much more useful in practice in place of the MTF curve in Fig. 10.1(a) in determining $f_{r\ \max}$. The reason is that image quality is determined by modulation contrast and threshold contrast required in the image plane. Hence, we use Fig. 10.1(b). Both figures become identical if contrast stretching methods are used, as described in Section 18.9, which effectively increase MCO. As seen From Figs. 10.1, information at spatial frequencies higher than $f_{r\ \max}$ is transmitted by the system. However, contrast at such high spatial frequencies is so low that information at such high spatial frequencies is essentially useless. Hence, $f_{r\ \max}$ is the spatial frequency of limiting resolution [10.1]. Here, we assume SNR is not a limitation. Such situations are called *contrast limited.* In contrast-limited imaging, an object or target is resolved only if the contrast between it and its background or clutter (unwanted background) is greater than the required threshold contrast. Since $f_{r\ \max}$ is a quantitative measure of blur extent, contrast-limited imaging is *also limited by blur.* Noise-limited imaging is characterized by time- and/or space-varying "snow" on the image display. In noise-limited imaging, an object or target is resolved against such random snow. Such noise, rather than background or clutter, limits target acquisition for noise-limited imaging. This situation is considered in Chapter 11.

Photographic systems involve graininess. A threshold contrast curve for film has been suggested to be [10.3]

$$T_f(f_r) = 0.034[dD/d(\log_{10}E)]^{-1}(0.033 + \sigma_{24}^2 f_r^2 S^2)^{1/2} \qquad (10.2.1)$$

where f_r is in lines·mm^{-1}, D is mean film density, E is exposure [Eq. (3.1.15)], σ_{24} is rms granularity for a 24 μm scanning aperture, and S is SNR for threshold viewing (about 4.5).

Generally, 2% used to be considered a feasible threshold contrast for the human visual system, until more recent measurements indicated a shape like that shown in Figs. 10.1. For simplicity, exercises here use 2% threshold contrast. Constant threshold

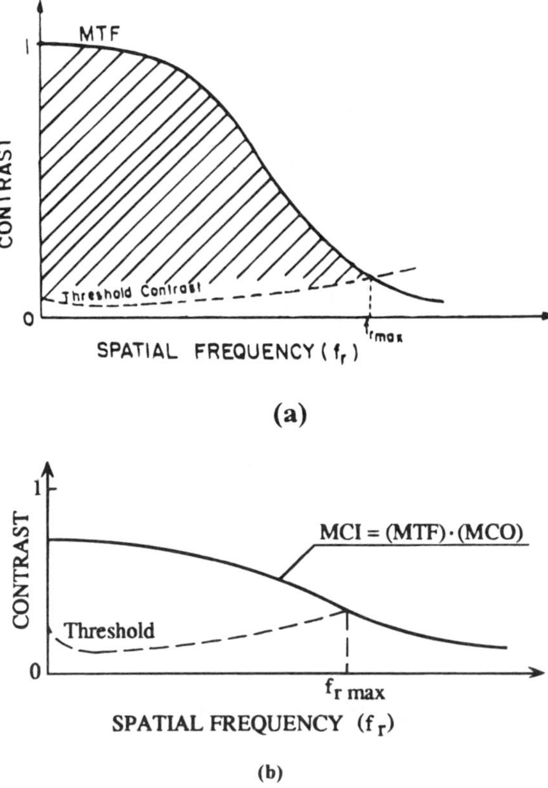

Fig. 10.1 (a) Limiting resolution $f_{r\ \max}$ defined by intersection of MTF and required threshold contrast curves (after [10.1–10.3]). (b) Limiting resolution $f_{r\ \max}$ defined by intersection of image plane modulation contrast (MCI) and required threshold contrast curves.

contrast is applicable to *automatic target acquisition* by instrumentation. Examples with actual human vision threshold contrast are presented in Chapter 19.

The system MTF curve in Fig. 10.2 is a cascade of all independent subsystem MTFs, including imager, optics, electronics, atmosphere, image motion and vibrations, etc. It is dominated by the weakest MTF, usually from vibrations or atmospheric blur; limiting resolution is determined principally from that effect, and improvements in other aspects such as aberrations or sensors have little effect on limiting resolution. This consideration is basic to cost effectiveness. However, image restoration as discussed in Chapter 18 can improve limiting and overall resolution considerably, so that improvements in any parts of imaging systems can then improve resolution even further under such circumstances.

In principle, the preceding definitions of $f_{r\ \max}$ and spatial-frequency bandwidth hold only for sine wave rather than bar resolution charts. However, as discussed in the previous chapter, sine wave and square wave response are quantitatively very similar. As a result, the same definitions are applied to both sine wave and square wave resolution charts. Both yield high correlations of image quality with $f_{r\ \max}$ [10.2, 10.3].

The role of $f_{r\ \max}$ or spatial-frequency bandwidth in determining the size of detail that can be resolved is illustrated in the following example.

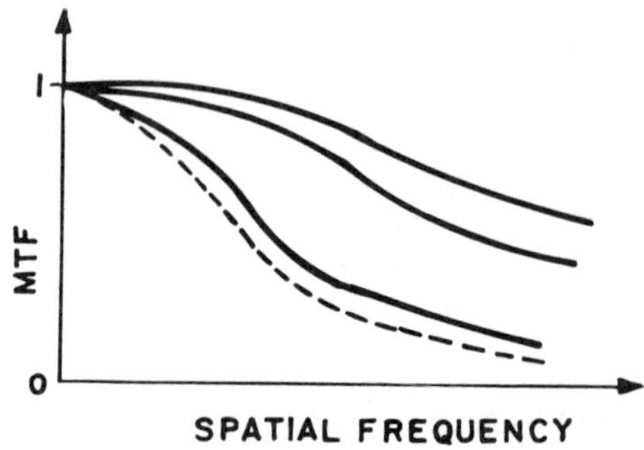

Fig. 10.2 Cascade of various component MTFs (solid curves) leads to system MTF (dotted curve) dominated by weakest MTF.

Consider the simplest type of imaging system: a thin lens. Similar triangles used in the derivation of lateral magnification in Eq. (2.6.8) indicate

$$\frac{x}{s} = -\frac{x'}{s'} \tag{10.2.2}$$

As shown in Fig. 10.3, the same relationship holds for the size of minimum resolvable detail in the object and image planes, labeled here Δx and $\Delta x'$, respectively, that is, resolution angle is

$$\frac{\Delta x}{s} = -\frac{\Delta x'}{s'} \tag{10.2.3}$$

where the minus sign simply refers to an inverted image. Experience with resolution charts indicates that if $f_{r\ \max}$ is the highest spatial frequency in the x direction at which a black-white line pair can be resolved, and if $\Delta x'$ is the width of either the black or white line of that line pair, then $2\Delta x' \approx f_{r\ \max}^{-1}$. Therefore, from Eq. (10.2.3)

$$\Delta x \simeq -(1/2)[f_{x\ \max}(s'/s)]^{-1} = -(2f_{x\ \max}m)^{-1} \tag{10.2.4}$$

where m is lateral magnification as defined in Eq. (2.6.9). When MTF is an isotropic

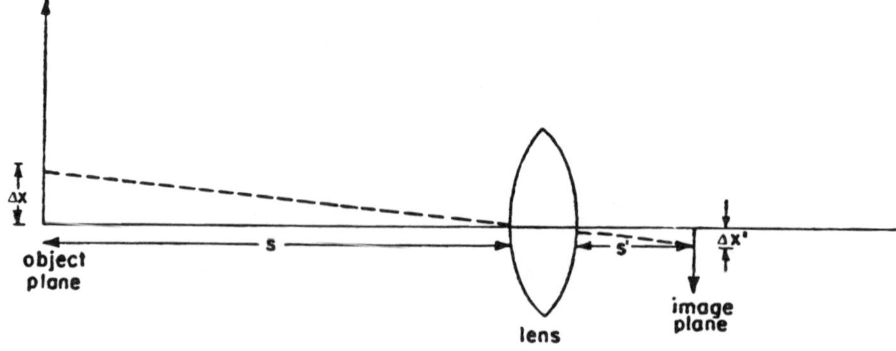

Fig. 10.3 Geometry for determining size of detail resolvable in object plane.

function of spatial frequency, then f_r is used here to denote spatial frequency rather than f_x. Detail size less than Δx^{-1} in the image plane is *blurred.*

Equation (10.2.4) tells us that the object plane detail detectable in direction r is inversely proportional to the product of spatial-frequency bandwidth $f_{r\ \max}$ and lateral magnification. If magnification is increased, then as target size increases, finer and finer detail can be resolved until a limit is reached. That limit is the blur extent inherent to the system, and it is defined by $(1/2)f_{r\ \max}^{-1}$. In other words, for low magnification resolution is limited by magnification itself and resolvable spatial detail Δr decreases as m increases, thereby improving resolution. If magnification is constant but $f_{r\ \max}$ increases, target image size is the same but more detail can be resolved. At high magnification, if image quality is already limited by spatial-frequency bandwidth, increased magnification merely increases blur radii by magnifying the blur, but without permitting resolution of finer detail. In this situation, resolution is limited by the system MTF as exemplified by $f_{r\ \max}^{-1}$. Thus Eq. (10.2.4) is valid for low magnification only. In general, the greater the spatial-frequency bandwidth $f_{r\ \max}$, the smaller the object plane detail that can be resolved. The larger the image size of the object, as compared to detail size $\Delta x'$, the greater the clarity with which the object can be seen. *Contrast-limited imaging is thus also blur-limited imaging.*

10.3 MTFA AND N_e

It should not be inferred, however, from Eq. (10.2.3) that determination of $f_{r\ \max}$ is *the* chief significance of OTF or MTF in determining image quality. Three MTF curves yielding identical $f_{r\ \max}$ are compared in Fig. 10.4. The imaging system whose MTF is curve 1 yields better contrast for large objects and poorer contrast for small objects than does the imaging system whose MTF is curve 2. Clearly, the system exhibiting the best image quality is that of MTF curve 3. It yields the best contrast for both large and small objects. This suggests that another criterion with which to quantitatively compare overall image quality is the area between the system MTF and required threshold contrast curves. This area is called the *MTF area*, designated MTFA, and psychophysical experiments with large numbers of independent observers do indeed support use of MTFA as a quantitative criteria of image quality [10.3]. An important advantage of this criterion is that image quality is reduced to a single number (MTFA) rather than a function (OTF or MTF).

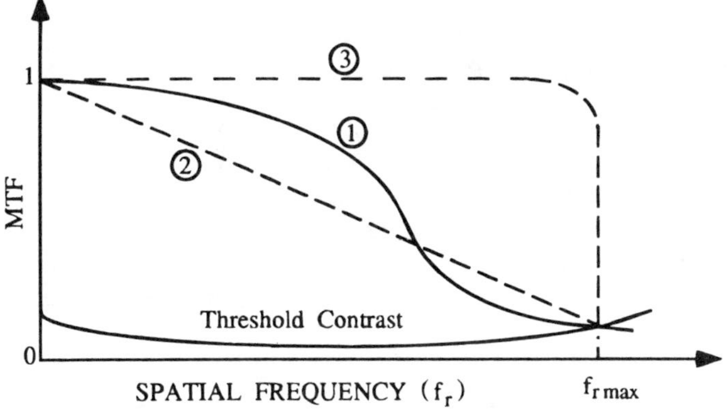

Fig. 10.4 MTFA concept.

Another criterion of image resolution is equivalent passband N_e, defined as the equivalent number of cycles or line pairs having unity modulation from zero frequency up to N_e and having zero modulation at all spatial frequencies above N_e that encloses an area equal to the square of the MTF area. The MTF area here is not reduced by the threshold contrast curve in Fig. 10.1 but includes all areas under the MTF curve. The concept is similar to that of noise bandwidth in electronics, defined in Eq. (5.1.10) and Fig. 5.1. Mathematically,

$$N_e = \int_{-\infty}^{\infty} [\tau(f_r)]^2 \, df_r \tag{10.3.1}$$

As in the case of MTFA, N_e is a convenient measure of resolution because it is a single number.

Similarly, equivalent noise spatial-frequency bandwidth is similar to that in the temporal frequency domain [Eq. (5.1.10)], that is,

$$\text{NBW} = \int_{0}^{\infty} [\tau(f_r)]^2 \, df_r = N_e/2 \tag{10.3.2}$$

For example, if there is little "dead" space between detectors, charge-coupled device (CCD) TV cameras are often limited in resolution by CCD pixel size. If $2a$ and $2b$ are pixel dimensions in the x and y directions, it has already been shown (Section 9.7.1) that pixel size MTF is

$$\tau_{\text{pixel}}(f_x, f_y) = \text{sinc}(2\pi f_x a)\,\text{sinc}(2\pi f_y b) \tag{10.3.3}$$

where cutoff frequencies $\pm f_{cx}$ and $\pm f_{cy}$ are $\pm 1/(2a)$ and $\pm 1/(2b)$, respectively, in the plus and minus f_x and f_y directions. Total spatial-frequency bandwidths in the f_x and f_y directions are $1/a$ and $1/b$, respectively. (A more accurate model of CCD MTF, including dead space and wavelength dependence, is given in Chapter 13.)

It can be shown (Exercise 10.8) that equivalent passband is approximately half of that, that is,

$$N_{ex} \lesssim 1/(2a), \qquad N_{ey} \lesssim 1/(2b) \tag{10.3.4}$$

10.4 THE JOHNSON CHART

Equation (10.2.4) considers only the minimum grade of resolution, namely, detection. Image detail of size $\Delta x' \approx (2f_{x\,\text{max}})^{-1}$ is "seen" just barely.

The probability of the object being resolved can be related to the relationship between image detail dimension $\Delta x'$, the imaged object size x', and image system resolution parameter $f_{x\,\text{max}}$. The treatment in Section 10.2 refers to *limiting* resolution where $\Delta x \approx -1/(2f_{x\,\text{max}}m)$ or $\Delta x' \approx -1/(2f_{x\,\text{max}})$. However, if the object size x is larger, say,

$$|x| = 2n|\Delta x| = 2n(2f_{x\,\text{max}}m)^{-1}, \qquad n \geq 1 \tag{10.4.1}$$

where n is number of line pairs (of width $2\Delta x$ each) whose width *altogether* is equivalent to object size x, then more detail can be seen. The $2n$ "equivalent" bars of width Δx fill up the object dimension x. The greater the value of n, the more small-size detail of size Δx that can be seen in the image of that object. Our question now is: Can the quality of resolution be quantified according to the value of the number n?

Johnson [10.4, 10.5] suggested four degrees of resolution: *detection, orientation, recognition*, and *identification*. He further suggested values of n with which to characterize each stage of resolution. His work was carried out for contrast-limited imaging with visible and near-infrared image intensifiers. Noise was not a limitation in his experiments, which involved large numbers of observers. His work was concerned with square wave response rather than MTF. Therefore, his model refers to the number of line *pairs* required to fill the minimum dimension of the image of the target. His psychophysical experiments showed that for the target geometries involved the minimum rather than maximum dimension was critical and is called the *critical dimension*. The Johnson chart is shown in Fig. 10.5 for different type of objects, and the numbers of line pairs refer to 50% probabilities of detection, orientation, recognition, and resolution.

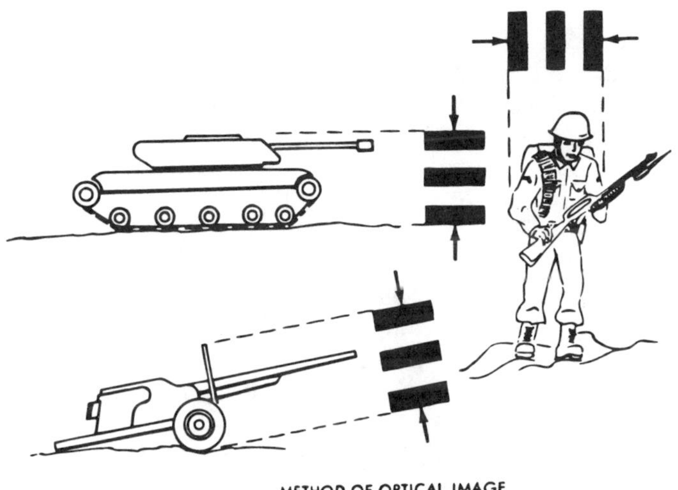

METHOD OF OPTICAL IMAGE TRANSFORMATION

TARGET	RESOLUTION PER MINIMUM DIMENSION IN LINE PAIRS			
BROADSIDE VIEW	DETECTION	ORIENTATION	RECOGNITION	IDENTIFICATION
TRUCK	0.90	1.25	4.5	8.0
M-48 TANK	0.75	1.20	3.5	7.0
STALIN TANK	0.75	1.20	3.3	6.0
CENTURION TANK	0.75	1.20	3.5	6.0
HALF-TRACK	1.00	1.50	4.0	5.0
JEEP	1.20	1.50	4.5	5.5
COMMAND CAR	1.20	1.50	4.3	5.5
SOLDIER (STANDING)	1.50	1.80	3.8	8.0
105 HOWITZER	1.00	1.50	4.8	6.0
AVERAGE	1.0 ± 0.25	1.4 ± 0.35	4.0 ± 0.8	6.4 ± 1.5

Fig. 10.5 Johnson chart (after [10.6]).

Qualitatively, Johnson's division of resolution into four classes involves the following concepts. *Detection* refers to the mere fact that the object is detected. No other detail can be resolved. No object shape is observed. All that is known is that there is some object imaged on the display screen. In this case, in Eq. (10.4.1), only one line pair on the average fills up a given image dimension. (This is consistent with our definition following Eq. (10.2.3) that $2\Delta x = (f_{x\ \max}m)^{-1}$.) In this case, $n \approx 1$. *Orientation* means that the image of the object fills up additional line pairs (1.4 on the average) and now some shape can be observed; that is, the object is seen to be *oriented* in a given direction. The term *recognition* means that the number n is sufficiently large so that the object is *recognizable* as a truck, tank, car, person (soldier), etc. This requires n to be on the average 4 line pairs per minimum dimension in the image. *Identification* refers to the fact that there are enough line pairs (6.4 on the average) filling up minimum image dimension that not only is the object recognizable as a truck, tank, etc., but the specific kind of truck or tank is identifiable. By designating Johnson chart numbers of line pairs as n_{50}, and considering image inversion, Eq. (10.4.1) is amended to

$$x = 2n_{50}\Delta x \approx -n_{50}(f_{\max}m)^{-1} \tag{10.4.2}$$

where, for simple detection $n_{50} = 1$ and $x = 2\Delta x$, in which case Eq. (10.4.2) reduces to Eq. (10.2.4). The more stringent the category of resolution, the greater the value of n_{50} and the larger the object size required to be resolved. The 50% probability associated with the Johnson chart implies that 50% of the population would detect such a target if the minimum dimension is filled by one line pair. The reason the consistency between Eq. (10.4.2) for $n_{50} = 1$ and Eq. (10.2.4) is limited to 50% of the population is that the human visual threshold contrast curve varies from person to person as shown in Chapter 12 in Fig. 12.17. Therefore, $f_{x\ \max}$ varies for different observers. Hence the Johnson chart can be representative only for an *average* human visual system. Nevertheless, from the comparison of the Johnson chart for detection and our postulation following Eq. (10.2.3) that $2\Delta x' = f_{\max}^{-1}$ it should be clear that the Johnson chart is consistent with resolution implications for detection deriving from MTF.

Equation (10.4.2) is appropriate to imaging with an analog device such as film or, for the horizontal direction, an electron tube camera such as a vidicon. With film the image is sensed continuously in both directions rather than sampled. In a vidicon, sampling takes place in the vertical direction, but not in the horizontal direction. Hence, for electron tube TV cameras, resolution in the vertical direction is limited generally by dead space between the horizontal raster lines. However, for semiconductor imagers such as CCDs, the image is quantized too in the horizontal direction. This means that for such devices target image dimension x' is equal to pixel width ($2a$ in Section 9.7.1) times the number of lines ($2n$) to be resolved that fill the target dimension, where n again is in line pairs. The minimum value of $\Delta x'$ here is pixel width $2a$. Therefore, for 50% resolution probability,

$$|x|' = (2a)(2n_{50}) \tag{10.4.3}$$

where n_{50} is the number of line pairs from the Johnson chart for the desired resolution criterion, that is, detection, orientation, recognition, or identification.

As shown in Section 9.7.1, pixel MTF is sinc $(2\pi f_x a)$. For the zero threshold contrast limit, $f_{x\ \max}$ for pixel-limited resolution is equal to $(2a)^{-1}$.

Plugging into Eq. (10.4.3),

$$|x'| = (f_{x\ \max})^{-1} 2n_{50} \tag{10.4.4}$$

or

$$|x| = 2n_{50}(f_{x\ \max} m)^{-1} \tag{10.4.5}$$

The equations are general results for 50% probability of resolution where $f_{x\ \max}$ is defined by threshold contrast in Figs. 10.1 and is therefore in practice somewhat less than $(2a)^{-1}$.

By comparing the preceding equations to Eq. (10.4.2), we can see that image sampling requires twice the number of equivalent line pairs to resolve an object of size x than if there were no sampling because a pixel can resolve only one line rather than a line pair. This means that for the same size dimension x the spatial frequency bandwidth of the sampling imager must be twice that of the nonsampling imager. In view of the high-resolution CCD imagers that are available, this is usually not a practical problem. Equation (10.4.5) is consistent with Nyquist sampling.

In Eqs. (10.4.2) and (10.4.5) the object resolution is seen to be proportional to the ratio of effective line pairs that fill the critical imaged target dimension to the number of effective line pairs that fill the display width. The latter number of TV line pairs is simply $f_{r\ \max}$ times the display width.

For vertical resolution in all TV cameras, including electron tube devices, a raster scan requires target image vertical dimension to be

$$y' = 2n\Delta y = (2b)(2n) \tag{10.4.6}$$

where $2b$ is pixel vertical width and $2n$ is the number of line pairs filling target image vertical dimension y', an example of which is shown in Fig. 10.6. This yields

$$y = 2n(f_{y\ \max} m)^{-1} \tag{10.4.7}$$

where $f_{y\ \max}$ cannot be greater than the raster scan spatial frequency. If $2b'$ is the

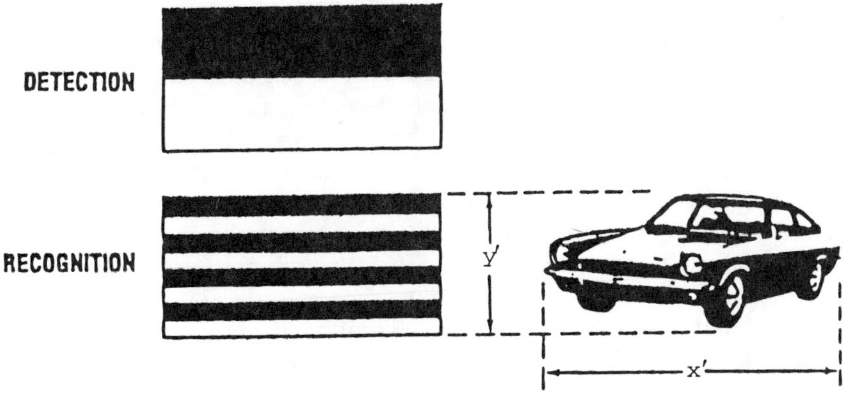

Fig. 10.6 Image and corresponding bar patterns for detection and recognition (after [11.1]).

distance from raster line center to adjacent raster line center, including dead space between raster lines, then

$$f_{y\,\max} \leq (2b')^{-1} \tag{10.4.8}$$

Here, horizontal minimum detail equivalent bar width in the vertical direction must be wider than the dead space between raster lines.

The last three equations refer to vertical resolution as limited by raster lines. However, vertical resolution can sometimes still be limited even more by pixel size (as in horizontal resolution). Hence, both raster scan limits and pixel size should be considered in determining which yields the smaller value of $f_{y\,\max}$. (See Exercise 10.18 and its solution.)

The above models are also appropriate for application to noise-limited imaging. The capability of adapting the Johnson chart to probabilities other than 50% is developed in the next section for contrast-limited imaging and in the next chapter for noise-limited imaging. The dependence of the Johnson chart on the minimum dimension of the object is understandable. If an object is of long length but near-zero width, it clearly will be difficult to resolve. However, the longer dimension and the ratio of long-to-short dimension also influence target recognition probabilities. The navy, for example, is interested in resolving ships at sea. Most ships have similar minimum dimensions, namely, height. The navy, therefore, is quite interested in use of the longer dimension to identify a ship's type. Hence, the Johnson chart is essentially useless in such efforts because it relates only to minimum dimension. Consequently, navy models depend on target area (Section 10.6).

In general, for several objects of the same minimum dimension, experiments with large numbers of observers indicate that small objects with larger maximum dimension have a higher probability of detection [10.7]. However, the reverse is true for large objects of minimum dimension subtending more than about 0.5 degrees [10.7]. It has been found that the human visual system is efficient in integrating a full image area only up to angular subtenses of about 0.5 degrees in both lateral dimensions. This angle is consistent with the angular spatial frequency at which human visual threshold contrast reaches a minimum in Fig. 10.1 (2 cycles·degree^{-1}) and may explain why required threshold contrasts increase for larger angular sizes as discussed in Chapter 12 for the human visual system. It seems that for large objects the human visual system actually integrates spatial signals from the perimeter area rather than from total area. Therefore, objects of large angular width do not yield larger detection probabilities [10.7].

The Johnson chart is usually applicable for four-bar patterns. The four bars and the three equal-sized spaces form a square. Therefore, the object (length-to-width) ratio of each bar and space is 7:1. Johnson assumed that the object or target and the equivalent bar pattern are of the same size and contrast at the same distance. Rosell and Wilson [10.7] considered aspect corrections for targets whose shapes are significantly different from that of a square. Because this affects primarily spatial noise statistics, aspect ratio correction is described in the next chapter where effects of photon statistics and noise are considered. For contrast-limited imaging, the Johnson chart is usually applied to minimum dimension without aspect ratio correction, although two-dimensional approaches are being considered, as described in Section 10.6.

Example 10.1

What focal length optics are required for 50% recognition probability of an 8-km-distant object whose minimum dimension is horizontal and is 2 m? Assume a spatial-frequency bandwidth of 50 line pairs per millimeter for a CCD system and consider hardware only without atmospheric effects.

Solution

From Eq. (10.4.5), a lateral magnification of

$$|m| = \frac{2n_{50}}{|x|f_{\max}} = \frac{2(4)}{|2|50 \times 10^3} = 8 \times 10^{-5}$$

is required. For the 8-km distance,

$$f_l = |m|s = 8 \times 10^{-5}(8 \times 10^3) = 640 \text{ mm}$$

If environmental effects such as atmospheric blur and image motion blur are considered, the situation changes considerably as shown in Parts 4 and 5 of this book.

10.5 TARGET ACQUISITION PROBABILITIES

An interesting application of the Johnson chart is the probability of detecting, recognizing, or identifying an object or target. The numbers n_{50} in the Johnson chart must first be modified to account for *clutter*. Clutter is irradiance from nonuniform backgrounds that interferes with target acquisition. For low clutter the numbers n_{50} for the various resolution categories remain unchanged. However, for medium or high clutter they are multiplied by *2* or *3*, respectively. The number n of resolvable cycles across target minimum dimension is calculated from Eq. (10.4.1) and is equal to $f_{x\ \max}x'$. A rough estimate of target acquisition probability over an infinite amount of time that is used when target is within field of view is found empirically to be [10.8]

$$P_\infty = \frac{(n/n_{50})^{x_0}}{1 + (n/n_{50})^{x_0}} \tag{10.5.1}$$

where [10.8]

$$x_o = 2.7 + 0.7\,(n/n_{50}) \tag{10.5.2}$$

and the factor (n/n_{50}) from the Johnson chart applies to the appropriate degree of resolution (detection, recognition, or identification) for which target acquisition probabilities are being considered, and has been multiplied already by the factor appropriate to clutter level. Equation (10.5.1) is known as the *target transfer function* (TTF). For convenience it is plotted in Fig. 10.7. Note that for $n = n_{50}$, P_∞ is equal to 50% as expected from the Johnson chart, which refers to target acquisition by 50% of the observer population.

Equation (10.4.2) can now be modified for different probabilities by using the appropriate value of n instead of n_{50} that is, $x = 2n\Delta x \approx -n(f_{x\ \max}m)^{-1}$. Similar modifications apply also to Eqs. (10.4.3), (10.4.4), and (10.4.5).

The time constant in seconds is determined empirically also as

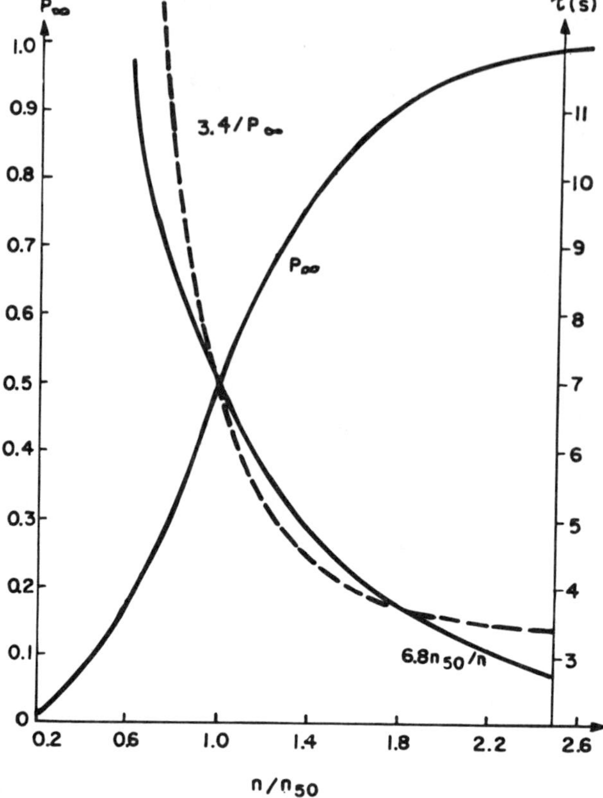

Fig. 10.7 Plots of Eq. (10.5.1) and (10.5.3).

$$\tau = 6.8n_{50}/n \approx 3.4/P_\infty \qquad (10.5.3)$$

where the approximation is most appropriate for $1 \leq n/n_{50} \leq 2$. Finally, target acquisition probability during time period t is

$$P(t) = P_\infty[1 - \exp(-t/\tau)] \qquad (10.5.4a)$$

This last equation refers to detection only. Recognition and identification search times are based on the assumption that the target has already been detected, and are characterized by

$$P(t) = P^*_\infty[1 - \exp(-t/\tau)] \qquad (10.5.4b)$$

where P^*_∞ is P_∞ for recognition or identification and τ in Eq. (10.5.3) uses n_{50} and P_∞ for detection. Thus τ is always time constant for detection, which is shorter than values that would be obtained if n_{50} and P_∞ in Eq. (10.5.3) were to refer to recognition or identification. The idea is that once an object has been detected, recognition or identification is essentially instantaneous if there is sufficient resolution.

Note that in general for a given geometry, as f_{rmax} increases so too does n, in which case $P(t)$ increases at a much faster rate, as shown in Fig. 10.8.

To understand Eqs. (10.5.4), let $P_o(t)$ be the probability that the object has not been acquired or detected during the time t. The incremental probability that the object will be acquired during the interval Δt and not before or after is

$$\Delta P_g(t) = (P_g/t_f)\Delta t P_o(t) \tag{10.5.5}$$

where, similar to the deviation of Poisson statistics for quantum noise in Section 5.1, $\Delta t << t_f$. Here, P_g is the probability that the object is found, and is called the *glimpse probability.* Glimpses occur at the rate t_f^{-1}, where t_f is called the *fixation time.*

Now, the sum of the probabilities of finding and not finding the object must be unity. Therefore, in terms of incremental probabilities

$$\Delta P_o(t) = -\Delta P_g(t) = -(P_g/t_f)\Delta t P_o(t) \tag{10.5.6}$$

or

$$\frac{\Delta P_o(t)}{P_o(t)} = -(P_g/t_f)\Delta t \tag{10.5.7}$$

which leads to

$$P_o(t) \propto \exp[(-P_g/t_f)t\} \tag{10.5.8}$$

The probability that the object is acquired is therefore

$$P(t) = P_\infty[1 - \exp(-t/\tau)] \tag{10.5.9a}$$

where

$$\tau = t_f/P_g \tag{10.5.9b}$$

and represents expected time of detection from Poisson processes.

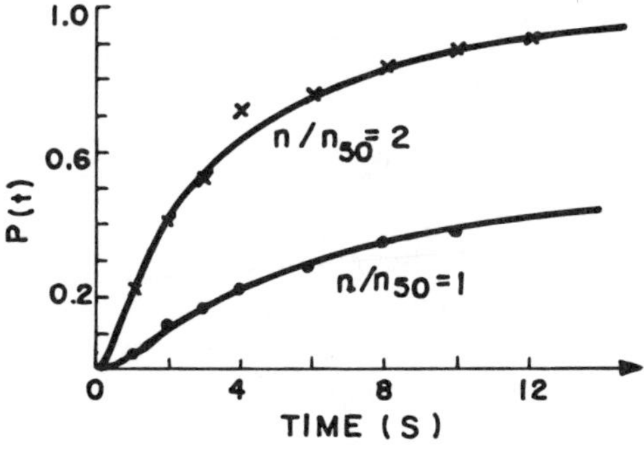

Fig. 10.8 Plot of Eq. (10.5.4a) for two resolution situations.

It is not intuitively obvious that the term P_∞ in Eq. (10.5.9a) is usually less than unity, although experimentally it is well established. For any particular ensemble of observers, there is assumed to exist a subset of observers who will never find a target with the given image size and contrast. Therefore, P_∞ is the fraction of observers who can find the target. Even among the subclass of observers who can find the object, there are those with higher than average glimpse probability and those with lower than average glimpse probability. However, the model uses a constant glimpse probability, P_g, for this group [10.8].

Generally, mean fixation time t_f is ≈ 0.3 s. Independent glimpses occur at the rate t_f^{-1}. Mean acquisition time is therefore given by Eqs. (10.5.9). From here, using Eq. (10.5.3),

$$P_g = t_f/\tau \approx \frac{0.3}{6.8}(n/n_{50}) \approx 0.044(n/n_{50}) \tag{10.5.10}$$

This static model, known as the CNVEO model [10.9] (Center for Night Vision and Electro-Optics of the U.S. Army where Johnson worked), has a number of significant limitations. It cannot be used for advanced systems for which the Johnson chart and the clutter coefficients are unreliable.Also, one might expect τ and P_∞ to depend on field of view (FOV), since the larger the FOV, the longer the time to be expected for target acquisition. Therefore, devices whose FOV/resolution ratios are significantly different from those considered in model development would be inappropriate for model implementation, although it does appear to be appropriate for current electro-optical systems in static situations. In addition, the model does not consider multiple targets nor multiple observers but only single targets and single observers. A more recent version of CNVEO modeling is FLIR 92, described in detail by Holst [10.10]. However, it pertains to noise-limited imaging, which is discussed in the next chapter.

The preceding model can also be adapted to search for targets over some large area (*field of regard*, FOR) which is larger than that seen instantaneously over one FOV. It is assumed that the FOR is systematically searched and the scan rate is such that each FOV is examined for t_o seconds per scan of FOR. This means each point remains in the FOV for t_o seconds for each scan of the FOR. P_∞ and τ were determined previously. The number n_{FOV} of FOVs required to cover the FOR is calculated, where $n_{\text{FOV}} = FOR/FOV$. Time t_s to search FOR once is then equal to $n_{\text{FOV}}t_o$. The probability that the target is found in one scan of the total FOR is [10.11]

$$P_s = 1 - \exp(-t_o/\tau) \tag{10.5.11}$$

Probability of acquisition in p complete scans over the time interval pt_s is

$$P_p = P_\infty[1 - (1 - P_s)^p] = P_\infty[1 - \exp(-pt_o/\tau)] \tag{10.5.12}$$

since $(1 - P_s)^p$ is the probability of not finding the target in p scans, where P_s is the probability of finding the target in one scan. The term P_p is the product of two probabilities. The first is the relative number of observers P_∞ who find the targets. The expressions in brackets are the second probability, which is that of each of these observers to actually find the target.

Let t_1 be the expected value of time required to find the object during m sweeps. The expected number of sweeps required is P_s^{-1}. Therefore, the expected time for the total number of sweeps is $t_s/P_s = n_{\mathrm{FOV}}t_o/P_s$. However, the object is not necessarily found at the end of the last sweep. On the average, it can be expected to be found in the middle of the last sweep, so that

$$t_1 = (t_o n_{\mathrm{FOV}}/P_s) - (t_0 n_{\mathrm{FOV}}/2) = t_o n_{\mathrm{FOV}}[(2 - P_s)/2P_s] \qquad (10.5.13)$$

When one considers that only a fraction P_∞ of the normal observer ensemble can find the target, then the probability of finding the object during time t is, from Eqs. (10.5.4),

$$P(t) = P_\infty[1 - \exp(-t/t_1)] \qquad (10.5.14)$$

If $t_s >> t_o$ then a search over many FOVs is required and $t_1 \approx \tau_o n_{FOV}/P_s \approx n_{FOV}\tau$ and

$$P(t) \approx P_\infty\{1 - \exp[-t/(n_{\mathrm{FOV}}\tau)]\} \qquad (10.5.15)$$

The preceding equation is true as long as $t_o << \tau$, which often is the case (see Exercise 10.17).

Equation (10.5.15) is similar in form to Eqs. (10.5.4) except that the time constant τ appropriate for one FOV has now been multiplied by the number of FOVs required to scan the FOR area.

The FOR search model inherits, of course, all the limitations of the FOV model. Neither has been sufficiently validated. Recent experiments indicate t_o for thermal imagers is about 1.25 s, and for TV systems it is about 1.8 s. The larger the value of t_o, the greater the probability P_p for constant τ, as seen from Eq. (10.5.12). However, the time constant τ cannot be expected to be identical for all observers even for a fixed value of n/n_{50}. The effort to describe image quality quantitatively and, as a result, target acquisition probabilities quantitatively as well can lead to order of magnitude calculation, but cannot be expected to be an exact science because of the great variability of human observer visual systems.

10.6 NAVY MODEL

As described in Section 10.4, one problem with reference to the Johnson chart is that critical object dimension is minimum object dimension. This is irrelevant to long objects such as ships. The U.S. Navy has found that target image area is a much more relevant criterion than minimum dimension. If Johnson's criteria are applied to a hypothetical target of square detected area, then since two pixels are required for each line pair the number of pixels required for detection, orientation, recognition, and identification would be $2 \times 2 = 4$, $2.8 \times 2.8 = 7.8$, $8 \times 8 = 64$, $12.8 \times 12.8 = 164$ pixels, respectively. Experience has shown that from 60 to 70 pixels per ship image are adequate to distinguish warships from commercial ships with a high probability of being correct. Generally, 66 pixels are required for ship "classification" [10.12]. This is close to Johnson's criterion for "recognition." For "identification," however, the required pixel number is on the order of 400 [10.12]. This criterion for beam aspect views of ships yields results different from Johnson's criteria for identification. The Navy model implies that for a square object about 10 line pairs are required for each direction, which is more stringent than Johnson's average of 6.4 ± 1.5 in Fig. 10.5 for the visible and infrared. (The thermal imaging community uses $n_{50} = 8$ for identification.)

On the basis of Eq. (10.4.2), for 50% resolution probabilities,

$$xy = n_{x50}n_{y50}(f_{x\,\max}f_{y\,\max})^{-1}(m)^{-2} \tag{10.6.1}$$

For sampled imaging in both horizontal and vertical directions as with a CCD or focal plane array, on the basis of Eq. (10.4.5),

$$xy = 4n_{x50}n_{y50}(f_{x\,\max}f_{y\,\max})^{-1}(m)^{-2} \tag{10.6.2}$$

Now n_{x50} and n_{y50} are line pairs while the Navy model deals with pixels. Two pixels are required for each line pair, so that application of Eq. (10.6.1) to the navy model requires that 66/4 = 16.5 line pairs^2 be the product of n_xn_y for classification and that 400/4 = 100 line pairs^2 be the product of n_xn_y for identification. The preceding two equations relate required resolution to object area.

Again, when TV systems are used, it is important to consider that the image is sampled in the vertical direction. Raster scans consist of 525 horizontal lines in America and 625 horizontal lines elsewhere, corresponding to frame rates of 60 and 50 s^{-1}, respectively, which are consistent with electrical power line frequencies. Vertical resolution is limited by the raster line process with which the image is scanned, since the space between raster lines is "dead space," meaning that no information is found there. Therefore n_{y50} is limited to the number of raster lines; in practice, it is even less, because some of the time over which the image is sampled or displayed is used for synchronization purposes to indicate the end of one line and the return of the raster scan process to the beginning of the next line, so that $f_{y\,max}$ is typically about 490 lines or 245 line pairs in America and about 590 lines or 295 line pairs elsewhere per unit effective vertical dimension of the camera tube.

CCD visible and thermal infrared cameras are often used with computers, in which case the image is displayed without raster scan, but with possible dead space between pixels.

10.7 TEXTUAL RESOLUTION

While the CNVEO and navy models usually refer to distant objects, it has been found that for textual resolution [10.13], such as in copying machines, *excellent* reproduction requires 8 line pairs per height of a lowercase letter *e*. *Legible* reproduction requires 5 line pairs per letter height, and *decipherable* resolution requires 3 line pairs per letter height.

Example 10.2

For an ideal circular lens, what f-number is required in order to image 4 μm size alphanumeric characters onto a CCD camera with 10 × 10 μm pixels using 0.6 μm wavelength radiation? Use "legibility" as a criterion for resolution.

Solution

Legibility requires $n = 5$ line pairs per character height. This implies the image height of the character must be 10 pixels. The magnification required is therefore $m = 10$ (10 μm)/4 μm = 25, which must also be the ratio of image distance s' to object distance s. Therefore,

$$\frac{1}{s} + \frac{1}{s'} = \frac{1}{f_l} = \frac{m}{s'} + \frac{1}{s'} = \frac{m+1}{s'}$$

The spatial-frequency bandwidth required for a circular aperture in diffraction-limited imaging (ideal lens) is

$$2\alpha_0 = \frac{D}{\lambda s'} \geq f_{x\ \max}$$

From Eq. (10.4.5), using Eqs. (10.5.1) and (10.5.2) to determine n,

$$x = \frac{2n}{f_{x\ \max} m}$$

Therefore,

$$f_{x\ \max} = \frac{2n}{mx} \leq \frac{D}{\lambda s'} = \frac{D}{\lambda(m+1)f_l}$$

which leads to

$$\frac{f_l}{D} \leq \frac{mx}{2n(m+1)\lambda} = \frac{25(4\ \mu\text{m})}{(2)5(26)(0.6\ \mu\text{m})} = 0.64$$

This f-number is small, but is obtainable by using combinations of large f-number lenses in parallel, as shown in Eq. (2.7.4), so as to produce an equivalent lens with the desired small f-number. For example, two 1.28 f-number lenses cemented together are equivalent to an 0.64 f-number lens.

10.8 EXAMPLES OF CONTRAST-LIMITED RESOLUTION

In Section 10.2 the reciprocal relationship of $f_{r\ max}$ with resolvable detail Δr is described. In this section pictorial examples are presented.

An unfiltered scene is shown in Fig. 10.9. This scene was Fourier transformed digitally using a personal computer and applying mathematically the techniques described in Chapter 7 for optical processing. A low spatial frequency filtered version resulting from inverse Fourier transformation is shown in Fig. 10.10. Removal of the high spatial-frequency content implies removal of fine detail. Examples are lack of detail concerning eyes and fingers and knobs on the display. The spatial-frequency cutoff frequency is lowered further by a factor of 3 in Fig. 10.11, and clearly much larger detail has been washed out, leaving only the largest detail. Figure 10.12 displays Fig. 10.9 after high spatial frequency filtering. Removal of low spatial-frequency content implies washing out of large spatial detail. Therefore, areas of large detail are of very low intensity ("blackish") because that irradiance has been removed. Only small details are of relatively bright irradiance because that is all that remains. As a result, high spatial-frequency filtering tends to emphasize borders, but clearly adds distortion.

Fig. 10.9 Unfiltered scene. (Courtesy of Y. Kadmon)

Fig. 10.10 Low spatial frequency filtered version of scene of Fig. 10.9. (Courtesy of Y. Kadmon)

Fig. 10.11 Lower spatial frequency filtered version of scene of Fig. 10.9. (courtesy of Y. Kadmon)

Fig. 10.12 High spatial frequency filtered version of scene of Fig. 10.9. (courtesy of Y. Kadmon)

A toy tank and soldier are shown in Fig. 10.13. Figure 10.14 is the same scene but with every 5 pixels in each direction averaged to the same irradiance. Consequently, the number of line pairs over each dimension has been reduced by 5. Figure 10.15 is the same scene as in Fig. 10.13 but with every 20 pixels in each direction averaged to the same irradiance, thereby reducing the number of line pairs over each dimension by a factor of 20. This can help illustrate the problems involved with Johnson's

Fig. 10.13 Unfiltered scene.

Fig. 10.14 Same as Fig. 10.13 but with averaging of every five pixels in each direction to the same irradiance.

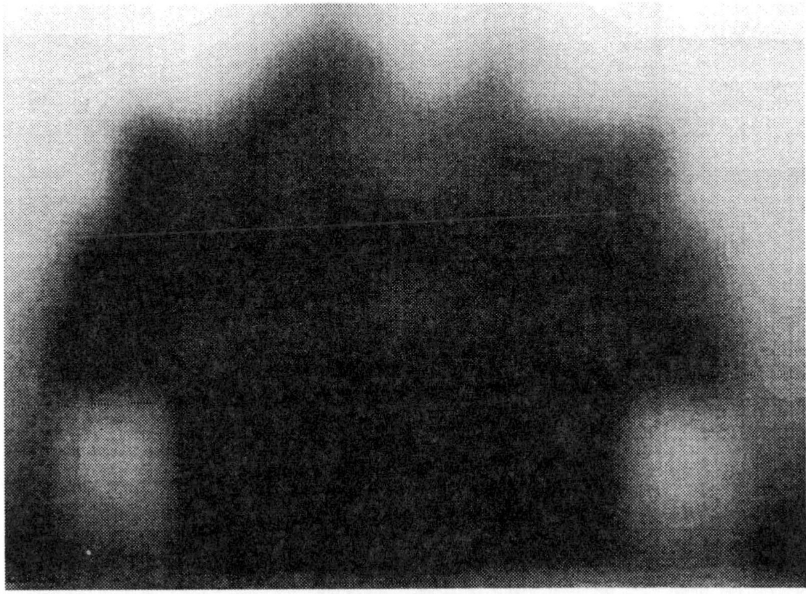

Fig. 10.15 Same as Fig. 10.13, but with every 20 pixels in each direction averaged to the same irradiance.

categories of resolution. For example, if Fig. 10.14 corresponds to identification, then Fig. 10.15 can correspond to orientation at best.

Finally, Figs. 10.16 and 10.17 demonstrate the effects of vertical sampling. Both involve Fig. 10.13, but in Fig. 10.16 the scene is reduced to only 8 lines in the vertical direction and in Fig. 10.17 to only 4 lines in the vertical direction. Horizontal resolution is unchanged.

Target acquisition probabilities, ranges, and times play a large role in image system design. The exercises at the end of the chapter integrate such modeling with geometrical and physical optics considerations in order to permit system design according to given target acquisition criteria. Until this point, and in the exercises at the end of the chapter, we have not considered contrast reduction prior to receipt of the image into the imaging system. However, contrast in the object plane itself is often less than unity. This has been considered in Section 9.6, and involves a uniform reduction in contrast at all spatial frequencies. In addition, path luminance glare over the propagation channel also causes a contrast reduction uniformly over the whole spatial-frequency domain. The net result is a uniform damping of system MTF, which also affects resolution adversely, as considered next.

10.9 PATH RADIANCE

Atmospheric background radiation is described in Chapter 4, and measurements are shown in Fig. 4.4. This radiation if recorded in the image can affect contrast quite adversely and thus limit noticeably target acquisition probabilities and ranges. The atmospheric background irradiance H_A, at wavelengths of less than 2 to 4 μm, consists of scattered background light such as sunlight during the day. This atmospheric background radiation is imaged over the same image space as the received irradiance

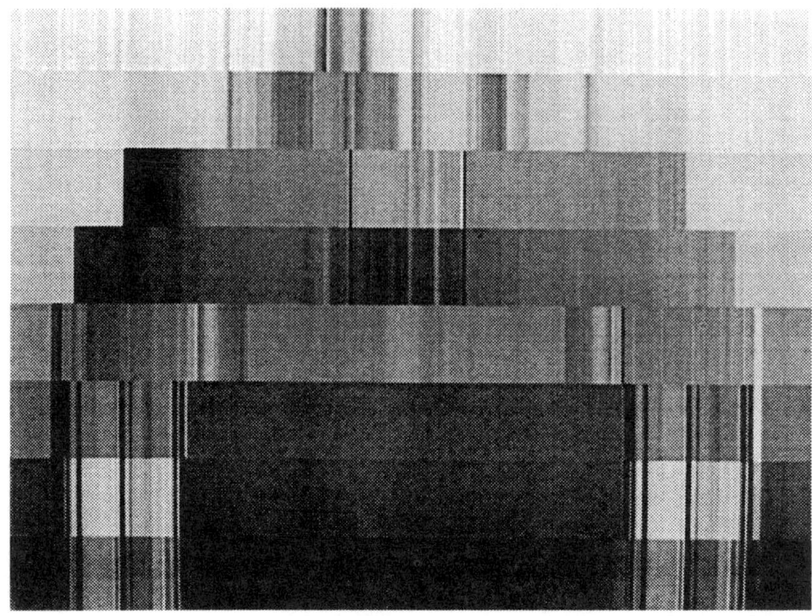

Fig. 10.16 Same as Fig. 10.13, but with only 8 lines in the vertical direction.

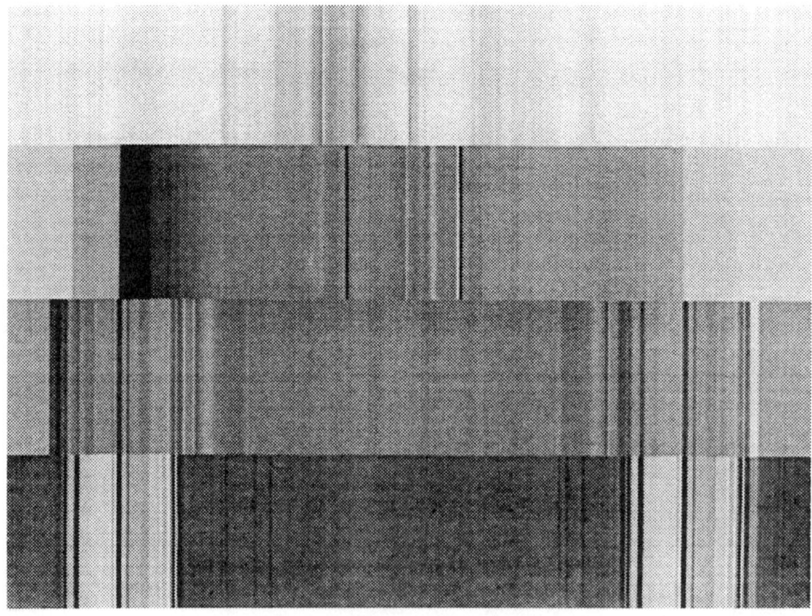

Fig. 10.17 Same as Fig. 10.13, but with only 4 lines in the vertical direction.

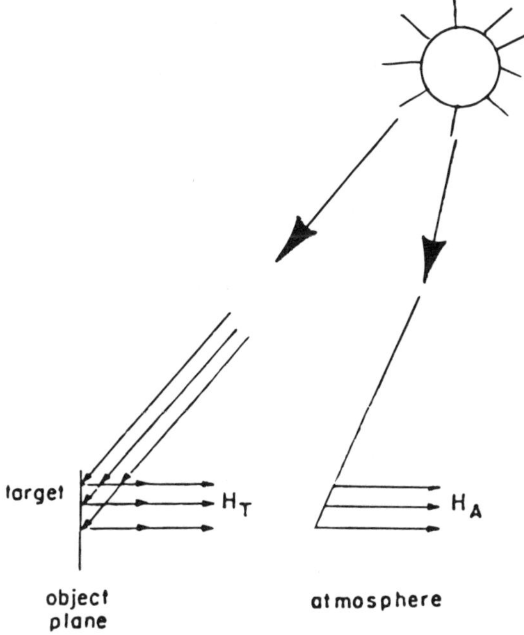

Fig. 10.18 Imaging of atmospheric path luminance irradiance H_A together with object or target irradiance H_T over identical image space.

H_T from the target plane, as shown in Fig. 10.18. The result is decreased contrast of the target plane scene as a result of atmospheric glare. The effect is similar to turning on the lights in a movie theater. In the dark, almost all light reaching the eye emanates from the movie screen, thus providing good contrast. When lights are turned on, the percentage of overall light reaching the eye from the movie screen is diminished because of the background light. The result is decreased contrast on the movie screen. Similarly, atmospheric background light or path radiance gives rise to decreased contrast of the target plane scene because the percentage of radiation emanating from the target plane relative to overall radiation reaching the receiver is diminished by the intervening atmosphere background radiation, or path radiance. This makes it more difficult to resolve small detail. At wavelengths larger than 4 μm, most of the path radiance is not scattered sunlight, as in Fig. 10.18, but rather thermal emission of the atmospheric constituents, gases in particular. This thermal path radiance decreases contrast in thermal imaging systems, particularly if they are dc coupled. This effect is very important in the 8–13 μm atmospheric window because the atmospheric emission there is high.

The reason for the glare is that the atmospheric background is imaged throughout the image plane, thus contributing to the irradiance of the images of both the target and the target-plane background. The greater the atmospheric background, the smaller the percentage of photons received that derive from the target. Examples of such effects in infrared imagery are presented by Watkins [10.14] and on ultraviolet and visible imagery by Livingston [10.15]. If ρ_t and ρ_b are target and target-plane background reflection coefficients, then from Eq. (9.6.1) the modulation contrast in the target plane is, for short wavelengths ($\lambda < 2$–4 μm)

$$\text{MCO} = \frac{H_o - H_b}{H_o + H_b} = \frac{H_s(\rho_t - \rho_b)}{H_s(\rho_t + \rho_b)} \tag{10.9.1}$$

where H_s is solar irradiance incident on the object plane, and H_o and H_b are target and object plane background irradiances, respectively, which comprise H_T in Fig. 10.18. Equation (10.9.1) reduces to Eq. (9.6.1) and describes the modulation contrast in the object plane prior to propagation through the atmosphere. This analysis is based on Fig. 10.18, but with the added complication that the object plane irradiance in Fig. 10.18 now consists not only of target, but also of background in the target plane, such as, for example, a vehicle (H_o) with vegetation or sky background (H_b). The apparent irradiances reaching the imager are $\tau' H_o$ and $\tau' H_b$, where τ' is atmospheric transmission defined in Eqs. (3.5.7) and (3.5.8). In addition, the atmospheric background irradiance or path radiance (H_A in Fig. 10.18) is assumed to fall uniformly over both target and object plane background spaces in the image plane. As a result, modulation contrast at receiver optics input is [10.16]

$$\text{MCI} = \frac{(\tau' H_o + H_A) - (\tau' H_b + H_A)}{(\tau' H_o + H_b) + (\tau' H_b + H_A)} = \frac{\tau' H_s(\rho_t - \rho_c)}{\tau' H_s(\rho_A + \rho_o) + 2H_A} \tag{10.9.2}$$

$$= \frac{\rho_t - \rho_b}{\rho_t + \rho_b + 2H_A/(\tau' H_s)}$$

By definition, the modulation contrast function is equal to MCI/MCO. Atmospheric background irradiance contrast reduction is [10.16], in view of the above two equations,

$$M_B = \frac{\text{MCI}}{\text{MCO}} = \frac{\rho_t + \rho_b}{\rho_t + \rho_b + 2H_A/(\tau' H_s)} \tag{10.9.3a}$$

Were there no path luminance ($H_A = 0$), then Eq. (10.9.3a) would be unity and there would be no degradation of contrast resulting from background atmospheric radiation. This last equation considers system effects of path radiance alone. Equation (9.6.1) or (10.9.1) must also be considered for object plane modulation contrast. Multiplication of Eq. (10.9.1) with Eq. (10.9.3a) yields Eq. (10.9.2) for the *image* plane, which represents the *combined* system effects of both object plane modulation contrast and path radiance.

At longer wavelengths, atmospheric irradiation is primarily thermal in nature, rather than transmitted and scattered sunlight. In this case [10.16],

$$M_B = \frac{H_o + H_b}{H_o + H_b + 2H_A/\tau'} \tag{10.9.3b}$$

where H_o, H_b, and H_A are essentially thermal emissions.

Note that the atmospheric background modulation contrast function, Eqs. (10.9.3a) and (10.9.3b), is not a function of spatial frequency. However, for a sine wave target, modulation contrast function also is an MTF. It represents a constant damping of the overall system MTF, thus causing higher spatial-frequency components of the overall imaging system MTF to be at contrast below the threshold contrast required at the output, as shown in Fig. 10.19.

This uniform damping across the spatial-frequency spectrum leads to a reduction in $f_{r\ \max}$ or increases in Δx and x in Eq. (10.4.2), thus impairing resolution and target acquisition probabilities. This impairment is very sensitive to wavelength [10.16–10.18]. However, since the glare is spatially uniform, it is often easy to remove it by removing the dc component in the optical Fourier transform of the object scene, and then inversely Fourier transforming the spatially filtered scene [10.19].

In the previous treatments, noise effects have been ignored. In the next chapter, resolution is considered for situations where the resolution limit is determined by noise.

Example 10.3.

Consider a target with reflection coefficient of 0.8 at 900-nm wavelength, with a background reflection coefficient of 0.5 at the same wavelength. If solar irradiance at this wavelength is 500 $\mathrm{W \cdot m^{-2} \cdot \mu m^{-1}}$ and atmospheric background irradiance is 10^{-1} $\mathrm{W \cdot m^{-2} \cdot \mu m^{-1}}$, and atmospheric transmission of sunlight to the target is 8% at this wavelength, then determine both object plane and atmospheric path luminance modulation contrasts at 900-nm wavelength.

Solution

From Eq. (10.9.1),

$$\mathrm{MCO} = \frac{0.8 - 0.5}{0.8 + 0.5} = \frac{0.3}{1.3} = 0.23$$

From Eq. (10.9.3a)

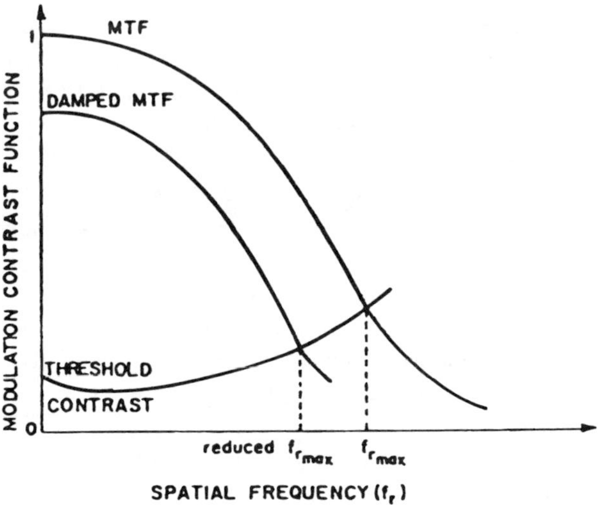

Fig. 10.19 Reduction in $f_{r\ \max}$ imposed by uniform damping of system MCF by atmospheric background.

$$M_B = \frac{0.8 + 0.3}{0.8 + 0.3 + 2(0.1)/[(0.08)500]} = 0.996$$

$$\text{MCO} \cdot M_B = 0.229$$

The overall MTF is multiplied by M_B to obtain image plane modulation contrast. The major image plane modulation contrast reduction in this example is that of the object scene itself rather than path radiance. Effects on resolution are considered in Exercise 10.19.

REFERENCES

10.1. H. P. Lavin, "System analysis," in *Photoelectronic Imaging Devices*, L. M. Biberman and S. Nudelman, Eds., Plenum Press, New York, 1971, pp. 333–374.

10.2. H. L. Snyder, "Image quality and face recognition in a television display," *Human Factors*, Vol. 16(3), 1974, pp. 300–307.

10.3. H. L. Snyder, "Image quality and observer performance," in *Perception of Displayed Information*, L. M. Biberman, Eds., Plenum Press, New York, 1973, pp. 87–118.

10.4. J. Johnson, "Analysis of image forming systems," paper presented at Image Intensifier Symposium, Ft. Belvoir, VA, October 6–7, 1958, AD 220160s, pp. 244–273.

10.5. J. Johnson, "Analytical description of night vision devices," in *Proc. Seminar on Direct-Viewing Electro-Optical Aids to Night Vision*, L. M. Biberman, Ed., Institute for Defense Analysis Study, 52254, October 1966, pp. 177–200.

10.6. L. M. Biberman, Ed., *Perception of Dipslayed Information*, Plenum Press, New York, 1973.

10.7. F. A. Rosell and R. H. Wilson, "Recent psychophysical experiments and the display signal-to-noise ratio concept," in *Perception of Displayed Information*, L. M. Biberman, Ed., Plenum Press, New York, 1973.

10.8. S. R. Rotman, E. S. Gordon, and M. L. Kowalczyk, "Modeling human search and target acquisition performance: I. First detection probability in a realistic multitarget scenario," *Opt. Eng.*, Vol. 28, November 1989, pp. 1216–1222.

10.9. J. A. Ratches, "Static performance model for thermal imaging systems," *Opt. Eng.*, Vol. 28, November 1989, pp. 1223–1226.

10.10. G. C. Holst, *Electro-Optical Imaging System Performance*, JCD Publishing, Winter Park, Florida, and SPIE Optical Engineering Press, Bellingham, WA, 1995.

10.11. S. R. Rotman, E. S. Gordon, and M. L. Kowalczyk, "Modeling human search and target acquisition performance: II. Simulating multiple observers in dynamic scenarios," *Opt. Eng.*, Vol. 28, November 1989, pp. 1223–1226.

10.12. P. M. Moser, "Mathematical model of FLIR performance," Technical Memorandum NADC-20203: PMM, Naval Air Development Center, Warminster, PA, October 1972.

10.13. W. J. Smith, *Modern Optical Engineering* (2nd ed.), McGraw-Hill, New York, 1990, p. 355.

10.14. W. Watkins, "Environmental bugs invade EO systems," in *Infrared Imaging Systems: Design, Analysis, Modeling and Testing IV*, G. C. Holst, Ed., *Proc. SPIE*, Vol. 1964, 1993, pp. 42–53.

10.15. W. Livingston, "Landscape as viewed in the 320-nm ultraviolet," *J. Opt. Soc. Am.*, Vol. 73, December 1983, pp. 1653–1657.
10.16. N. S. Kopeika, "Spectral characteristics of image quality for imaging horizontally through the atmosphere," *Appl. Opt.*, Vol. 16, September 1977, pp. 2422–2426.
10.17. N. S. Kopeika, "The general wavelength dependence of imaging through the atmosphere," *Appl. Opt.*, Vol. 20, May 1981, pp. 1532–1536.
10.18. N. S. Kopeika, S. Solomon, and Y. Gencay, "The wavelength variation of visible and near IR resolution through the atmosphere: dependence on aerosol and meteorological conditions," *J. Opt. Soc. Am.*, Vol. 72, July 1981, pp. 897–901.
10.19. L. Levi, *Applied Optics Vol. 2*, Wiley, New York, 1980.

EXERCISES

In these exercises, human visual system threshold contrast and MTFs for the atmosphere and image motion and vibrations are neglected. They appear in later chapters. Assume object plane modulation contrast is unity unless specified otherwise.

10.1 A 25-mm-diameter and 50-mm focal length lens is covered with a 22-mm (horizontal), 16-mm (vertical) rectangular aperture. Assuming a required threshold contrast of 2%, find spatial-frequency bandwidths in the horizontal and vertical directions for 500-nm radiation, for an object at a 1-km distance from the lens. Assume limiting MTF is determined by aperture diffraction limit. What is the approximate best case detail that can be detected in each direction for diffraction-limited imaging?

10.2 Consider Exercise 10.1 but without the rectangular aperture.

10.3 A 1/2-in. vidicon system is to be used to recognize a 2-m-wide vehicle at a 4-km distance. Effective dimensions of the TV tube are 11.8 mm (horizontal) × 8.8 mm (vertical). If the optics focal length is 400 mm, how many vertical TV lines are required for at least 50% probability of vehicle recognition with uniform (low) clutter? Assume TV square wave response is dominant in system MCF, and that one line pair equals 2 TV lines (one white and one black).

10.4 What size detail can be identified at a 5-km range with 50% probability in medium clutter with an imager consisting of a 500-mm focal length lens and a 1/2-in. vidicon featuring 500 TV lines, assuming resolution is limited essentially by the TV system? Objects of which size can be recognized?

10.5 Repeat Exercise 10.3 but for 90% instead of 50% recognition probability.

10.6 Repeat Exercise 10.4 but for 90% instead of 50% identification probability.

10.7 What is the equivalent passband for the optical system of Exercise 10.1?

10.8 What is the approximate passband of a CCD TV camera whose MTF is limited by a 20- × 20-μm pixel size? Assume CCD effective front surface dimensions are 11.8 mm horizontal and 8.8 mm vertical.

10.9 If the CCD camera of Exercise 10.8 is used with the lens of Exercise 10.1, which element limits overall MTF?

10.10 Repeat Exercises 10.1 and 10.2 using the CCD TV camera of Exercise 10.8. This illustrates the difference between diffraction-limited imaging and instrumentation-limited imaging.

10.11 A CCD TV camera whose MTF is limited by a 20-μm × 20-μm pixel size is used with a 50-mm-diameter zoom (variable focal length) lens to image a 4-m (horizontal) × 6-m (vertical) object at a distance of 2 km. What should be the focal length so that probability of object recognition in medium clutter is at least 50%? Assume 2% required threshold contrast. What must focal length be for 90% recognition probability?

10.12 Consider the geometry of Exercise 10.11 but with the constraint that 50% detection probability should take place within 8 s. What is typical glimpse probability for this case?

10.13 Using the same CCD-zoom lens combination as above, what should focal length be to identify in low clutter a 2-m (horizontal) × 3-m (vertical) object at a 5-km distance with 90% probability? How much time is expected to be needed for the average observer to identify the object?

10.14 In Exercises 10.11 and 10.12, what are average acquisition times at the required focal lengths?

10.15 A 4- × 6-m object is viewed through a 1.2-m focal length telescope having an 80-mm-diameter diffraction-limited objective through medium clutter and imaged with the CCD of Exercise 10.11. What is the probability of identifying the object within 5 s? What are glimpse and expected acquisition times?

10.16 If the effective width for the CCD of Exercise 10.15 is 11.8 mm, and its imaging system is to scan a field of regard (FOR) that is 98.3 mrad wide, answer these questions: (a) How many fields of view is the total scan area? (b) What is the probability of identifying a 4- × 6-m target in one scan of the total FOR? (c) What is the overall expected time required to acquire the target? (d) What are the probabilities of acquisition within 15 s? Within 70 s?

10.17 Show that for $t_s = n_{\mathrm{FOV}} t_o >> t_o$, t_i is approximated by $n_{\mathrm{FOV}}\tau$.

10.18 (a) Using the naval model, what focal length must be used so that there is a 50% probability that an 8-m × 50-m ship can be "classified" at a 10-km distance using the CCD imaging system of the above exercises? Neglect, as in the other exercises, atmospheric and image motion MTFs. Assume effective CCD dimensions are 11.8 mm horizontal and 8.8 mm vertical. (b) Would the ship be "recognized" by Johnson's 50% probability criteria? (c) What focal length would permit recognition according to the Johnson chart for low clutter? Assume in all cases that resolution is limited by the CCD.

10.19 A parabolic index optical fiber is to be used to image a 5-cm-diameter picture into the eye. Assuming a numerical aperture of 0.30 for the fiber and an eye pupil size of 0.3 cm, if 130 line pair resolution in each direction is required for the final image seen by the eye, determine the following: (a) the distance from the picture to the fiber input, (b) the distance from the fiber output to the eye, (c) the focal length required for the optical fiber, and (d) the minimum fiber diameter for an imaging wavelength of 0.55 μm. Assume an aberration-free fiber.

10.20 Repeat Exercise 10.10 considering the object plane and atmospheric channel modulation contrast conditions of Example 10.3.

CHAPTER

11

Noise-Limited Imaging and Target Acquisition

Basic noise concepts were developed in Chapter 5 and applied to individual "point" detectors in Chapter 6. Here, these concepts are extended to the image as a whole, where object or target area must be considered. Noise manifests itself when irradiance received from the object plane scene is weak, for example, as random "snow" when weak TV signals are received. This stems from random time and space fluctuations in image irradiance giving rise to random nonuniformities in image brightness, and also from electronic noise. In noise-limited imaging, the object or target contrast is imaged against noise, and the latter rather than required threshold contrast limits the ability of the observer to resolve the object or target. The poorer the signal-to-noise ratio (SNR), the poorer the image quality and resolution. Quantitatively, system modulation transfer function (MTF) is incorporated into the expression for signal power, thus giving rise to SNR as a function of spatial frequency. Intersection of system SNR curve with required SNR threshold defines $f_{r\,\max}$ for *noise-limited imaging*, as shown in Fig. 11.1.

The required threshold value varies according to resolution requirement. SNR thresholds required for "identification" are greater than those required for "detection."

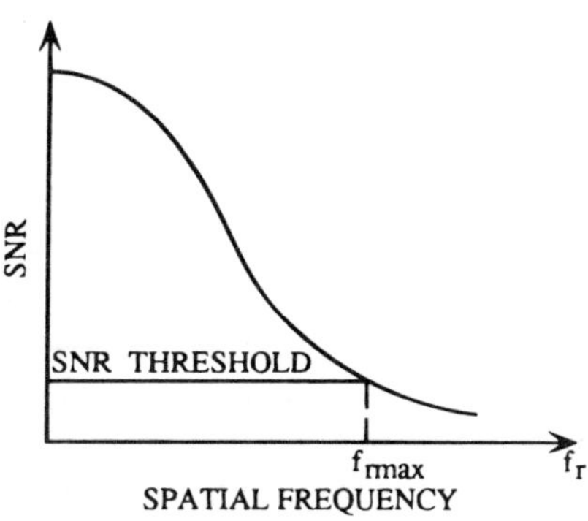

Fig. 11.1 Spatial frequency bandwidth ($f_{r\,\max}$) for noise-limited resolution.

Spatial frequency bandwidth ($f_{r\ \max}$) is thus determined separately by system MTF intersection with required contrast threshold for contrast-limited imaging as in Chapter 10 and by system SNR intersection with required SNR threshold here. The smaller of the two values for $f_{r\ \max}$ is thus essentially the limiting resolution. In general, "snowy" images are noise limited, while images that are not snowy are contrast limited (Chapter 10). Transition from contrast- to noise-limited imaging is shown in Fig. 11.2.

The noise-limited imaging resolution described here distinguishes between systems and imagers of reflected irradiance as in the visible and near-infrared regions and between systems and imagers of emitted irradiance as in thermal imaging. The former is considered next.

11.1 SIGNAL-TO-NOISE CURRENT RATIO

A photoelectronic system involving optics, imager, signal processor, and display is assumed. It is further assumed that the various components MTFs are independent, so that they can be cascaded so as to form overall system MTF. Although specific analyses in the literature have been made for various types of imagers or sensors such as electron tube imagers [11.1 and references therein] or CCD imagers [11.2], the following analysis is generalized so that in principle it can be applied to any square law imager (current proportional to received irradiance) with proper substitution of specific device properties.

Consider background irradiance H_b, which is incident uniformly across the entire imager or sensor photodetecting surface A. Object or target irradiance H_T is focused onto an incremental area a of the imager. The incremental signal over the area a is the difference between peak target current and average background current. In terms of photoelectron rates n_T and n_b for target and background, respectively, the signal is

$$\Delta s = (\dot{n}_T - \dot{n}_o)a\tau_i \tag{11.1.1}$$

where

$$\dot{n}_T = \Delta i_T/(qa) \tag{11.1.2}$$

$$\dot{n}_b = \Delta i_b/(qa) \tag{11.1.3}$$

where τ_i is integration time (usually about 0.1s for the human visual system). The incremental image plane area a considered here is that of the target if the target fills up the entire field of view (FOV) and is imaged against path luminance only. If the target is imaged against background scenery such as trees, sky, or hills surrounding part or all or the target, then a is an incremental area of the surrounding background of the same size as the target incremental area. In this case two adjacent incremental areas of the same size are considered, and irradiance in each is to be compared with that in the other.

In the preceding discussion, Δi_T and Δi_b are incremental currents deriving from H_T and H_b, respectively, that is,

Fig. 11.2 Transition from contrast-limited image (top) to noise-limited image (bottom) of same scene. (Courtesy of Nir Corse and Ofer Hadar.)

$$\Delta i_T = (\eta q/h\nu)GH_Ta \tag{11.1.4}$$

$$\Delta i_b = (\eta q/h\nu)GH_ba \tag{11.1.5}$$

where G represents the internal gain of the sensor. Equations (11.1.4) and (11.1.5) refer only to incremental target and background currents over the incremental area a only. Overall background current, for example, is given by $i_b = (\eta q/h\nu)(H_bA)$.

The signal in Eq. (11.1.1) can be reexpressed as

$$\Delta s = 2(\text{MCI})\dot{n}_{av}a\tau_i \tag{11.1.6}$$

where MCI is image plane modulation contrast (the photodetecting surface of the imager is located in the image plane) equal to

$$\text{MCI} = \frac{\dot{n}_T - \dot{n}_b}{\dot{n}_T + n_b} \tag{11.1.7}$$

and the average photoelectron rate is approximately

$$n_{av} \approx (\dot{n}_T + \dot{n}_b)/2 \tag{11.1.8}$$

In this way, the signal is related to modulation contrast, which is related to system MTF through

$$\text{MCI} = (\text{MTF})_s\text{MCO} \tag{11.1.9}$$

where MCO is object plane modulation contrast and $(\text{MTF})_s$ refers only to that part of the imaging system relating contrast transfer from object plane to the imaging sensor. This includes MTFs for optics, atmosphere, and image motion and vibration. It does not include imager or display MTFs which are added later.

Having found the incremental signal, we now consider noise. As shown in Section 5.1, quantum noise is governed by Poisson statistics. For the case of an object imaged against background, total mean square noise is the average of the object and background photoelectrons summed in quadrature, or

$$\Delta n_n = \{[(\dot{n}_T + \dot{n}_b)/2]a\tau_i\}^{1/2} = (\dot{n}_{av}a\tau_i)^{1/2} \tag{11.1.10}$$

Thus, mean SNR over incremental target area a at imager output is, from Eqs. (11.1.6) and (11.1.10),

$$\text{SNR}_I = \Delta s/\Delta n_n = 2(\dot{n}_{av}a\tau_i)^{1/2}(\text{MCI}) \tag{11.1.11}$$

Next, in addition to the photoconversion noise considered above, additional noises such as amplifier or Johnson noise generated internally within the signal processor or display must be considered. These noises, referenced to the output of the imager photosurface, result in mean square photoelectron rate $\dot{n}_{spd}$ during the integration time of the entire image. This mean is independent of space. Therefore,

$$\Delta n_n = [(\dot{n}_{av} + \dot{n}_{spd})a\tau_i]^{1/2} \tag{11.1.12}$$

and, from Eq. (11.1.6),

$$\text{SNR}_I = \frac{2(\text{MCI})\dot{n}_{av}(a\tau_i)^{1/2}}{(\dot{n}_{av} + \dot{n}_{spd})^{1/2}} \tag{11.1.13}$$

The minimum *detection* threshold value (50% probability) for Eq. (11.1.13) is usually around 5 for photographs and about 2.9 for dynamic images as in TV systems, including thermal imaging, where about three images are typically presented during a 0.1-s integration period. This means the observer coherently sums the signal while incoherently summing noise, thus reducing threshold by about $\sqrt{3}$.

To consider the image of the entire scene, rather than that over incremental area a, note that for each current, $\dot{n} = i/(qA)$. Therefore, for the total scene, in view of Eq. (11.1.9),

$$\text{SNR}_{IT} = \frac{2(a/A)^{1/2}(\text{MCO})i_{av}(\tau_i)^{1/2}(\text{MTF})_s}{(i_{av} + i_{spd})^{1/2}q^{1/2}} \tag{11.1.14a}$$

where $i_{av} = (i_T + i_b)/2$ and i_{spd} is noise current deriving from $\dot{n}_{spd}$. The numerator and denominator of Eq. (11.1.14a) are now multiplied by $(2B)^{1/2}$, where B is noise bandwidth, resulting in

$$\text{SNR}_{IT} = \frac{2^{3/2}(\text{MCO})(a/A)^{1/2}i_{av}(B\tau_i)^{1/2}(\text{MTF})_s}{[2qB(i_{av} + i_{spd})]^{1/2}} \tag{11.1.14b}$$

The denominator here is in the form of shot noise or quantum noise, while the expression $(B\tau_i)^{1/2}$ represents SNR enhancement through integration, as described in Section 6.1.6. The SNR is now a function of spatial frequency according to system MTF, designated $(\text{MTF})_s$.

The preceding discussion refers to a continuous image. However, when arrays of detectors are used to form a spatially sampled image, as in the case of most photoelectronic imagers, a different detector senses each portion of the image, and each detector has its own responsivity. The different responsivities lead to spatial noise that degrades performance. This *spatial noise* is simply [11.3]

$$\overline{(\delta i_{spn})^2} = (i_{av}\sigma_\eta/\overline{\eta})^2 \tag{11.1.15}$$

where σ_η is quantum efficiency standard deviation and $\overline{\eta}$ is average quantum efficiency across the array. The ratio $\sigma_\eta/\overline{\eta}$ is the relative nonuniformity. For low incident irradiances, shot noise dominates spatial noise and SNR improves with increasing quantum efficiency and irradiance. However, for high irradiance levels, SNR is limited by nonuniformity of detector response [11.3] as seen in Exercise 11.1. In such situations SNR reaches an asymptotic limit imposed by detector nonuniformity and does *not* increase with increased irradiance.

To consider SNR at the display (SNR_D), note that gains or amplifications of i_{av} and i_{spd} are not equal. The former current is amplified by both internal amplification in the imager, G_I, as well as that in the signal processing, G_s. The latter current actually originates in the processing; hence, it is amplified by G_s only. Therefore, in view of the Poisson nature of the spatial noise,

$$\mathrm{SNR}_D = \frac{2^{3/2}(\mathrm{MCO})(a/A)^{1/2} i_{av} G_I G_s (B\tau_i)^{1/2} (\mathrm{MTF})_{sT}}{\{[2qi_{av}B + (i_{av}\sigma_\eta/\overline{\eta})^2]G_I^2 G_s^2 + 2qi_{spd}G_s^2 B\}^{1/2}} \tag{11.1.16}$$

or

$$\mathrm{SNR}_D = (2aB\tau_i/A)^{1/2}(\mathrm{MCO})(\mathrm{MTF})_{sT}\mathrm{SNR}_v \tag{11.1.17}$$

where the video SNR is

$$\mathrm{SNR}_v = \frac{2i_{av}G_I G_s}{\{[2qi_{av}B + (i_{av}\sigma_\eta/\overline{\eta})^2]G_I^2 G_s^2 + 2qi_{spd}G_s^2 B\}^{1/2}} \tag{11.1.18}$$

and $(\mathrm{MTF})_{sT}$ in Eqs. (11.1.16) and (11.1.17) is total system MTF including imager and display MTF, as well as $(\mathrm{MTF})_s$.

If the imager output signal is broadcast to the processor and display, then G_I must include transmission losses. Radiometric or diffraction losses of the transmission through the atmosphere can convert contrast-limited to noise-limited situations.

The preceding equations are derived in principle for an isolated rectangular image. It is assumed, however, that detection of a bar pattern depends on detection of single bars. This is justified since it has been found [11.1] that threshold SNRs required to detect a bar in the presence of a number of bars do not differ significantly from those required to detect a single bar with a uniform background. Consider, therefore, for vertical critical target image dimension, as shown in Fig. 10.6 in Chapter 10, an isolated horizontal bar of width $\Delta x'$ and height (vertical) $\Delta y'$. Let ϵ_a be bar length-to-width $(\Delta x'/\Delta y')$ ratio. Therefore, incremental area

$$a = \Delta x'\Delta y' = \epsilon_a(\Delta y')^2 \tag{11.1.19}$$

Similarly, consider the whole imager surface in the image plane upon which the image irradiance is incident. Let the horizontal width of the imager be W_x and let its vertical extent be V_y so that

$$A = W_x V_y = \alpha_a V_y^2 \tag{11.1.20}$$

where $\alpha_a = W_x/V_y$ is horizontal-to-vertical-picture aspect ratio. Therefore,

$$\frac{a}{A} = \frac{\epsilon_a}{\alpha_a N_{sf}^2} \tag{11.1.21}$$

where $N_{sf} = V_y/\Delta y$ and represents *equivalent* spatial frequency of a vertical periodic bar pattern representing the target *image where each bar provides detail required for detection, orientation, recognition or identification purposes*. The greater the value of N_{sf}, the better the resolution. Note that N_{sf} is *not* in bar *pairs* per picture height, nor is it a property of the imager, except that N_{sf} cannot be greater than the value of $2f_{r\,\max}$ permitted by the imaging system. If not limited by system $f_{r\,\max}$, N_{sf} is essentially a property of target or object geometry, as in Fig. 10.6. For vertical critical dimension, Eq. (11.1.17) becomes

$$\mathrm{SNR}_D = \mathrm{MCO}(2\epsilon_a/\alpha_a)^{1/2}(B\tau_i)^{1/2}(\mathrm{MTF})_{sT}\mathrm{SNR}_v/\mathrm{N}_{sf} \tag{11.1.22}$$

If the critical dimension is horizontal, then equivalent vertical bars are considered and N_{sf} is defined $W_x/\Delta x'$, resulting in

$$\frac{a}{A} = \frac{\alpha_a}{\epsilon_a N_{sf}^2} \tag{11.1.23}$$

and

$$\mathrm{SNR}_D = \mathrm{MCO}(2\alpha_a/\epsilon_\alpha)^{1/2}(B\tau_i)^{1/2}(\mathrm{MTF})_{sT}\mathrm{SNR}_v/\mathrm{N}_{sf} \tag{11.1.24}$$

Conventionally, aspect ratio α_a is 4/3, although this changes for high-definition television (HDTV), as described in Chapter 13. Aspect ratio is greater than unity, that is, the horizontal picture display dimension is larger than the vertical because our two eyes are in the horizontal direction, thus permitting us greater resolution in the horizontal direction than in the vertical direction.

SNR_D is the perceived SNR and, unlike SNR_v, it is influenced by the "smart" spatial and temporal noise filtering properties of human intelligence, which can even discern objects when object SNR is as low as 0.05.

Although SNR for nonimaging applications is defined usually in terms of signal-to-noise current power ratios, for imaging applications it is defined usually as signal-to-noise current ratio as shown here, or as signal to effective noise wave irradiance at imager input. In practice, SNR_D may be slightly less than Eqs. (11.1.22) or (11.1.24) because of finite aperture effects, which increase noise slightly [11.1].

Since SNR_D is a function of $(\mathrm{MTF})_{sT}$, display SNR is a function of spatial frequency as depicted in Fig. 11.1. A critical question to be considered is the value of the threshold SNR, the intersection of which with SNR_D determines $f_{r\,\max}$. Information in the open literature is limited.

In experiments involving noise-limited imaging of vehicles [11.1], it was found that for 50% probability of correct *recognition* a threshold SNR_D of 3.3 is required for uniform background. However, for a road background, threshold SNR_D for 50% probability rose to 3.89. Grass backgrounds required a threshold SNR_D of 4.1, and a grass and tree background caused threshold SNR_D to increase to 5.0. As expected, increase of clutter also caused required SNR_D threshold to increase.

Similar variations in threshold SNR_D were also found in experiments involving vehicle *identification*.

The Johnson chart in Chapter 10 is based on both contrast and SNR_D. Treatment here assumes contrast is no problem but noise is. Threshold SNR_D values (SNR_{D50}) required for various levels of resolution obtained empirically are summarized in Table 11.1 for 50% probabilities. Since increased values of N_{sf} imply better resolution, it is not surprising that SNR_{D50} decreases as N_{sf} increases. Note that the number of TV lines or equivalent bars across minimum dimension corresponds to the Johnson chart since two TV lines (1 black and 1 white) correspond to one line pair. These SNR_D thresholds can be adjusted for other levels of probability with Fig. 11.3 involving "normalized" SNR_D. One multiplies the value of threshold SNR_D by the

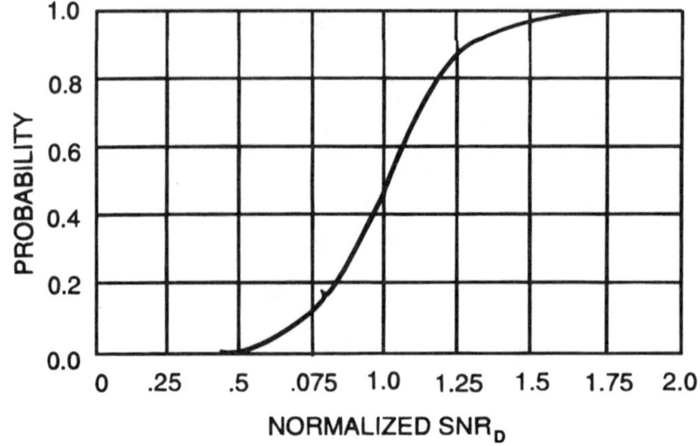

Fig. 11.3 Probability of target acquisition versus normalized SNR_D (from [11.1].

normalized SNR_D corresponding to the desired probability. For example, from Fig. 11.3 target acquisition probabilities of 90% correspond to a normalized SNR_D value of about 1.30. Therefore, threshold SNR_{D50} values in Table 11.1 should be multiplied by 1.30 to obtain threshold SNR_D values for 90% acquisition probabilities.

It has been suggested that overall acquisition probabilities can also be approximated [11.4] using a log-normal distribution

$$P(\mathrm{SNR}_D) = \frac{1}{\sqrt{2\pi}\log\sigma}\exp -\frac{1}{2}\left[\frac{\log \mathrm{SNR}_D - \log \mathrm{SNR}_{D50}}{\log\sigma}\right]^2 \qquad (11.1.25)$$

where $\log\sigma = 0.198$ and SNR_{D50} values are the threshold values listed in Table 11.1. For example, if background is uniform then for detection $SNR_{D50} = 2.8$. If 90% detection probability is required, then from normal distribution tables

$$\frac{\log \mathrm{SNR}_D - \log 2.8}{0.198} = 1.282 \qquad (11.1.26)$$

Since $\log 2.8 = 0.447$, then $\log SNR_D = 0.700$ and $SNR_D = 5.01$. This approach, however, is more stringent than that of Fig. 11.3, which suggests

TABLE 11.1 *Best Estimates of Threshold of SNR_D for Detection, Recognition, and Identification of Images (after [11.1])*

Discrimination Level	Background	Equivalent Bars per Minimum Dimension from Johnson Chart	Threshold SNR_{D50} for a Single Equivalent Target Bar of Spatial Frequency N_{sf} lines per Picture Dimension Equal to			
			100	*300*	*500*	*700*
Detection	Uniform	1	2.8	2.8	2.8	2.8
Detection	Clutter	2	4.8	2.9	2.5	2.5
Recognition	Uniform	8	4.8	2.9	2.5	2.5
Recognition	Clutter	8	6.4	3.9	3.4	3.4
Identification	Uniform	13	5.8	3.6	3.0	3.0

$$\mathrm{SNR}_D = (2.8)(1.30) = 3.64 \tag{11.1.27}$$

The threshold SNR_D values above are recommended only if the angular subtense of the displayed image is between 4 and 30 ft of arc (~1.1 and 8.5 mrad). Otherwise required SNR_D increases [11.1].

Sensors today are so sophisticated that most images, even at great distances, are not "snowy." Hence, they are not noise limited but rather contrast- or blur-limited (Chapter 10). This trend should be strengthened in the future as cameras are improved even further.

Example 11.1

An object of 0.5-m horizontal critical dimension at a 10-km distance is viewed through a telescope. If minimum display signal-to-noise ratio for 50% probability object *identification* in the presence of uniform clutter is to be at least 5.8, what should be the maximum focal length? Assume an 11.8-mm effective width for an imager with 500 TV lines per picture width, and neglect atmospheric effects.

Solution

To use Table 11.1, image size for critical dimension must be found. In the image plane, a 0.5-m critical object dimension is reduced to $0.5(s'/s) = 0.5f_l\,10^{-4}$, where f_l is focal length. This can be divided into $12.8 \approx 13$ equivalent vertical bars of width $(0.5/13)10^{-4}f_l$ m in the image plane. The horizontal width of the imager is 11.8 mm. The number of equivalent vertical target bars per picture width is $N_{sf} =$ 11.8 mm/$(0.5/13)10^{-4}f_l$ mm $= 3068/f_l$ where f_l is in millimeters. If $\mathrm{SNR}_D \geq 5.8$ then, from Table 11.1, $N_{sf} \geq 100$. Therefore, $f_l \leq 30.68 \approx 31$ mm. (The 500 TV lines characterize the TV system rather than the image of the target.)

If focal length is reduced, so too is magnification and, as a result, the width of each of the 13 equivalent bars comprising the target image is reduced. This increases N_{sf} and can permit reduced threshold SNR_D as shown in Table 11.1. From Section 3.4, reduced *f*-number also increases image plane irradiance, thus improving SNR_D. Therefore, reduced focal length and magnification can result in conversion of noise-limited imaging to contrast-limited imaging (Chapter 10).

11.2 THERMAL IMAGING SYSTEMS

For forward-looking infrared (FLIR) sensors, the noise equivalent parameter of interest is the noise equivalent temperature difference (NETD), derived in Chapter 6 [Eq. (6.2.6)]. It is the temperature difference between adjacent large objects which would provide an SNR equal to unity. It also permits video SNR_v to be predicted directly for a given signal ΔT according to [11.1]

$$\mathrm{SNR}_v = \Delta T/\mathrm{NETD} \tag{11.2.1}$$

where ΔT is temperature difference between adjacent objects. Using Eq. (11.1.17) SNR_D can be computed from SNR_v, and then use of Table 11.1 and Fig. 11.3 allows one to predict noise-limited resolution.

It is customary, however, to correct NETD according to aspect ratio α_T of the *target* [11.5]. The Johnson chart, for example, is based on psychophysical experiments involving four bars and three equal spaces between them as shown in Fig. 11.4. They form a square aspect ratio (length:width) of each bar and space equal to 7:1. Let α_T be equivalent bar length-to-width ratio where length is not confined to the x-direction (definition of ϵ_a) but is in whatever direction the object is longest. Psychophysical experiments [11.1, 11.5] show that NETD should be divided by $\sqrt{\alpha_T/7}$ in order to obtain an aspect corrected value of NETD. For example, Fig. 10.6 shows an automobile with major dimension, or length, equal approximately to twice the minor or critical dimension (height) and the corresponding bar patterns for detection (1 line pair) and recognition (4 line pairs) according to the Johnson chart. Here, aspect ratio of a single bar (α_T) for detection is 4:1 and for recognition is 16:1. Therefore, for detection $\mathrm{SNR}_v = \Delta T/[\mathrm{NETD}/\sqrt{4/7}]$, and for recognition $\mathrm{SNR}_v = \Delta T/[\mathrm{NETD}/\sqrt{16/7}]$. Hence SNR_v for recognition is only 1/2 of that for detection for the same ΔT/NETD ratio, that is, recognition requires twice the ΔT/NETD ratio as detection. The greater the aspect ratio of the target, the smaller the signal-to-noise ratio. This means, for example, a long and very thin target provides a much smaller SNR than a square target of the same area. The thinner the first target, the harder it is to acquire.

In addition to aspect ratio, another correction normally implemented is to use minimum resolvable temperature (MRT) as a noise parameter rather than NETD. MRT is the smallest change in blackbody equivalent temperature that can be detected clearly with a thermal imaging system. MRT is a function of spatial frequency since smaller target images require larger temperature differences in order to be observed. Therefore, MRT increases with spatial frequency. In view of these corrections, *normalized signal-to-noise ratio* is often used instead of Eq. (11.2.1), and is defined [11.5] for thermal imaging as

$$\mathrm{SNR}_N = \frac{\Delta T\sqrt{7/\alpha_T}}{\mathrm{MRT}(f_a)} \tag{11.2.2}$$

where f_a is angular spatial frequency and MRT is a function of it. Normalized signal-to-noise ratio is proportional to SNR_D.

Furthermore, contrast between target at a distance z and path luminance background is degraded according to atmospheric transmission. The assumption is that the apparent differential between target and path radiance background, ΔT, is degraded to $(\Delta T)\tau'_{\mathrm{atm}}$, where τ'_{atm} is atmospheric transmission, while received path radiance

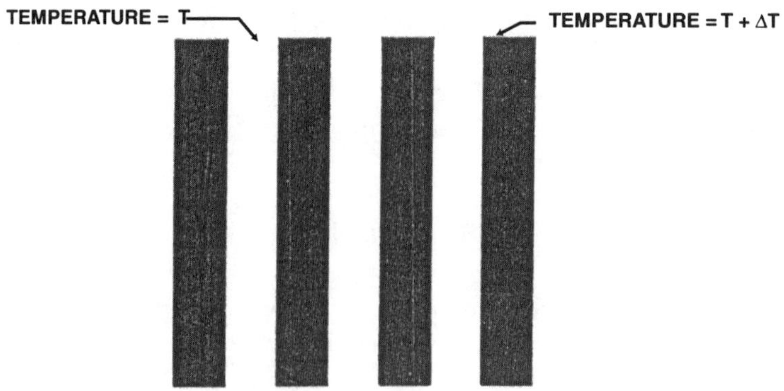

Fig. 11.4 Standard four-bar test pattern.

background irradiance is not degraded but is independent of receiver position. This assumption is often justified except for cloud conditions, which cause received path radiance to vary along the line of sight. Accordingly [11.5],

$$\mathrm{SNR}_N(z) = \frac{\Delta T\sqrt{7/\alpha_T}}{\mathrm{MRT}(f_a)} \exp\left[-\int_0^z \alpha(\lambda, z', T)\, dz'\right] \tag{11.2.3}$$

where the exponential term is atmospheric transmission and α is attenuation or extinction coefficient.

This result assumes atmospheric attenuation of contrast transfer is not spatial-frequency dependent. This assumption is shown in Part 5 to be *false*, not only in the visible and near-infrared regions, but also in the thermal infrared. It is not atmospheric transmission but rather *atmospheric blurring* that limits target acquisition. The spatial-frequency dependence of contrast transfer and blur through the atmosphere at visible and near-infrared wavelengths is described by the atmospheric MTF, which should be included in $(\mathrm{MTF})_s$ in Eq. (11.1.9) and $(\mathrm{MTF})_{sT}$ in Eq. (11.1.17). For thermal imaging Eq. (11.2.3) suggests an atmospheric MTF equal to atmospheric transmission at all spatial frequencies. Data described in Part 5 show this can be approximately true at high spatial frequencies only and under conditions of weak turbulence only. For the present we will continue in the traditional manner in which atmospheric MTF blur is neglected. We will reconsider this shortly and add atmospheric MTF.

Under fairly homogeneous weather conditions, Eq. (11.2.3) can be approximated by

$$\mathrm{SNR}_N(z) = \frac{\Delta T\sqrt{7/\alpha_T}}{\mathrm{MRT}(f_a)} \exp[-\alpha(\lambda, T)z] \tag{11.2.4}$$

As shown shortly, this permits determination of resolution range z for a given required SNR_N, shown in Fig. 11.5. Note that probability of target acquisition is 50% when

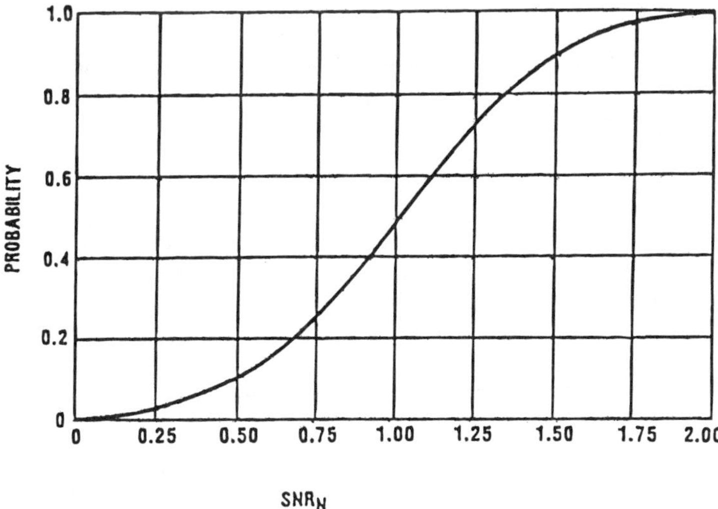

Fig. 11.5 Probability of target acquisition versus normalized signal-to-noise ratio (from [11.5]).

SNR_N is unity. Other values of SNR_N yield different probabilities of target acquisition. For example, 90% probability requires an SNR_N of at least 1.5.

The probabilities shown in Fig. 11.5 are equivalent to P_∞ in Section 10.5 for target acquisition probabilities but here they are for noise-limited imaging.

The increase of MRT with spatial frequency is often exponential and expressed as [11.5]

$$\mathrm{MRT}(f_a) = \mathrm{MRT}_0 \exp(\beta_{sys} f_a) \tag{11.2.5}$$

where f_a is angular spatial frequency in cycles·mrad^{-1}, MRT_0 is extrapolated zero spatial frequency MRT as indicated in Fig. 11.6, and β_{sys} characterizes in mrad·cycle^{-1} the system resolution limitations. The value of β_{sys} varies with optics magnification and imager quality.

To determine CNVEO target model acquisition ranges for noise-limited thermal imaging, angular spatial frequency f_a should be equal to at most $f_{a\ \max}$, where, analogous to $f_{r\ \max}$ in the previous chapter, $f_{a\ \max}$ is the maximum usable angular spatial frequency of the imaging system. Since, from Eq. (10.2.4), if there is no angular magnification one equivalent resolution bar equals

$$\Delta x = -\frac{1}{2 f_{r\ \max} m} = -\frac{z}{2 f_{r\ \max} s'}$$

then, ignoring the minus sign for image inversion, for a target acquisition probability requiring n line pairs across the critical target dimension from Eq. (10.4.2),

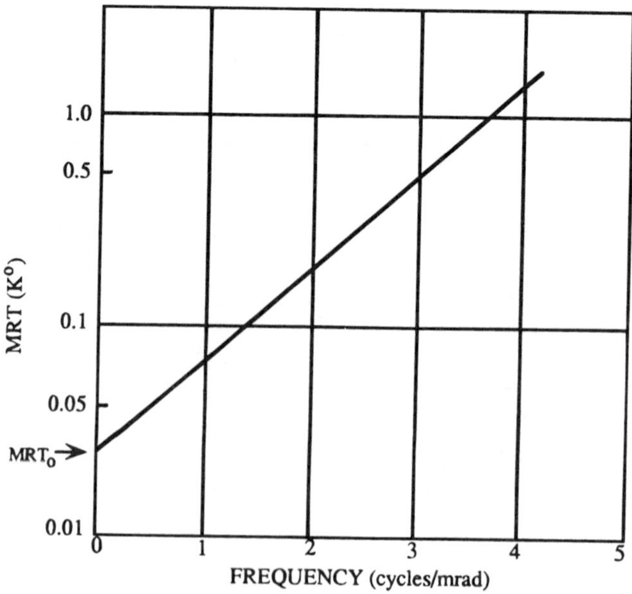

Fig. 11.6 Sample MRT(f_a).

$$f_{r\,\max} = \frac{n}{|xm|} = \frac{n}{|x'|} \tag{11.2.6}$$

for lateral magnification m where, for large range z, image distance s' = focal length f_l. Therefore,

$$f_{a\,\max} = \frac{n}{x'} f_l = \frac{nz}{x} \tag{11.2.7}$$

where n again is the number of line pairs required to fill x in object space or its image x' in image space. Then, from Eq. (11.2.5)

$$\mathrm{MRT}(f_{a\,\max}) = \mathrm{MRT}_0 \exp(\beta_{sys} nz/x) \tag{11.2.8}$$

and, from Eq. (11.2.4), for a fairly homogeneous atmosphere,

$$\mathrm{SNR}_N(z, f_{a\,\max}) = \frac{\Delta T \sqrt{7/\alpha_T} \exp[-\alpha(\lambda, T)z]}{\mathrm{MRT}_0 \exp(\beta_{sys}\, nz/x)} \tag{11.2.9}$$

Solving for acquisition range, the above yields target acquisition range

$$z = [\alpha(\lambda, T) + \beta_{sys} n/x]^{-1} \ln\left(\frac{\Delta T \sqrt{\alpha_T/7}}{\mathrm{SNR}_N \mathrm{MRT}_0}\right) \tag{11.2.10}$$

where z varies with the resolution criterion (detection, orientation, recognition, or identification) used to determine the number n of line pairs filling the critical target dimension. From Eq. (10.4.5) for a thermal imager that also samples in the horizontal direction, such as a focal plane array, n should be replaced by $2n$.

In Part 5 of this book, optical properties of the atmosphere are described. These include both attenuation and blurring effects. One form of attenuation is absorption. This limits thermal imaging to two infrared wavelengths bands in particular, the 3- to 5-μm wavelength (called *short wave*) and the 8- to 14-μm wavelength (called *long wave*). Absorption by water vapor is so strong between 5 to 7 μm, for example, that radiation at those wavelengths is essentially extinguished entirely over path lengths of only a few meters. As a result, thermal imaging systems are made for either long-wave (LW) or short-wave (SW) application. For LW application, the dc value of MRT, called here MRT_0, is usually on the order of 0.01 K, while β_{sys} for 1978–79 devices was on the order of 0.633 mrad·cycle^{-1} [11.5] but has been improved considerably in recent devices.

For SW devices, MRT_0 is also typically on the order of 0.01 K, but β_{sys}, in 1974 technology devices, was [11.5] on the order of 1.0 mrad·cycle^{-1}. This too has been improved considerably in recent devices which can extend out to spatial frequencies at least an order of magnitude greater than did 1974 devices. In recent SW focal plane arrays of InSb (256 × 256 pixels), MRT_0 is typically on the order of 0.003°K while β_{sys} is on the order of 2.4 mrad·cycle^{-1} for 50 mm focal length lenses and about an order of magnitude smaller for 500 mm focal length lenses. For 960 × 4 pixel focal plane arrays of PtSi, MRT_0 is on the order of 0.001°K while β_{sys} is approximately 0.78 mrad·cycle^{-1} for 50-mm focal length optics.

Equation (11.2.10) is included in many target acquisition models based on theories and experiments that evolved from use of thermal imagers developed in the 1970s. Such devices had very limited spatial-frequency bandwidths and their performances were not greatly affected by atmospheric blur, which affects image quality primarily at higher spatial frequencies. However, in view of more recent development of high-resolution thermal imagers such as focal plane array devices, experiments indicate atmospheric blur effects should be included because they are very evident for long-range imaging with high-resolution hardware. In this case, Eq. (11.2.9) should include atmospheric MTF [$MTF_a(f_a)$] instead of atmospheric transmission and is amended to [11.6]

$$SNR_N(z, f_{a\,\max}) = \frac{\Delta T\sqrt{7/\alpha_T}\,MTF_A(f_a)}{MRT_0\exp(\beta_{sys}nz/x)} \tag{11.2.11}$$

As shown in an expanded discussion in Chapter 19, Eq. (11.2.10) can still be approximately correct but for weak turbulence only. Since atmospheric MTF concepts are considered in Part 5, we will limit ourselves here temporarily to the previous model of Eq. (11.2.10), and consider Eq. (11.2.11) later in Section 19.3. The more recent CNVEO version for Eq. (11.2.10) is FLIR 92, which has been summarized by Holst [11.7].

As described in Section 10.6 for contrast-limited imaging, the U.S. Navy also prefers, for noise-limited thermal imaging, to consider total object area rather than just critical dimension. Experiments regarding detection and recognition of ships indeed support the Navy approach [11.8].

Thermal imaging systems today are so quiet and sensitive that most images, even at long distances, are not "snowy." It is incorrect to use the CNVEO approach described in this section for thermal imagery that is not snowy. The MTF approach of Chapter 10 should be used instead since such images are limited by blur rather than by noise. This trend should be strengthened in the future as technology improves even more.

Example 11.2

While a tank exhibits a front aspect (view) that is approximately square, the side aspect involves a length-to-height ratio of approximately 3, where tank height is approximately 3 m. Assuming an atmospheric extinction coefficient of 0.50 km^{-1} for LW imaging, find the maximum range for 90% "recognition" probability, assuming $\beta_{sys} = 0.633$ mrad·cycle^{-1} and $MRT_0 = 0.01$ K. The temperature differential of the tank is 2 K. How does maximum range change if a focal plane array is used?

Solution

For recognition, 4 *horizontal* line pairs or 8 bars (or lines) are required to fill an object area of critical vertical dimension 3-m × 9-m horizontal dimension. Since the overall aspect of the side view is 3:1, the aspect ratio of each of 8 bars is 24:1. Therefore, from Eq. (11.2.10), with the requirement that $SNR_N = 1.50$ for 90% probability,

$$z = [0.50 \text{ km}^{-1} + 0.63 \text{ mrad}\cdot\text{cycle}^{-1}(4 \text{ line pairs}/3 \text{ m})]^{-1} \cdot \ln\left[\frac{2\sqrt{24/7}}{1.5(0.01)}\right]$$

$$= (0.75 \text{ km}) \ln 247 = 4.1 \text{ km}$$

If a focal plane array is used, we have to double n and α_T. Consequently $z = 0.46 \ln 349 = 2.69$ km.

REFERENCES

11.1. F. A. Rosell and R. H. Wilson, "Recent psychophysical experiments and the display signal-to-noise ratio concept," in *Perception of Displayed Information*, L. M. Biberman, Ed., Plenum Press, New York, 1973.

11.2. R. H. Wight, "First order performance prediction techniques for charge-coupled device scanning imagers," *Opt. Eng.*, Vol. 23, May/June 1984, pp. 269–274.

11.3. J. M. Mooney and E. L. Dereniak, "Comparison of the performance limit of Schottky-barrier and standard infrared focal plane arrays," *Opt. Eng.*, Vol. 26, March 1987, pp. 223–227.

11.4. G. C. Holst, "Applying the log-normal distribution to target detection", in *Infrared Imaging Systems*, G. C. Holst, Ed., *Proc. SPIE*, Vol. 1689, 1992, pp. 213–215.

11.5. L. N. Seekamp, "Field manual to determine detection or recognition range of a FLIR sensor," IDA Paper P-1419, Institute for Defense Analyses, September 1979. Reprinted by courtesy of the Institute for Defense Analyses.

11.6. D. Sadot, N. S. Kopeika, and S. R. Rotman, "Incorporation of atmospheric blurring effects in target acquisition modeling of thermal images," *Infrared Phys. Technol.*, Vol. 36, February 1995, pp. 551–564.

11.7. G. C. Holst, *Electro-Optical Imaging System Performance*, JCD Publishing, Winter Park, Florida, and SPIE Optical Engineering Press, Bellingham, WA, 1995.

11.8. A. W. Cooper, P. L. Walker, E. A. Milne, and B. J. Cook, "Evaluation of tactical decision aid code predictions of FLIR range performance," in *Characterization, Propagation, and Simulation of Sources and Backgrounds II*, D. Clement and W. Watkins, Eds., *Proc. SPIE*, Vol. 1687, 1992, pp. 147–157.

EXERCISES

11.1 Determine limit of SNR_D for large irradiances incident on a camera including a detector array.

11.2 Determine the optics MTF coefficient in SNR at output for an amplifier noise-limited imager of high spatial uniformity with a 25-s^{-1} frame rate, assuming (1) the target image occupies 10% of FOV, and that the rest of the FOV is atmospheric background path luminance; (2) equivalent noise current referenced to input is 3 mA; (3) object is observed at 800-nm wavelength through F8 optics with a 200-nm-wide optical filter; (4) object reflection coefficient at 800-nm wavelength is 0.7; (5) optical density of the atmosphere at 800 nm is three; and (6) solar irradiance and atmospheric irradiance at 800-nm wavelength are 1000 $\text{W}\cdot\text{m}^{-2}\cdot\mu\text{m}^{-1}$ and 100 $\text{W}\cdot\text{m}^{-2}\cdot\mu\text{m}^{-1}$, respectively. Assume further that imager responsivity is 0.1 $\text{A}\cdot\text{W}^{-1}$ at 800-nm wavelength.

Imager effective dimensions are 11.8 × 8.8 mm. (*Hint*: Refer to Sections 3.4 and 10.7.)

11.3 (a) If in Exercise 11.2 imager gain is unity and amplifier gain is 100, what is display SNR?
(b) Does the preceding result indicate image quality is likely to be noise limited or contrast limited?.

11.4 (a) Consider again Exercise 11.3, assuming the imager output signal is broadcast to a TV receiver, and that antenna diffraction losses limit G_I to 0.5%.
(b) What happens to image quality if G_I is limited to 0.1%?

11.5 Consider Exercise 11.4(a) assuming that combined optics, imager, signal processing, and display MTFs are characterized by $\exp(-0.04 f_r)$, where f_r is in cycles·mm^{-1}.
(a) What is $f_{r\max}$ for 50% probability of detection through uniform clutter?
(b) Assuming unity modulation contrast incident on the receiver, what is $f_{r\ \max}$ if the system is contrast limited? Assume 2% required contrast.
(c) What is the smallest image plane detail detectable?
(d) What is the smallest image plane detail identifiable with 50% probability? Assume a rectangular target of image plane dimensions $0.1W_x$ and $0.1V_y$, where W_x and V_y are defined following Eq. (11.1.19).

11.6 Compare required threshold $\mathrm{SNR_D}$ values for 90% probability of identification through uniform clutter using both the techniques of Fig. 11.3 and of Eq. (11.1.25) for $N_{sf} = 300$.

11.7 The imager in Example 11.2 is required to identify the side aspect of a tank at an 8.2-km distance. Optics FOV and focal length must be changed to accommodate a new required value of β_{sys}. What must be the new required value of β_{sys}? Assume atmospheric conditions and required acquisition probability as in Example 11.2.

11.8 In Example 11.2, what is maximum range if the tank is viewed from the front?

11.9 In Example 11.2, what is the probability of recognizing the tank within 2 s?

11.10 Using the system and atmospheric conditions of Example 11.2, what is the probability of recognizing the tank within 4 s, when it is viewed from the front at a range of 3 km?

CHAPTER

12

Human Visual System MTF and Threshold Contrast

In the study of the behavior of optical systems it is essential to know how the human eye interacts with the output of those systems. The ultimate quality of an image depends directly on the modulation transfer function (MTF) and on the threshold contrast function of the human visual system. These also determine the optimum distance of the observer from the display so as to view a target of given size (Exercise 12.4). To begin to understand these, a knowledge of the physiology of the eye is essential.

12.1 PHYSIOLOGY OF THE EYE

A diagram of the eye is presented in Fig. 12.1. The eye is an almost spherical (24 mm long by 22 mm across) mass contained within the tough outer shell, the *sclera*. The sclera is continuous with the *cornea*, which bulges slightly outward from the body of the sphere. Light enters through the cornea. Most of the bending imparted to a bundle of rays entering the eye takes place at the air/cornea interface. One of the reasons that one cannot see very well underwater is that the index of refraction of the water (1.33) is too close to that of the cornea (1.376) to allow for adequate refraction. Light passing through the cornea enters the aqueous humor, a chamber filled with clear watery fluid with an index of refraction of 1.366. A ray that is strongly refracted toward the optical axis at the air/cornea interface will be only slightly redirected at the cornea/aqueous humor interface because of the similarity in their indices.

Between the cornea and the lens lies the *iris* diaphragm, a heavily pigmented opening that serves to limit the amount of light entering the eye through the pupil. The iris is what gives the eye its characteristic color. The circular and radial muscles of the iris can expand or contract the pupil from about 2 mm in bright light to 8 mm in darkness. In addition, the iris aids in focusing by contracting when doing close-up work to increase image sharpness.

Immediately behind the iris lies the *lens*. The lens is a convex and transparent membrane containing no nerves or blood vessels. In structure the lens is somewhat like a transparent onion, formed of roughly 22,000 very fine layers. It contains more

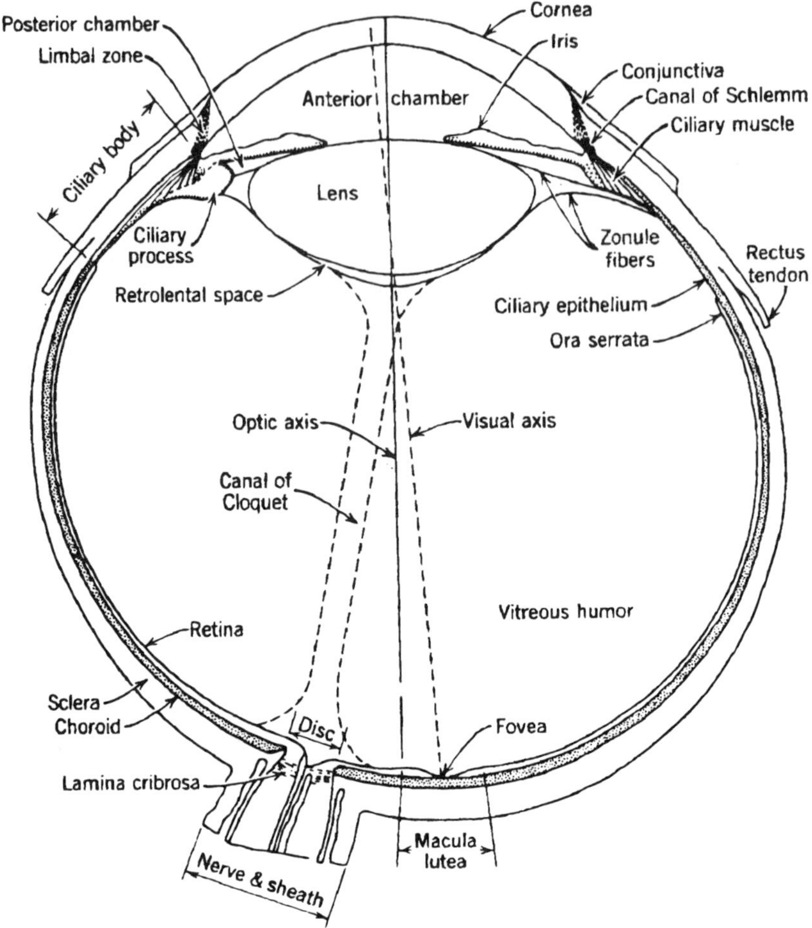

Fig. 12.1 Anatomical cross section of the eye (after [12.1]). © John Wiley & Sons, Inc.

protein than any other tissue. Because of this laminar structure, rays entering the lens follow paths made up of minute, discontinuous segments. The lens is very flexible, and becomes less flexible with age. The index of refraction of the lens varies from about 1.406 at the inner core to 1.386 at the less dense cortex. Since the lens is pliable, it has a variable focal length, and can therefore focus on near or far targets, through a process called *accommodation*. Generally focal length is on the order of 17 mm. As can be seen in Fig. 12.1, the radius of curvature of the anterior surface of the lens is considerably greater than that of the posterior surface. It absorbs approximately 8% of the visible spectrum, with relatively higher absorption at shorter wavelengths. Both infrared and ultraviolet radiation is absorbed within the lens, and can cause damage to the lens in excessive amounts.

The next structure behind the lens is a chamber filled with a transparent gelatinous substance known as the *vitreous humor*, which has an index of refraction of 1.337.

Within the tough wall of the sclera is an inner shell, the *choroid*. It is a dark layer richly supplied with blood vessels, and heavily pigmented with melanin. The choroid serves to absorb stray light, much as a thin coat of black paint on the inside of a camera. A thin layer (about 0.5 to 1 mm thick) of light receptor cells, analogous to detectors, covers much of the inner surface of the choroid. This is the *retina*. A beam of light coming through the eye is focused onto the retina, and through electrochemical reactions, which are discussed in detail later, an image is formed and sent to the brain. There are two types of receptor cells in the retina called rods and cones, after their characteristic shapes. The *cones* are for photopic vision under normal light levels and give us color vision. *Rods* are for scotopic vision under low light levels and are responsible for night vision. They give no color information at all. (Photopic and scotopic spectral responses are described in Fig. 3.12).

In the center of the retina is a small depression from 2.5 to 3 mm in diameter called the *macula*. In the center of the macula is a tiny, rod-free region called the *fovea centralis*. In this region the cones are packed more densely than anywhere else in the retina, and the fovea provides the most detailed information about the image. The eyeball is constantly moving in order to keep in this region light from the object of interest since the high cone density there permits maximum resolution (about 0.002 nm between image points). At normal daytime brightness cone vision predominates and irradiance from objects within a field of view of 2 deg or less falls in the fovea. This is called *photopic* or foveal vision. Night vision is attributed to rods that are more peripheral in location and is called *scotopic* or peripheral (extrafoveal) vision. Cone concentration in the fovea and rod concentration in the periphery are illustrated in Fig. 12.2(a). The normal human retina contains about 120 million rods and 6 million cones.

The *optic nerve* is the central carrier of information from the eye to the brain. The nerve fibers from the output of the retina all converge at the optic nerve. The optic nerve sits in a section of the eye that has no retinal cover, and is called the disc or *blind spot*. Since the eye is in constant motion, the perception of a scene is constructed from the time-varying motions of the image on the retina, and so the blind spot does not present a problem. From the optic nerve the signal goes into the visual cortex of the brain, and by a process still not fully understood, vision results.

As mentioned previously, the eye focuses on objects using a process called accommodation, which is performed by the lens in conjunction with the structures and muscles surrounding the lens. The lens is suspended by the ciliary zone. When the muscles of accommodation are at rest, the zonule is under tension and the lens tends to be relatively flattened. Under these conditions, the eye is accommodated for distant objects. Contraction of the annular or circular fibers within the ciliary body reduces the tension in the zonule and allows the lens to become thicker so that the images of near objects are focused on the retina. In this state the eye is accommodated for near objects. In the normal eye, the variation in lens shape permits a range of accommodation from 6 m to about 10 to 20 cm. Rays beyond 6 m are so parallel that variations in focus beyond this distance are unnecessary and ineffective in discriminations of distance [12.1, 12.3].

12.1.1 The Structure of the Retina

The retina is a complex structure made up of many layers of different types of cells. The light receptor cells (analogous to detectors), composed of rods and cones, are the last of these layers to see a light ray as it passes through the retina. Figure 12.2(b) is a semi-schematic drawing of the layers of the retina.

The outermost layer, farthest from the lens of the eye, contains the receptor rod and cone cells. The next layer includes the bipolar cells, and the horizontal and amacrine cells. Bipolar cells transmit messages from the receptors to cells in the

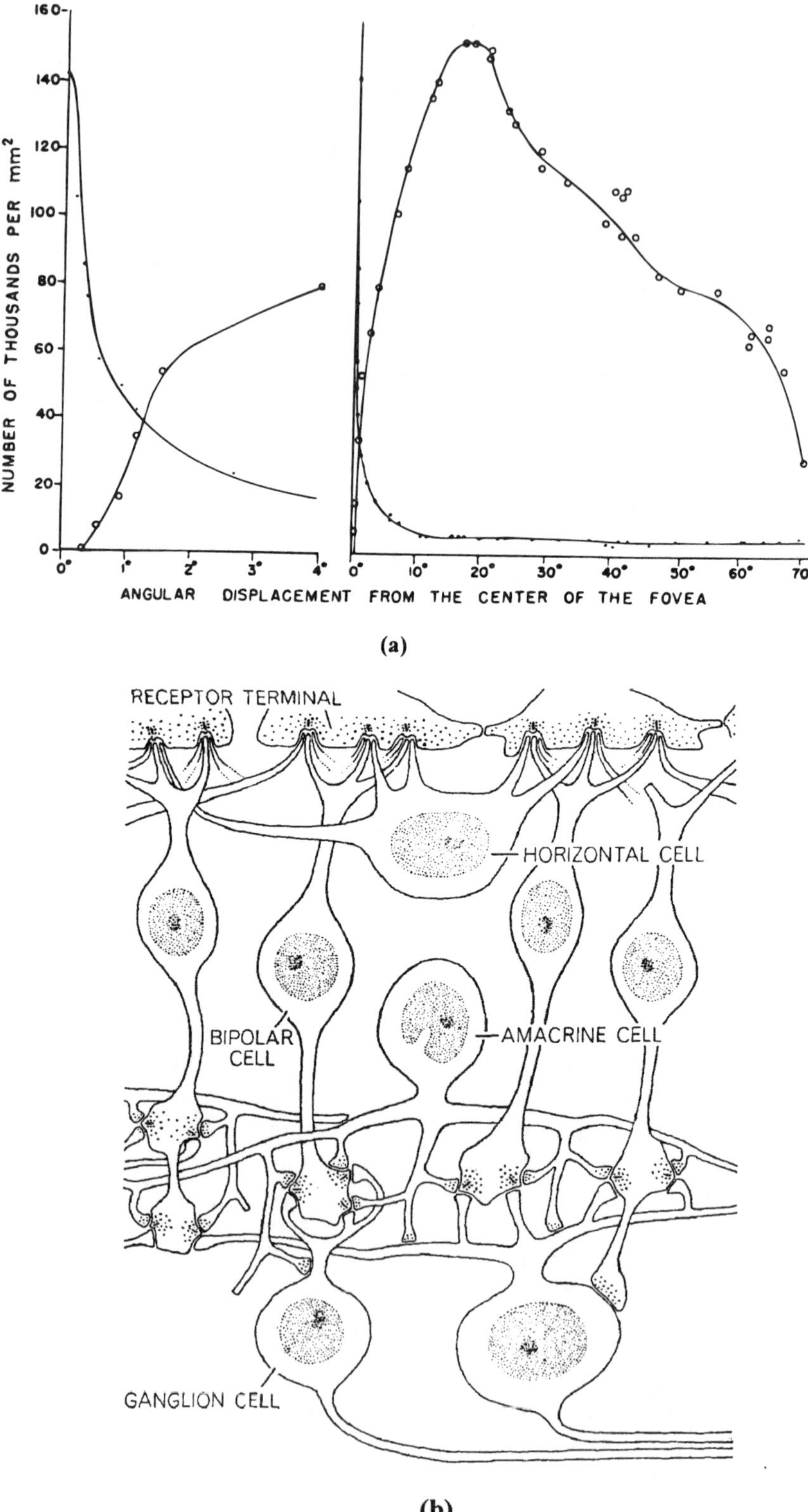

Fig. 12.2 (a) Concentration of rods (circles) and cones (dots) with relation to the center of the fovea along the visual axis (horizontal direction for person standing). (b) Structural schematic of the retina (after [12.2]).

third layer, while amacrine and horizontal cells appear to be involved in the lateral transmission of the information. This lateral transmission of information is central to the understanding of the origins of the MTF of the eye. The third layer of the retina contains the *ganglion* cells, whose axons come together to form the *optic nerve*, which is the sole output of the retina. It is the highly complex network of synaptic connections between the various types of nerve cells that makes possible a wide variety of integrative mechanisms necessary for visual perception.

As can be seen in Fig. 12.2(b), the terminals of the bases of the receptor cells form *synapses* (neural connections) with bipolar cell dendrites, and horizontal cell processes. The horizontal cells connect neighboring receptors while the bipolar cells send information on to the inner synaptic layer (the layer closest to the lens).

In the inner synaptic layer of the retina, each bipolar cell forms synapses with two types of cells, the *ganglion* cell and an amacrine cell. Amacrine cells act on ganglions just as horizontal cells do on receptors, transmitting information between ganglions. A given retinal ganglion cell receives information from a small population of receptor cells. The area covered by these receptor cells is called the receptive *field* of that ganglion cell.

The receptive field is that set of receptor cells that, when stimulated for photopic vision, influences the electrical activity in the ganglion cell in either an *excitory* or *inhibitory* manner. These influence strongly the MTF of the known visual system as shown later in Fig. 12.14, and cause maximum photopic response to be at a nonzero spatial frequency. Experiments to determine the receptive field have been done, and it is possible to identify the receptive field of a given ganglion through a painstaking process of trial and error.

In 1938, H. K. Hartline of the University of Pennsylvania mapped such receptive fields in the frog for the first time, and received the Nobel prize for this pioneering work in 1967. In 1953 Steven W. Kuffler found that the receptive fields of ganglion cells for almost all vertebrates were organized in a concentric manner, with a circular central area surrounded by a ring-shaped outer zone. If the central zone of certain ganglia was stimulated (illuminated), the ganglion cell was excited, whereas light on the outer ring inhibited the cell. The reverse situation is also possible, with light falling on the central region resulting in an inhibited response. This on/off response is the active mechanism in sensing of contrast differences, and also senses movement. As an object moves across the image plane of the retina causing many on/off responses, a series of on/off pulses is generated and sent to the visual cortex for processing. This on/off sequence not only enables us to see stationary objects, but also allows us to perceive motion when an object crosses our field of view, sending a series of on/off pulses in rapid sequence to the brain [12.2].

12.1.2 Detection of Quanta in Photoreceptors

Now let us consider the process involved in the detection of light by the rods and cones in the retina. Figure 12.3 shows an electron microscope picture of a typical rod cell. It is evident from the picture that the rod is filled with a membrane folded back on itself hundreds of times which looks like a stack of coins. The cone membrane is similar in structure to that of a rod, and consists of an elaborately folded sheet of a different topology. These membranes in booth rods and cones contain a molecule called the 11-*cis* retinal, described in Fig. 12.4.

The retina is embedded in a protein that determines the spectral responsivity of the pigment within the receptor. The pigment contained in the rods is called *rhodopsin*,

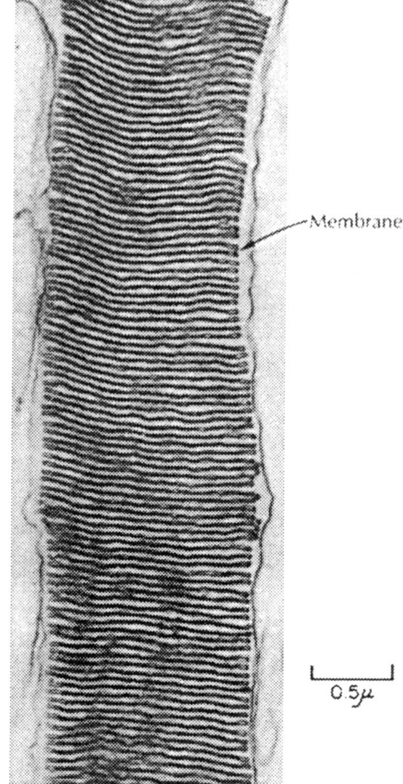

Fig. 12.3 Electron micrography of the receptor area of a typical rod cell [after (12.4)].

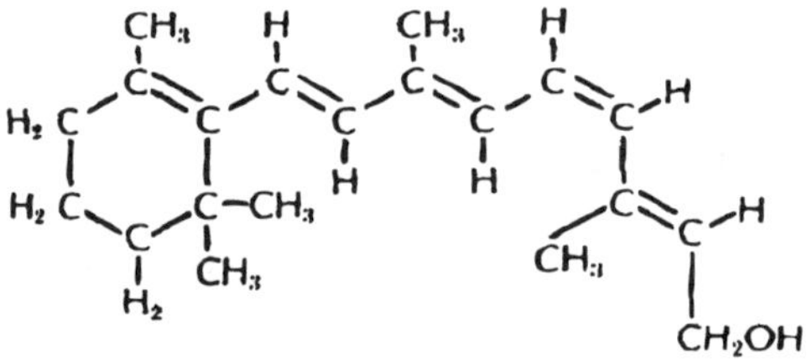

Fig. 12.4 Retinal molecule.

and has a wide spectral response, thereby giving no color information. Cone cells can be classified into one of three categories—green absorbing, blue absorbing, or red absorbing—depending on the type of pigment contained within the cell. Because pigments within the cones are tuned to receive a certain color, the cones are responsible for color vision. The spectral responses of rhodopsin and the color pigments in the cones are shown in Fig. 12.5. Together they span the entire visible spectrum and even beyond.

When a quanta of light strikes the retinal molecule, the energy from the quantum causes the bonds of the retinal molecule to be rearranged, and the molecule changes its shape by untwisting and straightening. Such a change in state is called a *cis-trans* isomerization, and the various stages are shown in Fig. 12.6.

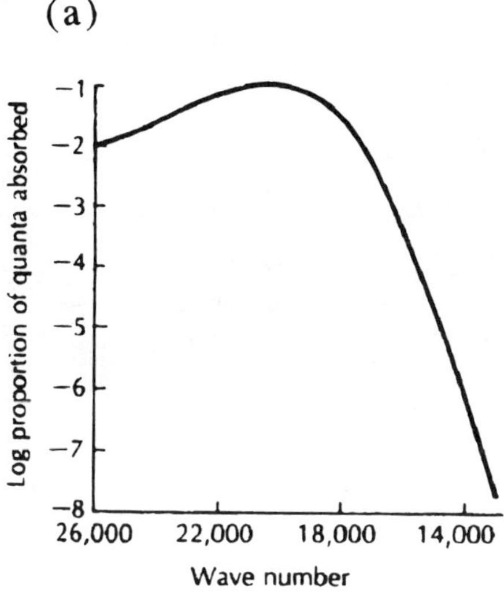

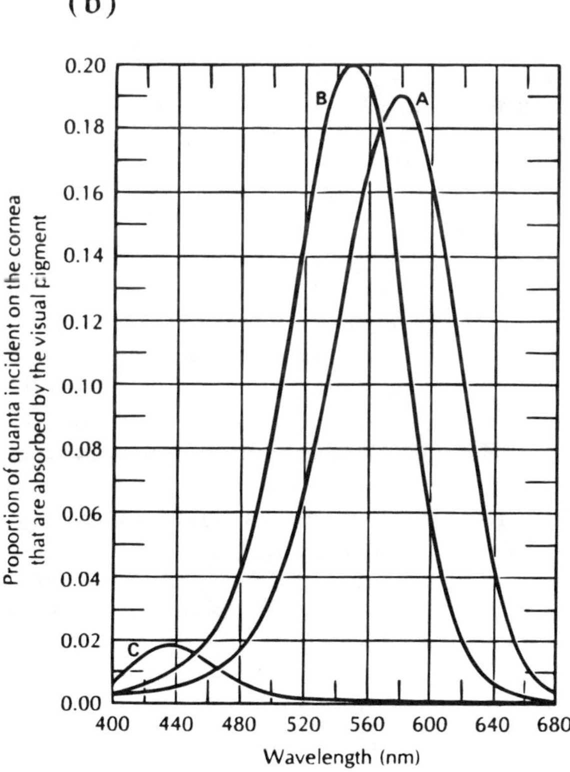

Fig. 12.5 (a) Spectral response of rhodopsin. (b) Cone color pigments; A (red), B (green), and C (blue). (Note that wavenumbers of 14,000 cm^{-1} and 26,000 cm^{-1} equal wavelengths of 714 and 385 nm, respectively, where wavenumber is λ^{-1} in cm^{-1}).

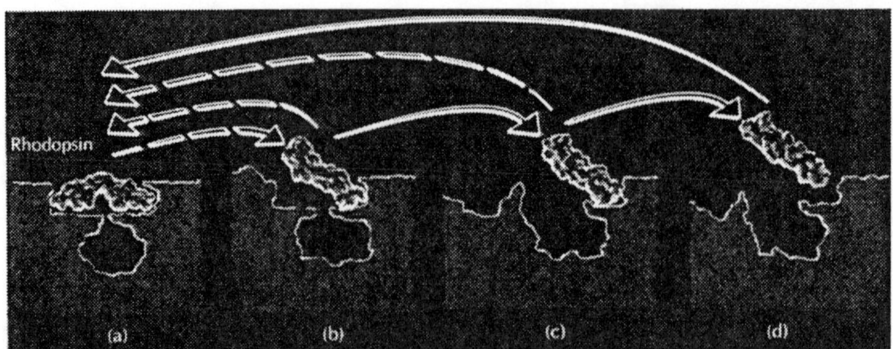

Fig. 12.6 The four stages of the *cis-trans* isomerization of a molecule of rhodopsin. The states are labeled (a) through (d) (after [12.4]).

In Fig. 12.6, state *b* is unstable at body temperature. If incoming light were turned off so that no new quanta could be absorbed, the molecule would quickly undergo a series of changes as shown in *c* and *d*. Note that the change from *b* to *c* represents a change in the protein molecule, and between *c* and *d* the retinal molecule floats free of protein. In rods, the pigment molecule in state *a* is purple in color, whereas a molecule in state *d* is yellowish in color. For that reason, the molecule in state *d* is called *bleached*. A molecule in state *d* will spontaneously regenerate, that is, return to state *a*. The time factor involved in these reactions is important to the understanding of the phenomenon of *dark adaption*. Once a quanta has been absorbed and a molecule of retinal has transited into state *b*, it takes on the average about 1 s to go all the way to state *d*. However, once the molecule has transited into state *d*, the spontaneous regeneration time back to state *a* is about 5 min.

The 1-s and 5-min times require some explanation. These are average times, meaning that the probability of any given state *b* molecule to change to a state *d* molecule in 1 s is 0.5. Similarly, a state *d* molecule has a 0.5 probability to go to state *a* within 5 min. If a rod is exposed to a flash that isomerizes every molecule of retinal in the rod (every molecule goes to state *d*), then the concentration of state *d* molecules as a function of time is plotted in Fig. 12.7.

It is known that the energy contained in a nerve pulse is much greater than that in a photon by some eight orders of magnitude [12.5]. How a photon of light is converted into a nerve impulse is still not fully understood, but clearly a biological amplifier of some type is at work. Somewhere between state *a* and state *d*, a biochemical reaction is catalyzed that causes the membrane of the receptor cell to block sodium flow into the cell. When this happens, the cell becomes hyperpolarized (typically the membrane goes from about -40 to -70 mV when the cell is exposed to a flash of light). This hyperpolarization starts the action potential down the nerve path. The enzymatic reaction acts as a photomultiplier, enabling one transformed rhodopsin molecule to create hundreds of molecules, which block the entry of millions of atoms of sodium into the cell, causing the cell to become hyperpolarized. This hyperpolarization causes an action potential to arise within the cell, starting the nerve pulse.

The speed in which the action potential arises is not the same for rods (scotopic vision) and cones (photopic vision). The rod's quantal response time is nearly four times that of the cone. A rod takes about 300 ms to finish signaling the absorption of a photon. Although the rods are far more sensitive to absolute light levels than the cones, they are much less effective for temporal resolution. Because of their faster response to movement, cones are better at encoding rapidly changing temporal stimuli. This results in a trade-off between sensitivity (photon counting) and temporal resolution (motion detection) [12.6].

The exact relation between the intensity of the nerve pulse and the number of quanta absorbed has not yet been determined, but there is much experimental evidence to suggest that the dependence is *logarithmic*. This logarithmic response explains a great many phenomena related to the eye, such as the log-normal distribution of Eq. (11.1.25) as well as the TTF curve of Eqs. (10.5.1) and (10.5.2), which can also be approximated by a log-normal distribution [11.4]. It is therefore important to keep in mind this logarithmic response when examining experimental results. It is also important in the understanding of the MTF of the human visual system.

12.1.3 Dark Adaptation

All are familiar with what happens when one enters a dark room from a lighted space. Virtual blindness persists for a few minutes until the eye gets adapted to the dark. After about 30 min have passed, details that were all but invisible on entry into the room can be clearly seen. The time dependence is shown in Fig. 12.8. This process is reversible as well. Going into a well-lighted area after sitting in a dark room always entails some time to adapt to the light, but light adaption is always much quicker than dark adaption. In addition, it is possible to destroy night vision with just one intense flash of light. After the flash, another 30 min is required to fully adapt.

The large increase in sensitivity as one passes from daylight into night vision is primarily due to rod vision taking over from cone vision. Physically, when illumination levels are reduced, the pupil of the eye expands and admits more light. The retina

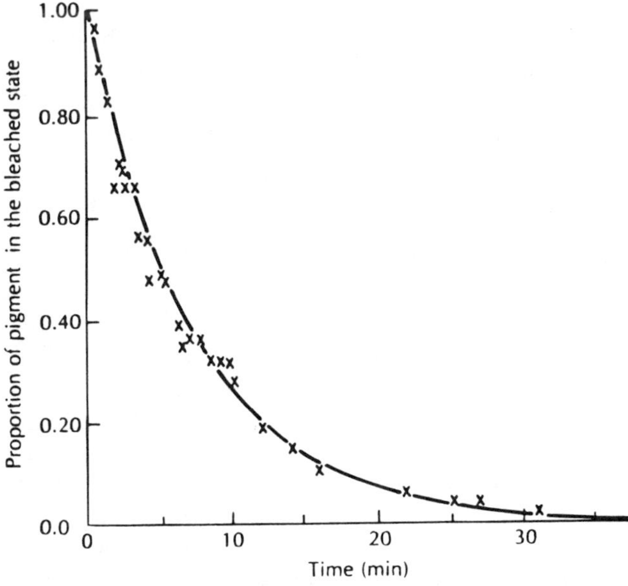

Fig. 12.7 Concentration of bleached (state *d*) rhodopsin versus time (after [12.4]).

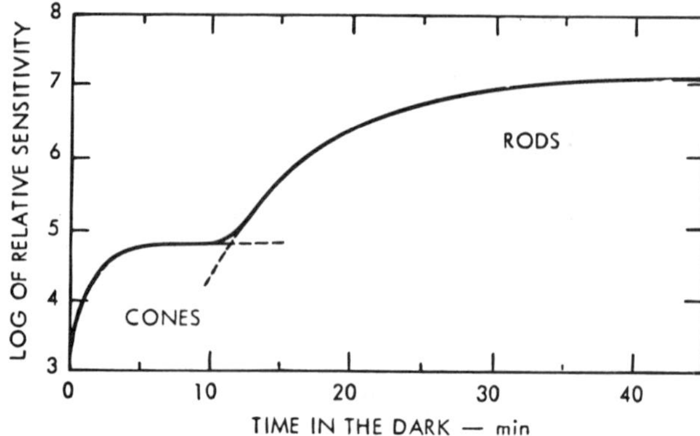

Fig. 12.8 Adaption of the eye to darkness.

becomes more responsive by switching from cone vision to rod vision. During daylight, the rods are saturated and contribute no useful information to the visual cortex, and vision is dominated by cones [12.6]. During the day virtually all of the pigment inside of the typical rod is bleached, or in state *d* in Fig. 12.6. As time passes, more and more pigment inside the rods transitions into state *a*, according to the graph of Fig. 12.7, which assumes a 5-min half-life for molecules in state *d*. Changing from state *d* to *a* corresponds to the eye adapting to the dark, for this enables the rhodopsin molecule to absorb a quanta of light. As more rhodopsin molecules change from state *d* to state *a* the sensitivity of the eye increases. Likewise, the change from state *a* to *d* takes about 1 s, hence the short time needed for light adaption and for destruction of one's night vision.

Experimental results point out an inconsistency, however. If a dark-adapted eye is exposed to a flash of light, which results in a 25% bleaching of the pigment, the decrease in sensitivity is not 25% but rather a 100,000 fold decrease is observed [12.5]. The threshold is not simply inversely proportional to the proportion of pigment in state *a*. Rather the decrease in sensitivity seems to go as a logarithmic function of the fraction of pigment in the bleached state. Put simply, a small increase in the amount of bleached pigment results in a very large decrease in sensitivity.

12.2 THE SPECTRAL RESPONSE OF THE HUMAN VISUAL SYSTEM

The spectral response of the human eye is common to most vertebrates, and extends over wavelengths from 400 to 700 nm as shown in Fig. 3.12 in Chapter 3. Figure 12.9 shows photopic spectral response as plotted against the solar radiation spectrum, where solar radiation is in watts per unit area per unit wavelength (The solar spectrum is shown in more detail in Fig. 4.3.).

From Fig. 12.9(a) the peak spectrum of the eye and that of the solar spectrum coincide nicely. However, since the eye responds to photons and not to irradiance, it would be better to use the plot in Fig. 12.9(b), which plots the solar spectrum in terms of photons per unit area per second per electron volt versus photon energy $h\nu$.

We can see from Fig. 12.9(b) that the photon energy most abundant is around 1 eV, but the eye is most efficient at counting 2.2 eV for both photopic and scotopic vision. If the human visual response peaked at 1.0 eV, the thermal excitations of the night sky, indicated by curve 3, would be visible to us and we would have very poor

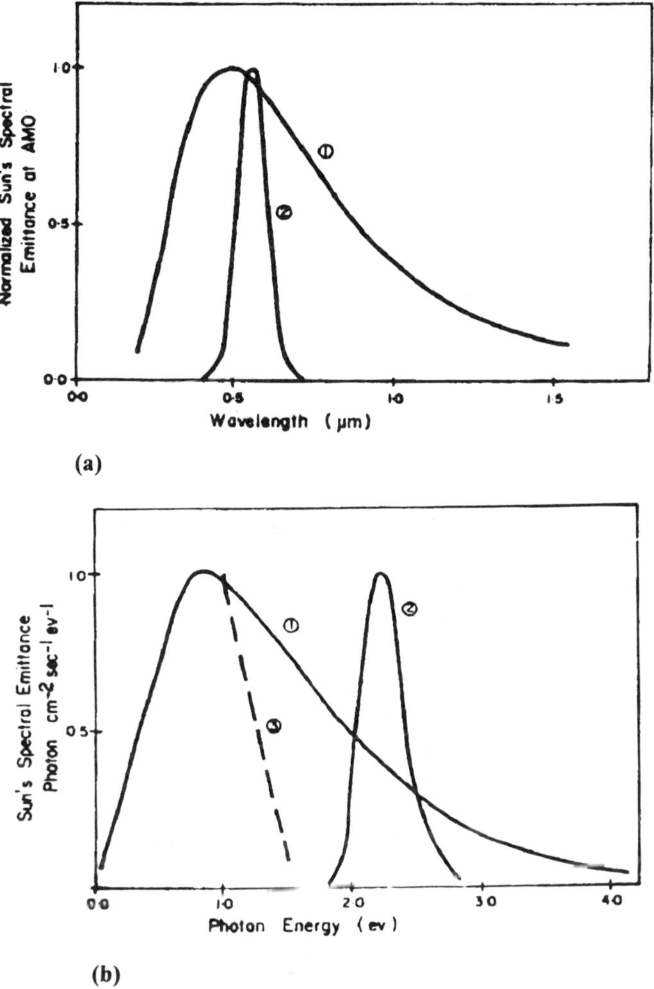

Fig. 12.9 (a) Spectral emittance of (1) sun and (2) eye. (b) (1) Spectral emittance of sun. (2) Spectral response of eye. (3) Spectral emittance of night sky after [12.7].

contrast at night because we would see arrays of flashes caused by such thermal excitations of the retina [12.7].

Although human visual spectral response peaks at around 511 and 556 nm for scotopic and photopic vision, respectively, as shown in an earlier chapter (Fig. 3.12), bright laser radiation has been known to be seen by observers at wavelengths below 312.5 nm and beyond 1 μm [12.7]. Radiation beyond these wavelength limits is strongly absorbed by protein of the eye lens and by liquid between the lens and retina, respectively. This means that the spectral response curves of Fig. 3.12 can essentially be extended on a log-log scale to these ultraviolet and infrared wavelength limits without changing the slope, thus permitting sufficiently bright ultraviolet and infrared radiation to be seen. For example, visual responsivity at 1.1 μm is about 10^{-11} of that at 556 nm. However, it is important to prevent eye damage in such cases because of the necessity of using high irradiances at such wavelengths. At such wavelengths, peripheral response becomes greater than foveal and the radiation appears colorless (rods only).

Quantum efficiency of the eye is typically on the order of 10%. This low value is attributed to reflections. Figure 12.5(b) indicates how quantum efficiency varies with wavelength for photopic vision.

In general, vision at levels below 10^{-4} millilambert (mL) is scotopic, while vision at 1 levels above 10^{-1} mL is photopic. The limit of perception of a point source of white light is about 0.1 to 0.2 $\mu lm \cdot m^{-2}$ for photopic vision and on the order of 0.004 $\mu lm \cdot m^{-2}$ for scotopic vision. For colored light the threshold is about the same as for white for photopic vision. However, for scotopic vision the bluer the light, the lower the threshold. For blue light the threshold is about 4×10^{-4} $\mu lm \cdot m^{-2}$ while for red light it is the same as for white light (about $4 \times$ 10-3 $\mu lm \cdot m^{-2}$). These are against a black background.

12.3 BRIGHTNESS CONSTANCY

The human visual system reacts not only to irradiance but also to relative irradiance. For example, a piece of white paper appears brighter than a piece of gray paper because it reflects more photons per second per unit area. If illumination on the piece of gray paper is increased such that it gives off the same number of photons per unit area per second as the white paper, common experience tells us that the paper will still be gray. Despite the fact that the white paper and the gray paper have the same irradiance, they do not look the same to us.

Consider the example in Fig. 12.10. Despite the fact that all four inner gray squares are the same absolute irradiance (they all are radiating the same number of photons) they appear quite different.

It seems that the apparent brightness of an object is a function too of the brightness of its surroundings, and not the number of photons coming from the object. The apparent brightness of an object depends not only on how many photons from the object are striking the retinal receptors, but also on how many photons from *around* the object are striking the retinal receptors. The strength of a signal coming from a receptor on the retina is proportional somehow to the strength of the signal from the receptors immediately surrounding it.

This phenomenon of relative brightness between two objects is called *brightness constancy*: The brightness of an object depends on its surroundings, rather than on absolute illumination.

There is a limit to brightness constancy, however, and it breaks down at very high contrasts. The ratio of reflectances of two objects needs to be no more than 9

Fig. 12.10 An example of simultaneous contrast. The small squares all have identical irradiances, but their apparent brightnesses are different because they lie on backgrounds that differ in intensity (after [12.4]).

or 10 to 1 in order for brightness constancy to break down. Since an ordinary sheet of white paper is 90% reflective and black paper is 10% reflective, brightness constancy holds for most ordinary objects but breaks down at very high contrasts [12.4].

12.4 VISUAL ACUITY

Visual acuity is a measure of the ability to recognize small and fine details. It is defined and measured in terms of the angular size of the smallest character that can be recognized. For example the letter *E* has three bars and two spaces. Visual acuity is then reciprocal of the angular size in minutes of arc of one element of the letter. If each element of the letter *E* subtends one minute of arc then visual acuity is unity. The height of the whole letter would then subtend five minutes of arc.

Acuity is often expressed as the ratio between target distance (usually 20 ft) and the distance at which one target element would subtend one minute of arc. Therefore, a visual acuity of 20/40 indicates one element of a letter of minimum recognizable size subtends 2 minutes while the letter itself subtends 10 minutes.

Normal acuity of one minute, or 20/20, is the value of resolution for the human visual system conventionally assumed for design of optical instruments. This is equivalent to 0.291-mrad resolution. However, this value pertains only to that part of the field of view which corresponds to the fovea. Outside the fovea acuity drops rapidly. For example, a displacement of 10 deg reduces acuity to only 20% of that with zero displacement. A displacement of only 2.5 deg is enough to reduce acuity by 50%.

Acuity also depends on illumination levels. As scene brightness diminishes, the iris opens wider and vision is determined mostly by cones rather than rods. One result of this process is a drop in visual acuity, as shown in Fig. 12.11. Normal illumination is about 1 mL. Also shown there is the effect of illumination on pupil size. Uniform illumination maximizes acuity.

Figure 12.12 shows how acuity is reduced according to reduced target contrast until a minimum is reached at around 2% contrast [12.8]. Smaller objects require greater contrast to be discernible.

Normally, visual acuity is given for white light. However, because of chromatic aberrations, it does display a wavelength dependence and improves slightly at yellow and yellow-green wavelengths where photopic response peaks. Acuity is slightly lower at red wavelengths and noticeably lower at short wavelengths. Reductions of 10% to 20% are typical in the blue and 20% to 30% in the violet [12.8]. There are simply fewer blue cones in the retina and particularly in the fovea. *Hence, while diffraction-limited aperture resolution improves at short wavelengths, that of the human visual system does not.*

Dynamic visual acuity differs from static. The human visual system is capable of detecting angular motion to the order of 10 seconds of arc. The slowest motion normally detectable is one to two minutes of arc per second. At the other extreme, motion faster than 200 deg/s appears as a blurred streak [12.8]. However, the wavelength dependence of dynamic visual acuity does not resemble that of static visual acuity. In dark-adapted viewing conditions and in the transition from photopic to scotopic vision moving blue targets are much more easily resolved than are other colors of equivalent photopic brightness. However, under dynamic photopic viewing conditions there is no clear dependence on target wavelength [12.9].

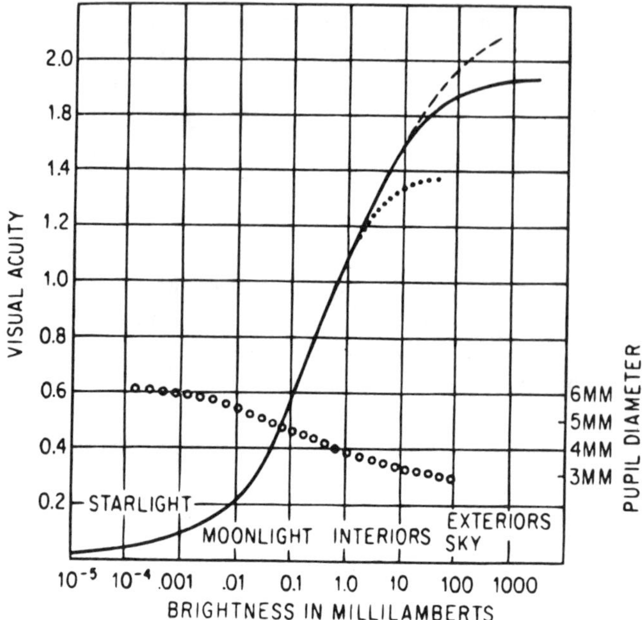

Fig. 12.11 Normalized visual acuity as a function of object brightness. The dashed and dotted lines show the effect of increased and decreased, respectively, surround brightness. The open circle curve indicates the diameter of the pupil; pupil diameters are larger in the young and smaller in the old, especially at lower brightness (after [12.8]). © 1991 The McGraw-Hill Cos.

12.5 MTF OF THE HUMAN VISUAL SYSTEM

Various techniques have been used to measure the spatial frequency dependence of the human visual system. Many effects such as accommodation and brightness constancy have to be considered, as well as the logarithmic rather than linear response of the human visual system to brightness. The relative MTF as measured by Davidson [12.4] is shown in Fig. 12.13 for photopic vision.

We can see from Fig. 12.13 that the human visual system MTF peaks at a spatial frequency of about 6 cycles·deg^{-1}. This is consistent with concepts of normal visual acuity. One minute of arc is 2.91×10^{-4} radians. Normal visual acuity, or 20/20 vision, requires a resolution angle five times larger, of 1.455×10^{-3} rad (0.083 deg). From Section 10.2, we have seen that an image plane detail $\Delta x'$ corresponds to a spatial frequency equal to $(2f_x)^{-1}$. Therefore, an image plane resolution angle $\Delta\alpha'$ corresponds to the angular spatial frequency $f_a = (2\Delta\alpha')^{-1}$. For normal visual acuity, the resolution angle of 0.083 deg corresponds to the spatial frequency $[2(0.083)]^{-1} = 6$ cycles·deg^{-1}. Therefore, it is not surprising that human visual MTF peaks at 6 cycles·deg^{-1} as shown in Fig. 12.13. On the other hand, as is shown in the next section, threshold contrast of the human visual system reaches a minimum normally at about 2 cycles·deg^{-1}, thus indicating that optimum sensitivity under low contrast conditions typically shifts to the lower frequency. This trend should not be completely surprising since it indicates that under low contrast conditions larger targets are easier to spot than small targets. However, Fig. 12.13 indicates that under high contrast conditions targets about one-third that size are easiest to spot. In other words, at minimum contrast targets subtending angles on the order of 4.365 mrad

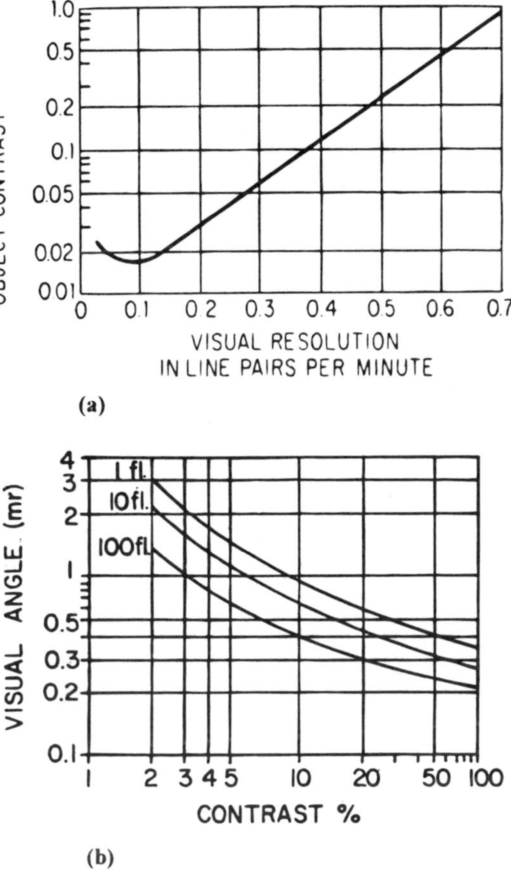

Fig. 12.12 (a) The object contrast ($\Delta H/H_{max}$) necessary for the eye to resolve a pattern of alternating bright and dark bars of equal width. For this plot bright bars had a brightness of H_{max} = 23 foot-lamberts, whereas ΔH varied with regard to dark bars (after [12.8]). (b) Visual acuity as a function of contrast for different background luminosities. © 1991 The McGraw-Hill Cos.

or $[2(2 \text{ cycles}\cdot\text{deg}^{-1})]^{-1} = 0.25$ deg are normally easiest to spot, while under high contrast conditions targets subtending angles on the order of 1.455 mrad or 0.083 deg are detected most easily.

The low photopic vision MTF values at low spatial frequencies in Fig. 12.13 are associated with inhibitory activity, described in Section 12.1.1, as if there is a negative MTF region in addition to a positive MTF that peaks at zero spatial frequency as is expected from a normal MTF. The sum of these positive and negative MTFs, as shown in Fig. 12.14, then yields the human visual MTF shown in Fig. 12.13. Physiologically, it is believed each ganglion cell acts as an impulse response with the surrounding larger area inhibiting it or yielding a negative impulse response as shown in Fig. 12.14. Generally, target acquisition models, such as that of the Night Vision Lab, do not include this inhibitory effect and assume eye MTF is constant at spatial frequencies from zero to 6 angles$\cdot$deg^{-1}. This is justified for scotopic vision, but not for photopic vision.

It has also been suggested that the MTF of the human visual system may not be a single continuous function but rather a composite of narrowband responses,

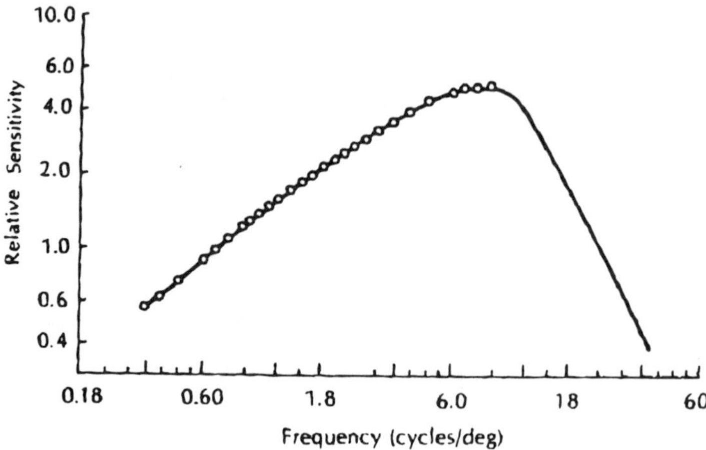

Fig. 12.13 Relative (unnormalized) MTF typical of the human visual system (after [12.4]). For normalization to maximum value of unity, relative sensitivities should be divided by 5.

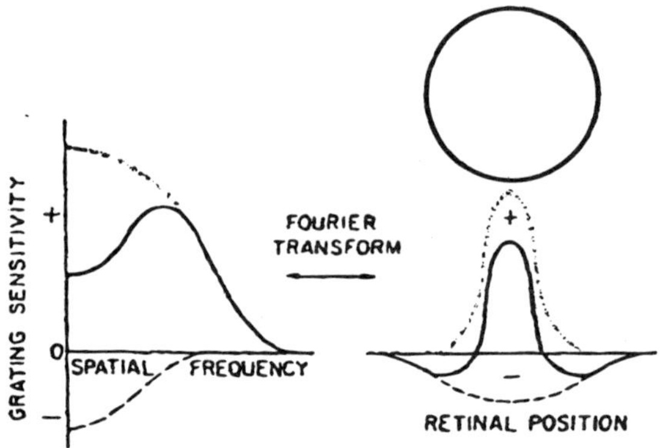

Fig. 12.14 Fourier transform of the MTF of the eye and that of a normal lens. Note that the point spread function for the eye indicates inhibitory activity, hence the negative region of the point spread function.

called *channels*, as shown in Fig. 12.15 [12.10] for photopic vision. In Fig. 12.15, *contrast sensitivity* is defined as the reciprocal of threshold contrast (see Fig. 12.17) for the human visual system. This measure of low contrast response peaks, as expected, at about 2 cycles·deg^{-1}.

Threshold contrast is measured with sine wave gratings displayed on a TV monitor. Contrast is decreased until threshold detection is obtained with large numbers of observers. These channels are almost independent of one another. High sensitivity for one channel does not mean high sensitivity for the others. Therefore, if in general 20/20 vision implies a certain visual sensitivity for 18 to 30 cycles·deg^{-1}, such acuity says nothing regarding channel sensitivity below 18 or above 30 cycles·deg^{-1}. This means that people with similar acuity, as manifested by reading the last line on eye charts, may react quite differently to targets of angular size below 18 or above 30 cycles·deg^{-1}. An example for three U.S. Air Force pilots is shown in Fig. 12.16.

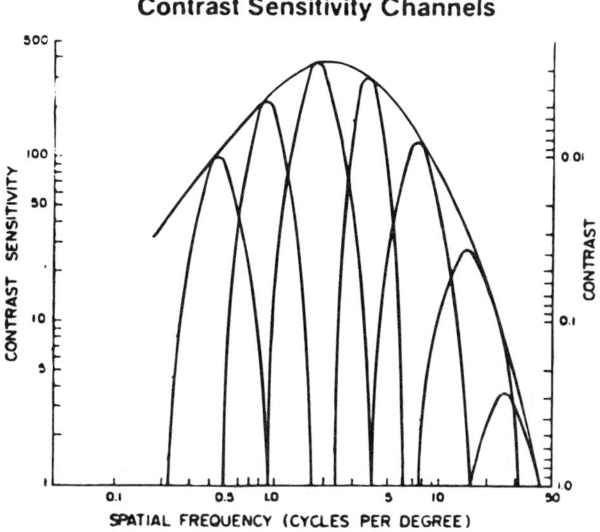

Fig. 12.15 Typical contrast sensitivity versus spatial frequency (after [12.10]).

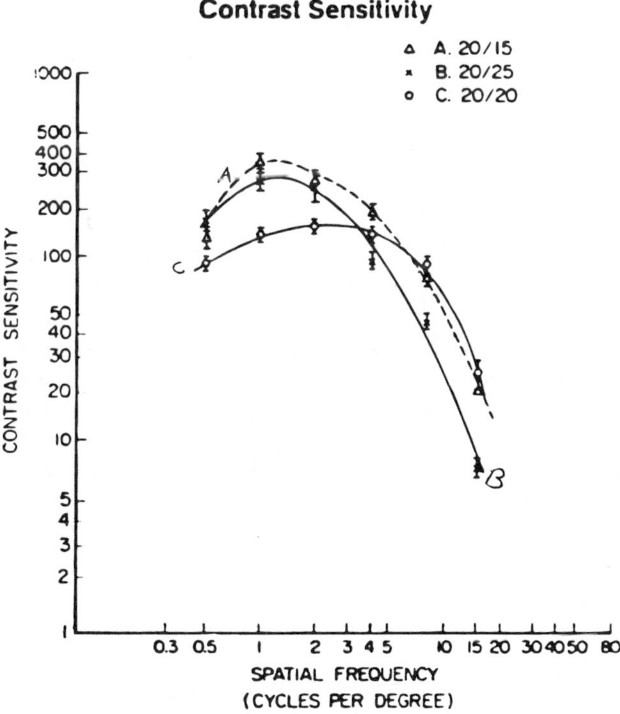

Fig. 12.16 Contrast sensitivities for three pilots (after [12.11]).

Although pilot B had lower visual acuity than the other two, his contrast sensitivity below 4 cycles·deg^{-1} is quite a bit higher than that of pilot C. Further testing indicated that contrast sensitivity (or threshold contrast) rather than visual acuity or eye MTF predicts a pilot's detection range [12.11], and that there is little correlation between acuity and contrast sensitivity. For example, two pilots displayed similar acuity (or eye MTF) under normal light conditions but exhibited peak contrast sensitivities differing by factors of 1.4 and 2.2 under normal and low light conditions. The pilot with higher contrast sensitivity (or lower threshold contrast) detected a MIG target at a distance 2.4 times greater than did his colleague. This difference in detection range led to target acquisition time differences of 21 and 10 s for clear and fog visibility conditions, respectively.

While contrast sensitivity correlated significantly with detection range, visual acuity did not. Average target acquisition difference in range and time for all visibility conditions was 2.2 miles and 56 s, which can be important [12.11]. This means that for *low contrast conditions threshold contrast or contrast sensitivity is much more important than eye MTF for the human visual system.* People with 20/20 vision see high contrast targets equally well, but can perform significantly differently with low contrast targets at all spatial frequencies.

12.6 THRESHOLD CONTRAST CURVE

The purpose of measuring the MTF of the human visual system is to obtain a measure as to how well an optical system will perform when it interacts with our eyes. The output of the optical system will be an input into the eye, and the MTF of the eye will be the final stage in the visual process. In other words, it does not matter if there is a great deal of small detail in a photograph; what matters is if the human visual system can see it.

The minimum contrast that must be present at each spatial frequency in order for information content at that spatial frequency to be observed is minimum threshold contrast, as described in Section 10.1. A typical such function for the human visual system is shown in Fig. 12.17. Examples for specific people can be obtained by taking the reciprocals of the curves in Fig. 12.16. Figure 12.17 is often called the *J curve.* Since in measuring and developing the J curve, the human visual system MTF has already been included, the system MTF should not be multiplied by the human visual system MTF when the J curve describes threshold contrast and its intersection with system MTF defines $f_{r\,\max}$ as in Fig. 10.1(a). Data for Fig. 12.17 were obtained by showing subjects different spatial frequencies, and changing the percent modulation for each frequency until the subject could just barely distinguish a grating. This study in threshold sensitivity was completed by Van Ness and Bouman in 1965, and the data still stand as valid [12.4].

An analytical approximation [12.12] for the "average" threshold curve in Fig. 12.17 which is fairly accurate for higher spatial frequencies is

$$C_{\text{threshold}}(f_a) = \frac{\nu_e}{\exp(-c_1 f_a) - \exp(-c_2 f_a)} \tag{12.6.1}$$

where $\nu_e = 0.001033$, $c_1 = 0.1138°$, $c_2 = 0.3250°$, and f_a is in cycles·deg^{-1}. This is shown in Fig. 12.18 on a scale of cycles·rad^{-1}.

Figures 12.17 and 12.18 refer to photopic vision, which is relevant to bright displays. However, threshold contrast does change with mean level of retinal illumi-

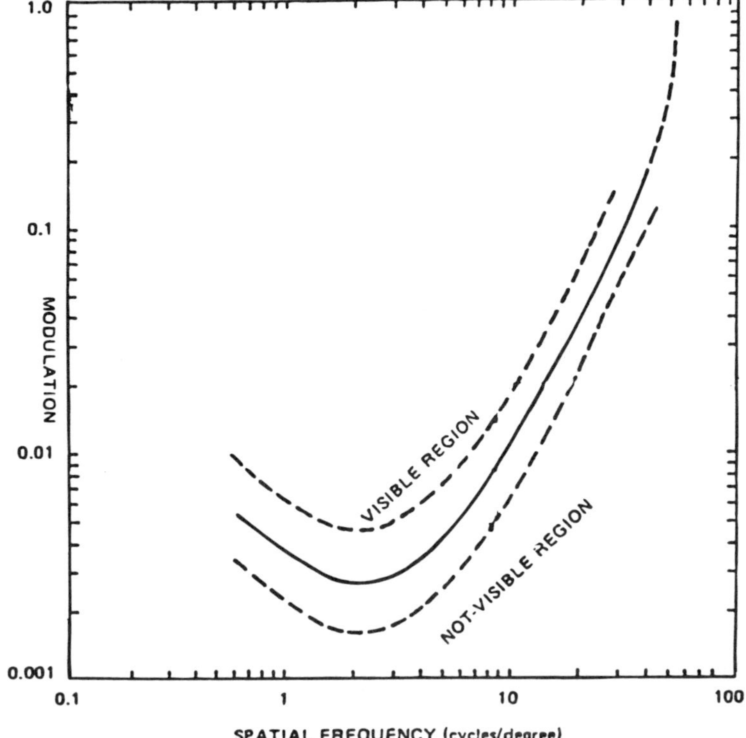

Fig. 12.17 Threshold contrast curves for the human visual system. The broken lines define limits estimated to be applicable to 90% of the population (after [12.4]). Luminance $\geq 50\ \text{cd}\cdot\text{m}^{-2}$ or 15 foot-lamberts.

nance, as shown in Fig. 12.19. Figure 12.19 shows how threshold contrast varies for the transition from photopic to scotopic vision. The inhibitory effect that causes maximum sensitivity to be at nonzero spatial frequency gradually disappears as illumination level decreases. As retinal luminance decreases, the threshold contrast not only rises but also exhibits a shift of peak sensitivity to lower spatial frequencies, that is, at medium and low luminances image detail at minimum contrast must increase in both size and contrast to be seen by the human visual system.

To evaluate an optical system as a whole, the MTF of the whole imaging system must be coupled with the threshold contrast function required at the output of the imager in order to get a complete picture of how the optical system will behave. In Fig. 10.1(a) the MTF of a typical optical system is plotted against the required threshold function of the system. The area in between the top curve (system MTF), and the threshold detectability curve is called the *MTFA* and represents the space of all frequencies that can be detected by the combination of imaging system (top curve) and observer or machine at system output (bottom curve).

It is clear that the effective bandwidth of an imaging system is limited by the bandwidth of the human or machine receiver of the image. As shown in Fig. 12.17, for human vision it is difficult to absorb information at spatial frequencies greater than about 53 cycles$\cdot$deg^{-1}, ($\approx$3037 cycles$\cdot$rad^{-1}). When a high-resolution sensor such as the 4-in. return beam vidicon ($\approx$10,000 TV lines) is used, as is often the case with satellites, normal size images can contain detail too fine to be observed

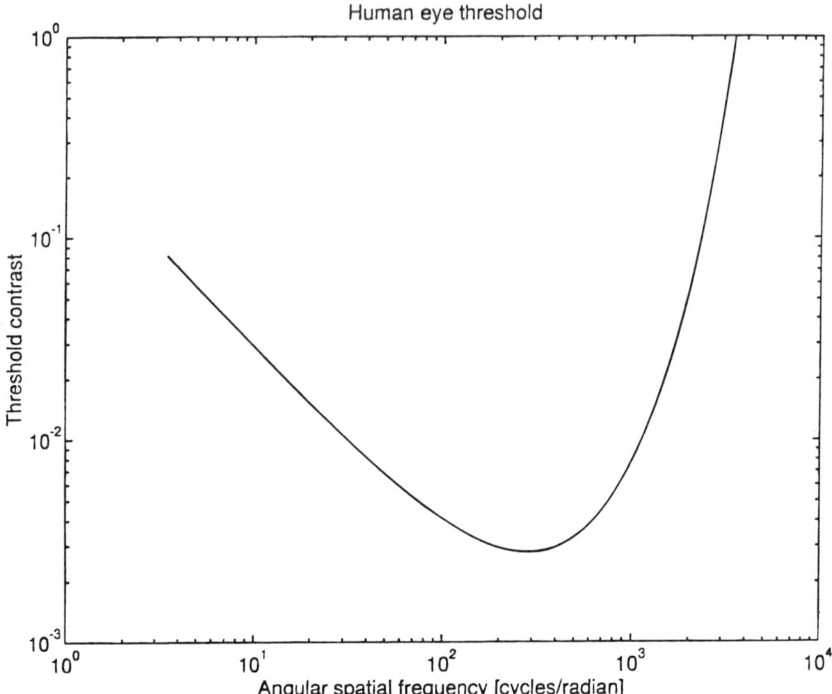

Fig. 12.18 Plot of Eq. (12.6.1).

with human vision. To see such detail, the images must be enlarged considerably. Angular magnification can decrease spatial frequencies toward those of reduced threshold contrast around the minimum of the J curve.

Eye response curves are valid only over a limited field of view at the eyes, which is essentially the angle occupied by the fovea for photopic vision. From Fig. 12.2(a) this is about ± 10 deg in azimuth from the visual axis and ± 7.5 deg in elevation, yielding an *aspect ratio* of 4/3. Thus, useful sensor field of view (FOV) when displayed is limited to 20×15 deg. As magnification is increased, usable FOV decreases. However, this may often be worthwhile in order to locate angular spatial frequencies of fine detail around the J curve minimum.

Example 12.1

An image subtending $\Delta\alpha' = 1$ mrad at the observer's eye is to be detected with a background luminance of 15 foot-lamberts. What is the optimum angular spatial frequency required for detection and what is the minimum system MTF required at that spatial frequency? What magnification should be used in order to obtain minimum threshold contrast for the average observer at the spatial frequencies of interest?

Solution

The target image angular size corresponds to the angular spatial frequency $f_a = (2\Delta\alpha')^{-1} = (2 \times 10^{-3})^{-1} = 500$ cycles·rad^{-1} or 500 $(\pi/180) = 8.73$ cycles·deg^{-1} based on Section 10.4. From the J curve in Fig. 12.17 an image plane modulation contrast of at least 0.009 is required for the average observer at that spatial frequency.

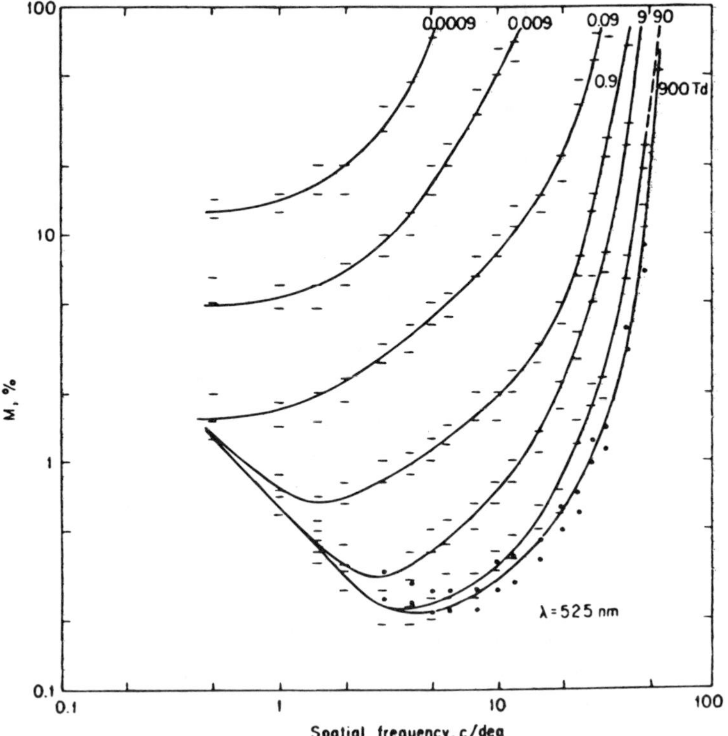

Fig. 12.19 Threshold modulation transfer functions for green light (λ = 525 nm) at seven mean levels of retinal illuminance (2.0-mm-diameter artificial pupil). The short horizontal lines denote the modulations at which the frequencies in question were perceived or not perceived. (These are the average results of three measurements.) For 900 Trolands (Td) retinal luminance they are replaced by dots. No difference was found between the modulation transfer function for 900 Td and that for 5900 Td. (Results of van Nes, 1968, in [12.4].) 1 Troland of retiral luminance equals 4.44×10^{-15} lumens · mm^{-2} for an eye primary focal length of 15 mm.

The J curve reaches a minimum of 0.003 modulation contrast at spatial frequencies slightly larger than 2 cycles·deg^{-1}. Therefore, an *angular* magnification of about 8.73/2 = 4.4 would optimize low contrast detection probability for the objects of interest.

12.7 MOVING IMAGES

The fovea integrates all received radiation at each point within a time interval of about 0.1 s. For shorter exposures, more contrast is required to achieve the same probability of detection, recognition, or identification. The exposure time presented by a particular object can be considered as the length of time the image of that object remains on any one point of the retina. When there is relative motion between a displayed target and the eye, effective exposure time is the target angular size divided by the angular rate of motion. If effective exposure time is less than the eye integration time of 100 ms, then motion effects must be considered. Examples showing variation of threshold contrast with brightness and exposure time are shown in Fig. 12.20. If, for example, threshold contrast for a stationary object is 10%, then threshold contrast for effective exposure time of 50 ms is 20%, for 20 ms it is 50%, etc. The product

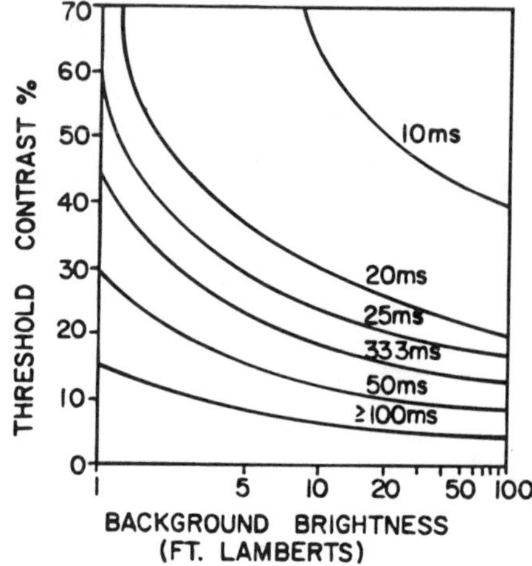

Fig. 12.20 Threshold contrast versus background brightness for various exposure times.

of exposure time and required threshold contrast for effective exposures of less than 100 ms is a constant. As a result, when system MTF is multiplied by image motion MTF (Part 4), $f_{a\ \max}$ is often affected very significantly.

12.8 NEARSIGHTEDNESS AND FARSIGHTEDNESS

Myopia or *nearsightedness* results from an excessive amount of positive power in the eye. The image of a distant object falls in front of rather than on the retina and cannot be focused sharply. It is corrected by placing a negative lens before the eye so that the resulting divergence spreads the incoming beam sufficiently so that the lens of the eye can then focus it on the retina. A person with 1 diopter of myopia cannot see clearly beyond 1 m, and a −1 diopter lens is used to correct this. Myopia often coincides with adolescence when growth is rapid.

Hyperopia or *farsightedness* is the reverse of myopia and occurs when the eye lens exhibits too long a focal length. This causes images to be formed behind rather than on the retina. It can be corrected by placing a positive or converging lens before the eye, thus decreasing the effective focal length of the eye.

REFERENCES

12.1. C. H. Graham et al., *Vision and Visual Perception*, Wiley, New York, 1966. Reprinted by permission of John Wiley & Sons, Inc.

12.2. R. Held and W. Richards, "Perception: mechanisms and models," in *Readings from Scientific American*, W. H. Freeman, San Francisco, 1972.

12.3. E. Hecht, *Optics*, Addison-Wesley, Reading, MA, 1987.

12.4. T. N. Cornsweet, *Visual Perception*, Academic Press, New York, 1970.

12.5. A. Rose, *Vision: Human and Electronic*, Plenum Press, New York, 1973.

12.6. J. L. Schnapf and D. A. Baylor, "How photoreceptor cells respond to light," *Sci. Am.*, April 1987.

12.7. N. Ben-Yosef and A. Rose, "Spectral response of the human eye," *J. Opt. Soc. Am.*, Vol. 68(7), 1978, pp. 935–936.

12.8. W. J. Smith, *Modern Optical Engineering* (2nd ed.), McGraw-Hill, New York, 1991.
12.9. G. M. Long and P. M. Garvey, "The effects of target wavelength on dynamic visual acuity under photopic and scotopic viewing," *Human Factors*, Vol. 30, 1988, pp 3–13.
12.10. A. P. Ginsburg, "Specifying relevant spatial information for image evaluation and display design: an explanation of how we see certain objects," *Proc. SID*, Vol. 21(3), 1980, pp. 219–227.
12.11. Arthur P. Ginsburg, et al., "Contrast sensitivity: relating visual capability to performance," *USAF Medical Service Digest*, Summer 1983, pp. 15–19. Permission for reprint, courtesy Society for Information Display.
12.12. T. J. Schulze, "A procedure for calculating the resolution of electro-optical systems," in *Airborne Reconnaissance XIV*, P. A. Henkel, F. R. La Gesse, and W. W. Schuster, Eds., *Proc. SPIE*, Vol. 1342, 1990, pp. 317–327.

EXERCISES

12.1 What magnification is necessary to enable a person with "normal" visual acuity to read letters 1 mm high at a distance of 250 m, neglecting atmospheric blur?

12.2 What magnification in Exercise 12.1 would be optimum using as a criterion contrast threshold of the "normal" photopic human visual system?

12.3 Explain the discrepancy between the answers to Exercises 12.1 and 12.2. Which answer is appropriate for high contrast objects with very little path luminance and large spatial frequency bandwidth of the imaging equipment?

12.4 A 1-in.-diameter CCD camera system is to be used with a 1-m focal length telescope to view a 2- × 2-m object 5 km away. Assuming low contrast because of (background) clutter, what would be the best observer distance from a 12- × 9-in. display so as to optimize target acquisition probability for normal human vision? What would happen for no clutter? Assume background luminance of 15 foot-lamberts in the display room.

12.5 A nearsighted person who cannot focus clearly on an object more than 20 cm away is in need of corrective lenses. What power lenses should be considered for his glasses?

CHAPTER

13

Imaging Devices

The previous chapters have considered system resolution and, in particular, the size of object detail resolvable in the image. The basic approach has been based on system optical transfer function (OTF), or its modulation transfer function (MTF) approximation. A real system includes hardware, particularly the imager or camera and accompanying display, environmental effects such as motion and vibration, and channel effects such as atmospheric turbulence and particulate light scattering and absorption. Each phenomenon can be characterized by an MTF, which can be incorporated into the resolution models of the previous chapters.

This chapter is concerned with imagers or cameras, particularly charge-coupled devices (CCDs). The following two parts are devoted to environmental and channel MTF phenomena, respectively.

13.1 INTRODUCTION

Images can be sensed in different ways. Film cameras, for example, sense the entire image plane simultaneously. Television cameras sample the image. As such, TV systems must obey the *Nyquist criterion* that the spatial sampling frequency f_s be equal to at least the total spatial frequency bandwidth of the image BW_s, that is,

$$f_s > BW_s \tag{13.1.1}$$

In Eq. (13.1.1) BW_s is equal to twice the positive spatial-frequency bandwidth if the image spectrum is symmetrical. If Eq. (13.1.1) does not hold, and the sampling frequency is too low, there is an overlap between the high and low spatial-frequency tails and image distortion takes place. High spatial-frequency detail then appears at low spatial frequencies. This is called *aliasing*.

TV cameras typically sample the image in the vertical direction. In North America, 525 horizontal lines are used to sample the image vertically according to specifications developed by the National Television Systems Committee (NTSC) in the United States. In Europe, the Middle East, and most of the rest of the world, images are sampled by TV systems over 625 horizontal lines, compatible with PAL (Phase Alternation Line Rate) and SECAM (Sequential Couleur avec Memoire) systems. These differences are associated with line frequencies (60 and 50 Hz). In North

America 30 images or frames per second are obtained; in PAL and SECAM systems, the frame rate is 25 per second. To obtain images that seem continuous in time rather than discrete, there are two fields per frame, so that the eye sees 60 images per frame in North America and 50 images per frame in PAL and SECAM systems. The dead space between horizontal sampling lines limits resolution in the vertical direction. Spot wobble in the display is often used to make the TV image appear more continuous in the vertical direction, but in terms of actual detail this is illusory since no new detail between lines is added. Such sampling processes are called *raster scan*, and the image is read out sequentially line by line. A field is one of two scans interlaced line by line so as to make up a frame. This means that each field contains an image sampled by alternating horizontal lines. One field usually consists of even TV lines, the other of odd TV lines. Because of the need for synchronization pulses and limited commercial TV bandwidth (6 MHz), the number of active raster lines is reduced by another 7.6% of the 6 MHz per TV channel. The video portion is 4.2 to 4.5 MHz.

Horizontal resolution is determined by system MTF and is usually characterized by the number of TV lines or line pairs resolvable per sensor width. It is more convenient to use TV lines or line pairs than spatial-frequency bandwidth since the latter depends on image dimension, which, in turn, varies with display dimension, whereas the number of horizontal TV lines or line pairs is usually constant for a given camera.

At present *high-definition television* (HDTV) is being considered commercially and is likely to be available in the not too distant future. The HDTV image has approximately twice as much definition horizontally and vertically as do the 525-line NTSC and 625-line PAL and SECAM systems. The total number of pixels is therefore about four times as great, and the wider screen adds about one-third more. Typically there are about 1920 pixels/line.

The increased vertical resolution is obtained by employing more than 1000 lines in the scanning patterns. Values currently proposed are 1050, 1125, and 1250 sampling lines. The increased horizontal detail and color require a video or temporal bandwidth from six to eight times greater than that used in conventional TV [13.1].

Because HDTV images are to be about 33% wider than in conventional TV, the aspect ratio (image width to height) increases from 4/3 to 16/9. This follows a trend of motion pictures from a ratio of 4/3 prior to 1953 to very wide screen formats [13.1].

TV cameras developed initially were based on electron tubes as the imager. A wide variety of such electron image tubes have developed, including orthicons, vidicons, plumbicons, epicons, newvicons, etc., each with different spectral responses, dynamic ranges, gains, sensitivities, etc., as summarized in the literature [13.2–13.4]. Examples of spectral responsivities for some tubes are shown in Fig. 13.1.

In recent years, semiconductor devices such as CCDs have been developed with improved resolution and are gradually supplanting electron tube image cameras. The advantages of CCDs include very small size, very small weight, very small power consumption, and greater ease in implementing digital image processing. Their effect on TV cameras is gradually becoming similar to that of transistor radios on electron tube radios. Hence, in view of the limited scope of this chapter, we will concentrate here on semiconductor imagers.

13.2 SEMICONDUCTOR IMAGE SENSORS

The CCD was developed in 1970 [13.6]. It can be used in purely electronic applications such as shift registers but is used most commonly for imaging.

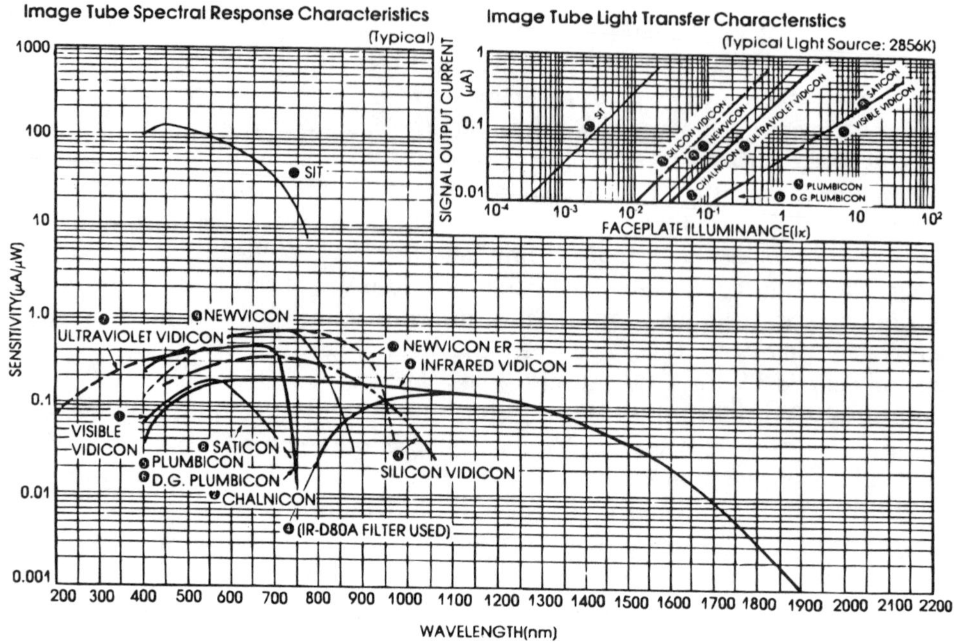

Fig. 13.1 Spectral responsivities of typical TV electron image tubes (from [13.5]).

The basic component of a CCD is the metal-oxide semiconductor (MOS) capacitor shown in Fig. 13.2. Its purposes are photodetection, charge storage, and information transfer. Arrays of such MOS capacitors form a CCD, in which each such capacitative device gives rise to an individual pixel in the overall CCD-sensed image.

In Fig. 13.2, a conducting electrode is separated by a thin insulating layer of SiO_2 from the *p*-doped silicon substrate. Application of a positive voltage V_g to the electrode generates a depletion layer in the substrate, typically a few micrometers thickness. Although the depletion region is characterized by an electric field, the remaining part of the substrate is field free. As described in Section 6.2.1.2, photons absorbed in the silicon create electron–hole pairs. If the absorption occurs within the depletion region (case A in Fig. 13.2) the electron and hole are separated by the electric field. The hole is attracted downward to the base electrode while the electron is trapped at the SiO_2/Si interface. Each trapped electron reduces the depletion region.

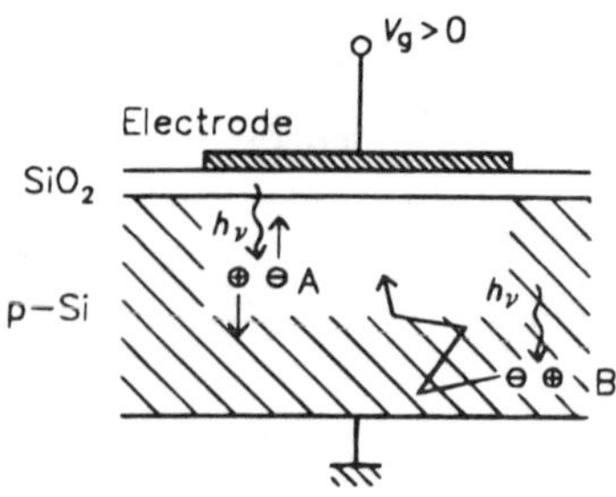

Fig. 13.2 MOS capacitor as the basic building block for a CCD.

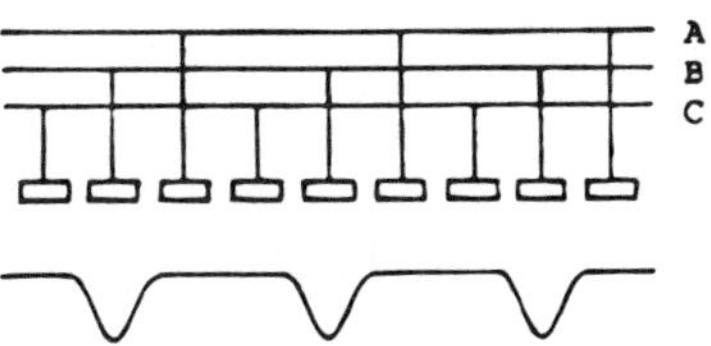

Fig. 13.3 CCD potential wells formed by a high voltage on line B.

Such photoelectrons fill potential wells, with electron content of a well equal to the time and space integral of the incident radiation as modified by the quantum efficiency of the CCD material. A "full well" occurs when the total charge at the SiO_2/Si interface corresponds to the positive charge at the electrode. Any further absorbed photons can still generate electron-hole pairs but they either recombine or the electron is trapped at a different site, such as the potential well of a neighboring yet "unfilled" MOS capacitor.

Photons absorbed outside the depletion region (case B in Fig. 13.2) give rise to electron-hole pairs that diffuse randomly. They may recombine before reaching the depletion region and thus do not contribute to the detected signal. (This is similar to what occurs in photodiodes as described in Section 6.2.1.2 with regard to spectral response).

To perform the readout or information transfer, the array of MOS capacitors forming the CCD is connected in a three-phase arrangement to bus lines A, B, and C as shown in Fig. 13.3. In the figure, line B has a higher voltage than lines A and C, thus giving rise to one potential well in the center of each pixel. A clocking scheme controls the signal current. During the charge integration interval or exposure time, electrons are collected and stored in the potential wells. During the charge transfer interval, photoelectrons are transferred to the CCD output. By attaching bus lines to different phases of a high-frequency clock, potential wells and their charge packets are quickly moved across the CCD as shown in Fig. 13.4.

The repetitive sequence of voltage corresponds to a shifting potential well. The individual stored charge packets follow the electrode of highest potential. The efficiency of charge transfer is extremely high. However, *crosstalk* can result from photons absorbed outside the depletion region.

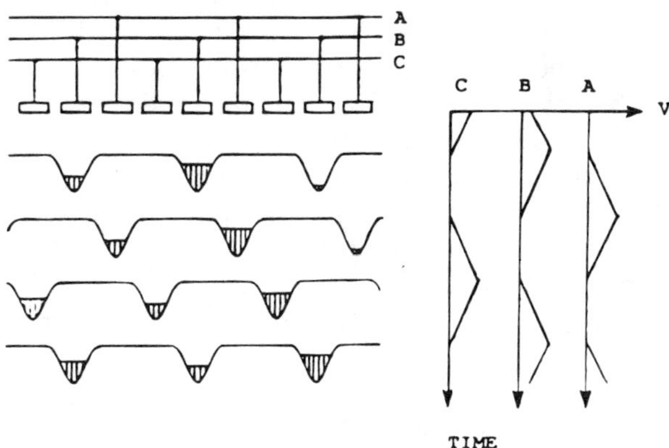

Fig. 13.4 Charge transfer process.

Two of the more common geometries used in CCDs are *frame transfer* (FT CCD) and *interline transfer* (IT CCD). In the former, pixel width and spacing between pixel centers (pixel pitch) are approximately equal along both array axes. IT CCDs have approximately equal pixel width and pitch along one CCD axis, but have pixel width equal approximately to half the pixel pitch along the other axis [13.7].

13.2.1 Noise

Temporal noise, described in Chapter 5, is the variation of signal from a given pixel in successive images and is defined there as the variance of the signal. Its main sources are temporal photon statistics, that is, quantum noise, dark current shot noise, and readout noise.

Spatial noise defined in Eq. (11.1.15) is the random variation over many pixels in a single image within a uniformly illuminated field. It derives from spatial photon statistics, readout noise, and fabrication tolerances or fixed-pattern noise. For example, variations in pixel size are a source of spatial noise, which increases proportionally to signal. Variations in dark current from pixel to pixel, resulting from slight variations in material properties, may be another source of fixed-pattern noise. In principle, all fixed-pattern noise can be removed by calibration and computation using a digital frame store [13.8].

As shown in Eq. (11.1.16), temporal and spatial noise add in quadrature.

As shown previously in Exercise 11.1 and here in Exercise 13.3, for low light levels temporal noise dominates spatial noise. In such cases, high detector responsivity or D^* or low NEP are important, particularly for limited atmospheric transmittance imaging. However, for higher incident light levels, signal-to-noise ratio (SNR) is limited by spatial nonuniformity.

13.2.2 Dynamic Range and Grayscale

Dynamic range is defined as the ratio of maximum output (*saturation*) signal over dark current noise. This determines the number of gray levels. Shades of gray represent steps of $\sqrt{2}$ in pixel luminances. Two shades of gray require a dynamic range of $(\sqrt{2})^2 = 2{:}1$, whereas eight shades of gray require a dynamic range of $(\sqrt{2})^8 = 16{:}1$. The range of available gray levels is called *grayscale*. In an 8-bit system the grayscale contains values from 0 to 255. This corresponds to 16 shades of gray.

The maximum number of electrons a pixel can hold is determined by the size of the depletion region and is proportional to pixel area. Typical saturation values are 10^4 electrons$\cdot\mu\text{m}^{-2}$. Both dark current noise and dynamic range increase with pixel dimension. Therefore, for large dynamic range, devices with large pixel area are required. This conflicts with resolution requirements as described by pixel MTF in Section 9.7.1 which require small pixel size.

Typical behavior of signal and noise as functions of incident light exposure is shown in Fig. 13.5. At low light levels temporal noise dominates. Within the linear range the expected square root dependence of quantum temporal noise is observed. At high illumination levels spatial noise dominates.

Adequate picture quality for visual inspection requires a dynamic range of at least 50 [13.9].

13.2.3 Modulation Transfer Function

The pixel MTF, which is a limit to best possible resolution, was described in Section 9.7.1. Exercises using it are given at the end of Chapters 9 and 10. The basic concept

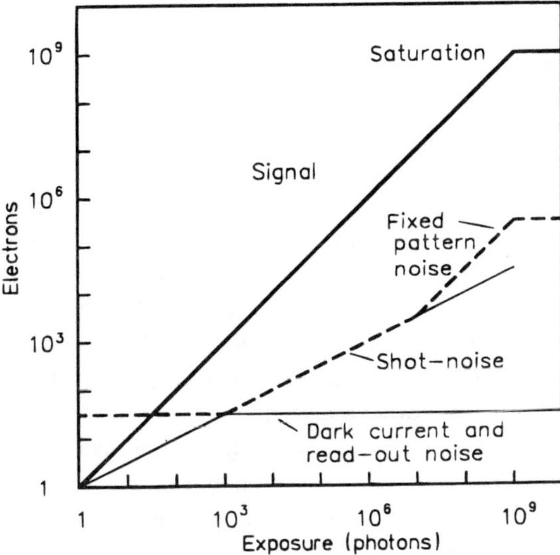

Fig. 13.5 Signal and noise electrons as functions of exposure for quantum efficiency of unity (signal equal to exposure until saturation).

is that under ideal conditions, and very little dead space between pixels (FT CCD), impulse response is one pixel. Hence, pixel optical transfer function is simply the Fourier transform of pixel shape. A rectangular pixel of dimensions $2a \times 2b$ yields a pixel OTF equal to $\text{sinc}2\pi f_x a\ \text{sinc}2\pi f_y b$. The smaller the pixel dimensions, the larger the pixel bandwidths of $1/a$ and $1/b$ and the better the resolution. The OTF has to be cutoff at the frequencies $-1/a$ and $1/a$ in the ω_x direction and $-1/b$ and $1/b$ in the ω_y direction since higher radian spatial frequencies would imply impulse response to be over an area smaller than pixel dimensions $2a \times 2b$. That would be false resolution that physically does not exist.

Here, qualifications to the pixel OTF and MTF are presented:

1. OTF theory assumes linear space invariance. TV cameras in general, including CCDs, are not space invariant because the image is sampled at discrete locations in the image plane. The image therefore is not continuous. Hence, periodic bar images of equal irradiance and size as shown in Fig. 13.6 can be sampled unequally by different pixels. An image of a sufficiently narrow bar may even fall between pixels and not be sampled, or fall on part of a pixel and be sampled. Hence, there is no unique CCD OTF or MTF, particularly at sufficiently high

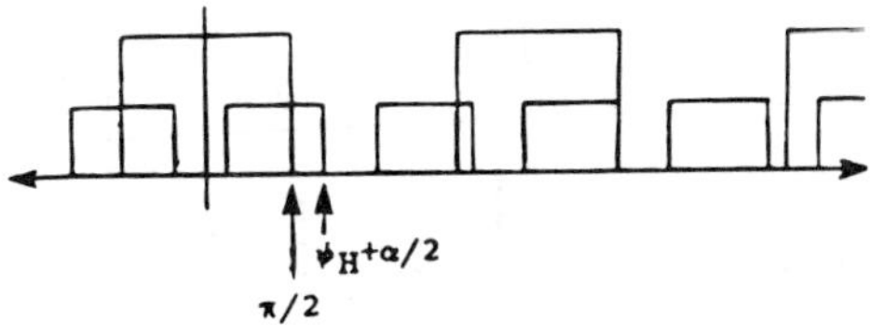

Fig. 13.6 Images of periodic bars of equal irradiance and size (upper pattern) sampled by pixels (lower pattern).

spatial frequencies. Nevertheless, detailed approximations have been developed [13.7]. In any event, to prevent aliasing of higher spatial frequency content to lower spatial frequencies, CCD OTF and MTF should be limited to the Nyquist spatial frequency limit of $1/(2p)$, where p is pixel pitch or distance between adjacent pixel centers. This can be done electronically or optically with appropriate MTF [13.9]. The Nyquist limit can cause point spread function to be considerably larger than a pixel (see Exercise 13.2).

2. In derivation of the pixel OTF or MTF it is assumed that the whole pixel responds uniformly to the incident light so that for a rectangular pixel, for example, impulse response is a rectangle identical with pixel shape. In a real CCD, however, several effects distort this convenient concept. Absorption and scattering of light by additional structures (such as electrodes), diffusion of photogenerated charge carriers, and readout imperfections result in a smearing of the pixel point spread function as shown in Fig. 13.7. The term *fill factor* is often used to describe the percent of usable pixel area. Diffusion MTF is considered below. Such nonideal behavior may also be characterized in terms of crosstalk, defined as the signal generated in neighboring pixels while illuminating exclusively one pixel only. Referring to Fig. 13.7, crosstalk corresponds to the area below the response curve which lies outside the pixel. Crosstalk of 30% to 50% is typical for a 10-μm pixel, for example [13.9].
3. Pixel OTF or MTF is more dominant at shorter wavelengths, but less so at longer wavelengths. The reason can be ascertained by referring again to Fig. 13.2 and considering the wavelength dependence of the silicon absorption coefficient. As explained in Section 6.2.1.2 with regard to photodiode spectral response and quantum efficiency, longer wavelength or lower energy photons penetrate longer distances into silicon before being absorbed. Hence, the low absorption coefficient at longer wavelengths ($\lambda > 800$ nm for visible wave CCDs) implies deeper concentration of incident radiation into the silicon bulk beyond the depletion region into the field-free region B in Fig. 13.2. Hence, at longer wavelengths there is more signal current carrier diffusion and degraded resolution. This results in an often noticeably wavelength-dependent MTF. In general, CCD diffusion MTF can be modeled as [13.10]

$$\mathrm{MTF}_{\mathrm{diff}} = \frac{1 - [\exp(-\alpha L_{depl})/(1 + \alpha L(f))]}{1 - [\exp(-\alpha L_{depl})/(1 + \alpha L_{diff})]} \tag{13.2.1}$$

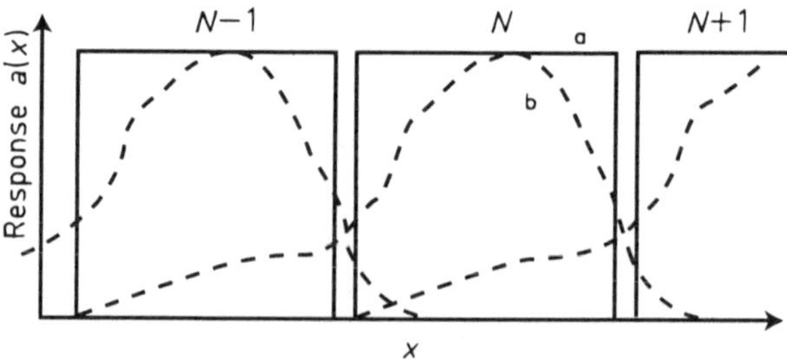

Fig. 13.7 Typical pixel response functions (dashed line) for pixels $N-1$, N, and $N+1$. Solid line represents ideal response within active pixel area (after [13.9]).

where α is silicon absorption coefficient at wavelength λ, L_{depl} is depletion width, L_{diff} is diffusion length of minority carriers such as electrons in p-type Si, and L is effective diffusion length, which is spatial frequency dependent and given by [13.10]

$$L(f) = \frac{L_{diff}}{\sqrt{1 + 4\pi^2 L_{diff}^2 f^2}} \tag{13.2.2}$$

Diffusion length is

$$L_{diff} = \sqrt{D\tau} \tag{13.2.3}$$

where τ is carrier lifetime and D is diffusion constant (0.0031 $m^2 \cdot s^{-1}$ for room-temperature silicon).

At short wavelengths where α is large, the image plane irradiance is absorbed near the surface and most of the photoelectrons are collected in depletion region A in Fig. 13.2. Because there is little diffusion $MTF_{diff} \approx 1$. However at wavelengths of more than 800 nm for visible-wave devices, the device MTF may often be affected strongly by Eq. (13.2.1), as illustrated in Exercise 13.4. The depletion region width determines the wavelength at which intermediate results between pixel and diffusion MTF extremes may be obtained.

4. If dead space is not negligible, then pixel MTF should be multiplied by $\text{sinc}\pi f_x p_x \ \text{sinc}\pi f_y p_y$ where p_x, p_y are pixel pitch in the x and y directions, respectively.

13.2.4 CCDs for the Ultraviolet

As described in Section 6.2.1.2, the spectral response of Si detectors is limited at shorter wavelengths by surface recombination losses. A method of extending CCDs into the ultraviolet is to use a thin sheet of fluorescent material on the surface of the device. The incident UV photons are of higher energy than visible wavelength photons and their absorption by the fluorescent layer gives rise to visible photons for CCD detection and imaging. For the near UV, plastic materials have been used with a quantum efficiency on the order of 25% or more. At shorter wavelengths down to soft x-rays, crystalline scintillator materials yield similar effects. For very short wavelengths below 1 μm (x-rays), Si is again well suited as the photodetector since photons can penetrate sufficiently into the active region. Because one high-energy photon can generate several electron–hole pairs, a quantum efficiency greater than unity is possible.

13.2.5 Thermal Imaging Focal Plane Arrays

Because of its band gap E_g the long-wavelength cutoff of Si is about 1.1 μm. For longer wavelengths more elaborate schemes must be used. A major effort has been concentrated on hybrid structures using low band-gap materials such as HgCdTe instead of Si to detect the radiation and then connecting such photodiodes via metal bonds to standard Si CCD or MOS readout structures. These are for long-wave (8-

to 14-μm wavelength) thermal imaging where previously only one or only a few detectors were used in mechanical scan schemes in order to sample the entire image.

The HgCdTe FPAs do have high D^* but also relatively high nonuniformity [13.11]. They are advantageous in low photon flux situations but not for low contrast because of their spatial noise.

For short-wave (3- to 5-μm) thermal imaging, platinum silicide (PtSi-Si) Schottky barriers are used as the detecting element, as described at the end of Section 6.2.1.2. Photons enter the device through the Si substrate and are not absorbed because of their low energy. Instead they are absorbed in the metallic PtSi. An excited charge carrier may cross the boundary into the Si and become trapped in the depletion region near the boundary. Subsequently, the charge can be read with the normal CCD techniques.

The disadvantages of such devices is their low quantum efficiency (a few percent) at these longer wavelengths. This can be important for longer range imaging where the image plane irradiance is reduced according to atmospheric transmittance (but not according to geometric radiometry as shown in Section 3.4). However, at longer wavelengths the photon rate ($P/h\nu$) increases, that is, there are more available photons for a given irradiance. Hence, despite the low quantum efficiency, platinum silicon devices are often quite successful even for long range imaging.

The advantage of PtSi-Si devices is their high degree of uniformity, thereby reducing spatial noise considerably.

To overcome problems of low photon flux imaging in the short-wave infrared, InSb focal plane arrays have been developed. These have much greater quantum efficiencies: ≈65% without an antireflection coating, ≈90% with an antireflection coating) than platinum silicide, and 640 × 640 device arrays have been developed.

REFERENCES

13.1. K. B. Benson and D. G. Fink, *HDTV Advanced Television for the 1990s*, McGraw-Hill, New York, 1991.

13.2. L. M. Biberman and S. Nudelman, Eds., *Photoelectronic Imaging Devices*, Vol. 2, *Devices and Their Evaluation*, Plenum Press, New York, 1971.

13.3. *RCA Electro-Optics Handbook*, RCA, Harrison, NJ, 1974.

13.4. L. J. Pinson, "Robot-vision: an evaluation of imaging sensors," *J. Robotics Syst.*, Vol. 1, 1984, pp. 263–314.

13.5. *The Photonics Design and Application Handbook, 1991, Book 3*, 1991. Reprinted with permission from Laurin Publishing Co., Inc.

13.6. W. S. Boyle and G. E. Smith, "Charge coupled semiconductor devices," *Bell System Tech. J.*, Vol. 49, 1970, pp. 587–593.

13.7. J. C. Feltz, "Development of the modulation transfer function and contrast transfer function for discrete systems, particularly charge-coupled devices," *Opt. Eng.*, Vol. 29, 1990, pp. 893–904.

13.8. T. S. Lomheim and L. S. Kalman, "Analytical modeling and digital simulation of scanning charge-coupled device imaging systems," in *Electro-Optical Displays*, M. A. Karim, Ed., Marcel Dekker, New York, 1992.

13.9. K. Knop, "Image sensors," in *Sensors*, W. Goepel, J. Hesse, and J. N. Zemel, Eds., Vol. 6, *Optical Sensors*, VCH Verlagsgesellschaft (Weinheim), 1992, pp. 233–252. Reprinted with permission from Wiley-VCH Verlag GmbH, Weinheim, Germany.

13.10. L. W. Schumann and T. S. Lomheim, "Modulation transfer function and quantum efficiency correlation at long wavelengths, (greater than 800 nm) in linear charge coupled imagers," *Appl. Opt.*, Vol. 28, May 1989, pp. 1701–1709.

13.11. P. R. Norton, "Infrared image sensors," *Opt. Eng.*, Vol. 30, November 1991, pp. 1649–1663.

EXERCISES

13.1 Compare expected observer distances for conventional and HDTV displays in order to make use of the higher resolution of HDTV systems.

13.2 For ideal 100% fill factor and short-wavelength imaging, compare horizontal resolution in terms of point spread function for FT CCD devices of 10-, 20-, and 30-μm pixel dimensions for an array of 512 pixels × 512 pixels, assuming effective detecting areas of 8.8 mm (vertical) and 11.8 mm (horizontal). What are the MTF values at the maximum usable spatial frequencies?

13.3 Equation (11.1.15) describes spatial noise current. Assuming constant photon flux across a CCD array, derive an expression for spatial noise in terms of variance in number of photoelectrons. Do the same for temporal or quantum noise, and then consider SNR for high and low incident photon fluxes [11.3].

13.4 Because Exercise 13.2 refers to short-wavelength imaging, compare MTF values at the maximum usable spatial frequency in light of possible MTF degradation via diffusion. Consider 410- and 820-nm wavelengths, assuming diffusion and depletion lengths equal to 75 and 5 μm, respectively. Note that for Si, the absorption coefficient at room temperature is approximately 7×10^4 cm^{-4} for 3-eV photons and 10^3 cm^{-1} for 1.5-eV photons. At which wavelength does the CCD produce a better quality image?

Part Four

OPTICAL TRANSFER FUNCTIONS FOR IMAGE MOTION AND VIBRATION

CHAPTER

14

Optical Transfer Functions for Image Motion and Vibration

14.1 INTRODUCTION

Image blurring caused by vibrations is a factor whose influence on resolution is often significant in imaging systems that involve mechanical motion. For example, an unstabilized, airborne system mounted on vibration isolators experiences an angular velocity typically on the order of 60,000 μrad/s. On the other hand, a stabilized system, which obtains image motion corrections by moving optical elements so as to counteract sensor motions, has a residual error typically on the order of 2500 μrad/s involving pitch, roll, yaw, and forward motion compensation errors. The greatest errors result usually from rotational vibration. A stabilized mount-type system has a combined error typically on the order of 100 μrad/s [14.1]. Even where no motion is involved, as in large fixed-position astronomical telescopes, vibrations deriving from thermal gradients in the walls and other parts of the telescope can be the limiting factor in image resolution.

As a consequence of such image motion, in many high-resolution systems, despite the use of high-quality sensors, resolution is limited by image motion and, as a result, the high-resolution capability of the sensor may be wasted. It can be a significant waste of means to use an expensive high-resolution sensor in such situations unless the blur can be corrected with image restoration. In imaging system design a convenient engineering tool is optical transfer function (OTF). The overall system OTF is generally limited by the OTF of the weakest link. In systems involving image vibration or motion, this weakest link is often the blur caused by the image vibration or motion, rather than that resulting from optical or electronic components. The formulation of such image blur into an OTF-type format is thus very convenient for system design and system analysis purposes, as well as for image restoration. It is the subject of Part 4, which consists of this chapter. High-resolution image restoration based on motion OTF is presented in Part 6.

Image motion can take many forms. Here, OTFs to describe image quality will be considered for uniform linear motion, sinusoidal vibrations at high vibration frequencies, sinusoidal vibrations at low vibration frequencies, and acceleration. This is followed by a discussion on the effects of quadratic motion, which becomes relevant in uniform linear motion compensation systems because of the nature of the compensation.

The OTF for any type of image motion can be calculated numerically in real time provided the motion function is known [14.2]. This is usually determined easily via a gyroscope or microelectronic accelerometer.

We should point out that, in general, the motion of interest here is not the actual relative motion between object and imager, but rather the relative motion between imager and image plane. The latter motion is that observed in the image plane. Both of these motions are related to each other by the image system magnification according to the similar triangles shown in Fig. 14.1. If V is actual relative velocity between object plane and image sensor, and if v is relative velocity between image sensor and image plane, then

$$V/s = -v/s' \tag{14.1.1}$$

or

$$v = -(s'/s)V \tag{14.1.2}$$

where the expression in parentheses is lateral magnification [Eq. (2.6.9)], which derives from the same similar triangles. For long-range imaging, the image distance s' is approximated by optics focal length f_l. For vertical imaging such as downward through the atmosphere s is image system height H and

$$v \approx -(f_l/H)V \tag{14.1.3}$$

Hence, image plane relative motion is usually much less than the actual relative motion. However, any motion during an exposure gives rise to blurring of image plane detail. One question to be considered in this part is how much blur can take place as a result of motion and still not be noticeable because of inherent blur deriving from the point spread function of the imaging system. In other words, assuming that image motion and vibration blur cannot be removed completely, how much is tolerable before resolution and target acquisition capability are affected? Examples and exercises are presented that integrate such questions with system design considerations.

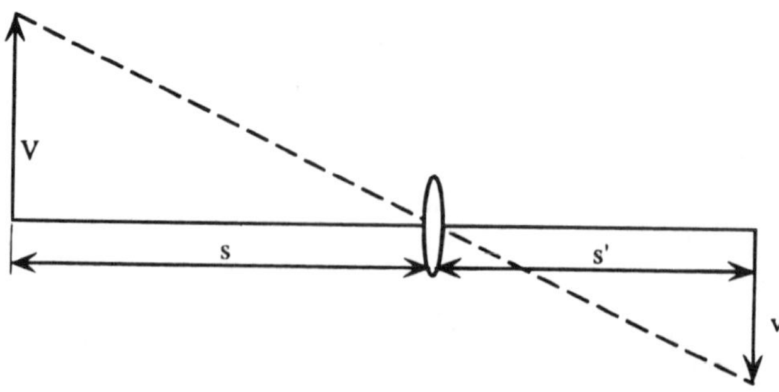

Fig. 14.1 Relationship between actual relative velocity V and apparent relative velocity v.

14.2 GENERAL METHOD OF OTF CALCULATION

In this section, the line spread function (LSF) derived from image motion transverse to the optical axis is obtained. The modulation transfer function (MTF) will be derived as the modulus of the OTF or Fourier transform of the LSF, and the phase transfer function (PTF) will be derived as the phase of the OTF. All graphs plotted for PTF are normalized by 2π.

Let $x(t)$ be the relative transverse displacement between object and sensor beginning at time t_s and ending at time $t_s + t_e$ where t_s is measured from the instant the sensor is first exposed and t_e is exposure time. The LSF of the motion is the PDF (probability density function) or the histogram of $x(t)$. The intuitive explanation for this determination is the following. Image motion causes the system line spread image response to move spatially. These displacements are integrated during the exposure. Displacement values of higher occurrence rate are values at which the line spread function is higher. Such occurrence can be described by a histogram of displacement $x(t)$, where frequency of occurrence is depicted as a function of x during the time interval $(t_s, t_s + t_e)$. This histogram is the LSF itself [14.2].

The quantity t_s is a random variable representing initial exposure time and it is uniformly distributed according to $f_t(t) = 1/t_e$.

By decomposing the relative displacement into n monotonic parts existing in one exposure time t_e, that is,

$$\Delta t_1 + \Delta t_2 + \Delta t_3 + \cdots + \Delta t_n = t_e \tag{14.2.1}$$

where $\Delta t_n = t_n - t_{n-1}$, the probability density function (PDF) for image motion is of the form

$$\begin{aligned} f_x(x) &= f_t(t) \cdot [1/|x'(t_1)| + \cdots + 1/|x'(t_n)| + \cdots] \\ &= 1/t_e \cdot [1/|x'(t_1)| + \cdots + 1/|x'(t_n)| + \cdots] \\ &\text{for } t_s < t < t_s + t_e \quad \text{and} \quad d_{\min} < x < d_{\max} \end{aligned} \tag{14.2.2}$$

where $x'(t)$ is the derivative of $x(t)$ and $f_x(x)$ is the PDF.

The lower and the upper limits, respectively, for x are results of minimum ($d_{\min}$) and maximum ($d_{\max}$) displacement between object and sensor. The PDF (or histogram) $f_x(x)$ represents the line spread function. The LSF is equal to $f_x(x)$ and the OTF is the one-dimensional fourier transform of the LSF

$$\text{OTF}(f) = \int_{-\infty}^{\infty} f_x(x) \exp(-j2\pi f_x)\, dx \tag{14.2.3}$$

where f_x is spatial frequency.

Thus, LSF for image motion can be determined from a histogram of the motion, and the resulting OTF for such motion is given by Eq. (14.2.3). This concept is illustrated now with examples.

14.3 LINEAR (UNIFORM) MOTION OTF

If the motion is linear at a constant velocity v_o in the image plane, the resulting displacement in that direction is given by

$$x = v_o t \tag{14.3.1}$$

where v_o is uniform relative velocity between object and sensor. Hence, from Eq. (14.2.2),

$$f_x(x) = 1/(v_o t_e) = 1/d \quad \text{for} \quad 0 < x < d \tag{14.3.2}$$

where d is the spatial extent of the blur and is equal to $v_o t_e$.

Therefore, from Eq. (14.2.3) the MTF for linear motion is

$$\begin{aligned} M_l(f) &= \left| \frac{1}{d} \int_0^d \exp(-j2\pi f_x)\, dx \right| = |\text{sinc}(\pi f_x d) \exp(-j\pi f_x d)| \\ &= |\text{sinc}(\pi f_x d)|, \qquad f_x \le d^{-1} \\ &= 0, \qquad \text{otherwise} \end{aligned} \tag{14.3.3}$$

and

$$\begin{aligned} &\text{PTF}(f) = \text{phase}[\text{sinc}(\pi f_x d) \exp(-j\pi f_x d)] = \pi f_x d \\ &\text{for} \quad n \cdot \frac{1}{d} < f_x < (n+1)\frac{1}{d} \quad \text{and} \quad n = 0, 1, 2, 3, \ldots \end{aligned} \tag{14.3.4}$$

Note that from Eq. (14.3.3) the smaller the blur extent d the wider the spatial-frequency bandwidth. This means it is desirable to decrease image plane velocity and exposure time so as to improve resolution.

Equation (14.3.3) is a well-known result (Fig. 14.2) previously derived [14.3, 14.4] in the spatial-frequency domain rather than in the spatial domain shown here. Since detail of size smaller than blur radius d is unresolvable and undergoes black–white phase or color reversal, MTF at frequencies higher than $1/d$ is labeled *false* or *spurious resolution*.

The PTF is an additional factor shown in Fig. 14.2(c). The PTF given by Eq. (14.3.4) is asymmetric because it is due to motion on one side of the image. The result is a phase factor that is a linear function of frequency. This implies that the mean image position is shifted by $d/2$ in the positive displacement direction due to linear smear. Linear PTF is not considered to be a degradation process because the image can be regarded as shifted bodily.

Uniform linear motion blurring effects can be largely offset by motion compensation systems. However, smaller blurring by quadratic motion often results from the compensation system itself. This is described in Section 14.5.

Example 14.1

Derive the MTF for uniform velocity, Eq. (14.3.3), on the basis of modulation contrast function [14.3, 14.4] or sine wave response defined in Section 9.2.

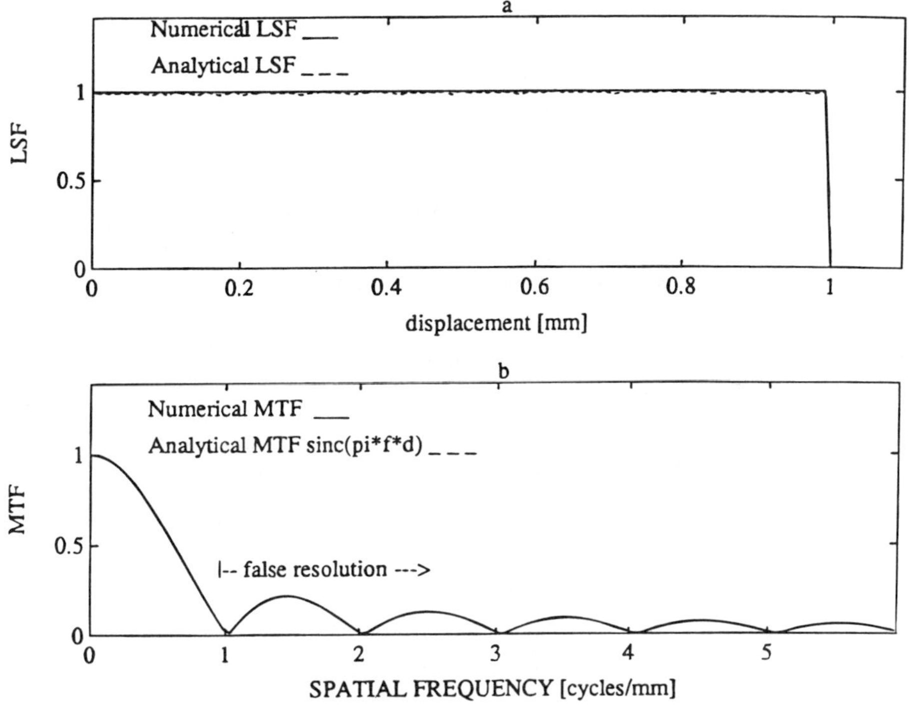

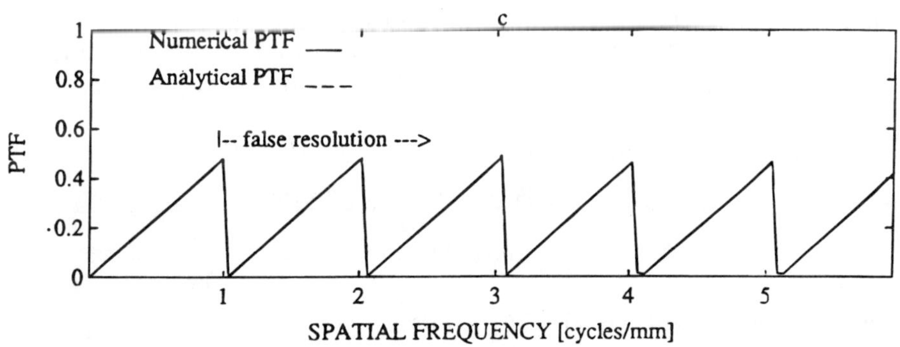

Fig. 14.2 (a) LSF calculated numerically and analytically for linear motion ($d = v_o \times t_e = 1$ mm). (b) MTF calculated numerically and analytically for linear motion. (c) PTF calculated numerically and analytically for linear motion (from [14.2]).

Solution

As a simple mathematical model, an image with a sinusoidal luminance pattern

$$i(x') = B_o + B_m \cos 2\pi f'_x \tag{14.3.5}$$

will be considered, where B_o and B_m are constants.

The modulation contrast of the image without image motion is, thus,

$$\text{MCO} = \frac{(B_o + B_m) - (B_o + B_m)}{(B_o + B_m) + (B_o - B_m)} = \frac{B_m}{B_o}$$

If the *image* moves at a constant velocity v, the new luminance distribution is

$$i(x', t) = B_o + B_m \cos 2\pi f_x(x' + vt)$$

The exposure of any point is proportional to the average of the intensity over the interval of the exposure time t_e

$$\overline{i(x't)} = \frac{1}{t_e}\int_0^{t_e} [B_o + B_m \cos 2\pi f_x(x' + vt)]\, dt$$

$$= B_o + \frac{B_m}{2\pi f v t_e} \sin 2\pi f_x(x' + vt)|_0^{t_e}$$

or

$$\overline{i(x', t)} = B_o + \frac{B_m}{2\pi f v t_e}[\sin 2\pi f_x(x' + vt_e) - \sin 2\pi f_x x'] \qquad (14.3.6)$$

To find the MTF for this image motion, we must know the modulation of the intensity pattern give by Eq. (14.3.6). This is probably easiest to do by expanding the above into exponential functions. By letting $A = 2\pi f_x$ and $B = 2\pi v t_e$, we see that

$$\begin{aligned}\sin(A + B) - \sin A &= (1/2j)[e^{jA}e^{jB} - e^{-jA}e^{-jB} - e^{jA} + e^{-jA}]\\ &= (1/2j)(e^{jB/2} - e^{-jB/2})(e^{jA}e^{jB/2} + e^{-jA}e^{-jB/2})\\ &= \sin(B/2)2\cos(A + B/2)\end{aligned}$$

Using this relation with Eq. (14.3.6) results in

$$\overline{i(x', t)} = B_o + B_m \frac{\sin \pi f_x v t_e}{\pi f_x v t_e} \cos 2\pi f_x\left(x + \frac{vt_e}{2}\right)$$

Thus the new pattern has the same shape as the original but with a phase lead given by $\pi f_x v t_e$ and a new modulation contrast of

$$\text{MCI} = \left(\frac{B_m \sin \pi v t_e}{B_o \pi f v t_e}\right)$$

By definition [Eq. (9.3.1)], the modulation contrast function is the ratio of modulation contrast with image motion to that without, or

$$\text{MCF}(f) = \left|\frac{\text{MCI}}{\text{MCO}}\right| = \left|\frac{\sin \pi f_x v t_e}{\pi f_x v t_e}\right| \tag{14.3.7}$$

which agrees with result (14.3.3).

Example 14.2

A target is moving horizontally at 20 m·s^{-1} velocity perpendicular to the line of sight of an imaging system at a distance of 5 km. The imaging system consists of a 1/2-in. vidicon camera system with a zoom lens. Effective dimensions of the camera tube are 8.8 × 11.8 mm. The square wave response of the TV system is essentially constant at unity until it descends sharply to zero at 295 line pairs or 590 TV lines. (a) What is the approximate maximum focal length to be used so as to not cause image motion blur on the display? In other words, we want resolution to be limited by the TV system and not by the image motion. Assume 25 frames per second and, for simplicity, zero required threshold contrast.

(b) How does the resolution vary with focal length?

Solution

(a) Motion here is transverse along the horizontal axis of the camera system. TV resolution therefore is limited by TV camera MTF and not by raster scan. For the TV camera itself,

$$f_{r\,\max} = 295 \text{ line pairs} \cdot (11.8 \text{ mm})^{-1} = 25 \text{ lp} \cdot \text{mm}^{-1} \tag{14.3.8}$$

For uniform image motion, Eq. (14.3.2) indicates that for 25 frames per second, there is a different $f_{r\,\max}$ which is from uniform motion and it is equal to

$$f_{r\,\max} = (vt_e)^{-1} = 25v^{-1} \tag{14.3.9}$$

In order for the image motion to be large and not degrade image quality, the spatial-frequency bandwidth of the motion MTF should be no smaller than 25 lp·mm^{-1}. Consequently, from Eqs. (14.3.9) and (14.3.8),

$$v = 25 \text{ seconds}^{-1}/25 \text{ lp·mm}^{-1} = 1 \text{ mm·s}^{-1}$$

From Eq. (14.1.1), in view of the 5-km distance,

$$|V/s| = v'/s' \approx v/f_l \tag{14.3.10}$$

As a result,

$$f_l = vs/V = (10^{-3} \text{ m·s}^{-1})(5 \times 10^3 \text{ m})/(20 \text{ m·s}^{-1}) = 1/4 \text{ m} = 250 \text{ mm}$$

(b) The longer the focal length, the smaller the field of view and the greater the lateral magnification. This means the blur deriving from image motion is magnified too at longer focal lengths.

The MTF relationships are depicted in Fig. 14.3 for the spatial-frequency domain. To generalize the concept, spatial frequency in the figure is symbolized by f, so as to keep it independent of direction. The MTF of the TV camera, $M_{TV}(f)$, is independent of optics parameters such as focal length. However, the MTF for image motion is very dependent on focal length, as illustrated in Eq. (14.3.10). For longer focal lengths, image plane velocity and blur deriving from it are increased by the higher magnification. Hence, the image motion MTF is narrower. Cascading the TV camera MTF and the image motion MTF for both long and short focal lengths is shown in Fig. 14.3. For short focal lengths (<250 mm) the combination is TV camera MTF limited, and the spatial-frequency cutoff is essentially that of the TV camera (295 line pairs/11.8 mm = 25 lp·mm^{-1}). From Eq. (10.2.4), the detectable detail is of size $2|\Delta r| = |(f_{r\,\max}\, m)^{-1}|$, where m is lateral magnification equal to f_l/s. Hence, the resolvable detail (excluding atmospheric effects) in the small focal length region is

$$
\begin{aligned}
|\Delta r| &= (1/2)[25(\text{lp}\cdot\text{mm}^{-1}) \cdot (f_l/5 \times 10^3 \text{ m})]^{-1} \\
&= (0.1/f_l) \text{ m}
\end{aligned}
\tag{14.3.11}
$$

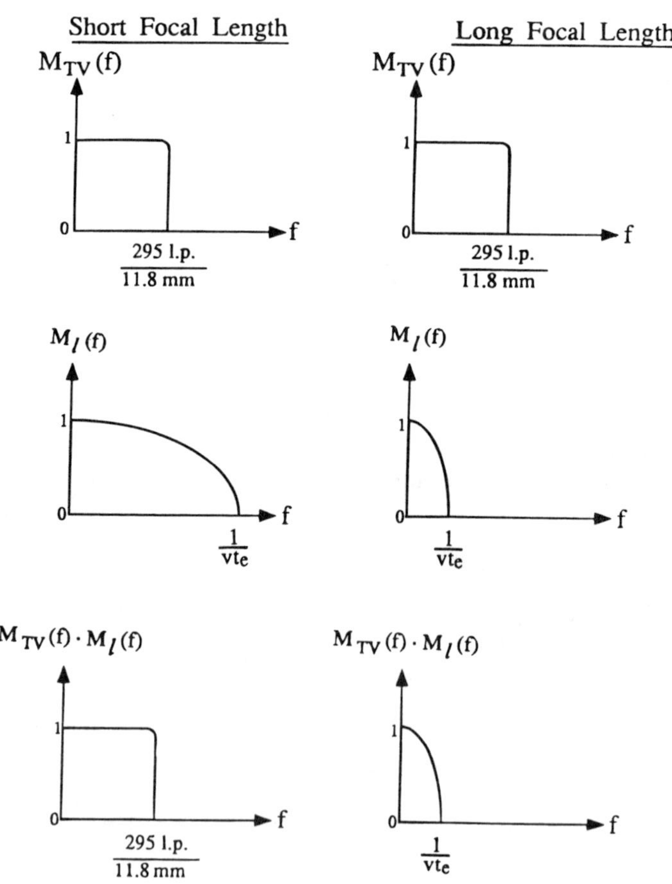

Fig. 14.3 MTF analysis for Example 14.2. M_{TV} stands for the TV camera's MTF and $M_l(f)$ is uniform motion MTF.

The greater the focal length and magnification, the better the resolution, as long as the image motion blur is small and is not seen. This holds for small focal lengths.

For long focal lengths (>250 mm), resolution is limited by image motion, since the cutoff spatial frequency of that MTF {which is $|(vt_e)^{-1}| = s/(f_l V t_e)$} is less than that of the TV camera. In this case

$$|\Delta r| = 1/2[s/(f_l V t_e)]^{-1}(f_l/s)^{-1} \tag{14.3.12}$$

where the last set of parentheses represents the lateral magnification. This simplifies to

$$|\Delta r| = Vt_e/2 = 20\ \text{m}\cdot\text{s}^{-1} \cdot \left(\frac{1}{25}\right)(1/2) = 0.4\ \text{m} \tag{14.3.13}$$

In this case, *resolution is independent of focal length*. The image motion blur is seen and is magnified together with the rest of the scene so that increased magnification does not produce smaller resolvable detail but does decrease field of view.

The largest useful magnification occurs when Eqs. (14.3.8) (TV MTF) and (14.3.9) (motion MTF) are equal to each other, or when Eqs. (14.3.11) and (14.3.13) are equal to each other. Both equalities occur at 250-mm focal length. Resolution is shown graphically in Fig. 14.4. For long focal lengths there is an asymptotic limit to resolvable detail set by the motion {14.3.13}. Optimum field of view occurs therefore when $f_l \approx 250$ mm.

14.4 SINUSOIDAL MOTION OTF

Sinusoidal motion usually results from mechanical vibrations and is a very critical factor in dynamic imaging systems as in robots, tanks, aircraft, boats, etc., where turbines and motors can give rise to mechanical vibrations {14.2}. In robotics and machine vision, linear motion is almost always accompanied by vibrations that are often close to being sinusoidal. Sinusoidal motion can be prevented in principle by proper design; in practice, however, it is often the most serious image motion.

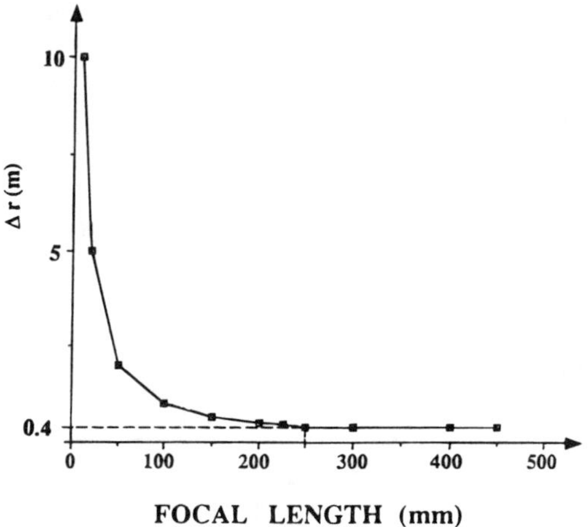

Fig. 14.4 Variation of resolution with focal length in Example 14.2. Optimum field of view for resolution occurs at 250 mm focal length.

Degradation of image quality as a result of sinusoidal motion [14.4–14.6] depends on the ratio of exposure time t_e to the period of the sinusoidal motion T_o. In this case, it is necessary to distinguish between two categories:

1. High-frequency vibration where the exposure period is long compared to the period of the simple harmonic motion ($t_e \geq T_o$).
2. Low-frequency mechanical vibrations in which exposure takes place during only a portion of the vibrational cycle ($t_e < T_o$). This situation is illustrated in Fig. 14.5 where it is seen that blur radius d is a random variable dependent on the portion of the sine wave vibrational cycle in which the exposure takes place, as well as on t_e itself.

Each type of motion is now considered in more detail.

14.4.1 High-Frequency Vibrations

In this case the blur radius is not a random variable but is a constant equal to the peak-to-peak displacement of the sine wave motion since one or more complete vibration cycles of time duration T_o fall within the exposure period t_e. Let the motion function be

$$x(t) = D\cos(\omega_o t) \qquad \text{for } \omega_o = 2\pi/T_o \tag{14.4.1}$$

It is assumed for simplicity that $t_e = nT_o$, $t_s = 0$ where $n \geq 1$ and t is a random variable distributed in the interval $(0, nT_o)$. As a result $f_t(t) = 1/t_e = 1/nT_o$ in Eq. (14.2.2). Note that the same blur of 2D occurs even if n is not an integer because $t_e > t_0$.

Since from Eq. (14.4.1)

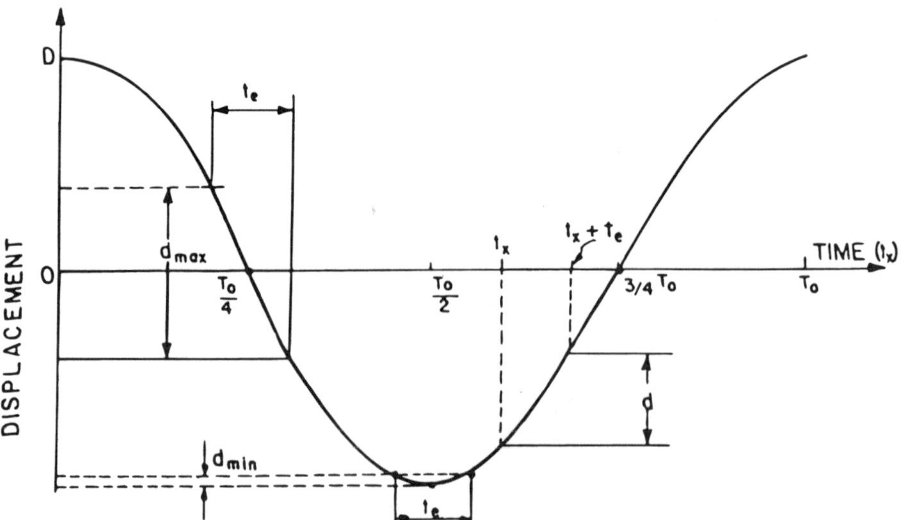

Fig. 14.5 Blur radius (d) as a function of exposure time relative to sine wave vibrational cycle for short exposures ($t_e < T_o$).

$$|x'(t_n)| = |-D\omega_o \sin(\omega_o t)| = \omega_o\sqrt{(D^2 - x^2)} \qquad (14.4.2)$$

then, by substituting this result into Eq. (14.2.2),

$$f_x(x) = \frac{1}{\omega_o\sqrt{D^2 - x^2}} \sum f_t(t_n) \qquad \text{for } |x| < D \qquad (14.4.3)$$

In this case Eq. (14.2.2) has exactly two solutions for $x(t)$ in one period T_o. Therefore, in the interval $(0,nT_o)$ it has $2n$ solutions, and $f_t(t) = 1/nT_o$, resulting in

$$f_x(x) = \frac{1}{\omega_o\sqrt{D^2 - x^2}} \frac{2n}{nT_o} = \frac{1}{\pi} \frac{1}{\sqrt{D^2 - x^2}} = \text{LSF} \qquad \text{for } |x| < D \quad (14.4.4)$$

The MTF for this motion is obtained from Eq. (14.2.3):

$$\text{MTF}(f) = \left| \frac{1}{\pi} \int_{-D}^{D} \frac{\exp(-j2\pi fx)}{\sqrt{D^2 - x^2}} \, dx \right| \qquad (14.4.5)$$

The solution for this integral is the known result [14.4]:

$$\text{MTF}(f) = \text{M}_s(f) = J_0(2\pi fD) \qquad (14.4.6)$$

where D is maximum vibration displacement and J_0 is the zero-order Bessel function.

The line spread function [Eq. (14.4.4)] is shown in Fig. 14.6. The two peaks at the extrema rather than the center result from the fact that image motion velocity

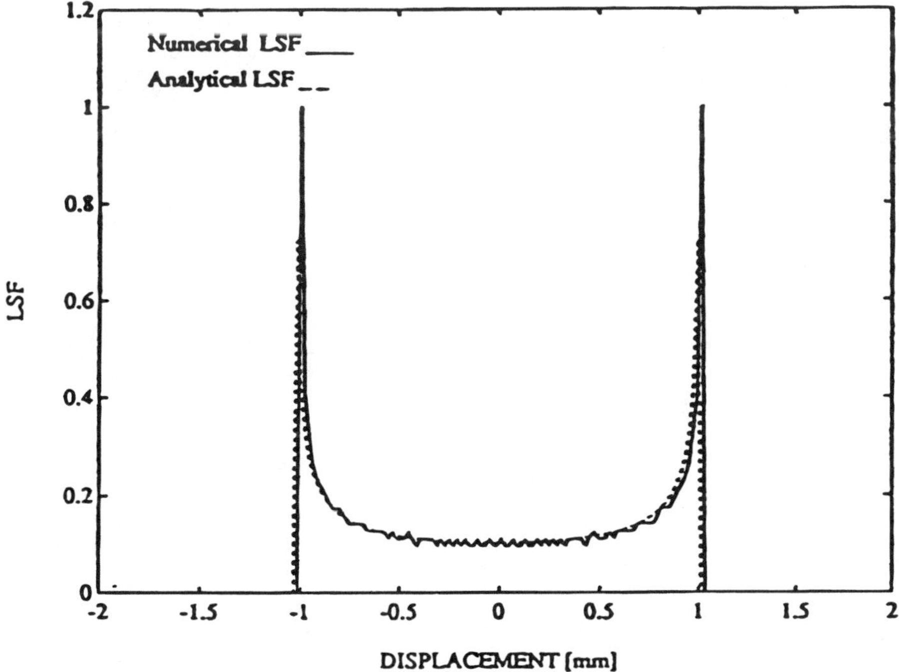

Fig. 14.6 Line spread function for high mechanical temporal vibration frequencies ($t_e \geq T_o$). D = 1 mm (after [14.2]).

is smallest and in fact is zero at the two extrema of vibrational motion. Hence the camera spends more time there than on the optical axis where displacement is zero, resulting in a spread function as depicted in Fig. 14.6.

The PTF is zero ideally since the LSF is an even function. Hence, Eq. (14.4.6) is not only MTF but also OTF. Assuming long-range imaging, then

$$D = f_l\theta_{pp}/2 \tag{14.4.7}$$

where θ_{pp} is peak-to-peak angular displacement, result (14.4.6) is depicted in Fig. 14.7. Note that at low contrasts spatial frequency in Eq. (14.4.6) hardly changes.

MTF for high temporal frequency vibrations can also be derived from the standpoint of modulation contrast function (see Exercise 14.2). The approach is similar to that for linear motion in Example 14.1.

14.4.2 Low-Frequency Vibrations

This type of image motion is characterized by a relatively long vibrational period T_o that is *longer* than the time exposure. This means image blur takes place only during a portion of the vibration period rather than the whole vibration period as in the previous case. Assume image motion is given by

$$x(t) = d \cos \frac{2\pi t}{T_o} \tag{14.4.8}$$

Image blurring at low vibration frequency ($t_e < T_o$) is a random process. It is a minimum when the exposure is entered either at a peak or at the bottom of the sine wave vibration curve as shown in Fig. 14.5. In this case, blur extent is

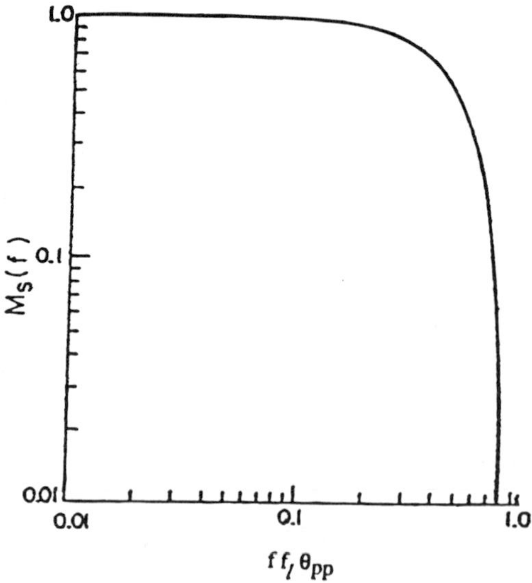

Fig. 14.7 OTF for high mechanical vibration frequencies.

$$d_{\min} = D\left\{1 - \cos\left[\left(\frac{2\pi}{T_o}\right)\left(\frac{t_e}{2}\right)\right]\right\} \tag{14.4.9}$$

When the exposure is centered between these points $[x(t) = 0]$, the image blur is maximum, that is, blur extent is

$$d_{\max} = 2D \sin\left[\left(\frac{2\pi}{T_o}\right)\left(\frac{t_e}{2}\right)\right] \tag{14.4.10}$$

In general, since this type of blur occurs only during a portion of low frequency vibration, then image blur $d < 2D$. However, this can be misleading since the spatial amplitude D of low-frequency vibrations is not necessarily constant with vibration frequency but can often be greater and perhaps even much greater than that of high-frequency vibrations [14.7]. Thus $d_{\max}$ for low-frequency vibrations is often considerably greater than D in Eqs. (14.4.1) and (14.4.6) for high-frequency vibrations. An example of both low and high temporal frequency vibrations is shown in Fig. 14.8 [14.1]. It is clear that the low-frequency vibration is of much greater amplitude than the high-frequency vibration. Hence, even though blurring takes place only over a portion of the low-frequency vibration period it is often much more serious than blur for high-frequency vibration. The randomness of the low-frequency blur requires special treatment in order to characterize it quantitatively.

Quantification of the low-frequency vibrational image blur extent d is much more complicated than that of high-frequency vibration because low-frequency vibrational blur depends on the initial phase of the oscillatory motion as well as on the instant and duration of the time exposure as shown in Fig. 14.5, both of which are often random processes. The influence of the MTF degradation is much more severe than that of the PTF. Therefore, the following discussion will concentrate on the MTF only.

Statistics on lucky shot probabilities, best case and average case MTF are detailed elsewhere [14.5, 14.6]. Numerical MTF calculation in the spatial-frequency domain for this case is described later [14.8]. The most important conclusion from these references is that the MTF is a random process.

The method of calculating the LSF for low-frequency vibrations is based on the PDF of $x(t)$ for high-frequency vibration in Eq. (14.4.3). This equation is applied

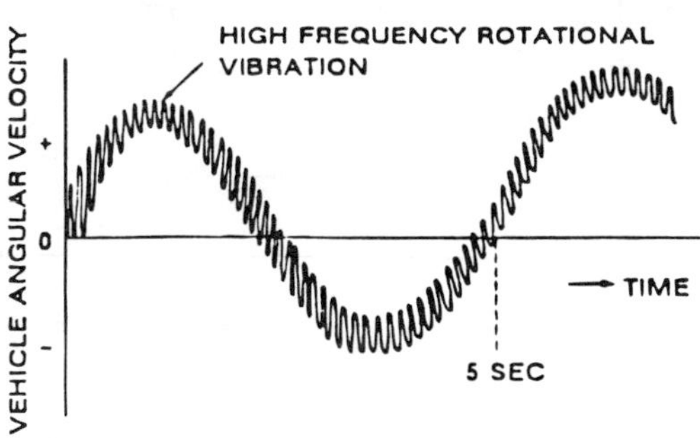

Fig. 14.8 Typical aircraft angular velocity combining both low and high frequency vibrations (after [14.1]). Low frequency vibration is envelope for high frequency vibrations.

here by evaluating $f_x(x)$ only over the particular displacement beginning at t_s and ending at $t_s + t_e$. In Fig. 14.9 a method of numerically calculating the MTF for two extreme cases of minimum (upper curve on right) and maximum (middle curve on right) blur extent is schematically demonstrated, and compared to the high-frequency vibration MTF (lower curve on right) which considers the LSF over the entire period T_o [14.2]. The possibility of a lucky shot (minimum blur radius) exists at the extreme points of the sine wave where the velocity of the image is minimum. Hence the LSF is characterized by a very narrow function so the MTF tends to be constant. The other extreme case (maximum blur radius) occurs around the zero crossing points of the sine wave where the velocity of the image is maximum and quite constant. Hence, the LSF approaches a rectangular pulse function that leads to the MTF formulation sinc ($\pi f d_{\max}$) for uniform velocity similar to Eq. (14.3.3). This would be depicted by the middle right MTF curve in Fig. 14.9 if exposure time and hence blur extent were smaller. All the other exposures fall between these two extreme cases and, as t_e/T_o increases, the variance of the MTF decreases so that in the extreme situation when $t_e/T_o = 1$ the MTF is the Bessel function described in Eq. (14.4.6) and depicted in the lower right portion of Fig. 14.9.

Figure 14.10 shows an example of nonlinear blur radius [14.8]. The LSF for this case contains two different parts of the LSF of high-frequency vibration. The first one shown at the bottom right is $f_x(x)$ in the interval $x(t_s + t_e) < x(t_s)$ and the second to the right of it is $2f_x(x)$ in the interval $x(t_s) < x < D$ because in the upper curve the absolute value of the derivative of x appears *twice* in this time interval—once for motion outward from $x(t_s)$ to D and then once for motion inward from D to $x(t_s)$. Therefore in this case the derivative of the LSF is discontinuous in $x(t_s)$ and there is a jump at this point. This discontinuity causes the MTF to be higher at all spatial frequencies and the MTF *never reaches zero*. The meaning of this result is that increasing the size of the jump in the spatial domain improves the MTF over all spatial frequencies. This is demonstrated later Fig. 14.12(c).

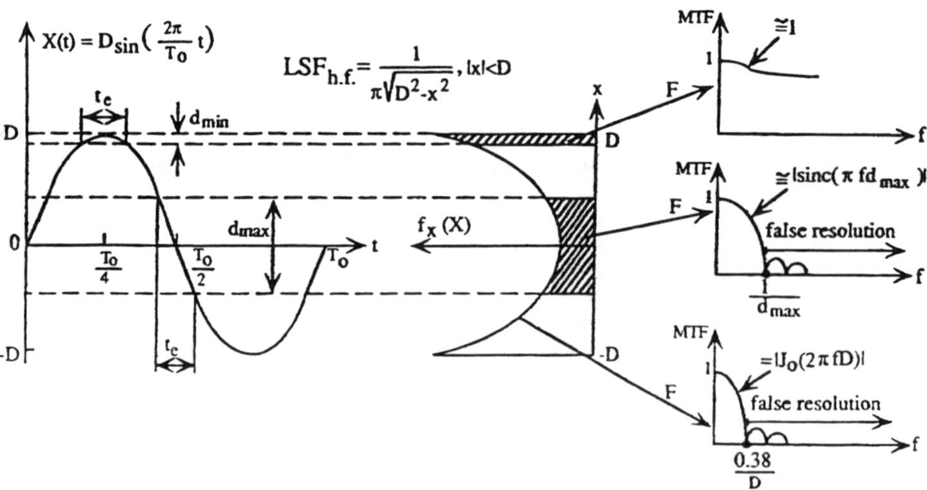

Fig. 14.9 Calculation of LSF and MTF for low-frequency vibrations based on the PDF of $x(t)$ for high-frequency vibration. PDF is $f_x(x)$ (after [14.2]).

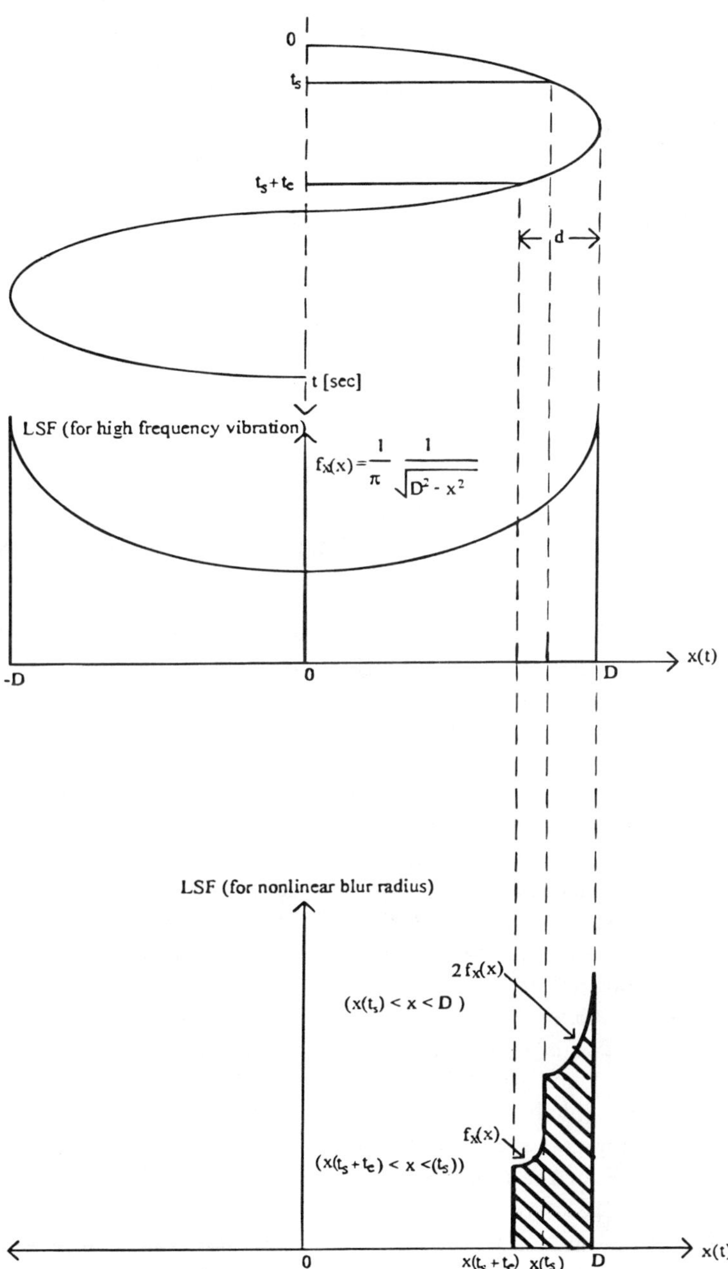

Fig. 14.10 Nonlinear motion example of obtaining LSF for low-frequency vibrations.

This method of obtaining the OTF for low-frequency vibrations can be used in real practical imaging systems. From the known mechanical vibration frequency (f) $= 1/T_o$ and the time exposure t_e, the OTF can be obtained for many different initial t_e values in one period T_o. Extraction of the blur radius is possible by tracking the amount of movement of a certain point in the image during two consecutive exposures. For each blur radius there is a unique OTF.

The general method of numerical calculation can be obtained from the following. From Eq. (14.2.2), since $t_e < T_o$,

$$\mathrm{LSF}(x) = \frac{1}{t_e} \cdot \left[\frac{1}{|x'(t_e)|} + \cdots + \frac{1}{|x'(t_n)|} \right] \qquad \text{for } (t_s < t < t_s + t_e)$$

$$\text{and } (d_{\min} < t < d_{\max}) \qquad (14.4.11)$$

However, for the same reason Eq. (14.2.3) must be amended to

$$\mathrm{OTF}(f) = \int_{x(t_s)}^{x(t_s+t_e)} \mathrm{LSF}(x) \exp(-j2\pi f x)\, dx \qquad (14.4.12a)$$

$$= \frac{1}{\pi} \int_{x(t_s)}^{x(t_s+t_e)} \frac{\exp(-j2\pi f x)}{\sqrt{D^2 - x^2}}\, dx \qquad (14.4.12b)$$

The method of calculating the LSF for low-frequency vibrations is again based on the PDF of $x(t)$ for high-frequency vibration in Eq. (14.4.4). Equation (14.4.4) is applied here by evaluating $f_x(x)$ only over the particular displacement beginning at t_s and ending at $t_s + t_e$. The function $f_x(x)$ is limited by the maximum and minimum displacement $d_{\max}$ and $d_{\min}$, respectively, during the exposure time, although the histogram is obtained over the entire exposure time. Since displacement or blur varies, $f_x(x)$ is not normalized to unity maximum value, as LSF is normally defined. After MTF is calculated, it must be normalized.

Analytical determination of MTF for $d_{\min}$ (Eq. 14.4.9) and $d_{\max}$ (Eq. 14.4.10) is considered in Exercises 14.4 and 14.5. For exposures between these two extremes, blur radius, LSF, and MTF are random processes that can be calculated numerically with the method shown here.

As presented in Table 14.1, there are six different cases for calculating blur radius and the LSF. Two of these cases contain blur for $x(t)$ increasing or decreasing monotonically (cases 1 and 4), and four of them contain nonmonotonic blur (cases 2, 3, 5, and 6) which involve maxima or minima of location coordinate $x(t)$ in Fig. 14.5. In the monotonic case $d_{\max}$ and $d_{\min}$ occur at t_s and $t_s + t_e$, not necessarily respectively. The LSF for the monotonic blur is calculated only in one interval of x, $d_{\min} < x < d_{\max}$, and LSF $= f_x(x)$, an example of which is given in Fig. 14.11 [14.9].

For the nonmonotonic blur cases $d_{\max}$ and $d_{\min}$ do not both or individually occur at t_s and $t_s + t_e$. Also, as in Fig. 14.10 which represents Case 6 in Table 14.1, the blur can be divided into two components, one of which represents LSF $= 2f_x(x)$ because there are two solutions to Eq. (14.2.2) and the absolute derivative is equal for these two solutions. The other blur component involves only one solution to Eq. (14.2.2) so that its LSF $= f_x(x)$ as in the case of monotonic blur. The overall LSF is then a superposition of both components, as in Figs. 14.11 [14.9] and 14.12 [14.9]. Two extreme cases are presented for relative exposure time $(t_e/T_o) = 0.1$. In Figs.

TABLE 14.1 Displacements and Line Functions for Exposures at Different Positions of Low Temporal Frequency Vibrations

Time	$d^* \equiv \frac{d(t_s)}{D}$	LSF	Blur Radius
1) $0 < t_s < \frac{T_o}{2} - t_e$	$\cos \frac{2\pi}{T_0} t_x - \cos \frac{2\pi}{T_0} (t_s + t_e)$	$f_x(x)$, $d_{min} \leq x < d_{max}$	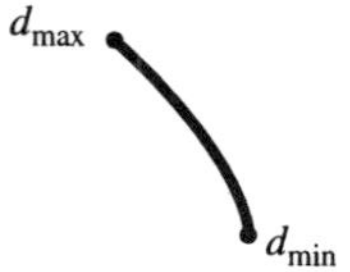 d_{max} d_{min}
2) $\frac{T_o}{2} - t_e < t_s < \frac{T_o}{2} - \frac{t_e}{2}$	$1 + \cos \frac{2\pi}{T_o} t_s$	$2f_x(x)$, $d_{min} \leq x < x(t_s + t_e)$ $f_x(x)$, $x(t_s) \leq x < d_{max}$	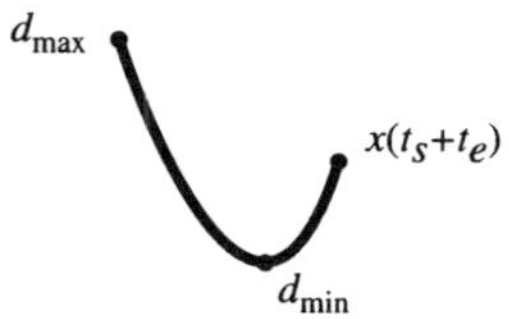 d_{max} $x(t_s+t_e)$ d_{min}
3) $\frac{T_o}{2} - \frac{t_e}{2} < t_s < \frac{T_o}{2}$	$1 + \cos \frac{2\pi}{T_o} (t_s + t_e)$	$2f_x(x)$, $d_{min} \leq x < x(t_s)$ $f_x(x)$, $x(t_s) \leq x < d_{max}$	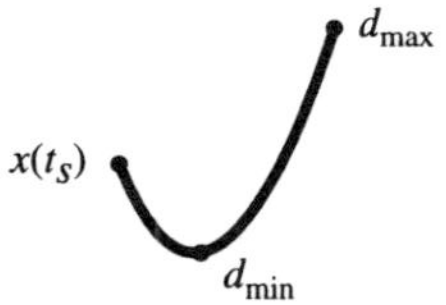 d_{max} $x(t_s)$ d_{min}
4) $\frac{T_o}{2} < t_s < T_o - t_e$	$\cos \frac{2\pi}{T_o} (t_s + t_e) - \cos \frac{2\pi}{T_o} t_s$	$f_x(x)$, $d_{min} \leq x \leq d_{max}$	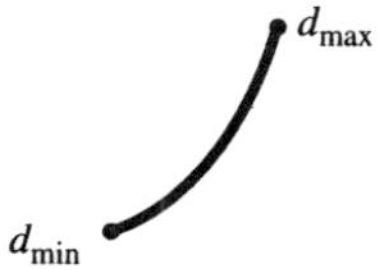 d_{max} d_{min}
5) $T_o - t_e < t_s < T_o - \frac{t_e}{2}$	$1 - \cos \frac{2\pi}{T_o} t_s$	$f_x(x)$, $d_{min} \leq x < x(t_s + t_e)$ $2f_x(x)$, $x(t_s + t_e) \leq x < d_{max}$	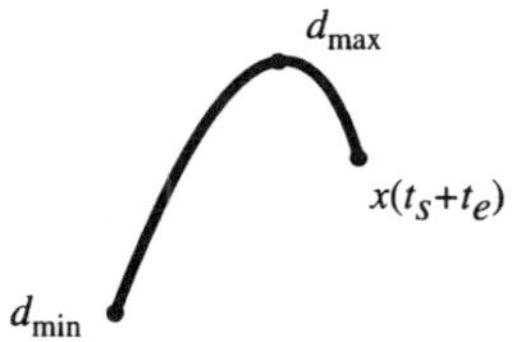 d_{max} $x(t_s+t_e)$ d_{min}
6) $T_o - \frac{t_e}{2} < t_s < T_o$	$1 - \cos \frac{2\pi}{T_o} (t_s + t_e)$	$f_x(x)$, $d_{min} \leq x \leq x(t_s)$ $2f_x(x)$, $x(t_s) \leq x < d_{max}$	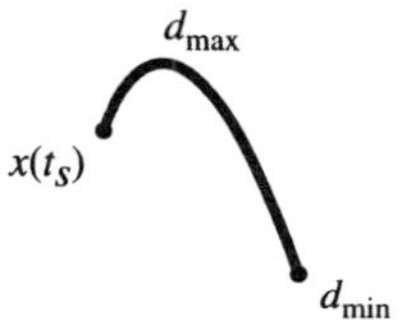 d_{max} $x(t_s)$ d_{min}

$x(t_s)$ = position at beginning of exposure, $x(t_s + t_e)$ = position at end of exposure.

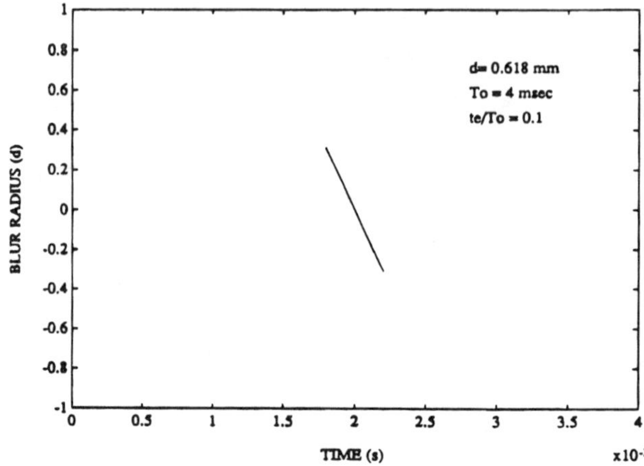

Fig. 14.11(a) Maximum blur radius obtained for t_s = 2.8 ms (after [14.9]).

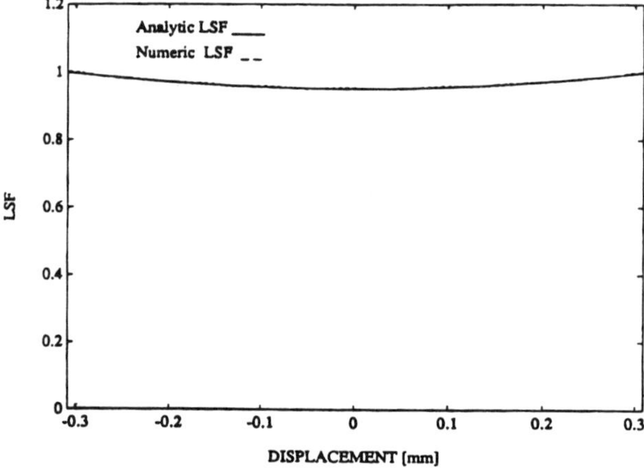

Fig. 14.11(b) Numerical and analytical LSF (see Exercise 14.4) calculated for maximum blur radius (after [14.9]).

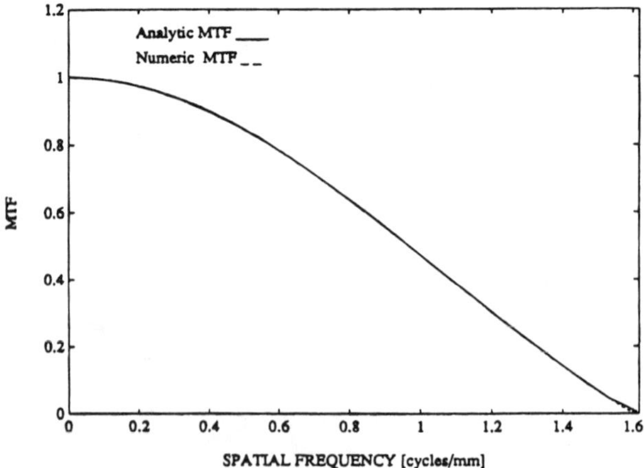

Fig. 14.11(c) Numerical and analytical MTF (see Exercise 14.4) calculated for maximum blur radius (after [14.9]).

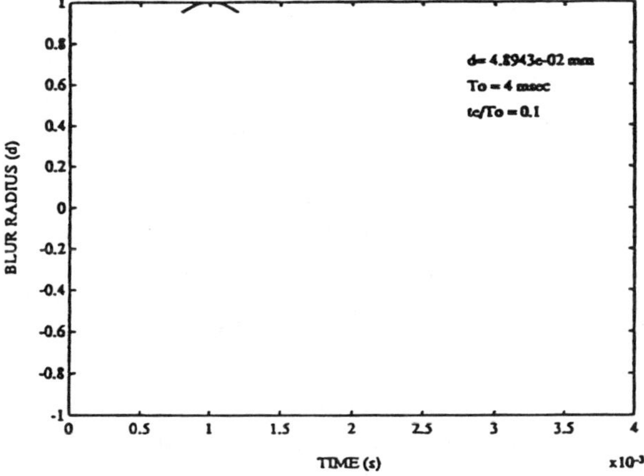

Fig. 14.12(a) Minimum blur radius for $t_s = 0.8$ ms (after [14.9]).

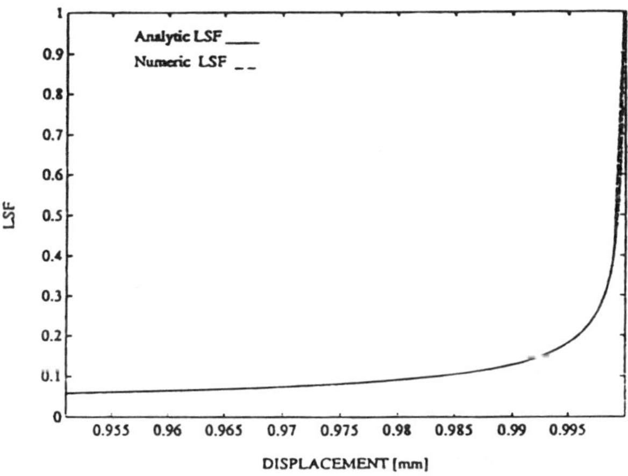

Fig. 14.12(b) Numerical and analytical LSF (see Exercise 14.5) calculated for minimum blur radius (after [14.9]).

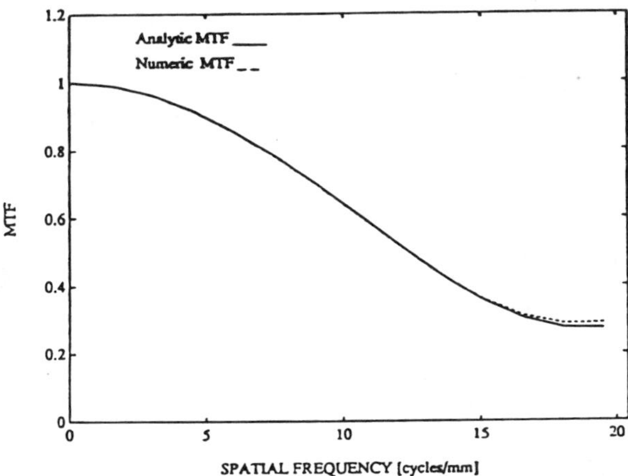

Fig. 14.12(c) Numerical and analytical MTF (see Exercise 14.5) calculated for minimum blur radius (after [14.9]).

14.11 blur shape is almost linear and monotonic so LSF turns out to be essentially a step function and the MTF is close to the sinc function derived in Eq. (14.3.3) as MTF for linear motion. In Figs. 14.12 the blur shape is not linear and is nonmonotonic, so the LSF approaches a one-sided delta function since peak displacement in one direction only occurs during the exposure, unlike Fig. 14.6 for high-frequency vibrations. The reason is that, at extremes of the sine wave vibration, image motion is slow, thereby maximizing frequency or probability of occurrence at such points in the histogram that form the LSF. The resulting LSF for such temporal vibration is very narrow for nonmonotonic displacement at points corresponding to extrema of the sine wave motion, thus giving rise to a broad MTF in Fig. 14.12(c), which falls off more slowly than the sinc MTF for uniform velocity, thereby yielding better contrast. This result has been suggested previously on the basis of theoretical considerations [14.10], but without the physical explanation and without consideration of nonmonotonic blur. Thus, *exposure during a nonlinear motion interval of the vibration cycle yields a double benefit*: (1) $f_{r\,\max}$ is *larger* because the blur radius is narrower than for a linear motion ($x_s \approx 0$) portion of the vibrational cycle and (2) *the MTF is higher and never reaches zero*.

The property of MTF never reaching zero in Fig. 14.12(c) is typical for other time intervals of sinusoidal vibration as long as they display nonuniform motion [14.8, 14.9]. It is also typical of nonuniform motion in general, such as acceleration [14.2].

In summary, low-frequency vibrational MTF is a random process that can be calculated numerically in real time. There is no general analytical MTF expression. The method of MTF calculation is valid for any type of one-dimensional motion in any given direction, including random motion.

14.5 QUADRATIC MOTION

14.5.1 Analytic Approach

As a general rule, whenever the image motion compensation varies with time, the blur is a quadratic equation in time. The effect of this quadratic motion can be found in the same way as in the case of linear motion. The approach first is that of modulation contrast function or sine wave response, followed by numerical calculation procedure. Consider a cosine wave object

$$i(x, t) = B_o + B_m \cos 2\pi f(x + v_0 t + at^2/2) \tag{14.5.1}$$

where v_0 is image plane velocity at $t = 0$ and a is acceleration. Therefore,

$$\overline{i(x, t)} = \frac{1}{t_e} \int_0^{t_e} [B_o + B_m \cos 2\pi f(x + v_0 t + at^2/2)]\, dt \tag{14.5.2}$$

Now,

$$x + v_0 t + at^2/2 = x - \frac{v_0^2}{2a} + \frac{a}{2}\left(t + \frac{v_0 a}{a}\right)^2 \tag{14.5.3}$$

Consequently we can rewrite Eq. (14.5.2) in the form

$$\overline{i(x,t)} = \frac{1}{t_e}\int_0^{t_e}\left\{B_o + B_m\left[\cos 2\pi f\left(x - \frac{v_0^2}{2a}\right)\cos \pi fa\left(t + \frac{v_0}{a}\right)^2\right.\right.$$

$$\left.\left. -\sin 2\pi f\left(x - \frac{v_0^2}{2a}\right)\sin \pi fa\left(t + \frac{v_0^2}{a}\right)^2\right]\right\} dt \quad (14.5.4)$$

Unfortunately, these integrals cannot be evaluated analytically. These are Fresnel integrals and are tabulated.

$$\int_{u_1}^{u_2} \cos\frac{\pi}{2}u^2\,du = 1_1 \quad (14.5.5)$$

$$\int_{u_1}^{u_2} \sin\frac{\pi}{2}u^2\,du = 1_2 \quad (14.5.6)$$

where $u^2 = 2fa(t + v_0/a)^2$ and $du = \sqrt{2fa}\,dt$.

The limits on the integrals are given by

$$u_1(\text{at } t = 0) = 2v_0\sqrt{f/a}$$

$$u_2(\text{at } t = t_e) = \sqrt{2fa}(t_e + v_0/a) \quad (14.5.7)$$

These integrals can be evaluated from the tables by noting that

$$\int_{u_1}^{u_2} f(u)\,du = \int_0^{u_2} f(u)\,du - \int_0^{u_1} f(u)\,du$$

Equation (14.5.4) can now be written as

$$\overline{i(x,t)} = B_o + \frac{B_m}{\sqrt{2fat_e}}\left[1_1 \cos \pi f\left(x - \frac{v_0^2}{2a}\right) - 1_2 \sin 2\pi f\left(x - \frac{v_0^2}{2a}\right)\right]$$

$$= B_o + \frac{B_m}{2fat_e}\sqrt{1_1^2 + 1_2^2}\cos(2\pi fx - \theta') \quad (14.5.8)$$

where θ' is the phase lag. The MTF is consequently [14.3]

$$M_Q(f) = \frac{B_m\sqrt{1_1^2 + 1_2^2}}{B_0 2fat_e} \div \frac{B_m}{B_0} = \frac{\sqrt{1_1^2 + 1_2^2}}{2fat_e} \quad (14.5.9)$$

This transfer function is shown in Fig. 14.13 [14.3] for the special case of $v_0 = 0$, that is, for motion that is proportional only to the square of the time, or parabolic motion. This applies to a situation where uniform linear motion has been compensated for entirely. A given amount of parabolic movement reduces the modulation less than an equal amount of linear movement. This is because most of the parabolic movement

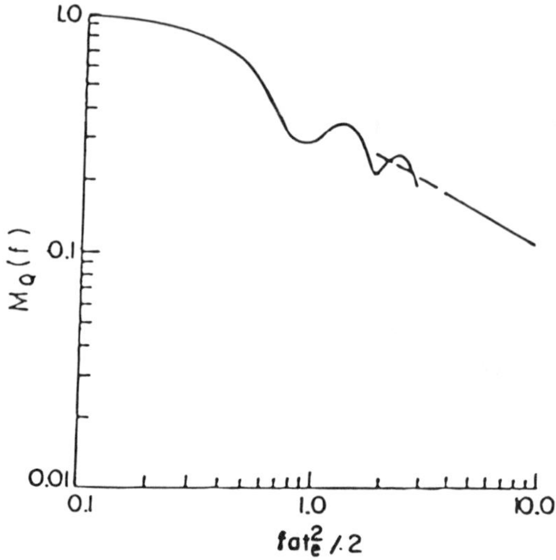

Fig. 14.13 MTF for parabolic motion without linear motion (after [14.3]). © John Wiley & Sons, Inc.

occurs in a relatively short time interval at the end of the exposure when the largest increase in t^2 takes place. Consequently, for parabolic motion the image blur is not spread as much. However, when $v_0 \neq 0$, the parabolic movement is in addition to rather than in place of the linear motion, thus increasing the blur and decreasing the modulation contrast function beyond that of linear motion alone.

14.5.2 Numerical Calculation

In general, OTF can be calculated numerically by following again the general approach of Eq. (14.2.2), including cases where v_0 is not zero. Since

$$x(t) = v_0 t + \frac{1}{2} a t^2 \qquad \text{for } t_s < t < t_s + t_e \tag{14.5.10}$$

The blur radius d is obtained by substituting $t = t_e$.

The image plane velocity at time t is

$$v(t) = v_0 + at \qquad \text{for } t_s < t < t_s + t_e \tag{14.5.11a}$$

In Eq. (14.5.10), $x(t)$ is a monotonic function for any t and therefore it is possible to obtain the velocity $v(t) = x'(t)$ as function of x instead of time t according to the solution of the following equation:

$$at^2 + 2v_0 t - 2x = 0 \qquad \text{for } t_s < t < t_s + t_e \tag{14.5.11b}$$

The positive solution for t in this equation is

$$t = -\frac{v_0}{a} + \frac{1}{a}\sqrt{v_0^2 + 2ax} \qquad \text{for } 0 < x < d \tag{14.5.12}$$

and hence, by substitution of this result into Eq. (14.5.11), $v(x)$ is equal to

$$v(x) = \sqrt{v_0^2 + 2ax} \qquad \text{for } 0 < x < d \tag{14.5.13}$$

According to Eq. (14.4.11) the line spread function is equal to

$$\mathrm{LSF}(x) = \begin{cases} \dfrac{1}{t_e\sqrt{v_0^2 + 2ax}} & 0 < x < d \\ 0 & \text{otherwise} \end{cases} \tag{14.5.14}$$

The OTF result can be obtained via the Fourier transform of Eq. (14.5.14):

$$\mathrm{OTF}(f) = \frac{1}{t_e}\int_0^d \frac{1}{\sqrt{v_0^2 + 2ax}} \exp(-2j\pi fx)\, dx \tag{14.5.15}$$

This can be obtained numerically by means of the fast Fourier transform.

An interesting analysis is comparison of OTF for linear motion and acceleration motion for the same blur radius. Acceleration is non linear motion. The modulus of the transfer function due to constant acceleration is equal to or greater than that due to constant velocity. This is because most of the parabolic movement occurs in a short time interval at the end of the exposure. Consequently the image luminance is not spread as much. Also, there is a slight difference in the phase domain where, for linear motion, the OTF is a linear function with spatial frequency as in Eq. (14.3.4) while for linear acceleration the PTF is a nonlinear function. The meaning of this is that for linear phase there is no degradation since the target image is displaced uniformly. For acceleration motion, phase distortion is an additional image degradation effect that has to be tolerated. The comparison results appear in Fig. 14.14.

Now, let us define R as [14.11]

$$R = \frac{v_0^2}{a} \tag{14.5.16}$$

This ratio expresses the relation between the image plane acceleration and the constant velocity of the relative motion. The LSF in Eq. (14.5.14) can be normalized with the parameter R [14.10]:

$$\mathrm{LSF}(x) = \begin{cases} d^{-1}[1 + 2x/R)]^{-0.5} & 0 \le x \le d \\ 0 & \text{otherwise} \end{cases} \tag{14.5.17}$$

As R increases v_0^2 becomes much greater than a and $2x/R$ becomes negligible compared to 1. Hence the dominant part of the motion is the constant velocity. The spread function will be almost constant and will differ very little from that due to constant velocity over the course of each exposure. However, acceleration causes the"constant"-velocity to be different for each exposure. On the other hand when R decreases, $2x/R >> 1$ and the dominant part of the motion is the acceleration component [14.10].

Now we consider two situations. The first is for constant blur radius d with R as a parameter. This implies variable mean exposure time t_e. The second one is for constant exposure time with different R values. The latter represents variable blur radii.

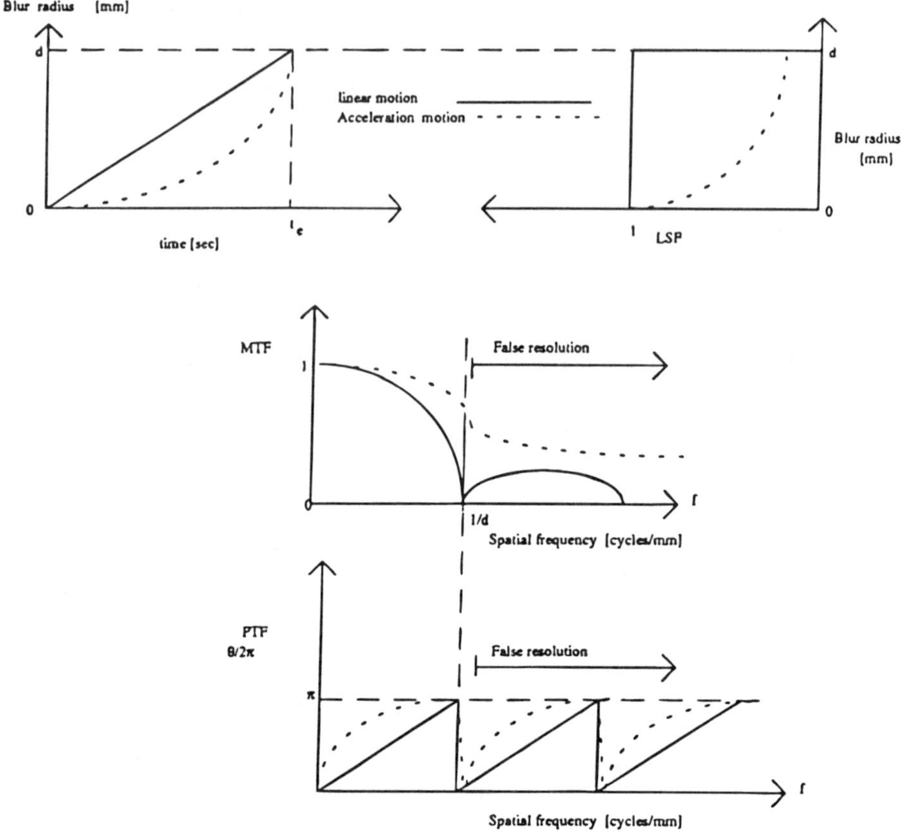

Fig. 14.14 OTF comparison for linear (solid lines) and acceleration (dotted curves) motion (after [14.10]).

14.5.2.1 CONSTANT BLUR RADIUS

Here, a presentation of the OTF for constant blur radius is considered, with R as a parameter. R ranges here from 0.001 up to 0.2 with d equal to 0.1 mm and v_0 equal to 1 mm·s^{-1}. The results are presented in Fig. 14.15. The difference in exposure time is calculated from Eq. (14.5.12) and the value of the MTF at the spatial frequency $f_{r\ \max} = 1/d_{\text{cutoff}}$ is considered with R as a parameter. This discrete value of MTF is called MTF$_{\text{-cutoff}}$.

All the graphs in Fig. 14.15 are plotted as functions of the parameter R. In Fig. 14.15(a), $x(t)$ is plotted as a function of time. In Fig. 14.15(b) LSF is plotted as a function of displacement x and in Figs. 14.15(c) and (d), MTF and PTF are plotted as functions of spatial frequency, respectively.

Several important conclusions [14.10] can be derived from the graphs of Fig. 14.15:

1. For constant blur radius, acceleration yields better image quality than does linear motion. This "surprising" result derives from nonlinearity of acceleration motion and is similar to that given in Fig. 14.12(c) for short-exposure sinusoidal motion.
2. As R increases the blur radius shape becomes similar to that of linear motion.
3. To obtain the same blur radius, longer time exposures are needed as R increases.

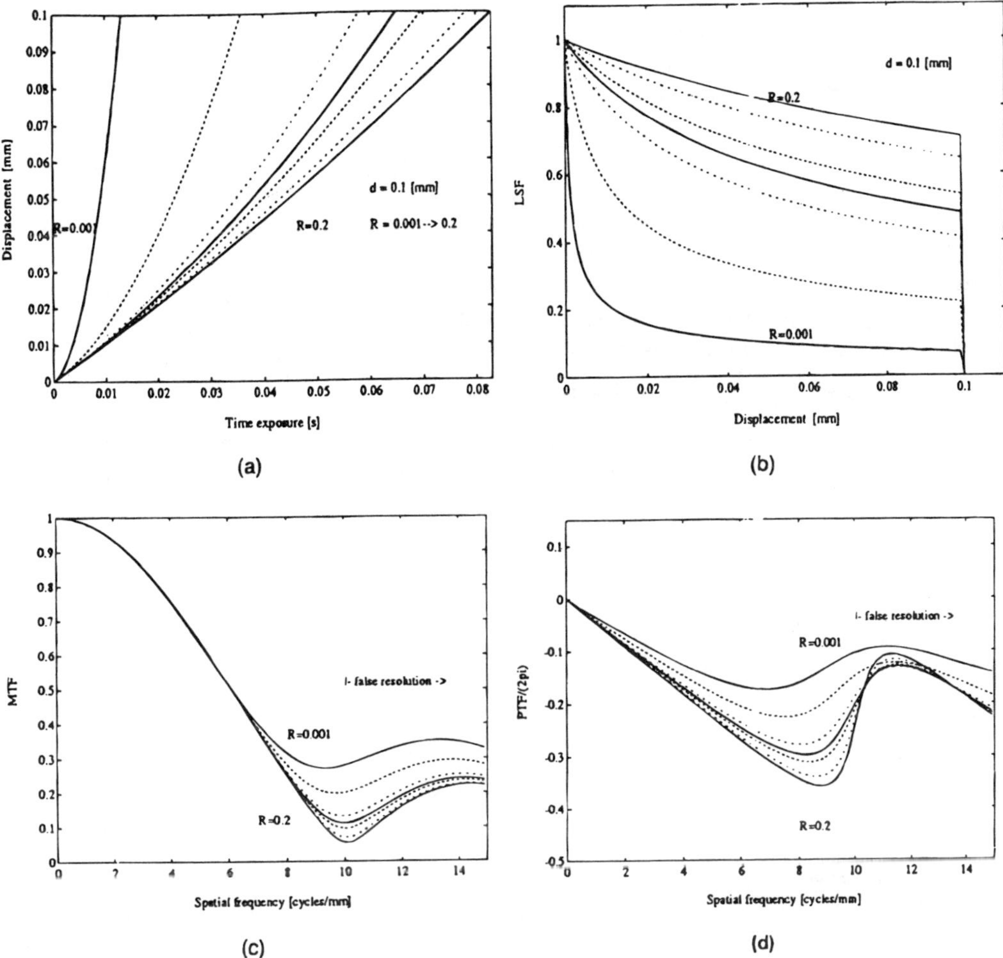

Fig. 14.15 Example of linear acceleration image motion with R as a parameter for constant blur radius $d = 0.1$ mm: (a) displacement; (b) LSF; (c) MTF; (d) normalized PTF (after [14.10]).

The reason is that v_0 here is constant while a is decreasing. This causes R to increase as defined in Eq. (14.5.16).

4. The $\text{MTF}_{\text{-cutoff}}$ values are reciprocal functions of R. As R increases, the MTF falls off more quickly and reaches lower contrast values at $f_{r\ \max}$. In the limit when R is sufficiently large ($R > 0.2$) the contrast at $f_{r\ \max}$ can be considered to be essentially zero.
5. The PTF can be considered a linear function of spatial frequency over large ranges as R increases. In the limit it approaches that of linear (constant) velocity as shown in Figs. 14.2 and 14.11(a).

14.5.2.2 CONSTANT TIME EXPOSURE

Now we consider the more familiar problem in mobile imaging systems where exposure time is constant but blur radius is a variable function of v_0 or a according to Eq. (14.5.10).

As in the previous case it is assumed here that v_0 is constant and equal to $1\ \mathrm{mm{\cdot}s^{-1}}$. However the exposure time is also constant at $t_e = 20$ ms, which is typical for a CCD camera. For this simulation different values of R were used over the range of 0.001 to 0.06. Smaller values of R were used because when R increases the change of the blur radius is not that noticeable. Figure 14.16 includes the same graphs as Fig. 14.15 but for the case of constant exposure time and variable blur radius [14.10].

The variations in all parameters are more noticeable for small values of R. For $R > 0.01$ there is only a small increase of $f_{r\ \max}$ and decrease of d and $\mathrm{MTF}_{\text{-cutoff}}$. On the other hand, for small R (high relative acceleration) image quality is greatly degraded. Also, compared to the case of constant blur radius here the decline of $\mathrm{MTF}_{\text{-cutoff}}$ as a function of R is much less noticeable [14.10].

14.6 Target Acquisition in the Presence of Vibrations

Image blur caused by mechanical vibrations was shown earlier to be a serious limitation in resolution. This limitation can also be expressed quantitatively in terms of a limitation on target acquisition probabilities and ranges.

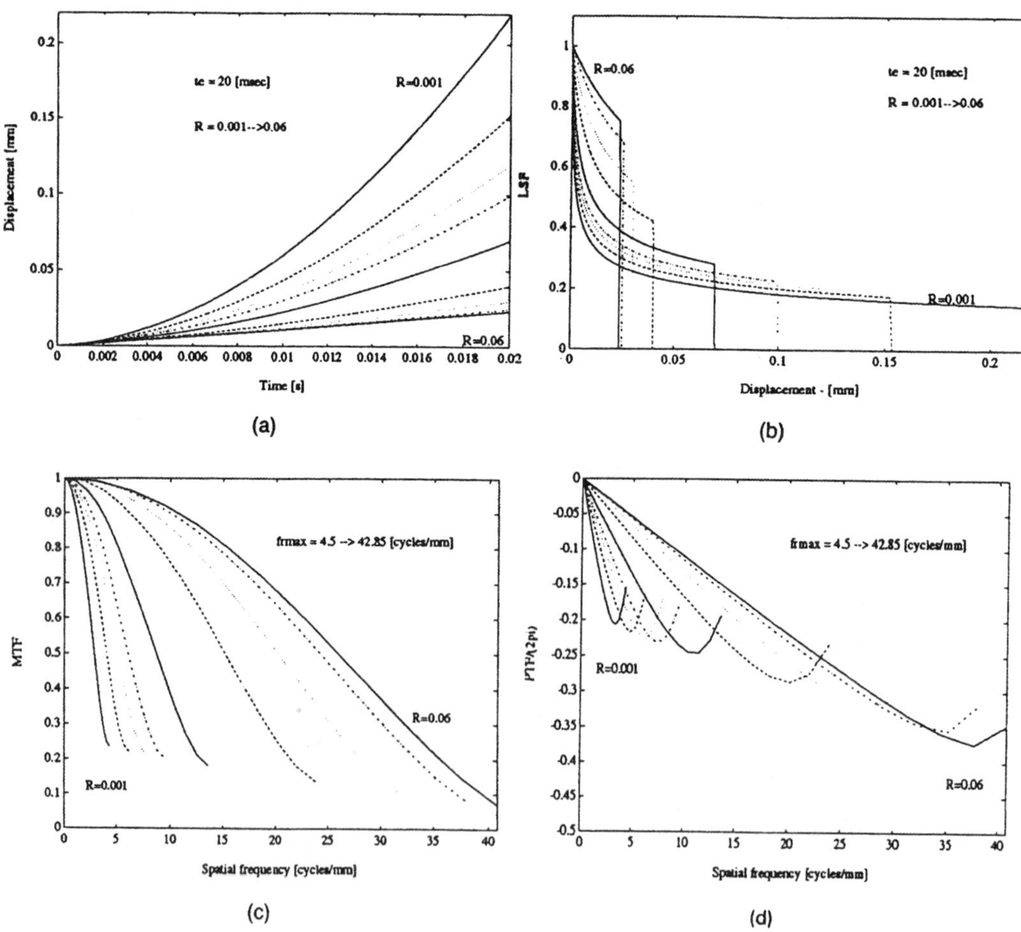

Fig. 14.16 Example of uniform acceleration image motion for constant exposure time $t_e = 20$ ms, with R as a parameter: (a) displacement; (b) LSF; (c) MTF; (d) normalized PTF (after [14.10]).

For contrast-limited imaging situations, the MTF of the vibrations is included in the overall system MTF in a multiplicative manner as described in Section 8.10. If resolution is limited by such image motion, as is often the case, the vibrational motion MTF will cause the overall system MTF to intersect the required threshold contrast curve at a reduced value of $f_{r\,\max}$ shown in Fig. 10.1(a) in Chapter 10. The reduction in $f_{r\,\max}$, or in spatial-frequency bandwidth, gives rise to a reduction in the number of line pairs n across critical target dimension described in Eqs. (10.4.2) and (10.4.5). The reduction in n/n_{50} caused by such vibrational blur then gives rise to a reduction in target acquisition probabilities calculated from the target transfer function curve in Fig. 10.7.

For noise-limited imaging in the visible and near-infrared regions, as in low light level situations, the vibrational motion MTF is also included in the overall system MTF, labeled $(\mathrm{MTF})_{ST}$ in Eq. (11.1.16), which gives rise to a reduction in display signal-to-noise ratio SNR_D. This reduces the probability of target acquisition as determined from Fig. 11.3.

For noise-limited thermal imaging, the vibrational motion MTF is incorporated into the parameter β_{sys} in Eq. (11.2.5) in the following way [14.12]. The parameter β_{sys} is altered to a new extinction coefficient β^*_{sys} equal to

$$\beta^*_{sys} = \beta_{sys} + \frac{1}{f_a}\, l_n\left[\frac{1}{\mathrm{MTF}_v(f_a)}\right] \qquad \text{for } f_a < f_{a\,\max} \tag{14.6.1}$$

where $\mathrm{MTF}_v(f_a)$ is the overall vibrational motion MTF. Often, both high- and low-frequency motions take place simultaneously, as shown in Fig. 14.8, and the overall vibrational motion MTF must consider all of them. Equation (14.6.1) is based on the exponential relationship of MRT to β_{sys} defined in Eq. (11.2.5), in which β_{sys} refers to thermal imaging hardware resolution limitations. The new parameter β^*_{sys} is larger than β_{sys} and implies a higher or worse MRT, which decreases SNR_N in Eqs. (11.2.9) and (11.2.11) and decreases target acquisition range in Eq. (11.2.10). In all of these equations, β^*_{sys} replaces β_{sys}.

An example of the effect of low-frequency vibrations only on target acquisition probabilities for low atmospheric turbulence is shown in Fig. 14.17 [14.12] for 2 ranges. Here, atmospheric attenuation is assumed to be 0.15 km^{-1}; critical target dimension is assumed to be 3 m; and 1974 FLIR parameters are $\mathrm{MRT}_0 = 0.0254°\mathrm{C}$ and $\beta_{sys} = 0.996$ mrad·cycle^{-1} for the 8- to 12-μm wavelength region. Vibration amplitude in the object plane is equivalent to only 2.5 cm. At the larger range, vibrational blur has less of a relative effect because of the poor SNR caused by lower atmospheric transmission. Nevertheless, the degradation in target acquisition probability is clearly evident here, and even more so for the shorter range where SNR_N is higher.

The great improvements in modern thermal imager hardware MRT and β_{sys} values imply that vibrational blur has an even larger relative degradation effect on target acquisition probabilities using such sophisticated devices.

The preceding analyses refer to vibrational motion of the camera, which means target and background images suffer from identical vibrations. However, if the target is actually moving relative to the background, such motion can actually improve target acquisition probability despite the increased blur in the target image.

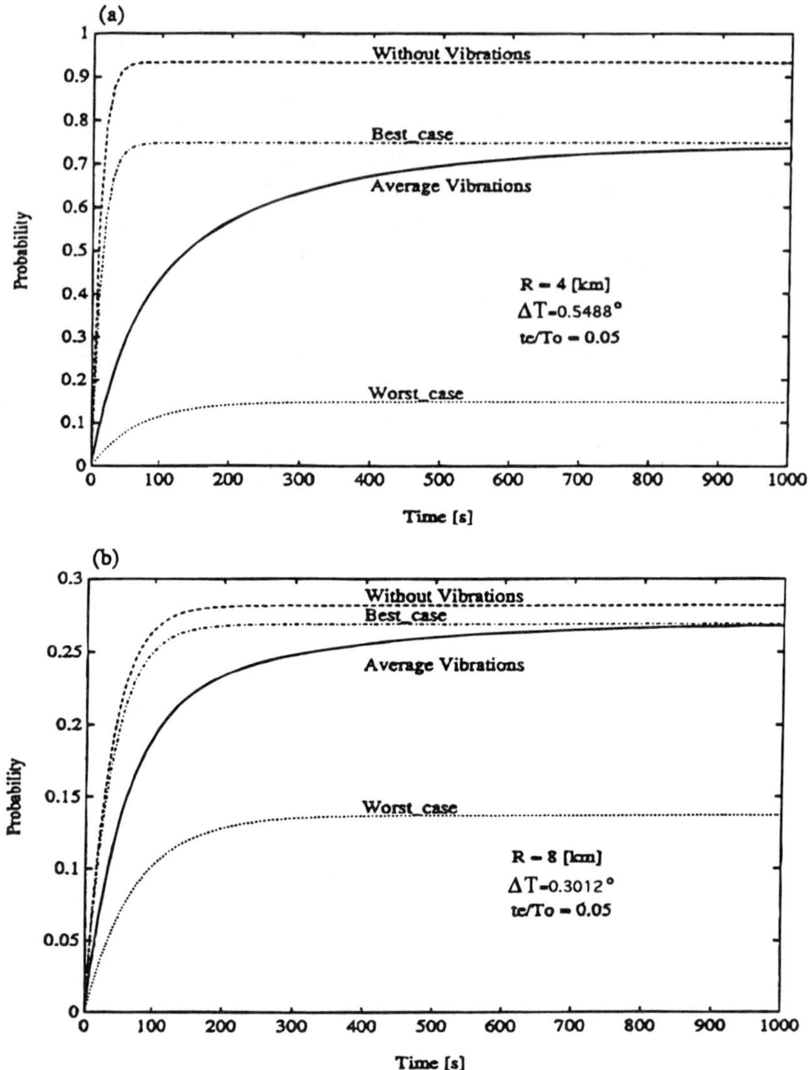

Fig. 14.17 Target acquisition probability results for $t_e/T_o = 0.05$: (a) 4-km range; (b) 8-km range (after [14.12]).

14.7 CONCLUSIONS

Optical transfer functions and examples involving their application in system design have been presented for uniform velocity, sinusoidal vibrations, and acceleration. For uniform velocity and high temporal frequency vibrational motion during exposure, closed-form analytical solutions are obtained to describe image blur, LSF, and OTF. For other types of motion, including random motion, a method of numerical calculation has been presented that is applicable to every type of motion in real time. MTF and PTF are obtained if the function of motion is known. This is usually easily obtainable with a miniature or microelectronic accelerometer attached to the image sensor, or often via a gyroscope. Methods of incorporating vibrational motion blur into target acquisition probabilities have been presented, based on motion MTF.

Knowledge of OTF permits high-resolution image restoration, as presented in Part 6.

AUTHOR'S NOTE

Recently a method of calculating motion OTF analytically has been developed for any type of motion [14.13]. Also, methods to calculate motion OTF without a motion sensor have been derived for a sequence of consecutive images [14.14] and even from a single motion-blurred image [14.15]. All three methods have been used successfully in image restoration.

REFERENCES

14.1. E. O'Donnell, "Image steadiness: a critical factor in system performance," in *Electron Optical Imaging System Integration*, R. Wright, Ed., *Proc. SPIE*, Vol. 762, 1987, pp. 51–56.

14.2. O. Hadar, I. Dror, and N. S. Kopeika, "Image resolution limits resulting from mechanical vibrations, part IV: real-time numerical calculation of optical transfer functions and experimental verification," *Opt. Eng.*, Vol. 33, February 1994, pp. 566–578.

14.3. N. Jensen, *Optical and Photographic Systems*, Wiley, New York, 1968, pp. 116–124.

14.4. T. Trott, "The effects of motion in resolution," *Photogramm. Eng.*, Vol. 26, 1960, pp. 819–827.

14.5. D. Wulich and N. S. Kopeika, "Image resolution limits resulting from mechanical vibrations," *Opt. Eng.*, Vol. 26, June 1987, pp. 529–533.

14.6. S. Rudoler, O. Hadar, M. Fisher, and N. S. Kopeika, "Image resolution limits resulting from mechanical vibrations, part II: experiment," *Opt. Eng.*, Vol. 30, May 1991, pp. 577–589.

14.7. R. A. Becker, *Introduction to Theoretical Mechanics*, McGraw-Hill, New York, 1954, p. 139.

14.8. O. Hadar, M. Fisher, and N. S. Kopeika, "Image resolution limits ranging from mechanical vibrations, part III: numerical calculation of modulation transfer functions," *Opt. Eng.*, Vol. 31, March 1992, pp. 581–589.

14.9. O. Hadar, I. Dror, and N. S. Kopeika, "Numerical calculation of image motion and vibration modulation transfer functions—a new method," in *Optomechanics and Dimensional Stability*, R. A. Paquin and D. Vukobratovich, Eds., *Proc. SPIE*, Vol. 1533, 1991, pp. 61–74.

14.10. O. Hadar, S. R. Rotman, and N. S. Kopeika, "Target acquisition modeling of forward-motion considerations for airborne reconnaissance over hostile territory," *Opt. Eng.*, Vol. 33, 1994, pp. 3106–3117.

14.11. S. C. Som, "Analysis of the effect of linear smear on photographic images," *J. Opt. Soc. Am.*, Vol. 61, 1971, pp. 859–864.

14.12. O. Hadar, S. R. Rotman, and N. S. Kopeika, "Thermal image target acquisition probabilities in the presence of vibrations," *Infrared Phys. Tech.*, Vol. 36, March 1995, pp. 691–702.

14.13. A. Stern and N. S. Kopeika, "Analytical method to calculate optical transfer functions for image motion and vibrations using moments," *J. Opt. Soc. Am. A.*, Vol. 14, Feb. 1995, pp. 388–396.

14.14. O. Hadar, M. Robbins, Y. Novogrozky, and D. Kaplan, "Image motion restoration from a sequence of images," *Opt. Eng.*, Vol. 35, Oct. 1996, pp. 2898–2904.
14.15 Y. Yitzhaky and N. S. Kopeika, "Identification of blur parameters from motion blurred images," *CVGIP: Graphical Models and Image Processing*, Vol. 59(5), pp. 321–332, Sep. 1997.

EXERCISES

14.1 Assume a certain imaging system is capable of 50 lines·mm^{-1} resolution with high contrast, that is, at spatial frequencies less than 50 lines·mm^{-1} modulation contrast function (MCF) is almost unity. At 50 lines·mm^{-1}, MCF drops suddenly to zero. The ratio of actual velocity to object plane distance is such that image motion v on the film is 4 mm·s^{-1}. What is the maximum exposure $t_{e\ \max}$ that does not involve image degradation beyond that inherent in the film itself? Assume a required threshold contrast of 3%.

14.2 Derive Eq. (14.4.6) based on the modulation contrast function.

14.3 Consider a stabilized TV tube with a zoom lens in which peak-to-peak high temporal frequency vibrations are limited to 100 μrad. The TV system is assumed to yield 700 TV lines, with an effective camera dimension of 11 mm in the direction of vibrations. Assume 10% required threshold contrast.

(a) What should zoom lens focal length be so that maximum resolution is achieved without degradation of image quality as a result of the vibrations?

(b) What is the angular resolution obtained at the above focal length? Plot resolution as a function of focal length.

(c) How is field of view affected by the resolution considerations.

14.4 Obtain an analytical expression of MTF for very short exposures centered at the center of sine wave vibration.

14.5 Obtain an analytical expression of MTF for very short exposures centered at the extrema of sine wave vibration.

Part Five

IMAGING THROUGH THE ATMOSPHERE

CHAPTER

15

Optical Properties of the Atmosphere

15.1 INTRODUCTION [15.1]

Many properties of the atmosphere affect the quality of images propagating through it. There are atmospheric phenomena that give rise to attenuation of the irradiance of the propagating image, thus reducing the contrast of the final image. There are also atmospheric phenomena that cause blurring of detail. Both types of phenomena prevent small detail from being resolved in the final image, thus degrading image quality.

Phenomena that give rise to attenuation are electromagnetic (EM) wave absorption and scattering by the constituent gases and particulates of the atmosphere and airborne particulates. Scattering of photons by airborne particulates is manifested as deflections of the photons to directions other than that of original propagation. If such scattering causes the deflected photons to miss the imaging receiver, then the scattering is manifested as attenuation. The received irradiance of the image propagated through the atmosphere is correspondingly diminished from that in the object plane. However, if the light scattering is at very small angles with respect to the original directions of propagation, and several such small-angle scattering events take place, then forward-scattered radiation can take round-about paths and still be received by the imaging system together with the unscattered radiation. The net effect is image blurring caused by a multitude of angles of arrival at the imaging receiver of radiation, emanating from the same point in the object plane, as illustrated in Fig. 15.1 for one multiply forward-scattered ray and one unscattered ray. Many such multiple-scattering paths give rise to a relatively large blurred-point image rather than a fine sharp-point image. Several adjacent object plane points can then appear as a single blurred image plane point, thus degrading image resolution.

Another effect of light scattering, particularly at large angles, is increased path radiance. This has been discussed in Section 10.9 and illustrated in Figs. 10.18 and 10.19. The atmospheric background irradiance H_A, at wavelengths less than 2 to 4 μm, consists of scattered background light such as sunlight through the day, as illustrated in Fig. 4.4. This atmospheric background radiation is imaged over the same image space as the received irradiance H_T from the target plane. The result is decreased contrast of the target plane scene. The effect is similar to turning on the

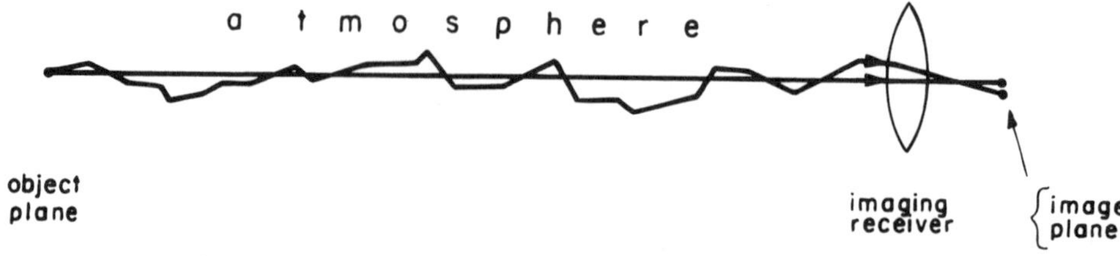

Fig. 15.1 Unscattered radiation (straight line) and multiply forward-scattered radiation from the same object point imaged at two different locations in the image plane. (after [15.1].)

lights in a movie theater. In the dark, almost all light reaching the eye emanates from the movie screen, thus providing good contrast. When lights are turned on, the percentage of overall light reaching the eye from the movie screen is diminished because of the background light. The result is decreased contrast from the movie screen. Similarly, atmospheric background light or path luminance gives rise to decreased contrast of the target plane scene because the percentage of radiation emanating from the target plane relative to overall radiation reaching the receiver is diminished by the intervening atmosphere background radiation, or path luminance. This makes it more difficult to resolve small detail. At wavelengths larger than 4 μm, most of the path radiance is not scattered sunlight, as in Fig. 10.18, but rather thermal emission of the atmospheric constituents, gases in particular, as shown in Fig. 4.4. This thermal path radiance decreases contrast in thermal imaging systems, particularly in the longwave (8–13 μm) infrared.

Although multiple forward scatter gives rise to image blur, another significant blur mechanism caused by the atmosphere is turbulence. Turbulence results from variations in the atmospheric refractive index, which are random in both time and space. These refractive index fluctuations are caused by local fluctuations in atmospheric temperatures, pressure, humidity, etc., all of which are affected by wind speed, which in itself is also random in time and space. Generally, the larger the temperature and humidity gradients, the more serious the image resolution degradation caused by turbulence. The lower the wind speed, the stronger the impact of turbulence whose limiting case is refractive bending such as mirages. Physically, the blur resulting from turbulence can also be described by Fig. 15.1, if it is assumed that deflections of radiation direction are now caused by random refractions instead of random scatterings. The net effect is still a variety of angles of arrival at the imaging receiver, thus causing image blur. However, scattering and turbulence are basically two very different physical mechanisms, each exhibiting quite different properties. Time dynamic properties such as scintillations and image dancing derive from turbulence, while aerosol scatter of light changes only slowly with time.

A point that must be strongly emphasized is that there is no such thing as a purely turbulent or purely scattering or purely absorbing atmosphere. All three properties occur simultaneously in the real-world atmosphere. Modeling and interpretation of experimental results must relate to all three processes. Any attempt to pretend otherwise is not realistic. In general, blur is caused by turbulence and small-angle scatter of light by aerosols and particles. Attenuation is caused by large-angle scatter and by absorption.

Each of the various optical properties of the atmosphere just described are now considered in turn.

15.2 ABSORPTION

Absorption of radiation is a mechanism by which the atmosphere is heated. The solar irradiance intercepted at the mean earth–sun distance is about 1365 to 1372 $W\cdot m^{-2}$, most of which is in the visible wavelength range with a peak at about 0.49-μm wavelength (Exercise 4.2). This corresponds to blue-green color. Part of this radiation is absorbed in the atmosphere, thus heating it, and part is transmitted to the ground. Part of the energy reaching the ground is absorbed, thus heating it, and part is reflected, thus further heating the atmosphere. For example, at a temperature of 288 K, the surface emits 390 $W\cdot m^{-2}$. Only 237 $W\cdot m^{-2}$ penetrates the atmosphere upward to space. The energy absorbed in the atmosphere is the 153 $W\cdot m^{-2}$ difference between surface emission and total energy loss. Because the atmosphere is generally colder than the ground during the day, atmospheric gases absorb more energy than they emit. The infrared absorption by the atmosphere is due primarily to water vapor and CO_2, with a smaller contribution from gases such as O_3, N_2O, and CH_4.

The atmospheric absorption is very wavelength selective, as shown in Fig. 15.2. Below about 0.2 μm, absorption by O_2 and O_3 is so great that there is essentially no propagation. This spectral region is referred to as the *vacuum ultraviolet* because at such wavelengths propagation is virtually possible only under vacuum conditions. Imaging through the earth's atmosphere here is practically impossible. (It is also for this reason that destruction of O_3 in the upper atmosphere is dangerous, for it permits ionizing radiation to reach the earth's surface.) On the other hand, there is hardly any absorption at visible wavelengths. This permits imaging using the human visual system. This absorption window continues out to about 1.3 μm.

As shown in Fig. 15.2, absorption windows also exist at 1.5 to 1.7 μm and 2.0 to 2.5 μm. The major thermal imaging infrared absorption windows are 3.4 to 4.2 μm, 4.5 to 5.0 μm, and 8.0 to 13 μm. In between are absorption bands where

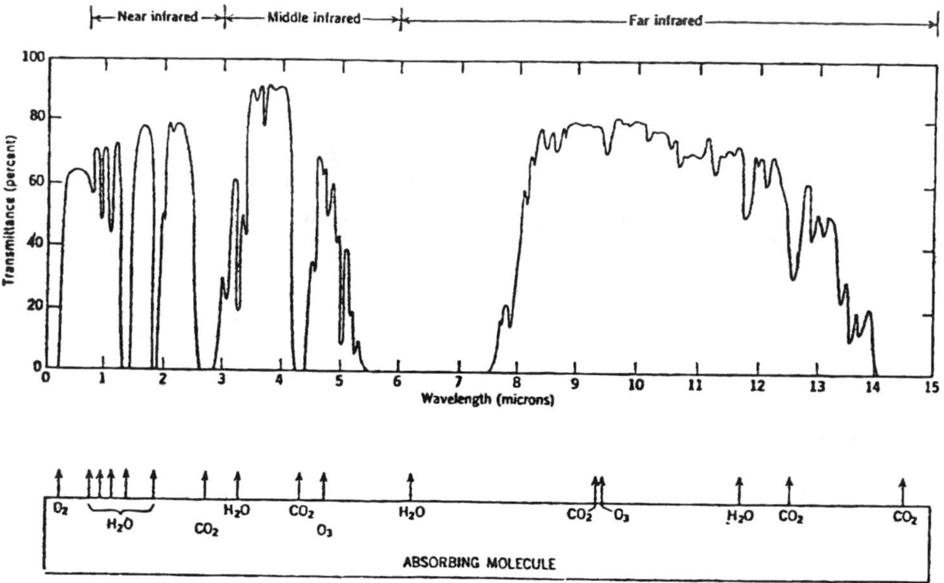

Fig. 15.2 Horizontal atmospheric transmission for 2-km path near ground (scattering not considered).

transmission is essentially negligible and, as a result, imaging through the atmosphere is essentially impossible. Absorption attenuation decreases over the 16- to 22-μm range, permitting limited transmission there. However, from 22-μm down to nearly 100-μm wavelength (300 GHz), water vapor absorption of radiation is so strong that imaging and transmission through the atmosphere are almost nonexistent. This is described in terms of electromagnetics in Section 1.3.2.

15.3 SCATTERING

Light scattering is very much wavelength dependent, according to the ratio of particulate radius to wavelength. Those particles that are small compared to the wavelength of the energy will produce Rayleigh scattering. The scattering coefficient in Rayleigh scattering is inversely proportional to the fourth power of the wavelength. For air molecules, scattering is negligible at wavelengths greater than about 3 μm. At a wavelength of 1 μm such scattering is no longer negligible, and in the visible band it is quite pronounced. The classical demonstration of its effect is the blue color of the sky on a clear day, caused by selective atmospheric scattering of the broad solar spectrum so that nearly four times more blue light is scattered to the observer than red light.

As shown in Section 3.5, the relationship between the scattering coefficient S_a, the absorption coefficient A_a, and transmission τ for a homogeneous medium is

$$\tau = \exp[-(A_a + S_a)]z = \exp(-\sigma z) \tag{15.3.1}$$

for a propagation path of length z, where σ is the extinction or attenuation coefficient. It, A_a, and S_a all vary with wavelength as described below. The atmosphere is essentially nonhomogeneous, so that

$$\tau = \exp\left[-\int_0^z [S_a(z') + A_a(z')]\, dz'\right] \tag{15.3.2}$$

The exponent in the above two equations is called *optical depth*. The *visual range* or "*visibility*" corresponds to the range at which 550-nm radiation is attenuated to 0.02 times its transmitted level and is equal to $3.912/\sigma_{550}$, where σ_{550} is atmospheric attenuation coefficient at 550-nm wavelength. Here Rayleigh scattering by molecules implies a visual range of about 340 km (213 miles). This situation is shown on the left in Fig. 15.3. Thus, Rayleigh scattering theory applies only to extremely clear air. Such a condition seldom, if ever, exists near the surface of the earth. The Rayleigh scattering coefficient for molecular scattering is therefore an extreme limit that can never be exceeded in practice. Rayleigh scattering applies when particulate radius r is much smaller than wavelength. Thus at optical wavelengths the atmospheric molecules themselves produce Rayleigh scatter.

The primary reason why Rayleigh, or molecular, scattering theory has limited applicability to visible and infrared energy transmission is that the air normally carries in suspension many particles that are comparable to or larger than the wavelength of the transmitted energy. The effect of these particles on the energy is explained by the Mie theory of scattering. Mie theory predicts that, because of diffraction effects, the scattering cross section of a particle relatively large compared to wavelength will be equal to about two times its physical cross section for all wavelengths much smaller

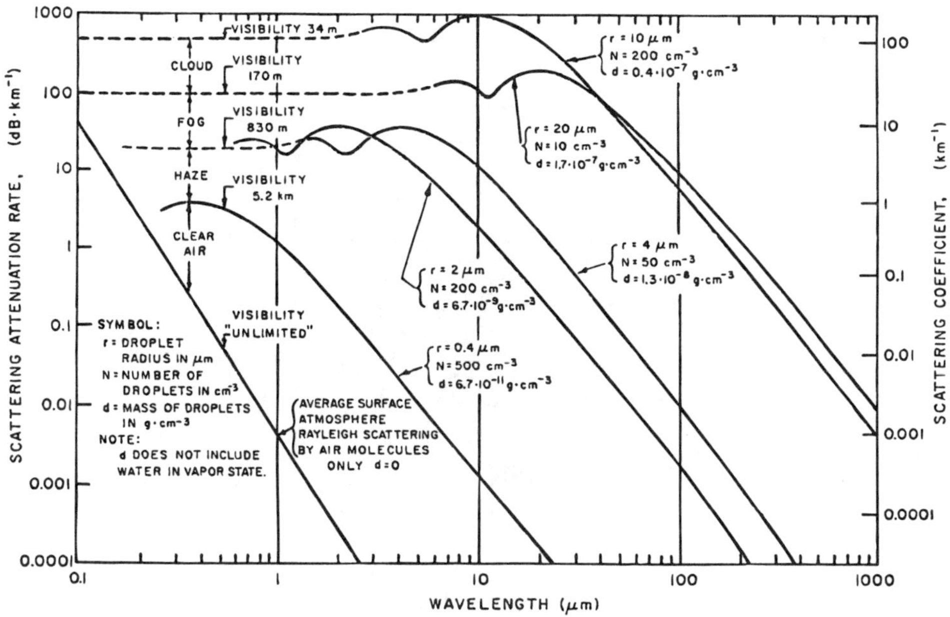

Fig. 15.3 Scattering coefficients for a variety of weather conditions (after [15.2]).

than the effective radius. Scattering losses decrease rapidly with increasing wavelength, approaching the Rayleigh scattering situation. The effects of Mie scattering are shown in Fig. 15.4, which illustrates the manner in which Mie scattering from water droplets affects scattering losses. Rayleigh scattering occurs at longer wavelengths which are much larger than particulate size. In this case S_a is proportional to λ^{-4}.

Some additional qualifications regarding Fig. 15.3 are in order. First, remember that these data apply only in the "window" bands for which absorption due to water vapor and CO_2 in the propagation path is negligible. Second, water droplets in haze, fog, and clouds are seldom uniform in size, so that the true characteristics will have less clearly defined "ripples" and "knees" than the curves shown in the figure, and will tend to be leveled out in most real situations. Third, no consideration has been given here to particles such as pollen, smoke, and combustion products that may be present in the air. Both scattering and absorption are involved in such conditions, and the absorption will tend to be selective according to wavelength.

Notice that in Fig. 15.4 the maximum scatter occurs at wavelengths essentially equal to particulate radius. As wavelength increases beyond that point, the scattering coefficient decreases with wavelength until for a given particulate size Rayleigh scattering conditions are obtained. The directionality of the scatter is very important in imaging applications. For Rayleigh scattering, the scattering phase function that describes the direction θ in which light is deflected is essentially proportional to $(1 + 0.9324 \cos^2\theta)$. Such scattering is significant in all directions and is manifested in imaging as attenuation and resultant loss of contrast. It is described in Fig. 15.5(a). However, as wavelength decreases toward maximum scatter conditions, and even beyond that point, the scatter becomes more and more directed into small angles in the forward direction. It is under these conditions that multiple scattering events significantly affect image quality by blurring, as shown in Fig. 15.1. Thus, image quality also depends largely on the ratio of particulate size to wavelength.

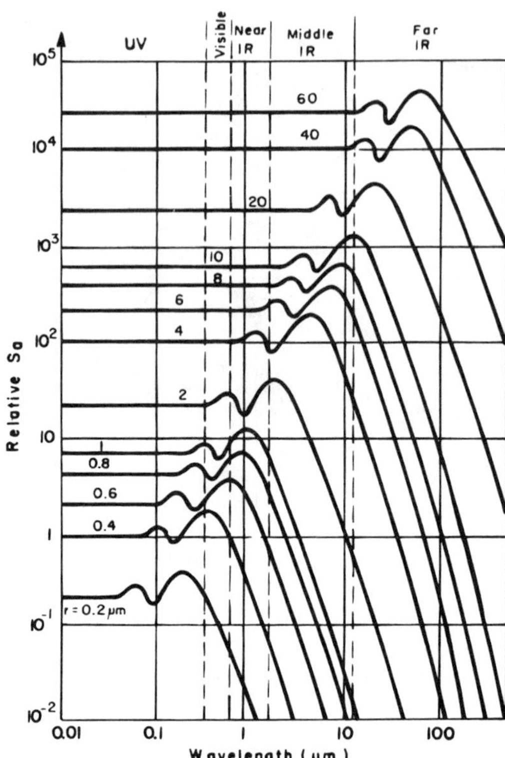

Fig. 15.4 Relative values of scattering coefficient as a function of wavelength for different particulate radii *r*.

Detailed data on values of particulate size and resulting scattering and absorption coefficients for various weather and climatic conditions can be obtained from the literature [15.2, 15.3] and from computer models such as the Air Force Geophysics Laboratory LOWTRAN 7 and MODTRAN computer programs. Often aerosol size distribution is modeled as bimodal or trimodal log-normal distributions.

Although scattered sunlight determines the color of the sky, it is the transmitted sunlight that determines the apparent color of the sun. Thus, even though, as shown in Fig. 4.3 the solar spectrum peaks at blue-green wavelengths, in clear weather the sun appears to be orangish or reddish rather than blue-green because of improved atmospheric transmission (or greatly reduced scatter) at the longer wavelengths.

In haze and fog conditions particulates increase in size because of water vapor adsorption and absorption. At visible wavelengths, the scattering becomes much less wavelength dependent and more neutral. As a result, both the sky (scattered light) and the sun (transmitted light) appear more and more "whitish" in color than in clear weather. This can be seen in Fig. 15.3.

The scattered versus transmitted irradiances play a significant role in overall image contrast in imaging through the atmosphere because the path luminance (Fig. 10.18) is determined by the background radiation in passive imaging systems. Atmospheric background radiance peaks strongly at around 0.5-μm wavelength for scattered sunlight and 10-μm wavelength for thermal emission of atmospheric gases for low elevations. A minimum between these two spectra exists at 2- to 4-μm

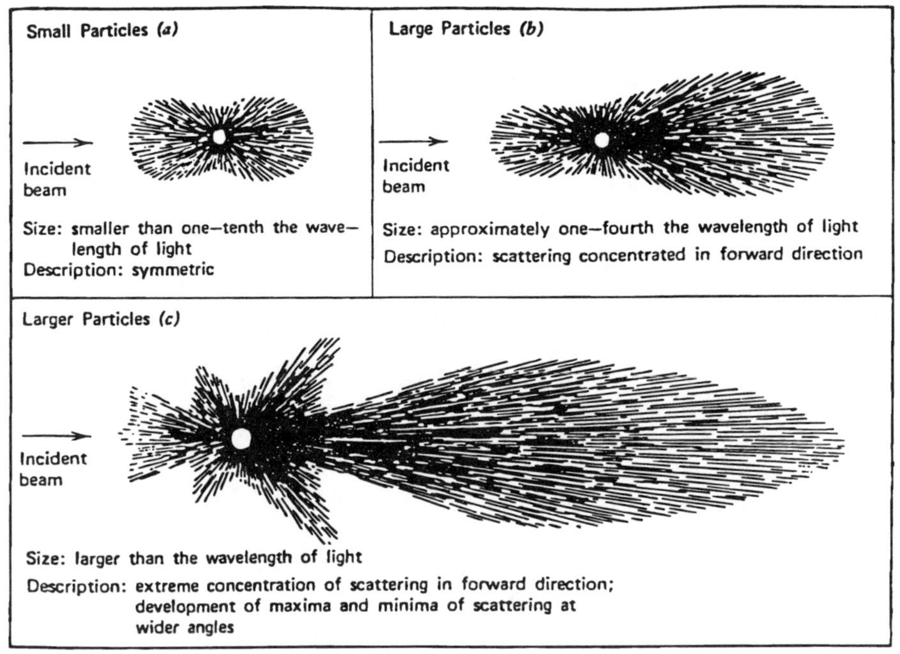

Fig. 15.5 Scatter diagrams for particulates of various sizes.

wavelength as shown in Fig. 4.4, and it is here that atmospheric background degradation of contrast is least. At visible and near-infrared wavelengths this atmospheric background radiance is a strong function of turbidity, with wavelength variation depending on the size distribution of airborne particle scatterers. Generally, particulate size distributions peak at radii on the order of submicrometer size, which is often due to pollution and reactions between various atmospheric gases. These both decrease the solar irradiance reaching the earth's surface and increase the atmospheric background radiance, primarily at near-UV and visible wavelengths.

Changing aerosol mass in the atmosphere can cause considerable turbidity fluctuations. This is true particularly on warm days when, as the ground heats up, thermal-induced convective mixing causes aerosol particles to be carried to higher altitudes and to higher concentrations. Such heating effects can cause large variations in turbidity between different seasons of the year and between different locations. For example, farming communities are usually much more turbid in the spring, summer, and fall when bare soil is exposed and such thermal activity is relatively high. Since the size of such particulates is relatively small compared to wavelength (except in the short end of the visible), scattering by them is generally at large angles and thus tends to decrease transmission and increase background radiance of the atmosphere. This is also characteristic of the early stages of fog formation, when there is a predominance of smaller droplets. However, as ambient humidity increases, water vapor may be adsorbed by aerosol particles suspended in the atmosphere. After ambient relative humidity has reached a threshold value such as 70%, which depends on the deliquescent property of the aerosol compound, the condensed water increases the size of the aerosol particles and tends to make them more spherical in shape. The size increase causes a greater proportion of the scattering to be small angle, or in the forward direction, thus increasing transmission when wavelengths are small compared to particle size [15.4]. Wind speed increases the impact and size of soil-derived particles that are airborne.

This increase in particle size and forward directionality of scattering with increased relative humidity is characteristic not only of common rural and urban aerosols but also of maritime salt-particle aerosols, desert dust, and fog particulates. Visibility is very closely linked to variations in relative humidity through these physical changes in aerosol size as the relative humidity changes. There is a general association of low transmittances with high relative humidity, and low relative humidity with high transmittances. The effects of ambient humidity and temperature changes are negligibly small when relative humidity is low. However, for a humid environment, a few percent increase in relative humidity or a few degrees decrease in temperature (which brings the atmosphere closer to 100% relative humidity) may cause a significant increase in scattering coefficient from aerosol particles of even small sizes at even middle-infrared (3.4- to 5-μm and 10-μm) wavelengths [15.4].

The main differences between infrared (IR) and visible light scattering can be explained by aerosol size distributions. The IR scattering is influenced mostly by aerosols with large sizes. When particulates are very large compared to optical wavelengths, the geometrical optics approximation suggests that scattering attenuation should be independent of wavelength, as seen on the left side of Fig. 15.4. However, numerous measurements have indicated that under rain, snow, and dust conditions there is less extinction in the visible than in the infrared. This difference of 10% to 20% has been attributed to scattering phase functions or diagrams being more sharply peaked in the forward direction for shorter wavelengths, thus increasing the forward directionality of their scatter. The increase in forward directionality of scatter is so pronounced that increase in field of view of optical communication receivers during snowstorms has been known to generate order-of-magnitude increases in the signal-to-noise ratio (SNR) at visible wavelengths [15.5]. In the calculation of scatter from aerosols, the Mie theory assumes that the particles are spherical. However, many atmosphere-borne particles are not spherical. Nevertheless, the assumption is made that because of random orientation of the particles, one can assume some "average" particle size. In many cases, this is a reasonable assumption. There are cases in which it is not [15.6]. One particular case is the one in which polarization properties are of interest. Another is in measuring backscatter. Another qualification of the Mie theory is that each particle is treated as being isolated from its neighbors. That is, it is assumed that the particles are in the far field of the scatter from each other. With this assumption it becomes possible to calculate the scatter and absorption properties of an ensemble of particles as the linear superposition of the scatter from each. The assumption of independent scattering is not always valid.

The Mie solution for scatter from a spherical particle is a lengthy but straightforward process that results in infinite series expressions for the field components polarized perpendicular and parallel to the scattering plane. Although the formal solution is lengthy, the series solutions for the field components are computed easily on a personal computer. The procedure begins by solving the vector wave equation in a spherical coordinate system, and obtaining a formal solution in terms of infinite series of spherical Bessel functions. By expanding the incident plane wave, the scattered field, and the internal field in vector spherical harmonics, and matching boundary conditions at the surface of the sphere, infinite series solutions for the field components are obtained [15.7–15.9], and are given in the Appendix at the end of this book. A detailed step-by-step procedure for how to calculate scattering (S_a), absorption (A_a), and extinction coefficients ($\sigma = S_a + A_a$) is presented in the Appendix, as well as a method for calculating the scattering diagram, called *phase function* [$P(\theta)$]. The solution for the scattered electric field is given in terms of two scalar components (S_1) and (S_2) perpendicular and parallel to the scatter plane in which θ is measured. These two scalar components are then used to determine the scattered intensities i_1 and i_2, from which $P(\theta)$ is obtained. The volume scattering and absorption coefficients, as well as the scattering phase function, depend strongly on the particulate size distribution and on the scattering size parameter since, as shown in Figs. 15.3 and 15.4, the scattering coefficient depends strongly on the ratio of particulate radius to wavelength. For particulates very small compared to wavelength, the Mie scattering solution reduces to Rayleigh scatter.

For vertical or slant path imaging, it is important to consider the vertical variability of particulate size distributions. Models for the vertical variability of atmospheric aerosols are generally broken into a number of distinct layers. In each of these layers a dominant physical mechanism determines the type, number density, and size distribution of particles. Figure 15.6 is an example of such a description. Generally accepted layer models consist of the following: a *boundary* layer that goes from 0 to 2 km, a region running from 2 to 6 km in which the number density displays an exponential decay with altitude (free troposphere), a stratospheric layer from 6 to 30 km, and layers above 30 km composed mainly of particles that are extraterrestrial in origin. Note that from 15- to 25-km altitudes there is usually an increase in aerosol light scatter caused by meteoric dust.

The average thickness of the aerosol-mixing region is approximately 2 km. This is the boundary layer. Within this region, one would expect the aerosol concentration

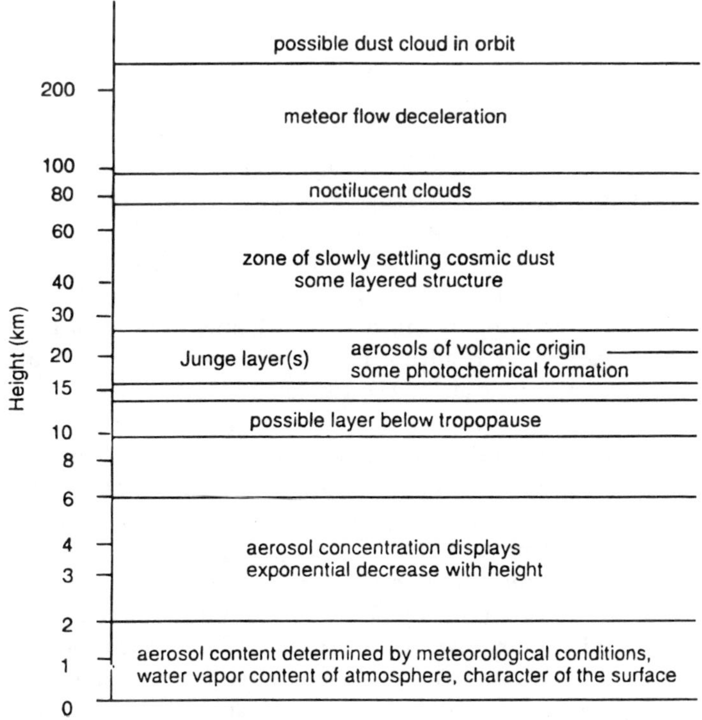

Fig. 15.6 Aerosol layers in the atmosphere (after [15.9]).

to be influenced strongly by conditions at ground level. Consequently, aerosols in this region display the highest variability with meteorological condition, climate, etc. In this layer particles are either transported from their point of origin into the atmosphere or are formed by gas-to-particle conversion or photochemical processes. In the maritime environment they are composed largely of sea-salt solutions and sulfates. Wind-blown dust from desert regions is an important constituent of continental aerosols. Generally, aerosol size distribution hardly decreases with height up to the boundary layer. Extant models describe environments for rural, urban, maritime, and desert areas and for fogs [15.10–15.13].

In the free troposphere layer, which extends from 2 to 6 km, an exponential decay is observed. Specifically, one often sees total number densities vary as follows:

$$N(z) = N(0)\exp\left(-\frac{z}{z_s}\right) \tag{15.3.3}$$

where z is height in km and the scale height z_s ranges from 1 to 1.4 km [15.10].

15.4 TURBULENCE

The refractive index n of the atmosphere is a function of the various meteorological parameters [15.14]. For example, for a marine atmosphere [15.15]

$$n \approx 1 + \frac{77p}{T} \times \left[1 + \frac{7.53 \times 10^{-3}}{\lambda^2} - 7733\,\frac{q}{T}\right] 10^{-6} = 1 + N \tag{15.4.1}$$

where p is air pressure (millibars), T is temperature (K), q is specific humidity ($g \cdot m^{-3}$), and λ is wavelength. By plugging in typical values for p, T, and q, one can see that $N << 1$. Nevertheless, the *fluctuations* in N, which depend largely on temperature and humidity gradients, strongly affect image quality because of the large number of random refractions that a light beam undergoes while propagating a relatively large distance through the atmosphere. The randomness in both time and space cause the incident light to be received at a large variety of angles of incidence, similar to the scattered light in Fig. 15.1. An example of such effects can be obtained by looking along a line of sight over an operating toaster at home. Because of the temperature fluctuations above the toaster resulting from the *moving* heated air, and because of the thermal gradient in the air above the toaster, the fluctuations in N are so great that even nearby objects appear to be dancing randomly and are somewhat blurred.

Since the ground is generally warmer than the surrounding air during the day and colder at night, it is analogous to the toaster example. Temperature fluctuations arise because of the moving heated air near the ground and a thermal gradient exists in the atmosphere that is usually greatest near the ground. Hence, image dancing and wavefront tilt result from random changes in N arising from random changes in air temperature and pressure in Eq. (15.4.1). Humidity fluctuations, particularly at low relative humidity, also give rise to similar effects. Such random refractions cause random wavefront tilt, which, when integrated over many tilt angles during the cause of an exposure, gives rise to image blur caused by optical turbulence.

Air temperature generally decreases with height, thus continuing the thermal gradients, albeit with smaller values, to increasing altitude. Water vapor is generally confined for the most part to regions near the ground. Thus, a humidity gradient also exists and basically decreases with increasing altitude.

Temperature and water-vapor fluctuations are known to be the primary mechanisms mediating the effects of atmospheric turbulence on the index of refraction. Pressure fluctuations can be neglected in the real atmosphere. Recent correlation coefficient studies at microwave frequencies, where the refractive index is much more sensitive to humidity than at optical frequencies, indicated, however, that wind speed was the dominant meteorological parameter affecting turbulence. In fact, increasing wind speed, because of the homogenizing effect on the index of refraction, was found to be the best meteorological predictor of reduced monthly accumulated microwave fading time. Higher wind speed can thus be expected to give improved imaging resolution as affected by turbulence, within certain limits. When wind speed is too high (above 8 $m \cdot s^{-1}$), it decreases air mixing and therefore increases turbulence strength. It also gives rise to increased aerosol loading as soil-derived particulates are uplifted by the wind. This increases both attenuation and blur. Nighttime clear skies have been associated with increased atmospheric degradation because of reduced vertical air mixing. The presence of clouds, on the other hand, tends to increase the mixing of air (decrease vertical stabilization) and thus to decrease the degradation of atmospheric imaging [15.14].

After sunrise and before sunset, ground and surrounding air temperatures are closest. At these times, when temperature gradients are least, turbulence has the least degradation effect on image quality.

15.4.1 Fluctuations in Refractive Index: Power Spectral Density

In modeling atmospheric turbulence it is customary to assume stationarity, homogeneity, and isotropism of atmospheric refractive index fluctuations [N in Eq. (15.4.1)]. Although these properties are not likely to be strictly true in the real world, they are utilized on the assumption that they are approximately true. Over some limited volume and time, such assumptions are reasonable. The key to the validity of such assumptions is in terms of local behavior [15.14].

A basic statistical parameter is the spatial cross-correlation of the refractive index, which is defined as

$$\Gamma_n(\bar{r}_1, \bar{r}_2) \equiv \langle N(\bar{r}_1) N(\bar{r}_2) \rangle \tag{15.4.2}$$

where $\langle \rangle$ signifies expected value. When N is wide sense stationary in three dimensions, then

$$\Gamma_n(\bar{\rho}) = \langle N(\bar{r}_1) N(\bar{r}_1 - \bar{\rho}) \rangle \tag{15.4.3}$$

where

$$\rho = \bar{r}_1 - \bar{r}_2 = (\Delta x, \Delta y, \Delta z) \tag{15.4.4}$$

The power spectral density of the fluctuations in N is the three-dimensional Fourier transform of $\Gamma_n(r)$, that is,

$$\Phi_n(\overline{K}) = \Phi_N(\overline{K}) = \frac{1}{(2\pi)^3} \int_{-\infty}^{\infty} \int_{-\infty}^{\infty} \int_{-\infty}^{\infty} \Gamma_n(\bar{r}) e^{j\bar{k}\cdot\bar{r}} d^3\bar{r} \tag{15.4.5}$$

where $\overline{K} = (K_x, K_y, K_z)$ is vector wave number and represents spatial frequencies. If N is statistically isotropic, the power spectrum will depend on the magnitude of $\overline{K}$ only, in which case, after integrating out the angular spherical coordinates,

$$\Phi_n(K) = \frac{1}{2\pi^2 K} \int_0^{\infty} \Gamma_n(r) r \sin(Kr)\, dr \tag{15.4.6}$$

$$\Gamma_n(r) = \frac{4\pi}{r} \int_0^{\infty} \Phi_n(K) K \sin(Kr)\, dk \tag{15.4.7}$$

where

$$K = [K_x^2 + K_y^2 + K_z^2]^{1/2} \tag{15.4.8}$$

$$r = [(\Delta x)^2 + (\Delta y)^2 + (\Delta z)^2]^{1/2} \tag{15.4.9}$$

When isotropicity is assumed, the power spectral density can be considered for different spatial frequency regions of turbulence fluctuations. To do so it is important to define the inner and outer scales. When $K < K_o = 2\pi/L_o$, where L_o represents the *outer scale* of turbulence, stationarity is essentially absent. Fortunately, such cases are not that frequent. The outer scale, L_o, represents the distance in Eq. (15.4.3) at which cross correlation goes to zero. It is the largest scale size at which homogeneity and isotropy are reasonable. Representative values of L_o are from tens to a few hundred meters. Near the surface it scales with height h above ground according to $L_o = 0.4h$. At the boundary layer it is typically hundreds of meters. It increases with strong turbulence [15.14].

The *inner scale* l_o is the small-scale size at which wind viscosity effects become dominant and energy is dissipated into heat [15.14]. Kolmogorov considered wind fluctuations and assumed that wind speed fluctuations can be represented by a locally homogeneous and isotropic random field for scales less than L_o. Between inner and outer scale distances a cascade of energy from large to small scales is assumed. The source of the energy may be convection or wind shear. At small scale sizes such energy is dissipated by viscosity. In between the inner and outer scales inertial forces dominate and fluctuations are assumed to be spatially homogeneous and isotropic. The inner scale ranges from a few millimeters near the surface to centimeters or more in the troposphere and stratosphere. In strong turbulence l_o decreases. Since simultaneously L_o broadens, the inertial range between them increases in strong turbulence [15.14].

The spatial frequency K_m is defined as being equal to $5.92/l_o$. For spatial frequencies between K_o and K_m, called the inertial ranges, by definition

$$\Phi_n(K) = 0.033 C_n^2 K^{-11/3} \qquad \text{for } K_o < K < K_m \tag{15.4.10}$$

where C_n^2 is called the *refractive index structure constant*, and is the parameter most commonly used to describe the strength of atmospheric turbulence. This is called the Kolmogorov spectrum. For large wave numbers (dissipation range), Tatarski introduced an empirical form of spectrum that is a reasonable approximation [15.16, 15.17], that is,

$$\Phi_n(K) = 0.033 C_n^2 K^{-11/3} \exp(-K^2/K_m^2) \qquad \text{for } K > K_m \qquad (15.4.11)$$

However, Eqs. (15.4.10) and (15.4.11) exhibit singularities at the origin in K space. This is not physically realizable since the size of the atmosphere is finite and therefore the turbulence power spectral density cannot be infinite. Experience indicates that Eqs. (15.4.10) and (15.4.11) should be amended to

$$\Phi_n(K) = 0.033 C_n^2 (K^2 + K_o^2)^{-11/6} \qquad \text{for } K_o < K < K_m \qquad (15.4.12)$$

$$\Phi_n(K) = 0.033 C_n^2 (K^2 + K_o^2)^{-11/6} \exp(-K^2/K_m^2) \qquad \text{for } K > K_m \qquad (15.4.13)$$

Equation (15.4.12) is called the *von Karman spectrum* and Eq. (15.4.13) is called the *modified von Karman spectrum.* A typical spectrum of turbulence fluctuations is shown in Fig. 15.7. Note that for spectral frequencies between K_o and K_m, the power spectrum decreases with wave number according to $K^{-11/3}$. Little is known of its behavior for $K < K_o$. These forms of the power spectrum are used for computational purposes and are not based on physical models. If results of calculations depend strongly on scales outside the inertial range these models must be used with caution. The numerical factor is empirical.

15.4.2 Fluctuations in Refractive Index: Structure Function

Treatment of spatial or temporal fluctuations of refractive index and accompanying power spectral density depends on assumptions of homogeneity or stationarity. Of particular importance is the assumption that the mean refractive index is a constant

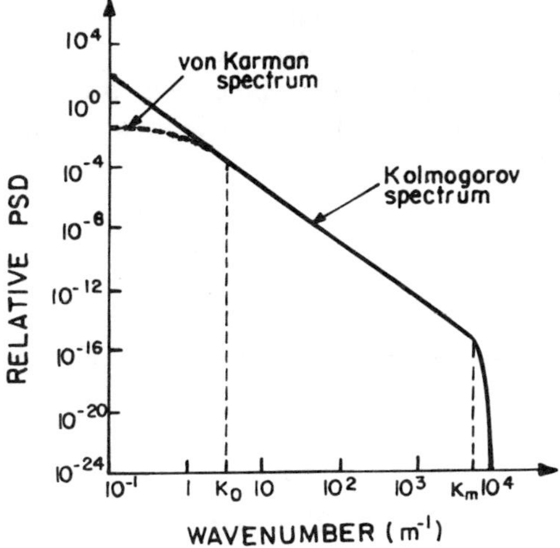

Fig. 15.7 Typical spectrum of turbulence power spectral density (PSD).

over all space and time. Such constancy is inconsistent with the dependence of refractive index on wind (air pressure), temperature, and humidity in Eq. (15.4.1). In the case of wind, for example, there is a mean that varies in space. *Structure functions* provide a framework for treatment of random processes with slowly varying means. The structure function for refractive index fluctuations is

$$D_n(\bar{\rho}) = \langle [n(\bar{r}_1) - n(\bar{r}_1 - \bar{\rho})]^2 \rangle \tag{15.4.14}$$

where $\bar{r}_1$ and $\bar{\rho}$ are defined in Eqs. (15.4.2), (15.4.3), and (15.4.4). Assuming stationarity so that $\langle \Delta n^2(r_1) \rangle = \langle \Delta n^2(\bar{r}_1 - \bar{\rho}) \rangle = \langle \Delta n^2 \rangle = \langle N^2 \rangle$, where Δn is fluctuations in refractive index, then using Eq. (15.4.1) this simplifies to

$$D_n(\bar{\rho}) = \langle [N(\bar{r}_1) - N(\bar{r}_1 - \bar{\rho})]^2 \rangle \tag{15.4.15}$$

which shows the utility of the structure function approach since for small enough separations it decomposes into the mean square *fluctuations* at two points spaced ρ apart. Using the Wiener-Khinchin theorem, for a locally homogeneous random field

$$D_n(\bar{\rho}) = 2 \int_{-\infty}^{\infty} \int_{-\infty}^{\infty} \int_{-\infty}^{\infty} \phi(\bar{K})[1 - \cos(\bar{K} \cdot \bar{\rho})] \, d\bar{K} \tag{15.4.16}$$

If the refractive index fluctuations are also locally isotropic, then [15.18]

$$D_n(\rho) = 8\pi \int_0^{\infty} \phi(K) \left[1 - \frac{\sin(Kr)}{Kr} \right] K^2 \, dK \tag{15.4.17}$$

The term in square brackets acts as a high-pass filter and removes spatial frequencies less than ρ^{-1}. In other words, contributions from scale size much larger than the separation between the two points are removed in the structure function so that D_n is affected mostly by inhomogeneities of dimensions ρ or less. This means that D_n is stationary over a small region. Kolmogorov showed by dimensional analysis that [15.16] in the inertial range

$$D_n(\rho) = C_n^2 \rho^{2/3} \qquad \text{for } l_o < \rho < L_o \tag{15.4.18}$$

This is the well-known $\rho^{2/3}$ power law.

Again assuming stationarity, from Eq. (15.4.15) D_n is related to Γ_n in Eq. (15.4.3) by

$$D_n(\rho) = 2[\Gamma_n(0) - \Gamma_n(\rho)] \tag{15.4.19}$$

where $\Gamma_n(0)$ is simply $\langle N^2 \rangle$ or $\langle \Delta n^2 \rangle$. The refractive index structure function and cross-correlation function are compared in Fig. 15.8. Note that $\Gamma_n(\rho)$ must decay to zero for $\rho \geq L_o$. Therefore, from the last two equations

$$C_n \approx \frac{\langle \Delta n^2 \rangle}{L_o^{2/3}} \tag{15.4.20}$$

Over very short distances [15.14],

$$D_n(\rho) = C_n^2 l_o^{-4/3} \rho^2 \tag{15.4.21}$$

In dealing with the correlation function under stationary conditions, it is sometimes useful to define a correlation length L_c beyond which refractive index fluctuations become nearly independent. If we consider this length to be the point where Γ_n is reduced to $1/e$ of its initial value, then $L_c \approx 0.5 L_o$ [15.19].

The basic parameter used to characterize turbulence MTF is C_n^2 defined by the structure function in Eq. (15.4.18) as well as by the power spectral density of turbulence fluctuations in Eqs. (15.4.12) and (15.4.13). Hence, discussion now centers on this parameter. Good image quality requires C_n^2 to be as small as possible. Since the refractive index itself varies according to meteorological parameters, it is natural to expect the same of C_n^2. Indeed, it has been shown that for a marine atmosphere [15.14, 15.20]

$$C_n^2 \approx \left(79 \times 10^{-6} \frac{P}{T^2}\right)^2 (C_T^2 + 0.113\, C_{Tq} + 3.2 \times 10^{-3} C_q^2) \tag{15.4.22}$$

where C_T^2 and C_q^2 are the air temperature and water vapor structure coefficients while C_{Tq} is the combined temperature–water vapor structure coefficient or covariance. In general [15.14],

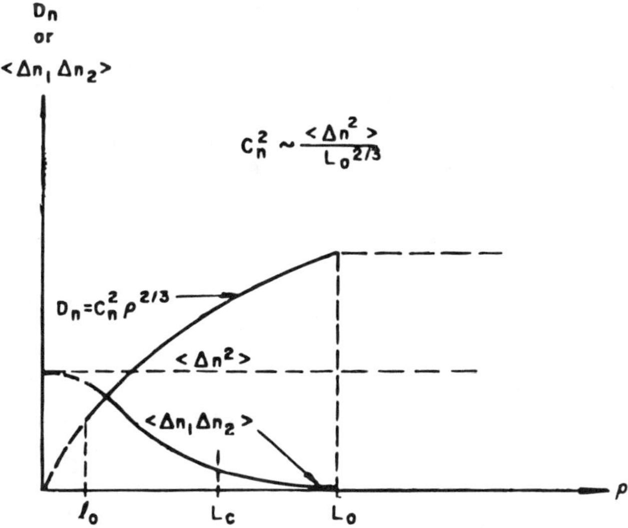

Fig. 15.8 Refractive index structure and cross-correlation functions (after [15.19]). © 1968 IEEE.

$$D_n(\rho) = \left(\frac{\partial T}{\partial n}\right)^2 D_T(\rho) + \left(\frac{\partial n}{\partial q}\right)^2 D_q(\rho) + 2\left(\frac{\partial n}{\partial T}\right)\left(\frac{\partial n}{\partial q}\right) \times \langle [T(\bar{r}_1) - T(\bar{r}_1 - \bar{\rho})][q(\bar{r}_1) - q(\bar{r}_1 - \bar{\rho})]\rangle \tag{15.4.23}$$

where D_T and D_q have 2/3 power law fluctuations similar to that for refractive index fluctuations in Eq. (15.4.18). Furthermore, since the fluctuations in T and q are locally homogeneous and isotropic, so is their cross variance, and the time-ensemble term in Eq. (15.4.23) can also be shown to satisfy the $\rho^{2/3}$ power law, which is used to determine C_{Tq} in Eq. (15.4.22). A general form of Eq. (15.4.22) is

$$C_n^2 = \left(\frac{\partial n}{\partial T}\right)^2 C_T^2 + \left(\frac{\partial n}{\partial q}\right)^2 C_q^2 + 2\left(\frac{\partial n}{\partial T}\right)\left(\frac{\partial n}{\partial q}\right) c_{Tq} \tag{15.4.24}$$

Often, the C_T^2 term dominates the expression for C_n^2, particularly over land. However, near a marine surface or saturated soil, humidity effects can seriously alter the overall value of C_n^2. Note that by definition of mean square surface fluctuations C_T^2 and C_q^2 are always positive. However, the C_{Tq} term can be negative, and has been known to essentially cancel both the C_T^2 and C_q^2 so as to reduce C_n^2 to almost zero [15.20]. Had one relied on the C_T^2 term alone under such conditions, the turbulence strength would have been seriously overestimated. Because D_T^2 and C_T^2 can be measured easily at discrete locations with fast response temperature sensors, it is convenient to use the first term only in Eq. (15.4.22) to measure C_n^2 at a given discrete location. However, caution is advisable with regard to neglect of the C_{Tq} term.

15.4.3 Path-Integrated Measurements of C_n^2

Since imaging is generally over a line of sight, a path-integrated average value of C_n^2 is more useful for turbulence MTF calculations than values of C_n^2 at discrete points. One way of measuring this path-integrated parameter is via angle-of-arrival fluctuations. For a plane wave, the variance in angle-of-arrival fluctuation is given by

$$\begin{aligned}\langle \alpha^2 \rangle &= 2.914\rho^{-1/3} \int_0^L C_n^2(z')\, dz' \\ &= 2.914\rho^{-1/3} C_{n\,\mathrm{eq}}^2 L \qquad \text{for } \sqrt{\lambda L} << D << L_o\end{aligned} \tag{15.4.25}$$

where L is path length and $C_{n\,\mathrm{eq}}^2$ is the path-integrated average value of C_n^2. Because it is primarily the larger scales that are responsible for these angle-of-arrival fluctuations, since phase variations smaller than the aperture are averaged it is customary to replace ρ with aperture diameter D, so that

$$C_{n\,\mathrm{eq}}^2 = \langle \alpha^2 \rangle D^{1/3}/(2.914L) \qquad \text{for } \sqrt{\lambda L} << D << L_o \tag{15.4.26}$$

For a spherical wave,

$$\langle \alpha^2 \rangle = 2.914 D^{-1/3} \int_0^L C_n^2(z)(z/L)^{5/3}\, dz \qquad \text{for } \sqrt{\lambda L} << D << L_o \tag{15.4.27}$$

where the spherical wave originates at $z = 0$. This is usually the object plane. For apertures approaching meters, which typify outer scale dimensions, $D^{-1/3}$ should be replaced by $(D^{1/3} - 1.4L_o^{-1/3})$ for both plane and sperical waves. Integration of Eq. (15.4.27) yields a value of $C^2_{n\,\mathrm{eq}}$ that is 3/8 of that in Eq. (15.4.25) for a plane wave. These formulations relating C^2_n to angle-of-arrival fluctuations are one dimensional. Thus, by considering the image dancing of a point object in a given direction (such as horizontal or vertical) or image dancing of a line object, Eqs. (15.4.26) and (15.4.27) can be used to determine the path-integrated average of C^2_n over the line of sight. For two-dimensional angle-of-arrival fluctuations the numerical coefficient is doubled, assuming isotropicity. This latter assumption may not necessarily hold, particularly on hot summer days when turbulence strength is high. In such cases, image dancing in the vertical direction may be slightly greater than in the horizontal direction. Since calculation of image dancing usually takes place in the focal plane of the system (for relatively large object distances) the mean square displacement δ of the image of the point or line can be obtained using the approximation $\delta = f_l\alpha$, in which case for a plane wave

$$C^2_{n\,\mathrm{eq}} = \langle\delta^2\rangle D^{1/3}/(2.914 f_l^2 L) \qquad \text{for } \sqrt{\lambda L} << D << L_o \qquad (15.4.28)$$

It is instructional to consider the integrand in Eq. (15.4.27) for a spherical wave. The factor $(z/L)^{5/3}$ indicates that image dancing is much stronger when it is caused by turbulence near the imager than when it derives from turbulence near the object plane where z is small. The physical cause has been shown to be that for spherical waves, large atmospheric refraction angles originating near the receiver can be still recorded in the image. This increases image blur. On the other hand, only for small atmospheric refraction angles originating near the object plane can the incoming radiation fall within the receiver field of view or even be incident on the aperture. This reduces blur. (Since a plane wave does not diverge, such angles of wavefront tilt caused by random refractions that enter the receiver field of view are independent of the relative location of the turbulence [15.21].) Hence, for a spherical wave $(z/L)^{5/3}$ is a spatial weighting function that modifies the effective value of C^2_n [15.21], and increases the effect of turbulence near the receiver on overall path-integrated average value of C^2_n. As explained at the beginning of this section, turbulence is most severe near the earth's surface. Hence, turbulence affects image quality much more severely when the imager looks upward than when it looks downward. Over long distances, a spherical wave can be approximated by a plane wave. However, use of Eq. (15.4.26) instead of Eq. (15.4.27) must be limited to situations where the turbulence is fairly homogeneous, such as over horizontal rather than vertical lines of sight. Use of Eq. (15.4.27) requires knowledge of C^2_n height profiles or slant paths [see Eqs. (15.4.35) and (15.4.36) and ensuing discussion].

When $l_o << D << \sqrt{\lambda L}$, the numerical coefficient in image dancing calculations should be 1.46 instead of 2.914 [15.22].

Another method of measuring path-averaged C^2_n is through scintillations. When an EM beam propagates through a turbulent medium, its amplitude through a fixed aperture fluctuates as a result of random partial interference at the receiver. This results from simultaneous reception of rays traversing through different physical paths caused by the random atmospheric refractions. An example is "twinkling" of stars, which is characterized by constructive and destructive interferences as functions of time at the receiver. When the turbulence is relatively weak, or the optical path is short, the amplitude fluctuations can be related to C_n^2 to yield [15.14, 15.16, 15.22, 15.23]

$$\sigma_x^2(L) = 0.56k^{7/6} \int_0^L C_n^2(z)(L - z)^{5/6}\, dz \tag{15.4.29}$$

for a plane wave, and

$$\sigma_x^2(L) = 0.56k^{7/6} \int_0^L C_n^2(z)(z/L)^{5/6}(L - z)^{5/6}\, dz \tag{15.4.30}$$

for a spherical wave. In these equations k is the wave propagation constant ($2\pi/\lambda$), x is defined as the logarithm of the field amplitude normalized to its free-space value, and σ_x^2 is the variance of x. It can be seen that scintillation from a spherical wave is less than that from a plane wave as shown by the weighting factor $(z/L)^{5/6}$. Scintillation favors turbulence far from the receiver. For constant C_n^2 as along a horizontal path,

$$\sigma_x^2(L) = 0.307k^{7/6}L^{11/6}C_{n\,\mathrm{eq}}^2 \tag{15.4.31}$$

for a plane wave, and

$$\sigma_x^2(L) = 0.124k^{7/6}L^{11/6}C_{n\,\mathrm{eq}}^2 \tag{15.4.32}$$

for a spherical wave. However, since detectors typically measure irradiance H it is important to relate amplitude fluctuations to irradiance fluctuations σ_H^2 through [15.22]

$$\sigma_x^2 = (1/4)\ln[1 + \sigma_H^2/\langle H^2\rangle] \tag{15.4.33}$$

In this way, measurement of average irradiance and irradiance fluctuations can be used to obtain path-integrated values of C_n^2. However, this method is valid only if (1) the irradiance fluctuations are log-normally distributed (weak turbulence or short optical path) rather than saturated and (2) if $D < \sqrt{\lambda L}$. The first condition implies that as the optical path or C_n^2 increases, σ_H^2 also increases. However, saturation occurs when irradiance fluctuations are so large that the irradiance fluctuates to zero much of the time. For this method of measurement, σ_H^2 should be less than 0.3 to 1.0. The latter condition implies scintillations are measured at one point only.

The angle-of-arrival method of $C_{n\,\mathrm{eq}}^2$ measurement, unlike that for scintillations, is valid also for strong turbulence.

Examples of irradiance fluctuations or scintillations are shown in Figs. 15.9 and 15.10. Figure 15.11 shows values of C_n^2 during the course of a typical summer day, at about a 15-m elevation. The diurnal minima near sunrise and sunset are quite obvious, as well as the maximum around midday. The sudden drops in C_n^2 at about 10:30 A.M., and 1 P.M., and midnight are a result of passing clouds.

15.4.4 Environmental Effects on C_n^2

Environmental effects of C_n^2 can be summarized as follows:

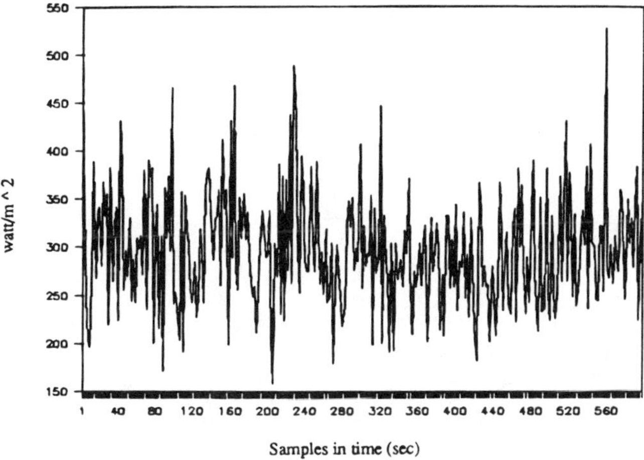

Fig. 15.9 CO_2 laser beam irradiance fluctuations over 4-km path (after [15.24]).

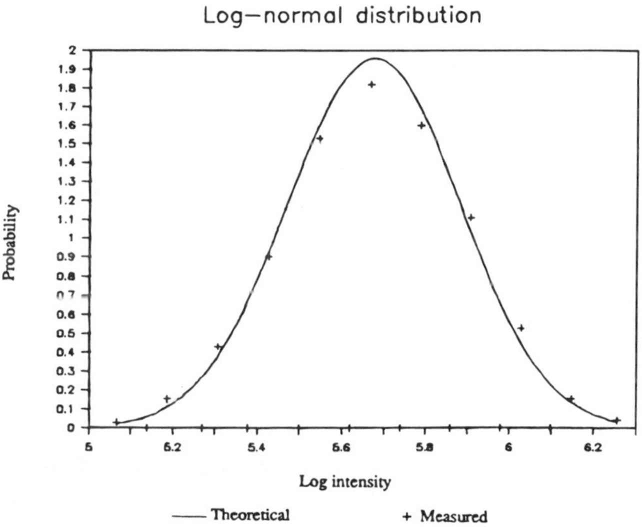

Fig. 15.10 Irradiance log-normal distribution of CO_2 laser beam scintillations in Fig. 15.10 (after [15.24]).

1. *Time of day*: C_n^2 is minimum at sunrise and sunset when air temperature is closest to ground temperature. Temperature gradient is generally greatest at midday and midnight when ground is warmer than overlying air, thus increasing C_n^2.
2. *Cloud cover*: During daytime, cloud cover limits surface heating by the sun, thus decreasing C_n^2. At night, cloud cover limits ground cooling, thereby increasing temperature gradient and thus C_n^2.
3. *Wind*: Wind produces more air mixing, thus decreasing temperature and humidity gradients, except for very strong winds on the order of 15 knots and higher.

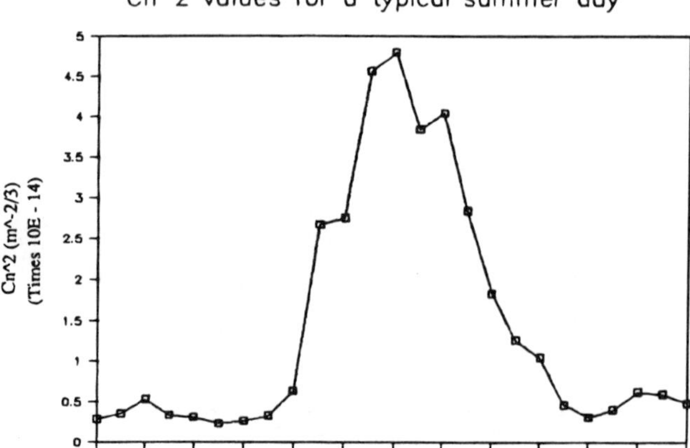

Fig. 15.11 C_n^2 values for a typical summer day as determined from He-Ne beam-of-arrival fluctuations (after [15.24]).

4. *Elevation*: C_n^2 is maximum near ground.
5. *Latitude*: Length of day and amount of solar heating are determined by latitude. There is less turbulence at higher latitudes.
6. *Surface conditions*: Surface moisture limits the humidity gradient, and thus C_n^2. On the other hand, surface roughness increases temperature gradient and thus C_n^2.

15.4.5 Prediction of C_n^2

It is useful to predict C_n^2 according to weather conditions, since this is almost equivalent to predicting seeing conditions if aerosol scatter MTF is predicted too. Furthermore, prediction of atmospheric MTF is important for cost-effective design, since it permits sensor selection according to atmospheric MTF, which often is the weakest link in imaging systems through the atmosphere. It makes no sense to use an expensive high-resolution sensor when the atmosphere does not permit such resolution, unless the turbulence is mitigated as discussed in Sections 16.7 and 18.5. In addition, knowledge of atmospheric MTF permits image restoration based on it, as described in Part 6 of this book. Therefore, many efforts have been made to connect atmospheric MTF to weather conditions. To predict C_n^2 notable computer programs such as IMTURB and PROTURB have been developed by scientists at the U.S. Army Atmospheric Sciences Laboratory [15.25, 15.26]. In addition, a simple empirical model [15.24] to predict C_n^2 has yielded excellent results in independent validation carried out by the U.S. Army Night Vision Laboratory through experiments in France. This model is based on standard meteorological parameters measured the world over. It is also based on the concept of temporal hours or relative part of day [15.27, 15.28].

Temporal hours are not fixed 60-min hours. A temporal hour is 1/12 of the time between sunrise and sunset. It is a sun clock hour. In the winter it is less than 60 min and in the summer it is greater. This concept provides a means of normalization so that, for example, three temporal hours mean one-quarter of a day after sunrise in both summer and winter.

Examination of the relations between C_n^2 values and the *th* (temporal hour) parameter have shown similar behavior as described in Fig. 15.11: high values of C_n^2 at noon (*th* = 06:00 h) and low values near sunrise (*th* = 00:00 h) and sunset (*th* = 12:00 h). That fact led to construction of a weight function with that kind of shape: For each temporal hour, a normalized weight was specified, and the weight vector was included in the regression model. The shape of the weight function was constructed from averaging several graphs of the type of Fig. 15.11 from different meteorological conditions. Table 15.1 contains all temporal hours and their respective weights. Values of *th* are obtained in the following way: one *th* is obtained by subtracting the hour of sunrise from that of sunset and dividing by 12. The current *th* is obtained by subtracting the hour of sunrise from the current hour and dividing by the value of 1 *th*. Note that the minimum and maximum values of the *th*, consequently, are not constants, and *th* can be negative (for time before sunrise).

Based on this concept of temporal hour, a simple but fairly reliable way to predict C_n^2 in units of $m^{-2/3}$ is

$$\begin{aligned} C_n^2 = {} & 3.8 \times 10^{-14}W + 2 \times 10^{-15}T - 2.8 \times 10^{-15}(\mathrm{RH}) + 2.9 \times 10^{-17}(\mathrm{RH})^2 \\ & - 1.1 \times 10^{-19}(\mathrm{RH})^3 - 2.5 \times 10^{-15}(\mathrm{WS}) + 1.2 \times 10^{-15}(\mathrm{WS})^2 \\ & - 8.5 \times 10^{-17}(\mathrm{WS})^3 - 5.3 \times 10^{-13} \end{aligned} \tag{15.4.34}$$

where W is temporal hour weight (Table 15.1), T is air temperature (K), RH is relative humidity (%), and WS is wind speed (m·s^{-1}). For elevations other than 15 m, the previously calculated value of C_n^2 can be scaled according to various models of the height profile of C_n^2. Although many models have been suggested, experiments [15.29] up to 100-m elevation support primarily the models of Tatarski [15.16] and Brookner [15.30] which are, respectively,

$$C_n^2(h) = C_{no}^2 h^{-4/3} \tag{15.4.35}$$

$$C_n^2(h) = C_{no}^2 h^{-5/6}\exp(-h/h_o) \tag{15.4.36}$$

TABLE 15.1 *Weight Function (W)*

	Temporal Hour Interval			Relative Weight (*W*)
		until	−4	0.11
	−4	to	−3	0.11
	−3	to	−2	0.07
	−2	to	−1	0.08
	−1	to	0	0.06
Sunrise	0	to	1	0.05
	1	to	2	0.10
	2	to	3	0.51
	3	to	4	0.75
	4	to	5	0.95
	5	to	6	1.00
	6	to	7	0.90
	7	to	8	0.80
	8	to	9	0.59
	9	to	10	0.32
	10	to	11	0.22
Sunset	11	to	12	0.10
	12	to	13	0.08
		over	13	0.13

where C_{n0} is refractive index structure coefficient at the surface and h_o is 320 m [15.30]. It has been suggested to replace the exponent in Eq. (15.4.35) with $-2/3$ for nighttime but this is often contradicted experimentally [15.14]. These height profiles would no longer be valid at the boundary level where C_n^2 suddenly decreases rapidly as elevation increases. Height profiles of C_n^2 including elevations above the boundary layer are summarized by Beland [15.14].

At higher altitudes, such as 10 to 25 km, C_n^2 increases significantly [15.14, 15.31] and exhibits a "belly" in its height profile similar to the "belly" exhibited in aerosol height profiles as a result of meteoric dust. These two "bellies" may not be coincidental. Indeed, a statistical connection has been observed between increases in C_n^2 and increases in aerosol loading [15.24]. This may be attributed to increases in temperature and humidity gradients produced by aerosol loading [15.32, 15.33]. A slightly more accurate empirical model [15.24] than Eq. (15.4.34) to predict C_n^2 includes aerosol loading as represented by the total cross-sectional area (TCSA) of the aerosol particulates in square centimeters per cubic meter, where

$$\begin{aligned}\text{TCSA} = {} & 9.69 \times 10^{-4}(\text{RH}) - 2.75 \times 10^{-5}(\text{RH})^2 + 4.86 \times 10^{-7}(\text{RH})^3 \\ & - 4.48 \times 10^{-9}(\text{RH})^4 + 1.66 \times 10^{-11}(\text{RH})^5 - 6.26 \times 10^{-3}\ln(\text{RH}) \\ & - 1.34 \times 10^{-5}(\text{SF})^4 + 7.30 \times 10^{-3}\end{aligned} \tag{15.4.37}$$

and

$$\begin{aligned}C_n^2 = {} & 5.9 \times 10^{-15}W + 1.6 \times 10^{-15}T - 3.7 \times 10^{-15}(\text{RH}) + 6.7 \times 10^{-17}(\text{RH})^2 \\ & - 3.9 \times 10^{-19}(\text{RH})^3 - 3.7 \times 10^{-15}(\text{WS}) + 1.3 \times 10^{-15}(\text{WS})^2 \\ & - 8.2 \times 10^{-17}(\text{WS})^3 + 2.8 \times 10^{-14}(\text{SF}) - 1.8 \times 10^{-14}(\text{TCSA}) \\ & + 1.4 \times 10^{-14}(\text{TCSA})^2 - 3.9 \times 10^{-13}\end{aligned} \tag{15.4.38}$$

In these equations, SF is solar flux in units of $\text{kW}\cdot\text{m}^{-2}$. Again, this model for C_n^2 applies to about 15 m of elevation. Equation (15.4.38) was obtained using Eq. (15.4.37). This second C_n^2 model is preferable for general use at night, and also in the day if aerosol loading is relatively high.

Although turbulence strength is known to depend on surface characteristics such as moisture and roughness, the preceding models have been validated over both desert sand as well as vegetation surfaces. The rationale is that macroscale weather parameters such as air temperature, relative humidity, wind speed, etc., are also affected by surface properties. For example, intensive vegetation increases humidity and also affects air temperature because of decreased absorption by CO_2 but increased absorption by H_2O. Desert sand with high albedo increases temperature gradient, decreases humidity, and affects air temperature as well. Therefore, such weather parameters contain information not only about the atmosphere, but also about the environment, including surface. Hence, this is the rationale for using them to predict C_n^2, in addition to the convenience, since such weather data are usually easily available. In these models, dynamic range for temperature is from 9 to 35°C, for relative humidity from 14% to 92%, and for wind speed from 0 to 10 $\text{m}\cdot\text{s}^{-1}$.

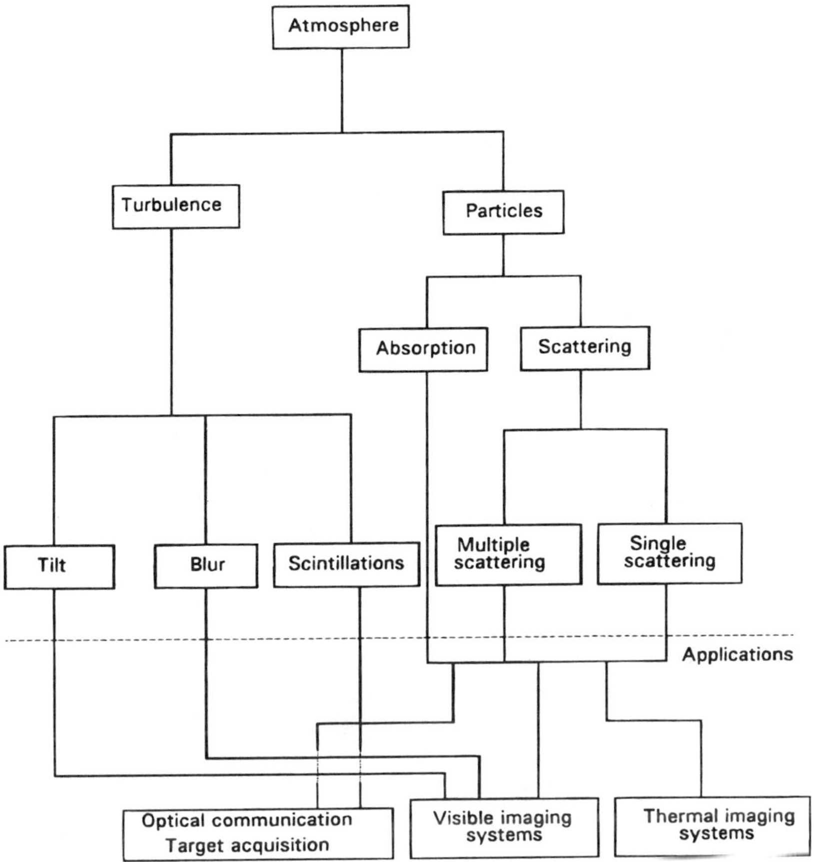

Fig. 15.12 Summary of optical properties of the atmosphere.

15.5 CONCLUSIONS

General properties of atmospheric absorption, scattering, and turbulence have been introduced here. For molecular absorption, more specific and fairly accurate data is available using computer programs such as LOWTRAN [15.12], FASCOD, and MODTRAN [15.9]. Absorption and large-angle scattering cause attenuation. The main mechanisms of image blur through the atmosphere are (1) forward small-angle scatter by airborne particulates and (2) optical turbulence. Modulation transfer functions for both are discussed in the next two chapters. Time dynamic properties such as scintillations and image dancing are attributed almost entirely to turbulence. Aerosol loading, however, indirectly affects such processes because of its effect on C_n^2. Aerosol scatter of light generally changes much more slowly with time. Historically, because turbulence has been considered to be the dominant and even only source of atmospheric blur, turbulence MTF is considered first. However, recent measurements do indicate that blur arising from forward light scatter by aerosols is often more dominant particularly at elevations more than several meters above the surface [15.34]. As shown in Chapter 17, absorption also indirectly enhances the relative blur caused by light scattering.

Figure 15.12 is a block diagram summarizing the optical properties of the atmosphere. These are elaborated on further in the next two chapters.

AUTHOR'S NOTE

There are indications of an increase or bump, called the "Hill bump", around wavenumber K_m in Fig. 15.7 [15.35, 15.36]. Analytic approximations have been developed for it [15.37].

REFERENCES

15.1. N. S. Kopeika, "Imaging through the atmosphere," in *Encyclopedia of Physical Science and Technology*, Vol. 7, Academic Press, New York, 1992, pp. 811–827.

15.2. N. S. Kopeika and J. Bordogna, "Background noise in optical communication systems," *Proc. IEEE*, Vol. 58, October 1970, pp. 1571–1577.

15.3. I. Dror and N. S. Kopeika, "Prediction of aerosol distribution parameters according to weather forecast: status report," in *Atmospheric Propagation and Remote Sensing*, A. Kohnle and W. B. Miller, Eds., *Proc. SPIE*, Vol. 1688, 1992, pp. 123–131.

15.4. N. S. Kopeika, "Imaging through the atmosphere for airborne reconnaissance," *Opt. Eng.*, Vol. 26, November 1987, pp. 1146–1154.

15.5. J. R. Clark, J. R. Baird, and R. S. Rearden, Jr., "Low visibility optical communications: received signal level as a function of receiver field-of-view," *Appl. Opt.*, Vol. 15, 1976, pp. 314–316.

15.6. D. W. Schuerman, Ed., *Light Scattering by Irregular Shaped Particles*, Plenum Press, New York, 1979.

15.7. C. F. Bohren and D. R. Huffman, *Absorption and Scattering of Light by Small Particles*, Wiley, New York, 1983.

15.8. D. Deirmendjian, *Electromagnetic Scattering on Spherical Polydispersions*, American Elsevier, New York, 1969.

15.9. M. E. Thomas and D. D. Duncan, "Atmospheric transmission," in *Atmospheric Propagation of Radiation*, F. G. Smith, Ed., in *The Infrared and Electro-Optical Systems Handbook*, Vol. 2, J. S. Accetta and D. L. Shumaker, executive editors, Environmental Research Institute of Michigan, Ann Arbor, and SPIE Optical Engineering Press, Bellingham, WA, 1993.

15.10. E. J. McCartney, *Optics of the Atmosphere*, Wiley, New York, 1976.

15.11. R. W. Fenn, S. A. Clough, W. O. Gallery, R. E. Good, F. X. Kneizys, J. P. Mill, L. S. Rothman, E. P. Shettle, and F. E. Voltz, "Optical and infrared properties of the atmosphere," Chapter 18, in *Handbook of Geophysics and Space Environment*, A. S. Jeursa, ed., Air Force Geophysics Laboratory, 1985.

15.12. F. X. Kneizys, E. P. Shettle, L. W. Abreu, J. H. Shetwynd, G. P. Anderson, W. O. Gallery, J. E. A. Selby, and S. A. Clough, "Users Guide to LOWTRAN 7," Report AFGL TR-88-0177, Environmental Research Papers No. 1010, Air Force Geophysics Laboratory, Hanscom AFB, MA, August 16, 1988.

15.13. R. A. McClatchey, R. W. Fenn, J. E. A. Selby, F. E. Volz, and J. S. Garring, "Optical properties of the atmosphere," in *Handbook of Optics*, W. Driscoll, Ed., McGraw-Hill, New York, 1978, pp. 14–54.

15.14. R. R. Beland, "Propagation through atmospheric optical turbulence," in *Atmospheric Propagation of Radiation*, F. G. Smith, Ed., in *The Infrared and Electro-Optical Systems Handbook*, Vol. 2, J. S. Accetta and D. L. Shumaker, executive editors, Environmental Research Institute of Michigan, Ann Arbor, and SPIE Optical Engineering Press, Bellingham, WA, 1993, pp. 157–232.

15.15. C. W. Fairall, K. L. Davidson, and G. E. Schachter, "Meteorological models for optical properties in the marine atmospheric boundary layer," *Opt. Eng.*, Vol. 26, 1982, pp. 847–857.

15.16. V. I. Tatarski, *Wave Propagation in a Turbulent Medium*, McGraw-Hill, New York, 1961.

15.17. V. I. Tatarski, "The effects of the turbulent atmosphere on wave propagation," NOAA Report TT68-50464, U.S. Department of Commerce, Springfield, VA, 1971.

15.18. S. F. Clifford, "The classical theory of wave propagation in a turbulent medium," in *Laser Beam Propagation in the Atmosphere*, J. W. Strohbehn, Ed., Springer Verlag, New York, 1978, Chap. 2.

15.19. H. Hodara, "Effects of a turbulent atmosphere on the phase and frequency of optical waves," *Proc. IEEE*, Vol. 56, December 1968, pp. 2130–2136.

15.20. C. A. Friehe, J. C. La Rue, F. H. Champagne, C. H. Gibson, and G. F. Dreyer, "Effects of temperature and humidity fluctuations on the optical refractive index in the marine boundary layer," *J. Op. Soc. Am.*, Vol. 65, 1975, pp. 1502–1511.

15.21. D. Sadot, D. Shemtov, and N. S. Kopeika, "Theoretical and experimental investigation of image quality through an inhomogeneous turbulent medium," *Waves in Random Media*, Vol. 4(2), 1994, pp. 177–189.

15.22. R. L. Fante, "Electromagnetic beam propagation in a turbulent media: an update," *Proc. IEEE*, Vol. 68, 1980, pp. 1424–1443.

15.23. A. Ishimaru, *Wave Propagation and Scattering in Random Media*, Academic Press, New York, 1978.

15.24. D. Sadot and N. S. Kopeika, "Forecasting optical turbulence strength on the basis of macroscale meteorology and aerosols: models and validation," *Opt, Eng.*, Vol. 31, February 1992, pp. 200–212.

15.25. W. B. Miller and J. C. Ricklin, "IMTURB: a module for imaging through optical turbulence," Report ASL-TR-0221-27, U.S. Army Atmospheric Sciences Laboratory, 1990.

15.26. W. J. Stewart, "A propagation model for Gaussian beam waves in clear air turbulence," Report ASL-CR-88-0001-2, U.S. Army Atmospheric Sciences Laboratory, 1988.

15.27. Babylonian Talmud, Berachot, Vilna edition, p. 9b.

15.28. Shulchan Aruch Orach Haim, 58:1.

15.29. N. Ben-Yosef, E. Tirosh, A. Weitz, and E. Pinsky, "Refractive-index structure constant dependence on height," *J. Opt. Soc. Am.*, Vol. 69, November 1979, pp. 1616–1618.

15.30. E. Brookner, "Improved model for the structure constant variations with altitude," *Appl. Opt.*, Vol. 10, August 1971, pp. 1960–1962.

15.31. G. R. Ochs, T. Wang, R. S. Lawrence, and S. F. Clifford, "Refractive-turbulence profiles measured by one-dimensional spatial filtering of scintillations, *Appl. Opt.*, Vol. 15, October 1976, pp. 2504–2510.

15.32. G. de Leeuw, "Profiling of aerosol concentrations, particle size distributions and relative humidity in the atmospheric surface layer area over the North Sea," *Tellus*, Vol. 43B, February 1990, pp. 342–354.

15.33. J. R. Hummel and E. P. Shettle, "Effects of solar heating by aerosols and trace gases on the temperature structure constant," Report GL-TR-90-0349, U.S. Air Force Geophysics Laboratory, Hanscom AFB, MA, August 9, 1990.

15.34. I. Dror and N. S. Kopeika, "Experimental comparison of turbulence MTF and aerosol MTF through the open atmosphere," *J. Opt. Soc. Am. A.*, Vol. 12, May 1995, pp. 970–980.

15.35 R. J. Hill, "Models of the scalar spectrum for turbulent advection," *J. Fluid Mech.*, Vol. 88, 1978, pp. 541–562.

15.36 R. J. Hill and S. F. Clifford, "Modified spectrum of atmospheric temperature fluctuations and its application to optical propagation," *J. Opt. Soc. Am.*, Vol. 68, 1978, pp. 892–899.

15.37 R. Frehlich, "Laser scintillation measurements of the temperature spectrum in the atmospheric surface layer," *J. Atm. Sci.*, Vol. 49, 1992, pp. 1494–1509.

CHAPTER

16

Turbulence Modulation Transfer Function

16.1 LONG AND SHORT EXPOSURES

The MTF for long-exposure atmospheric turbulence can be described by [16.1, 16.2]

$$M_{TL}(f_a) = \exp\left[-57.53 f_a^{5/3} \lambda^{-1/3} \int_0^L C_n^2(z)\, dz\right] \tag{16.1.1a}$$

for a plane wave, and

$$M_{TL}(f_a) = \exp\left[-57.53 f_a^{5/3} \lambda^{-1/3} \int_0^L C_n^2(z)(z/L)^{5/3}\, dz\right] \tag{16.1.1b}$$

for a spherical wave propagating through an inhomogeneous medium with object plane at $z = 0$. In the preceding equations, f_a is angular spatial frequency (cycles·rad^{-1}), λ is wavelength, and L is path length. Angular spatial frequency is equal to the product of spatial frequency (f_r) and focal length (f_l), and C_n is the refractive index structure coefficient defined in Eqs. (15.4.18), (15.4.12), and (15.4.13). The greater the value of C_n, the greater the degradation imposed by the turbulence. For short exposures (<<10 ms) the degradation is less. For aperture diameter D that MTF is [16.2]

$$M_{TS}(f_a) = M_{TL}(f_a)\exp\left[1 - \frac{1}{b}\left(\frac{f_a\lambda}{D}\right)^{1/3}\right] \tag{16.1.2}$$

where $b = 2$ at the image center and unity at the edges. Here, the center is defined as encompassing radial coordinates in the image plane much less than $\sqrt{L\lambda}$. The situation $b = 2$ thus corresponds to relatively long or far-field path lengths. These MTF expressions, including the 5/3 power law dependence for spherical waves, have been confirmed experimentally [16.3]. The 5/3 power law or weighting factor affects spherical rather than plane waves because the former are diverging while the latter are not [16.3] (see Fig. 1.1 in Chapter 1). Because of such spherical wave divergence, turbulence near the sensor gives rise to larger tilt angles and therefore blur radii actually recorded in the image than those caused by turbulence further away. The reason is that the angle subtended by the receiver is larger at short distances to the receiver than at larger ones. Hence, large tilt angles emanating from more distant turbulence may not be intercepted by the receiver [16.3].

We can see from the preceding equations that long-exposure turbulence, unlike diffraction MTFs, favors imaging at longer wavelengths [16.4]. The improvement is very small but is increased when imaging with short exposures.

Because of the vertical profile of $C_n^2(z)$ it is not correct to approximate a spherical wave by a plane wave for vertical or slant paths (see Exercise 16.1). As with the angle-of-arrival fluctuations described in the previous chapter, imaging upward is much more degraded by turbulence than imaging downward for a spherical wave because C_n^2 is greatest near the surface.

A typical comparison of long- and short-exposure turbulence MTFs for imaging horizontally is presented in Fig. 16.1. The difference between long- and short-exposure degradations is that short exposures involve fewer angles of arrival, thus improving the image quality [16.5]. When the incoming light interacts with large turbulence eddies, the light tends to be deflected, whereas for small turbulence eddies objects tend to appear broadened. The longer the exposure, the greater the number of deflections or beam wander, thus causing further blurring of the short-exposure image [16.6].

A limitation involved in short-exposure imaging is that although there are fewer and perhaps even only one angle of arrival at a given location, *that angle itself is apt to vary randomly over the receiver aperture.* Thus the probability of obtaining a good

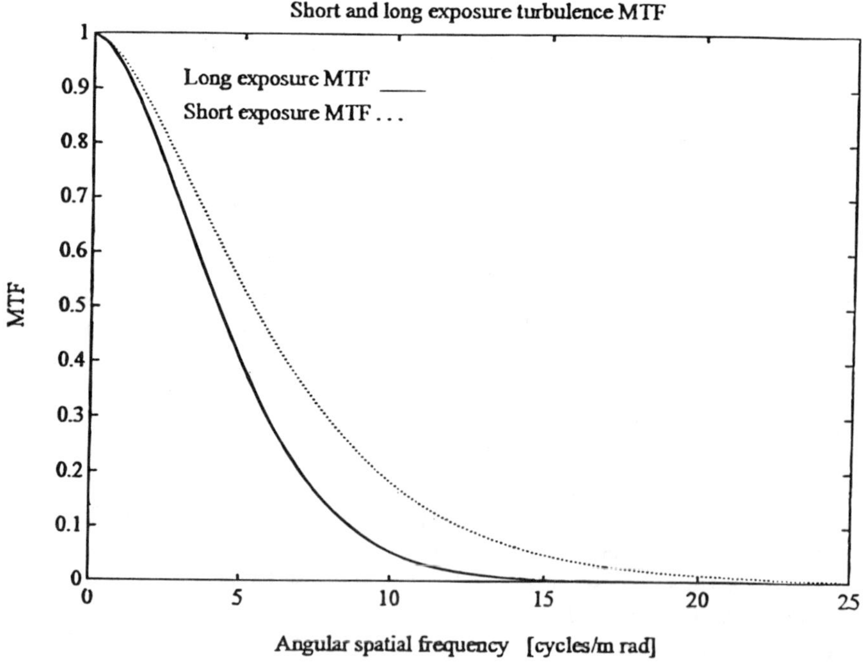

Fig. 16.1 Short- and long-exposure turbulence MTF calculations at image center for 633-nm wavelength, where $D = 89$ mm and $L = 5$ km. This is applicable to imaging horizontally where C_n^2 is representative of a path-integrated average value of 5×10^{-14} m$^{-2/3}$.

quality "lucky shot" image decreases as receiver area increases, since there can be more multiple images of small detail, even though each may be of good quality. An example of this can be seen in Fig. 16.2 where a star is imaged with different exposure times through a large telescope. Although it may be desirable to use large apertures to compensate for reduced energy received during short exposures, there is no guarantee of a "lucky" shot. The probability of a lucky shot varies [16.8] as $\exp(-D^2)$. Equation (16.1.2) indicates that the short-exposure MTF improvement over long exposures decreases for larger apertures.

16.2 EFFECT OF FOCAL LENGTH

In Eqs. (16.1.1), since angular spatial frequency f_a is equal to $f_r f_l$, optics focal length can be a factor in system design. A large focal length decreases field of view but increases magnification. Increased magnification can either permit resolution of smaller detail or simply magnify the turbulence-derived blur without any resolution improvement, depending on whether or not it is turbulence that is limiting the resolution. Two cases are considered here. In the first, effect of focal length on turbulence alone is considered. In the second, effect of focal length on a TV system imaging through optical turbulence is considered.

Example 16.1

Compare turbulence MTF and size of resolvable detail over a 10-km horizontal path through long-exposure turbulence characterized by C_n^2 equal to 1.3×10^{-15} cm$^{-2/3}$. Assume 0.7-μm wavelength and 2% threshold contrast and consider focal lengths equal to 1 and 5 m.

Solution

For a horizontal path, a constant path-integrated value of C_n^2 can be considered, in which case Eq. (16.1.1b) simplifies, after integration, to

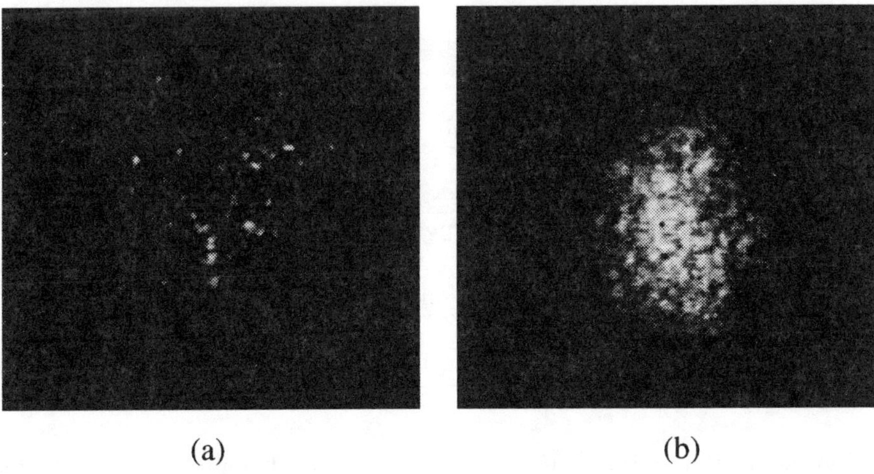

(a) (b)

Fig. 16.2 Long- (4-s) and short- (1/30-s) exposure images of a star using the 4-m telescope at Kitt Park Observatory with field of view just under 3 s of arc. (a) Short-exposure image, and (b) long-exposure image (after [16.7]).

$$M_{TL}(f_r) = \exp[-(3/8)57.53 f_r^{5/3} f_l^{5/3} \lambda^{-1/3} C_n^2 L] \tag{16.2.1}$$

since the atmosphere at a given time is near homogeneous for constant elevation. By changing all units to centimeters, the above for $f_l = 100$ cm is equal to

$$\begin{aligned} M_{TL}(f_r) &= \exp[-(3/8)57.53 f_r^{5/3}(100)^{5/3}(7 \times 10^{-5})^{-1/3} \cdot 1.3 \times 10^{-15} \cdot 10^6] \\ &= \exp(-1.50 \times 10^{-3} f_r^{5/3}) \end{aligned} \tag{16.2.2}$$

If limiting resolution occurs at $M_{TL} = 2\%$, then

$$\ln(0.02) = -3.912 = -0.015\, f_{r\,\max}^{5/3} \tag{16.2.3}$$

$$f_{r\,\max} = \left(\frac{3.912}{0.0015}\right)^{3/5} = 112.15 \text{ cycles}\cdot\text{cm}^{-1} \tag{16.2.4}$$

where $f_{r\,\max}$ was defined in Figs. 10.1.

For a 5-m focal length,

$$M_{TL}(f_r) = \exp(-0.0219 f_r^{5/3}) \tag{16.2.5}$$

$$f_{r\,\max} = \left(\frac{3.912}{0.0219}\right)^{3/5} = 22.45 \text{ cycles}\cdot\text{cm}^{-1} \tag{16.2.6}$$

Equations (16.2.2) and (16.2.5) are plotted in Fig. 16.3. It is clear that increased focal length gives rise to decreased $f_{r\,\max}$ for turbulence by itself. Our question is how does focal length increase affect resolution as limited by turbulence. To answer this question, consider Eq. (10.2.4), which is the conceptual basis of the Johnson chart, that is,

$$\frac{\Delta s}{s} = -\frac{\Delta x'}{s'} = -\frac{1}{2 f_{r\,\max} f_l} \tag{16.2.7}$$

Substituting data for 1-m focal length, the resolution angle is

$$\frac{\Delta x}{s} \approx -\frac{1}{2(112.15)(100)} = -\frac{1}{22430} = -4.4583 \times 10^{-5} \text{ rad} \tag{16.2.8}$$

while for a 5-m focal length, the resolution angle is

$$\frac{\Delta x}{s} \approx -\frac{1}{2(22.45)(500)} = -\frac{1}{22448} = -4.4548 \times 10^{-5} \text{ rad} \tag{16.2.9}$$

It is clear that resolution is the same in both cases, and is not affected by focal length. When resolution is limited by turbulence, increasing optics focal length simply magnifies the turbulence-induced blur without permitting resolution of any smaller detail. At the same time, the increased focal length also decreases field of view, which is often undesirable. Therefore, when image quality is limited by turbulence, there

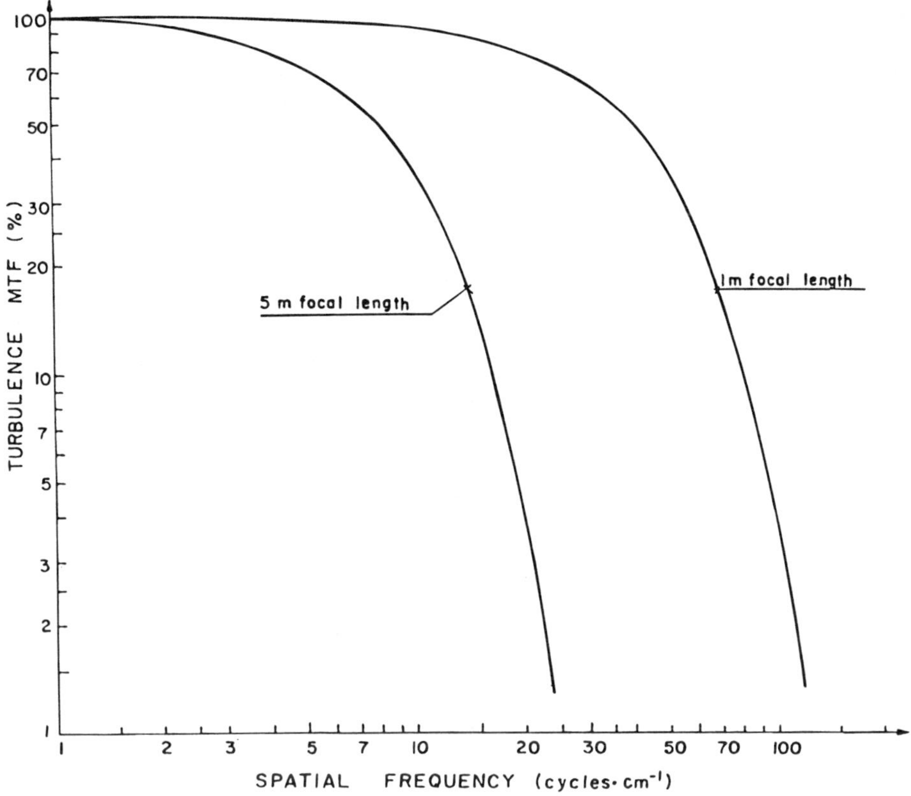

Fig. 16.3 Long-exposure turbulence MTFs for Example 16.1

is no advantage to using large focal length optics. Focal length can be increased to that value where resolution is limited by turbulence, but it is generally not worthwhile to increase focal length any further. This is demonstrated in the next example. (However, as shown in Section 18.5, image restoration appropriate to turbulence limits or even eliminates turbulence blur, in which case longer focal lengths can be useful.)

Example 16.2

A target 5-km distant is viewed horizontally through a turbulence atmosphere. The imaging system consists of a 1/2-in. vidicon camera system with a high-quality zoom lens. Effective dimensions of the camera tube are 8.8 × 11.8 mm. Square wave response of the TV system is essentially constant at unity until it descends sharply to zero at 295 line pairs or 590 TV lines. What is the approximate maximum focal length to be used so as to not generate image blur by turbulence? In other words, how long can focal length be so that resolution is limited by the TV system rather than turbulence? As seen from the previous example, increased focal length beyond this value would only serve to decrease field of view without any improvement in resolution. Assume 5% threshold contrast, and assume $C_n^2 = 1.3 \times 10^{-15}\ \text{cm}^{-2/3}$ and $\lambda = 0.7\ \mu\text{m}$ as before. Assume lens MTF is too good to limit resolution.

Solution

Turbulence MTF is required to approach 5% contrast at the limiting TV spatial frequency which is 295 line pairs/11.8 mm = 250 lp·cm^{-1}. Therefore, substituting into Eq. (16.2.1) for λ, Cn^2, L, and $f_{r\ \max}$ of the TV system yields

$$M_{\mathrm{TL}}(f_{r\ \max}) = 0.05 = \exp(-0.00346 f_l^{5/3}) \tag{16.2.10}$$

Therefore,

$$\ln 0.05 = -2.996 = -0.00346 f_l^{5/3} \tag{16.2.11}$$

which results in

$$f_l = 57.9\ \mathrm{cm} \tag{16.2.12}$$

If focal length is less than 57.9 cm, resolution is limited by the TV system. Otherwise, resolution is limited by turbulence.

For $f_l << 57.9$ cm, the overall MTF is

$$\mathrm{MTF} = M_{\mathrm{TV}} \cdot M_{\mathrm{TL}} \approx \begin{cases} M_{\mathrm{TV}}, & f_r < 250\ \mathrm{lp{\cdot}cm^{-1}} \\ 0, & f_r > 250\ \mathrm{lp{\cdot}cm^{-1}} \end{cases} \tag{16.2.13}$$

where M_{TV} is the TV system's MTF.

In this case, angular resolution is

$$\frac{\Delta x}{s} \approx = \frac{1}{2 f_{r\ \max} f_l} = -\frac{1}{2(250\ \mathrm{lp{\cdot}cm^{-1}}) f_l(\mathrm{cm})} = -0.002/f_l \tag{16.2.14}$$

which improves as focal length increases. For $f_l = 57.9$ cm, $\Delta x/s$ is 3.45×10^{-5} rad.

For $f_l >> 57.9$ cm, overall MTF is

$$\mathrm{MTF} = M_{\mathrm{TV}} \cdot M_{\mathrm{TL}} \approx M_{\mathrm{TL}} \tag{16.2.15}$$

and the situation is similar to that of turbulence-limited imaging in Example 16.1.

In general, since $f_{r\ \max}$ can be less than 250 lp·cm^{-1} for turbulence-limited imaging, substitution for λ, Cn^2, and L into Eq. (16.2.1) yields

$$\mathrm{MTL} = \exp[-3.49 \times 10^{-7} (f_{r\ \max} f_l)^{5/3}] \tag{16.2.16}$$

For 5% limiting contrast,

$$0.05 = M_{\mathrm{TL}} = \exp[-3.49 \times 10^{-7} \cdot 10^{-7} (f_{r\ \max} f_l)^{5/3}] \tag{16.2.17}$$

This yields

$$f_{a\,\max} = f_{r\,\max} f_l = \left(\frac{2.996}{3.49 \times 10^{-7}}\right)^{3/5} = 1.45 \times 10^4 \qquad (16.2.18)$$

The accompanying resolution angle is

$$\frac{\Delta x}{s} \approx -\frac{1}{2 f_{r\,\max} f_l} = -3.45 \times 10^{-5}\ \text{rad} \qquad (16.2.19)$$

Again, for turbulence-limited imaging ($f_l >> 57.9$ cm), increased focal length does not improve resolution. Both cases, TV-limited resolution ($f_l << 57.9$ cm) and turbulence-limited resolution ($f_l >> 57.9$ cm), are depicted schematically in the spatial frequency domain in Fig. 16.4. Resolution as a function of optics focal length is shown in Fig. 16.5. As long as resolution is limited by the TV system, it can be improved by increasing the focal length. However, *once focal length is sufficiently long so that resolution is limited by turbulence, resolution cannot be improved further by increasing the focal length.* Increasing the focal length will only decrease the field of view and magnify the turbulence-caused blur, without permitting resolution of smaller detail. Optimum field of view for resolution occurs when $f_l \approx 57.9$ cm.

The approach developed here permits evaluation of atmospheric turbulence effects on target acquisition probabilities, as shown in Exercise 16.4.

16.3 WAVELENGTH DEPENDENCE

The wavelength dependence of image resolution through turbulence is very weak. Long exposures do favor imaging at longer wavelengths, but only slightly. For a given contrast threshold independent of spatial frequency, as in machine vision or automatic target recognition (ATR), the exponents in Eqs. (16.1.1a) and (16.1.1b) would each have to be equal at different wavelengths. For wavelengths λ_1 and λ_2 this requires

$$f_{a\,\max 1}^{5/3} \lambda_1^{-1/3} = f_{a\,\max 2}^{5/3} \lambda_2^{-1/3}$$

or

$$\frac{f_{a\,\max 1}}{f_{a\,\max 2}} = \left(\frac{\lambda_1}{\lambda_2}\right)^{1/5} \qquad (16.3.1)$$

The same relationship would hold for spatial frequencies ($f_{r\,\max}$) instead of angular spatial frequencies. The above relationship means, for example, that for comparison of image qualities at 0.5- and 10-μm wavelengths, through the same turbulence

$$\frac{f_{a\,\max,10\mu\text{m}}}{f_{a\,\max,0.5\mu\text{m}}} = \left(\frac{10}{0.5}\right)^{1/5} = 1.8 \qquad (16.3.2)$$

For a 20 times longer wavelength the improvement in limiting spatial frequency is only 1.8. Since the variation of C_n^2 itself with wavelength is only very slight in the optical spectral region such comparisons are representative of the wavelength depen-

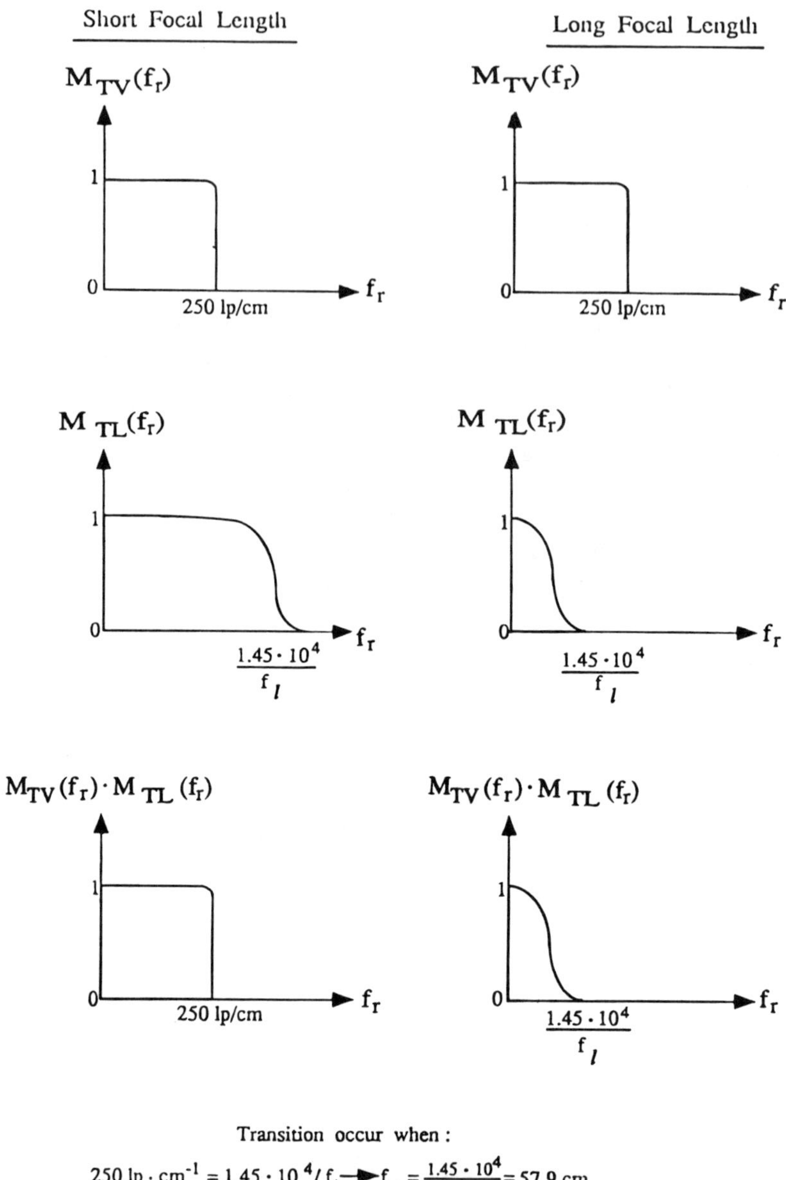

Fig. 16.4 Spatial frequency domain representation for TV-limited (left side) and turbulence-limited (right side) resolution.

dence of image quality through turbulence. (However, C_n^2 does vary much more when compared at optical and microwave wavelengths, for example.)

If contrast threshold is not constant with spatial frequency but varies as shown in Figs. 12.17 or 12.18 for human vision, the longer wavelength advantage of imaging through optical turbulence is diminished even further.

Relative turbulence MTF improvements associated with short exposures versus long exposures are diminished at longer wavelengths as can be seen from Eq. (16.1.2).

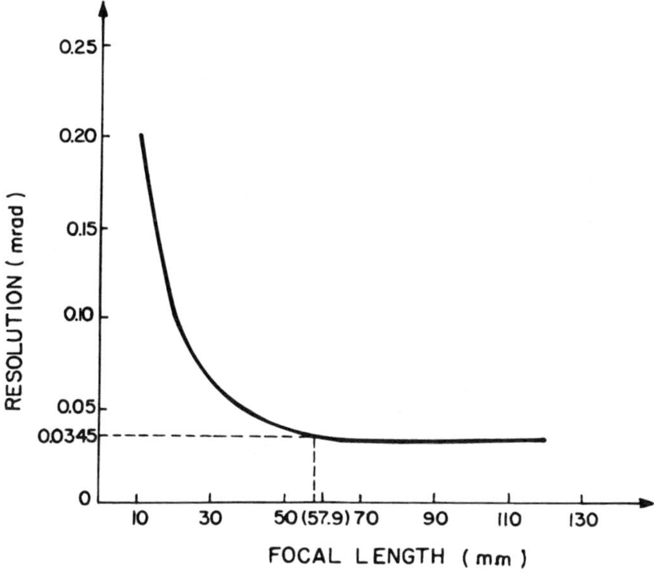

Fig. 16.5 Resolution angle as a function of a focal length for Example 16.2. Optimum field of view for resolution occurs at 57.9 mm focal length.

16.4 VERTICAL OR SLANT PATH VIEWING

The dependence of turbulence MTF on altitude is determined basically by that of C_n^2 with altitude, which has been discussed previously in terms of Eqs. (15.4.35) and (15.4.36). Vertical profiles of C_n^2 have been observed to change considerably even over a few hours. Nevertheless, to aid in system design several mathematical models such as Eqs. (15.4.35) and (15.4.36) have been suggested. These models are most useful from the surface up to only the boundary or inversion layer, typically 1- to 3-km elevation. It is at such low elevations that turbulence is strongest. To use Eqs. (16.1.1) for vertical paths, let z be the vertical coordinate. For slant paths, incremental path length is then (dz) secθ, where θ is viewing angle relative to a vertical path. For $z = 0$ at the surface, when imaging downwards Eq. (16.1.1b) becomes, using profile (15.4.35),

$$
\begin{aligned}
M_{\mathrm{TL}}(f_a) &= \exp\left[-57.53 f_a^{5/3} \lambda^{-1/3} \sec\theta \int_0^L (C_{no}^2 z^{-4/3})(z/L)^{5/3}\, dz\right] \\
&= \exp\left[-57.53 f_a^{5/3} \lambda^{-1/3} \sec\theta\, C_{no}^2 L^{-5/3} \int_0^L z^{1/3}\, dz\right] \\
&= \exp[-43.15 f_a^{5/3} \lambda^{-1/3} \sec\theta\, C_{no}^2 L^{-1/3}]
\end{aligned}
\tag{16.4.1}
$$

For example, consider viewing downward from a height of 2 km (assuming that this elevation is not above the boundary layer), at a wavelength of 0.5 μm and a depression angle of 40° ($\theta = 50°$). In this case the long-exposure turbulence MTF is

$$M_{\mathrm{TL}} = \exp(-3.60 \times 10^{-12} f_a^{5/3}) \tag{16.4.2}$$

for $C_{no}^2 = 9 \times 10^{-14}$ cm$^{2/3}$, which is fairly strong turbulence. Consider a rather high angular spatial frequency of 50 cycles·mrad^{-1}. At such a high frequency, the exponent in Eq. (16.4.2) becomes equal to -2.45×10^{-4} and long-exposure turbulence MTF has a value of 0.9998. Hence, except for small depression angles turbulence generally has only a very weak effect on the image quality when imaging downward and aerosol blur is much more dominant. Note that the value of the integral in the first line of Eq. (16.4.1) is

$$\begin{aligned}\int_0^L C_{no}^2 z^{-4/3}(z/L)^{5/3}\, dz &= C_{no}^2 L^{-5/3} \int_0^L (z^{1/3})\, dz = \frac{3}{4} C_{no}^2 L^{-5/3}(L^{4/3}) \\ &= \left(\frac{3}{4}\right) C_{no}^2 L^{-1/3} = \left(\frac{3}{4}\right)(9 \times 10^{-14})(2 \times 10^5)^{-1/3} \\ &= 1.15 \times 10^{-15}\ \mathrm{cm}^{1/3}\end{aligned} \tag{16.4.3}$$

where path length L is converted to centimeters.

This value is almost on the order of the integrals in Eq. (16.1.1) for viewing downward through the entire atmosphere [16.9, 16.10], rather than from a 2-km elevation. Hence, for downward viewing over vertical or slant paths, even through the entire atmosphere, turbulence is usually not too serious a factor (even if the turbulence is very strong, as in Exercise 16.5). Clearly, the contributions of atmospheric layers above the boundary layer to the turbulence strength integrals in Eqs. (16.1.1) are often small relative to those contributions below the boundary layer.

For viewing upward from the surface of the earth, the turbulence height profile must be revised to correspond to the coordinates. In Eq. (16.1.1b) the coordinate $z = 0$ at the object plane which is now at elevation L. The value of C_n^2 at this height, according to the Tatarski height profile [Eq. (15.4.35)] is $C_o^2 L^{-4/3}$. Now C_n^2 must be made to increase as z increases (downward) so that at $z = L - 1$ (near earth's surface), $C_n^2 = C_{no}^2$. The profile $C_n^2 = C_{no}^2\,[(z + 1)/L]^{4/3}$ addresses these needs. Hence, using this height profile for C_n^2 and letting z be zero at 2-km elevation, which is assumed to be lower than the boundary layer, if we look upward at an object at this elevation Eq. (16.1.1b) is modified to

$$M_{\mathrm{TL}}(f_a) = \exp\left\{-57.53 f_a^{5/3} \lambda^{-1/3} \sec\theta\, C_{no}^2 \times \int_0^L \left[\frac{z+1}{L}\right]^{4/3} \left(\frac{z}{L}\right)^{5/3} dz\right\} \tag{16.4.4}$$

where values of L , λ, and θ are as in the previous example. The height profile of C_n^2 for looking upwards requires C_{no}^2 to be of identical units. Hence, C_{no}^2 here is assumed to be equal to 9×10^{-14} cm$^{-2/3}$. In this way, both the upward-looking and downward-looking height profiles of C_n^2 exhibit identical values of C_n^2 at each elevation. The integral in Eq. (16.4.4) is evaluated numerically to approach $L/4$ (or 5×10^4 cm in this example) except for very short paths. This implies that, in terms of turbulence, imaging upward for a 2-km distance is equivalent to imaging horizontally for an almost 0.5-km distance at an elevation of 1 cm. The long-exposure turbulence MTF for imaging upward is calculated to be

$$M_{\mathrm{TL}}(f_a) = \exp(-1.4 \times 10^{-5} f_a^{5/3}) \tag{16.4.5}$$

This MTF expression for imaging upward drops off orders of magnitude more quickly with increasing angular spatial frequency than does Eq. (16.4.2) for imaging downward. This is a result of the strong turbulence in the vicinity of the receiver near the surface of the earth. The nearness of the strong turbulence to the receiver implies that larger tilts of the image or portions of it are recorded in the image than when the strong turbulence layers are distant from the receiver [16.3]. The larger refraction angles increase the blur radii in the image. This is the implication of the 5/3 power law for spherical wave MTF through turbulence. The integral in Eq. (16.1.1b) for imaging upward from near the ground is equal, in this example, to $(9 \times 10^{-14}) \cdot (5 \times 10^4) = 4.5 \times 10^{-9}\ \mathrm{cm}^{1/3}$. This is more than six orders of magnitude greater than the result in Eq. (16.4.3) for imaging downward. For imaging upward from near the surface the exponent of Eq. (16.4.5) at the high angular spatial frequency of 50 cycles·mrad^{-1} is equal to -950, and the long-exposure turbulence MTF is essentially zero.

The examples here can help us understand why images from satellites can have very high resolution, while astronomical images recorded near the surface of the earth have such limited resolution. Observatories are located generally on mountain tops of high elevation to help alleviate this problem, as well as similar ones deriving from blur and attenuation caused by aerosols.

16.5 IMAGING HORIZONTALLY

For imaging horizontally, a path-integrated average value of C_n^2 can be assumed that is removable from the integrals in Eqs. (16.1.1). In this case, long-exposure turbulence MTF after integration becomes

$$M_{\mathrm{TL}}(f_a) = \exp(-a 57.53 f_a^{5/3} \lambda^{-1/3} C_n^2 L) \tag{16.5.1}$$

where a equals unity for a plane wave and 3/8 for a spherical wave. Equation (16.5.1) can then be substituted into Eq. (16.1.2) to determine short-exposure turbulence MTF for horizontal imaging. Examples for a specific situation are shown in Fig. 16.1. Turbulence effects for the same path length are much more severe for low elevation horizontal rather than vertical or slant path imaging because of the strong turbulence near the earth's surface. Generally the spherical wave formulation is more accurate for passive horizontal imaging than is the plane wave approximation.

16.6 EFFECTIVE APERTURE SIZE

Because of the blurring effects introduced by atmospheric turbulence, actual resolution through the atmosphere is less than that obtainable with diffraction-limited optics. In the discussion on diffraction-limited imaging in Chapter 8, MTF spatial-frequency bandwidth is seen to increase with increasing aperture diameter D. However, even if the optics performance itself is diffraction limited, optical turbulence does not permit such high resolution to be realized. Consequently, an effective aperture size was derived by Fried [16.5] to describe the maximum imaging system diameter at which image resolution received through the atmosphere is the best or nearly the best permitted by the atmosphere. The diameter of such an aperture is called *atmospheric coherence diameter*, designated r_o by Fried. Use of larger size apertures improves signal-to-noise ratio but there is only negligible image resolution improvement because Rayleigh criterion resolution angle through the turbulent atmosphere is $\approx 1.22\lambda/r_o$ provided $D \geq r_o$. In the next chapter we will see that the atmospheric coherence diameter is limited also by airborne particulate scattering. However, here we will limit ourselves to turbulence effects.

Fried's mathematical definition of imaging system resolution is

$$\mathfrak{R}_s = 2\pi \int_0^\infty f_a M(F_a)\, df_a \tag{16.6.1}$$

where $M(F_a)$ is the system MTF which, for simplicity, is assumed to be the product of optics MTF and atmospheric turbulence MTF. Fried did not consider aerosol MTF. His definition of r_o applies only to simple viewing through a telescope without aid of photoelectronic imagers, since their MTF is not included. Consequently, for a diffraction-limited circular lens aperture as in Eq. (8.5.8), assuming a plane wave, for simplicity,

$$\begin{aligned} M(f_a) = {} & (2/\pi)[\cos^{-1}u - u(1 - u^2)^{1/2}] \\ & \times \exp\left[-57.53 f_a^{5/3}\lambda^{-1/3} \int_0^L C_n^2(z)\, dz\right] \\ & \times \left[1 - \frac{1}{b} u^{1/3}\right] \end{aligned} \tag{16.6.2}$$

where u equals $\lambda f_a/D$ and $(1/b)$ equals zero for a long exposure or, for a short exposure, 1/2 at the image center (far field) or unity at the image edge (near field). Substitution of Eq. (16.6.2) into Eq. (16.6.1) yields

$$\begin{aligned} \mathfrak{R}_s = {} & 4(D/f_l)^2 \int_0^1 [u(\cos^{-1}u) - u(1 - u^2)^{1/2}] \\ & \times \exp[-3.44(Du/r_o)^{5/3}\, du] \cdot (1 - u/b) \end{aligned} \tag{16.6.3}$$

where f_l again is focal length (equal to f_a/f_r) and, for a plane wave,

$$r_o \triangleq 0.185\left[\lambda^2 / \int_0^L C_n^2(z)\, dz\right]^{3/5} \tag{16.6.4a}$$

For a spherical wave

$$r_o \triangleq 0.185\left[\lambda^2 / \int_0^L C_n^2(z)(z/L)^{5/3}\, dz\right]^{3/5} \tag{16.6.4b}$$

The greater the turbulence strength (C_n^2), the smaller the effective aperture diameter imposed by turbulence. Typical values of r_o are about 20 cm for exceptionally good vertical seeing conditions and about 5 cm for moderately poor vertical imaging conditions through the entire atmosphere. For horizontal imaging, r_o may be much less, depending on elevation and path length. Maximum value of $\Re_s$ is obtained for infinite diameter optics. This condition implies that u approaches zero. Hence,

$$\Re_{s\,\max} = \lim_{D\to\infty} \Re_s = (\pi/4)[r_o(\lambda f_l)] \tag{16.6.5}$$

The ratio $\Re_s/\Re_{s\,\max}$ as a function of normalized optics diameter D/r_o is plotted in Fig. 16.6. It is clear that by increasing optics diameter to values much larger than r_o there is very little improvement to be gained in resolution for long exposures. There is some improvement in resolution obtainable for short exposures when D is on the order of three or four times the size of r_o, $D \approx 3.8r_o$ being optimum for turbulence.

From Eq. (16.6.4) it is clear that r_o increases with wavelength according to $\lambda^{6/5}$. This means that for the same C_n^2, imaging at 10-μm wavelength results in a value of r_o that is 36.4 times larger than that for imaging at 0.5 μm. This suggests that resolution through the atmosphere in the thermal infrared can be improved considerably over that in the visible. However, this potential improvement is mitigated very considerably by blur resulting from forward scatter by atmospheric aerosols, as shown in the next chapter where the effects of such scatter on overall atmospheric coherence diameter are considered.

Atmospheric coherence diameter is also connected with atmospheric MTF. The connection is through the atmospheric mutual coherence function. By defining the radial coordinate $\rho = \lambda f_a$ in the entrance pupil plane and substituting into Eqs. (16.1.1) and (16.1.2), turbulence MTF is converted to turbulence mutual coherence function. Lutomirski and Yura [16.11] define a transverse coherence length as the

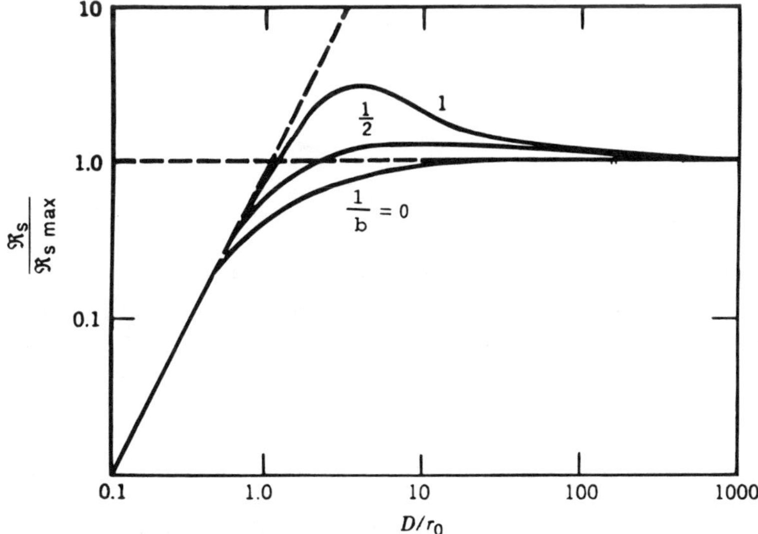

Fig. 16.6 Normalized resolution versus normalized optics imaging diameter (after [16.5]).

transverse length at which mutual coherence decays to its e^{-1} value. From Eq. (16.1.1a) the angular spatial frequency at which turbulence MTF decreases to its e^{-1} value for a plane wave is

$$f_{a\,1/e} = \lambda^{1/5} / \left[57.53 \int_0^L C_n^2(z)\, dz \right]^{3/5} \tag{16.6.6}$$

Consequently, for the turbulence mutual coherence function

$$\rho_{1/e} = \lambda f_{a\,1/e} = \lambda^{6/5} / \left[57.53 \int_0^L C_n^2(z)\, dz \right]^{3/5} \tag{16.6.7}$$

By comparing Eqs. (16.6.4) and (16.6.7),

$$r_o / \rho_{1/e} = 0.185(57.53)^{3/5} = 2.1 \tag{16.6.8}$$

Hence, r_o and $\rho_{1/e}$ can both be obtained from C_n^2 and from each other.

The preceding discussion applies to plane waves. However, similar relationships hold too for spherical waves. For spherical waves,

$$r_o = 0.185\lambda^{6/5} / \left[\int_0^L C_n^2(z)(z/L)^{5/3}\, dz \right]^{3/5} \tag{16.6.9}$$

$$\rho_{1/e} = \lambda^{6/5} / \left[57.53 \int_0^L C_n^2(z)(z/L)^{5/3}\, dz \right]^{3/5} \tag{16.6.10}$$

and, again,

$$r_o = 2.1\rho_{1/e} \tag{16.6.11}$$

For both spherical and plane waves, r_o can be measured directly by placing different size apertures D in front of an imaging system and calculating $\mathfrak{R}_s$ according to Eq. (16.6.1) for each aperture [16.12]. Saturation is approached for long exposures when $D \approx r_o$. Alternatively, r_o can be obtained indirectly by measuring turbulence MTF and finding the spatial frequency $f_{a\,1/e}$ at which the turbulence MTF is reduced to its $1/e$ value. This determines $f_{a\,1/e}$ which is equal to $\rho_{a\,1/e}/\lambda$. Having determined transverse coherence length $\rho_{1/e}$, r_o is obtained by simply multiplying $\rho_{1/e}$ by the factor 2.1 [16.12].

It should be clear that r_o represents a limit on effective aperture diameter of an imaging system, provided the image is contrast limited rather than noise limited. However, increasing apertures beyond r_o can be desirable to improve SNR.

Turbulence MTF is also described in terms of r_o or $\rho_{1/e}$, that is,

$$M_{TL}(f_a) = \exp[-3.44(\lambda f_a/r_o)^{5/3}] \quad (16.6.12)$$

$$= \exp[-(\lambda f_a/\rho_{1/e})^{5/3}] \quad (16.6.13)$$

Example 16.3

Consider a telescope of aperture diameter 9 cm and focal length 2.8 m used to image horizontally an object 6.5 km distant through turbulence described by $C_n^2 = 10^{-13}\ \text{m}^{-2/3} = 4.64 \times 10^{-15}\ \text{cm}^{-2/3}$ at 0.5 μm = 5×10^{-5} cm wavelength. Compare resolution with and without long-exposure optical turbulence. Assume 2% threshold contrast.

Solution

For diffraction-limited optics alone, the MTF cutoff radian frequency as shown in Eq. (8.5.1), is $2\alpha_o = k_o D/s'$. Therefore, $f_{a\ \max} \approx D/\lambda = 1.8 \times 10^5$ cycle·rad^{-1}. As in Example 12.1, based on considerations in Section 10.4, resolution angle ≥ $(2f_{a\ \max})^{-1} = \lambda/2D = 2.8$ μrad. At 6.5 km, object plane detail on the order of $(2.8 \times 10^{-6})(6.5 \times 10^3) = 1.8$ cm could begin to become distinguishable were it not for atmospheric distortions and degradation.

Equation (16.6.6) indicates $f_{a1/e}$ through optical turbulence is approximately $f_{a\ 1/e} \approx \lambda^{1/5}/(3/8)(57.53 C_n^2 L)^{3/5} = 2830$ cycles·rad^{-1} after substituting for C_n^2 and L.

If 2% threshold contrast is considered, the turbulence spatial frequency bandwidth is larger, that is,

$$\ln 0.02 = -3.912 = -(3/8)57.53\lambda^{-1/3} f_{a\ \max}^{5/3} C_n^2 L$$

or

$$f_{a\ \max} = \left(\frac{3.912\lambda^{1/3}}{21.57 C_n^2 L}\right)^{3/5} = 2.27 f_{a\ 1/e} = 6420\ \text{cycles·rad}^{-1} > f_{a\ 1/e}$$

This result is not surprising since $f_{a\ 1/e}$ represents 37% contrast instead of 2% contrast. These calculations assume a spherical wave. The turbulence resolution angle $(2f_{a\ \max})^{-1}$ is about 140 μrad. This is 50 times worse or larger than that deriving from diffraction-limited optics alone (2.8 μrad). The same ratio occurs from comparing the values of $f_{a\ \max}$ without and with turbulence. At the distance involved, object plane detail must be at least 91 cm in order to be distinguishable because of the turbulence. This is about 50 times larger than the 1.8 cm resolvable detail that can be seen if there were no atmospheric turbulence. (The actual resolution through the atmosphere is actually worse due to aerosols.)

In this example, the transverse coherence length is $\rho_{1/e} = \lambda f_{a\ 1/e} = (5 \times 10^{-5}$ cm)(2830 cycles·rad^{-1}) = 14 cm, and $r_o = 2.1\rho_{1/e} = 30$ cm. The latter may also be calculated directly from Eq. (16.6.4b).

Example 16.4

Assume angle of arrival variance is used to determine path-integrated C_n^2. Express long exposure turbulence MTF as a function of such variance. Does plane wave propagation change the turbulence MTF formulation from that of spherical waves?

Solution

From Eqs. (15.4.26) and (15.4.27), the C_n^2 integral equals

$$a\, C_{n\,\mathrm{eq}}^2 L = \langle \alpha^2 \rangle D^{1/3}/2.914, \qquad \sqrt{\lambda L} << D << L_0$$

where a is unity for a plane wave and 3/8 for a spherical wave. From Section 16.1, long exposure turbulence MTF can be written as

$$M_{\mathrm{TL}}(f_a) = \exp[-57.53\, f_a^{5/3} \lambda^{-1/3} (a\, C_n^2 L)]$$
$$= \exp[-57.53\, f_a^{5/3} \lambda^{-1/3} < \alpha^2 > D^{1/3}/2.914].$$

Since a cancels out, the above expression is valid for both spherical and plane waves. In it, turbulence MTF is expressed directly as a function of one-dimensional angle-of-arrival variance which is pleasing since angle-of-arrival fluctuations cause the blur. As long as one is consistent in using plane or spherical wave formulation for both determination of C_n^2 and turbulence MTF, the *same turbulence MTF is obtained for both plane and spherical waves*.

16.7 TECHNIQUES TO CORRECT FOR TURBULENCE

There are basically two techniques used to correct for turbulence degradation. One is adaptive optics, whose purpose is to prevent turbulence-derived distortions from being recorded in the image. The other technique is image restoration using typically digital computers. In the latter case, turbulence-derived blur has already been recorded in the image but is removed through image processing.

Adaptive optics [16.13] is an expensive process not readily available, applicable primarily to astronomical imaging. Briefly, the concept in adaptive optics is to sense the wavefront tilt and distortion using an incoming laser beam or the image of a distant point source such as a star. These make it most convenient to apply when looking upward. Telescopes used in adaptive optics use mirrors as objectives. The thickness of each area of the mirror is varied electronically using transducers such as piezoelectric elements. The information obtained from the wavefront sensors is used to control piezoelectric elements so as to alter the surface of the mirror to conform to the incoming wavefront shape. In this way, if the wavefront from a distant point source such as a star would be planar because of the long distance but is distorted by turbulence, the mirror surface distortions by the piezoelectric elements are such that the wavefront recorded is now planar despite the fact that turbulence causes the incident wavefront to be nonplanar. Adaptive optics have been used successfully with large telescopes to produce very fine quality images through the atmosphere. To work successfully, adaptive optics requires each element in the mirror to be no larger than the isoplanatic patch size incident to the mirror. Adaptive optics are intended to prevent turbulence blur from being recorded in the image. This improves the measurement SNR at high spatial frequencies, and can then allow postprocessing techniques such as digital image restoration [16.13, 16.14].

However, as shown in Chapter 18, digital image restoration using an atmospheric Wiener filter can also be amazingly successful in removing turbulence blur. Restoration also has the advantage of being much cheaper and can be done with readily available personal computers. Techniques based on atmospheric MTF are discussed in Section

18.5.3 of this book. Essentially, all atmospheric blur, including that derived from both turbulence and aerosols, as well as path radiance is removed. A high-resolution image is obtained as if there were no atmospheric channel. Such image restoration is applicable to any contrast-limited imaging situation, including terrestrial imaging, in *any* imaging direction.

REFERENCES

16.1. R. E. Hufnagel and N. R. Stanley, "Modulation transfer function associated with image transmission through turbulent media," *J. Opt. Soc. Am.*, vol. 54, pp. 52–61January 1964.

16.2. D. L. Fried, "Optical resolution through a randomly inhomogeneous medium for very long and very short exposures," *J. Opt. Soc. Am.*, vol. 56, pp. 1372–1379, October 1966.

16.3. D. Sadot, D. Shemtov, and N. S. Kopeika, "Theoretical and experimental investigation of image quality through an inhomogeneous turbulent medium," *Waves in Random Media*, vol. 4(2), pp. 177–189 April 1994.

16.4. N. S. Kopeika, "Spectral characteristics of image quality for imaging horizontally through the atmosphere," *Appl. Opt.*, vol.16, 2422–2426, September 1977, pp. vol. 17, August p. 1162 Aug. 1978.

16.5. D. L. Fried, "Limiting resolution looking down through the atmosphere," *J. Opt. Soc. Am.*, vol. 56, pp. 1380–1384, October 1966.

16.6. R. L. Fante, "Electromagnetic beam propagation in turbulent media," *Proc. IEEE*, vol. 63, pp. 1669–1692, December 1975.

16.7. Harold A. Alister, "Seeing stars with speckle interferometry," *Amer. Scientist*, vol. 76, pp. 166–173. March–April 1988.

16.8. D. L. Fried, "Probability of getting a lucky short-exposure image through turbulence," *J. Opt. Soc. Amer.*, vol. 68, pp. 1651–1658 (1978).

16.9. C. Roddier, "Measurements of the atmospheric attenuation of the spectral components of astronomical images," *J. Opt. Soc. Amer.*, vol. 66, pp. 478–482, May 1976.

16.10. N. S. Kopeika, "Imaging through the atmosphere for airborne reconnaissance," *Opt. Eng.* vol. 26, pp. 1147–1154, November 1987.

16.11. R. F. Lutomirski and H. T. Yura, "Wave structure function and mutual coherence function of an optical wave in a turbulent atmosphere," *J. Opt. Soc. Amer.*, vol. 61, pp. 482–487, April 1971.

16.12. D. Sadot, A. Melamed, N. Dinar, and N. S. Kopeika, "Effects of aerosol forward scatter on long and short exposure atmospheric coherence diameter," *Waves in Random Media*, vol. 4(4), pp. 487–498 Oct. 1994.

16.13. R. K. Tyson, *Principles of Adaptive Optics*, Academic Press, New York, 1991.

16.14. M. C. Roggemann, E. L. Caudill, D. W. Tyler, M. J. Fox, M. A. Von Bokern, and C. L. Matson, "Compensated speckle imaging: theory and experimental results," *Appl. Opt.*, Vol. 33, 10 May 1984, pp. 3099–3110.

EXERCISES

16.1 In the examples of vertical imaging considered in Section 16.4, would it have been justified to use the plane wave approximation for spherical wave propagation for long propagation paths?

16.2 In the examples considered in Section 16.4, determine atmospheric coherence diameter and transverse coherence lengths for both downward and upward viewing.

16.3 Compare the results of the previous question to those for imaging horizontally as in Example 16.2.

16.4 Consider Example 16.2. For an optics focal length of 60 cm, what size object has a 50% probability of recognition under the turbulence conditions and distance involved? How does this result change for a 50-cm focal length?

16.5 Evaluate Eq. (16.4.2) for strong turbulence, where C_{no}^2 is two orders of magnitude larger. Is turbulence then a serious factor in downward viewing?

CHAPTER

17

Aerosol Modulation Transfer Function

17.1 INTRODUCTION

As described in the previous two chapters, turbulence gives rise to image degradation as a result of wavefront tilt as well as random image detail displacement deriving from random changes in refractive index of the propagation channel. Because of the randomness, a statistical rather than deterministic approach characterizes turbulence. However, forward light scatter by aerosols and atmospheric particulates in general cause relatively very little wavefront distortion and tilt. Rather, they cause broad diffusion of details in the propagating image, thus generating image blur in a different fashion. These latter processes are rather steady and lack the time dynamic properties of turbulence.

A general configuration describing the problem of imaging through the atmosphere is illustrated in Fig. 17.1. In the object plane is the source of radiation, which, in this example, is a point source producing a spatial delta function. The propagation channel is the atmosphere, which is a random medium containing both turbulence, and scattering and absorbing particles. At the other end of the atmospheric path is the imaging system, including both the optics and electronics. The main effect of the turbulent medium is to produce a wavefront tilt as illustrated in Fig. 17.1, which causes image shifts in the image plane. Those tilts are random and their temporal power spectra are usually limited to several tens up to a few hundred hertz under ordinary atmospheric conditions. The image distortions caused by the wavefront tilt (typically on the order of tens or maybe hundreds of microradians) can be compensated either by adaptive optics techniques or by means of software (Section 18.5) if the exposure time is short enough, less than the characteristic fluctuation time (usually a few milliseconds). There is also a blur effect related to turbulence (inner scale wavefront distortions that displace image plane detail) not illustrated in this figure, which usually affects very high spatial frequencies. This blur effect can be characterized by the turbulence short-exposure MTF, described in the previous chapter. In addition, scintillations cause noticeable degradation of image quality as well, particularly for horizontal imaging near the ground. Current adaptive optics techniques have proven inadequate for horizontal imaging unless the turbulence strength is very low. However, digital image restoration, described in Section 18.5.3, has proven to be quite effective in image restoration.

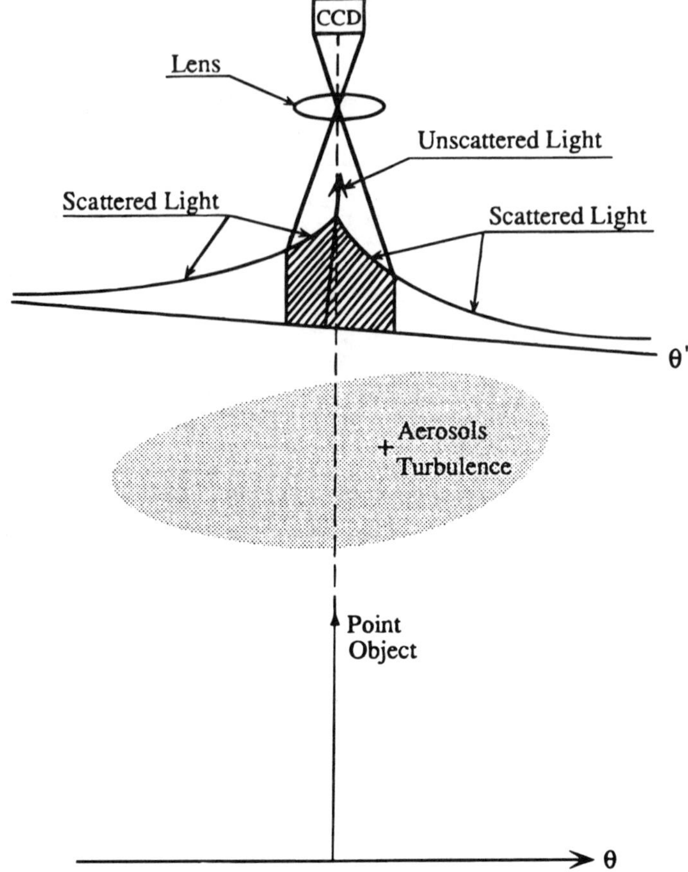

Fig. 17.1 Propagation of angular point object through turbulent and scattering atmospheric channel to image sensor such as a CCD.

In addition to turbulence, scattering and absorption effects are produced by molecules and aerosols. Scattering gives rise to very wide diffusion of the point object radiation as shown in Fig. 17.1. This causes a wide variance in angle of arrival according to the particulate scattering phase function, and thus also produces blur at the image plane [17.1–17.5]. Unlike the short-exposure turbulence case, the blur due to aerosols is very wide, usually on the order of radians which is typically orders of magnitude wider than the imaging system field of view.

An example would be the large blurred diffuse image of the moon through clouds. Even in clear weather, if one searches for the moon through a telescope, it will be found by moving the instrument in the direction of increasing skylight or moonlight even if the moon is not within the field of view. Such moonlight consists of photons reflected or emitted from the moon that should have been part of the image of the moon but have been scattered or "misdirected" by atmospheric aerosols to be imaged outside of the image of the moon. Particulate scattering in the atmosphere is over a very broad angular region. The whole "sky" is simply scattered light—from the sun during the day and principally from the moon during the night. The image of a point is broadened greatly by such scatter. This implies a wide point spread function (PSF). However, some practical instrumentation-based limitations should be applied to the scattered light actually recorded in an image [17.5]. The angular spread of the incident scattered light, according to diffraction theory [Eq. (7.4.14)], is approximately on the order of λ/a, where a is the dominant particulate radius and λ is the radiation wavelength. In the atmosphere in the optical spectral region, this angular spread of scattered light is on the order of several *radians*, and will usually be truncated by the imaging system field of view, which may be typically on the order of *milliradians*. Therefore, as shown in Fig. 17.1 not all the scattered light will reach the sensor in the image plane. Furthermore, even if it does reach the sensor, the whole quantity of received scattered radiation will not necessarily be detected by it [17.5]. The reason is that every sensor has a limited dynamic range, and scattered radiation of very low irradiance compared to the unscattered light will not be detected at all.

As shown in Fig. 17.1, the weak intensity scattered light is at larger scatter angles. Thus, dynamic range limitations of the imaging system sensor also limit scatter angles recorded in the image, often even more so than field of view [17.5]. In addition, the dynamic range might be limited by the object-to-background irradiance ratio of the scene itself. If the scattered light is at irradiance levels lower than that of the background irradiance, its contribution to the blur will be minor and usually negligible. Hence, dynamic range truncates larger scattering angles where scattered light is of much less irradiance than unscattered light. Similar effects arise too with the human visual system. If one looks at the moon one does not see the scattered moonlight in clear weather simultaneously because of the limited dynamic range of the human visual system.

In addition, practical limitations must also be applied to the *unscattered* component of the radiation too. The finite angular spatial-frequency bandwidth of the imager truncates the high angular spatial-frequency radiation. This is because unscattered light from a point object (delta function) is equivalent to a constant of infinite spectrum in the angular spatial-frequency plane. Limited spatial-frequency bandwidth of the sensor limits the amount of unblurred or unscattered light actually recorded in the image from that incident to the imager, because the higher spatial-frequency components are not within the spatial-frequency bandwidth of the hardware. This causes broadening of the unscattered light PSF. Because of conservation of energy, such widening of the unscattered light PSF gives rise to a decrease in amplitude of unscattered light relative to scattered light in the image, often very significantly. The scattered light PSF is so broad that it is hardly—if at all—affected by the spatial-frequency bandwidth limitations of the instrumentation [17.5].

Absorption should be taken into account too [17.4]. In the thermal infrared range molecular and aerosol absorption is very pronounced, unlike the visible range. The light reaching the imaging system of Fig. 17.1 contains both unscattered and scattered components. In the far-infrared region [17.6, 17.7], for moderate atmospheric conditions most of the scattered light to a good approximation is due to single scattering events. This is in view of the long wavelength and typically smaller scatterers, giving rise to smaller scattering coefficients than in the visible. This assumption leads to the conclusion that most of the *scattered* light does not experience absorption, since only a single interaction can occur between each photon and particulate—either absorption or scattering. Therefore, light that was scattered was not absorbed. Only the light that is unscattered is subject to absorption [17.4].

Consider Figs. 17.2 and 17.3, which represent a typical particulate scattering pattern and its corresponding MTF at optics input before instrumentation effects are considered. The unscattered light is a delta function (Fig. 17.2) with respect to the

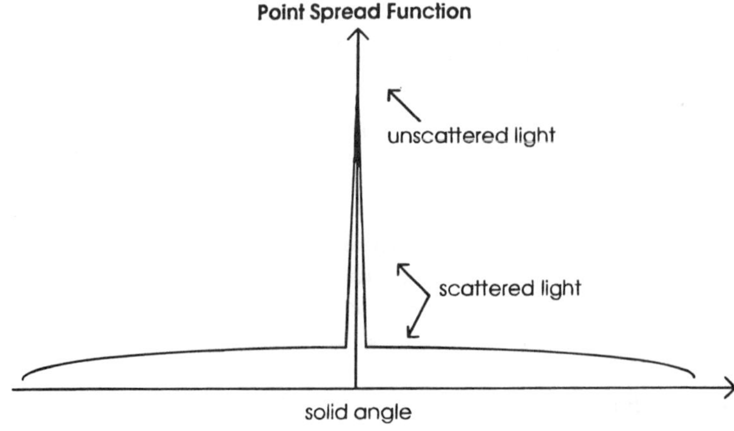

Fig. 17.2 Point spread function of atmospheric scatter channel incident to optics.

angular spatial domain, since it has no angular spread arising from scatter. Therefore, in the spatial-frequency domain (Fig. 17.3) in accordance with its Fourier transform it is constant for all spatial frequencies up to, in principle, infinite spatial frequency. The scattered light of Fig. 17.2 has wide angular spread and in the spatial-frequency domain (Fig. 17.3) is a narrow function with cutoff angular frequency f_{ac}. In its "classical" form, aerosol MTF is described by [17.1–17.3]

$$M_a(f_a) = \begin{cases} \exp\{-A_aL - S_aL(f_a/f_{ac})^2\} & f_a < f_{ac} \\ \exp\{-(A_a + S_a)L\} & f_a > f_{ac} \end{cases} \tag{17.1.1}$$

where, as before, L is path length, A_a and S_a are absorption and scattering coefficients, and f_a is angular spatial frequency. Note that for $f_a > f_{ac}$ classical aerosol MTF equals atmospheric transmission [Eq. (15.3.1)]. The angular spatial-frequency cutoff f_{ac} is given by a/λ, where a is aerosol radius. Although Eq. (17.1.1) was developed for large optical depths, that is, $\int_0^L \alpha(z)\,dz \approx S_aL >> 1$, its form has been shown to apply also to very small optical depths as low as 1 or 2 [17.8], provided instrumentation effects are considered [17.5]. Equation (17.1.1) would apply to Fig. 17.3 if there

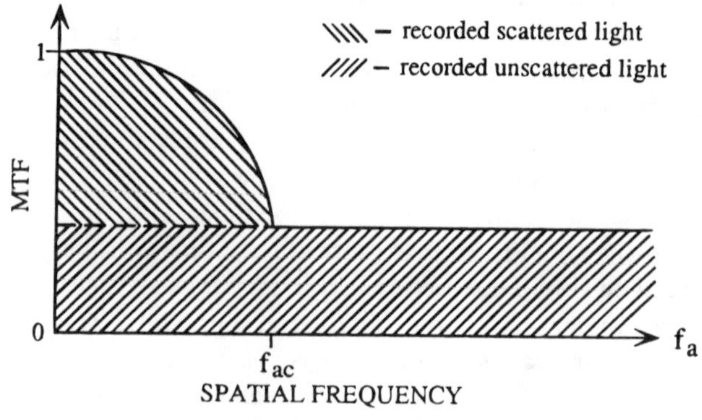

Fig. 17.3 Shape of aerosol MTF.

were no instrumentation limitations on the angles of scattered light actually recorded in the image. In Eq. (17.1.1) it is assumed that light can be scattered out to the diffraction limit angle of λ/a and still be recorded in the image. This classical aerosol MTF can apply, however, only up to the optics input plane. In the image, field of view and dynamic range truncate the angles of light scatter actually recorded in the image by orders of magnitude, thereby reducing blur radius correspondingly. These instrumentation effects are treated in Section 17.2. Currently, we must consider further the effects of absorption in the atmospheric channel since it precedes the optics input.

As mentioned earlier, a correction must be made to this "classical" aerosol MTF, which treats absorption as a "simple" multiplicative factor to the aerosol MTF. In this classical approach, absorption is introduced as a constant attenuation to the aerosol MTF at all spatial frequencies. However, as explained earlier, the absorption must be applied only to the relative part of the *unscattered* light in Fig. 17.3 and not to the total incident radiation, which includes also scattered light. This is demonstrated in Fig. 17.4a. This analysis can be interpreted in the spatial domain in the following way: Consider the radiation at the image plane of Fig. 17.1. All the *blur*, which is related to the *scattered* light, was not attenuated by particulate absorption. This gives rise to a blurred image. On the other hand, the unscattered radiation that would give rise to an unblurred image for infinite spatial-frequency bandwidth was subject to absorption attenuation. This causes attenuation of the relatively unblurred image deriving from the unscattered light but not to the blurred image deriving from the scattered light. Thus, the blur caused by scattering is relatively enhanced by absorption. This is particularly important in the thermal infrared, where absorption is very pronounced. The final form of the "classical" aerosol MTF then becomes [17.4]

$$M_a(f_a) = \begin{cases} \exp\left\{-S_aL\left(\dfrac{f_a}{f_{ac}}\right)^2\right\} \cdot \exp\left\{\left[\exp\left\{-S_aL\left[1 - \left(\dfrac{f_a}{f_{ac}}\right)^2\right]\right\}\right.\right. & \\ \qquad\qquad \left.\left. -\exp(-S_aL)\right](-A_aL)\right\} & f_a < f_{ac} \\ \exp(-S_aL)\exp\{[1 - \exp(-S_aL)] - A_aL\} & f_a > f_{ac} \end{cases} \tag{17.1.2}$$

as illustrated in Fig. 17.4a. When absorption is significant, Eq. (17.1.2) rather than Eq. (17.1.1) must be used if there are no instrumentation limitations on scatter angles recorded in the image. If A_a is zero, this equation reduces to Eq. (17.1.1).

Even though the "classical" aerosol MTF format of Eq. (17.1.2) does not include the instrumentation limitations on angles of light scatter actually recorded in the image, it can still be used to determine the limiting value of aerosol MTF at very high angular spatial frequencies, that is,

$$T_{\text{eff}} = -\ln M_a(\infty) \tag{17.1.3}$$

where T_{eff} is effective optical depth for $f_a > f_{ac}$.

17.2 PRACTICAL INSTRUMENTATION-BASED AEROSOL MTF

While Fig. 17.4(a) considers atmospheric phenomena such as scattering and absorption, instrumentation parameters strongly affect the aerosol MTF actually recorded

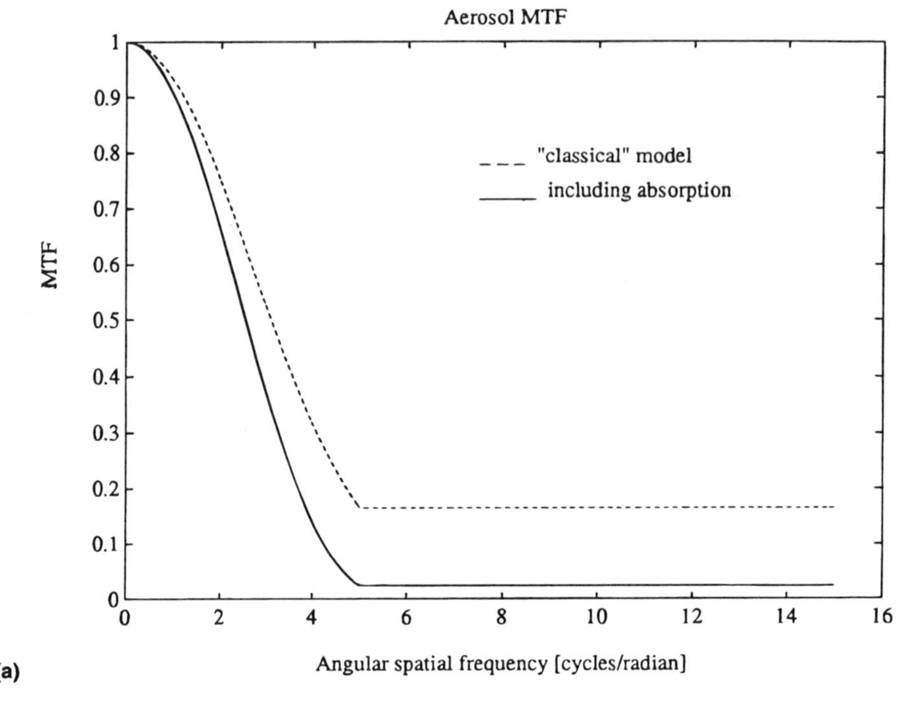

(a)

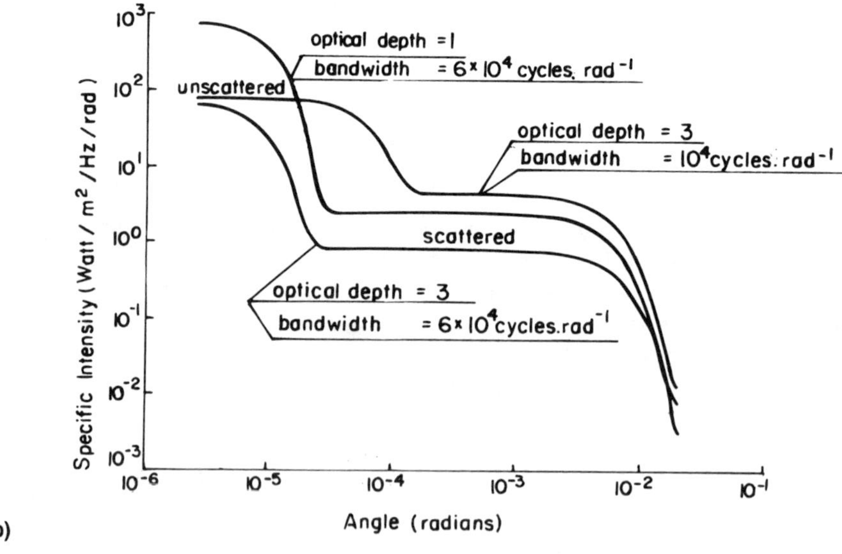

(b)

Fig. 17.4 (a) Effect of absorption on typical aerosol MTF for single interaction case when $\lambda >> a$. (b) Typical point spread functions from aerosol scatter actually recorded in images for various optical depths and angular spatial frequency bandwidths. Contributions from unscattered and scattered light, respectively, comprise the left and right sides of the point spread functions (after [17.9]).

in the image. This occurs because field of view and dynamic range usually limit the maximal angle of light scatter recorded in the image, and because the spatial frequency bandwidth alters significantly the point spread function actually recorded in the image from that shown in Fig. 17.2, which is incident to the optics. Such alteration is shown in the form of typical spread functions or specific intensities depicted in Fig. 17.4(b) for various optical depths and spatial frequency bandwidths. The unscattered light, which is a delta function in angle space in Fig. 17.2, is affected most by the limited spatial frequency bandwidth of any imaging system, since it causes a broadening of the image of the delta function. Consider the upper curve on the right-hand side and the lower curve in Fig. 17.4(b), which refer to the same optical depth of 3. Since the angular spatial frequency bandwidth of the upper curve is smaller (10 cycles·mrad^{-1})than that for the lower curve, the contribution to the point spread function deriving from the received unscattered light is broadened more for the upper curve than for the lower. Because energy is conserved, such broadening also causes a decrease in apparent irradiance which can be by orders of magnitude, so that *for lower-resolution systems* (smaller angular spatial frequency bandwidth) the *irradiance of the unscattered light is much closer to that of the scattered light* than for higher-resolution systems. This is important when considering the truncation on the scattering diagram caused by limited system dynamic range. For example, if dynamic range is two orders of magnitude, then only scattered light of irradiance greater than 1% of that of the unscattered light is recorded in the image. For the right-side upper curve in Fig. 17.4(b) discussed here, only scattering angles less than about 10 mrad are recorded in the image. Larger scattering angles are of too little irradiance relative to the unscattered light to pass the dynamic range threshold of the imaging system. Thus, dynamic range limits the light scatter angle recorded in the image and therefore decreases the size of blur extent. For the lower curve in Fig. 17.4(b), which represents a spatial frequency bandwidth six times larger, such a two-order-of-magnitude dynamic range limitation limits angles of light scatter actually recorded in the image to the order only of a milliradian, thus reducing blur extent even further.

It is useful to consider higher-resolution (60 cycles·mrad^{-1} bandwidth) imaging system response for different optical depth weather conditions. Compare the upper curve on the left side of Fig. 17.4(b) with the lowest. When optical depth is only unity, there is so little scattering taking place that a two-order-of-magnitude dynamic range limitation would eliminate essentially the entire scattered light image, leaving only the unscattered light image whose point spread function is broadened by spatial frequency bandwidth only so that there is essentially no blur caused by the scattered light. A reduction of spatial frequency bandwidth, which broadens and reduces the irradiance of the unscattered light portion of the aerosol scatter spread function, can permit recording of an aerosol MTF at such low optical depths. High-resolution systems are therefore less affected by aerosol MTF. In general, an optical depth of at least unity is required in order for aerosol MTF to have a noticeable effect on image quality. This depends on dynamic range and spatial frequency bandwidth. For elevations a few meters and higher above the surface, aerosol light scatter is often the dominant source of atmospheric blur (but not scintillations or image dancing which are caused by turbulence) in clear weather. In haze, fog, and duststorms, aerosol MTF is dominant over turbulence at even lower elevations.

To calculate actual aerosol MTF deriving from the angles of light scatter actually recorded in the image, it is necessary to know the maximal scatter angle limits deriving from receiver field of view, dynamic range limitation, and target-to-background ratio in the scene itself. The latter is important because if this object scene signal-to-noise ratio (SNR) is poor, scattered light from only very small scatter angles with larger irradiances can be recorded as part of the video signal since irradiance from larger scatter angles yields a target signal too weak to be recognizable above background energy or noise. Hence, angles of light scatter recorded in the image can be limited by object scene irradiances, rather than by instrumentation. *The smallest of these scatter angles is the angle at which the scattering diagram, or phase function, is truncated.* Incoming light at larger scatter angles is removed from the recorded image by the instrumentation. This is summarized schematically in Fig. 17.5.

Figure 17.1 illustrates the case when field of view limits the scatter angles recorded in the image. However, usually the scatter angle limitation stems from instrumentation dynamic range or object scene SNR. Once the scatter angle limitation is known, then the aerosol MTF is calculated numerically from the aerosol size distribution and refractive index as shown step by step in the Appendix. A few comments on the calculation procedure follow. The phase function or scattering diagram is usually assumed to be spatially Gaussian. Therefore, in Eq. (16) in the Appendix α_p is evaluated in a best fit manner from the particulate size distribution [17.3, 17.5]. Also, since photon scatter (as distinguished from absorption) is at the outer layer of the aerosol, it is the refractive index there that is of interest. For relative humidities (RHs) above about 75%, that outer aerosol layer is often water adsorbed to the aerosol. Therefore, aerosols grow in size significantly under extended high humidity conditions. An imaginary component to refractive index signifies absorption, as derived in Section 1.3.2. Such absorption is particularly significant in the infrared and it affects adversely the high spatial-frequency dependence of the aerosol MTF as in Eq. (17.1.2). Since it is dynamic range that often limits light scatter angles actually recorded in the image, whether limited by the object scene itself or limited by the hardware instrumentation, it follows that increased dynamic range is at the expense of spatial resolution [17.5]. Limited dynamic range implies smaller angles of light scatter are actually recorded in the image, thus improving spatial resolution. Hence, often there is a trade-off between dynamic range and spatial resolution. If one wishes to detect both high- and low-intensity objects within the same scene, spatial resolution suffers. If one wishes to obtain high angular resolution so as to resolve objects that appear close to one another, dynamic range should be limited.

Cascading classical aerosol MTF with hardware MTF is not legitimate because of the dependence of aerosol MTF on hardware parameters. It is only after such

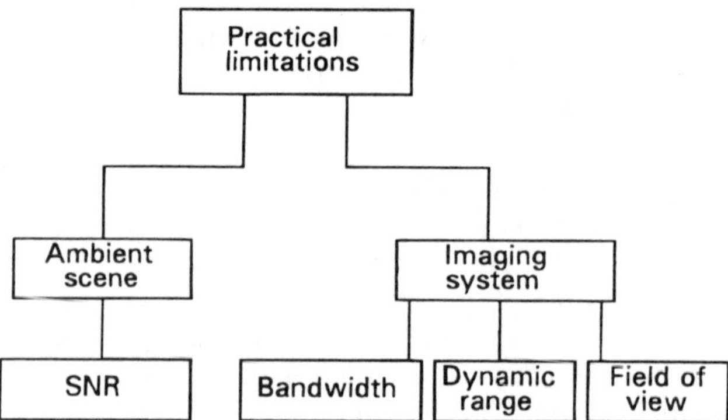

Fig. 17.5 Summary of practical limitations on maximal angle of light scatter actually recorded in image. Smallest of angles shown is used in computations.

hardware limitations have been used in the calculation of aerosol MTF that the latter is independent of hardware MTF. Therefore, cascading of *practical instrumentation-based* aerosol MTF with hardware MTF is entirely proper. The same reasoning applies to cascading of short exposure turbulence MTF [Eq. (16.1.2)] with hardware MTF, which is also quite proper.

Following the "classical" aerosol MTF formulation of Eq. (17.1.1), for limited absorption, a version of the practical instrumentation-based aerosol MTF approximate to the numerical calculation in the Appendix is

$$M_a(f_a) = \begin{cases} \exp[-A_aL - S_a^*L(f_a/f_{ac}^*)^2] & f_a \leq f_{ac}^* \\ \exp[-(A_a + S_a^*)L] & f_a > f_{ac}^* \end{cases} \tag{17.2.1}$$

where $\exp[-(A_a + S_a^*)L]$ is the high spatial-frequency asymptote of the practical aerosol MTF obtained either experimentally or by calculation [17.5–17.7]. Coefficient S_a^* is an effective scattering coefficient that can be evaluated from the high spatial-frequency practical aerosol MTF asymptote, the absorption coefficient A_a, and path length L. The practical aerosol MTF cutoff angular spatial-frequency f_{ac}^* is approximately reciprocal to the limiting angle of light scatter actually recorded in the image. Classical aerosol MTF in Eq. (17.1.2) and Fig. 17.4 starts out at unity at zero spatial frequency [or near unity at zero spatial frequency in Eq. (17.1.1)] and decreases with increasing spatial frequency until the cutoff f_{ac}, at which point it levels off to its high spatial-frequency asymptotic value. Practical aerosol MTF behaves similarly with regard to cutoff spatial frequency f_{ac}^*. However, since f_{ac} equals a/λ, which is typically milliradians, f_{ac}^* is usually several orders of magnitude greater than f_{ac}. While f_{ac} may be a few tens of cycles per radian, f_{ac}^* may be a few tens of cycles per milliradian, depending on the limiting angle of light scatter actually recorded in the image [17.5, 17.8]. Another difference between the practical instrumentation-based and classical aerosol MTF expressions is that the high spatial-frequency asymptote for the practical aerosol MTF is slightly greater than that for the classical aerosol MTF. Hence, S_a^* is used in Eq. (17.2.1) instead of S_a. This means the high frequency asymptote in Eq. (17.2.1) is at an MTF value slightly higher than atmospheric transmission.

The reason $S_a > S_a^*$ is that the received unscattered light energy is limited by instrumentation to finite instead of infinite spatial-frequency extent. This limited spectral range for the same energy causes the high spatial-frequency asymptote [Eq. (17.1.3)] for practical aerosol MTF to be greater than its value (atmospheric transmission) for infinite spatial-frequency bandwidth. Another way of looking at it is as follows: Because of the limited maximal scattering angle recorded in the image, part of the scattered light is not detected by the imager. This results in a lower unnormalized MTF at low spatial frequencies, especially at zero frequency, since the detected scattered light influences aerosol MTF primarily at low frequencies as shown in Fig. 17.3. Because the MTF is normalized to its dc value, which is reduced because of the manner in which instrumentation reduces light scatter angles recorded in the image, the MTF values at high spatial frequencies are increased slightly.

Approximation (17.2.1) yields a constant MTF value for high spatial frequencies ($f_a > f_{ac}^*$). In reality, the practical aerosol MTF is not constant at high spatial frequencies, but displays a certain amount of "waviness" or "ringing" for high spatial-frequency values as shown in Figs. 17.9 and 17.10 in the next section. Another difference between Eq. (17.1.1) and Eq. (17.2.1) is that the former considers a single discrete aerosol size, whereas the latter is relevant to real-atmosphere size distributions as shown in the Appendix. For a given discrete aerosol size Eq. (17.1.1) can be considered as aerosol MTF in the optics plane, while Eq. (17.2.1) is aerosol MTF actually influencing image quality in the image plane according to the effects of instrumentation on light scatter angles actually recorded in the image.

To calculate aerosol MTF it is imperative to know the particulate size distribution, as shown in the Appendix. Although many models have been suggested [17.10], it is often convenient to use bimodal or trimodal log-normal approximations to model size distributions, that is,

$$\frac{dn(r)}{dR} = \sum_{i=1}^{n} \frac{N_i}{\ln(10)\sqrt{2\pi}\,R\sigma_i} \exp - \left[\frac{(\log R - \log R_i)^2}{2\sigma_i^2}\right] \qquad (17.2.2)$$

where R is particulate radius, and $i = 2$ or 3 depending on whether the model is bimodal or trimodal. The terms R_i and σ_i are mean radius and standard deviation for each mode, while N_i is number density. By convention, increasing i implies larger size particulates. Generally, as size increases particulate concentration decreases. This is illustrated in Fig. 17.6, which depicts a typical measurement of "coarse"particulate size distribution ($R > 0.16$ μm). The"fine"particulate ($R < 0.16$ μm) distribution increases with decreasing particulate radius and peaks typically at 0.01 to 0.03 μm radius. As radius decreases further, the particulate size distribution falls off sharply. Although there are fewer large particles, each one does more scattering than do smaller particulates. Since the probability of photon/aerosol interaction increases with aerosol cross-sectional area, a more accurate picture of scattering phenomena can be ascertained from aerosol distribution according to cross-sectional area rather than number. This too is shown in Fig. 17.6. Since each particulate is multiplied by πR^2, the aerosol area distribution emphasizes the greater role played by the larger particulates. Also, the three modes composing the size distribution begin to become more distinct.

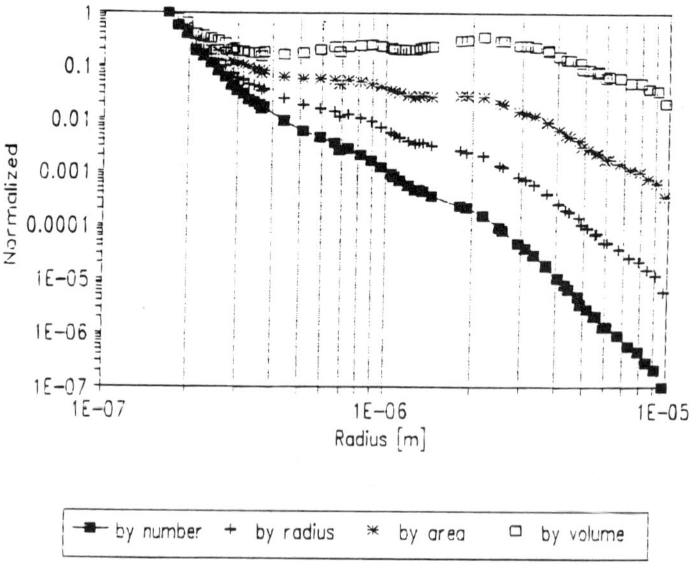

Fig. 17.6 Typical size distribution measurement of "coarse" aerosols at about 30-m elevation.

Particulate size distribution changes strongly with weather, particularly humidity. Under extended relatively high humidity conditions, water adsorbs to particulate surfaces and is even absorbed. These processes cause aerosol size to increase and to become spherical. Tables 17.1, 17.2, and 17.3 show typical values to substitute in Eq. (17.2.2) for continental aerosols for a trimodal distribution. An example of the accuracy of this model to actual measurement is shown in Fig. 17.7. Marine aerosols also are modeled with a trimodal distribution [17.12]. Often continental aerosols are modeled according to a bimodal size distribution, as in MODTRAN and LOWTRAN produced by the Air Force Geophysics Laboratory. However, the more modes, generally the more accurate the size distribution model.

17.3 EXPOSURE TIME

Although turbulence MTF was seen in the previous chapter to improve noticeably with decreased exposure time, the same is not true of aerosol MTF. One of the properties of turbulence is wavefront tilt. For longer exposures, there are a larger number of wavefront tilts integrated into the image. Each wavefront tilt produces a given PSF in a different location in the image, thus increasing blur radius until saturation is reached, as shown in Fig. 16.2. Particulate light scatter diffuses and broadens the photon beam comprising the image as it propagates through the atmosphere. However, it causes little, if any, wavefront tilt. Consequently, aerosol light scatter does not give rise to the time dynamic properties of turbulence. This concept is illustrated in Figs. 17.8, 17.9, and 17.10 [17.13].

Overall atmospheric PSF was evaluated by imaging a HeNe laser beam at 633-nm wavelength over a horizontal 5.5-km path of about 25-m elevation. This was done for short time exposures of 1/1200 s using a CCD TV camera with a telescope having a 3200-mm equivalent focal length. To obtain long-exposure PSF, 1200 short-

TABLE 17.1 *Dependence of σ_i on Relative Humidity [17.11]*

RH (%)	σ_1	σ_2	σ_3
0–40	0.28	0.18	0.27
40–60	0.23	0.21	0.25
60–70	0.31	0.1	0.23
70–80	0.27	0.1	0.28

TABLE 17.2 *Dependence of N_i on Relative Humidity [17.11]*

RH (%)	N_1	N_2	N_3
0–40	1.14×10^7	8.5	2.7
40–60	1.02×10^5	15.5	2.6
60–70	3.33×10^5	4.8	2.25
70–80	0.17×10^5	11.6	3.85

TABLE 17.3 *Dependence of R_i on Relative Humidity [17.11]*

RH (%)	R_1	R_2	R_3
0–40	0.01	0.3	1
40–60	0.03	0.28	1.1
60–70	0.016	0.6	1.3
70–80	0.043	0.66	1.22

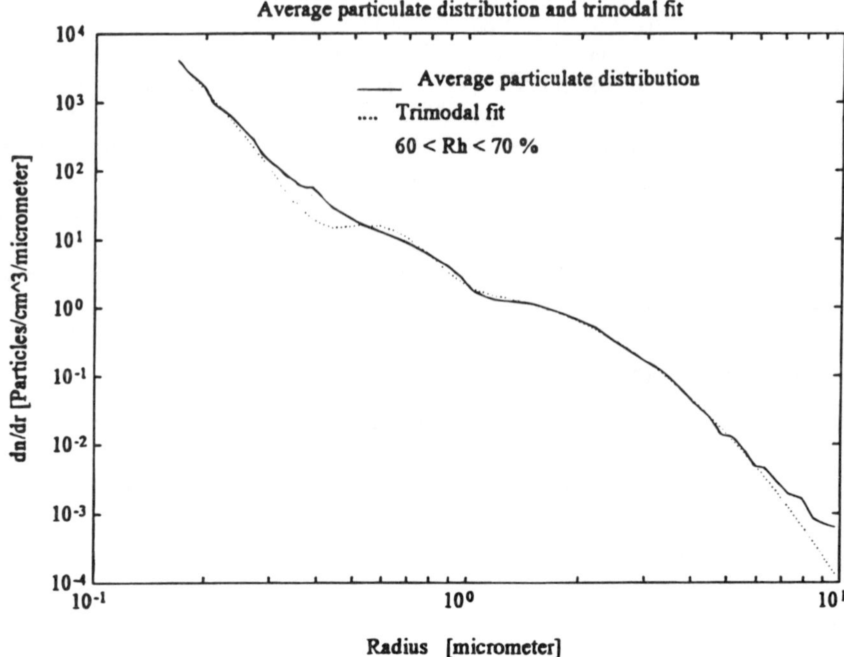

Fig. 17.7 Comparison of trimodal log-normal aerosol size distribution to measurement (after [17.11]).

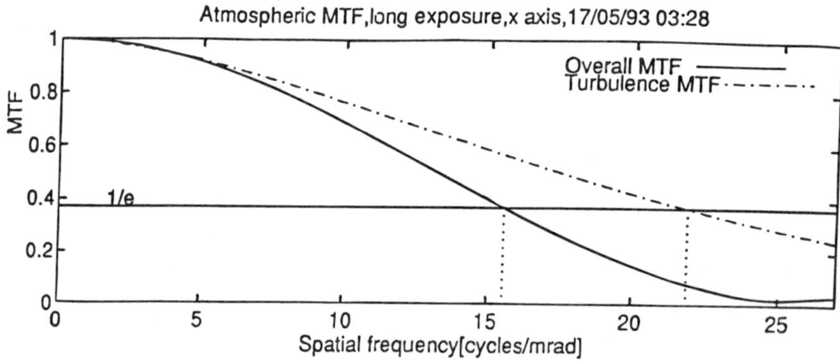

Fig. 17.8 Overall atmospheric and turbulence long-exposure MTFs measured over 5.5-km horizontal path of 25-m average elevation ($C_n^2 = 1.3 \times 10^{-15}$ m$^{-2/3}$). Vertical dotted lines designate $f_{a\,1/e}$ (after [17.12]).

exposure images of the focused laser beam comprising the PSF were added. The shapes of 1200 short-exposure PTFs were averaged to obtain an average 1/1200 second short-exposure PSF. Short- and long-exposure PSFs were Fourier transformed to obtain short- and long-exposure MTFs, which were divided by equipment MTF so as to obtain short- and long-exposure overall atmospheric MTFs.

The refractive index structure coefficient C_n was obtained simultaneously from angle-of-arrival variance according to Eq. (15.4.26). The angles of arrival were mea-

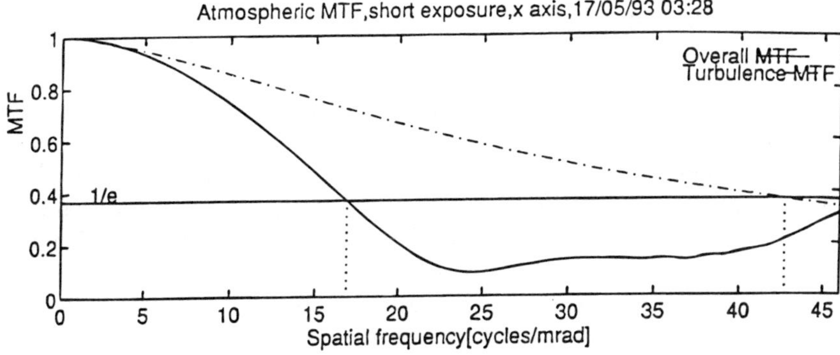

Fig. 17.9 Same as Fig. 17.8 but for short exposures (after [17.12]).

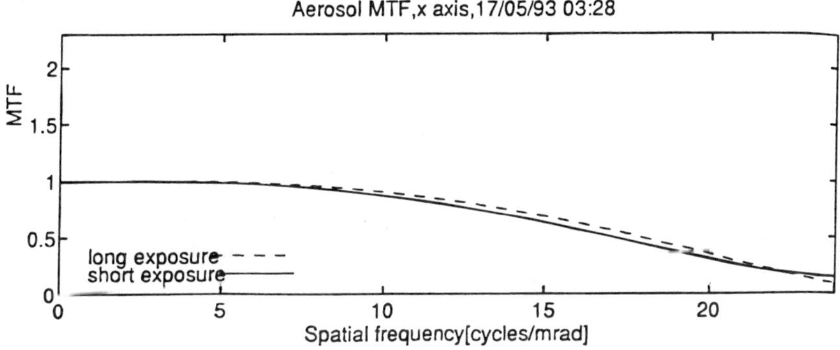

Fig. 17.10 Short- and long-exposure aerosol MTFs for the experiment of Figs. 17.8 and 17.9 (after [17.12]).

sured over the same pixels as the PSF measurements, since both measurements involve imaging of the focused laser beam. From C_n^2, both short- and long-exposure turbulence MTF curves were obtained according to Eq. (16.5.1) for a plane wave. (As shown in Example 16.4 the exact same results apply to a spherical wave since the plane wave formulation was used to obtain C_n^2.) In Figs. 17.8 and 17.9, the lower MTF curves are overall atmospheric MTF. The upper curves are turbulence MTF. The ratio of overall atmospheric MTF to turbulence MTF yields aerosol MTF. Note how overall atmospheric MTF levels off at $f_{ac}^* \approx 24$ cycles·mrad^{-1} for both long and short exposures. The shape of the aerosol MTF curve measured in this way conforms to theoretical shape. The long and short exposure aerosol MTF curves are shown in Fig. 17.10. They essentially coincide, showing that over exposure times typical of short and long exposures for turbulence, aerosol MTF remains essentially unchanged [17.12]. Note also the waviness or ringing typical of practical aerosol MTF in Fig. 17.9. (In many applications, it may be desirable to minimize such artifacts through use of Hanning or Blackman window functions.) Overall atmospheric MTF is dominated more by aerosol MTF at this elevation than by turbulence MTF, particularly for short exposures where turbulence MTF improves (increases) but aerosol MTF does not.

For active imaging systems involving short pulses in particular, and for high-speed imaging systems, the much shorter time scales involved may show aerosol effects deriving from multipath phenomena caused by light scattering.

17.4 EFFECTIVE APERTURE SIZE

The role of turbulence in limiting atmospheric coherence diameter r_o was discussed in Section 16.6. The concept is that increasing aperture diameter to much beyond r_o can improve SNR, but not resolution. Here, we are concerned with effects of blur deriving from light scatter by aerosols on image resolution. Is there a value of aperture diameter beyond which resolution does not improve because of such forward scatter by aerosols? Although for diffraction-limited imaging, resolution increases with increased aperture size, as shown in Sections 8.4 and 8.5, do atmospheric aerosols impose an upper limit to useful aperture diameter, as does turbulence? To consider this question analytically, it is more convenient to use approximation (17.2.1) than the actual practical aerosol MTF, which must be calculated numerically.

Until now, aerosol and turbulence MTFs have been considered individually. However, the real-world atmosphere gives rise to both effects. To a first approximation, at least, both effects are independent of each other. Hence, their MTFs can be cascaded. As was done for Eqs. (16.6.1) and (16.6.2), we define total resolution as follows:

$$\Re_T = 2\pi \int_0^\infty f_a M_0(f_a) M_T(f_a) M_a(f_a)\, df_a \tag{17.4.1}$$

where $M_0(f_a)$ is diffraction-limited optics MTF, $M_T(f_a)$ is turbulence MTF, and $M_a(f_a)$ is given by Eq. (17.2.1). In view of the difference in aerosol MTF for $f_a < f_{ac}$ and $f_a > f_{ac}$, Eq. (17.4.1) can be decomposed into

$$\Re_T = I_1 + I_2 \tag{17.4.2}$$

where, defining $u = \lambda f_a/D$ as in Section 16.6 and $u_c = \lambda f_{ac}^*/D$,

$$\begin{aligned} I_1 = 4\left(\frac{D}{\lambda}\right)^2 \int_0^{u_c} & u[\cos^{-1}u - u(1-u^2)^{1/2}] \\ & \times \exp\left[-57.4 \int_0^L C_n^2(x)\, dx \lambda^{-1/3} u^{5/3} \left(\frac{D}{\lambda}\right)^{5/3}\right] \\ & \times \exp\left[-\int_0^L A_a(x)\, dx - \int_0^L S_a^*(x)\, dx \cdot u^2 \left(\frac{D}{f_{ac}^*\lambda}\right)^2\right] du \end{aligned} \tag{17.4.3}$$

$$\begin{aligned} I_2 = 4\left(\frac{D}{\lambda}\right)^2 \int_{u_c}^{1} & u[\cos^{-1}u - u(1-u^2)^{1/2}] \\ & \times \exp\left[-57.4 \int_0^L C_n^2(x)\, dx\, \lambda^{-1/3} u^{5/3} \left(\frac{D}{\lambda}\right)^{5/3}\right] \\ & \times \exp\left[-\int_0^L (A_a(x)\, dx + S_a^*(x))\, dx\right] du \end{aligned} \tag{17.4.4}$$

We introduce a practical aerosol atmospheric coherence diameter (r_{oa}) in a way analogous to the turbulence-derived atmospheric coherence diameter defined by Fried in Eqs. (16.6.4), such that r_{oa} is defined as [17.14]

$$r_{oa} \triangleq f_{ac}^{*}\lambda / \left[\int_0^z S_a^*(x)\,dx\right]^{1/2} \tag{17.4.5}$$

Using those parameters in Eqs. (17.4.3) and (17.4.4) results in

$$\begin{aligned} I_1 = 4(D/\lambda)^2 \int_0^{u_c} & u[\cos^{-1}u - u(1 - u^2)^{1/2}] \\ & \times \exp\{-3.44u^{5/3}(D/r_o)^{5/3}\} \\ & \times \exp\left\{-\left[\int_0^L A_a(x)\,dx + u^2(D/r_{oa})^2\right]\right\} du \end{aligned} \tag{17.4.6}$$

$$\begin{aligned} I_2 = 4(D/\lambda)^2 \int_{u_c}^{1} & u[\cos^{-1}u - u(1 - u^2)^{1/2}] \\ & \times \exp\{-3.44u^{5/3}(D/r_o)^{5/3}\} \\ & \times \exp\left\{-\left[\int_0^L (A_a(x) + S_a^*(x))\,dx\right]\right\} du \end{aligned} \tag{17.4.7}$$

The resolution function can be calculated by evaluating the integrals of Eqs. (17.4.6) and (17.4.7) and adding the results as in Eq. (17.4.2). Typical values of r_o for turbulence are between a few centimeters under conditions of moderately poor seeing to 20 cm under conditions of exceptionally good seeing. In examples here, the value of r_o was chosen to be 10 cm, which corresponds to moderate turbulence conditions. The measured practical aerosol MTF cutoff frequencies are on the order of thousands to tens of thousands of cycles per radian [17.15, 17.16], depending on optics field of view and imager dynamic range. The integral in Eq. (17.4.5) is close to the optical depth (replacing S_a with S_a^*), and thus on the order of 1 to 5 in the open atmosphere [17.6–17.8, 17.13, 17.16] for path lengths of several kilometers. The optical wavelength λ considered is a 0.5 μm wavelength, in the visible range. With these parameters, the practical aerosol atmospheric coherence diameter can be evaluated using Eq. (17.4.5). Equation (17.4.2) was evaluated numerically as a function of aperture diameter D and results are presented for different atmospheric conditions corresponding to different values of r_{oa}. Figure 7.11 shows results for $f_{ac}^* = 20$ cycles·mrad^{-1} and an optical depth of 5. The dashed line is the resolution as evaluated originally by Fried, considering the atmospheric MTF as turbulence MTF only. The continuous line is the resolution evaluated considering atmospheric MTF as *both* turbulence and practical aerosol MTF. As can be seen, there is a significant difference between the curves, and the overall atmospheric coherence diameter is clearly close to r_{oa}.

On the other hand, Fig. 17.12 compares the resolution achieved by considering the atmospheric MTF as practical aerosol MTF only (dashed line), and by considering the atmospheric MTF as deriving from both turbulence and aerosols (continuous line, as in Fig. 17.11). An excellent fit between both curves is clear for aperture diameters

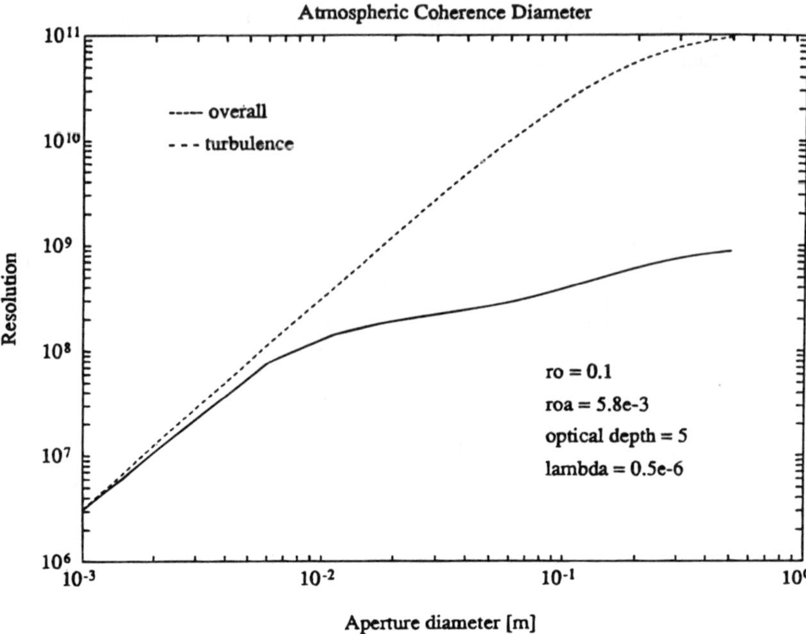

Fig. 17.11 Simulated comparison between overall and turbulence atmospheric coherence diameters. Optical depth is 5, r_o is 10 cm, and wavelength is 0.5 μm. As seen from the first bend in the overall atmospheric coherence diameter curve, $r_{oa} \approx 5.8$ mm $<< r_o$ (after [17.14]).

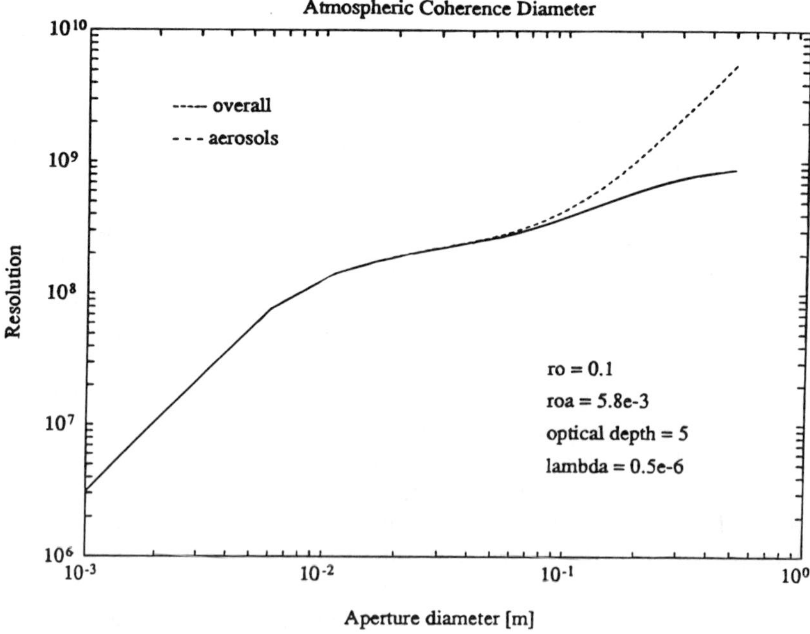

Fig. 17.12 Simulated comparison between overall and practical aerosol atmospheric coherence diameters for conditions of Fig. 17.11 (after [17.14]).

on the order of r_{oa}, in contradiction to the comparison between the overall and turbulence-derived resolutions in Fig. 17.11. For aperture diameters on the order of r_o and larger, the aerosol-derived resolution continues to grow but the overall atmospheric resolution is bounded by the turbulence MTF, and thus remains constant [17.14].

The growth of the practical aerosol-derived resolution for $D >> r_{oa}$ is as the square of D as can be seen from Eq. (17.4.4). It is similar to the growth of the resolution derived from either turbulence or practical aerosol MTF for $D << r_{oa}$ as can be seen in Eq. (17.4.3). Therefore, for $D << r_{oa}$ the contribution of the practical aerosol MTF to resolution is negligible, as is the contribution of turbulence MTF, and the dominant limitation is of the optics diffraction limit MTF. For $D >> r_{oa}$, the practical aerosol MTF contribution is negligible too since the turbulence MTF is dominant. However, in the intermediate zone $r_{oa} < D < r_o$ the *practical aerosol* MTF plays an important role in the atmospheric resolution, and it is in this intermediate zone in which the atmospheric coherence diameter is derived as the diameter for which the knee in the resolution curve occurs. Therefore, the practical aerosol effect on the overall atmospheric coherence diameter is quite important since its main effect is on the knee of the resolution curve, usually shifting it significantly from r_o toward r_{oa}, which, for typical atmospheric conditions, is up to an order of magnitude less [17.14].

Another consideration in choosing aperture diameter is near-field scattering effects. For optical depths on the order of unity or more, such near-field scattering typically produces a uniform path radiance background irradiance that reduces contrast independent of spatial frequency as described in Section 10.9.

We should point out that when atmospheric conditions are those of a small particulate size distribution and optical depth is less than 1, the practical aerosol MTF contribution to overall atmospheric coherence diameter is less important as can be seen in Fig. 17.13, which was created under the same conditions as Fig. 17.11 but for an optical depth of unity. In that case, the knee that occurs in the resolution function due to aerosols is of much less significance, and the overall resolution is improved beyond r_{oa} up to r_o, though at a somewhat slower rate due to practical aerosol scattering effects actually recordable in the image [17.14].

As shown following Eq. (16.6.11),

$$r_o \approx 2.1\rho_{1/e} = 2.1\lambda f_{a\,1/e} \tag{17.4.8}$$

Figures 17.8 and 17.9 illustrate that $f_{a1/e}$ is different for overall atmospheric and turbulence MTF curves for long exposures and particularly for short exposures. These differences in coherence diameters indicate the significance of r_{oa} defined in Eq. (17.4.5). This concept is particularly important to thermal imaging [17.14]. At such long wavelengths, the $\lambda^{6/5}$ dependence of r_o in Eq. (16.6.4) suggests that the coherence diameter at 10 μm wavelength is 36.4 times larger than at 0.5 μm wavelength. This presumption would be grossly in error if applied to overall atmospheric coherence diameter rather than turbulence coherence diameter. The reason is that thermal imaging atmospheric MTFs are usually affected strongly by aerosol rather than turbulence MTF [17.6, 17.7] because of particulate absorption [17.4], although, with high-resolution thermal imaging focal plane arrays, turbulence can still affect high spatial-frequency atmospheric MTF, particularly for short-wave (3 to 5 μm wavelength) thermal imaging [17.17].

17.5 HORIZONTAL AND VERTICAL IMAGING

The effect of imaging geometry on aerosol MTF is determined primarily by aerosol size distribution and concentration along the line of sight. In regions where soil is

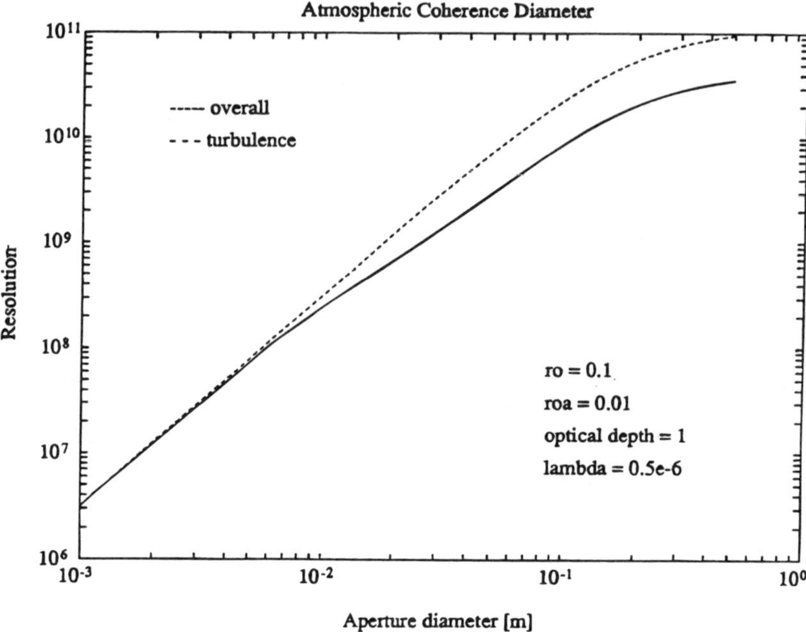

Fig. 17.13 Simulated comparison between overall and turbulence atmospheric coherence diameters for conditions of Fig. 17.11, except that optical depth is reduced to unity after [17.14].

dry, such as beaches and deserts in the absence of rain, winds give rise to large particulate sizes and concentrations near the surface. The same holds for marine atmospheres, where winds pick up salt particulates and snowcaps, thus giving rise to very large particles at low elevations. However, over elevations from a few meters to the boundary layer, particulate sizes and concentrations generally do not vary much. Above the boundary layer, aerosol concentration decreases rapidly, although relative size distribution usually varies little up to about 5 km. At higher elevations aerosol effects diminish considerably until meteoric dust particles are encountered at the 10- to 25-km elevations. Above 25 km, there are very few aerosols.

As shown with regard to turbulence, inhomogeneity over vertical or slant paths causes large differences in aerosol MTF between imaging upward and imaging downward. This has been called the "shower curtain" effect [17.18]. If aerosol scattering takes place near the receiver, light scattered at large angles can still be recorded in the image, thus giving rise to larger blur radii. However, if the light scattering takes place far away from the receiver, such as near the object plane, only small-angle light scatter can be incident on the sensor. Hence, blur radii are smaller. For the same reason, one can see with better clarity through a shower curtain if one is further away than if nearer. With regard to imaging geometry, aerosol MTF—like turbulence MTF—is better for imaging downward than imaging upward, according to the vertical profile of aerosol size distribution and concentration. The shower curtain approach is to divide the line of sight into layers and then integrate over the different layers [17.18]. More data are needed in order to model more accurately the manner in which scattering coefficient varies with elevation under various climatic conditions.

17.6 WAVELENGTH DEPENDENCE

Unlike turbulence, aerosol MTF displays very strong dependence on wavelength according to the wavelength dependencies of scattering and absorption coefficients and scattering phase functions (scatter diagrams). In general, the narrower the forward-scattering patterns and the greater the optical depth, the greater the image degradation as a result of particulate light scatter. In clear weather, aerosol MTF generally improves as wavelength increases from shorter wavelengths into the near-infrared region because, as shown in Fig. 15.4, scattering coefficients decrease. However, if larger particulates in large quantities are present, as in desert and marine atmospheres, the opposite result frequently occurs [17.15]. Even small changes in wavelength, such as from 0.55 to 0.7 μm, can sometimes generate significant improvement in image quality. The amount of improvement depends on aerosol size distribution. As pointed out in scattering discussions in Sections 15.3 and 17.1, the ratio λ/a plays a critical role in determining scattering coefficient and density. This ratio is very weather dependent [17.15]. Figure 17.14, for example, is a photograph of a closed-circuit TV screen image of a large-resolution chart 4.15 km distant (a fluorescent lamp is reflected off the upper right-hand corner of the TV display). This image was obtained in the late afternoon, when turbulence is not too great, at about 0.55 μm wavelength and average elevation of 4- to 5-m. Figure 17.15 is a photograph taken near midday the same day and at the same wavelengths. The difference between the qualities of the two images is primarily turbulence. Figure 17.16, on the other hand, was taken near midday just after Fig. 17.15. The only difference is wavelength. The latter figure was taken in the near infrared at about 0.75 μm wavelength. For Figs. 17.15 and 17.16 turbulence is essentially the same. The difference in sharpness derives from the wavelength dependences of aerosol MTF and path radiance. The advantage of the slightly longer wavelengths is clear even when turbulence is maximum at midday.

Fig. 17.14 Photograph of TV display showing late afternoon image of 4.15-km distant resolution chart at 0.55-μm wavelength (after [17.15]).

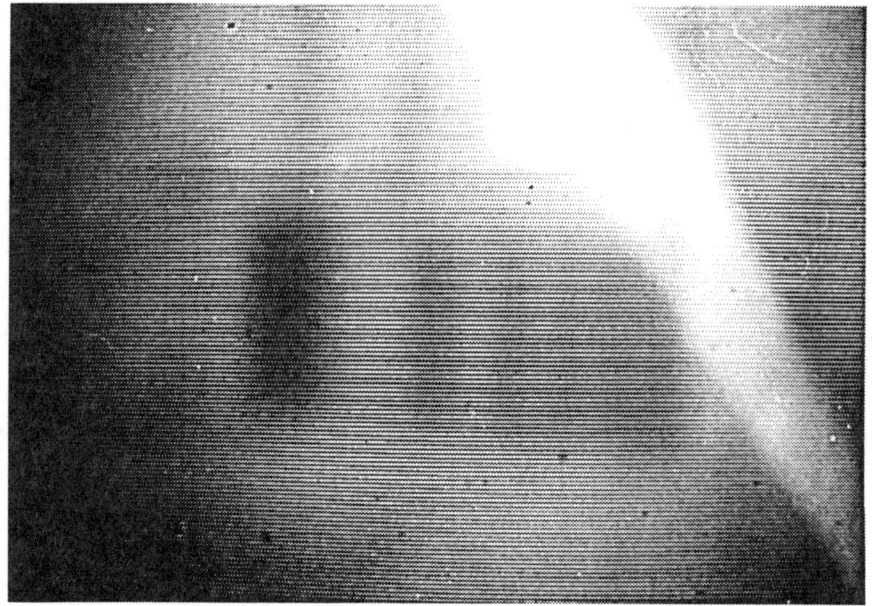

Fig. 17.15 Same as Fig. 17.14 but at around midday (after [17.15]).

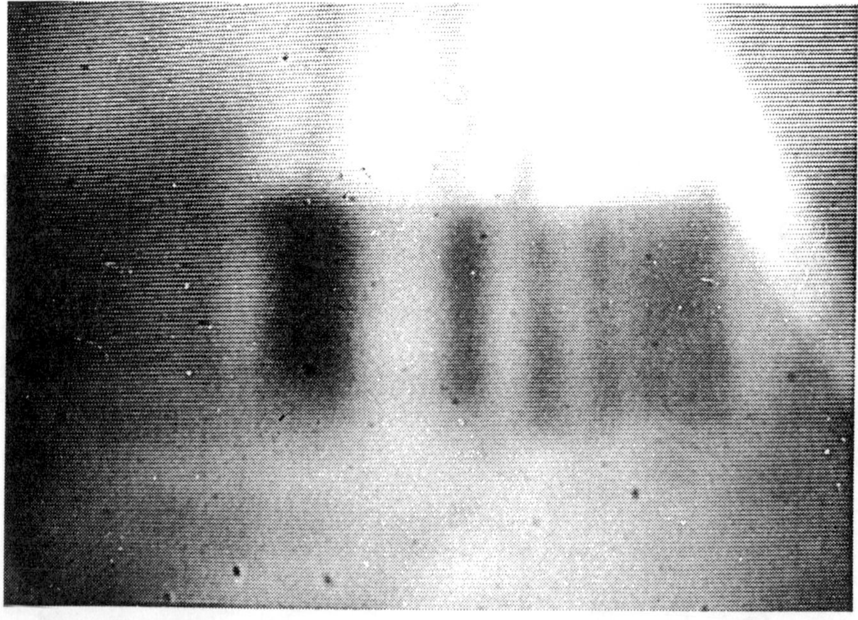

Fig. 17.16 Same as Fig. 17.14 but at 0.75-μm wavelength instead of 0.55-μm wavelength (after [17.15]).

In some weather conditions, aerosol MTF even permits imaging ranges of 40 to 50 km at near-infrared wavelengths, whereas the visible imaging range may be limited to less than 1 km. The key parameter is the ratio λ/a, which changes strongly with weather conditions and with climatic conditions since the presence of many *large* soil-derived particulates in the air can cause visible wavelengths to be more advantageous than near infrared ones for both horizontal [17.15] and vertical [17.19] imaging.

This section and the last suggest that improved modeling of aerosol size distribution and concentration and their dependence on weather conditions [17.20] and elevation are crucial to good design of systems for imaging through the atmosphere.

17.7 IMAGING THROUGH THE ATMOSPHERE: COMPARISON OF TURBULENCE TO AEROSOL MTFS

Historically, blur deriving from atmospheric optics has almost always been attributed to optical turbulence. However, experiments during the last 20 years have shown that, even for optical depths as low as about unity [17.6, 17.7, 17.9, 17.13, 17.16] or close to it [17.17, 17.20], the image degradation deriving from particulate scatter is very significant and, in fact, is often even more significant than that deriving from turbulence. Indeed, measurements at about 15-m average elevation indicate that even in the middle of summer days when turbulence is at a maximum (see Fig. 15.11) aerosol MTF is often more significant [17.16]. An example is shown in Fig. 17.17 for midday. The lower curve is overall atmospheric MTF as measured by edge response (Section 9.7.2). The edge separating the wide white and black bars is imaged horizontally at a distance of 5.5 km. The image of the edge is not a step function because of atmospheric blur. The gradient of the edge image is the line spread function (LSF), from which we obtain overall atmospheric MTF after the Fourier transform of the LSF is divided by the hardware MTF. The upper curve is turbulence MTF calculated

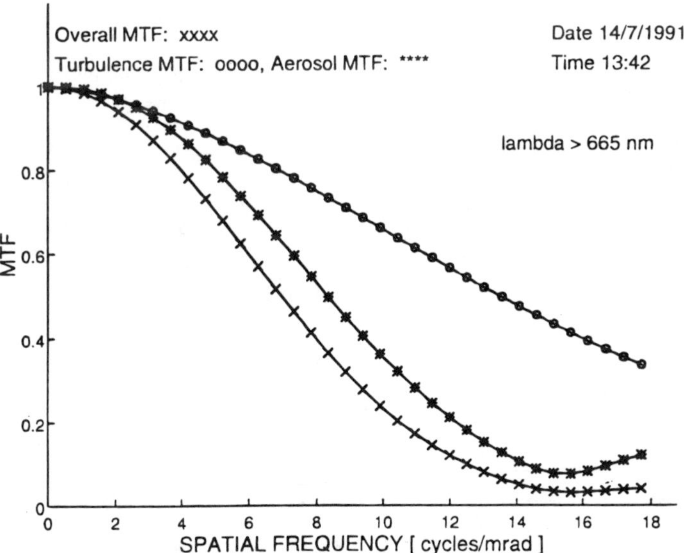

Fig. 17.17 Comparison of overall atmospheric MTF to turbulence MTF and aerosol MTF. Multiplication of the latter two yields overall atmospheric MTF (after [17.16]). Spectral range is limited by silicon CCD to about 1-μm wavelength. Lambda is wavelength.

according to Eq. (16.5.1) for a spherical wave where the value of C_n^2 is obtained from the angle-of-arrival variance of the edge location according to Eq. (15.4.27). Both the upper and lower curves in Fig. 17.17 are long-exposure measurements which are based on 1600 individual measurements of angle of arrival and edge response, respectively. The measurements are simultaneous, independent of each other, and over the same pixels. Division of overall atmospheric MTF (lower curve) by turbulence MTF (upper curve) yields aerosol MTF on the assumption that turbulence blur and aerosol blur are independent of one another. This assumption is largely true, although there may be some small interaction between such phenomena. Figure 17.18 shows the coarse or larger size particulate distribution at the time of aerosol MTF determination in Fig. 17.17. By adding the fine aerosols from MODTRAN, aerosol specific intensity or point spread function was calculated using the technique in the Appendix. The resulting aerosol MTF well resembled that in Fig. 17.17 [17.16]. Figure 17.17 indicates that, in that particular measurement, the aerosol MTF dominated the turbulence MTF, even though it was around midday during a hot summer day. Such a situation is actually typical [17.16] for a 15-m elevation. At shorter wavelengths aerosol MTF was even more dominant [17.16].

From a broader conceptual standpoint, the value of C_n^2 near the surface would have been much larger than that at 15-m elevation. Hence, turbulence MTF is much more dominant near the surface. However, at low elevations aerosol MTF begins to become much more dominant and this usually continues several kilometers up to the boundary layer, since aerosol size distribution changes little up to that point and sometimes even beyond. Hence, for vertical or slant path imaging, the aerosol MTF can usually be expected to be dominant as regards blur, but not image dancing or scintillations which are caused by turbulence. Indeed, aerosol blur is often very significant for imaging of earth from satellites [17.20, 17.21], where it is generally

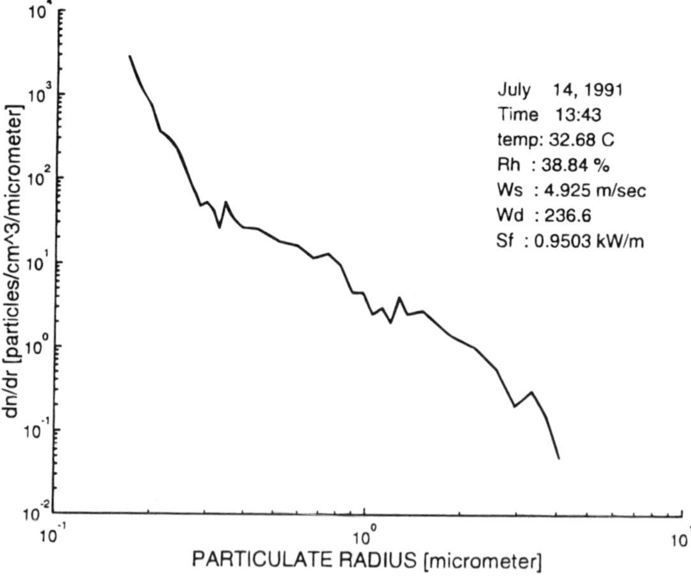

Fig. 17.18 Coarse aerosol size distribution at time at which Fig. 17.17 was obtained in Beer-Sheva, Israel. *Rh* is relative humidity, *Ws* is wind speed, *Wd* is wind direction relative to north, and *Sf* is solar flux (after [17.16]).

referred to as the *adjacency effect* since aerosol scatter causes photons to be imaged in pixels adjacent to those where they would have been imaged otherwise, as summarized elsewhere [17.22].

The preceding discussion refers to visible and near-infrared wavelengths. For thermal imaging, particulate absorption as well as the wavelength dependence of turbulence MTF cause aerosol MTF to usually be more dominant [17.6, 17.7].

In general, the larger the optical depth, the more significant the aerosol MTF [17.9, 17.16, 17.23].

Example 17.1

Consider Example 16.2 but add aerosol MTF. For this purpose assume $f_{ac}^* = f_r f_l = 2 \times 10^4$ cycles·rad^{-1} and that $S_a^* L = 2.5$. Assume no absorption at 0.7-μm wavelength.

Solution

From Eq. (17.2.1), practical aerosol MTF is $\exp[-2.5f_a^2/(2 \times 10^4)^2]$ for $f_a \leq f_{ac}^*$, and $\exp(-2.5) = 0.082$ for $f_a > f_{ac}^*$. To a good first approximation, atmospheric MTF is the product of turbulence and aerosol MTF where, in Example 16.2, turbulence MTF is $\exp(-3.49 \times 10^{-7} f_a^{5/3})$. A key spatial frequency that determines whether resolution is limited by the atmosphere or the TV system is the limiting TV resolution of 295 line pairs per 11.8 mm = 250 lp·cm^{-1}. At that frequency, atmospheric MTF is

$$M_A = M_{TL} M_a = \exp(-3.49 \times 10^{-7} f_r^{5/3} f_l^{5/3}) \exp[-2.5(f_r f_l / 2 \times 10^4)^2]$$
$$= \exp(3.46 \times 10^{-3} f_l^{5/3}) \exp(-3.9 \times 10^{-4} f_l^2)$$

Calculations using the preceding result show that, assuming a 5% contrast threshold, focal length must be less than 47.1 cm for the resolution to be limited by the TV system. For larger focal lengths, resolution is limited by the atmosphere.

For $f_l << 47.1$ cm, overall MTF is

$$\text{MTF} = \text{MTF}_{\text{TV}} \cdot M_A \approx \text{MTF}_{\text{TV}}, \qquad f_r < 250 \text{ lp·cm}^{-1}$$
$$= 0, \qquad f_r > 250 \text{ lp·cm}^{-1}$$

In this case angular resolution is the TV-limited value

$$\frac{\Delta x}{s} \approx -\frac{1}{2 f_{r\,\text{maxTV}} f_l} = -\frac{1}{2(250) f_l} = -0.002/f_l$$

which improves as focal length increases. This results from the increased magnification. For $f_l = 47.1$ cm, angular spatial frequency at the TV spatial-frequency resolution limit is (250)(47.1) = 11.8 cycles·mrad^{-1}, which is less than f_{ac}^*. Hence, our consideration of aerosol MTF here only for $f_a < f_{ac}^*$ has been justified. At this TV resolution-limiting angular spatial frequency, angular resolution is $\Delta x/s = 0.002/47.1 \approx 4.24 \times 10^{-5}$ rad because of the reduced optimal focal length caused by aerosol MTF considerations. This result is 1.23 times worse than that in Eq. (16.2.19) when aerosol MTF was neglected.

For focal length much greater than 47.1 cm, resolution is limited by the atmosphere rather than by the TV systems, that is,

$$\begin{aligned}\text{MTF} &= \text{MTF}_{\text{TV}} \cdot M_A \approx M_A = \text{M}_{\text{TL}} \cdot M_a \\ &= \exp(-3.49 \times 10^{-7} f_a^{5/3}) \exp(-6.25 \times 10^{-9} f_a^2), \qquad f_a \le f_{ac}^* \\ &= 0.082 \exp(-3.48 \times 10^{-7} f_a^{5/3}), \qquad f_a > f_{ac}^*\end{aligned}$$

A 5% limiting contrast yields $f_{a\,\max} = 11.8$ cycles·rad^{-1}. Again, this is less than f_{ac}^*. The accompanying resolution angle is

$$\frac{\Delta x}{s} \approx -(2f_{a\,\max})^{-1} = -4.24 \times 10^{-5} \text{ rad}$$

This is the same as that at the TV limiting spatial frequency since $f_{a\,\max} = f_{r\,\max} f_l$. Increased optical magnification via longer focal length does not yield smaller resolvable detail. As in the case of turbulence by itself, when resolution is limited by the atmosphere, including aerosol MTF, there is no improvement to resolution deriving from increased focal length and magnification. The only result of increased focal length (beyond 47.1 cm here) is magnified blur with no improvement in image detail, but with decreased field of view. The widest field of view without sacrificing resolution is that obtained with 47.1 cm focal length.

Because of aerosol MTF, resolution in this example is more than 1.23 times worse than it would be on the basis of turbulence alone.

However, improved resolution of small detail—theoretically possible with higher resolution hardware (including long focal length)—is achievable using image restoration techniques described in the next chapter. In many cases, the atmospheric blur can be completely removed, particularly for contrast-limited imaging even on the basis of weather data [17.24].

17.8 TECHNIQUES FOR CORRECTION OF AEROSOL MTF BLUR

Blur deriving from particulate scatter does not exhibit the time dynamic properties of turbulence. The scatter does not give rise to significant scintillation or wavefront tilt, but does give rise to radiation being incident from different directions, as shown in Fig. 15.1. Hence, since light from *different* directions can be incident on the same point in the objective mirror plane, adaptive optics offers little correction for aerosol MTF. However, since the aerosol MTF can be estimated fairly well [17.23], it can be well corrected for with proper image processing, even in real time [17.7, 17.24]. This is developed in the next chapter. Another possibility is active imaging.

17.8.1 Active Imaging

Active imaging systems usually involve scene illumination with a laser beam that scans the scene. Often, the laser beam is pulsed. At the receiving end, a computer then assigns each return pulse to its relevant image pixel. There are several advantages to active imaging:

1. Increased transmitter power can be used to overcome poor atmospheric transmission.
2. By using pulsed radiation, range information from the propagation time is obtained as well. Thus, the resulting computer image is three dimensional.
3. By using range gating techniques in which the detector is "closed" during

times corresponding to returns from ranges closer than the object scene, path radiance background radiation (Section 10.9) is greatly reduced. This improves contrast, often considerably.

4. By using optical heterodyne detection if possible, essentially all scattered light is removed from the image because, as shown in Section 6.1.4, the efficiency of IF conversion is drastically reduced as the angle between the received signal and local oscillator radiant powers increases. This means that aerosol MTF, deriving from scattered light recorded in the image, is of greatly reduced significance if heterodyne detection is feasible [17.14]. Thus, active imaging can greatly improve image quality if the number of image pixels is large. At present, much effort is being invested in this direction, particularly with regard to increasing the spatial-frequency bandwidth.

A potential disadvantage of active imaging with lasers is multiple images of edges stemming from diffraction effects in the object plane, as well as speckle arising from diffraction effects by atmospheric particulates. The latter can produce both noise and temporal effects. However, such coherence effects are negligible if the atmospheric coherence diameter of the return beam is much smaller than the receiving aperture diameter [17.7, 17.13]. Using Eq. (17.4.8), the former can be determined from the atmospheric MTF, which can be estimated from expected weather conditions using Section 15.4.5 and the Appendix. In this way, since the atmosphere affects coherence adversely, lower bounds for the receiving aperture diameter can be established so as to produce a better quality image deriving from incoherent light. If multimode lasers are used, which comprise the great bulk of commercial lasers, coherence length is typically only tens of centimeters, so that this too should pose no problem in obtaining incoherent active images.

REFERENCES

17.1. R. F. Lutmokirski, "Atmospheric degradation of electrooptical system performance," *Appl. Opt.*, Vol. 17, December 15, 1978, pp. 3915–3921.

17.2. H. T. Yura, "Small angle scattering of light by ocean water," *Appl. Opt.*, Vol. 10, January 1971, pp. 114–118.

17.3. A. Ishimaru, "Limitation on image resolution imposed by a random medium," *Appl. Opt.*, Vol. 17, February 1978, pp. 348–352.

17.4. D. Sadot and N. S. Kopeika, "Effects of aerosol absorption on image quality through a particulate medium," *Appl. Opt.*, Vol. 33(30), October 20, 1994, pp. 7107–7111.

17.5. D. Sadot and N. S. Kopeika, "Practical instrumentation-based theory and verification of aerosol MTF," *J. Opt. Soc. Am. A*, Vol. 10, January 1993, pp. 172–179.

17.6. D. Sadot, G. Kitron, N. Kitron, and N. S. Kopeika, "Thermal imaging through the atmosphere: atmospheric MTF theory and verification," *Opt. Eng.*, Vol. 33, March 1994, pp. 880–887.

17.7. D. Sadot, A. Dvir, I. Bergel, and N. S. Kopeika, "Restoration of thermal images distorted by the atmosphere, based upon measured and theoretical atmospheric modulation transfer functions," *Opt. Eng.*, Vol. 33, January 1994, pp. 44–53.

17.8 I. Dror and N. S. Kopeika, "Aerosols and turbulence modulation through modulation transfer functions: comparison measurements in the open atmopshere," *Opt. Lett.*, Vol. 7, November 15, 1992, pp. 1532–1534.

17.9. N. S. Kopeika and D. Sadot, "Imaging through the atmosphere: practical instrumentation-based theory and verification of aerosol MTF: reply to comment," *J. Opt. Soc. Am. A*, Vol. 12, May 1995, pp. 1017–1023.

17.10. M. F. Thomas and D. D. Duncan, "Atmospheric transmission," in *Atmospheric Propagation of Radiation*, F. G. Smith, Ed., Vol. 2 of *The Infrared and Electro-Optical Systems Handbook*, J. S. Accetta and D. L. Shumaker, executive editors, Environmental Research Institute of Michigan, Ann Arbor, and SPIE Optical Engineering Press, Bellingham, WA, 1993.

17.11. I. Dror and N. S. Kopeika, "Prediction of aerosol distribution parameters according to weather forecast: status report," in *Atmospheric Propagation and Remote Sensing*, A. Kohnle and W. B. Miller, Eds., *Proc. SPIE*, Vol. 1688, 1992, pp. 1232–131.

17.12. J. H. Richter and H. G. Hughes, "Marine atmospheric effects on electro-optical system performance," *Opt. Eng.* Vol. 30, November 1991, pp. 1804–1820.

17.13. D. Sadot, A. Melamed, N. Dinur, and N. S. Kopeika, "Effects of aerosol forward scatter on long and short exposure atmospheric coherence diameter," *Waves in Random Media*, Vol. 4(4), October 1994, pp. 487–498.

17.14. D. Sadot and N. S. Kopeika, "Effects of aerosol forward scatter of infrared and visible light on atmospheric coherence diameter: theory and validation," *Opt. Eng.*, Vol. 34, January 1995, pp. 144–153.

17.15. N. S. Kopeika, S. Solomon, and Y. Gencay, "Wavelength variation of visible and near-infrared resolution through the atmosphere: dependence on aerosol and meteorological conditions," *J. Opt. Soc. Am.*, Vol. 71, July 1981, pp. 892–901.

17.16. I. Dror and N. S. Kopeika, "Experimental comparison of turbulence MTF and aerosol MTF through the open atmosphere," *J. Opt. Soc. Am. A*, Vol. 12, May 1995, pp. 970–980.

17.17. D. Sadot, G. Lorman, R. Lapardon, and N. S. Kopeika, "Restoration of thermal images distorted by the atmosphere using predicted atmospheric modulation transfer function," *Infrared Phys. Technol.*, Vol. 36, February 1995, pp. 565–576.

17.18. Y. Kuga and A. Ishimaru, "Modulation transfer function of an inhomogeneous and layered random medium using the small angle approximation," *J. Opt. Soc. Am. A*, Vol. 3, January 1986, pp. 115–119.

17.19. N. S. Kopeika, "Spatial frequency and wavelength-dependent effects of aerosols on atmospheric modulation transfer function," *J. Opt. Soc. Am.*, Vol. 72, August 1982, pp. 1092–1094.

17.20 E. F. Vermote, N. El Saleous, C. D. Justice, Y. J. Kaufman, J. L. Privette, L. Remer, J. C. Roger, and D. Tanre, "Atmospheric correction of visible to middle-infrared EOS-MODIS data over land surfaces: background, operational algorithm, and validation," *J. Geophys. Res.*, Vol. 102 (D14), July 1997, pp. 17,131–17,141.

17.21 Y. J. Kaufman, D. Tanre, H. R. Gordon, T. Nakajima, J. Lenoble, R. Frouin, H. Grassl, B. M. Herman, M. D. King, and P. M. Teillet, "Passive remote sensing of tropospheric aerosol and atmospheric correction for the aerosol effect," *J. Geophys. Res.*, Vol. 102(D14), July 1997, pp. 16,815–16,830.

17.22. N. S. Kopeika, I. Dror, and D. Sadot, "Causes of atmospheric blur: comment on scattering effect on spatial resolution of imaging systems," *J. Opt. Soc. Am. A.*, Vol. 15, Dec. 1998, pp. 3097-3106.

17.23. D. Sadot, S. Shamriz, I. Sasson, I. Dror, and N. S. Kopeika, "Prediction of overall atmospheric MTF with standard weather parameters: comparison with measurements with two imaging systems," *Opt. Eng.*, Vol. 34, November 1995, pp. 3239–3248.

17.24. Y. Yitzhaky, I. Dror, and N. S. Kopeika, "Restoration of atmospherically-blurred images according to weather-predicted atmospheric modulation transfer function," *Opt. Eng.*, Vol. 36, November 1997, pp. 3064–3072.

EXERCISES

17.1 Consider Example 17.1. For an optics focal length of 60 cm, what size object has a 50% probability of recognition under the atmospheric conditions and distance involved? How does this change for 90% probability of recognition?

17.2 A. thermal imaging system yields 140 pixel resolution over 2.5°. For $C_n^2 = 10^{-13}\ \text{m}^{-2/3}$ (or $4.64 \times 10^{-15}\ \text{cm}^{-2/3}$), which describes fairly strong turbulence, is turbulence a serious limitation on system resolution for horizontal imaging? Assume a 10-μm wavelength and 2-km range. If hardware PSF is a single pixel, consider resolution limitations of hardware versus turbulence.

17.3 If hardware resolution is improved to 512 pixels over the same field of view as in Exercise 17.2, is such high turbulence a serious problem?

17.4 For the imaging system of Exercise 17.2, assume $f_{ac}^* \approx 1$ cycle·mrad^{-1}, and that atmospheric scattering coefficient is effectively $S_a^* = 0.5\ \text{km}^{-1}$ and absorption coefficient is $A_a = 0.2\ \text{km}^{-1}$ at 10-μm wavelength. Compare relative effects of particulate-derived blur to turbulence-derived blur.

17.5 For the thermal imager of Exercise 17.3, assume S_a^* decreases to $0.3\ \text{km}^{-1}$. Compare aerosol and turbulence MTF effects, assuming f_{ac}^* has been increased to 2 cycles·mrad^{-1}.

Part Six

IMAGE PROCESSING AND EFFECTS ON RESOLUTION

CHAPTER

18

Image Restoration

18.1 INTRODUCTION

Image processing, both digital and optical, is an extremely broad field. Here, we will concentrate on digital image restoration on the basis of the optical transfer function (OTF) or even modulation transfer function (MTF) alone describing the image degradation. Generally, resolution is limited not by the electronics and not by the optics, but often by image motion and vibration. For long-range imaging, the image quality may be limited by atmospheric effects. These environmental effects are emphasized.

First, some basic general concepts on image restoration filters are presented. This is followed by descriptions of some specific techniques that can be used to restore images degraded by image motion and vibration and by atmospheric effects, including pictorial results illustrating corrections. Such restorations improve resolution by broadening overall system MTF. The chapter concludes with a description of noise-reduction techniques and resolution.

By image restoration filters we refer primarily but not exclusively to spatial-frequency filters placed in the spatial-frequency plane (P_2) in the optical data processor of Fig. 7.17. Although data processing is done optically there, the same mathematical concepts can be carried out with a digital computer, including personal computers. In addition, it is also possible to do restoration in the spatial domain. We will consider first spatial-frequency filters. The restorations here are digital.

The simplest image resoration filter is the inverse filter introduced in Section 7.6.

18.2 INVERSE FILTERS

Although this filter is not too effective in the presence of noise compared to others, it does illustrate some of the basic concepts.

An image $i(x, y)$ derives from convolution of the object $o(x, y)$ and the point spread function $s(x, y)$, that is,

$$i(x', y') = \int_{-\infty}^{\infty} \int_{-\infty}^{\infty} o(x, y)s(x' - x, y' - y)\, dx\, dy \qquad (18.2.1)$$

In the Fourier plane, again using capital letters **I** and **O** to designate Fourier transforms of the image and object, respectively,

$$\mathbf{I}(f_x, f_y) = \mathbf{O}(f_x, f_y)\tau(f_x, f_y) \tag{18.2.2}$$

If $s(x, y)$ and $\tau(f_x, f_y)$ designate, respectively, the point spread functions (PSFs) and OTFs of processes degrading image quality, such as atmospheric effects and image motion and vibration, then an inverse filter

$$M(f_x, f_y) = 1/\tau(f_x, f_y) \tag{18.2.3}$$

in the Fourier transform plane P_2 of Fig. 7.17 would yield an output equal to

$$\mathbf{I}(f_x, f_y) = \mathbf{O}(f_x, f_y)\tau(f_x, f_y)/\tau(f_x, f_y) = \mathbf{O}(f_x, f_y) \tag{18.2.4}$$

and thus it is possible, in principle, to obtain the original undistorted image $o(x, y)$ in plane P_3 after inverse Fourier transformation. The processing can be carried out with a personal computer using a frame grabber card. The received image is Fourier transformed mathematically with the computer. The image spectrum is then multiplied in the computer by the inverse transform in Eq. (18.2.3). The product is then inverse Fourier transformed with the computer to yield the restored image.

However, a number of practical problems have to be considered. They involve noise $n(x', y')$, which can be spatially or even temporally changing. If we assume the noise is only spatially varying (frozen in time), then Eq. (18.2.2) must be corrected to

$$\mathbf{I}(f_x, f_y) = \mathbf{O}(f_x, f_y)\tau(f_x, f_y) + \mathbf{N}(f_x, f_y) \tag{18.2.5}$$

where $N(f_x, f_y)$ is the Fourier transform of the image noise $n(x', y')$. (In a noisy image such noise would be manifested as "snow" as in Fig. 11.2.) Consequently, if an inverse filter is used, Eq. (18.2.4) must be modified to [18.1]

$$\mathbf{I}(f_x, f_y) = \frac{\mathbf{O}(f_x, f_y)}{\tau(f_x, f_y)} + \frac{\mathbf{N}(f_x, f_y)}{\tau(f_x, f_y)} \tag{18.2.6}$$

This means that at spatial frequencies where $\tau(f_x, f_y)$ is small, the effects of noise are enhanced signficantly. Furthermore, in practice images often display a spatial-frequency spectrum that is ideally "white," but which decreases in amplitude at higher spatial frequencies, while $N(f_x, f_y)$ decreases with increasing spatial frequency at a slower rate than does τ. Consequently, the last ratio on the right-hand side of Eq. (18.2.6) typically increases with increasing spatial frequency, thus enhancing the relative effects of noise at higher spatial frequencies.

One solution is to define a band-limited inverse filter such that [18.1]

$$M(f_x, f_y) = \begin{cases} 1/\tau(f_x, f_y) & f_x^2 + f_y^2 \le W^2 \\ 1 & f_x^2 + f_y^2 > W^2 \end{cases} \tag{18.2.7}$$

Here, W is determined so that the denominator in Eq. (18.2.6) never approaches zero. Although this technique does bring about improved restorations, other techniques described later are often better. However, an advantage of inverse filter techniques is their simplicity and, in low-noise situations, they can bring about significant image restoration with short computer run times.

When noise becomes a problem, Wiener filters are often used.

18.3 LEAST SQUARES OR WIENER FILTERS

One approach with which to obtain good fidelity or resemblance between the restored image $\hat{o}(x, y)$ and the original object scene $o(x, y)$ is to minimize any discrepancies between the two by minimizing the mean-squared error between them. This filter is called the *least squares* or *Wiener filter*. Our purpose is to find an $\hat{o}(x, y)$ such that the mean square error

$$e^2 = E\{[o(x, y) - \hat{o}(x, y)]^2\} \tag{18.3.1}$$

is minimum. For a given unrestored or degraded image $i(x', y')$, $\hat{o}(x, y)$ is called the least squares estimate of $o(x, y)$. The problem is simplified mathematically if the estimate $\hat{o}(x, y)$ is assumed to be a linear function of the unrestored image such that

$$\hat{o}(\bar{r}) = \int_{-\infty}^{\infty}\int_{-\infty}^{\infty} m(\bar{r} - \bar{r}')i(\bar{r}')\, d\bar{r}' \tag{18.3.2}$$

where the position vector $\bar{r}$ is used and $m(\bar{r})$ is a weighting function that describes the impulse response of the image restoration process. In Eq. (18.3.2) the unrestored image $i(\bar{r}')$ is the input and the restored image $\hat{o}(r)$ is the output. Our purpose is to determine a function $m(\bar{r})$ that yields the minimum mean square error in Eq. (18.3.1), which is then used to restore the image. To do so, Eq. (18.3.2) is substituted into Eq. (18.3.1), yielding

$$e^2 = E\left\{\left[o(\bar{r}) - \int_{-\infty}^{\infty}\int_{-\infty}^{\infty} m(\bar{r} - \bar{r}')i(\bar{r}')\, d\bar{r}'\right]^2\right\} \tag{18.3.3}$$

The results of an involved derivation [18.1] indicate that minimum error is obtained with the following spatial-frequency domain filter:

$$\mathbf{M}(f_x, f_y) = \mathbf{S}_{oi}(f_x, f_y)/\mathbf{S}_{oo}(f_x, f_y) \tag{18.3.4}$$

where the numerator is the cross spectral density of the degraded and undegraded images and the denominator is the spectral density of the original object scene $o(x, y)$ or $o(\bar{r})$.

When the original object $o(\bar{r})$ and the noise $n(\bar{r})$ are uncorrelated and at least one of these two functions exhibits zero mean so that

$$E\{o(\bar{r})n(\bar{r})\} = E\{o(\bar{r})\}E\{n(\bar{r})\} = 0 \tag{18.3.5}$$

then the restoration filter becomes [18.1]

$$M(f_x, f_y) = \frac{1}{\tau(f_x, f_y)} \cdot \frac{|\tau(f_x, f_y)|^2}{|\tau(f_x, f_y)|^2 + [\mathbf{S}_{nn}(f_x, f_y)/\mathbf{S}_{oo}(f_x, f_y)]} \tag{18.3.6}$$

Note that if there is no noise, then the noise spectral density $\mathbf{S}_{nn} = 0$ and the Wiener filter reduces to the ideal inverse filter $1/\tau(f_x, f_y)$. The square brackets term in the denominator of Eq. (18.3.6) is a modification function whose purpose is to provide optimum restoration in the mean square sense in the presence of noise. The high spatial-frequency enhancement is concentrated at those spatial frequencies where the square bracket term in the denominator of Eq. (18.3.6) is minimum. At these spatial frequencies the noise enhancement is least.

Often the noise is assumed to be spectrally white so that $S_{nn}(f_x, f_y) = S_{nn}(0, 0) =$ a constant. Such an assumption is approximately correct if the signal spectral density $S_{oo}(f_x, f_y)$ falls off much faster as a function of increasing spatial frequency than the noise spectral density $S_{nn}(f_x, f_y)$. If no statistical properties of the random processes are known it is not uncommon to assume a constant signal-to-noise power density ratio Γ^{-1} at all spatial frequencies so that

$$M(f_x, f_y) = \frac{1}{\tau(f_x, f_y)} \cdot \frac{|\tau(f_x, f_y)|^2}{|\tau(f_x, f_y)|^2 + \Gamma} \tag{18.3.7}$$

Under such assumptions the Wiener filter is no longer necessarily optimum since such assumptions may be questionable.

Methods to estimate the signal term $S_{oo}(f_x, f_y)$ are discussed in Section 18.5.

Wiener filter restoration and the techniques to follow are relevant in particular to contrast-limited blur imaging (Chapter 10) because with these techniques high spatial-frequency noise is typically enhanced. If the image were to be noise limited to begin with, then since the noise enhancement is identical to the signal enhancement there would be no net improvement in image resolution. However, if the noise is low and the resolution is limited by blur, the restored image is usually still limited by contrast and the noise increase is not too noticeable so that resolution after restoration is usually still limited by blur rather than by noise. What restoration does is to broaden the MTF curve and thus increase $f_{r\,\max}$ as shown later in Fig. 18.10. This improves resolution as shown in Sections 10.4 and 10.5. The higher MTF at higher spatial frequencies after restoration, again as shown in Fig. 18.10, also improves contrast of small detail. This is important in contrast-limited imaging.

18.4 CONSTRAINED LEAST SQUARES FILTER

A limitation to use of the optimum Wiener filter is the necessity of knowing the power spectra of signal and noise. However, if all that is known is the variance of noise as discussed later, then the constrained least squares (CLS) filter is often useful [18.1, 18.2]. The CLS filter is based on two mathematical requirements. The first is minimum mean square error between the original and restored image, as in Eq. (18.3.1). The second involves smoothing of the noise. The mathematical criterion for this is described elsewhere [18.1, 18.2], as well as the pixel matrix development and manipulation. Only the main concept and examples are considered here.

Based on Eq. (18.2.1), the "residue" in the restoration process is defined as

$$\rho = i - s * \hat{o} \tag{18.4.1}$$

where * signifies convolution. A parameter γ is defined as a convergence parameter of the iterative algorithm and a restoration filter similar in form to the Wiener filter is derived:

$$M(f_x, f_y) = \frac{\mathbf{t}^*(f_x, f_y)}{|\tau(f_x, f_y)|^2 + \gamma|\mathbf{C}(f_x, f_y)|^2} \tag{18.4.2}$$

where raised * implies complex conjugate and $C(f_x, f_y)$ is the Laplacian operator used for error minimization. The condition for convergence is

$$n^t \cdot n = \rho^t \cdot \rho \tag{18.4.3}$$

where t symbolizes the transpose matrix operator. The purpose of the iterative algorithm is to find a value of γ that will satisfy Eqs. (18.4.2) and (18.4.3) with an error boundary often suggested to be on the order of $\alpha = 0.01(n^t \cdot n)$. The convergence process is determined by the content of noise in the image. For Gaussian noise it is sufficient to know the noise variance. One method of measuring the spatial noise variance in an imaging system is to use a black board in place of the target. Since the irradiance of a "black" pixel is zero, each received value different from zero is noise. Noise variance can be calculated from an imaged reference black board for several images.

The calculation process starts with the input of the blurred image and the noise variance into the computer. A first value of γ is assumed. A matrix method [18.1, 18.2] is used to estimate $\tau(f_x, f_y)$ and $C(f_x, f_y)$, leading to an estimated filter $\mathbf{M}(f_x, f_y)$ in Eq. (18.4.2). A restored Fourier transformed image $\hat{\mathbf{O}}(f_x, f_y) = \mathbf{M}_e(f_x, f_y)\mathbf{I}(f_x, f_y)$ is calculated, where subscript e means "estimated." This result is inverse Fourier transformed to obtain the estimated image $\hat{o}(x, y)$. From it, the residue $\rho(\gamma)$ is calculated from Eq. (18.4.1) according to $i_e - s * \hat{o}_e$. If $\rho^*\rho < n^*n_e - \alpha$, then γ must be decreased. Otherwise, γ must be increased. The process is repeated until $\rho^*\cdot\rho = n_e^t n_e \pm \alpha$, where typically $\alpha = 0.01n$, thus yielding the restored image.

Figures 18.1 and 18.2 illustrate CLS filter restoration of low-frequency vibration blur based on vibration OTF calculation as in Section 14.4.2. In Fig. 18.1 blur radius is linear as in Fig. 14.11(a). In Fig. 18.2, blur radius is nonlinear, as in Fig. 14.10. Figure 18.3 illustrates CLS filter restoration for high-frequency vibrations, based on vibration OTF calculation as in Section 14.4.1. In all three figures, motion during exposure causes widening of the target image, which gives rise to reduced irradiance, SNR, and contrast, so that raster lines in the video display become evident rather than target shape. In all these examples, the degradation is extreme in that blur extent is much larger than the width or thickness of the letters and numbers. Nevertheless, the restoration is quite clear.

18.5 ATMOSPHERIC WIENER FILTER FOR CORRECTION OF ATMOSPHERIC DEGRADATION

To use the standard Wiener filter the power spectral density (PSD) of the original object scene $S_{oo}(f_x, f_y)$ is required in order to apply the restoration filter to the corrupted image. One way of estimating this term is by rigorously assuming that it equals the image's PSD. This, of course, is not true, but in many cases yields useful results. To check this issue, we can do restoration iteratively until the restored images converge to a stable solution. This often happens after one or two iterations. Another way of

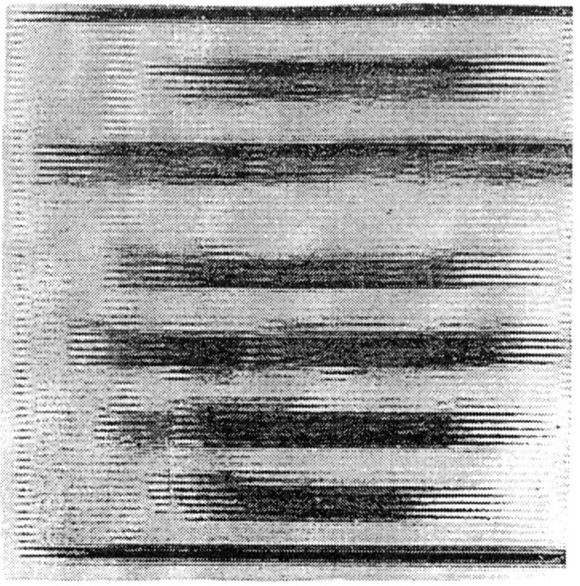

Fig. 18.1 Low vibration frequency (21-pixel linear motion blur extent) degraded and restored images (after [18.3]). Width of numbers is about 5 pixels. In the restoration process, $\gamma = 0.2667$. (The English is a translation of the Hebrew.)

Fig. 18.2 Low vibration frequency (26-pixel nonlinear blur extent) motion degraded and restored images (after [18.3]) Letter width is on the order of 5 pixels. In the restoration process, $\gamma = 0.236$. (The English is a translation of the Hebrew.)

Fig. 18.3 High vibration frequency motion (40-pixel blur extent) degraded and restored images (after [18.3]). Letter width is on the order of 5 pixels. In the restoration process, $\gamma = 0.201$. (The English is a translation of the Hebrew.)

estimating object PSD is by assuming a fractal model. This often yields very good results for visible images of natural scenes [18.4], in contrast to infrared images, which tend to be modeled by Markov random fields [18.5].

After discussing fractal and Markovian techniques to estimate $S_{oo}(f_x, f_y)$, an improved Wiener filter applicable to atmospheric blur is presented.

18.5.1 Fractal Model

Average power spectral density of a fractal scene is modeled

$$E\{S_{oo}(f_r)\} = bf_r^{-a} \tag{18.5.1}$$

where a is fractal dimension and b is an irradiance parameter. Deviation from Eq. (18.5.1) is very little at low spatial frequencies. This equation is a straight line on a log-log plot. Hence, by Fourier transforming the image and using a best fit approach to Eq. (18.5.1), the constants a and b can be evaluated. The resulting expression for $S_{oo}(f_x, f_y)$ can then be used in Eq. (18.3.6) or (18.3.7) for Wiener filter restoration, or in an improved Wiener filter designed to overcome atmspheric blur discussed below. Alternatively, $S_{oo}(f_x, f_y)$ can be simply inverse Fourier transformed to produce a restored image. The restoration is manifest primarily at the higher spatial frequencies where information concerning small detail has been lost in the degradation process since blur affects higher frequencies more than lower ones. We assume the higher frequencies in the original natural object plane scene followed the same f_r^{-a} dependence as the lower frequencies. In this way the higher frequency information is restored [18.6].

Results of images restored by using least mean square fits between the object scene's power spectral density and Eq. (18.5.1) are shown in Fig. 18.4. It is clear that resolution is improved. However, when object plane scenes involve more artificial rather than natural objects, restoration quality decreases. In the thermal infrared, fractal models fail completely [18.7]. This is not unexpected since man-made and thermal images are more Markovian than fractal.

18.5.2 Scene Representation by a Wide Sense Markovian Model

This model is also recursive, using least mean square estimation techniques to estimate fields at points neighboring those at which the field is known. This technique can be used to estimate scene power spectra, which can be implemented in Wiener filters. It can also be used by itself for image restoration, as was the fractal method in Fig. 18.4. For this technique *a sampled image* is assumed, which can be described by a two-dimensional matrix.

When images are represented statistically, each pixel is considered to be a random variable. A given image is then considered a sample function of an ensemble of images. Each random variable is called a *discrete random field.* Here we consider *Markov* random fields. We define a collection of picture points satisfying

$$x_{m,n} = [(i, j) | i < m \text{ or } j < n] \tag{18.5.2}$$

Consider a discrete, random, homogeneous field $o(m, n)$ defined by an array of $m \times n$ points. Assume object scene gray levels (i, j) are known for points in $x_{m,n}$. Our purpose

Fig. 18.4 Restoration of two scenes at 6.5-km horizontal distance (visible wavelengths) using best fit fractal power spectral density for $S_{oo}(f_x, f_y)$ (after [18.7]).

is to estimate the gray levels in $x_{m,n}$ by means of a linear function f of gray levels in $x_{m,n}$. In other words, if $\hat{o}(m, n)$ is the optimum or least square estimate for $o(m, n)$, then

$$\hat{o}(m, n) = \sum_{\substack{\text{for all} \\ (i, j) \text{ such that} \\ (m - i, n - j) \in x_{m,n}}} c_{i,j} f(m - i, n - j) \tag{18.5.3}$$

where the constant $c_{i,j}$ is determined by the least mean square process:

$$e_{m,n} = E\{[\hat{o}(m, n) - o(m, n)]^2\} \tag{18.5.4}$$

in which error $e_{m,n}$ is minimum. Equation (18.5.3) is substituted into Eq. (18.5.4), which is differentiated with respect to each $c_{i,j}$. By substituting each derivative equal to zero, a set of orthogonality relationships governed by

$$E\{[o(m, n) - \hat{o}(m, n)]o(i, j)\} = 0 \tag{18.5.5}$$

is obtained which satisfied for all $(i, j) \in x_{m,n}$.

We define the collection of pairs (i, j):

$$D = \{(0, 1), (1, 1), (1, 0)\} \tag{18.5.6}$$

If $\hat{o}(m, n)$ defined in Eq. (18.5.5) can be shown [18.1] to equal

$$\hat{o}(m, n) = \sum_{i,j \in D} c_{i,j} o(m - i, n - j) \tag{18.5.7}$$

then the random field $o(m, n)$ is wide sense Markovian. This means that the least square estimate of $o(m, n)$in terms of $x(m, n)$is the same as that in terms of only the three immediate neighbors on the right and above. The orthogonality relationship can be applied to Eq. (18.5.7), yielding

$$E\left\{\left[o(m, n) - \sum_{i,j \in D} c_{i,j} o(m - i, n - j)\right] o(p, q)\right\} = 0 \tag{18.5.8}$$

For the Markovian model the preceding relationship holds for all $(p, q) \in x_{m,n}$ and particularly for the values of (p, q) equal to $(m - 1, n)$, $(m, n - 1)$, and $(m - 1, n - 1)$. Substitution of these values for (p, q) into Eq. (18.5.8) permits determination of $c_{1,0}$, $c_{0,1}$, and $c_{1,1}$ from the set of equations

$$\begin{aligned} c_{1,0} R_{00}(0, 0) + c_{1,1} R_{00}(0, 1) + c_{0,1} R_{00}(-1, 1) &= R_{00}(-1, 0) \\ c_{1,0} R_{00}(0, 1) + c_{1,1} R_{00}(0, 0) + c_{0,1} R_{00}(-1, 0) &= R_{00}(-1, 1) \\ c_{1,0} R_{00}(1, -1) + c_{1,1} R_{00}(1, 0) + c_{0,1} R_{00}(0, 0) &= R_{00}(0, -1) \end{aligned} \tag{18.5.9}$$

where

$$R_{00}(\alpha, \beta) = E\{o(m, n) o(m + \alpha, n + \beta)\}$$

The error at each point is designated

$$\zeta(m, n) = o(m, n) - \hat{o}(m, n) \tag{18.5.10}$$

It can be shown [18.1] that a random field is wide-sense stationary if and only if the following are true:

1. The difference equation

$$o(m, n) - \sum_{i,j \in D} c_{i,j} o(m - i, n - j) = \xi(m, n) \tag{18.5.11}$$

is satisfied for all points (m, n) where $m > 1$, $n > 1$.

Solutions for the above are $c_{i,j}$.

2. The relationships

$$o(1, 1) = \xi(1, 1) \tag{18.5.12}$$

$$o(m, 1) - \phi o(m - 1, 1) = \xi(m, 1), \qquad m > 1 \tag{18.5.13a}$$

$$o(1, n) - \eta o(1, n - 1) = \xi(1, n), \qquad n > 1 \tag{18.5.13b}$$

where

$$\phi = \frac{R_{00}(1, 0)}{R_{00}(0, 0)}, \qquad \eta = \frac{R_{00}(0, 1)}{R_{00}(0, 0)}$$

hold for the upper row and left-hand column.

3. $\xi(m, n)$ is a random discrete field of random orthogonal variables, that is,

$$E\{\xi(m, n)\zeta(p, q)\} = 0, \qquad m \neq p \text{ or } n \neq q \tag{18.5.14}$$

If given a random, homogeneous field of known autocorrelation that is not Markovian, it is possible to determine $c_{i,j}$ from Eqs. (18.5.9) and to represent the field with a Markovian model. The accuracy of approximating the field with a Markovian model depends on the degree of correlation of $\zeta(m, n)$. If there is no correlation, then the Markovian approximation is accurate.

18.5.3 Atmospheric Wiener Filter

We have thus far presented two models (fractal and Markovian) for estimating $S_{oo}(f_x, f_y)$ in Eq. (18.3.6). Use of the Wiener filter for correction of atmospheric blur is often not effective because, although aerosol MTF is rather deterministic, turbulence MTF is random. The improved Wiener filter [18.8] presented here is one method for overcoming turbulence jitter, which is random changes of turbulence MTF from its average value. In this approach, the atmospheric MTF variance, which derives essentially from turbulence, is treated as an additional noise source since noise by definition is random. Aerosol MTF, on the other hand, since it is fairly constant as long as atmospheric conditions do not vary too much, contributes primarily to an average atmopsheric MTF. Turbulence MTF changes with time due to its tilt jitter characteristic. These tilts are random and their temporal power spectra are usually limited to several tens up to a few hundred hertz under ordinary atmospheric conditions. The image distortions caused by overall atmospheric MTF can thus be regarded as the sum of a deterministic and a random filter [18.9, 18.10]. The deterministic filter includes aerosol and average turbulence MTFs, whereas the random filter includes the noise component induced on the imaging system, both by the turbulence MTF variance and the hardware [18.8]. Stated mathematically, the atmospheric Wiener filter is determined by

$$s' = s + n_1 \tag{18.5.15}$$

where s' is the instantaneous atmospheric PSF, s is the average atmospheric PSF, and n_1 is an additive random component with zero expectation. Using this model, Eq.

(18.2.1) is altered to yield the image received at the imaging system after propagating through the atmosphere as

$$i(x', y') = [s(x, y) + n_1(x, y)] * o(x, y) + n_2(x, y) \qquad (18.5.16)$$

where $*$ signifies convolution and n_2 is an additive nose imposed by the instrumentation, including camera, digitization, electronics etc., described in Section 18.2, but not by the atmosphere. In other words, n_1 describes turbulence MTF random jitter, whereas n_2 describes the usual hardware noise, primarily electronic. Fourier transforming of Eq. (18.5.16) yields:

$$\mathbf{I}(f_x, f_y) = [\tau(f_x, f_y) + \mathbf{N}_1(f_x, f_y)] \cdot \mathbf{O}(f_x, f_y) + \mathbf{N}_2(f_x, f_y) \qquad (18.5.17)$$

where $\mathbf{I}$, τ, $\mathbf{O}$, $\mathbf{N}_1$, and $\mathbf{N}_2$ are Fourier transforms of i, s, o, n_1 and n_2, respectively. The received image thus is a sum of a deterministic part $\mathbf{I}_1$ and a random part $\mathbf{N}$:

$$\mathbf{I} = \mathbf{I}_1 + \mathbf{N} \qquad (18.5.18)$$

where

$$\mathbf{I}_1(f_x, f_y) = \tau(f_x, f_y) \cdot \mathbf{O}(f_x, f_y) \qquad (18.5.19)$$

and

$$\mathbf{N}(f_x, f_y) = \mathbf{O}(f_x, f_y) \cdot \mathbf{N}_1(f_x, f_y) + \mathbf{N}_2(f_x, f_y) \qquad (18.5.20)$$

The improved atmospheric Wiener filter is defined similarly to the standard Wiener filter, differing by the noise component, which includes an additional term imposed by the turbulent atmosphere [18.8]

$$\mathbf{M}(f_x, f_y) = \frac{|\tau(f_x, f_y)|^2}{\tau(f_x, f_y) \cdot \{|\tau(f_x, f_y)|^2 + [\mathbf{S}_{n_1n_1}(f_x, f_y) + \mathbf{S}_{n_2n_2}(f_x, f_y)/\mathbf{S}_{oo}(f_x, f_y)]\}} \qquad (18.5.21)$$

where $\mathbf{M}$ is the restoring filter, $\tau(f_x, f_y)$ is the average atmospheric MTF, $\mathbf{S}_{n_1n_1}(f_x, f_y)$, $\mathbf{S}_{n_2n_2}(f_x, f_y)$, and $\mathbf{S}_{oo}(f_x, f_y)$ are the power spectral densities of $\mathbf{N}_1$, $\mathbf{N}_2$ and $\mathbf{O}$, and n_1 and n_2 are the inverse Fourier transforms of $\mathbf{N}_1$ and $\mathbf{N}_2$ in Eq. (18.5.20).

Assuming independence between aerosol and turbulence effects, the term $\tau(f_x, f_y)$ can be measured or calculated by a multiplication of the turbulence MTF (for either the short- or long-exposure case) and aerosol MTF. Turbulence MTF can be evaluated with a knowledge of standard meteorological parameters using a C_n^2 prediction model in Section 15.4 and verified independently by U.S. Army Night Vision Laboratory or IMTURB or PROTURB (U.S. Army Atmospheric Sciences Laboratory). The aerosol MTF can be evaluated from the Appendix according to knowledge of particle size distribution, which can also be predicted via LOWTRAN, MODTRAN, or other models [18.12]. The term $\mathbf{S}_{oo}(f_x, f_y)$ can be estimated either by using the Fourier transform of the received image $\mathbf{I}(f_x, f_y)$ or by using estimation models of the object's PSD, which has been shown to obey a fractal model in the visible range, and a Markovian model in the thermal range [18.5]. As with the standard Wiener filter, the term $S_{n_2n_2}(f_x, f_y)$ can be assumed to be constant for all spatial frequencies since the additive noise N_2 is assumed to be white noise. This assumption is commonly used and very practical, and it has a relatively weak effect on the Wiener filter. The term $S_{n_1n_1}(f_x, f_y)$ is very important, since it includes the random part of the atmospheric distortions. One way of estimating $S_{n_1n_1}(f_x, f_y)$ is by a direct measurement. By using the relation

$$S_{n_1n_1}(f_x, f_y) = E\{N_1^2(f_x, f_y)\} \tag{18.5.22}$$

and the Fourier transform of Eq. (18.5.15)

$$N_1(f_x, f_y) = \tau'(f_x, f_y) - \tau(f_x, f_y) \tag{18.5.23}$$

it follows that $S_{n_1n_1}$ equals the variance of the instantaneous atmospheric MTF

$$S_{n_1n_1}(f_x, f_y) = E\{\tau'^2(f_x, f_y) - \tau^2(f_x, f_y)\} \tag{18.5.24}$$

Here, τ' and τ represent instantaneous and average overall atmospheric MTFs.

Since the contribution to the random part of the atmospheric MTF is due mainly to turbulence rather than aerosols, in Eqs. (18.5.22), (18.5.23), and (18.5.24), atmospheric MTF $|\tau(f_x, f_y)|$ refers to turbulence only. However, in Eq. (18.5.21), $|\tau(f_x, f_y)|$ includes aerosol MTF in addition to turbulence MTF since it refers to the average atmospheric MTF which is the product of the two. The variance of overall atmospheric MTF can be evaluated by calculating both terms on the right side of Eq. (18.5.23). This can be carried out by measuring a series of instantaneous atmospheric MTFs, and evaluating the average of both the MTF and its square. For example, in high-resolution astromonical imaging it has been found that short exposures which lack the temporal smear of long exposures can be combined in such a way that a useful signal-to-noise ratio (SNR) is available at higher spatial frequencies [18.11]. This technique is called *speckle interferometry.* It is, however, not a very practical method for *real-time* image restoration.

An alternative way of obtaining $S_{n_1n_1}(f_x, f_y)$ is to evaluate both terms inside the brackets on the right side of Eq. (18.5.24). The second term τ is the square of the turbulence MTF, which can be predicted or measured. The first term $E\{\tau'^2\}$ was evaluated analytically [18.8] to yield

$$E\{\tau'^2(f_x, f_y)\} \propto \tau^2(f_x, f_y) \times \int_{-\infty}^{\infty}\int_{-\infty}^{\infty} \tau^2(f'_x, f'_y)\tau(f'_x + f_x, f'_y + f_y) \times \tau(f'_x - f_x, f'_y - f_y)\, df'_x\, df'_y \tag{18.5.25}$$

which determines the expected value of the squared MTF, or in other words the PSF's power spectral density. The integral in Eq. (18.5.25) can be evaluated numerically with the use of the average turbulence MTF only. Hence, the input required is simply average turbulence and aerosol MTF at the time the image is recorded. These can be evaluated according to weather [18.12].

Examples of restored visible images using this technique, obtained over a 6.5-km horizontal path length through the atmosphere, are presented in Fig. 18.5. The restoration was carried out by implementing Eq. (18.5.21). The term $\tau(f_x, f_y)$ was the measured atmospheric MTF. The term $S_{oo}(f_x, f_y)$ was estimated by best fit to a fractal model. Obtaining the term $S_{oo}(f_x,f_y)$ from the received image iteratively yielded less impressive results. The term $S_{n_2n_2}(f_x, f_y)$ was assumed to be white noise and the term $S_{n_1n_1}(f_x, f_y)$ was evaluated via Eqs. (18.5.25) and (18.5.24). Restoration time was only about 2 s per frame using parallel processing. This can be shortened to a fraction of a second, using appropriate hardware. Therefore, restoration via this method can be in real time. There is a distinct improvement in the fine details of the images, even though the image's SNR is not degraded significantly. This is so in spite of the severe imaging conditions (long horizontal distance), where the turbulence isoplanatic patch was less than image size, and standard Wiener filters were not too successful in restoring the image [18.7]. In the standard Wiener filter of Eq. (18.3.6), $S_{nn}(f_x,f_y)$ refers to white noise only. The restoration of Fig. 18.5 is seen to be better than that of Fig. 18.4. Indeed, the restoration with this atmospheric Wiener filter deblurs essentially all atmospheric blur and resolution is limited by hardware only [18.8]. Whereas the conventional Wiener filter optimizes restoration at those spatial frequencies where signal-to-instrumentation noise is highest, the atmospheric Wiener filter optimizes restoration at those spatial frequencies where the *turbulence jitter is also minimum*. The atmospheric Wiener filter is thus *most advantageous when* $S_{n_1n_1}$ *is not negligible when compared to* $S_{n_2n_2}/S_{oo}$ *in Eq. (18.5.21)*.

original picture

original picture

restoration with improved Wiener

restoration with improved Wiener

Fig. 18.5 Restoration via atmospheric Wiener filter using the fractal PSD estimation of Fig. 18.4 ($\Gamma = S_{n_1n_2}/S_{oo} = 0.017$) (after [18.7, 18.8]).

The atmospheric Wiener technique yielded very good results in the thermal range too (8- to 12-μm spectral window), and restoration for a picture obtained over a 2-km horizontal path length through the atmosphere, using this technique, is presented in Fig. 18.6. The term $S_{oo}(f_x, f_y)$ was estimated from the received image, and no iteration was needed (restoration converged after the first iteration). Since the thermal images involved significantly lower resolution than those in the visible range (due to instrumentation limitations), the main atmospheric distortions were due to aerosol rather than turbulence MTF. Therefore, as explained previously, MTF variance was significantly weaker and excellent restoration was also possible using an average atmospheric MTF (rather than instantaneous), as demonstrated at the bottom right of Fig. 18.7. This is important when considering real-time restoration since atmospheric MTF can be estimated once to create an efficient restoration filter that can be used for a large set of images. The fact that atmospheric MTF in the thermal range is relatively static gives rise to relatively impressive restoration results using standard Wiener filters (which failed in the visible range). An example can be seen in Fig. 18.7.

The atmospheric Wiener filter for restoration of atmospherically blurred images is seen to yield significant improvements. Processing time takes typically 1 to 2 s using parallel processing, and can be reduced considerably by converting software to hardware [18.8].

18.6 MAXIMUM ENTROPY RESTORATION

This powerful method of image restoration requires a lot of computer power and time. It is useful not only for contrast-limited imaging but, unlike the other image restoration filters considered here, also for noise-limited imaging. The reason is that the maximum entropy method can also be used for noise reduction. Entropy is defined as $\ln(W_0)$ where W_0 is the number of substates in which a system can be found. The second law of thermodynamics states that systems strive toward maximum entropy. In imaging it is assumed that the original scene is of maximum entropy [18.1]. Hence, the restored image should also be of maximum entropy. The number of substates of the original scene is

original picture

restoration with atmospheric Wiener

Fig. 18.6 Restoration of 8- to 12-μm wavelength thermal image scene 2 km distant horizontally using the atmospheric Wiener filter ($\Gamma = \mathbf{S}_{n_1n_2}/\mathbf{S}_{oo} = 0.015$) (after [18.7]).

Fig. 18.7 Standard Wiener filter restoration of the thermal image of Fig. 18.6. "Average MTF" refers to averaging atmospheric MTF over both days and nights for several weeks. Because it is composed primarily of aerosol MTF, it does not necessarily vary much, unlike turbulence MTF ($\Gamma = 0.015$) (after [18.13]).

$$W_0(g_1, g_2, \ldots, g_p) = \frac{\left(\sum_{j=0}^{p} g_j\right)!}{g_1!g_2! \ldots ! \, g_p!} \tag{18.6.1}$$

where g_j is the gray level of the j'th pixel and p is the number of pixels. Similarly, the number of image noise substates is

$$N_0(n_1, n_2, \ldots, n_p) = \frac{\left(\sum_{j=0}^{p} n_j\right)!}{n_1!n_2! \ldots ! \, n_p!} \tag{18.6.2}$$

where n_j corresponds to each pixel g_j.

It is assumed then that the best reproduction of the original scene is the restored scene consistent with maximum entropy

$$\ln(W_0 N_0) = \text{maximum} \tag{18.6.3}$$

It is therefore necessary to determine the set of $g_1, g_2, \ldots, g_p$ that yields maximum entropy. Since a large number of variables and equations is required [18.1] a large amount of computation is also required.

For a received image, all received photons are considered indistinguishable, the order in which they arrive being unimportant. Therefore, the distribution of photons to the various pixels can be carried out in various ways. For some images the number of ways to form the image is much greater than for others. These images with greater multiplicity have a higher entropy. They can be formed in more ways, so they are more likely. The most likely "image" is for all pixels to receive the same average number of photons, producing a uniform gray scene. Such a result, however, is contrary to the recorded received image, which is viewed here as a constraint.

One typically assumes random values for each pixel. The difference between this assumed scene and the one actually recorded can be allowed to have a set upper limit, depending on the noise characteristics of the sensor, the received irrandiance, etc. The maximum entropy method involves finding the scene that has the highest multiplicity.

It is usually found that the solution having the maximum feasible entropy permitted by constraints is hard against the boundary formed by those constraints. If the constraints were relaxed, the solution would move higher to a more uniform image. For *noise removal*, a common constraint is the chi-squared value of the smoothed image as compared to the recorded one, where

$$\chi^2 = \sigma^{-2} \sum (g_j - g_j')^2 \tag{18.6.4}$$

where g_j refers to j'th pixel gray level of the original recorded image and g_j' to the assumed one; σ is the standard deviation of the values. An upper limit can be set on the value of χ^2 permitted in the calculation of a new set of g_j' values to maximize the entropy. A typical but arbitrary limit for χ^2 is p, the number of pixels in the array.

An iterated solution for values of g_j' produces a new image with smoothness and noise characteristics that often are an improvement over those of the original recorded image.

18.7 MODIFIED BACKUS-GILBERT FILTER

The modified Backus-Gilbert filter [18.14, 18.15] is a spatial-domain filter. It is assumed that the atmospheric PSF is a finite impulse response (FIR), and so is the restoring filter. Two constraints in the restored image are enforced. These are image sharpness in order to resolve details as fine as possible, and minimum noise variance. Image sharpness is obtained by narrowing the restored PSF as much as possible.

Defining the degraded PSF as $s(x, y)$ and the restoring spatial domain filter as $m(x, y)$, the restored PSF is defined as

$$\hat{s}(x, y) = s(x, y) * m(x, y) \tag{18.7.1}$$

To achieve perfect restoration, $\hat{s}(x, y)$ should be as close to a spatial delta function as possible. One possible measure of a narrow restored PSF is a minimized second moment, that is,

$$R = E\left\{\int r^2 \hat{s}(\bar{r})\bar{r}\right\} \rightarrow \min \tag{18.7.2}$$

The requirement of minimizing noise variance can be stated mathematically as

$$V = E\{\hat{n}^2(\bar{r})\} \rightarrow \min \tag{18.7.3}$$

where

$$\hat{n}(\bar{r}) = n(\bar{r})*m(\bar{r}) \tag{18.7.4}$$

is the noise of the restored image. Since minimizing average spread R and minimizing noise variance V cannot be achieved simultaneously, a trade-off between resolution and noise must be accomplished, and a compromise is reached by minimizing the weighted sum:

$$F = (1 - \lambda)R + \lambda V \rightarrow \min, \qquad 0 \leq \lambda \leq 1 \tag{18.7.5}$$

where parameter λ should be selected according to the restoration requirements, blur versus noise. A detailed mathematical description of solving Eq. (18.7.5) is presented in [18.14].

An important advantage of FIR filters such as the Backus-Gilbert filter is the speed of the computation time. Since the filter is based on the PSF, only a limited number of pixels are involved. This decreases restoration time significantly.

An example of visible image restoration for the same scene as in Fig. 18.5 is presented in Fig. 18.8. In this case the degrading PSF is that of the atmosphere, and the PSF's dimensions were limited to only 15 × 15 pixels. In the thermal infared (8- to 12-μm wavelength), this filter yielded very good atmospheric correction, as shown in Fig. 18.9 for the scene of Fig. 18.6.

18.8 RESOLUTION IMPROVEMENT VIA IMAGE RESTORATION

A variety of spatial-frequency and spatial-domain image restoration techniques have been presented. These are a few of many different techniques that exist. The pictorial examples do indicate that excellent corrections are achievable. The corrections for image motion and vibration blur are as if no image motion or vibration took place.

The atmospheric Wiener filter yields excellent correction for atmospheric blur, as if no atmospheric blur ever took place. Detail on the order of pixel size is resolved. For the visible images over a 6.5-km distance, pixel size in the object plane is on the order of 7 cm [18.7, 18.8]. Detail of this size, such as poles in fences, or elements of vegetation, can be seen in Fig. 18.5. For the thermal images at a 2-km distance, pixel size in the object plane corresponds to about 62 cm [18.3]. This is about the size of windows, which can be seen clearly as black dots in the restored images, since glass does not pass long wave infrared radiation. The key to the quality of the correction is knowledge as to the MTF or PSF of the phenomenon corrupting the image. The better this is known, the more accurate and faster the

Fig. 18.8 Restoration of visible image of scene of Fig. 18.4 using modified Backus-Gilbert FIR filter ($\lambda = 0.017$) (after [18.7]).

image restoration is since fewer iterations are required. Hence it is important to be able to evaluate the required MTF or PSF accurately. They need not be known exactly. Often MTF or PSF approximations are sufficient to obtain good quality corrections. What is critical is that the atmospheric MTF be of the correct shape. If aerosol MTF is neglected, the resulting shape for atmospheric MTF is wrong, and restoration for such cases will tend to be quite poor. In general, high spatial-frequency enhancement does improve contrast. However, blur radius reduction requires that the right shape be used for the degrading MTF [18.16]. Even if that MTF is not characterized exactly, but is of the right shape, excellent restorations are still achieved.

A parameter affecting speed of restoration is the number of pixels involved in the calculations. In this regard the FIR filters that deal with PSF can be advantageous since PSF involves only a small number of pixels. However, the correction for atmospheric blur may not be quite as good as that with the improved atmospheric Wiener filter, although it may be close, as seen in Figs. 18.5 and 18.8.

Since image motion and vibration and atmospheric blur are often the primary causes of image degradation, the examples shown here concentrate on correction of these two phenomena.

Fig. 18.9 Restoration of a thermal image using modified Backus-Gilbert FIR filter for scene of Fig. 18.6 (γ = 0.001) (after [18.13]).

The image restoration techniques presented here are basic and are intended to restore the image to its original unaltered form prior to distortions introduced by sources such as the atmospheric channel or image motion and vibration blur. Further processing can be considered according to application requirements. Much of the restoration treatment here is based on MTF, OTF, or PTF of the image degradation. Such treatments are fundamental and can yield essentially total deblurring of the blurring phenomena, without *a priori* knowledge of target characteristics or shape. Other forms of image processing, such as edge enhancement, can emphasize various features of an image but do not necessarily restore the image as if no blurring had taken place. They can even cause distortion of the image from its original contrast and spatial-frequency content. Such enhancement can be obtained through high-pass spatial-frequency filtering, for example, or by spatially differentiating an image. Examples of low- and high-pass spatial-frequency filtering are shown in Section 10.8. The purpose of the present chapter is fundamental restoration of the image scene to its original unblurred version.

An important advantage of image restoration is that for sufficient SNR, the system MTF is widened, thus increasing the spatial-frequency bandwidth as characterized by f_{rmax} in Fig. 10.1. The system MTF is also increased at higher spatial frequencies, thus improving contrast of small detail. Examples are shown here in Fig. 18.10 for the atmospheric Wiener filter restorations of Fig. 18.5 and standard Wiener filter restorations in Fig. 18.11. The system MTF broadening using the atmospheric Wiener filter is considerable, particularly at low contrasts, and is limited by hardware MTF and spatial-frequency bandwidth. Essentially, just about all atmospheric blur can be removed with atmospheric Wiener filter correction. The broadening indicates considerable increase in $f_{r\ max}$ and therefore decrease in the size of resolvable detail. To permit appreciation of the improved image quality, an expanded image of the building in Fig. 18.5 is shown in Fig. 18.12.

Since, as shown in Sections 10.4 and 10.5, target acquisition for contrast-limited imaging is based on $f_{r\ max}$, image restoration improves target acquisition probabilities and ranges considerably, which is the subject of the next chapter. However, before considering the effects of the improved resolution brought about by image restoration techniques such as the above on target acquisition, two other types of image processing should also be considered. These are improved contrast via contrast enhancement and noise reduction techniques. Both can improve image quality, even if the resulting image is not the initial unblurred object scene.

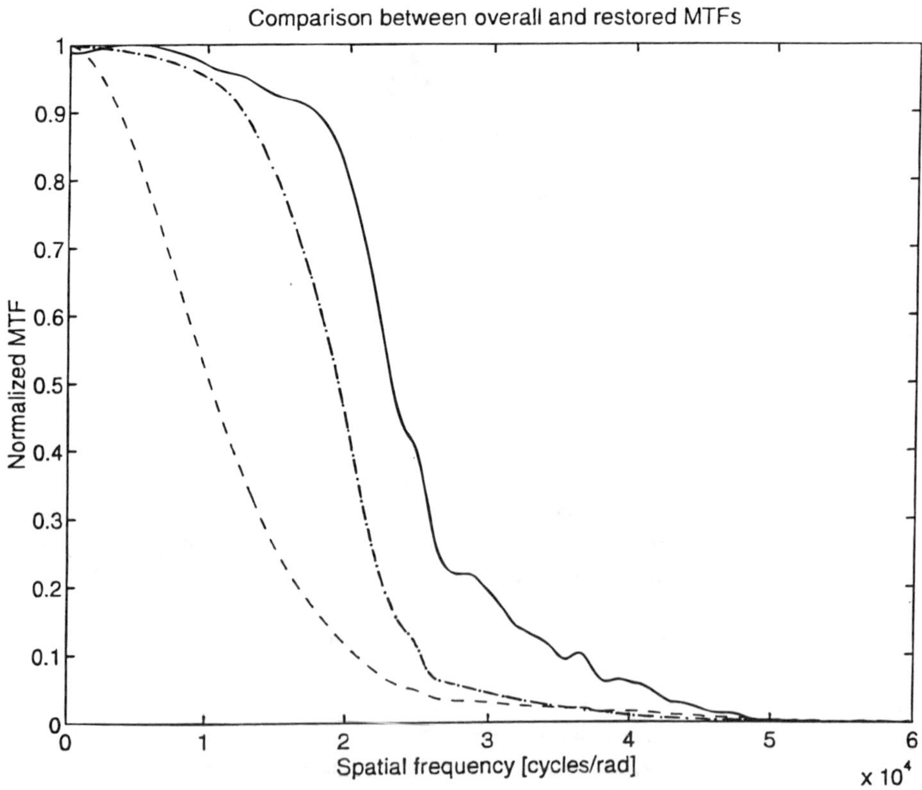

Fig. 18.10 Overall system (hardware and atmospheric) MTFs prior to image restoration of the visible scene of Fig. 18.5 (dashed curve); after restoration with standard Wiener filter (dashed-dotted curve); and after restoration with atmospheric Wiener filter (solid curve) (after [18.8]).

Fig. 18.11 Image restoration of visible scenes of Fig. 18.5 with standard Wiener filter (after [18.8]).

18.9 CONTRAST ENHANCEMENT

As described above, the overall system MTF broadening brought about by image restoration also improves contrast because the MTF is also increased in the vertical direction, especially at higher spatial frequencies. However, it is possible to improve contrast without improving resolution. This can occur in place of or in addition to the image restoration techniques described previously.

Typical digitizers convert analog voltage range from cameras or other sensors to gray-level numbers from 0 to 255. The 256 gray levels corresponds to 8 bits, as shown in Section 13.2.2. This dynamic range in brightness levels covers several orders of magnitude in numbers of photons. Twelve-bit systems are also available. However, often the inherent range of variation in brightness of an image is much smaller than that permitted by the hardware, including imager, electronics, digitizer, and computer. The contrast reduction may be inherent in the object domain scene itself or it may be imposed by the propagation channel, with the atmosphere (path radiance and aerosol and turbulence MTFs) being one example.

Contrast can be improved by stretching the received contrast so that brightness values of pixels are reassigned to cover the entire range permitted by the hardware. This means darker gray-level pixels are reassigned to appear even darker; lighter gray-

Original Image

Restored Image

Fig. 18.12 Expanded image of building viewed from 6.5 km distance shown in Fig. 18.5 using visible wavelengths.

level pixels are reassigned to appear white. Such mapping can be linear or nonlinear. All pixels in the original image that had a particular gray-level value continue to have one gray-level value in the contrast-expanded image, although it will probably be a different one. An example of a nonlinear manipulation of pixel brightness values can be a logarithmic scale. Such manipulation functions need not be mathematical. Often a lookup table is used [18.17]. A typical goal in such a pixel brightness manipulation function may be to achieve histogram equalization, where the number of pixels for each brightness level is equal. However, this criterion is rather arbitrary.

The contrast improvement with such techniques can be quite impressive. However, since the overall system MTF is not broadened, the size of discernible detail is largely unaffected. However, the small detail can be seen with much improved contrast. This speeds up target acquisition times. The image restoration techniques described previously can be even more impressive when contrast expansion is added too.

18.10 NOISE REDUCTION

As shown in Section 6.1.6, integration can improve SNR. The signal strength increases with integration time, while the noise is smoothed out because of its randomness. The same goal can be achieved by averaging a number of frames. The signal appears in each frame. The noise, because of its randomness, is present at some pixels during certain frames and absent from them or located at other pixels during different frames. Thus, averaging a number of frames achieves the same goal as integration. However, by their very nature neither technique is real time. Both techniques are elaborated on in the following discussion.

Charge integration characterizes "staring" mode systems, where the name refers to the fact that the camera is simply open to the incident light [18.17]. System cooling can reduce dark current and thermal noise, as described in Chapters 5 and 6. An infrared camera must be cooled to lower temperatures than a visible one because the infrared photons have less energy. This means the production of a signal charge carrier at longer wavelengths requires less energy so that more dark current is present at room temperature. Staring mode sytems are particularly applicable to low-irradiance situations since the integration improves the SNR. Having considered time averaging, we now consider space averaging.

A simple form of spatial averaging is to add the pixel brightness values in each small region of the image, divide by the number of pixels in each such region or "neighborhood," and use the resulting value to construct a new image. The resulting image is of the same size but with effectively fewer pixels. If, for example, neighborhoods of 5×5 pixels are used, the improvement in SNR is by a factor of $25/(25^{1/2}) = 5$ for small numbers of pixels since Poisson noise varies according to the square root of the signal. However, the cost of this SNR improvement is a possible reduction in this example of up to five in the resolution, depending on the signal-to-noise ratio. To be discernible, detail must be five times larger. For *very* noisy images where resolution anyhow is much poorer than a pixel such neighborhood averaging may still be advantageous. A more common way to accomplish neighborhood averaging is to replace each pixel with the average of itself and its neighbors. This is often described as a "kernel" operation and is discussed elsewhere [18.17]. In general, the larger the neighborhood, the greater the amount of noise smoothing as a result of increased SNR, but the decrease in resolution is also greater.

A widely used method for noise smoothing is based on ranking of neighborhood pixels according to brightness. Then, for example, the *median* value in this ordered list can be used as the brightness value for the central pixel. This is used to produce a new image and only the original pixel values are used for the neighborhood ranking around each pixel. This technique is called *median filtering.* The median filter is an excellent rejector of certain kinds of noise. For example, if a pixel is randomly changed by noise to an extreme value of brightness, it will be eliminated from the image and replaced by a "reasonable" value, that of the median brightness value in the neighborhood. An example is shown in Fig. 18.13. The computational effort rises quickly with the number of values to be sorted. The larger the neighborhood, the more effective the noise smoothing, but the larger the computation time. As far as contrast is concerned, note that the median filter method does not significantly reduce

Fig. 18.13 (a) Original image; (b) noise added by randomly selecting 10% of the pixels and setting them to black and another 10% and setting them to white; (c) application of median filter to part (b) using a 3 × 3 square pixel neighborhood; (d) same as part (c) but using a 5 × 5 pixel octagonal neighborhood (from [18.17]).

the brightness difference across edges, because the values available are only those present in the neighborhood rather than an average between those values. Also, median filtering does not shift boundaries as averaging might do, depending on the relative magnitude of brightness values in the neighborhood [18.17]. Furthermore, the minimal degradation to edges from median filtering allows repeated application of this method [18.17]. In such situations, fine detail is erased, and large neighborhoods take on the same brightness values. However, if the target dimension is more than one-half the neighborhood dimension, edges do remain in place and well defined. Such leveling of brightness as a result of repetitive median filtering is often described as *contouring* or *posterization*.

On the other hand, for a neighborhood dimension of n pixels, any target of dimension less than one-half the n pixels will be removed by a median filter. Such filters are therefore inappropriate for point source detection or where preserving small detail is important [18.18]. However, for larger targets, resolution loss is minimal, which makes the median filter preferable over many noise reduction applications.

Other noise reduction techniques are detailed in [18.1] and [18.17].

While the noise-reduction techniques described here do not improve resolution and can affect it adversely, the maximum entropy method in Section 18.6 has the advantage that it can improve resolution while simultaneously reducing noise.

REFERENCES

18.1. A. Rosenfeld and A. C. Kak, *Digital Picture Processing*, 2nd ed., Vol. 1, Academic Press, New York, 1982.

18.2. B. R. Hunt, "The application of constrained least squares estimation to image restoration by digital computer," *IEEE Trans. Comput.*, Vol. C-22, 1973, pp. 805–812.

18.3. O. Hadar, Z. Adar, A. Cotter, and N. S. Kopeika, "Restoration of images degraded by extreme mechanical vibrations," in *Optics and Laser Tech.,* Vol. 29, pp. 171–177, June 1997.

18.4. A. P. Pentland, "Fractal based description of natural scenes," *IEEE Trans. Patt. Anal. Machine Intel.*, Vol. PAMI-6, 1989, pp. 661–674.

18.5. G. Tidhar, S. R. Rotman, and M. L. Kowalczyk, "Clutter matrices for target detection systems," *IEEE Trans. Aerospace Electr. Syst.*, Vol. 30, January 1994, pp. 81–91.

18.6. G. Cohen, G. Reina, G. Tidhar, and S. R. Rotman, "Improved electro-optical target detection in a natural fractal environment," in *8th Meeting on Optical Engineering and Remove Sensing*, M. Oron, I. Shladov, and Y. Weissman, Eds., *Proc. SPIE*, Vol. 1971, 1993, pp. 78–92.

18.7. D. Sadot, S. R. Rotman, and N. S. Kopeika, "Comparison between high-resolution restoration techniques of atmospherically distorted images," *Opt. Eng.*, Vol. 34, January 1995, pp. 144–153.

18.8. D. Sadot, A. Rosenfeld, G. Shuker, and N. S. Kopeika, "High resolution restoration of images distorted by the atmosphere, based upon average MTF," *Opt. Eng.*, Vol. 34, June 1995, pp. 1799–1807.

18.9. L. Guan and R. K. Ward, "Restoration of randomly blurred images by the Wiener filter," *IEEE Trans. Acoustics, Speech, Signal Proc.*, Vol. 12, 1989, pp. 589–592.

18.10. L. Guan and R. K. Ward, "Deblurring random time-varying blur," *J. Opt. Soc. Am. A*, Vol. 6, November 1989, pp. 1727–1737.

18.11. F. Roddier, "The effects of atmospheric turbulence in optical astronomy," *Prog. Optics*, Vol. 19, 1981, pp. 283–376.

18.12. D. Sadot, S. Shamriz, I. Sasson, I. Dror, and N. S. Kopeika, "Prediction of overall atmospheric MTF with standard weather parameters: comparison with measurements with two imaging systems," *Opt. Eng.*, Vol. 34, November 1995, pp. 3239–3248.

18.13. D. Sadot, A. Dvir, I. Bergel, and N. S. Kopeika, "Restoration of thermal images distorted by the atmosphere based upon measured and theoretical atmospheric modulation transfer function," *Opt. Eng.*, Vol. 33, January 1994, pp. 44–53.

18.14. R. K. Ward and B. E. A. Saleh, "Deblurring random blur," *IEEE Trans. Acoustics, Speech, Signal Proc.*, Vol. 35, October 1986, pp. 1494–1498.

18.15. L. Guan and B. E. A. Saleh, "Restoration of images distorted by systems of random impulse response," *J. Opt. Soc. Am. A*, Vol. 2, August 1985, pp. 1254–1259.

18.16. D. Sadot, G. Lorman, R. Lapardon, and N. S. Kopeika, "Restoration of thermal images distorted by the atmosphere using predicted atmospheric modulation transfer function," *Infrared Phys. Technol.*, Vol. 36, February 1995, pp. 565–576.

18.17. J. C. Russ, *The Image Processing Handbook*, 2nd ed., CRC Press, Boca Raton, FL, 1992. Reprinted with permission from CRC Press. © 1992.

18.18. G. C. Holst, *Electro-Optical Imaging System Performance*, JCD Publishing, Winter Park, FL, and SPIE Optical Engineering Press, Bellingham, WA, 1995.

CHAPTER

19

Effects of Atmospheric Blur and Image Restoration on Target Acquisition

19.1 INTRODUCTION

Here we consider atmospheric blur effects on target acquisition for both contrast-limited and noise-limited imaging. There is an important difference between the two, in addition to the mathematical differences discussed in Chapters 10 and 11. The restoration techniques of the previous chapter involve spatial-frequency filtering. As such, both signal and noise are altered equally at each spatial frequency. Thus, there is no signal-to-noise improvement. This means that image restoration is essentially ineffective in noise-limited imaging. However, it can be quite effective in contrast-limited imaging. As shown in the previous chapter, the image usually still remains contrast-limited despite the increase in noise deriving from high spatial-frequency enhancement. In such situations, blur is reduced significantly by such image restoration and fine, smaller detail can now be resolved. Target acquisition is improved. *Target acquisition represents a quantitative measure of image quality.*

For both contrast-limited and noise-limited imaging, effects of noise can be reduced by use of conventional techniques such as recursive or median filters [19.1] discussed in Section 18.10. Environmental factors such as image motion and vibration and atmospheric blur can strongly effect image quality. Effects of the former on target acquisition have been considered in Section 14.6. Here, we consider effects of atmospheric blur.

In the first part of this chapter, atmospheric blur effects are considered in contrast-limited target acquisition, with and without image restoration. We will see that, quantitatively speaking, after imager restoration with the atmospheric Wiener filter of Section 18.5.3, target acquisition probabilities are essentially equivalent to those obtainable if there were no atmospheric distortion whatsoever. This means that essentially all atmospheric blur is removed. In the thermal infrared, where turbulence blur is much less significant than in the visible, similar target acquisition probability improvement is also obtained with conventional image restoration techniques applicable to deterministic blur rather than random blur, in accordance with the static properties of aerosol MTF.

In the second part of this chapter, atmospheric effects on noise-limited target acquisition are described. Here, image restoration is much less relevant.

In general, most imaging situations are contrast limited rather than noise limited, since the images are not "snowy" but are blurred and limited by poor contrast. Hence, image restoration techniques are usually applicable [19.2]. The question, however, as to whether an image is contrast or noise limited depends on whether the $f_{r\ \max}$ in Figs. 10.1 or in Fig. 11.1 is smaller. It is usually smaller in the former case. The analysis here is based on the Johnson chart concept of a critical dimension, but can be extended to two-dimensional acquisition, corresponding to the U.S. Navy model described in Section 10.6.

19.2 CONTRAST-LIMITED TARGET ACQUISITION

As described in Sections 10.4 and 10.5, contrast-limited target acquisition probabilities depend on the number of cycles or line pairs, n, resolvable across the critical target dimension. It is often more convenient to consider n from the standpoint of angular spatial frequency $f_{a\ \max}$ rather than planar spatial frequency $f_{r\ \max}$. The atmospheric MTF developments are as functions of angular spatial frequencies.

19.2.1 Angular Spatial-Frequency Approach

The finest target detail Δx equivalent to one bar (in the object plane) that can be resolved by the imaging system, that is, system resolution, determines the number of line pairs resolved over the target dimensions. Corresponding to the development in Section 10.4 for planar resolution, the corresponding resolvable angle $\Delta\theta$ is directly related to the target's distance z

$$\Delta\theta = 2\Delta x/z \tag{19.2.1}$$

where $\Delta\theta$ is the angle subtended by one bar *pair.*

This finest resolvable angle is related to the system's angular spatial-frequency bandwidth, as shown in Example 12.1, via

$$\Delta\theta = \frac{1}{f_{a\ \max}} \tag{19.2.2}$$

Combining Eqs. (19.2.1) and (19.2.2) yields the relation between minimum resolvable size at the target and system angular spatial-frequency bandwidth

$$\Delta x = \frac{z}{2f_{a\ \max}} \tag{19.2.3}$$

Minimum target dimension x is related to minimum angle subtended by the target

$$\theta = \frac{x}{z} \tag{19.2.4}$$

If the system is characterized by resolution better than *detection*, that is, the angle θ subtended by the entire target critical dimension is larger than the minimum resolution angle $\Delta\theta$, then more detail can be seen on the target. Generally, the number of cycles (line pairs) resolved over the target is given by

$$n = \frac{\theta}{\Delta\theta} \tag{19.2.5}$$

where n is the same parameter as in the Johnson criteria given in Sections 10.4 and 10.5. Combining Eqs. (19.2.1), (19.2.4), and (19.2.5) yields

$$n = \frac{f_{a\,\max} x}{z} \tag{19.2.6}$$

which determines the number of line pairs resolvable across the minimum target dimension as a function of target distance and system angular spatial-frequency bandwidth. In developing this result a simple type of imaging system is assumed—a thin lens. Therefore, it is assumed that target angle θ in both object and image planes is identical. Thus Eq. (19.2.6) is used to determine n, and n is used in Eq. (10.5.1) to determine P_∞, although the Johnson chart deals with the image plane rather than the object plane. Most imaging systems, however, include also angular magnification such as zoom lenses or telescopes, and electronic magnification relating display image size to image size at the camera input. In that case, a scaling factor is needed in order to permit implementation of Eq. (19.2.6) in target transfer function calculations [Eq. (10.5.1)].

19.2.2 Angular Magnification

If system spatial-frequency bandwidth $f_{a\,\max}$ is constant and not affected by the atmospheric or vibration environment, then resolution angle $\Delta\theta$ is constant. Target angle θ (subtended by the target) in the image plane, however, is magnified by the angular magnification m_a. Thus Eq. (19.2.5) is changed to include angular magnification:

$$(n)_{m_a} = \frac{m_a\theta}{\Delta\theta} = m_a n \tag{19.2.7}$$

where $(n)_{m_a}$ is the extended version of n including angular magnification. Consequently, Eq. (19.2.6) becomes modified to

$$(n)_{m_a} = \frac{f_{a\,\max} m_a x}{z} \tag{19.2.8}$$

where Eq. (19.2.8) determines a linear relationship between the number of resolvable line pairs versus angular magnification m_a. However, this relation holds only under the assumption of constant angular spatial-frequency bandwidth of the system, which is valid for an ideal imaging system that is diffraction limited and free of environmental effects. The next paragraph determines a way of evaluating the system's maximum frequency $f_{a\,\max}$ while dealing with practical nonideal imaging systems and human observers.

For a practical imaging system, the angular spatial-frequency bandwidth $f_{a\,\max}$ is determined as the frequency at which the system's MTF is reduced to the required threshold contrast value for contrast-limited imaging as shown in Figs. 10.1. Without image processing, information at frequencies higher than $f_{a\,\max}$, though transmitted by the system, is useless since contrast at such frequencies is lower than the observer's contrast threshold. Here we concentrate on the human eye as the detector, in which case the system's contrast threshold is determined by the human visual threshold curve. This curve is fitted with a closed-form formula in Fig. 12.18 and Eq. (12.6.1) which is fairly accurate particularly at higher angular spatial frequencies.

Up to this point, the imaging propagation channel considered here was free of environmental effects and the only limitation was the one imposed by the hardware (optics and electronics) as characterized by the imaging system MTF. However, when the optical path length is increased, atmospheric distortions become very significant, and this must be considered in the target search model.

19.2.3 Atmospheric MTF

In atmospheric imaging problems, the propagation channel is the atmosphere, which is a random medium containing both turbulence and scattering and absorbing particles. The main effect of the turbulent medium is to reproduce a wavefront tilt, which causes image shift at the image plane. Those tilts are random in both time and space and their temporal power spectra are usually limited to several tens up to a few hundred hertz under ordinary atmospheric conditions. There is a blur related to turbulence that usually affects very high spatial frequencies. Both blur and tilt effects can be characterized by the well-known turbulence long-exposure spherical wave MTF:

$$\mathrm{MTF}_{le} = \exp\{-(3/8)57.53 f_a^{5/3} C_{ne}^2 \lambda^{-1/3} z\} \tag{19.2.9}$$

where MTF_{le} is the long-exposure MTF, C_n is the refractive index structure coefficient, and λ is the radiation wavelength. For the short-exposure case, the turbulence first-order tilt effect is removed and turbulence spherical wave MTF is evaluated from

$$\mathrm{MTF}_{se} = \exp\left\{-(3/8)57.53 f_a^{5/3} C_{ne}^2 \lambda^{-1/3} z\left[1 - (1/b)\left(\frac{\lambda f_a}{D}\right)^{1/3}\right]\right\} \tag{19.2.10}$$

where MTF_{se} is the short-exposure MTF, $1/b$ is 1 in the near field (image edges) and 0.5 in the far field (image center), and D is aperture diameter. These equations were described in detail in Chapter 16.

In addition to turbulence, scattering and absorption effects are produced by molecules and aerosols. Scattering gives rise to very wide diffusion of point object radiation. This causes a wide variance in angle of arrival according to the particulate scattering phase function, and thus also produces blur at the image plane. Unlike the short-exposure turbulence case, the angular blur due to aerosols is very wide, usually on the order of radians, which is typically orders of magnitude wider than imaging field of view. Consequently, some practical limitations should be applied to the scattered light recorded in the image, as described in Chapter 17.

In addition, further practical instrumentation limitations must be applied also to the *unscattered* component of the radiation too. The finite angular spatial-frequency bandwidth of the sensor truncates the high angular spatial-frequency radiation. This is because unscattered light from a point object is equivalent to a constant requiring an infinite spectrum in the angular spatial-frequency plane. Limited spatial-frequency bandwidth of the receiver limits the maximum value of unscattered light actually recorded in the image. The limited spatial-frequency bandwidth smears sharp edges and small detail but much less so than does the received scattered light. The limitation imposed by spatial-frequency bandwidth on the recorded unscattered light enhances the relative blur deriving from the received scattered light, which has a significant adverse effect on image quality [19.3].

All those practical instrumentation limitations have been applied to the classical theory dealing with aerosol MTF in Chapter 17. In the thermal infrared (IR), absorption effects are considered too. All aerosol effects imposed on a practical imager can be characterized by the practical instrumentation-based aerosol MTF, which can be approximated by the Gaussian form

$$\mathrm{MTF}_a(f_a) = \begin{cases} \exp\{-A_a z - S_a^* z (f_a/f_{ac})^2\} & f_a \leq f_{ac} \\ \exp\{-(A_a + S_a)z\} & f_a > f_{ac} \end{cases} \tag{19.2.11}$$

where $\mathrm{MTF}_a(f_a)$ is the practical aerosol MTF, S_a^* and A_a are the effective scattering and absorption coefficients, and f_{ac} is the effective aerosol cutoff frequency. The term *effective* means that S_a and f_{ac} in the image actually recorded depend on instrumentation parameters in clear weather. For large optical depths, as in the case of fog or rain, they depend on the aerosol size distribution, in which case $f_{ac} \approx a/\lambda$, where a is dominant aerosol radius. The precise procedure of calculating the effective parameters S_a, A_a, and f_{ac} are described step by step in Chapter 17 and in the Appendix. Alternatively, these parameters can be evaluated experimentally by measuring the aerosol MTF. A procedure for calculating aerosol MTF numerically instead of by approximation is given in the Appendix.

Equation (19.2.11) indicates that for high spatial frequencies practical aerosol MTF is limited by atmospheric transmission. (In practice, the practical aerosol MTF high-frequency asymptote is slightly higher than atmospheric transmission τ because the finite spatial-frequency bandwidth of the sensor cannot pass a spatial delta function comprised of the received unscattered light. Hence, the instrumentation limitations cause this part of the spectrum to rise since energy is conserved.) This means that for relatively large optical depths, such as in fog or rain, the aerosol and overall atmospheric high-frequency asymptote is very low so that $f_{a\,\max}$ is limited by $f_{ac} \approx a/\lambda$, which is then a very small value for $f_{a\,\max}$ in Eq. (19.2.8). In clear weather, haze, and light fog, where the high spatial-frequency aerosol MTF asymptote is higher, further atmospheric MTF reduction at high angular spatial frequencies derives from turbulence. In such cases scatter angles actually recorded in the image are generally also limited by instrumentation and $f_{a\,\max}$ increases by orders of magnitude from a/λ as discussed in Chapter 17.

Both turbulence and aerosol MTF should be considered as part of the system MTF, in addition to the hardware MTF mentioned previously. Therefore, the overall system MTF, where "system" refers to hardware, atmosphere and motion effects together, is evaluated by a multiplication of the three component MTFs: turbulence, aerosol, and hardware MTFs. Here we assume statistical independence between all MTF components.

In the case of imaging through the atmosphere, with relatively large optical path lengths, the atmospheric MTF often becomes the weakest link in the imaging system.

TABLE 19.1 *Atmospheric, System, Target, and Task Characteristics for Moderate to Strong Turbulence, Slight Haze*

z (range)	6.5 [km]
x target minimum size	1 [m]
n_{50} (recognition)	4 line pairs
λ (radiation wavelength)	0.5×10^{-6} [m]
ν_o (hardware cutoff frequency $-e^{-1}$ value)	35,000 [cycles/rad]
m_a (angular magnification)	6
S_a^* (scattering coefficient)	0.1 [1/km]
A_a (absorption coefficient)	0.01 [1/km]
f_{ac} (aerosol MTF cutoff frequency)	10,000 [cycles/rad]
C_n^2 (turbulence strength parameter)	1.3×10^{-14} [$m^{-2/3}$]

The overall system is influenced mostly by the atmospheric MTF, and so is the system angular spatial-frequency bandwidth. Here, the evaluation of target acquisition probabilities and search time will be demonstrated, with the use of real-world data of both imaging system and atmospheric parameters.

19.2.4 Incorporating Atmospheric Effects into Target Acquisition Modeling

Table 19.1 contains the typical parameters of a static visible-near-infrared imaging system [19.2, 19.3], atmosphere [19.2–19.4], target, and search task [19.5] to be analyzed in this section. A similar analysis for a thermal infrared imaging system appears elsewhere [19.6]. In Fig. 19.1, all MTFs of the imaging system components

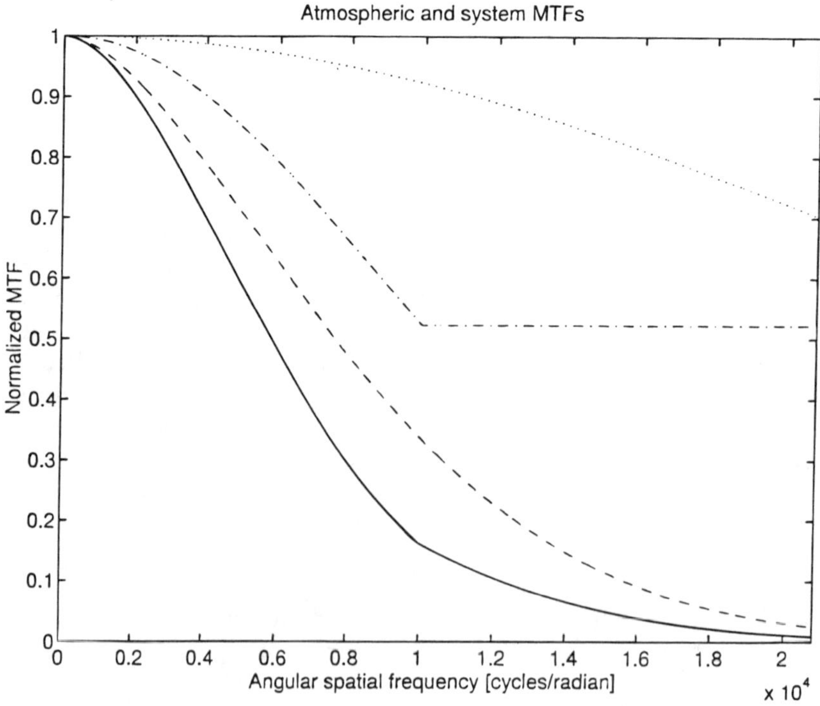

Fig. 19.1 Atmospheric and system MTFs for the data of Table 19.1: turbulence MTF (dashed line), aerosol MTF (dashed-dotted line), optics and electronics hardware MTF (dotted line), and overall system MTF (solid line) (after [19.2]).

with the parameters of Table 19.1 are presented.It is clear from this real-world example that the overall system MTF is limited mainly due to environmental limitations rather than hardware. Figure 19.2 illustrates the procedure of evaluation of the system angular spatial-frequency cutoff. This is the frequency corresponding to the intersection between the overall system MTF and detector's (eye) threshold contrast curve (~2200 cycles/rad in this example). Frequencies higher than $f_{a\ max}$, though transmitted through the system, will not contribute any information to the human observer since without image restoration they are not sensed by the human visual system. (For an *ideal* imaging system operating in free space without vibrations, the overall system MTF would be unity and $f_{a\ max}$ would depend only on the eye threshold contrast where it reaches unity.) With the knowledge of $f_{a\ max}$ from Fig. 19.2, the number of line pairs across the target's minimum dimension can be evaluated from Eq. (19.2.8). The probability of target recognition for an infinite search time can be calculated from Eq. (10.5.1). Figure 19.3 presents graphical evaluation of the probabilities of target recognition for the data of Table 19.1. Comparison is presented between the probability (0.48) calculated using the model incorporating environmental distortions, where $f_{a\ max}$ is evaluated in Fig. 19.2, and the probability (0.8) calculated by ignoring atmospheric effects where $f_{a\ max}$ is evaluated only with the consideration of the MTF of the hardware. From Figs. 19.1 and 19.2, when the atmosphere is included, $f_{a\ max} \approx 2.2$ cycles·mrad^{-1}. On the basis of hardware alone, $f_{a\ max} \approx 3.3$ cycles·mrad^{-1}.

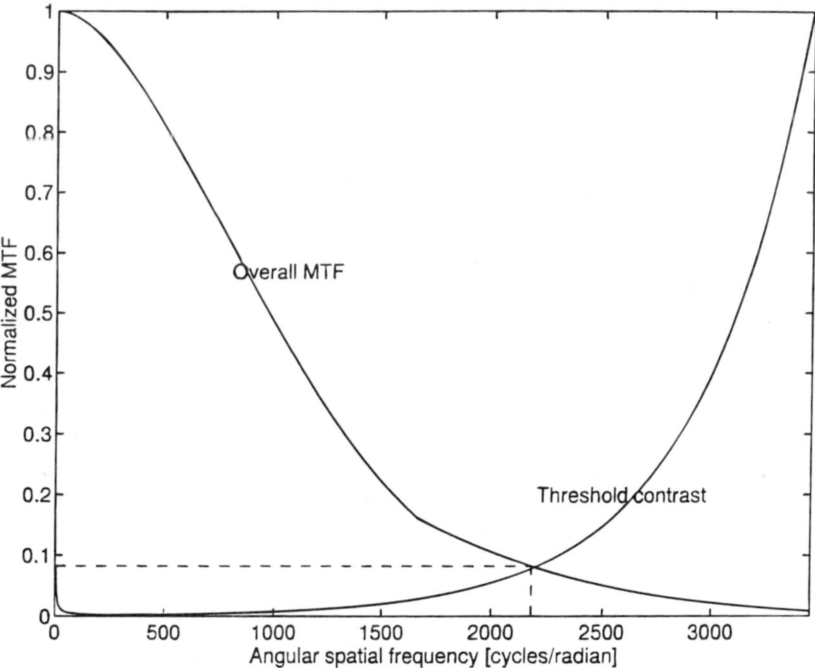

Fig. 19.2 Evaluation of the overall system bandwidth $f_{a\ max}$ for the data of Table 19.1. Angular spatial frequency corresponding to overall system MTF and eye contrast threshold intersection. The closed-form formula of Fig. 12.18 is used for the eye contrast threshold curve. We assume here MCO ≈ 1 or that contrast enhancement described in Section 18.9 has been used to effectively bring about this assumption (after [19.2]).

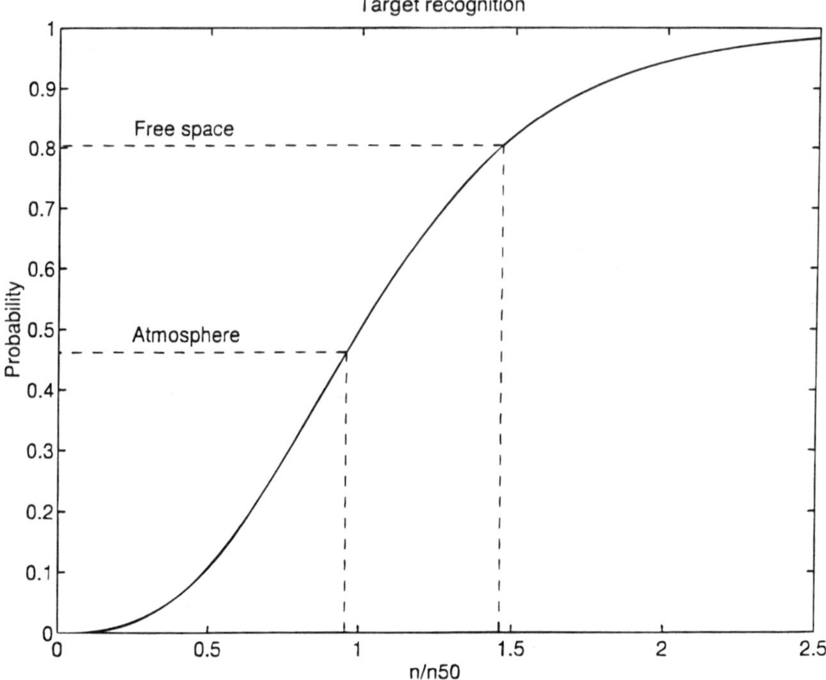

Fig. 19.3 Target recognition probability versus n/n_{50} for the data of Table 19.1. (after [19.2]).

This analysis can be expanded to estimate time-dependent performance incorporating human search. The temporal search model for detection proposed by CNVEO is presented in Chapter 10 {Eqs. (10.5.4a) and (10.5.4b)}.

In Fig. 19.4, the recognition search time is illustrated for the data of Fig. 19.3, using Eq. (10.5.4b). It is clear that *both rise time and asymptotic probabilities for infinite search time differ significantly with and without considering atmospheric blur.*

19.2.5 Angular Magnification Results

When dealing with long distances it is common to think that using improved imagers with stronger zoom lenses or telescopes with longer focal lengths will improve search probabilities, or, conversely, by the use of improved imagers, target recognition ranges can be increased with the same search probabilities. This is true in free space provided the hardware MTF does not decrease $f_{a\ max}$ significantly. From Eq. (19.2.8) we can see that as long as $f_{a\ max}$ remains constant, the number of line pairs n across the target will increase linearly with m_a. However, as seen in Fig. 19.2, note that $f_{a\ max}$ is determined by the intersection between the eye threshold contrast curve and overall system MTF. Since the eye threshold curve refers to angular spatial frequencies at the *image* plane, the angular magnification m_a should be considered, and an appropriate scaling done to the overall system MTF curve related to the object plane. The effect of angular magnification as shown in Section 2.9.1 is to increase the angle of resolution at the *image* plane because of the larger target image size. This decreases angular spatial frequencies. The angular spatial-frequency axis of the system MTF curve thus should be scaled via division by m_a, in order to make the intersection compatible

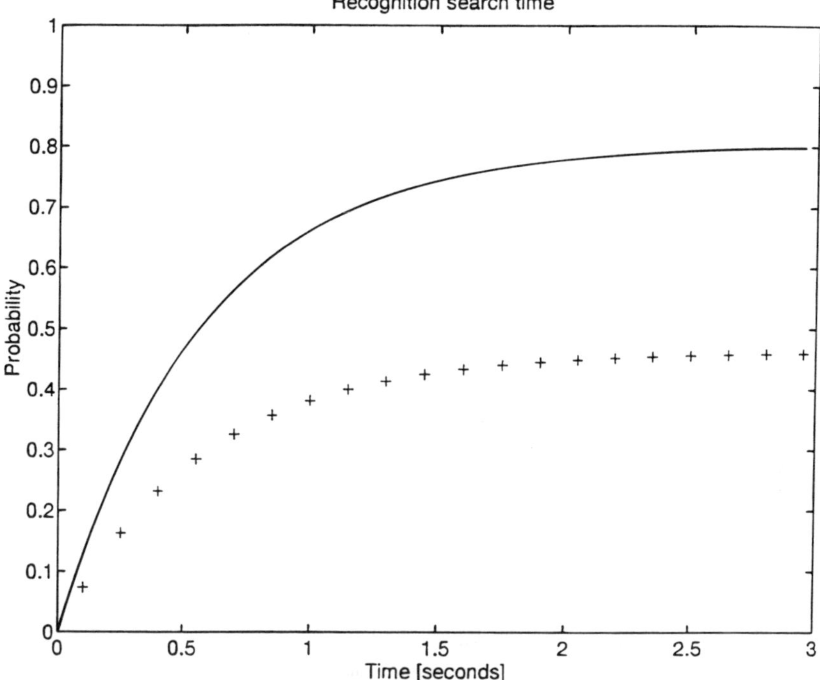

Fig. 19.4 Probability of recognition versus search time for the data of Table 19.1. (after [19.2]).

with the visual system contrast threshold. From this discussion it is clear that the effect of angular magnification is to compress the system MTF curve, thus decreasing the frequency of intersection between the MTF and threshold curves or, in other words, decreasing $f_{a\ max}$. Therefore, the number n/n_{50} will no longer increase linearly with zooming, since it causes an increase in m_a together with a decrease in $f_{a\ max}$, and the role of both effects will determine the changes imposed on n/n_{50} via Eq. (19.2.8).

The effect just discussed becomes much more dominant when considering environmental blur. When dealing with longer distances, environmental distortions are very dominant and, when zooming in on the target, the vibrations and atmospheric distortions are magnified as well. Thus, no significant improvement is achieved in target search times and probabilities by increased magnification. This is particularly true for the turbulence MTF, but not for the aerosol MTF, which (in its asymptotic form) remains constant for frequencies larger than the aerosol MTF cutoff frequency f_{ac}. Magnification will decrease f_{ac} but hardly decrease $f_{a\ max}$ deriving from aerosol MTF alone, since the spatial-frequency scaling of dividing angular spatial frequency by m_a will not change the MTF value for $f_a > f_{ac}$. The knowledge of the atmospheric MTF is crucial for the system designer and it is important to estimate when a possible saturation in zooming improvement occurs. It can be easily shown that for a linearly decreasing MTF and a constant threshold curve, there will be no improvement at all by angular magnification since $f_{a\ max}$ decreases linearly with m_a, and thus n remains constant in Eq. (19.2.8).

Investigation of the results of the effect of angular magnification for the data of Table 19.1 is presented in Fig. 19.5. Figure 19.5(a) presents the normalized number

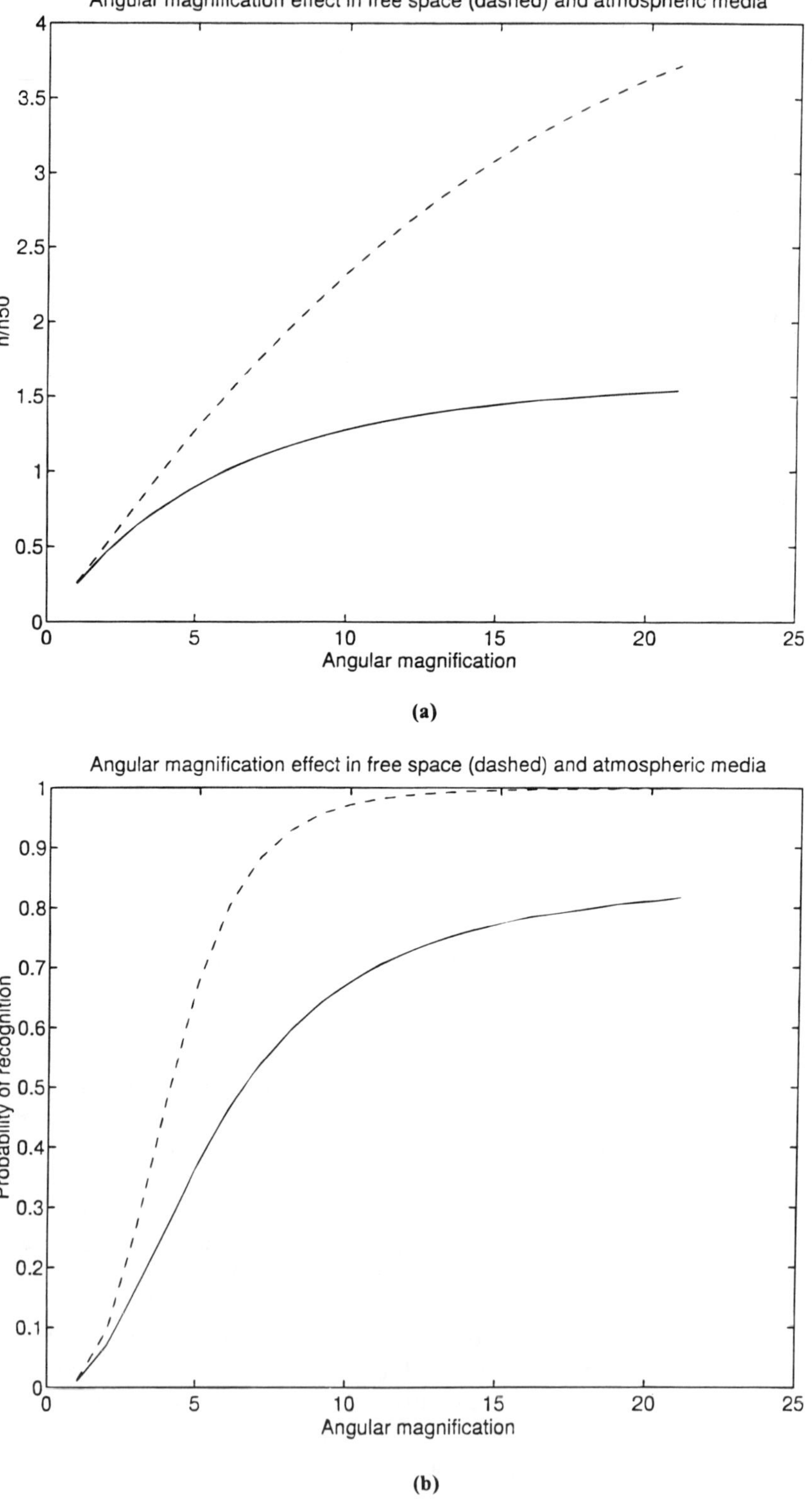

Fig. 19.5 (a) n/n_{50} versus angular magnification for the data of Table 19.1; free-space medium (dashed line) and atmospheric medium (solid line). (b) Probability of recognition versus angular magnification for the data of Table 19.1; free-space medium (dashed line) and atmospheric medium (solid line). (after [19.2]).

of line pairs across the target n/n_{50} versus angular magnification m_a, for free space and atmospheric media. While in free space n/n_{50} increases almost linearly with m_a, for an atmospheric medium the improvement here in n/n_{50} starts approaching an asymptote at angular magnification of about 10 because of turbulence. For larger magnification the atmospheric blur is magnified as much as is fine detail in the target and almost no additional information is added. Field of view is decreased because of the magnification. Because of turbulence the lack of smaller resolvable detail is explained by the decrease in $f_{a\ max}$, which exceeds the magnification increase. A comparison of probability of recognition versus m_a is presented in Fig. 19.5(b), indicating similar conclusions.

A similar investigation was carried out for shorter path lengths where atmospheric distortions are less dominant. Figure 19.6 presents an analog to the results of Fig. 19.5, with the same data of Table 19.1 and differing only by the target range, which was reduced to 2 km. In Fig. 19.6(a) it is clear that angular magnification improves n/n_{50} significantly, almost as in free space. From Fig. 19.6(b) it is clear that for this situation (of relatively short atmospheric path) zooming significantly improves search times and probabilities, almost as in free space.

On the other hand, for much larger distances (16 km), Fig. 19.7 demonstrates that zooming is almost useless. This saturation effect in zooming brought about by turbulence is analogous to the saturation in resolution while increasing the aperture diameter of the imaging system in order to increase the system diffraction limit. The largest aperture diameter that contributes to improvement in resolution is on the order of the atmospheric coherence diameter (Sections 16.6 and 17.4). For apertures larger than the atmospheric coherence diameter, system resolution is limited by the atmospheric MTF. Here, in a similar manner, atmospheric MTF limits $f_{a\ max}$, thus causing saturation in search probabilities. Hence, both saturations are caused by the same physical mechanism: atmospheric MTF limitations. This phenomenon is also illustrated by Examples 17.1 and 16.2, which describe focal lengths at which there is a transition from hardware- to atmospheric-limited resolution. The focal length determines optical magnification.

In cases where atmospheric distortions are relatively weak, target search tasks can be achieved for much longer distances by using stronger angular magnification.

19.2.6 Image Restoration Effects on Target Acquisition

It is interesting to consider the effects of image restoration on target acquisition. Most atmospheric blurring effects can be removed or minimized if image corruption is characterized accurately. Recent image restoration techniques, as shown in Chapter 18, have shown significant improvement in image resolution by including atmospheric aerosol MTF in addition to turbulence MTF. An *atmospheric Wiener filter* was introduced in Section 18.5 that considers jitter in the turbulence MTF as an additional source of noise, similar to electronic noise, since both are random. In this way, high spatial-frequency enhancement is concentrated only at those spatial frequencies where the sum of both the turbulence jitter "noise" and electronic noise is minimum. The required input is the average turbulence and practical aerosol MTFs, which can be predicted according to weather as described in Chapters 15 and 17. Resolution improvement was achieved in 1 or 2 s in both the thermal IR and visible spectral ranges using parallel processing. By converting software to hardware, this can be reduced to a fraction of a second. When using these image processing algorithms, image resolution is limited essentially by instrumentation only, as if there were no

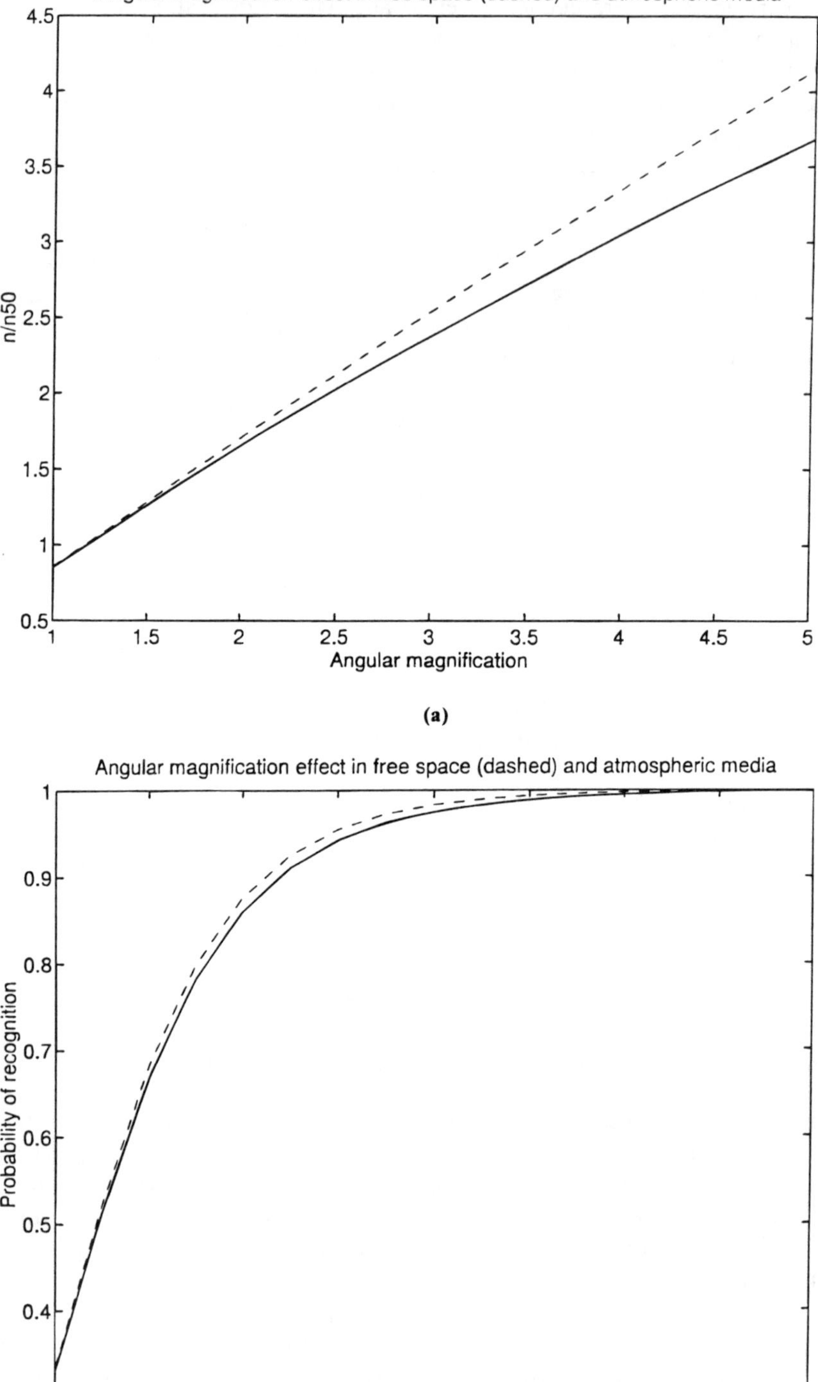

(b)

Fig. 19.6 (a) Same as Fig. 19.5(a) for 2-km path length (instead of 6.5 km). (b) Same as Fig. 19.5(b) for 2-km path length (instead of 6.5 km) (after [19.2]).

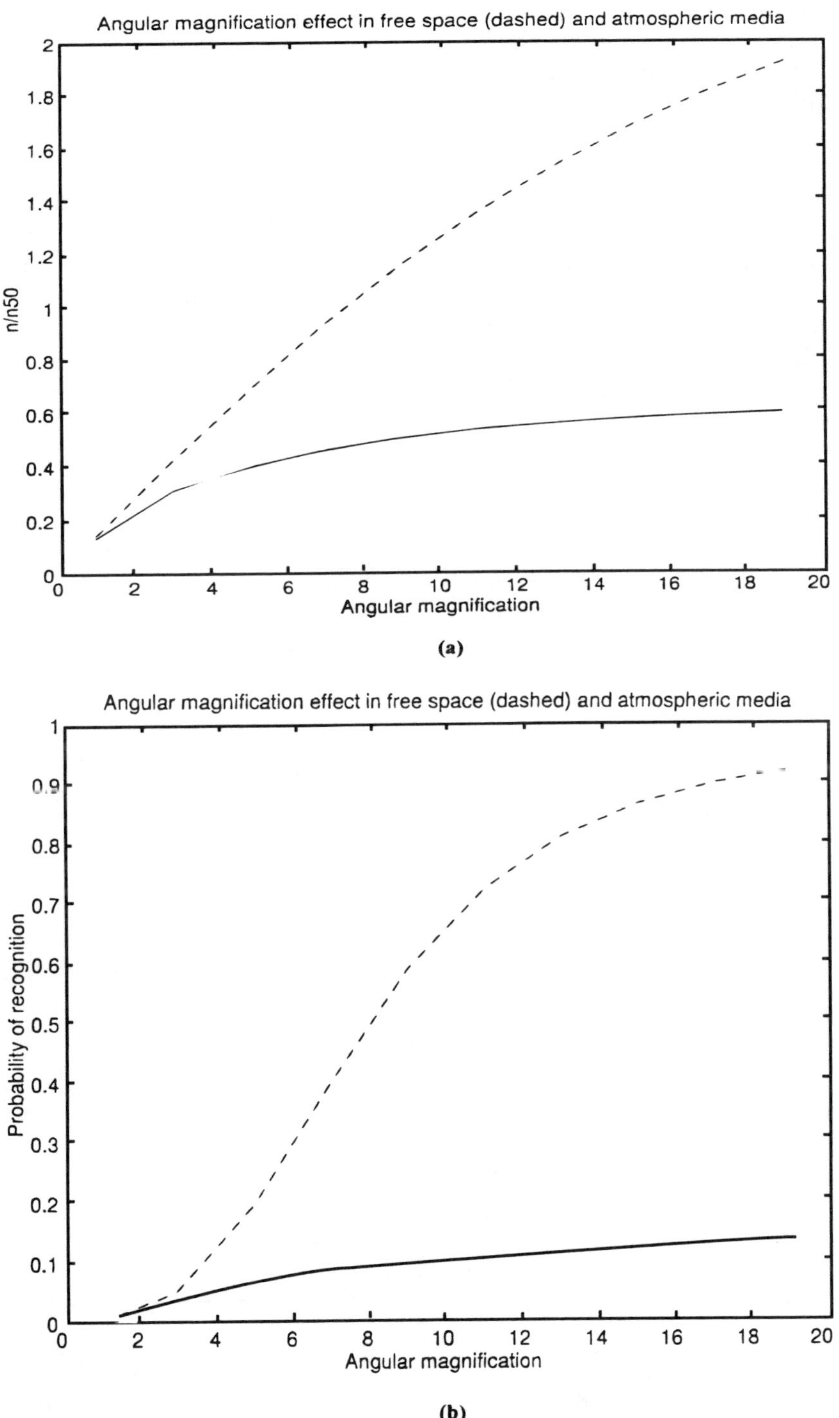

Fig. 19.7 (a) Same as Fig 19.5(a) for 16-km path length (instead of 6.5 km). (b) Same as Fig. 19.5(b) for 16-km path length (instead of 6.5 km) (after [19.2]).

atmosphere. In that case *restored* atmospheric MTF improves significantly and, for successful restoration, it equals approximately unity. This applied to contrast-limited rather than noise-limited imaging. Target acquisition ranges and probabilities then also increase correspondingly. Thus, in cases where the received image is contrast limited rather than noise limited, the increase in target detection and recognition probabilities will be related to the improvement in the restored MTF, and so too is the improvement in ranges of acquisition.

In Fig. 19.8 an example of restored (circles) system MTF and *actual* atmospheric Wiener filter image processing techniques with a computer is presented, and compared to the atmospheric and hardware MTFs. This example refers to Figs. 18.5 and 18.10. This system MTF improvement significantly increases $f_{a\ max}$, as shown in Fig. 19.9. Improved probabilities and search times after restoration are presented in Figs. 19.10 and 19.11, respectively. It is clear that after image restoration, probabilities essentially approach those for free space, that is, for hardware only as if there were no atmospheric blur. This means that, after using image processing techniques, system performance for target search tasks is almost as if there were no atmosphere at all, and large angular magnification can then be used to see smaller detail in selected portions of the image.

Although the treatment here is for the visible, it is equally applicable toward thermal imaging, where restoration is similarly effective and yields target acquisition probabilities as if there were no atmosphere [19.6].

However, if atmospheric transmission is *very* poor the image quality can become noise limited. This is typically a "snowy" image. For a noise-limited situation, target

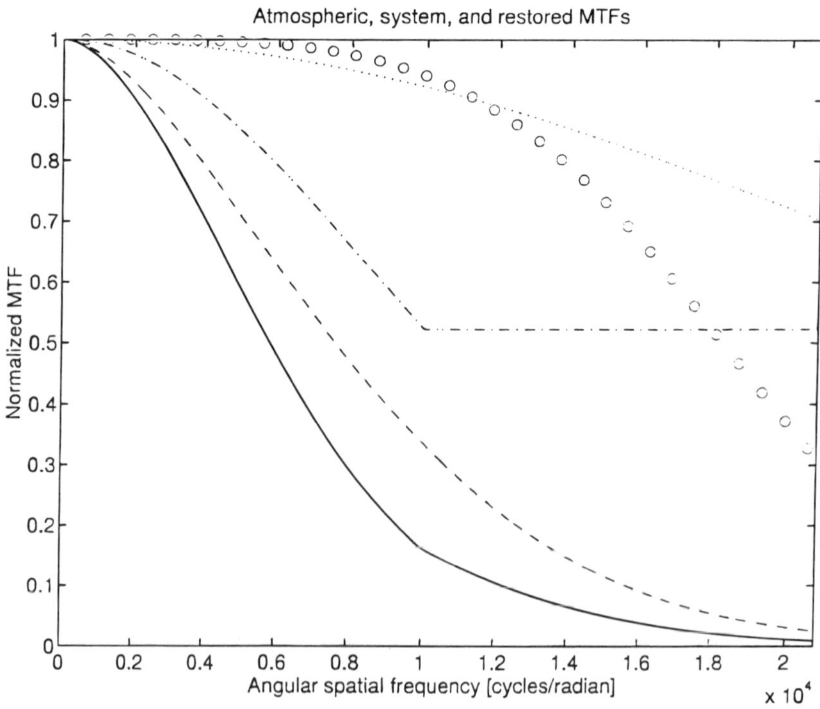

Fig. 19.8 Same as Fig. 19.1 with an additional curve (circles) representing the restored overall system MTF (after [19.2]).

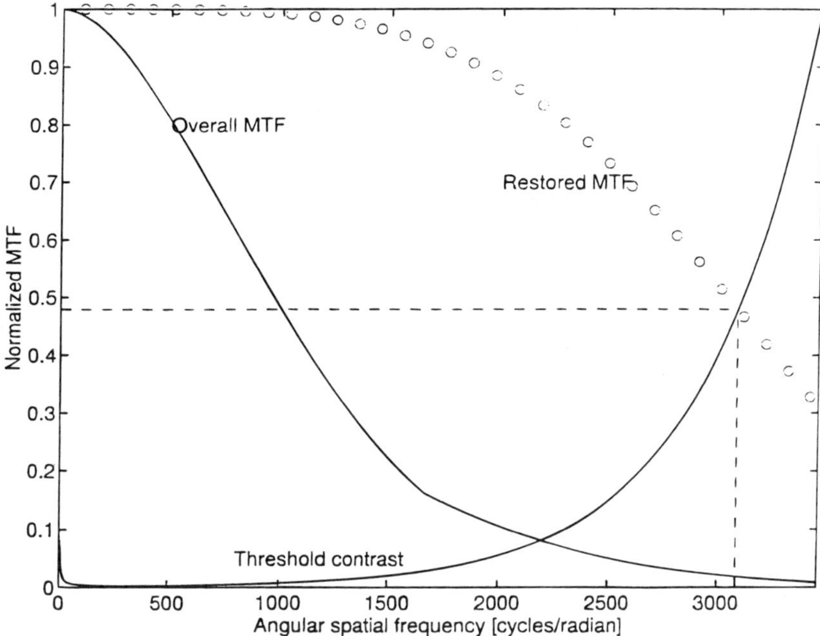

Fig. 19.9 Evaluation of the system bandwidth $f_{a\,max}$ for the data of Table 19.1, after applying image restoration, compared to that without restoration (after [19.2]).

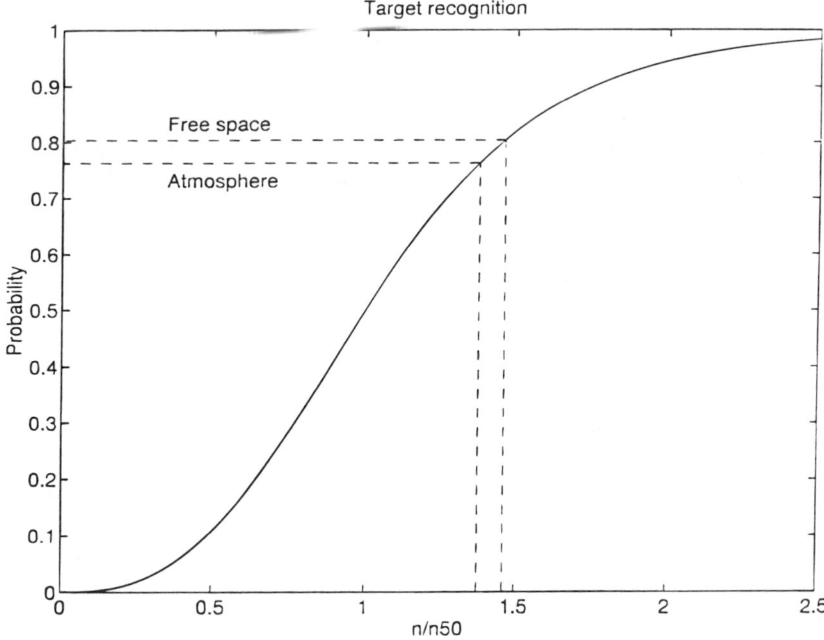

Fig. 19.10 Same as Fig. 19.3, after applying image restoration (after [19.2]).

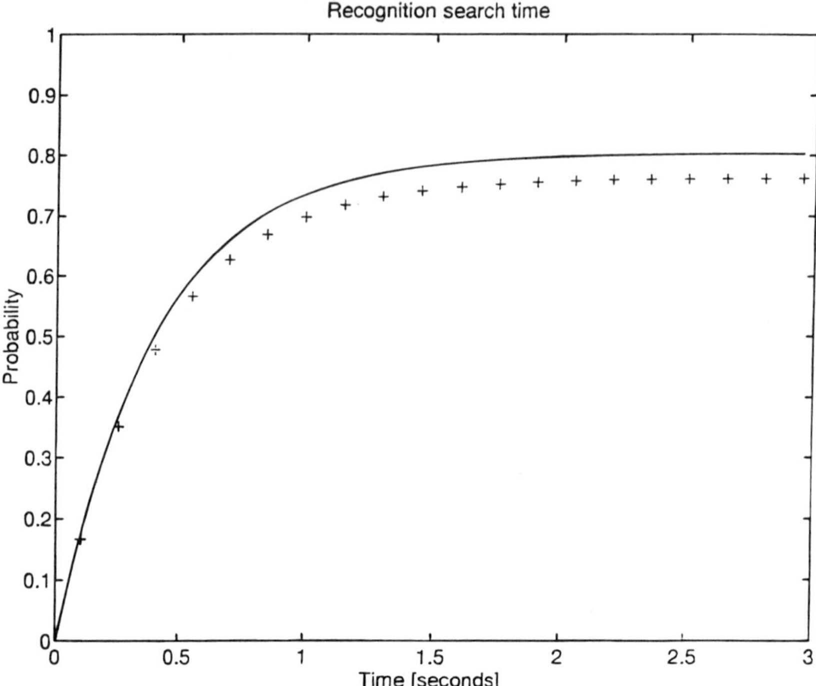

Fig. 19.11 Same as Fig. 19.4, after applying image restoration (after [19.2]).

acquisition is determined by display or normalized SNR as discussed in Chapter 11, rather than contrast. Because noise at each spatial frequency is enhanced by spatial-frequency filtering the same way as is the signal, such image restoration cannot improve the SNR at each spatial frequency.

The fact that image restoration is becoming more and more effective serves as an indication that more often contrast-limited, rather than noise-limited, performance is being achieved. As image sensors become quieter and more sophisticated, this trend to contrast-limited imaging can only be expected to become stronger. However, there are still situations when image quality is noise limited. Atmospheric effects in such situations are considered next.

19.2.7 Conclusions

The effect of atmospheric blur is included in contrast-limited target acquisition modeling. Atmospheric blur effects are based on recent investigations of atmospheric MTF, which have shown significant spatial-frequency dependence deriving from both aerosol and turbulence MTFs. This spatial-frequency dependence directly influences the intersection between the system MTF and human eye contrast threshold and, consequently, the probability of detection or recognition obtained with the imaging systems. When atmospheric conditions are of moderate or weak turbulence, and/or vibration displacement is small, angular magnification or zooming can significantly improve system performance. However, when atmospheric conditions get more severe (such as at midday on sunny days for example), the magnification of the image will not contribute any significant improvement to target search probabilities since such environmental blur will be magnified as well. This saturation effect in zooming is analogous to the saturation in resolution while increasing the aperture diameter of the imaging system, with the latter being related directly to the atmospheric coherence diameter. The atmosphere thus not only blurs but also strongly affects optimum field of view.

The zooming saturation effect was quantitatively modeled here and is important for the system designer. Knowledge of atmospheric MTF is essential to good system design. The improvement achieved by correcting for this blur automatically in real time is also discussed. If such environmental distortions are modeled accurately (including aerosol MTF), most image corruption can be corrected. In that case, for contrast-limited images, the restored atmospheric MTF is improved significantly and approaches unity. This improves greatly the probabilities and ranges of detection and recognition. However, under very poor atmospheric transmission conditions, the normalized SNR becomes noise limited and no improvement in probability of detection will be achieved.

19.3 ATMOSPHERIC EFFECTS ON NOISE-LIMITED TARGET ACQUISITION

For noise-limited imaging in the middle and near-infrared regions, as in low-light-level situations in the visible, the overall atmospheric MTF must be included in the overall system MTF, labeled $(\mathrm{MTF})_{ST}$ in Eq. (11.1.16), which gives rise to a reduction in display signal-to-noise ratio SNR_D. This alters the probability of target acquisition as determined from Fig. 11.2 for noise-limited imaging.

For noise-limited thermal imaging, we use Eq. (11.2.11) where atmospheric MTF replaces atmospheric transmission in Eq. (11.2.9). Overall atmospheric MTF is essentially the product of aerosol MTF approximated by Eq. (19.2.11) (or calculated numerically as described in the Appendix) and the turbulence MTF described in Eq. (19.2.9) or (19.2.10). The contrast reduction described by multiplication of temperature difference ΔT by atmospheric transmission τ_{atm} in Eq. (11.2.9) is also included in the aerosol MTF, since aerosol MTF approaches τ_{atm} at higher spatial frequencies. The need to revise Eq. (11.2.9) in order to substitute atmospheric MTF for atmospheric transmission arises from the fact that Eq. (11.2.9) implies that target acquisition is independent of target size, while atmospheric MTF size measurements indicate that not only in the visible but also in the thermal infrared, smaller detail is affected more adversely than is larger detail, that is, there is a strong infrared atmospheric MTF spatial-frequency dependence [19.6–19.11]. The existence of an atmospheric MTF spatial-frequency dependence was not appreciated in the 1970s when target acquisition theory began to be applied to thermal imaging because sensors then could not go out to spatial frequencies sufficiently high to sense atmospheric blur. Their low spatial-frequency instrumentation was sensitive to atmospheric transmission only, rather than atmospheric blur.

To include atmospheric MTF in Eq. (11.2.11), note that required values of angular spatial frequencies can be obtained from Eq. (19.2.8):

$$f_a = \frac{m_a n z}{x} \qquad (19.3.1)$$

and that the "cutoff" range z_c corresponding to aerosol cutoff angular spatial frequency f_{ac} is

$$z_c = \frac{f_{ac}x}{m_a n} \tag{19.3.2}$$

For ranges smaller than z_c, *apparent* atmospheric transmission is increased because of small-angle forward scattered light that is measured in addition to the unscattered light.

By substitution of these equations into Eq. (19.2.11), the aerosol MTF can be expressed as a function of range

$$\mathrm{MTF}_a(s) = \begin{cases} \exp(-A_a s - S_a^* z^3/z_c^2) & z \le z_c \\ \exp(-\alpha(\lambda, T)z) & z > z_c \end{cases} \tag{19.3.3}$$

where α is atmospheric attenuation coefficient, equal to $A_a + S_a^*$. Usually, aerosol MTF is more dominant than turbulence MTF [19.4], particularly at longer infrared wavelengths [19.8, 19.10] as seen from the wavelength dependencies of Eqs. (19.2.9) and (19.2.10). Hence, we consider aerosol MTF first and then add turbulence MTF.

In view of Eq. (19.3.3) and the preceding assumption, apparent or received temperature difference sensed at range z is, using Eq. (19.2.11),

$$\Delta T_a = \Delta T\ \mathrm{MTF}_a(z) \tag{19.3.4}$$

and normalized SNR, from Eq. (11.2.11), is

$$\mathrm{SNR}_N = \begin{cases} \dfrac{\sqrt{7/\alpha_T}\ \exp(-A_a z - S_a^* z^3/z_c^2)}{\mathrm{MRT}_o \exp[\beta_{sys} m_a n z/x\}} & z \le z_c \\[2ex] \dfrac{\sqrt{7/\alpha_T}\ \exp(-\alpha z)}{\mathrm{MRT}_o \exp(\beta_{sys} m_a n z/x)} & z > z_c \end{cases} \tag{19.3.5}$$

where angular magnification has been included through Eq. (19.3.1). Similar in form to Eq. (10.5.1) for contrast-limited target acquisition, it has been shown [19.12] that the probability P_∞ of carrying out a given task of detection or recognition is related to SNR_N via a cumulative Gaussian function approximated by [19.13]

$$P_\infty = \frac{(\mathrm{SNR}_N)^E}{1 + (\mathrm{SNR}_N)^E} \tag{19.3.6}$$

where

$$E = 2.7 + 0.7\mathrm{SNR}_N \tag{19.3.7}$$

Note that when $\mathrm{SNR}_N = 1$, $P_\infty = 50\%$.

For a given value of P_∞, expected range can be calculated by substituting Eq. (19.3.5) into Eq. (19.3.6). From Eq. (19.3.5) the dependence of acquisition range on SNR_N is

$$
\begin{aligned}
&A_a z + \frac{S_a^* z^3}{z_c^2} + \beta_{sys}\frac{nz}{x} + \ln\left\{\frac{\mathrm{SNR}_N(\mathrm{MRT}_o)}{(\epsilon/7)^{1/2}\Delta T_o}\right\} = 0 \qquad z \le z_c \\
&z = \ln\left\{\frac{\Delta T_o(\epsilon/7)^{1/2}}{\mathrm{SNR}_N(\mathrm{MRT}_o)}\right\} \Big/ (\alpha + \beta_{sys} m_a n/x) \qquad z > z_c
\end{aligned}
\tag{19.3.8}
$$

The SNR_N required for a probability P_∞ for the expected acquisition range depends on the value of the range itself. It depends on whether or not the range is smaller than the cutoff range z_c. For ranges smaller than z_c, we have an equation that must be solved numerically, whereas for ranges larger than z_c, z can be evaluated analytically. Therefore, to calculate the expected range of detection or recognition for a given probability, one has to estimate first whether this range is greater or smaller than the range z_c given in Eq. (19.3.2). Of course, it is easier to try first the second part of Eq. (19.3.8), assuming $z > z_c$, and either verify this assumption or contradict it and recalculate z from the first term ($z \le z_c$).

The search model should be expanded to include turbulence effects too, particularly when dealing with a relatively clear midday atmosphere for which turbulence is often relatively strong. In that case, the received or apparent temperature difference at the FLIR, ΔT, is affected by the spatial-frequency dependence of turbulence MTF (in addition to aerosol MTF), and Eq. (19.3.4) must be extended to include this effect. Following the analysis that led to Eq. (19.3.8), multiplying the received temperature difference in Eq. (19.3.4) by the turbulence MTF and using the change of variables in Eq. (19.3.1), the SNR at the FLIR for the long-exposure case is given by

$$
\mathrm{SNR}_N \begin{cases}
\dfrac{\Delta T(7/\alpha_T)^{1/2}\exp(-A_a z - S_a^* z^3/z_c^2)\mathrm{MTF}_{le}(f_a, z)}{\mathrm{MRT}_o \exp(\beta_{sys} m_a n z/x)} & z \le z_c \\[2ex]
\dfrac{\Delta T(7/\alpha_T)^{1/2}\exp(-\alpha z)\mathrm{MTF}_{le}(f_a, z)}{\mathrm{MRT}_o \exp(\beta_{sys} m_a n z/x)} & z > z_c
\end{cases}
\tag{19.3.9}
$$

for long exposures (much larger than about a millisecond), where $\mathrm{MTF}_{le}(f_a, z)$ is given by Eq. (19.2.9). For short exposures on the order of a millisecond or less,

$$
\mathrm{SNR}_N \begin{cases}
\dfrac{\Delta T(7/\alpha_T)^{1/2}\exp(-A_a z - S_a^* z^3/z_c^2)\mathrm{MTF}_{se}(f_a, z)}{\mathrm{MRT}_o \exp(\beta_{sys} m_a n z/x)} & z \le z_c \\[2ex]
\dfrac{\Delta T(7/\alpha_T)^{1/2}\exp(-\alpha z)\mathrm{MTF}_{se}(f_a, z)}{\mathrm{MRT}_o \exp(\beta_{sys} m_a n z/x)} & z > z_c
\end{cases}
\tag{19.3.10}
$$

where $\mathrm{MTF}_{se}(f_a, z)$ is given by Eq. (19.2.10).

Equations (19.3.9) and (19.3.10) can be substituted into Eq. (19.3.6) to obtain the effects of the atmosphere on target acquisition probability for various ranges or, given a required acquisition probability, to determine maximum range. Examples of numerical calculation are presented next.

19.3.1 Examples

Examples for actual systems and atmospheric conditions are presented in this section [19.14]. The data of FLIR and target characteristics taken from [19.15] are summarized in Tables 19.2, 19.3, and 19.4. [Data for more recent devices is summarized following Eq. (11.2.10).] Atmospheric optical parameters were calculated using actual meteorological data measurements at Ben-Gurion University of the Negev, located in a semidesert area in Israel. Calculations were similar to those in [19.8]. In the examples here, angular magnification is assumed to be unity. An example for a fairly "dusty" condition with "normal" turbulence strength is summarized in Table 19.5, while an example for relatively"clear"atmosphere with strong turbulence strength is summarized in Table 19.6.

A comparison between the "traditional" (IDA/CNVEO) model of Eq. (11.2.9) and Eq. (19.3.9) for normalized SNR is presented in Fig. 19.12. The "classical" normalized SNR in which atmospheric degradation is considered attenuation only and not blur, using the parameters in Tables 19.2 and 19.3, is represented by the pluses in Fig. 19.12. Eq. (19.3.9) which includes atmospheric blur, according to the parameters in Tables 19.2, 19.3, and 19.5 is represented by the solid line in Fig. 19.12. In this example, the cutoff frequency z_c from Eq. (19.3.2) equals 6 km. For all cases where the target is located nearer than the range of 6 km, the actual SNR_N received at the sensor is higher than the SNR_N calculated by the classical IDA/CNVEO model [19.15]. The reason for this is that the *measured* atmospheric transmission of the target's temperature difference ΔT_o is angular spatial-frequency dependent, and for ranges closer than z_c, that transmission is greater than the received atmospheric transmittance as shown in Fig. 17.3 for angular spatial frequencies less than f_{ac}. This

TABLE 19.2 *Target Characteristics*

x	3 [m]
ϵ	2
ΔT_o	2°C

TABLE 19.3 *Sensor Characteristics (1974 FLIR, 8–12 μm)*

MRT_o	0.0254°C
β_{sys}	0.996 [cycles/mrad]

TABLE 19.4 *Sensor Characteristics (1978 FLIR, 8–12 μm)*

MRT_o	0.0112°C
β_{sys}	0.663 [cycles/mrad]

TABLE 19.5 *Atmospheric Characteristics*

S_a^*	0.3 [1/km]
A_a	0.5 [1/km]
f_{ac}	2 [cycles/mrad]
C_n^2	2.7×10^{-14} [$m^{-2/3}$]

TABLE 19.6 *Atmospheric Characteristics*

S_a^*	0.05 [1/km]
A_a	0.05 [1/km]
f_{ac}	2 [cycles/mrad]
C_n^2	2.7×10^{-13} [$m^{-2/3}$]

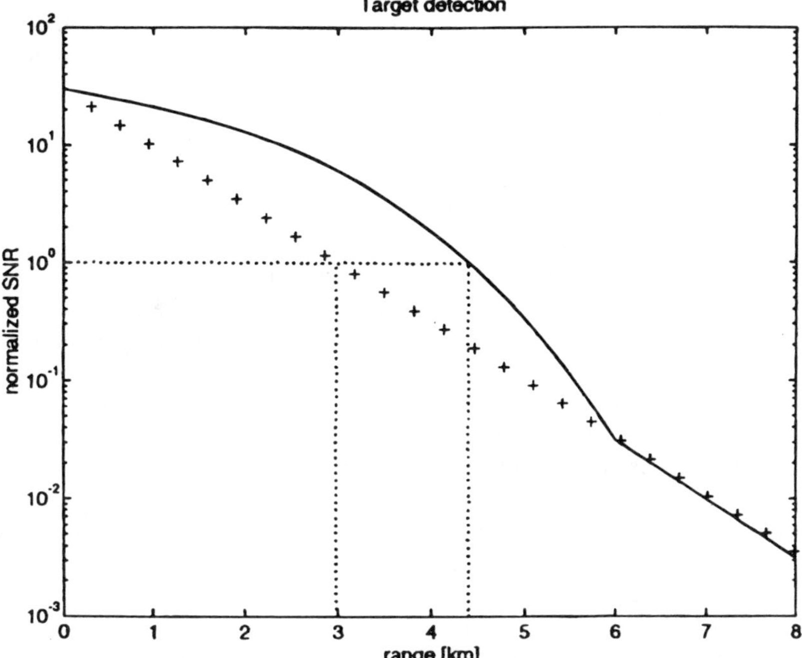

Fig. 19.12 SNR_N versus range for Eq. (11.2.9) (IDA/CNVEO) (plus signs) and Eq. (19.3.9) (solid line) models. Ranges for 50% detection probability shown by dotted lines (after [19.14]).

near-field increase in SNR_N is mainly due to contributions of received scattered light in addition to the received unscattered light.

For a range larger than z_c, the two curves converge. The reason is that for spatial frequencies corresponding to a range larger than z_c, the high spatial-frequency asymptote of the aerosol MTF equals approximately the atmospheric transmittance, in agreement with the IDA/CNVEO model. {This is an approximation rather than an equality since instrumentation causes the practical aerosol MTF value at high spatial frequencies to be slightly greater than that of atmospheric transmission [19.3], as discussed regarding Eq. (19.2.11). Therefore the agreement between Eq. (19.3.9) and the IDA/CNVEO model at high spatial frequencies is only approximate.} Note that in the thermal infrared, moderate turbulence hardly affects atmospheric MTF. In this example, the range for 50% static detection is larger than that expected by the IDA/CNVEO model. Figure 19.12 illustrates ranges (dotted lines) for 50% target detection probability predicted by the "classical" model and that shown here. The range increase is by almost 50%. Other probabilities can be found by substituting into Eq. (19.3.6) value of SNR_N appropriate to range in Fig. 19.12.

For the case of target recognition, differences between the models are much smaller. This is due to the inverse proportion between n and z_c, so that z_c is four times smaller for recognition than for detection. In that case, the two curves ["classical" and that of Eq. (19.3.9)] converge much earlier, and with significantly less difference. However, if dealing with larger targets (larger x), or instrumentation of wider spatial-frequency bandwidth that yields larger f_{ac} [19.10], z_c increases and the target recognition model should be corrected too. An example for target recognition analysis including atmospheric conditions similar to those in Fig. 19.12, but with f_{ac} equal to 3 instead of 2 cycles·mrad^{-1}, is presented in Fig. 19.13. Again, the range for 50% static recognition (dotted lines) is larger than that expected for the IDA/CNVEO model, this time by about 40%.

In the previous normal turbulence examples, aerosol MTF was assumed to be much more dominant than turbulence MTF, as is often the case [19.8, 19.10]. However, in cases where the atmosphere is relatively clear, and turbulence is strong [19.7], turbulence MTF has a significant effect on Eq. (19.3.9), particularly at very low elevations. This would apply to midday on hot days and would apply when using FLIRs with improved high-resolution performance. An example of *strong* turbulence long-exposure effects on target detection, according to target parameters from Table 19.2, FLIR 1978 parameters from Table 19.4, and atmospheric parameters from Table 19.6, is presented in Fig. 19.14. It is clear that in the conditions given in this example, atmospheric turbulence must be considered, and at longer ranges the received SNR at the FLIR is significantly lower than expected according to the IDA/CNVEO model.

The reason for the difference is turbulence blur, which is spatial frequency dependent and affects the contrast much more at high spatial frequencies corresponding to larger target ranges. In Fig. 19.15, an example of target recognition is presented, for the same parameters given in Fig. 19.14. A similar explanation follows for this result. A comparison of calculated SNR_N using all the search models that were presented in this section, for target FLIR and atmospheric data taken from Tables 19.2, 19.4, and 19.6, respectively, is given in Fig. 19.16. The solid line represents

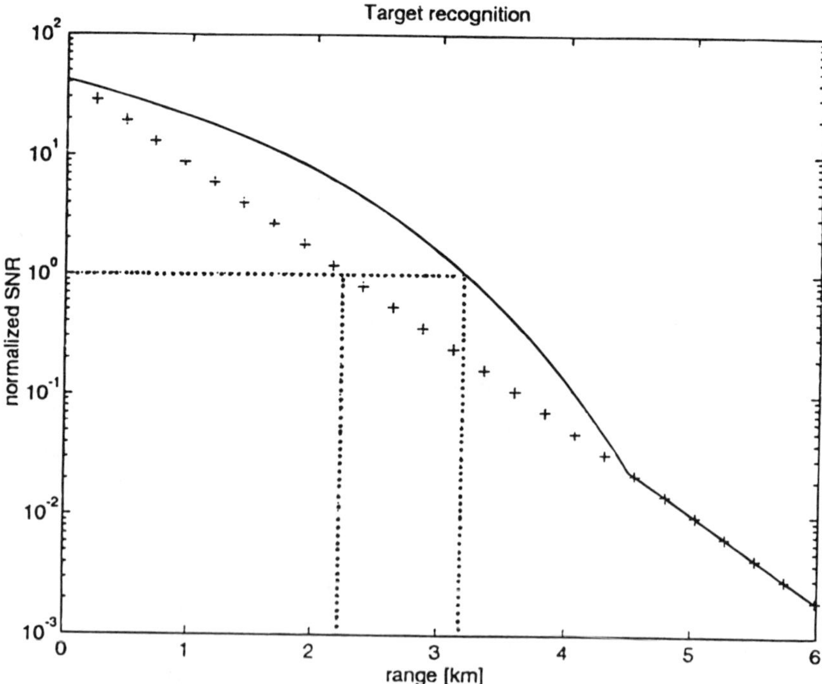

Fig. 19.13 SNR_N versus range for Eq. (11.2.9) (IDA/CNVEO) (plus signs) and target recognition models of Eq. (19.3.9) (solid line) which include atmospheric blur. Ranges for 50% recognition probabilities designed by dotted lines (after [19.14]).

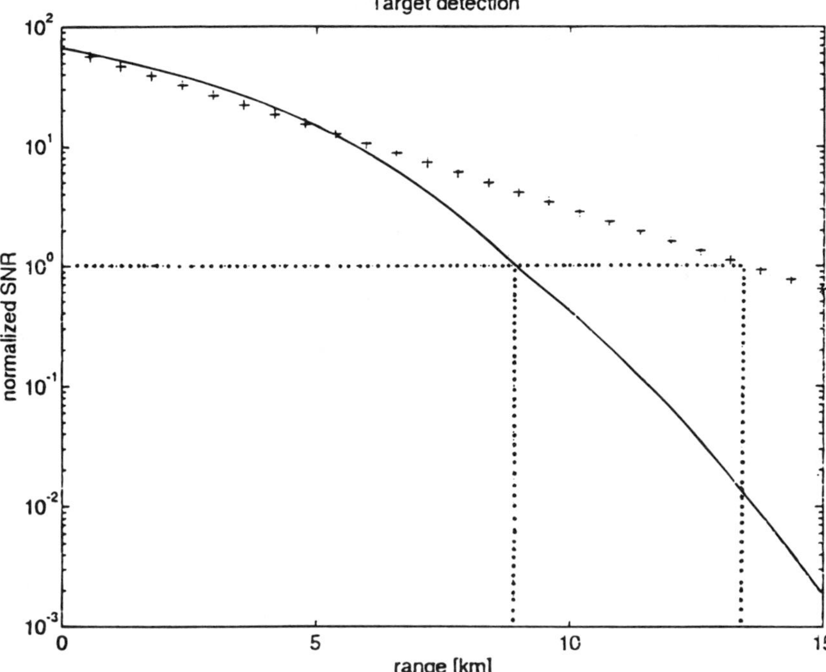

Fig. 19.14 Same as Fig. 19.12, but for a clear atmosphere and strong turbulence (after [19.14].

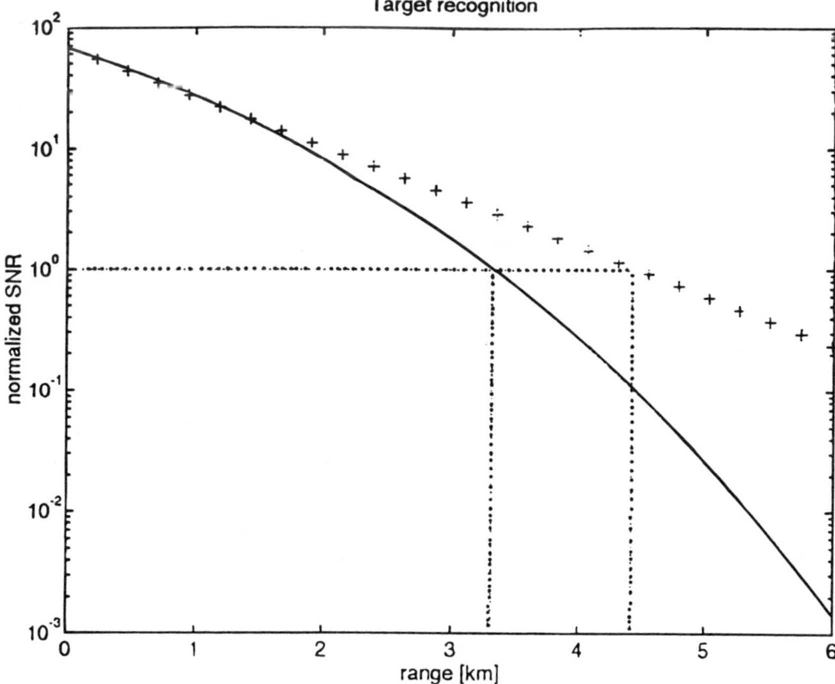

Fig. 19.15 Same as Fig. 19.13 but for a clear atmosphere and strong turbulence (after [19.14]).

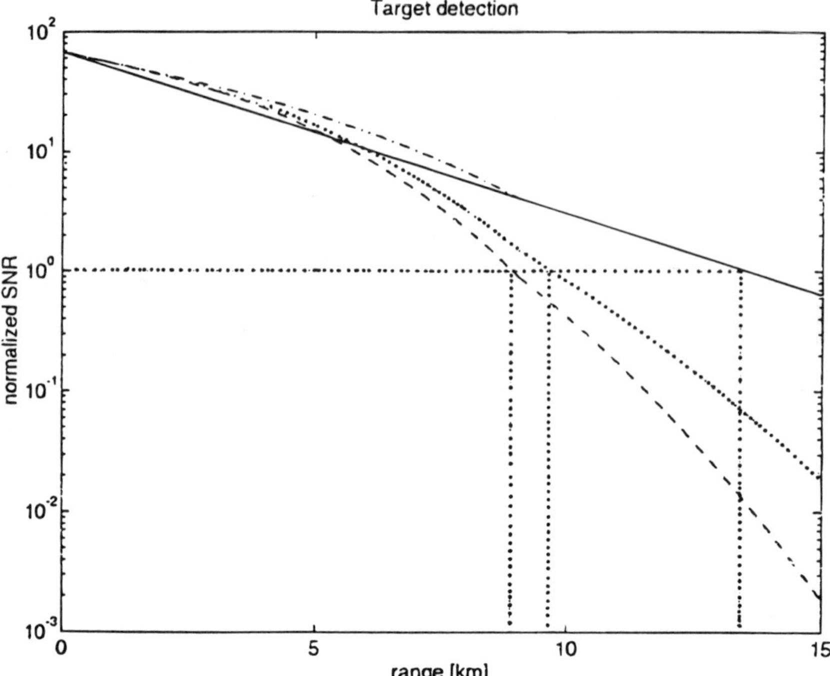

Fig. 19.16 Comparison between calculated SNR_N using different techniques: Eq. (11.2.9) (IDA/CNVEO) model (solid line); Eq. (19.3.5) incorporating aerosol MTF (dashed-dotted line); Eq. (19.3.9) incorporating aerosol MTF and long-exposure turbulence MTF (dashed line); and Eq. (19.3.10) incorporating aerosol MTF and short-exposure turbulence MTF (dotted line) (after [19.14]).

the IDA/CNVEO model. The aerosol MTF target acquisition model [Eq. (19.3.5)] is represented by the dashed-dotted curve and applied to medium and weak turbulence. The aerosol MTF and strong long-exposure turbulence MTF target acquisition model [Eq. (19.3.9)] is represented by the dashed curve, and the suggested aerosol MTF and short-exposure strong turbulence MTF target acquisition model [Eq. (19.3.10)] is represented by the dotted curve. For calculating the short-exposure turbulence MTF, the aperture diameter D was chosen to be 10 cm. Probabilities different from 50% for various ranges can be obtained by substituting appropriate value of SNR_N into Eq. (19.3.6).

This analysis can be expanded to estimate time-dependent performance incorporating human search [19.14]. The temporal search model proposed by IDA/CNVEO is also given in Eqs. (10.5.4) where P_∞ is defined here in Eq. (19.3.6) and again represents the fraction of the normal observer ensemble that can successfully find the target in infinite time.

19.3.2 Conclusions

The effect of atmospheric blur is included in thermal imaging noise-limited target acquisition modeling. Atmospheric blur effects are based on recent investigation of atmospheric MTF in the thermal range, which has shown significant spatial-frequency dependence deriving from a very strong aerosol MTF, and from a turbulence MTF that improves at long wavelengths and thus causes less image degradation than in the visible. This spatial-frequency dependence directly influences SNR_N and probability of detection or recognition obtained in FLIR systems. When atmospheric conditions are of moderate or weak turbulence, its main effect is to increase SNR for ranges smaller than the cutoff range z_c, mainly due to near-field contributions of scattered light to received apparent temperature difference. For ranges larger than z_c, aerosol forward scatter effects are manifested as constant attenuation (the atmospheric transmittance), in which case both the IDA/CNVEO and the new models that include atmospheric blur models approximately converge. However, when atmospheric conditions are those of a relatively clear atmosphere but with *strong* turbulence (as on middays on sunny days after rain for example), turbulence dominates mainly for larger target ranges corresponding to high spatial frequencies. In that case, received SNR and probability of detection are significantly poorer than expected by the IDA/CNVEO model.

If atmospheric distortions are modeled accurately (including aerosol MTF), most image corruptions can be corrected for contrast-limited imaging as shown here in Section 19.2. In that case, the restored atmospheric MTF is improved significantly and approaches unity. Apparent thermal signals are improved greatly, as are probabilities and ranges of detection and recognition. However, under very poor atmospheric transmission conditions normalized SNR becomes noise limited and very little improvement in probability of detection will be achieved even with restoration.

REFERENCES

19.1. G. C. Holst, *Electro-Optical Imaging System Performance*, JCD Press, Winter Park, FL, and SPIE Optical Engineering Press, Bellingham, WA, 1995, pp. 371–372.

19.2. D. Sadot, N. S. Kopeika, and S. R. Rotman, "Target acquisition modeling for contrast-limited imaging: effects of atmospheric blur and image restoration," *J. Opt. Soc. Am. A*, Vol. 12, November 1995, pp. 2401–2414.

19.3. D. Sadot and N. S. Kopeika, "Practical instrumentation-based theory and verification of aerosol MTF," *J. Opt. Soc. Am. A*, Vol. 10, January 1993, pp. 172–179.

19.4. I. Dror and N. S. Kopeika, "Experimental comparison of turbulence MTF and aerosol MTF through the open atmosphere," *J. Opt. Soc. Am. A*, Vol. 12, May 1995, pp. 970–980.

19.5. S. R. Rotman, E. S. Gordon, and M. L. Kowalczyk, "Modeling human search and target acquisition performance: III. Target detection in the presence of obscurants," *Opt. Eng.*, Vol. 30, June 1991, pp. 824–829.

19.6. D. Sadot and N. S. Kopeika, "Thermal imaging atmospheric effects and image restoration," in *Infrared Spaceborne Remote Sensing II*, M. Scholl, Ed., *Proc. SPIE*, Vol. 2268, 1994, pp. 366–385.

19.7. W. R. Watkins, S. B. Crow, and F. T. Kantrowitz, "Characterizing atmospheric effects in target contrast," *Opt. Eng.*, Vol. 33, October 1991, pp. 1563–1575.

19.8. D. Sadot, G. Kitron, and N. S. Kopeika, "Thermal imaging through the atmosphere: atmospheric MTF theory and verification," *Opt. Eng.*, Vol. 33, March 1994, pp. 880–887.

19.9. D. Sadot and N. S. Kopeika, "Effects of absorption on image quality through a particulate medium," *Appl. Opt.*, Vol. 33, October 20, 1994, pp. 7107–7111.

19.10. D. Sadot, A. Dvir, I. Bergel, and N. S. Kopeika, "Restoration of thermal images distorted by the atmosphere, based upon measured and theoretical atmospheric modulation transfer function," *Opt. Eng.*, Vol. 33, January 1994, pp. 44–53.

19.11. D. Sadot, G. Lorman, R. Lapardon, and N. S. Kopeika, "Restoration of thermal images distorted by the atmosphere using predicted atmospheric modulation transfer function" *Infrared Phys. Technol.*, Vol. 36, February 1995, pp. 565–576.

19.12. R. L. Legault, "Visual detection process for electrooptical images: man—the final stage of an electrooptical imaging system," in *Photoelectronic Imaging Devices, Vol. 1: Physical Processes and Methods of Analysis*, L. M. Biberman and S. Nudelman, Eds., Plenum Press, New York, 1971, pp. 69–87.

19.13. G. Kornfeld and W. Lawson, "Visual perception model," *J. Opt. Soc. Am.*, Vol. 61, July 1971, pp. 811–820.

19.14. D. Sadot, N. S. Kopeika, and S. R. Rotman, "Incorporation of atmospheric blurring effects in target acquisition modeling of thermal images," *Infrared Phys. Technol.*, Vol. 36, February 1995, pp. 551–564.

19.15. L. N. Seekamp, "Field manual to determine detection or recognition range of a FLIR sensor," IDA Paper P-1419, Institute for Defense Analyses, September 1979.

Appendix

Calculating Practical Aerosol MTF

Mie Scattering, Absorption, and Extinction Coefficients and the Mie Phase Function

1. Define the following parameters:

m = relative refractive index (complex)
λ = radiation wavelength (meters)
a = particulate radius (meters)
$x = 2\pi a/\lambda$—size parameter
$\rho = x \times m$
θ = scattering angle (steradians)
$\mu = \cos\theta$
$n(x)$ = particle size distribution.

2. Create the spherical Bessel functions, which satisfy the recurrence relations:

$$z_{n-1}(\rho) + z_{n+1}(\rho) = \frac{2n+1}{\rho} z_n(\rho) \tag{1}$$

$$(2n+1)\frac{d}{d\rho} z_n(\rho) = n_{z_{n-1}}(\rho) - (n+1)_{z_{n+1}}(\rho) \tag{2}$$

where z_n is either J_n or Y_n. From the first two orders,

$$J_0(\rho) = \frac{\sin\rho}{\rho} \qquad J_1(\rho) = \frac{\sin\rho}{\rho^2} - \frac{\cos\rho}{\rho} \tag{3}$$

$$Y_0(\rho) = -\frac{\cos\rho}{\rho} \qquad Y_1(\rho) = -\frac{\cos\rho}{\rho^2} - \frac{\sin\rho}{\rho} \tag{4}$$

higher order functions can be generated by recurrence.

3. Create the spherical Bessel functions of the third kind (Hankel functions):

$$H_n(\rho) = J_n(\rho) - iY_n(\rho) \tag{5}$$

4. Create the Riccati-Bessel functions:

$$\phi_n(\rho) = \rho J_n(\rho) \qquad \xi_n(\rho) = \rho H_n(\rho) \tag{6}$$

5. Calculate the scattering parameters a_n and b_n:

$$a_n = \frac{m\phi_n(\rho)\phi_n'(x) - \phi_n(x)\phi_n'(\rho)}{m\phi_n(\rho)\xi_n'(x) - \xi_n(x)\phi_n'(\rho)} \tag{7}$$

$$b_n = \frac{\phi_n(\rho)\phi_n'(x) - m\phi_n(x)\phi_n'(\rho)}{\phi_n(\rho)\xi_n'(x) - m\xi_n(x)\phi_n'(\rho)} \tag{8}$$

6. Calculate the Mie scattering, extinction, and absorption efficiency factors:

$$Q_{sc}(m, x) = \frac{2}{x^2}\sum_{n=1}^{\infty}(2n + 1)(|a_n|^2 + |b_n|^2) \tag{9}$$

$$Q_{ext}(m, x) = \frac{2}{x^2}\sum_{n=1}^{\infty}(2n + 1)\,\mathrm{Re}\{a_n + b_n\} \tag{10}$$

$$Q_{abs} = Q_{ext} - Q_{sc} \tag{11}$$

7. Calculate the angle-dependent functions that satisfy the recurrence relations:

$$\pi_n = \frac{2n - 1}{n - 1}\mu_{n-1} - \frac{n}{n - 1}\pi_{n-2} \tag{12}$$

$$\tau_n = n\pi_n - (n + 1)\pi_{n-1} \tag{13}$$

where $\pi_0 = 0$ and $\pi_1 = 1$.

8. Calculate the complex amplitudes of scattered electric fields:

$$S_1(m, x, \theta) = \sum_{n=1}^{\infty}\frac{2n + 1}{n(n + 1)}(a_n\pi_n + b_n\tau_n) \tag{14}$$

$$S_2(m, x, \theta) = \sum_{n=1}^{\infty}\frac{2n + 1}{n(n + 1)}(b_n\pi_n + a_n\tau_n) \tag{15}$$

where S_1 and S_2 are components perpendicular and parallel to the scatter plane.

9. Calculate the dimensionless intensity parameters (for scattered light only):

$$i_1(x, m, \theta) = S_1S_1^* \tag{16}$$

$$i_2(x, m, \theta) = S_2S_2^* \tag{17}$$

10. Use the measured or predicted atmospheric particle size distribution to calculate the medium's scattering coefficient (scattering cross section per unit volume):

$$S_a[\lambda, n(x)] = \pi k^{-3} \int_0^\infty x^2 n(x) Q_{sc}(x)\, dx \tag{18}$$

11. Calculate the medium's phase function:

$$P(\theta) = \frac{2\pi}{k^3 S_a} \int_0^\infty n(x)[i_1(\theta) + i_2(\theta)]\, dx \tag{19}$$

which obeys the following normalization condition:

$$\int_\Omega P(\theta)\, dw = 4\pi \tag{20}$$

where the integration is with respect to solid angle over all directions corresponding to all space around a point.

12. Calculate the medium's extinction coefficient (extinction cross section per unit volume):

$$\sigma[\lambda, n(x)] = \pi k^{-3} \int_0^\infty x^2 n(x) O_{est}(x)\, dx \tag{21}$$

13. Calculate the medium's absorption coefficient (absorption cross section per unit volume):

$$A_a[\lambda, n(x)] = \pi k^{-3} \int_0^\infty x^2 n(x) Q_{abs}(x)\, dx \tag{22}$$

Atmospheric Absorption: Spatial-Frequency Dependence

14. Calculate aerosol MTF including absorption spatial frequency dependence:

$$M_a(f_a) = \begin{cases} \exp\left\{-S_a z\left(\dfrac{f_a}{f_{ac}}\right)^2\right\} \cdot \exp\left\{\left[\exp\left\{-S_a z\left(1 - \left(\dfrac{f_a}{f_{ac}}\right)^2\right)\right\} - \exp\{-S_a z\}\right](-A_a z)\right\} & f_a < f_{ac} \\ \exp\{-S_a z\} \cdot \exp\{[1 - \exp\{-S_a z\}](-A_a z)\} & f_a > f_{ac} \end{cases} \tag{23}$$

15. Calculate the new MTF asymptote at high spatial frequencies $= M_a(\infty)$ (differing from the atmospheric transmittance), and calculate the effective optical depth:

$$T_{\text{eff}} = -\ln[M_a(\infty)] \tag{24}$$

Practical Aerosol MTF

16. Approximate the medium's phase function by a Gaussian form (not necessary):

$$P(\theta) = 4\sqrt{\pi\alpha_p}\,\exp\{-\alpha_p\theta^2\} \tag{25}$$

17. Calculate the "classical" aerosol MTF using the *effective optical depth*:

$$K(f_a) = \exp\left\{-\int_0^z T_{\text{eff}}\left[1 - \exp\left(-\frac{\pi^2 f_a^2}{\alpha_p}\right)\right] dz'\right\} \tag{26}$$

18. Determine the main instrumentation limitations:

(a) Finite angular spatial frequency bandwidth, $f_{a\ \max}$.

(b) Finite field of view, $\otimes'_{\max}$.

(c) Dynamic range limitation; minimal detected received irradiance, which is related to maximal angle of detected scattered light, $\otimes''_{\max}$.

(d) SNR limitation; minimal detected received irradiance, which is of higher intensity than the background noise or radiation, $\otimes'''_{\max}$.

19. Choose the worst case of (18b) through (18d), obtaining the instrumentation limitation of the maximal angle of recorded scattered light: $\otimes_{\max} = \min\{\otimes'_{\max}, \otimes''_{\max}, \otimes'''_{\max}\}$.

20a. Calculate the "classical" I (specific intensity) function for the case of an indefinite plane wave:

$$I(\theta) = \frac{1}{(2\pi)^2}\int_{-f_{a\ \max}}^{f_{a\ \max}} df_a \cdot \exp(-j\theta f_a) \cdot K(f_a) \tag{27a}$$

20b. In case where the hardware's MTF is known, Eq. (27a) can be replaced by

$$I(\theta) = \frac{1}{(2\pi)^2}\int_{-\infty}^{\infty} df_a \cdot \exp(-j\theta f_a) \cdot K(f_a) \cdot M_0(f_a) \tag{27b}$$

where $M_0(f_a)$ is the hardware MTF.

21. Calculate the practical aerosol MTF by inverse Fourier transforming the specific intensity function:

$$M_a(f_a) = \int_{-\otimes_{\max}}^{\otimes_{\max}} d\theta \cdot \exp(j\theta f_a) I(\theta) \tag{28}$$

CHAPTER 1

SOLUTIONS

1

1.1 From Eq. (1.3.27a)

$$\alpha = \omega\left(\frac{\mu\epsilon'}{2}\right)^{1/2}\left[\left(1 + \left(\frac{\sigma + \omega\epsilon''}{\omega\epsilon'}\right)^2\right)^{1/2} - 1\right]^{1/2}$$

Use binomial expansion

$$(1 + x)^n = 1 + nx + \frac{n(n-1)}{2!}x^2 + \frac{n(n-1)(n-2)}{3!}x^3 + \cdots$$

where $x = (\sigma + \omega\epsilon''/\omega\epsilon')^2 << 1$ for lossless media.

$$\therefore (1 + x)^{1/2} \approx 1 + \frac{1}{2}\left(\frac{\sigma + \omega\epsilon''}{\omega\epsilon'}\right)^2$$

$$\therefore \alpha \approx \omega\left(\frac{\mu\epsilon'}{2}\right)^{1/2}\left[1 + \frac{1}{2}\left(\frac{\sigma + \omega\epsilon''}{\omega\epsilon'}\right)^2 - 1\right]^{1/2}$$

$$\approx \omega\left(\frac{\mu\epsilon'}{2}\right)^{1/2}\left[\frac{1}{2}\left(\frac{\sigma + \omega\epsilon''}{\omega\epsilon'}\right)^2\right]^{1/2}$$

$$\approx \left(\frac{\mu}{\epsilon'}\right)^{1/2}\left(\frac{\sigma + \omega\epsilon''}{2}\right)$$

From Eq. (1.3.27b)

$$\beta = \omega\left(\frac{\mu\epsilon'}{2}\right)^{1/2}\left\{\left[1 + \left(\frac{\sigma + \omega\epsilon''}{\omega\epsilon'}\right)^2\right]^{1/2} + 1\right\}^{1/2}$$

$$= \omega\left(\frac{\mu\epsilon'}{2}\right)^{1/2}\left[1 + \frac{1}{2}\left(\left(\frac{\sigma + \omega\epsilon''}{\omega\epsilon'}\right)^2 + 1\right)^2\right]^{1/2}$$

SOLUTIONS

$$= \omega\left(\frac{\mu\epsilon'}{2}\right)^{1/2}\left[2 + \frac{1}{2}\left(\frac{\sigma + \omega\epsilon''}{\omega\epsilon'}\right)^2\right]^{1/2}$$

$$= \omega(\mu\epsilon')^{1/2}2^{1/2}\left[1 + \frac{1}{4}\left(\frac{\sigma + \omega\epsilon''}{\omega\epsilon'}\right)\right]^{1/2}$$

$$= \omega(\mu\epsilon')^{1/2}\left[1 + \frac{1}{8}\left(\frac{\sigma + \omega\epsilon''}{\omega\epsilon'}\right)^2\right]$$

1.2 (a) From Eqs. (1.3.31) and (1.3.17),

$$\beta = \omega(\mu\epsilon)^{1/2} = \frac{\omega}{v} = \frac{\omega n}{c} = \frac{\omega n' - j\omega n''}{c}$$

From Eq. (1.3.9),

$$\exp(-j\beta z) = \exp\left(-\frac{j\omega n'}{c}z\right)\exp\left(-\frac{\omega n''}{c}z\right)$$

where n' or real β imply propagation; n'' or imaginary β implies attenuation according to

$$\exp(-\gamma\beta z) = \exp(-j\beta z)\exp(-\alpha z)$$

$$\therefore \alpha = \omega n''/c$$

(b) From Eq. (1.3.18),

$n = \sqrt{\epsilon_r\mu_r} \approx \sqrt{\epsilon/\epsilon_o}$ for nonferromagnetic medium.

$$\therefore n = \left[\frac{\epsilon' - j\epsilon''}{\epsilon_o}\right]^{1/2} = \left(\frac{\epsilon'}{\epsilon_o}\right)^{1/2}(1 - j\epsilon''/\epsilon')^{1/2}$$

$$\approx \left(\frac{\epsilon'}{\epsilon_o}\right)^{1/2}\left(1 - \frac{j\epsilon''}{\epsilon_o}\right) = n' - jn'' \text{ for } \epsilon'' < \epsilon', \text{ where}$$

$$n' = (\epsilon'/\epsilon_o)^{1/2}; \qquad n'' = \left(\frac{\epsilon'}{\epsilon_o}\right)^{1/2}\left(\frac{\epsilon''}{2\epsilon'}\right) = n'\,\frac{\epsilon'}{2\epsilon'}$$

1.3

$$n = \sqrt{\epsilon_{r_{eq}}} = \left(\frac{\epsilon'}{\epsilon_o} - j\,\frac{\sigma}{\omega\epsilon_o}\right)^{1/2}$$

$$= \left(\frac{-j\sigma}{\omega\epsilon_o}\right)^{1/2}\left(1 + j\,\frac{\omega\epsilon'}{\sigma}\right)^{1/2} = \sqrt{\frac{-j\sigma}{\omega\epsilon_o}} \text{ for a good conductor.}$$

Now $j = e^{j\pi/2}$.

$$\therefore j^{-1/2} = e^{j\pi/2(-1/2)} = e^{-j\pi/4} = \cos\frac{\pi}{4} - j\sin\frac{\pi}{4} = \frac{1-j}{\sqrt{2}}$$

$$\therefore n = n' - jn'' = \sqrt{\frac{\sigma}{2\omega\epsilon_o}}\,(1-j)\,; \qquad \therefore n' = n'' = \sqrt{\frac{\sigma}{2\omega\epsilon_o}}$$

for a good conductor.

1.4 From Eqs. (1.3.23) and (1.3.13),

$$\eta = \sqrt{\frac{\mu}{\epsilon_{eq}}} = \sqrt{\frac{\mu}{\epsilon' - j(\sigma/\omega) + \epsilon'')}} = \left(\frac{\mu}{\epsilon'}\right)^{1/2}\left[1 - j\left(\frac{\sigma}{\omega\epsilon'} + \frac{\epsilon''}{\epsilon'}\right)\right]^{-1/2}$$

$$\approx \eta'\left[1 + \frac{j}{2}\left(\frac{\sigma}{\omega\epsilon'} + \frac{\epsilon''}{\epsilon'}\right)\right] \text{ where } \eta' = \left(\frac{\mu}{\epsilon'}\right)^{1/2}$$

1.5 From Eq. (1.3.34),

$$\frac{E_x}{A_x} = \exp(-\alpha z)\exp[j(\omega t - \beta z + \delta)]$$

$$\left(\frac{E_x}{A_x}\right)^2 = \exp(-2\alpha z)\exp[j2(\omega t - \beta z + \delta)]$$

From Eq. (1.3.10),

$$\left(\frac{E_y}{A_y}\right)^2 = \exp(-2\alpha z)\exp[j2(\omega t - \beta z)]$$

$$\therefore \left(\frac{E_x}{A_x}\right)^2 + \left(\frac{E_y}{A_y}\right)^2 = \exp(-2\alpha z)\exp[j2(\omega t - \beta z)][1 + \exp(j2\delta)]$$

$$= 2\exp(-2\alpha z)\exp[j2(\omega t - \beta z)]\exp(j\delta)\cos\delta$$

$$= 2\,\frac{E_x E_y}{A_x A_y}\cos\delta$$

This is the equation of an ellipse at an angle α to the x axis, where α is given by Eq. (1.3.36).

1.6

$$n = (1-j)\sqrt{\frac{\sigma}{2\omega\epsilon_o}} = \sqrt{\epsilon_r} = \sqrt{\frac{\epsilon}{\epsilon_o}}$$

$$\therefore \epsilon = (1-j)^2\frac{\sigma}{2\omega} = (1 - 2j + j^2)\frac{\sigma}{2\omega} = -\frac{j\sigma}{\omega}$$

SOLUTIONS

This expression indicates that the medium upon which the parallel-polarized wave is incident is a good conductor. In this case

$$\gamma_2 \approx \sqrt{j\omega\mu\sigma} = (1 + j)\sqrt{\frac{\omega\mu\sigma}{2}} = \alpha_2 + j\beta_2$$

$$\therefore \beta_2 = \sqrt{\frac{\omega\mu\sigma}{2}}$$

From Eq. (1.4.16),

$$\beta_0 \sin\theta_1 = \beta_2 \sin\theta_2$$

where β_o refers to free space and is equal to ω/c. The imaginary part of refractive index is irrelevant to direction, but influences only attenuation.

Here, $(\omega/c)\sin\theta_1 = \sqrt{\omega\mu_0\sigma/2}\,\sin\theta_2$ is the relevant form of Snell's law. Plugging in numbers,

$$\sin\theta_2 = \frac{\omega\sqrt{\mu_0\epsilon_0}}{\sqrt{\omega\mu_0\sigma/2}}\sin\theta_1 = \sqrt{\frac{2\omega\epsilon_0}{\sigma}}\sin\theta_1$$

$$= \sqrt{\frac{2 \cdot 2\,\pi \cdot 10^5 \cdot 8.85 \times 10^{-12}}{40}}\sin\theta_1$$

$$= \sqrt{8.85\pi \cdot 10^{-8}}\sin\theta_1 \approx 5.27 \times 10^{-4}\sin\theta_1$$

Therefore, no matter what θ_1 may be, where $R_e\,\theta_1 \leq \pi/2$, θ_2 is essentially zero. Waves incident from medium 1 into medium 2 propagate only perpendicular to the interface, regardless of the angle of incidence.

1.7 From Snell's law,

$$\sin\theta_2 = \frac{n_1}{n_2}\sin\theta_1$$

but $\sin\theta_1 = \sin\theta_{\text{Brewster}} = [n_2^2/(n_2^2 + n_1^2)]^{1/2}$ for a wave going from medium 1 to medium 2.

$$\therefore \sin\theta_2 = \frac{n_1}{n_2}\frac{n_2}{(n_2^2 + n_1^2)^{1/2}} = \frac{n_1}{(n_2^2 + n_1^2)^{1/2}} = \sin\theta_{\text{Brewster}}$$

for a wave going from medium 2 to medium 1. Therefore, 100% of incident electric field exits the window.

1.8 To obtain essentially total reflection, $\theta_2 > \theta_c = \sin^{-1}(n_2/n_1)$

$$\therefore \theta_1 < \frac{\pi}{2} - \sin^{-1}\left(\frac{n_2}{n_1}\right)$$

At the input, from Snell's law,

$$\sin\phi = \frac{n_1}{n_0}\sin\theta_1 = \frac{n_1}{n_0}\cos\theta_2 = \frac{n_1}{n_0}[1 - \sin^2\theta_2]^{1/2}$$

Since it is required that $\sin\theta_2 > n_2/n_1$,

$$\text{then } \sin\phi < \frac{n_1}{n_0}\left[1 - \left(\frac{n_2}{n_1}\right)^2\right]^{1/2} = \frac{[n_1^2 - n_2^2]^{1/2}}{n_0}$$

$$\therefore n_0 \sin\phi = (n_1^2 - n_2^2)^{1/2}$$

1.9 With reference to Section 1.4, all fields are presented there. The reflection coefficient from core (medium 1) to cladding (medium 2), from Eq. (1.4.25), is

$$r_\perp = \frac{n_1\cos\theta_1 - n_2\cos\theta_2}{n_1\cos\theta_1 + n_2\cos\theta_2}$$

$$= \frac{n_1\cos\theta_1 - n_2(1 - \sin^2\theta_2)^{1/2}}{n_1\cos\theta_1 + n_2(1 - \sin^2\theta_2)^{1/2}}$$

From Snell's law, $\sin\theta_2 = n_1/n_2 \sin\theta_1$. Substituting,

$$r_\perp = \frac{n_1\cos\theta_1 - n_2\left[1 - \left(\frac{n_1}{n_2}\right)^2\sin^2\theta_1\right]^{1/2}}{n_1\cos\theta_1 + n_2\left[1 - \left(\frac{n_1}{n_2}\right)^2\sin^2\theta_1\right]^{1/2}}$$

Since $\theta_1 > \theta_c = \sin^{-1}(n_2/n_1)$, θ_2 is imaginary and $(n_1/n_2)\sin^2\theta_1 > 1$. Therefore,

$$r_\perp = \frac{n_1\cos\theta_1 + jn_2\left[\left(\frac{n_1}{n_2}\right)^2\sin^2\theta_1 - 1\right]^{1/2}}{n_1\cos\theta_1 - jn_2\left[\left(\frac{n_1}{n_2}\right)^2\sin^2\theta_1 - 1\right]^{1/2}}$$

$$= (1)\cdot e^{j\phi}$$

where $\phi = 2\tan^{-1}\{[\sin^2\theta_1 - (n_2/n_1)^2]^{1/2}/\cos\theta_1\}$

As expected, for $\theta_1 > \theta_c$ the reflection coefficient has a magnitude of unity, since magnitude of the real and imaginary components of the numerator of $r_\perp$ are equal to those of the denominator. The transmission coefficient, from Eq. (1.4.26), is

$$t_{\perp} = \frac{2n_1 \cos\theta_1}{n_1 \cos\theta_1 + n_2 \cos\theta_2} = \frac{2n_1 \cos\theta_1}{n_1 \cos\theta_1 - jn_2\left[\left(\frac{n_1}{n_2}\right)^2 \sin^2\theta_1 - 1\right]^{1/2}}$$

$$= \frac{2 \cos\theta_1}{\cos\theta_1 - j\left[\sin^2\theta_1 - \left(\frac{n_2}{n_1}\right)^2\right]^{1/2}} = \frac{2 \cos\theta_1}{\left[\cos^2\theta_1 + \sin^2\theta_1 - \left(\frac{n_2}{n_1}\right)^2\right]^{1/2}} e^{j\phi/2}$$

In view of the definition above for ϕ,

$$t_{\perp} = 2 \cos(\phi/2) \exp j(\phi/2)$$

The coefficients $r_{\perp}$ and $t_{\perp}$ are used to determine the fields. Since $\theta_1 = \theta_3$ in Fig. 1.5a, the total electric field in the core (medium 1) is, from Eq. (1.4.2) and (1.4.7),

$$\begin{aligned}\overline{E}_{y1} &= \overline{E}_{yI} + \overline{E}_{yR} \\ &= \hat{y}A_{yI} \exp(j\omega t) \exp(-j\beta_{1z} \sin\theta_1)[\exp j\beta_{1x}x + r_{\perp} \exp(-j\beta_{1x}x)] \\ &= \hat{y}2A_{yI} \exp(j\omega t) \exp(-j\beta_{1z} \sin\theta_1 + \phi/2) \cos(\beta_1 x \cos\theta_1 - \phi/2)\end{aligned}$$

In view of Eq. (1.3.12), the total magnetic field in the x direction in the core is

$$\begin{aligned}\overline{H}_{x1} &= \overline{H}_{xI} + \overline{H}_{xR} = -\hat{x}\frac{j}{\omega\mu}\frac{\partial E_{y1}}{\partial z} = -\frac{jE_{y1}}{\eta_1} \cdot (-j\beta_1 \sin\theta_1) \\ &= -\hat{x}\frac{E_{y1} \sin\theta_1}{\eta_1}\end{aligned}$$

Similarly,

$$\begin{aligned}\overline{H}_{z1} &= \overline{H}_{zI} + \overline{H}_{zR} = \hat{z}\frac{j}{\omega\mu}\frac{\partial E_{y1}}{\partial x} \\ &= \hat{z}(-j)\frac{2A_{yI} \cos\theta_1}{\eta_1} \exp(j\omega t) \exp(-j\beta_1 z \sin\theta_1 + \phi/2) \\ &\quad \times \sin(\beta_1 x \cos\theta_1 - \phi/2).\end{aligned}$$

Note that H_{x1} is real, while H_{z1} is imaginary. That is because H_{xI} and H_{xR} are in the same direction (from core to cladding), or negative x direction, while H_{zI} and H_{zR} are in opposing directions, causing real parts to cancel.

With regard to time-averaged power flow, in the core,

$$\overline{P}_1 = \frac{1}{2} \mathrm{Re}\,\hat{y}\overline{E}_{y1} \times (\hat{x}\overline{H}^*_{x1} + \hat{z}\overline{H}^*_{z1}) = \frac{1}{2}(\hat{y}\overline{E}_{y1} \times \hat{x}\overline{H}^*_{x1})$$

because E_{y1} and H_{x1} are real, while H_{z1} is imaginary.

$$\overline{P}_1 = \hat{z}P_1 = \hat{z}2\frac{A_{yI}}{\eta_1}\sin\theta_1\cos^2(\beta_1 x\cos\theta_1 - \phi/2)$$

In the cladding (medium 2 in Fig. 1.5), using $t_\perp$ above and Eqs. (1.4.2) and (1.4.10),

$$E_{y2} = 2A_{yI}\cos(\phi/2)\exp(j\omega t)\exp(-j\beta_2 z\sin\theta_2 + \phi/2)\cdot\exp(j\beta_2 x\cos\theta_2)$$

In view of Snell's law and its effect on θ_2,

$$E_{y2} = 2A_{yI}\cos(\phi/2)\exp(j\omega t)\exp\left(-j\beta_2 z\frac{n_1}{n_2}\sin\theta_1 + \phi/2\right)$$

$$\times\exp\left\{\beta_2 x\left[\left(\frac{n_1}{n_2}\right)^2\sin^2\theta_1 - 1\right]^{1/2}\right\}$$

Note that at the $x = 0$ interface, in view of Eq. (1.4.14), $\overline{E}_{y1}$ and $\overline{E}_{y2}$ are continuous. The electric field in the cladding is composed of a traveling wave in the z direction and is exponentially damped in the x direction. That is, despite the imaginary nature of θ_2, because of the continuity of tangential electric fields, E_y must be identical on both sides of the core/cladding interface, thus causing an EM field to exist in the cladding as well, while decreasing exponentially with increasing penetration into the cladding. There is no propagation in the x direction because there is no imaginary component to γ_{2x} in the last exponential above. From Eq. (1.4.12a), applied to the cladding

$$\beta_{2x} = \beta_2\cos\theta_2 = \beta_2[1 - \sin^2\theta_2]^{1/2} = j\beta_2\left[\left(\frac{n_1}{n_2}\right)^2\sin^2\theta_1 - 1\right]^{1/2}$$

Since β_{2x} is imaginary, $j\beta_{2x}$ is real, and there is no imaginary component to γ_{2x}. However, in the z direction, β_{2z} is real and there is propagation in the z direction despite the unity magnitude of $r_\perp$. This propagation in the cladding results from the continuity of tangential electric fields.

As concerns magnetic fields in the cladding, from Eq. (1.4.11),

$$H_{x2} = -\frac{E_{y2}}{\eta_2}\sin\theta_2 = -\frac{E_{y2}}{\eta_2}\frac{n_1}{n_2}\sin\theta_1$$

$$H_{z2} = -\frac{E_{y2}}{\eta_2}\cos\theta_2 = -j\frac{E_{y2}}{\eta_2}\left[\left(\frac{n_1}{n_2}\right)^2\sin^2\theta_1 - 1\right]^{1/2}$$

In the cladding too, H_z is imaginary. As in the core, this produces real power flow in the z direction only

$$\overline{P}_2 = \frac{1}{2}\mathrm{Re}\left[\hat{y}\overline{E}_{y2}\times(\overline{H}^*_{x2} + \overline{H}^*_{z2})\right] = \frac{1}{2}\mathrm{Re}\,\hat{y}\overline{E}_{y2}$$

$$\times\left\{\hat{x}\frac{E^*_{y2}}{\eta_2}\frac{n_1}{n_2}\sin\theta_1 - \hat{z}j\frac{E_{y2}}{\eta_2}\left[\left(\frac{n_1}{n_2}\right)^2\sin^2\theta_1 - 1\right]^{1/2}\right\}$$

$$= \hat{z}\,\frac{1}{2}\,\frac{|E_{y2}|^2}{\eta_2}\,\frac{n_1}{n_2}\sin\theta_1$$

$$= \hat{z}\,\frac{A_{yI}^2}{\eta_2}\,\frac{n_1}{n_2}\sin\theta_1\,\cos^2(\phi/2)\,\exp\left\{2\beta_2 x\left[\left(\frac{n_1}{n_2}\right)^2\sin^2\theta_1 - 1^{1/2}\right]\right\}$$

Wave power flow in the cladding is in the z direction, while being attenuated exponentially in the x direction with increasing distance from the core/cladding boundary.

1.10

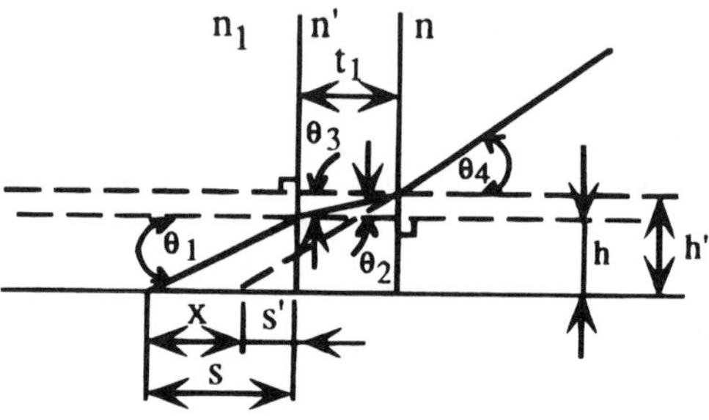

$$\frac{\sin\theta_1}{\sin\theta_2} = \frac{n'}{n_1}; \qquad \frac{\sin\theta_3}{\sin\theta_4} = \frac{n}{n'}$$

Since normal 1 and normal 2 and the optical axis are all normal to the glass plate, they are all parallel. Therefore, $\theta_2 = \theta_3$.

$$\therefore \frac{\sin\theta_3}{\sin\theta_4} = \frac{n}{n'} = \frac{\sin\theta_2}{\sin\theta_4} = \frac{n_1\sin\theta_1}{n\sin\theta_4}$$

$$\therefore \frac{\sin\theta_1}{\sin\theta_4} = \frac{n}{n_1} = \frac{1}{n_1} \qquad \text{for observing from air } (n = 1).$$

For paraxial rays, $\sin\theta_1 \approx \theta_1$ and $\sin\theta_4 \approx \theta_4$. Therefore, $\theta_1/\theta_4 \approx n/n_1 = 1/n_1$ where $\theta_4 > \theta_1$ because $n_1 > 1$. For a thin glass plate, $h \approx h'$ and $h = s\tan\theta_1 \approx h' = s'\tan\theta_4$.

$$\therefore s' \approx s\,\frac{\tan\theta_1}{\tan\theta_4}$$

$$x = s - s' \approx s\left(1 - \frac{\tan\theta_1}{\tan\theta_4}\right) \approx s\left(1 - \frac{\theta_1}{\theta_4}\right) = s\left(1 - \frac{n}{n_1}\right) = s\left(1 - \frac{1}{n_1}\right)$$

Ratio s'/s same as in Example 1.6.

1.11

$$d = AC\,\sin(\theta_1 - \theta_2)$$

$$AC = t/\cos\theta_2$$

$$\therefore d = \frac{t}{\cos\theta_2}\sin(\theta_1 - \theta_2)$$

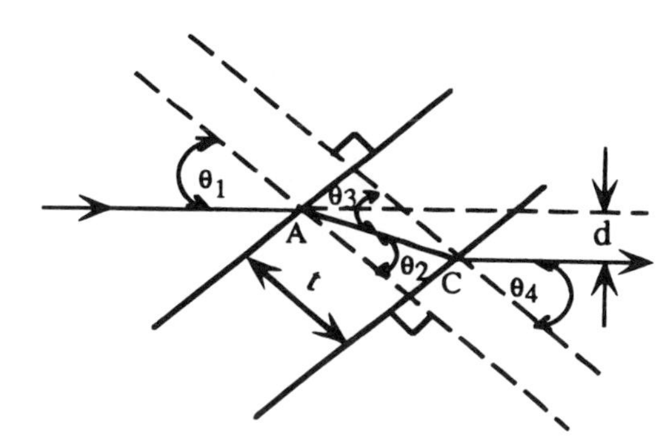

$$d = \frac{t(\sin\theta_1\cos\theta_2 - \cos\theta_1\sin\theta_2)}{\cos\theta_2}$$

$$= t\sin\theta_1 - t\sin\theta_2\cos\theta_1/\cos\theta_2$$

$$= t\sin\theta_1\left(1 - \frac{\sin\theta_2\cos\theta_1}{\sin\theta_1\cos\theta_2}\right)$$

$$= t\sin\theta_1\left(1 - \frac{n_1\cos\theta_1}{n_2\cos\theta_2}\right)$$

1.12 Starting from the right,

$$\eta_{T4} = \eta_4 = \sqrt{\frac{\mu}{\epsilon_{eq}}} = \sqrt{\frac{\mu}{\epsilon - j\sigma/\omega}} = \sqrt{\frac{j\mu\omega}{\sigma}} = 0.$$

As a result,

$$r_{34} = \frac{\eta_{T4}(-d_3) - \eta_3}{\eta_{T4}(-d_3) + \eta_3} = -1;$$

$$\gamma_3 = j\beta_3 = j\beta_0 = \frac{2\pi}{\lambda_0} = 2\pi \cdot 10^6 \text{ cycles}\cdot\text{m}^{-1}$$

$$\eta_{T3}(-d_3) = \eta_3\left\{\frac{\eta_{T4}(-d_3)\cos\beta_3\cdot 10^{-3} - j\eta_3\sin\beta_3\cdot 10^{-3}}{\eta_3\cos\beta_3\cdot 10^{-3} - j\eta_{T4}(-d_3)\sin\beta_3\cdot 10^{-3}}\right\}$$

$$= -j\eta_3\tan\beta_3\cdot 10^{-3} = -j\eta_0\tan 2\pi\cdot 10^3 = 0$$

$$r_{23} = \frac{\eta_{T3}(-d_2) - \eta_2}{\eta_{T3}(-d_2) + \eta_2} = -1; \qquad \gamma_2 = j\beta_2 = j\eta_2\beta_0 = j2\beta_0 = 4\pi\cdot 10^6$$

$$\eta_{T2}(-d_1) = \eta_2\left\{\frac{\eta_{T3}(-d_2)\cos\beta_2\cdot 10^{-3} - j\eta_2\sin\beta_2\cdot 10^{-3}}{\eta_2\cos\beta_2\cdot 10^{-3} - j\eta_{T3}(-d_2)\sin\beta_2\cdot 10^{-3}}\right\}$$

$$= -j\eta_2\tan\beta_2\cdot 10^{-3} = -j(\eta_0/2)\tan 4\pi\cdot 10^3 = 0$$

$$\therefore r_{12} = \frac{\eta_{T2}(-d_1) - \eta_1}{\eta_{T2}(-d_1) + \eta_1} = -1$$

As concerns the EM fields,

$$E_{x1}(z) = A_{i1}\left[e^{-j\beta_1(z+d_1)} + r_{12}e^{j\beta_1(z+d_1)}\right]$$

$$= A_{i1}\left[e^{-j\beta_o(z+d_1)} - e^{j\beta_o(z+d_1)}\right] = -j2A_{i1}\sin\beta_o(z + d_1)$$

$$H_{y1}(z) = \frac{A_{i1}}{\eta_1}\left[e^{-j\beta_o(j+d_1)} + e^{j\beta_o(z+d_1)}\right] = \frac{2A_{i1}}{\eta_o}\cos\beta_o(z + d_1)$$

where $d_1 = 2 \cdot 10^{-3}$m.

$$E_{x2}(z) = A_{i2}\left[e^{-j\beta_2(z+d_2)} + r_{23}e^{j\beta_2(z+d_2)}\right]$$

$$= A_{12}\left[e^{-j2\beta_o(z+d_2)} - e^{j2\beta_o(z+d_2)}\right] = -j2A_{i2}\sin 2\beta_o(z + d_2)$$

$$H_{y2}(z) = \frac{A_{i2}}{\eta_2}\left[e^{-j2\beta_o(z+d_2)} + e^{j2\beta_o(z+d_2)}\right] = \frac{2A_{i2}}{\eta_o/2}\cos 2\beta_o(z + d_2)$$

where $d_2 = 10^{-3}$ m, $\eta_2 = \sqrt{\mu_o/\epsilon_2} = \eta_0/2$

$$E_{x3}(z) = A_{i3}\left[e^{-j\beta_3 z} + r_{34}e^{j\beta_3 z}\right] = A_{i3}\left[e^{-j\beta_0 z} - e^{j\beta_0 z}\right] = -j2A_{i3}\sin\beta_o z$$

$$H_{y3}(z) = \frac{A_{i3}}{\eta_3}\left[e^{-j\beta_o z} + e^{j\beta_o z}\right] = \frac{2A_{i3}}{\eta_o}\cos\beta_o z$$

$$E_{x4}(z) = A_{i4}\left[e^{+j\beta_4 z} + r_{45}e^{-j\beta_4 z}\right]$$

$r_{45} = 0$ because there is no interface to the right of $z = 0$

$A_{i4} = 0$ because in a perfect conductor $E = 0$

$\therefore H_{y4}(z) = 0$

Boundary conditions:

$$E_{x1}(-d_1) = E_{x2}(-d_1) \rightarrow j_2A_{i1}\sin\beta_0(0) = 0 = -j2A_{i2}\sin\beta_2(d_2 - d_1)$$

$$\text{but } -j2A_{i2}\sin\beta_2(d_2 - d_1) = -j2A_{i2}\sin 4\pi \cdot 10^6(10^{-3} - 2 \cdot 10^{-3})$$
$$= j_2A_{i2}\sin(-4\pi \cdot 10^3) = 0$$

$\therefore$ no information relating A_{i1} to A_{i2}. However, by watching magnetic fields, all of which are tangential to the boundary,

$$H_{y1}(-d_1) = H_{y2}(-d_1) \rightarrow \frac{2A_{i1}}{\eta_o}\cos\beta_o(0) = \frac{2A_{i2}}{\eta_o/2}\cos 2\beta_o(d_2 - d_1)$$

$$\frac{2A_{i1}}{\eta_o}\cos(0) = \frac{4A_{i2}}{\eta_o}\cos 2\beta_0(-10^{-3}) = \frac{4A_{i2}}{\eta_o}$$

$$\therefore A_{i2} = A_{i1}/2$$

At $z = -d_2$,

$$E_{x2}(-d_2) = E_{x3}(-d_2)$$

As before, it turns out here that the values of both sine functions in the expressions for E_{x2} and E_{x3} are zero. Therefore, no information can be obtained here to relate A_{i3} to A_{i2}. However, by matching tangential magnetic fields,

$$H_{y2}(-d_2) = H_{y3}(-d_2) \rightarrow \frac{2A_{i2}}{\eta_o/2}\cos 2\beta_o(0) = \frac{2A_{i3}}{\eta_o}\cos\beta_o(-d_2)$$

$$\rightarrow \frac{2A_{i2}}{\eta_o/2} = \frac{2A_{i3}}{\eta_o} \rightarrow A_{i3} = 2A_{i2} = A_{i1}$$

As a result, in media 2 and 3 the fields are rewritten:

$$E_{x2}(z) = -j2A_{i2}\sin 2\beta_o(z + d_2) = -jA_{i1}\sin 2\beta_o(z + d_2)$$

$$H_{y2}(z) = \frac{2A_{i1}}{\eta_0}\cos 2\beta_o(z + d_2)$$

$$E_{x3}(z) = -j2A_{i1}\sin\beta_o z$$

$$H_{y3}(z) = \frac{2A_{i1}}{\eta_o}\cos\beta_0 z$$

Poynting vectors are

$$\overline{P}_1(z) = \frac{1}{2}\text{Re}\,\overline{E}_{x1}(z) \times \overline{H}^*_{y1}(z) = 0$$

$$\overline{P}_2(z) = \frac{1}{2}\text{Re}\,\overline{E}_{x2}(z) \times \overline{H}^*_{y2}(z) = 0$$

$$\overline{P}_3(z) = \frac{1}{2}\text{Re}\,\overline{E}_{x3}(z) \times \overline{H}^*_{y3}(z) = 0$$

Since reflection coefficients in each medium have a magnitude of unity, there is no time-averaged real power flow in any of the media.

1.13 Assume refractive indices are a function of x only. From Eq. (1.7.11), since $\partial n/\partial z = 0$

$$\frac{d}{ds}\left(n\frac{dz}{ds}\right) = 0$$

Integrating,

$$n\frac{dz}{ds} = \text{constant}$$

Letting s be in the direction of the light ray in Fig. 1.5, angles there in each medium are defined such that

$$\frac{dz}{ds} = \sin\theta, \qquad \text{so that } \frac{n\,dz}{ds} = n\sin\theta = \text{constant}$$

$$\therefore n_1\sin\theta_1 = n_2\sin\theta_2 = \text{constant}$$

1.14 For a homogeneous medium, $\nabla n = 0$ and the ray equation reduces to

$$n\frac{d^2\bar{r}}{ds^2} = 0$$

or

$$\frac{d\bar{r}}{ds} = \text{constant} = \bar{a}$$

Integrating,

$$\bar{r} = \bar{a}s + \bar{b}$$

which represents a straight line in three-dimensional space.

2

2.1 $$[S] = n[h^2 + (P - x)^2]^{1/2} + n[(h')^2 + x^2]^{1/2}$$

$$\frac{d\{S\}}{dx} = [h^2 + (P - x)^2]^{-1/2}(2)(P - x)(-1) + \frac{n}{2}[(h')^2 + x^2]^{-1/2}(2x) = 0$$

$$\therefore -\frac{h(P - x)}{\sqrt{h^2(P - x)^2}} + \frac{hx}{\sqrt{(h')^2 + x^2}} = 0$$

$$\therefore \frac{P - x}{\sqrt{h^2(P - x)^2}} = \frac{x}{\sqrt{(h')^2 + x^2}}$$

now

$$\frac{P - x}{\sqrt{h^2 + (P - x)^2}} = \sin\theta_1; \qquad \frac{x}{\sqrt{(h')^2 + x^2}} = \sin\theta_2$$

$$\therefore \theta_1 = \theta_2, \qquad \theta_1 < \frac{\pi}{2}$$

2.2 $$[S] \approx n_1(s^2 + r^2)^{1/2} + 2n_1 z + n_2(d - 2z) + n_1[(s')^2 + r^2]^{1/2}$$

$$\cong 2n_1 z + n_2(d - 2z) + n_1(s^2 + r^2)^{1/2} + n_1(s'^2 + r^2)^{1/2}$$

$$\cong n_2 d + 2z(n_1 - n_2) + n_1 s\left(1 + \frac{r^2}{2s^2}\right) + n_1 s'\left(1 + \frac{r^2}{2s'^2}\right)$$

$$\cong n_2 d + 2z(n_1 - n_2) + n_1(s + s') + \frac{n_1 r^2}{2}\left(\frac{1}{s} + \frac{1}{s'}\right)$$

Substituting $z(r)$ from Eq. (2.3.7) to this last equation yields:

$$[S] \approx n_2 d + \frac{1}{2f} r^2 \frac{n_1}{n_2 - n_1}(n_1 - n_2) + n_1(s + s') + \frac{n_1 r^2}{2}\left(\frac{1}{s} + \frac{1}{s'}\right)$$

$$\cong n_2 d - \frac{n_1 r^2}{2f} + n_1(s + s') + \frac{n_1 r^2}{2}\left(\frac{1}{s} + \frac{1}{s'}\right)$$

$$\cong n_2 d + n_1(s + s') + \frac{n_1 r^2}{2}\left(\frac{1}{s} + \frac{1}{s'} - \frac{1}{f}\right)$$

By Fermat's principle,

SOLUTIONS

$$\frac{d}{dr}[S] = n_1 r\left(\frac{1}{r} + \frac{1}{s'} - \frac{1}{f}\right) = 0 \Rightarrow \frac{1}{s} + \frac{1}{s'} = \frac{1}{f}$$

which is the Gaussian form of the classical thin lens formula.

2.3 $1/f = (n' - 1)(1/r_1 - 1/r_2) = (n' - 1)(1/15 - 1/20) > 0$
$\therefore$ positive (thicker at center than at edges).

2.4 From Eq. (2.4.9),

$$\frac{n}{s} + \frac{n''}{s''} = \frac{n' - n}{r_1} + \frac{n'' - n'}{r_2}$$

$$\frac{1.6}{s} + \frac{1.6}{s''} = \frac{(1.5 - 1.6)}{r_1} + \frac{(1.6 - 1.5)}{r_2}$$

$$= -(1.6 - 1.5)\left(\frac{1}{r_1} - \frac{1}{r_2}\right) = \frac{1}{f}$$

$\therefore$ negative focal length since $(1/r_1 - 1/r_2)$ is positive for a convex lens.

2.5 Use geometry in Fig. 2.7. Phase change from A to F is given by

$$\phi = \beta_o[2n_1 z + n_2(t - 2z) + n_1(f^2 + r^2)^{1/2}]$$

For paraxial rays incident on a thin lens this reduces to, using the binomial approximation,

$$\phi \approx \beta_o\left[2n_1 z + n_2(t - 2z) + n_1 f\left(1 + \frac{r^2}{2f^2}\right)\right]$$

$$= \beta_o\left[n_2 t + 2z(n_1 - n_2) + n_1 f + \frac{n_1 r^2}{2f}\right]$$

If all paraxial rays are in phase at F, then ϕ is not a function of r, that is, $d\phi/dr = 0$. In this case,

$$\frac{d\phi}{dr} = \beta_o\left[2(n_1 - n_2)\frac{dz}{dr} + \frac{n_1 r}{f}\right] = 0$$

$$\therefore \frac{dz}{dr} = -\frac{n_1 r}{2f(n_1 - n_2)} = \frac{r}{2f\left(\frac{n_2}{n_1} - 1\right)}$$

$$z(r) = \int_0^r \frac{r\,dr}{2f\left(\frac{n_2}{n_1} - 1\right)} = \frac{r^2}{4f\left(\frac{n_2}{n_1} - 1\right)}$$

This agrees with Eq. (2.3.7).

2.6 The lens separation d is adjusted to the following geometry. From similar triangles ABC and CDE,

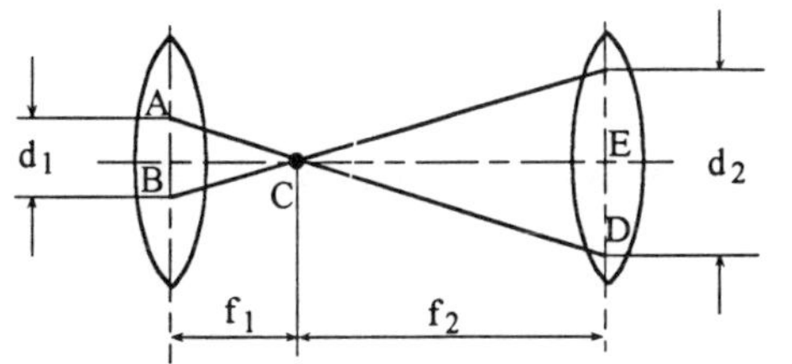

$$\frac{d_1/2}{f_1} = \frac{d_2/2}{f_2}$$

$$\therefore \frac{d_2}{d_1} = \frac{f_2}{f_1} \quad \text{and} \quad d = f_1 + f_2$$

2.7 By using the Newtonian form,

$$f^2 = zz' = 100 \text{ cm}^2 = 40z' \text{ cm}^2 \rightarrow z' = \frac{100}{40} = 2.5 \text{ cm}$$

The image is located 2.5 cm to the right of the back focal point, or 12.5 cm behind the lens if immersed in air. With regard to height, $m = -f/z = -10$ cm/40 cm $= -0.25$. Therefore, $x' = mx = (-0.25)(5)$ cm $= -1.25$ cm.

2.8
$$AT = f \tan\theta$$
$$AT = F'x = z' \tan\theta'$$

$$\therefore f \tan\theta = z' \tan\theta'$$

$$\therefore \frac{\tan\theta'}{\tan\theta} = \frac{f}{z'} = \frac{f}{f'/x} = -\left(\frac{f}{f''}\right)\left(\frac{x}{x'}\right) = \left(\frac{n}{n''}\right)\left(\frac{x}{x'}\right)$$

$$\therefore nx \tan\theta = n''x' \tan\theta' = H$$

2.9 From Eq. (2.7.4),

$$f = \frac{f_1 f_2}{f_1 + f_2 - d} = \frac{6(-9)}{6 - 9 - 4} \text{ cm} = \frac{-54}{-7} = 7.71 \text{ cm}$$

$$s'_{eq} = s'_2 + df/f_1 = s'_2 + \frac{4(7.71)}{6} = s'_2 + 5.14 \text{ cm}$$

Note that because of the negative focal length of L_2, the secondary principal plane is actually located behind L_2. The former is 5.14 cm behind L_1 while L_2 is 4 cm behind L_1. The negative focal length makes f_{eq} long.

$$s_{eq} = s_1 + df/f_2 = 16 + \frac{4(7.71)}{-9} = 16 - 3.43 \text{ cm} = 12.57 \text{ cm}$$

$$\frac{1}{s'_{eq}} = \frac{1}{f} - \frac{1}{s_{eq}} = \frac{1}{7.71} - \frac{1}{12.57} = 0.1297 - 0.796$$

$$= 0.0501 \rightarrow s'_{eq} = 19.96 \text{ cm} = s'_2 + 5.14 \text{ cm}$$

$$\therefore s'_2 = 19.96 - 5.14 = 14.8 \text{ cm}$$

$$m_{eq} = \frac{s'_{eq}}{s_{eq}} = -\frac{19.96 \text{ cm}}{12.57 \text{ cm}} = -1.59$$

$$\therefore x' = m_{eq}x = (-1.59)3 = 4.77 \text{ cm}$$

2.10 In Fig. 2.19, from similar triangles QMF and DHF

$$\frac{x}{x'} = \frac{z}{f}$$

From similar triangles C′H′F′ and Q′M′F′

$$\frac{x}{x'} = \frac{f'}{z'}$$

Therefore, $ff' = zz'$. This is the familiar Newtonian form. Since $z = s - f$, and $z' = s' - f'$, then $ff' = zz' = (s - f)(s' - f') = ss' - sf' - s'f + ff'$ $\therefore 0 = ss' - sf' - s'f = ss' - sf - s'f$ since $n = n''$. Divide by $ss'f$

$$\frac{1}{f} - \frac{1}{s'} - \frac{1}{s} = 0$$

2.11 The canopy acts as a lens of focal length

$$\frac{1}{f_1} = (n' - 1)\left(\frac{1}{r_1} - \frac{1}{r_2}\right)$$

$$= (1.49 - 1)\left(-\frac{1}{163} + \frac{1}{165}\right) \cdot 10^3 = -0.036438 \text{ m}$$

Therefore, the canopy and the zoom lens act as a two-lens combination. For 20-mm zoom lens focal length

$$\frac{1}{f_{20}} = \frac{1}{f_1} + \frac{1}{f_2} - \frac{d}{f_1 f_2} = -0.036438 + \frac{1}{0.02} - \left(-\frac{30}{20}\right)(-0.036438)$$

$$= 50 + \frac{1}{2}(0.036438) = 50.018219 \text{ m}$$

$$\therefore f_{20} = 19.993 \text{ mm}$$

For 150-mm zoom lens focal length

$$\frac{1}{f_{150}} = \frac{1}{f_1} + \frac{1}{f_2} - \frac{\text{d}}{f_1 f_2} = 0.036438 + \frac{1}{0.15} + \frac{30}{150}(0.036438)$$

$$f_{150} = 150.66 \text{ mm}$$

Now, magnification $= m_{eq20} = s'_{eq20}/s_{eq20}$ for 20-mm focal length and $m_{eq150} = s'_{eq150}/s_{eq150}$ for 150 mm focal length

$$S_{eq20} = s + \frac{df_{20}}{f_2}$$

$$= (1000 \text{ m} - 30 \text{ mm}) + \frac{30}{20}(19.993)\text{mm} = 999.999990$$

$$S_{eq150} = (5) + \frac{df_{150}}{f_2}$$

$$= (1000 \text{ m} - 30 \text{ mm}) + 30.13 \text{ mm} = 1000.00013 \text{ m}$$

$$\frac{1}{s'_{eq20}} = \frac{1}{f} - \frac{1}{s_{eq20}} = \frac{1}{0.0019993} - \frac{1}{999.999990} = 50.016506 \text{ m}^{-1}$$

$$s'_{eq20} = 19.993399$$

$$\therefore m_{eq20} = \frac{19.993399}{999.999990} = 1.9993399 \times 10^{-3}$$

Without canopy

$$s'^{-1}_{20} = \frac{1}{f_2} - \frac{1}{s} = \frac{1}{0.02} - \frac{1}{1000} = 50 - 0.001 = 49.9990$$

$$s'_{20} = 20.00004 \text{ mm}$$

$$m_{20} = \frac{s'_{20}}{s_{20}} = \frac{20.00004 \times 10^{-3}}{1000} = 2.00004 \times 10^{-5}$$

$$\frac{m_{eq20} - m_{20}}{m_{20}} = \frac{1.9993399 - 2.00004}{2.00004} = -0.0332\%$$

For 150-mm focal length

$$\frac{1}{s'_{eq150}} = \frac{1}{f_{150}} - \frac{1}{s_{eq150}} = \frac{1}{0.015066} - \frac{1}{1000.00013}$$

$$s'_{eq150} = 150.6827 \text{ mm}$$

$$m_{eq} = \frac{s'_{eq150}}{s_{eq150}} = \frac{150.6827 \times 10^{-3}}{1000.00013} = 1.5068 \times 10^{-4}$$

Without canopy

$$\frac{1}{s'} = \frac{1}{f} - \frac{1}{s} = \frac{1}{0.15} - \frac{1}{1000} \rightarrow s' = 0.1500225 \text{ m}$$

$$m = \frac{s'_{150}}{s_{150}} = \frac{1.500225}{1000} = 1.500225 \times 10^{-4}$$

$$\frac{m_{eq150} - m_{150}}{m_{150}} = \frac{1.5068 - 1.5002}{1.5002} = 0.44\%$$

To prevent distortion by the canopy, its lens action must be minimized. To let $f_1 \rightarrow \infty$, decrease canopy thickness and refractive index.

2.12 $r(L) = r_0 \cos(\sqrt{n_1/n_0}L) \approx r_0$

$$\Delta\phi = \frac{w}{c}\left[L\left(n_0 - \frac{1}{2}n_1 r_0^2\right) + (r_0^2 + f^2)^{1/2}\right]$$

$$\approx \frac{w}{c}\left[Ln_0 - \frac{L}{2}n_1 r_0^2 + f + \frac{r_0^2}{2f}\right]$$

$$\therefore \frac{L}{2}n_1 r_0^2 = \frac{r_0^2}{2f}$$

$$\therefore f = \frac{1}{n_1 L}$$

2.13 $1/s + 1/s' = -2/r = -2/-30 = 1/15$. For $s = 60$ cm, $1/s' = 1/15 - 1/60 = 3/60 = 1/20$

$\therefore s' = 20$ cm $> f = 15$ cm $\rightarrow$ real image (to left of mirror)

$m = -s'/s = 20/60 = -1/3$ (real and inverted image)
For $s = 10$ cm, $1/s' = 1/15 - 1/10 = -1/30$
$s' = -30$ cm (to right of mirror)
$m = -s'/s = -(-30/10) = 3$ (virtual and erect image)

2.14

$$m_a = \frac{s'}{s} = \frac{25 \text{ cm}}{f} - \frac{25 \text{ cm}}{s'} = \frac{25 \text{ cm}}{f} = \frac{25 \text{ cm}}{16 \text{ mm}} = 15.6$$

2.15 For objective, $1/s = 1/f - 1/s' = 1/16$ mm $- 1/160$ mm $\approx 1/16$ mm, magnification of the objective is

$$m_o = -\frac{s_1'}{s} = -\frac{160 \text{ mm}}{f_1} = -\frac{160 \text{ mm}}{16 \text{ mm}} = -10$$

Magnification of the eyepiece is

$$m_e = -\frac{s_2'}{s_2} = \frac{25.4 \text{ cm}}{f_2} = -\frac{25.4 \text{ cm}}{12.7 \text{ mm}} = -20$$

Overall magnification is $m_o m_e = 200$.

2.16 (a) $m_a = -f_o/f_e = -1.4/0.014 = -100$

(b) From above, the equivalent eyepiece focal length should be 7 mm.

$$\frac{1}{f_{eq}} = \frac{1}{0.007} = \frac{1}{f_e} + \frac{1}{f_B} - \frac{d}{f_e f_B} = \frac{f_B + f_e - d}{f_e f_B} = \frac{f_B + 0.014 - 0}{0.014 f_B}$$

where f_B = focal length of Barlow lens = 0.014 m.

2.17 Referring to Fig. 1.14, $r_o = 0$

$$\therefore r(z) = \sqrt{\frac{n_0}{n_1}}\, r_i' \sin \sqrt{\frac{n_1}{n_0}}\, z$$

$$r'(z) = r_i' \cos \sqrt{\frac{n_1}{n_o}}\, z$$

If fiber radius is a, maximum value of $r(z)$ is a. When $r(z) = a$, $r'(z) = 0$ so that the ray is bent toward the z axis. This occurs when

$$\sqrt{\frac{n_1}{n_0}}\, z = \frac{\pi}{2} \rightarrow z = \frac{\pi}{2}\sqrt{\frac{n_0}{n_1}}$$

$$\therefore r(z) = a = \sqrt{\frac{n_0}{n_1}}\, r_i' \sin\left(\sqrt{\frac{n_1}{n_0}}\frac{\pi}{2}\sqrt{\frac{n_0}{n_1}}\right) = \sqrt{\frac{n_0}{n_1}}\, r_i'$$

$$\therefore r_i' - a\sqrt{\frac{n_1}{n_0}}$$

let $2\theta = 2 \arctan r_i'$ be acceptance angle inside fiber, and let 2ϕ be acceptance angle outside fiber in air

$$(1)\ \sin\phi = n_0 \sin\theta = n_0 \sin \tan^{-1}\left(a\sqrt{\frac{n_1}{n_0}}\right)$$

From geometry, $(1)\ \sin\phi = \dfrac{n_0 a\sqrt{n_1/n_0}}{\sqrt{1 + a^2 n_1/n_0}}$

$$\approx n_0 a\sqrt{n_1/n_0}\left(1 - \frac{1}{2}a^2\frac{n_1}{n_0}\right) \text{ from the Taylor series expansion}$$

$$= a\sqrt{n_1/n_0}\left(n_0 - \frac{1}{2}a^2 n_1\right) \approx a\sqrt{n_1 n_0}$$

This is the numerical aperture at the fiber center.

$$\phi = \sin^{-1}(a\sqrt{n_1 n_0} \rightarrow 2\phi = 2\sin^{-1} a\sqrt{n_1 n_0}$$

2.18 $\alpha = 10^{-3}$ rad $= d/f$

$$\therefore f = \frac{d}{\alpha} = \frac{0.2 \times 10^{-3}\text{ m}}{10^{-3}\text{ rad}} = 20\text{ cm}$$

Minimum lens diameter D is such that all radiation incident on lens plane is collected by lens. That occurs when $(D/2)/f = \tan(\theta/2)$ where θ is beam width of LED. $D > 2f \tan \theta/2 = 2(20\text{ cm}) \tan(10°)$ $\therefore D > 7.05$ cm.

2.19 (a) The TV tube must be in the image plane to obtain sharp image. Here, $s' = f = 1.4$ m.

$$\therefore \alpha = \frac{d}{f} = \frac{0.01127\text{ m}}{1.4\text{ m}} = 9.0714\text{ mrad}$$

(b) $1/s' = 1/f - 1/s = 1/1.4 - 1/5000 = 0.714$ $\therefore s' = 1.4004$ m
From Eq. (2.13.11),

$$\begin{bmatrix} r_1 \\ r_1' \end{bmatrix} = \begin{bmatrix} 1 - \dfrac{s'}{f} & s + s'\left(1 - \dfrac{s}{f}\right) \\ -\dfrac{1}{f} & 1 - \dfrac{s}{f} \end{bmatrix} \begin{bmatrix} r_0 \\ r_0' \end{bmatrix}$$

Substituting

$$\begin{bmatrix} r_i \\ r_i' \end{bmatrix} = \begin{bmatrix} 1 - \dfrac{1.4004}{1.4} & 5000 + 1.4004\left(1 - \dfrac{5000}{1.4}\right) \\ -\dfrac{1}{1.4} & 1 - \dfrac{5000}{1.4} \end{bmatrix} \begin{bmatrix} r_0 \\ r_0' \end{bmatrix}$$

$$= \begin{bmatrix} -0.00027 & 0.04324 \\ -0.7143 & -3570.43 \end{bmatrix} \begin{bmatrix} r_0 \\ r_0' \end{bmatrix}$$

For rays proceeding from the top of the lens to the top of the TV camera in Fig. 2.25(c) (angle α_1 reverse arrows), $r_i = 1/4$ in. $= 0.00635$ m and

$$r_i' = -\frac{(D - d)/2}{s'} = -\frac{\left(\dfrac{0.089}{2} - 0.00635\right)}{1.4004} = -0.27243$$

Because of the reverse direction of the arrows in Fig. 2.23(c), r_i' is negative. Substituting into the matrix

$$0.00635 = -0.00027r_0 + 0.04324r_0'$$

$$-0.027243 = -0.7143r_0 - 3570.43r_0'$$

Consequently,

$$r_0 = \frac{\begin{vmatrix} 0.00635 & 0.04324 \\ -0.027243 & -3570.43 \end{vmatrix}}{\begin{vmatrix} -0.00027 & 0.04324 \\ -0.7143 & -3570.43 \end{vmatrix}} = -22.79 \text{ m}$$

$$r_0' = \frac{\begin{vmatrix} -0.00027 & 0.00635 \\ -0.7143 & -0.027243 \end{vmatrix}}{\begin{vmatrix} -0.00027 & 0.04324 \\ -0.7143 & -3570.43 \end{vmatrix}} = 0.00456$$

$\therefore\ \alpha_1$ = field of view (FOV) = $2 \tan^{-1}(0.00456) = 9.12$ mrad. This is slightly larger than for $s \to \infty$, as expected, since the TV camera tube is now at a distance $s' > f$.

If we consider chief rays from the lens center to the edges of the TV camera tube,

$$r_i' = \pm\frac{d - 0}{s'} = \pm\frac{0.00635}{1.4004} = \pm 4.545 \text{ mrad} = r_0'$$

Therefore, FOV here $= 2r_0' = 9.06$ mrad, which is slightly less than FOV formed by outside rays in the analysis above [α_1 in Fig. 2.25(c)] as also expected from Fig. 2.25(c). Hence, overall FOV is $\alpha_1 = 9.12$ mrad.

(c) If $s = 100$ m, then

$$\frac{1}{s'} = \frac{1}{1.4} - \frac{1}{100} = \frac{98.6}{140} = 0.704 \to s' = 1.42 \text{ m}$$

The matrix now becomes

$$\begin{bmatrix} r_i \\ r_i' \end{bmatrix} = \begin{bmatrix} -0.0142 & 0.0086 \\ -0.7143 & -70.43 \end{bmatrix} \begin{bmatrix} r_0 \\ r_0' \end{bmatrix}$$

For rays proceeding from top of lens to top of TV tube in Fig. 2.25(c) (reverse the arrows), again $r_i = 1/4$ in. $= 0.00635$ m, but

$$r_i' = \frac{-(D - d)/2}{s'} = -\frac{\left(\frac{0.089}{2} - 0.00635\right)}{1.42} = -0.0269$$

Substituting into the matrix,

$$0.00635 = -0.0142 r_0 + 0.0086 r_0'$$

$$-0.0269 = -0.7143 r_0 - 70.43 r_0'$$

$$\therefore r_0' = \frac{\begin{vmatrix} -0.0142 & 0.00635 \\ -0.7143 & -0.0269 \end{vmatrix}}{\begin{vmatrix} -0.0142 & 0.0086 \\ -0.7143 & -70.43 \end{vmatrix}} = \frac{0.000382 + 0.00454}{1.00012 + 0.006143} = 0.00489$$

This is larger than the previous result because s' has increased. The corresponding FOV $= \alpha_1 = 2\tan^{-1}(0.00489) = 9.78$ mrad, which is larger than before.

If we consider chief rays from the lens center to the TV tube edges,

$$r_i = \frac{d - 0}{s'} = \pm\frac{0.00635}{1.42} = 0.00447$$

Corresponding FOV $= 8.94$ mrad. Therefore, overall $\alpha_1 = 9.78$ mrad, corresponding to the analysis in Fig. 2.25(c).

Conclusion: This example shows that as s' increases (corresponding to shorter object ranges), FOV also increases.

Minimum FOV occurs when $s' = f$.

2.20

$$M_{sf} = \begin{bmatrix} \cos\sqrt{\frac{n_1}{n_0}}\,L & \sqrt{\frac{n_0}{n_1}}\sin\sqrt{\frac{n_1}{n_0}}\,L \\ -\sqrt{\frac{n_1}{n_0}}\sin\sqrt{\frac{n_1}{n_0}}\,L & \cos\sqrt{\frac{n_1}{n_0}}\,L \end{bmatrix}$$

2.21 Consider propagation from object plane to lens L_1, to lens L_2, and from there to the image plane. The matrix from object plane to L_2 is obtained from Eq. (2.13.11) where lens separation d is substituted for s' and focal length f_1 for L_1 is used, that is,

$$M_{0-L_2} = \begin{bmatrix} 1 - \frac{d}{f_1} & s + d\left(1 - \frac{s}{f_1}\right) \\ -\frac{1}{f_1} & 1 - \frac{s}{f_1} \end{bmatrix}$$

The matrix for input to output of lens L_2 is $[M_2]$ in Eq. (2.13.4), where f_2 is used for f, that is,

$$[M_2] = \begin{bmatrix} 1 & 0 \\ -\frac{1}{f_2} & 1 \end{bmatrix}$$

The matrix for propagation from the output of L_2 to the image plane is M_1 in Eq. (2.13.5), where s' is used in place of L, that is,

$$M_{L_2-i} = \begin{bmatrix} 1 & s' \\ 0 & 1 \end{bmatrix}$$

Accordingly, the overall matrix to describe propagation from object to image plane is

$$[M_{0-i}] = [M_{L_2-i}][M_2][M_{0-L_2}]$$

$$\left[M_{L_2-i}\right][M_2] = \begin{bmatrix} 1 - \dfrac{s'}{f_2} & s' \\ -\dfrac{1}{f_2} & 1 \end{bmatrix}$$

$$[M_{0-i}] = \begin{bmatrix} 1 - \dfrac{s'}{f_2} & s' \\ -\dfrac{1}{f_2} & 1 \end{bmatrix}\begin{bmatrix} 1 - \dfrac{d}{f_1} & s + d\left(1 - \dfrac{s}{f_1}\right) \\ -\dfrac{1}{f_1} & 1 - \dfrac{s}{f_1} \end{bmatrix}$$

$$= \begin{bmatrix} 1 - \dfrac{d}{f_1} - s'\left(\dfrac{1}{f_1} + \dfrac{1}{f_2} - \dfrac{d}{f_1 f_2}\right) & s + d\left(1 - \dfrac{1}{f_1}\right) + s'\left(1 - \dfrac{s}{f_1} - \dfrac{s}{f_2} + \dfrac{sd}{f_1 f_2}\right) \\ -\dfrac{1}{f_1} - \dfrac{1}{f_2} + \dfrac{d}{f_1 f_2} & 1 - \dfrac{d}{f_2} - \dfrac{s}{f_1} - \dfrac{s'}{f_2} + \dfrac{sd}{f_1 f_2} \end{bmatrix}$$

If this is compared to Eq. (2.13.11) for a single lens, then from the c_{21} term (lower left corner), using f for equivalent focal length,

$$\frac{1}{f} = \frac{1}{f_1} + \frac{1}{f_2} - \frac{d}{f_1 f_2}$$

The c_{11} term (upper left corner), when compared to matrix (2.13.11) implies, using s'_{eq} for equivalent image distance,

$$1 - \frac{s'_{eq}}{f} = 1 - \frac{d}{f_1} - s'\left(\frac{1}{f_1} + \frac{1}{f_2} - \frac{d}{f_1 f_2}\right) = 1 - \frac{d}{f_1} - \frac{s'}{f}$$

$$\therefore \frac{s'_{eq}}{f} = \frac{s'}{f} + \frac{d}{f_1} = \frac{1}{f}\left(s' + d\frac{f}{f_1}\right)$$

or

$$s'_{eq} = s' + \frac{df}{f_1}$$

The c_{22} term (lower right corner) when compared to Eq. (2.13.11) indicates

$$1 - \frac{s_{eq}}{f} = 1 - \frac{d}{f_2} - \frac{s}{f_1} - \frac{s}{f_2} + \frac{sd}{f_1 f_2} = 1 - \frac{d}{f_2} - \frac{s}{f}$$

or

$$\frac{s_{eq}}{f} = \frac{s}{f} + \frac{d}{f_2} = \frac{1}{f}\left(s + \frac{df}{f_2}\right)$$

$$\therefore s_{eq} = s + \frac{df}{f_2}$$

As a result, $\{M_{0-i}\}$ is reduced to a form similar to that of Eq. (2.13.11) for object to image plane for a single lens, that is, for a two-lens combination

$$\{M_{0-i}\} = \begin{bmatrix} 1 - \frac{s'_{eq}}{f} & s_{eq} + s'_{eq}\left(1 - \frac{s_{eq}}{f}\right) \\ -\frac{1}{f} & 1 - \frac{s_{eq}}{f} \end{bmatrix}$$

where f is equivalent focal length for the two-lens combination as derived in Eq. (2.7.4).

3.1 Area of source $= (\pi/4)d^2 = (\pi/4)(0.2)^2 = (0.01)\pi\ \text{cm}^2$ where d = mean source diameter. Equivalent solid angle $\Omega_L = \pi \sin^2\theta = 1.57$ sr. Therefore,

$$P = \pi N\, dA \sin^2\theta = \pi\left(20\, \frac{\text{W}}{\text{cm}^2\cdot\text{sr}}\right)(0.01\pi)\ \text{cm}^2 \sin^2\left(\frac{\pi}{4}\right) \text{sr}$$

$$= 0.2\pi^2\left(\frac{1}{\sqrt{2}}\right)^2 = 0.986\ \text{W}$$

3.2 $\theta = \arctan D/2s$ where D = lens diameter and s is object distance

$$= \arctan \frac{3\ \text{cm}}{2(10)\ \text{cm}} = 0.15 \rightarrow \theta = 0.149\ \text{rad}$$

Therefore, $\Omega_L = \pi \sin^2\theta = 0.069$ sr.

$$P = N\, dAW = \pi N\, dA \sin^2\theta$$

$$= \left(20\, \frac{\text{W}}{\text{cm}^2\cdot\text{sr}}\right)(0.01\pi)\ \text{cm}^2\ (0.069\ \text{sr}) = 0.043\ \text{W}$$

3.3 For a specular source, Eq. (3.2.10) becomes $P(\theta) = 2\pi N\, dA \int_0^\theta \sin\theta\, d\theta = 2\pi N\, dA(1 - \cos\theta)$. For radiation into a hemisphere ($\theta = \pi/2$), $P = 2\pi N\cdot dA$ and $W = P/dA = 2\pi N$. From above, in general, $\Omega_s = 2\pi \int_0^\theta \sin\theta\, d\theta = 2\pi(1 - \cos\theta)$. For hemisphere, $\Omega_s = 2\pi$.

3.4 For a specular source, $\Omega_s = 2\pi(1 - \cos\theta)$ where, for Exercise 3.2, $\theta = \arctan 0.15 = 0.149$ rad. Therefore, $\Omega_s = 2\pi(1 - 0.989) = 0.07$ sr. The solid angles are almost identical. Here, $P_s = N dA\Omega_s = 20(0.01\pi)(0.07) = 0.043$ W when rounded off. It can be seen that for radiation of small beam width there is little difference between specular and Lambertian sources.
For Exercise 3.1, $\theta = \pi/4$ (much larger) and $\Omega_s = 2\pi[1 - \cos(\pi/4)] = 1.84$ sr. $P_s = N\, dA\Omega_s = 20(0.01\pi)(1.84) = 1.156$ W.

3.5 Power incident on dA_r is $P = N\, dA_s\, (\cos\theta)\Omega$

$$\Omega = \frac{dA_r \cos\theta}{r^2}.$$

Since $r = z/\cos\theta$, $\Omega = dA_r \cos^3\theta/z^2$. Therefore,

$$P = \frac{N\, dA\, dA_r \cos^4\theta}{z^2} \rightarrow H = \frac{P}{dA_r} = \frac{N\, dA_s \cos^4\theta}{z^2}$$

S O L U T I O N S

3.6 Source radiance is $N = P/(\pi\, dA)$. Power P_r incident on the detector is $N\, dA \cos\theta \Omega \rho_{eq}$ where ρ_{eq} is equivalent reflection coefficient as manifested by an infinite number of reflections by the interior surface. Accordingly, $\rho_{eq} = \rho + \rho^2 + \rho^3 + \cdots = \rho/(1 - \rho)$. From the figure, $dA_p/d^2 = dA_r \cos\theta/d^2$. But by bisecting d, $\cos\theta = d/2R$. Substituting for d, $\Omega = dA_r/(4R^2 \cos\theta)$. Substituting into expression for P_r, one obtains

$$P_r = \frac{N\, dA \cos\theta\, dA_r}{4R^2 \cos\theta}\left(\frac{\rho}{1 - \rho}\right) = \frac{P\, dA\, dA_r \rho}{\pi dA\ 4R^2(1 - \rho)}$$

$$= \frac{P\, dA_r \rho}{4\pi R^2(1 - \rho)}$$

Irradiance on detector is

$$\frac{P_r}{dA_r} = \frac{P\rho}{4\pi R^2(1 - \rho)}$$

3.7 Radiance of light source is $N = P/(A\Omega)$, where A is emitting area of light source of radiant power P.

(1) $A_s \cos\theta/z^2 < \Omega$ (Illuminated area > scene area): Power incident on the scene is $P_s = N(\cos\theta_2)(A \cos\theta_2)(A_s \cos\theta/z^2)$, where θ_2 is angle of light source inclination with respect to scene, A is source emission area, and N is source radiance. Note $\cos\theta_2$ is unity. The scene now acts as a Lambertian source of mean radiance

$$N_s = \frac{\rho P_s}{A_s \pi} = \frac{\rho N A \cos\theta}{\pi z^2} = \frac{\rho P \cos\theta}{\Omega \pi z^2} \text{ since } N = P/(A\Omega).$$

Power incident on the imager is

$$P_r = N_s \cos\theta (A_s \cos\theta)(A_r/z^2) = \frac{\rho P \cos^3\theta A_s A_r}{\Omega \pi z^4}$$

where A_r is receiving area of imager.

$P_r = N_s \cos\theta(A_s \cos\theta)(A_r/z^2) = \rho P \cos^3\theta\ A_s A_r/\Omega\pi z^4$ where A_r is the receiving area of the imager.

Irradiance falls off with z^4 instead of z^2 because propagation is round-trip. Only part of light beam is intercepted by scene, and only part of scene-reflected beam is intercepted by imager. Mean irradiance is $P_r/A_r = \rho P \cos^3 A_s/\Omega\pi z^4$.

(2) $A_s \cos\theta/z^2 > \Omega$; scene area is larger than the cross-sectional area of the beam; only part of the scene is illuminated. Therefore, all radiant power reaching the scene is incident on the scene. Power incident on scene is $P_s = NA(\cos\theta_2)\Omega = P$. Scene acts as Lambertian reflector of radiance $N_s = \rho P_s/A_s\pi = \rho P/A_s\pi$. Power incident on imager is $P_r = N_s \cos\theta(A_s \cos\theta)A_r/z^2 = \rho P A_r \cos^2\theta/\pi z^2$. The inverse square law holds because all power transmitted by source illuminates scene and therefore the inverse square law is relevant only for radiant power transfer from scene to imager, where only part of this

reflected radiant power is intercepted by imager according to solid angle A_r/z^2 subtended by imager. Mean irradiance is $P_r/A_r = \rho P \cos^2\theta/\pi z^2$.

3.8 If the light source is inclined at angle ϕ_s, projected source area is diminished to $A \cos\phi_s$. Therefore, $N = P/(A \cos\phi_s \Omega)$. For case 1 ($A_s \cos\theta/z^2 < \Omega$), $P_s = N \cos\phi_s(A \cos\phi_s)(A_s \cos\theta)/z^2$;

$$N_s = \frac{\rho P_s}{A_s \pi} = \frac{\rho P \cos\phi_s \cos\theta}{\pi \Omega z^2}$$

$$P_r = N_s \cos\theta(A_s \cos\theta)(A_r \cos\phi_i)/z^2$$

$$= \frac{(\rho P A_r A_s \cos^3\theta \cos\phi_s \cos\phi_i)}{(\pi \Omega z^2)}$$

Irradiance on the imager is

$$\frac{P_r}{A_r \cos\phi_i} = \frac{\rho P A_s \cos^3\theta \cos\phi_s}{\pi \Omega z^4}$$

For case 2 ($A_s \cos\theta/z^2 > \Omega$), $P_s = N \cos\phi_s(A \cos\phi_s)\Omega$ where N is as above. Therefore, $P_s = P \cos\phi_s$ and

$$N_s = \frac{\rho P_s}{A_s \pi} = \frac{\rho P \cos\phi_s}{A_s \pi}$$

$$P_r = N_s \cos\theta(A_s \cos\theta)(A_r \cos\phi_i)/z^2 = \frac{\rho P \cos^2\theta \cos\phi_s \cos\phi_i}{\pi z^2}$$

Average irradiance at imager is

$$\frac{P_r}{A_r \cos\phi_i} = \frac{\rho P \cos^2\theta \cos\phi_s}{\pi z^2}$$

Note that in both cases P_r depends on tilt (ϕ_i) of imager, but irradiance on imager does not.

3.9 Solid angle of the laser beam divergence is approximately

$$\Omega_r = \pi \sin^2(2\alpha) \approx 4\pi\alpha^2$$

Therefore, radiance of laser is $N = P/(4\pi\alpha^2 A)$ where A is area of transmitter. Power incident on retroflector is $P_{rr} = NA \cdot A_r/z^2 = PA_r/(4\pi\alpha^2 z^2)$ for case (1) and $P_{rr} = NA\Omega$ for case (2). Radiance retroflected *converges* rather than diverges ideally back to the laser transmitting aperture. Because of tolerance ϕ, the solid angle beam width of retroflected beam is $\pi\phi^2$. Radiance of retroflector is $N_{rr} = \rho P_{rr}/[A_r(\pi\phi^2)]$. Power of retroflected beam in imager plane is $P_r = N_{rr}A_r(\pi\phi^2) = \rho P_{rr}$. Irradiance there is $H_r = P_r/(\pi\phi^2)z^2 = \rho P_{rr}/[\pi\phi^2 z^2]$ where the denominator is cross-sectional area of retroflected beam at imager. For case (1), $H_r = \rho NAA_r/(\pi\phi^2 z^4) = \rho PA_r/(4\pi^2\alpha^2 z^4\phi^2)$; for case (2), $H_r = \rho P/(\pi\phi^2 z^2)$.

3.10 P_{rr} same as previous. However, laser beam of planar beam divergence 2α incident on mirror is reflected with same divergence of $\Omega = 4\pi\alpha^2$. Radiance of specular target therefore is $N_s = \rho P_{rr}/(A_r 4\pi\alpha^2)$. For case (1), $H_r = \rho PA_r/(16\pi^2\alpha^4 z^4)$ and, for case 2, $H_r = \rho P/(4\pi\alpha^2 z^2)$.

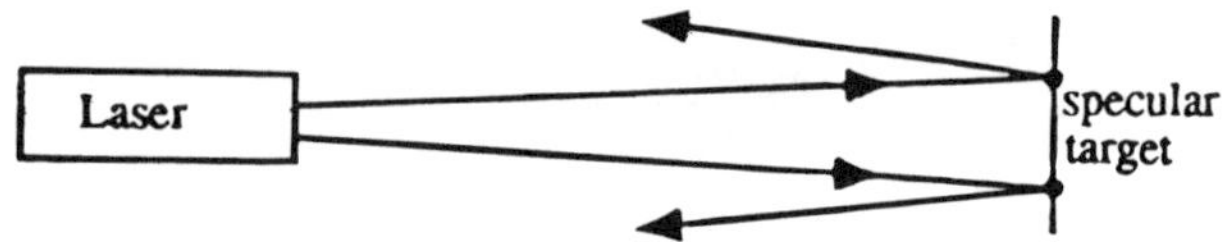

3.11 N_L = radiance of laser $= P\tau_t/A_L\Omega_L$. Here $A_L = (\pi/4)D^2$ and $\Omega_L = \pi(1.22\lambda/D)^2$, and τ_t = transmitter optics transmission. Power reaching satellite is $P_s = N_L A_L(A_s\tau_t/z^2)$. (Note that plugging in numbers reveals $A_s/z^2 < \Omega_L$, where A_s is satellite area.) Satellite radiance is $N_s = \rho P_s/\Omega A_s = \rho N_L A_L/(\pi z^2) = \rho P/\pi\Omega_L z^2$. Received power is $P_r = \tau_r N_s A_s(A_r/z^2) = \tau_r PA_s A_r\tau_t/(\pi\Omega_L z^4)$. Received pulse energy $E_r = P_r\tau$, where τ is pulse width. Therefore $E_r = \tau_r P\tau A_s A_r\tau_t/\pi\Omega_L z^4 \geq 1000\, h\, c/\lambda$. Therefore,

$$P_\tau \geq \frac{1000h\dfrac{c}{\lambda}\cdot\pi(1.22)^2 z^4}{\rho\tau_r\tau_t A_s A_r D^2}$$

$$\geq \frac{10^3(6.63\times10^{34})(3\times10^8)\pi(1.22)^2(1.06)^2 10^{12}(5\times10^5)^4}{(1.06\times10^{-6})0.7(0.95)^2(1)(\pi)(1)^2(2\times10^{-3})^2} = 2.47\text{ J}$$

3.12 Eq. (3.4.4):

$$H_i = \tau\pi N\left(\frac{n'}{n}\right)^2\sin^2\theta'_{\text{max}};\qquad \sin\theta'_{\text{max}} = \frac{D/2}{\sqrt{(D/2)^2+(s')^2}}$$

$$s' = \frac{sf}{s-f} = \frac{10(9)}{10-9} = 90\text{ cm.}\qquad \sin\theta'_{\text{max}} = \frac{1.5}{\sqrt{1.5^2+90^2}} = 0.0167$$

$$H_i = 0.9\pi(20)\left(\frac{1}{1}\right)^2(0.0167)^2 = 0.0157\text{ W}\cdot\text{cm}^{-2}$$

Eq. (3.4.5):

$$H_i = \tau W\left(\frac{n'}{n}\right)^2\left(\frac{D^2}{4(s')^2}\right) = 0.9\pi(20)\left(\frac{1}{1}\right)^2\frac{3^2}{4(90)^2} = 0.0157\text{ W}\cdot\text{cm}^{-2}$$

Eq. (3.4.7):

$$H_i = \frac{\tau W}{4F^2(1+m)^2}\left(\frac{n'}{n}\right)^2; \qquad F = \frac{f}{D} = \frac{9}{3} = 3; \qquad m = \frac{s'}{s} = \frac{90}{10} = 9$$

$$H_i = \frac{0.9\pi(20)}{4(3)^2(1+9)^2}\left(\frac{1}{1}\right)^2 = 0.0157\ \text{W}\cdot\text{cm}^{-2}$$

All of the above equations refer to an object not very distant. The results are in very close agreement here because s' is rather large compared to focal length. Lateral magnification is rather high, causing the image to be of relatively large size. The radiant power collected by the lens is spread over the large image, causing image irradiance to be low.

Eq. (3.4.8):

$$H_i = \frac{\tau W}{4F^2}\left(\frac{n'}{n}\right)^2 = \frac{0.9\pi(20)}{4\cdot 3^2} = 1.57\ \text{W}\cdot\text{cm}^{-2}$$

Eq. (3.4.9):

$$H_i = \frac{\tau W}{4F^2+1}\left(\frac{n'}{n}\right)^2 = \frac{0.9\pi(20)}{4(3)^2+1} = 1.53\ \text{W}\cdot\text{cm}^{-2}$$

These last two equations refer to an object infinitely distant. As a consequence the image is assumed to be at the focal point of the lens. Although there is a decrease in power collected by the lens, image size is virtually microscopic, thus causing image irradiance to be much higher. Equations (3.4.8) and (3.4.9) are thus *not* relevant to the problem at hand since in this exercise $s = 10$ cm and not infinity, and $s' = 90$ cm and not $f = 9$ cm.

3.13 If $s = 90$ cm, then $s' = sf/(s-f) = 90(9)/(90-9) = 10$ cm.

$$m = \frac{s'}{s} = \frac{10}{90} = 0.11; \qquad \left(\frac{n}{n'}\right) = 1; \qquad \sin^2\theta'_{max} = \frac{1.5^2}{1.5^2+10^2} = 0.022$$

Eq. (3.4.4):

$$H_i = \tau\pi N\left(\frac{n'}{n}\right)^2 \sin^2\theta'_{max} = 0.9\pi(20)1^2(0.022) = \frac{1.24}{\text{W}\cdot\text{cm}^{-2}}$$

Eq. (3.4.5):

$$H_i = \tau W\left(\frac{n'}{n}\right)^2 \frac{D^2}{4(s')^2} = 0.9\pi(20)1^2\,\frac{3^2}{4(10)^2} = 1.27\ \text{W}\cdot\text{cm}^{-2}$$

Eq. (3.4.7):

$$H_i = \frac{\tau W(n'/n)^2}{4F^2(1+m)^2} = \frac{0.9\pi(20)1^2}{4(3)^2(1+0.11)^2} = 1.27 \text{ W}\cdot\text{cm}^{-2}$$

Here too there is little difference despite the fact that the optical path assumed by the last two is different from that used for Eq. (3.4.4). Equations (3.4.8) and (3.4.9) yield identical calculations as in previous exercises since they assume $s \to \infty$ and are thus unaffected by the change in finite s here. Therefore $H_i = 1.57 \text{ W}\cdot\text{cm}^{-2}$ for Eq. (3.4.8) and $1.53 \text{ W}\cdot\text{cm}^2$ for Eq. (3.4.9). Eqs. (3.4.4), (3.4.5), and (3.4.7) consider the real value of s' which is 10 cm. Only when $s \to \infty$ does $s' \to f = 9$ cm, and only then do approximations (3.4.8) and (3.4.9) hold.

3.14 Eq. (3.4.15):

$$H_2 = W\left(\frac{n'}{n}\right)^2 \frac{D^2}{D^2 + 4z^2} = W\left(\frac{n'}{n}\right)^2 \left.\frac{D^2}{D^2 + 4\left(\frac{fD}{2r_s}\right)^2}\right|_{z=z_c}$$

$$= W\left(\frac{n'}{n}\right)^2 \frac{(r_s)^2}{r_s^2 + f^2} = \text{Eq. (3.4.14)}$$

Answer is yes.

3.15 From Eq. (3.4.16), $H_2 = W_s D^2/(4z^2)$ where W_s is source emittance. Now, $W = \rho H_2 = \rho W_s D^2/(4z^2)$. From Eq. (3.4.4),

$$H_i = \tau W \sin^2\theta'_{max} = \tau\rho W_s D^2 \sin^2\theta'_{max}/(4z^2)$$

$$= \frac{\tau\rho P D^2 \sin^2\theta'_{max}}{4z^2 A \cos\phi_s} \text{ for both cases.}$$

3.16 $H = \int N \, dA\Omega/A_r$ where A_r is cross-sectional area of incident beam and distance from source to attenuating medium is R. At a distance $R + z$ from the source, that is, in the attenuating medium, the observed spectral irradiance is $H' = N' \, dA\Omega'$ where $\Omega' =$ unit area$/(R + z)^2$. If

$$H' = H_o \exp\left[-\int_0^z \alpha(z')\, dz'\right]$$

then

$$N' \, dA\Omega' = N \, dA\Omega \exp\left[-\int_0^z \alpha(z')\, dz'\right]$$

But

$$\frac{\Omega}{\Omega'} = \frac{1/R^2}{1/(R+z)^2} = \left(\frac{R+z}{R}\right)^2 = 1 + \frac{2z}{R} + \frac{z^2}{R^2} = \left(1 + \frac{z}{R}\right)^2$$

Therefore, only if $z/R << 1$ does

$$N' = N \exp\left[-\int_0^z \alpha(z')\,dz'\right]$$

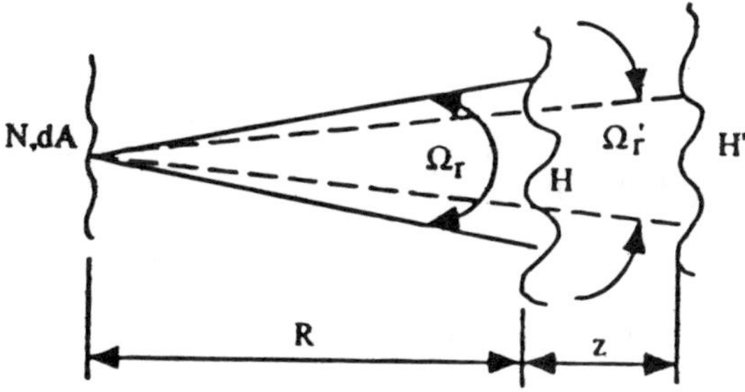

4

4.1 $W_\lambda = (2\pi hc^2/\lambda^5)\exp(hc/\lambda kT - 1)^{-1}$

$$\frac{dW_\lambda}{d\lambda} = 0 = 2\pi hc^2\{-5\lambda^{-6}\exp(hc/\lambda kT - 1)^{-1} + \lambda^{-5}\frac{hc}{\lambda^2 kT}\exp(hc/\lambda kT)\}$$

$$\times [\exp(hc/\lambda kT) - 1]^{-2}$$

Therefore,

$$5 = \frac{hc}{\lambda kT}\frac{\exp(hc/\lambda kT)}{\exp(hc/\lambda kT) - 1} \approx \frac{hc}{\lambda kT}$$

since at the peak of thermal emittance $hc >> \lambda kT$. Therefore,

$$\lambda_m T \cong \frac{hc}{5k} = \frac{(6.63 \times 10^{-34})(3 \times 10^8)}{5(1.38 \times 10^{-23})} = 2.883 \times 10^3\ \mu\text{m}\cdot\text{K}$$

A more exact solution can be obtained graphically or by using Taylor series expansions for the exponents until the solution converges to $\lambda_m T = 2897$ μm·K.

4.2 $\lambda_m = 2.897 \times 10^3/(5.9 \times 10^3) = 0.497\ \mu\text{m} \rightarrow$ sun is blue-green primarily

4.3 $W = \sigma T^4 = 5.67 \times 10^{-9}\,(5.9 \times 10^3)^4 = 6.9 \times 10^7\ \text{W}\cdot\text{m}^{-2}$

4.4 $N_\lambda = 2hc^2/\lambda^5\,[\exp(hc/\lambda kT - 1]^{-1}$

$$\lambda_m = \frac{2.897 \times 10^3}{T}\ \mu\text{m} = \frac{2.897 \times 10^{-3}}{T}\ \text{m}$$

$$N_\lambda(\lambda_m) = \frac{2hc^2}{\lambda_m^5}[\exp(hc/2.897 \times 10^{-3}k) - 1]^{-1}$$

$$\frac{1}{2}N_\lambda(\lambda_m) = \frac{hc^2}{\lambda_m^5}[\exp(hc/2.8897 \times 10^{-3}k) - 1]^{-1}$$

$$= \frac{6.63 \times 10^{-34}(3 \times 10^8)^2}{\lambda_m^5}$$

$$\times \left\{\exp\left[\frac{6.63 \times 10^{-34}(3 \times 10^8)}{2.897 \times 10^{-3}(1.38 \times 10^{-23})}\right] - 1\right\}^{-1}$$

$$= \frac{5.963 \times 10^{-17}}{\lambda_m^5}(143.03)^{-1} = \frac{4.17 \times 10^{-19}}{\lambda_m^5}$$

$$= \frac{4.17 \times 10^{-19} T^5}{(2.897 \times 10^{-3})^5}$$

Let $\lambda_{1/2}$ be wavelengths where $N_\lambda = 1/2 N_\lambda(\lambda_m)$:

$$N_\lambda(\lambda_{1/2}) = \frac{4.17 \times 10^{-19} T^5}{(2.897 \times 10^{-3})^5} = \frac{5.963 \times 10^{-17}}{\lambda_{1/2}^5}$$

$$\times \left\{ \left[\frac{(6.63 \times 10^{-34}) 3 \times 10^8}{\lambda_{1/2} T (1.38 \times 10^{-23})} \right] - 1 \right\}^{-1}$$

$$= \frac{5.963 \times 10^{-17}}{\lambda_{1/2}^5} \left[\exp \frac{1.44 \times 10^{-2}}{\lambda_{1/2} T} - 1 \right]^{-1}$$

$$\lambda_{1/2}^5 T^5 = \frac{(2.897 \times 10^{-3})^5 5.963 \times 10^{-17}}{4.17 \times 10^{-19}}$$

$$\times \left[\exp\left(\frac{1.44 \times 10^{-2}}{\lambda_{1/2} T} \right) - 1 \right]^{-1}$$

$$= (2.987 \times 10^{-3})^5 \frac{\left[\exp\left(\frac{1.44 \times 10^{-2}}{\lambda_m T} \right) - 1 \right]}{\left[\exp\left(\frac{1.44 \times 10^{-2}}{\lambda_{1/2} T} \right) - 1 \right]}$$

$$= \frac{(2.897 \times 10^{-3})^5 (143.03)}{\exp\left(\frac{1.44 \times 10^{-2}}{\lambda_{1/2} T} \right) - 1} = \frac{2.92 \times 10^{-11}}{\exp\left(\frac{1.44 \times 10^{-2}}{\lambda_{1/2} T} \right) - 1}$$

where λ is in meters.

Check the validity:

(1) $\lambda_{\text{short}} T = 1.8 \times 10^{-3}$ m·K

$$\lambda_{\text{short}}^5 T^5 = \frac{2.92 \times 10^{-11}}{\exp\left(\frac{1.44 \times 10^{-11}}{1.8 \times 10^{-3}} \right) - 1} = \frac{2.92 \times 10^{-11}}{e^8 - 1} = 9.80 \times 10^{-15}$$

$$\lambda_{\text{short}} T = 1.6 \times 10^{-3} \text{ m·K} = 1.6 \times 10^3 \ \mu\text{m·K} < 1.8 \times 10^3 \ \mu\text{m·K}$$

Discrepancy is attributed to 1.8×10^3 µm·K being rounded off which affects exponent strongly.

(2) $\lambda_{\text{long}} T = 5.1 \times 10^{-3}$ m·K

SOLUTIONS

$$\lambda_{\text{long}}^5 T^5 = \frac{2.92 \times 10^{-11}}{\exp\left(\frac{1.44 \times 10^{-11}}{5.1 \times 10^{-3}}\right) - 1} = \frac{2.92 \times 10^{-11}}{15.84} = 1.84 \times 10^{-12}$$

Therefore,

$$\lambda_{\text{long}} T = 4.50 \times 10^{-3}\ \text{m}\cdot\text{K} = 4.50 \times 10^3\ \mu\text{m}\cdot\text{K} < 5.1 \times 10^3\ \mu\text{m}\cdot\text{K}$$

The discrepancy is attributed to $5.1 \times 10^3\ \mu\text{m}\cdot\text{K}$ also being rounded off, which affects the exponent strongly.

4.5 From the previous exercise,

$$N_\lambda(\lambda_m) = \frac{hc^2}{\lambda_m^5}\left[\exp\frac{hc}{2.897 \times 10^{-3}k} - 1\right]^{-1}$$
$$= \frac{4.17 \times 10^{-19}}{\lambda_m^5} = \frac{8.34 \times 10^{-19} T^5}{(2.897 \times 10^{-3})^5}$$
$$= 4.09 \times 10^{-6} T^5$$

Therefore, $N_\lambda(\lambda_m, T_1) = 4.09 \times 10^{-6} T_1^5$; $N_\lambda(\lambda_m, T_2) = 4.09 \times 10^{-6} T_2^5$. Therefore,

$$\frac{N_\lambda(\lambda_m, T_2)}{N_\lambda(\lambda_m, T_1)} = \left(\frac{T_2}{T_1}\right)^5.$$

4.6 $U_f = 8\pi h\nu^3/c^3\ [\exp(h\nu/kT) - 1]^{-1}$
For low frequencies, a Taylor series expansion of exponent reduces to

$$\exp\frac{h\nu}{kT} \approx 1 + \frac{h\nu}{kT}$$

$$U_f \cong \frac{8\pi h\nu^3}{c^3}\left(1 + \frac{h\nu}{kT} - 1\right)^{-1} = \frac{8\pi h\nu^3}{c^3}\cdot\frac{kT}{h\nu} = \frac{8\pi\nu^2 kT}{c^3}$$

$$U_\lambda = \frac{8\pi hc}{\lambda^5}\left[exp\left(\frac{hc}{\lambda kT}\right) - 1\right]^{-1} \cong \frac{8\pi hc}{\lambda^5}\left(1 + \frac{hc}{\lambda kT} - 1\right)^{-1}$$
$$= \frac{8\pi hc}{\lambda^5}\left(\frac{\lambda Tk}{hc}\right) = \frac{8\pi Tk}{\lambda^4}$$

4.7 Assume the sun to be a Lambertian source of radiance $N = \sigma' T^4$. At the moon the sun appears as a uniform disk of radius R, so that the projected area of the sun is $dA = \pi R^2$. Power incident on moon is $P_s = N_{bb}\, dA\, A_r/s^2$ where A_r/s^2 is the solid angle subtended by the moon, A_r being the projected area of the moon and s the distance to the sun. Therefore, $H_s = N_{bb}\, dA/s^2$ equals

$$\sigma' T^4 \frac{\pi R^2}{s^2} = \left(\frac{6.9 \times 10^7}{\pi}\right) \pi \left(\frac{4.3 \times 10^5 \text{ miles}}{9.3 \times 10^7 \text{ miles}}\right)^2 = 1.48 \times 10^3 \text{ W}\cdot\text{m}^{-2}$$

where use has been made of the answer to Exercise 4.3. For the Nd:YAG laser,

$$H_L = \frac{P_r}{A_{rL}} = \frac{N_L \, dA \, A_{rL}}{A_{rL} \quad z^2} = \frac{P_L}{dA\Omega_L} \cdot \frac{dA}{A_{rL}} \cdot \frac{A_{rL}}{z^2} = \frac{P_L}{\Omega_L z^2}$$

where

$$\Omega_L = \pi(1.22\lambda/D)^2 \quad \text{and} \quad z = \text{distance to moon}$$

$$\cong 3.8 \times 10^5 \text{ km} = 3.8 \times 10^8 \text{m}.$$

$$H_L = \frac{10^9}{\pi[(1.22)^2(1.06 \times 10^{-6})^2/(10^{-2})^2](3.78 \times 10^8)^2} = 0.72 \text{ W}\cdot\text{m}^{-2}$$

where $H_L/H_s = 0.72/1480 = 4.9 \times 10^{-4}$.

Conclusion: Virtually impossible to see laser irradiance on moon with sunlight background.

4.8 The filter should be centered at the laser wavelength so as to pass the laser radiation into the receiver while removing sunlight background at other wavelengths. Let $\Delta\lambda$ be the linewidth of the filter. Since the filter is so narrow compared to the solar spectrum,

$$\int_{\lambda_1}^{\lambda_2} H_{s\lambda L}(\lambda) \, d\lambda \cong H_{s\lambda L}\Delta\lambda$$

where $H_{s\lambda L}$ is solar irradiance at $\lambda = 1.06$ μm. It is assumed that solar irradiance is virtually constant over this narrow passband.

To find $H_{s\lambda L}$, note that $H_{s\lambda L} = N_{s\lambda L} \, (dA/s^2) = N_{s\lambda L} \, \Delta\lambda(\pi R^2/s^2)$.

$$N_{s\lambda L} = \frac{2hc^2}{\lambda^5}\left[\exp\left(\frac{hc}{\lambda kT}\right) - 1\right]^{-1}$$

$$\text{where} \quad \lambda = 1.06 \ \mu\text{m and } T = 5900 \text{ K}$$

$$= 9.9 \times 10^{-12} \rightarrow N_{s\lambda L}\Delta\lambda = 9.9 \times 10^3$$

$$\text{instead of} \quad 6.9 \times 10^7/\pi$$

from Exercise 4.3. H_s is now reduced to 6.65×10^{-1} W·m^{-2} from 1480 W·m^{-2}.

$$\frac{H_L}{H_s} = \frac{0.72}{0.665} = 1.08$$

It is now possible to see the laser spot even against sunlight background.

Where laser spot is located, total irradiance is $H_L + H_s = 0.72 + 0.67 = 1.39\ \mathrm{W \cdot m^{-2}}$ as against $H_s = 0.67\ \mathrm{W \cdot m^{-2}}$ outside of laser spot.

4.9 Irradiance in the image plane depends on object emittance, *f*-number, magnification, etc., as shown in Chapter 3. The optical parameters are identical for the whole scene. What varies, therefore, is only emittances of various objects in the scene. This causes the difference in irradiance of images of those objects in the image plane. The question of which factor is responsible for contrast then is reduced to differences in emittances of different objects. Over the whole spectrum, for any object,

$$W = \epsilon\sigma T^4$$

$$dW = \sigma(T^4\,d\epsilon + 3\epsilon T^3\,dT) = \sigma T^3(T\,d\epsilon + 3\epsilon\,dT)$$

$$= \sigma T^3 \cdot \left[\left(272 + \frac{20 + 26}{2}\right)(0.95 - 0.05) + 3\left(\frac{0.95 + 0.05}{2}\right)(272 + 26 - 272 - 20)\right]$$

$$= \sigma T^3[(272 + 23)(0.90) + 3(0.5)6] = \sigma T^3(265.5 + 9)$$

It is clear that dW depends primarily on the first term $\sigma T^4\,d\epsilon$ rather than on the second term $3\epsilon T^3\,dT$. Hence, when different surfaces are in the field of view, emissivity differences are the primary contrast factor in thermal imaging unless temperature differences are *very* large. In general, however, emissivity differences can be decreased by environmental and meteorological factors such as moisture on all surfaces. Temperature differences can also be decreased by factors such as winds, clouds, etc., for passive objects (not heated artificially).

4.10 $H_i = W\sin^2\theta'_{max} = \sigma T^4 \sin^2(25°) = 5.67 \times 10^{-8}(272 + 37)^4 \cdot 1.9 \times 10^{-3} = 0.98\ \mathrm{mW \cdot m^{-2}} = 98\ \mu\mathrm{W \cdot cm^{-2}}$

4.11

$$1 = \rho + \alpha' + \tau'; \qquad \tau = 7/9 = \exp(-A_a z) = \exp(-2z)$$

Therefore, $z = (1/2)\ln(7/9) = 0.125$ cm.

4.12

$$P = \frac{\pi}{4} d_o^2 \epsilon\sigma T^4$$

$$\epsilon = \alpha' = 1 - \rho, \qquad \text{since } \tau' = 0$$

$$P = \frac{\pi}{4} d_o^2 (1 - \rho)\sigma T^4$$

$$\rho = 1 - \frac{4P}{\pi d_o^2 \sigma T^4}$$

4.13 The laser spectrum must contain one mode of bandwidth 4.77 MHz and may have spacing of $\Delta f_m = 150$ MHz on each side. Therefore,

$$LW < 2\Delta f_m + \delta f_m = 300 + 4.77 \text{ MHz} = 304.77 \text{ MHz}$$

4.14

$$\Delta f_m = \frac{c}{2l} = \frac{3 \times 10^8}{2(0.5)} = 300 \text{ MHz}$$

$$\Delta f = 2f_o\sqrt{\frac{2kT \ln 2}{Mc^2}} = \frac{2}{\lambda}\sqrt{\frac{2kT \ln 2}{M}}$$

$$= \frac{2}{0.488 \times 10^{-6}}\left[\frac{2(1.38 \times 10^{-23})300(0.7)}{40(1.66 \times 10^{-24})}\right]^{1/2} = 1.2 \text{ GHz}$$

$$df_m = \frac{f_m}{Q} = \frac{c\delta}{2\pi l} = \frac{3 \times 10^8(0.029)}{2\pi(0.5)} = 2.77 \text{ MHz}$$

$$LW = 2\Delta f_m + 3df_m = 608.3 \text{ MHz}$$

$$l_c = \frac{c}{LW} = \frac{3 \times 10^8}{6.083 \times 10^8} = 49.3 \text{ cm}$$

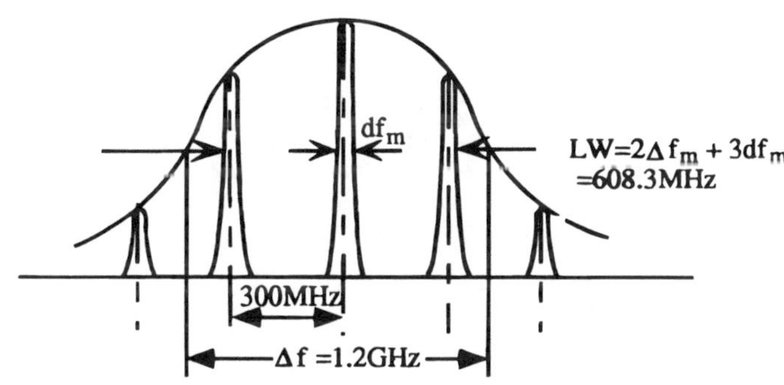

4.15 If $|b| > 1$, then

$$r_s = r_{01}e^{b-\sqrt{b^2-1}} + r_{02}e^{b+\sqrt{b^2-1}}$$

Since $|b| > 1$, the exponents are both positive. As s increases as a result of reflection back and forth between the mirrors, r increases and the ray eventually leaves the resonator → unstable resonator.

4.16 (a) $l/R_1 = 2/4/3 = 3/2$; $l/R_2 = 2$ → stable (see Fig. 4.13)
(b) $l/R_1 = 0$, $l/R_2 = 1/2$ → within stable region
(c) $l/R_1 = 0$, $l/R_2 = 1$ → on border of stable region

4.17

$$0 \le \left(1 - \frac{l}{R_1}\right)\left(1 - \frac{l}{R_2}\right) \le 1$$

$$0 \leq \left(1 + \frac{l}{250}\right)\left(1 - \frac{l}{50}\right) \leq 1$$

l(cm)	(1 + l/250)	(1 − l/250)	(1 + l/250)(1 − l/50)
0	1	1	1
10	1.04	0.8	0.832
30	1.12	0.4	0.448
50	1.20	0	0
60	1.24	−0.2	−0.25

stable for $0 < l < 50$ cm.

4.18

$$0 \leq (1 - l/R_1)(1 - l/R_2) \leq 1$$

$$0 \leq (1 - l/R_1)^2 \leq 1$$

$$0 \leq (1 - l/2f)^2 \leq 1$$

$$-1 \leq (1 - l/2f) \leq 1$$

4.19

$$\theta \cong \frac{\lambda}{\pi w_0} \rightarrow w_0 \cong \frac{\lambda}{\pi\theta} = \begin{cases} 0.2 \text{ mm for 1 mrad} \\ 20 \text{ cm for 1 } \mu\text{rad} \end{cases}$$

The latter requires a very wide laser.

4.20 $z_1 = 70$ cm, $w_{01} = 4.36 \times 10^{-2}$ cm, $f = 26$ cm

$$\left(\frac{w_{01}}{w_{02}}\right)^2 = \left(1 - \frac{z}{f}\right)^2 + \left(\frac{\pi w_{01}^2}{\lambda f}\right)^2 = 2.86 + 13.28 = 16.14$$

$$w_{02}^2 = \frac{w_{01}^2}{16.14} = \frac{(4.36 \times 10^{-4})^2}{16.14} = 1.18 \times 10^{-8} \text{ m}^2$$

$$w_{02} = 1.085 \times 10^{-2} \text{ cm} < w_{01}$$

$$z_2 = f + \frac{(z_1/f - 1)f}{(z_1/f - 1)^2 + \left(\frac{\pi w_{01}^2}{\lambda f}\right)^2}$$

$$= 2.6 \times 10^{-1} + 2.72 \times 10^{-2} \text{ m}$$

$$= 2.872 \times 10^{-1} \text{ m}$$

Output beam waist is 0.11 mm after lens, located 28.72 cm beyond the back focal plane.

4.21

$$w_{02} \cong \frac{f\lambda}{\pi w_{01}} = \frac{2.6 \times 10^{-1}(0.6328 \times 10^{-6})}{\pi(4.36 \times 10^{-4})} = 1.20 \times 10^{-2}\ \text{cm}$$

instead of 1.085×10^{-2} cm → error of 10.6%.

4.22 If the laser beam is not focused by a lens, then from Eq. (4.5.14),

$$w(z) = \frac{\lambda}{\pi w_0} z = \frac{1.06 \times 10^{-6}(4 \times 10^3)}{\pi(0.2)10^{-3}} = \frac{1.06}{0.2\pi} = 1.69\ \text{m}$$

which is too large. To decrease $w(z)$ in the object plane, it is necessary to use a lens after the laser so as to increase w_{02} in the lens output.

From the above equation,

$$w(z) = 1 = \frac{\lambda z}{\pi w_0} \rightarrow w_0 = \frac{\lambda z}{\pi} = 1.35\ \text{mm}$$

where $w_0 = w_{02}$ is the beam waist at the lens output, and w_{01} from the laser is known. Now, lens focal length must be found. The distance z_1 from beam waist inside the laser to the lens can be examined. For simplicity, let $z_1 = f$. Equation (4.5.16) reduces to

$$\frac{\pi w_{01}}{\lambda f} - \frac{1}{w_{02}}$$

From here,

$$f = \frac{\pi w_{01} w_{02}}{\lambda} = \frac{\pi(0.2 \times 10^{-3})(1.35 \times 10^{-3})}{1.06 \times 10^{-6}} = 80\ \text{cm}$$

Thus, an 80-cm focal length lens must be placed 80 cm from the beam waist in the laser resonator. At the lens input, laser beam radius is, from Eq. (4.5.13),

$$\begin{aligned} w(z) &= w_0\left[1 + \left(\frac{\lambda}{\pi w_0^2}\right)^2 z^2\right]^{1/2} \\ &= 0.2 \times 10^{-3}\left\{1 + \left[\frac{1.06 \times 10^{-6}}{\pi(0.2 \times 10^{-3})^2}\right]^2 0.8^2\right\}^{1/2} \\ &= 0.2 \times 10^{-3}(1.0045) = 0.2\ \text{mm} \end{aligned}$$

Therefore, lens diameter must be at least 0.6 mm, which is no problem. If these requirements are fulfilled then from Eq. (4.5.14) for free-space propagation at a distance 4 km from the lens

S O L U T I O N S

$$w(z) = \frac{\lambda z}{\pi w_{02}} = \frac{1.06 \times 10^6 \cdot 4 \times 10^3}{\pi \cdot 1.3 \times 10^{-3}} = 1 \text{ m}$$

as expected. Furthermore, laser irradiance on the object varies as

$$H = H_o e^{-2r^2/w^2(z)}$$

where $w(z) = 1$ m and, from (Eq. 4.5.26), $H_o = P/\pi w^2(z)$, P being laser radiant power.

4.23 From Eq. (3.4.8), since $n = n' = 1$, and assuming Lambertian reflections,

$$H_i = \frac{W}{4F^2} = \frac{WD^2}{4f^2} = \frac{\rho H_o e^{-2(r')^2}(0.089)^2 \cos\theta}{4(1400)^2}$$

where F is f-number and f is focal length of imager, and r' is now radial coordinate in image plane, which is related to r in object plane by magnification, that is,

$$r' = r\frac{s'}{s} = r\frac{f}{s} = r\frac{1.4}{2 \times 10^3} = 7 \times 10^{-4} r$$

$w(z)$ in the image is reduced correspondingly, so that in the image $w'(z) = 7 \times 10^{-4} w(z) = 7 \times 10^{-4}$ m, and

$$H_i = \frac{\rho P(0.089)^2}{4\pi(1)^1(1400)^2} \cos\theta \exp\left[-2\left(\frac{r'}{w'(z)}\right)^2\right]$$

where ρ is reflectivity of the object and θ is angle of reflection from object to imager, in accordance with Lambert's law. P is total power of the laser. If atmospheric extinction is considered,

$$H_i = \frac{\tau\sigma P(0.089)^2}{4\pi(1)^2(1400)^2} \cos\theta \exp\left[-2\left(\frac{2r'}{w'(z)}\right)^2\right]$$

where $\tau = \exp[-a(z_1 + z_2 + s + s']$ and $z_1 + z_2 = 4$ km and $s + s' = 2$ km.

4.24 Fourier transform is

$$F(f) = \int_{-\infty}^{\infty} f(t) \exp(-j2\pi ft)\, dt = \int_0^{\infty} \exp[j2\pi(f_o - f)t - t/\tau]\, dt$$
$$= [j2\pi(f_o - f) - \tau^{-1})]^{-1}$$

Power spectral density is

$$|F(f)|^2 = \{[2\pi(f_o - f)]^2 + \tau^{-2}\}$$

This peaks at $f = f_o$, and drops to one-half its value at the peak when

$$[2\pi(f_o - f)]^2 = \tau^{-2}$$

or

$$|(f_o) - f| = (2\pi\tau)^{-1}$$

Half-power linewidth is twice this, or

$$\mathrm{LW} = 2(f_o - f) = 2(\pi\tau)^{-1} = (\pi\tau)^{-1}; \qquad t_c = \pi\tau; \qquad l_c = ct_c = c\pi\tau$$

5

5.1 Let

$$\lim_{\Delta t \to 0} (1 - \bar{r}\Delta t)^{t/\Delta t} = y$$

$$\lim_{\Delta t \to 0} \frac{t}{\Delta t} \ln(1 - \bar{r}\Delta t) = \ln y$$

$$= \lim_{\Delta t \to 0} \frac{\ln(1 - \bar{r}\Delta t)}{\Delta t/t} = \lim_{\Delta t \to 0} \frac{-\bar{r}}{\left(\frac{1 - \bar{r}\Delta t}{1/t}\right)} \text{ by L'Hopital's rule}$$

$$= \lim_{\Delta t \to 0} \frac{-\bar{r}t}{1 - \bar{r}\Delta t} = -\bar{r}t$$

$$\ln y = -\bar{r}t \rightarrow y = e^{-\bar{r}t}$$

5.2 (a) From Eq. (5.1.17), $\Re = I/P = 5 \times 10^{-6}\ \text{A}/6 \times 10^{-6}\ \text{W} = 0.83\ \text{A}\cdot\text{W}^{-1}$

(b) $\Re = \eta q/h\nu \rightarrow \eta = h\nu\Re/q = hc\Re/\lambda q = 0.80$

5.3 $I_b = \Re P_b = 0.83(10^{-6}) = 0.83\ \mu\text{A}$

$$(\delta \overline{I_b^2})^{1/2} = (2qI_bB)^{1/2} = [2(1.6 \times 10^{-19})(0.83 \times 10^{-6})(10^6)]^{1/2}$$

$$= 5.15 \times 10^{-10}\text{A} << I_b$$

$$\frac{I_b}{(\delta \overline{I_b^2})^{1/2}} = 1.61 \times 10^3$$

5.4 (a) $\overline{N} = \eta/h\nu \int_0^\tau \underline{P}(t)\, dt = \eta E/h\underline{\nu}$

(b) $P(n) = (\overline{N})^n e^{-\overline{N}}/n!;\ P(0) = e^{-\overline{N}} = e^{-\eta E/h\nu}$

(c) $e^{-\eta E/h\nu} = 10^{-9}$

$$\therefore E \geq \frac{h\nu}{\eta}(9) \ln 10 \cong 20.7\ h\nu/\eta$$

(d) $\eta = 1 \therefore E \geq 20.7\ h\nu \cong 21$ photons

5.5 If the detected current is passed through a capacitor, the signal pulses and background radiation current fluctuations remain. The constant background radiation current is removed. Substituting into Eq. (5.1.36)

$$\text{SNR} = \frac{(0.1)(10^{-8})^2}{2(6.6 \times 10^{-34})(3 \times 10^8/10^{-6})10^6 \cdot 10^{-6}} = 25.1$$

The time-varying signal is detected despite the large background current because fluctuations in the background current are less than the signal pulses.

If the signal were not pulsed, both i_s and I_b would be constant and should not be put through a capacitor. In this case

$$\text{SNR} = \frac{i_s^2}{I_b^2} = \left(\frac{P_s}{P_b}\right)^2 = \left(\frac{10^{-8}}{10^{-6}}\right)^2 = 10^{-4}$$

The signal is *very* difficult to detect.

5.6

$$(\overline{\delta i_d^2})^{1/2} = (2qi_dB)^{1/2}$$

$$\frac{(\overline{\delta i_d^2})^{1/2}}{i_d} = \left(\frac{2qB}{i_d}\right)^{1/2}$$

As long as $i_d > 2qB$, the time-varying component of dark current is of average magnitude much less than the dark current itself. If $i_d < 2qB$, then both are so small that essentially neither one is measurable and both can be neglected compared to other noise sources.

5.7

$$RC = 50 \cdot 10 \cdot 10^{-12} = 5 \times 10^{-10}\ \text{s}$$

$$\frac{1}{f_s} = \frac{\lambda_s}{c} = \frac{10^{-6}}{3 \times 10^8} = 3.3 \times 10^{-14}\ \text{s}$$

Circuit time constant = 500 ps. Circuit cannot react to time changes on the order of 0.033 ps as required to follow carrier frequency of electric or magnetic field.

5.8 (a)

$$(\overline{i_{nj}^2})^{1/2} = (4kTB/R)^{1/2} = \left[\frac{4(1.38 \times 10^{-23})(3 \times 10^2)10^6}{10^6}\right]^{1/2}$$

$$= 1.29 \times 10^{-10}\ \text{A} < \overline{\delta I_b^2} > = 5.15 \times 10^{-10}\ \text{A}$$

(b) If $R_L = 10^3\ \Omega$, $(\overline{i_{nj}^2}) = 4.08\ \text{nA} > (\overline{\delta I_b^2})^{1/2}$

(c)

$$(\overline{\delta I_b^2})^{1/2} = (5.15 \times 10^{-10})100 = 5.15 \times 10^{-8}\ \text{A} > (\overline{I_{nj}^2})^{1/2}$$
$$= 4.08\ \text{nA}$$

Internal detector gain, without increasing Johnson noise temperature, tends to make Johnson noise negligible compared to quantum noise and the signal itself.

6

6.1

$$i_s(t) = \frac{\eta q G A_r}{2h\nu\eta_o} a^2(t)[1 + \cos 2(\omega_s t - B_z z)]$$

The time average of the cosine term is zero. Hence,

$$\bar{i}_s(t) = \frac{\eta q G A_r}{2h\nu\eta_o} \overline{a^2(t)} = K\Re G\overline{a^2(t)}$$

6.2 For a nonphotoconducting quantum detector, output

$$\mathrm{SNR} = \frac{(\eta q/h\nu)^2 \tilde{P}_s^2 G^2}{2qi_d BG^2}$$

where $\tilde{P}_s$ is the time-varying component of P_s.

$$\tilde{P}_{s\,\min} = (h\nu/\eta)(2i_d B/q)^{1/2}$$

$$\therefore\ \mathrm{NEP}_d = (h\nu/\eta)(2i_d/q)^{1/2}\ \mathrm{W\cdot Hz^{-1/2}}$$

For a photoconductor,

$$\mathrm{NEP}_d = (2h\nu/\eta)(i_d/q)^{1/2}\ \mathrm{W\cdot Hz^{-1/2}}$$

6.3 $\mathrm{SNR} = (\eta q/h\nu)^2 \tilde{P}_s^2/4kTB/R_L$

$$\tilde{P}_{s\,\min} = \frac{h\nu}{\eta q}\left(\frac{4kTB}{R_L}\right)^{1/2}$$

$$\therefore\ \mathrm{NEP} = \frac{2h\nu}{\eta q}\left(\frac{kT}{R_L}\right)^{1/2}\ \mathrm{W\cdot Hz^{-1/2}}$$

where it is assumed $G = 1$ in order for the device to be Johnson noise limited.

6.4

$$\overline{i_s^2(t)} = \frac{1}{T}\int_0^T i_s^2(t)\,dt = \frac{\omega_s}{2\pi}\int_0^{2\pi/\omega_s} \Re^2 G^2 P_s^2(t)\,dt$$

$$= \frac{\omega_s}{2\pi}\Re^2 G^2 A^2 \int_0^{2\pi/\omega_s} (1 + 2m\cos\omega_s t + m^2\cos^2\omega_s t)\,dt$$

$$= \frac{\omega_s}{2\pi} \Re^2 G^2 A^2 \left[t + \frac{2m}{\omega_s} \sin\omega_s t \right|_0^{2\pi/\omega_s} + \frac{m^2}{2} \int_0^{2\pi/\omega_s} (1 + \cos 2\omega_s t)\, dt$$

$$= \frac{\omega_s}{2\pi} \Re^2 G^2 A^2 \left[2\pi/\omega_s + 0 + \frac{m^2}{2} \cdot \frac{2\pi}{\omega_s} + 0 \right]$$

$$= \Re^2 G^2 A^2 (1 + m^2/2) \qquad \text{where } \Re = \eta q/h\nu$$

The modulation term is $m^2 \Re^2 G^2 A^2/2$ and this is to be used in the numerator of the SNR.

6.5 (a) Since P_s is pulsed, it is time varying. During a pulse,

$$i_s = \frac{\eta q}{h\nu} G P_s = \frac{0.5(1.6 \times 10^{-19})(1)10^{-9}}{6.63 \times 10^{-34}(3 \times 10^8/10^{-6})} = 0.4 \text{ nA}$$

Here, the signal pulse amplitude is much less than that of i_d. However, since i_s is pulsed, the relevant dark current noise is not i_d, but fluctuations of i_d. The detector current is passed through a capacitor, thus removing constant i_d. Therefore,

$$\text{SNR} = \frac{(\eta q/h\nu)^2 (\tilde{P}_s)^2}{2qi_d B + 4kTB/R_L}$$

$$= \frac{(4 \times 10^{-10})^2}{2(1.6 \times 10^{-19})10^{-7}(10^3) + 4(1.38 \times 10^{-23})(300)10^3/10^3}$$

$$= \frac{4^2 \times 10^{-20}}{3.2 \times 10^{-23} + 1.66 \times 10^{-20}} = 9.74$$

Dominant noise is Johnson noise.

(b) If $G = 100$, then

$$\text{SNR} = \frac{4^2 \times 10^{-20} \times 10^4}{3.2 \times 10^{-23}(10^4) + 1.66 \times 10^{-20}}$$

$$= \frac{4^2 \times 10^{-16}}{3.2 \times 10^{-19} + 1.66 \times 10^{-20}}$$

$$= 4.8 \times 10^3 >> 9.74$$

Dominant noise is dark current fluctuations.

6.6 (a)

$$\text{SNR} = \frac{(\eta q/h\nu)^2 \tilde{P}_s^2 G^2}{4kTB/R_L} = 3.92 \times 10^{-6}$$

Signal not detected. (Johnson noise power is $1.66 \times 10^{-16}\ A^2$.)

(b)

$$\text{SNR} = \frac{2(\eta q/h\nu)^2 \tilde{P}_s P_{LO}}{2q\Re B P_{LO} + 4kTB/R_L}$$

$$= \frac{2(\eta q/h\nu)^2 \tilde{P}_s P_{LO}}{(8.15 + 1.66)10^{-16}} = \frac{5.09 \times 10^{-14}}{9.81 \times 10^{-16}} = 51$$

Signal easily detected.

6.7 Noise current $i_n^2 = \Re(\text{NEP})B^{1/2} = 0.1(10^{-14})10^4 = (2qi_dB)^{1/2} = 10^{-11}$ A $\rightarrow i_d = 3.1 \times 10^{-12}$ A.

$$\overline{i_s^2} = (\Re P_s m)^2/2$$

$$\overline{i_n^2} = 2qB[\Re P_s(1 + m^2/2)^{1/2} + i_d]$$

$$\text{SNR} = \frac{(10^{-9})^2/8}{3.2 \times 10^{-19} \cdot 10^8(10^{-9}(9/8)^{1/2} + 3 \times 10^{-12}]}$$

$$\cong \frac{10^{-18}}{25 \times 10^{-20}} = 4 \qquad \text{for } m = 0.5$$

$$\text{If } m = 1,\ \text{SNR} \cong \frac{(10^{-9})^2}{3.2 \times 10^{-19} \cdot 10^8[10^{-9}(3/2)^{1/2}]} = 25.5$$

$$\text{If } m = 0.1,\ \text{SNR} \cong \frac{(10^{-9})^2(0.005)}{3.2 \times 10^{-19} \cdot 10^8[10^{-9}(1.005)^{1/2} + 3 \times 10^{-12}]}$$

$$= 0.156$$

6.8 Using the notation of Eq. (6.1.17a) and assuming beating or mixing involving background radiation is negligible,

$$i = \frac{\Re G A_r}{\eta_o} |E_s + \overline{E}_o|^2$$

$$= \frac{\Re G A_r}{\eta_o} |a \cos[\omega_o t + \phi_s(t)] + A_o \cos(\omega_o t + \phi_o)|^2$$

$$= \frac{\Re G A_r}{\eta_o} [[a \cos\phi_s(t) + A_o \cos\phi_o] \cos\omega_o t$$

$$- (a \sin\phi_s(t) + A_o \sin\phi_o) \sin\omega_o t]^2$$

$$= \frac{\Re G A_r}{\eta_o} [(a \cos\phi_s(t) + A_o \cos\phi_o)^2 + (a \sin\phi_s(t) + A_o \sin\phi_o)^2]$$

$$= \frac{\Re G A_r}{\eta_o} [a^2 \cos^2\phi_s(t) + A_o^2 \cos^2\phi_o + 2aA_o \cos\phi_s(t) \cos\phi_o$$

$$+ a^2 \sin^2\phi_s(t) + A_o^2 \sin^2\phi_o + 2aA_o \sin\phi_s(t) \cos\phi_o)]$$

SOLUTIONS

$$= \frac{\Re G A_r}{\eta_o} [a^2 + A_o^2 + 2aA_o \cos\phi_o[\cos\phi_s(t) \cos\phi_o + \sin\phi_s(t) \sin\phi_o]]$$

The third term is the original phase modulation information term, and its amplitude is exactly twice that of the IF terms in Eq. (6.1.19).

6.9 NEP for heterodyne detection is

$$h\nu/\eta = \frac{6.63 \times 10^{-34}(3 \times 10^8)}{(0.5)(0.8 \times 10^{-6})} = 4.97 \times 10^{-19}\ \text{W}\cdot\text{Hz}^{-1}$$

$$\text{SNR} = 10 = \frac{P_s}{\text{NEP}(B)} = \frac{P_s}{4.97 \times 10^{-15}} \rightarrow P_s \geq 4.97 \times 10^{-14}\ \text{W}$$

6.10 Assume one plane wave is propagating according to $\exp(-j\beta x)$ and the other according to $\exp(j\beta_x x - j\beta_z z) = \exp j(x \cos\theta - z \sin\theta)$, using the setup in Eq. (1.4.2) and Fig. 1.5, where x is normal (outward from) the detector and z is in the plane of the detector. Both beam centers are incident on the origin. There, phases of both beams are zero since $x = z = 0$. At $z = z_1$, however, the phases are zero for the normally incident beam (since $x = 0$ on the detector surface) and $-\beta z_1 \sin\theta$ for the diagonally incident beam.

The phase difference $0 - (-\beta z_1 \sin\theta) = \pi$ occurs where $z_1 = \pi/(\beta \sin\theta) \cong \pi\lambda/(2\pi\theta)$ for small θ, or $z_1\theta \cong \lambda/2$. If $z_1 = 1$ mm, then

$$\theta \cong \frac{6.33 \times 10^{-7}\ \text{m}}{2 \times 10^{-3}} = 3.17 \times 10^{-4}\ \text{rad}$$

Thus, heterodyne alignment is very critical.

6.11 Without integration

$$\text{SNR} = \frac{\eta}{\exp(h\nu/kT) - 1} = \frac{1}{\exp\left[\dfrac{(6.6 \times 10^{-34}(3 \times 10^8)}{10^{-5} \cdot (1.38 \times 10^{-23})(800)}\right] - 1}$$

$$= \frac{1}{\exp(1.8) - 1} = 0.198$$

With integration, $(B\tau_i)^{1/2} = (10^6)^{1/2} = 10^3$ and SNR = 198.

6.12 For the lower detector in Fig. 6.3 incident radiant local oscillator power is proportional to ρ, while for the upper detector it is proportional to τ'. Incident signal radiant power is proportional to τ' for the lower detector and to ρ for the upper detector. For each detector, $I_{IF} \propto (P_s P_o)^{1/2} \propto (\rho\tau')^{1/2} = [\rho(1 - \rho)]^{1/2}$, since $\rho + \tau' = 1$. Maximum IF current is obtained when $dI_{IF}/d\rho = 0 \rightarrow 1 - 2\rho = 0$. This occurs when $\rho = \tau' = 1/2$.

6.13 From Table 6.2, $\lambda_c = 0.87\ \mu\text{m}$. From Eq. (6.2.9b), $W = 1.24/\lambda_c = 1.43$ V. Therefore $\phi = 1.43$ eV. For 0.80-μm wavelength photons,

$$E_i = h\nu - \phi = \left(\frac{6.62 \times 10^{-34} \cdot 3 \times 10^8}{8 \times 10^{-7}}\right) - (1.43\ \text{eV})(q)$$

$= 2.48 \times 10^{-19} - 2.29 \times 10^{-19}\ \text{J} = 0.19 \times 10^{-19}$ J, while at the 0.40-μm wavelength

$$E_i = h\nu - \phi = \left(\frac{6.62 \times 10^{-34} \cdot 3 \times 10^8}{4 \times 10^{-7}}\right) - (1.43\ \text{eV})(q)$$

$= 4.97 \times 10^{-19} - 2.29 \times 10^{-19}\ \text{J} = 2.68 \times 10^{-19}$ J. From Eq. (6.2.12) at 0.80 μm, $v_i = (2E_i/m)^{1/2} = [2(0.19 \times 10^{-19})/9.1 \times 10^{-31}]^{1/2} = 2.04 \times 10^5\ \text{m}\cdot\text{s}^{-1}$, while at 0.40 μm, $v_i = 7.67 \times 10^5\ \text{m}\cdot\text{s}^{-1}$.

Substituting into Eq. (6.2.15), at 0.80 μm, $T = 1.04$ ns, while at 0.40 μm, $T = 0.983$ ns. This represents a decrease of 5.5% in time response at the shorter wavelength.

With regard to current response per photon, at 0.80 μm,

$$I_o = \frac{qv_i}{d} = \left(\frac{(1.6 \times 10^{-19})(2.04 \times 10^5)}{5 \times 10^{-3}}\right) = 6.53\ \text{pA}$$

while at 0.40 μm, $I_o = 24.5$ pA. Substituting into Eq. (6.2.16), at $t = T$, for $\lambda = 0.80$ μm, $i(T) = 0.299$ nA. For $\lambda = 0.40$ μm, $i(T) = 0.313$ nA. This represents an increase of 4.68% in response at the shorter wavelength. Thus, improvements in both responsivity and time response occur at shorter wavelengths because of greater initial electron energy after photoemission.

6.14 Quantum noise exiting the cathode is $\overline{i_{nq}^2} = 2qI_cB$, where I_c is total cathode current. It is multiplied by N dynodes, yielding an amplification $G = \delta^N$. This noise originates in the cathode and is equal to $2qI_cBG^2$ on reaching the anode. Current exiting the first dynode is δI_c. Fluctuations in secondary emission from the first dynode generate an average shot noise current power $2q\delta I_cB$ exiting the first dynode. This is multiplied by $N - 1$ dynodes, resulting in a gain δ^{N-1}. Therefore, noise power generated at the first dynode that reaches the anode is $2q\delta IB\delta^{2(N-1)} = 2qIB\delta^{2N-1}$. Current reaching the j'th dynode is δ^jI_c. Average shot noise power exiting the j'th dynode is $2q\delta^jI_cB$. It is amplified by $N - j$ dynodes, so that noise power generated at the j'th dynode is equal to $2q\delta^jI_cB\delta^{2(N-j)} = 2qI_cB\delta^{2N-j}$ at the anode. Summing the shot noise current power added at the cathode and at each of the N dynode results in

$$\overline{i_{nqN}^2} = 2qI_cB[\delta^{2N} + \delta^{2N-1} + \cdots + \delta^{2N-j} + \cdots + \delta^{2N-N}]$$

$$= 2I_cB\delta^{2N}[1 + \delta^{-1} + \cdots + \delta^{-j} + \cdots + \delta^{-N}]$$

$$= 2qI_cBG^2\left(\frac{1 - \delta^{-N}}{1 - \delta^{-1}}\right), \qquad \text{where } G = \delta^N$$

For a large number of dynodes, this is approximately

$$\overline{i_{nqN}^2} \cong 2qI_c BG^2\left(\frac{\delta}{\delta - 1}\right)$$

For photon-limited detection, signal current is much greater than dark current and background current, so that from Section 6.2 for a photomultiplier

$$\mathrm{SNR} = \frac{\left(\frac{\eta q}{h\nu}\right)^2 P_s^2 G^2}{2q\left(\frac{\eta q}{h\nu}\right) P_s BG^2\delta/(\delta - 1)} = \frac{\eta P_s}{2g\nu B}\left(\frac{\delta - 1}{\delta}\right)$$

Therefore photon-limited SNR is reduced by $(\delta - 1)/\delta$. For dark-current-limited detection,

$$\mathrm{SNR} = \frac{\left(\frac{\eta q}{h\nu}\right)^2 P_s^2 G^2}{2qi_d BG^2(\delta/\delta - 1)} = \frac{\eta^2 P_s^2 q}{2h^2\nu^2 i_d B}\left(\frac{\delta - 1}{\delta}\right)$$

which is reduced by the same factor from its value if a unity gain detector had been used. For background-limited detection,

$$\mathrm{SNR} = \frac{\left(\frac{\eta q}{h\nu}\right)^2 P_s^2 G^2}{2q\frac{\eta q}{h\nu} P_b BG^2(\delta/\delta - 1)} = \frac{\eta P_s^2}{2h\nu P_b B}\left(\frac{\delta - 1}{\delta}\right)$$

again reduced by the same factor from unity gain value.

6.15 Optimum gain occurs when quantum noise equals Johnson noise. In that case,

$$\mathrm{SNR} = \frac{(\eta q/h\nu)^2 G^2 P_s^2}{2(4)kTB/R_L}$$

so that

$$P_{s\,\mathrm{min}} = \frac{2h\nu}{\eta qG}(2kTB/R_L)^{1/2}$$

leading to

$$\text{NEP} = \frac{2h\nu}{\eta q G}(2kT/R_L)^{1/2}$$

The larger the value of optimum G, the smaller the value of signal power that can be detected.

7

7.1 Since the vertical edges are not illuminated, there is no diffraction by them. Therefore,

$$u(f_x, f_y) = C_2 A \int_{-b}^{b} e^{-j2\pi f_y y_s}\, dy_s = -\frac{C_2 A e^{-j2\pi f_y}}{j2\pi f_y}\Bigg|_{-b}^{b}$$

$$= C_2 A b \operatorname{sinc} 2\pi f_y b$$

$$H(f_x, f_y) = C_2^2 A^2 b^2 \operatorname{sinc}^2(2\pi f_y b)$$

Note the sinc is in the vertical direction.

7.2 See photograph below.

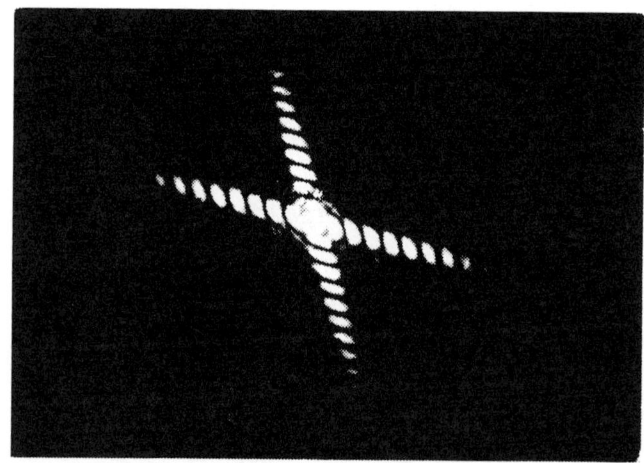

7.3 For a selfoc lens, from Eq. (2.10.3)

$$n = n_o - \frac{n_1}{2} r^2$$

Therefore, phase change is

$$\Delta\phi = k_o n L = k_o(n_o - 1/2 n_1 r^2)L$$

$$t(\rho) = \exp(-jk_o n_o L)\exp\left(\frac{jk_o n_1}{2} L r^2\right)$$

Phase transmission of a thin convex ordinary lens is from Eq. (7.3.3),

$$t(\rho) = \exp(-jk_o n d_o)\exp\left(\frac{jk_o\rho^2}{2f}\right)$$

Both exhibit a phase shift proportional to maximum path and to radial coordinate squared.

7.4 For simplicity we will consider x terms only. Identical treatment can be carried out for y terms. Consider the exponent in Eq. (7.3.13).

$$-\frac{jk_o}{2z_o}(\xi - x_s)^2 - \frac{jk_o}{2f_l}[(x_d - \xi)^2 - \xi^2]$$

$$= -\frac{jk_o}{2z_o}(\xi - x_s)^2 - \frac{jk_o}{2f_l}(x_d^2 - 2\xi x_d)$$

$$= \frac{-jk_o}{2z_o}(\xi^2 - 2\xi x_s + x_s^2) - \frac{jk_o}{2f_l}x_d^2 + \frac{jk_o\xi x_d}{f_l}$$

$$= -\frac{jk_o}{2z_o}[(\xi^2 - 2\xi x_s + x_s^2) + \frac{jk_o\xi x_d}{f_l} - \frac{jk_o x_d^2}{2f_l}$$

$$-\frac{jk_o x_s x_d}{f_l} + \frac{jk_o x_s x_d}{f_l} - \frac{jk_o z_o x_d^2}{2f_l^2} + \frac{jk_o z_o x_d^2}{2f_l^2}$$

$$= \frac{-jk_o}{2z_o}\left[\xi^2 - 2\xi\left(x_s + \frac{z_o x_d}{f_l}\right) + \left(x_s + \frac{z_o x_d}{f_l}\right)^2\right]$$

$$-\frac{jk_o}{2f_l}x_d^2 + \frac{jk_o z_o x_d^2}{2f_l^2} + \frac{jk_o x_s x_d}{f_l}$$

$$= -\frac{jk_o}{2z_o}\left[\xi - \left(x_s + \frac{z_o x_d}{f_l}\right)\right]^2 - \frac{jk_o x_d^2}{2f_l}\left(1 - \frac{z_o}{f_l}\right) + \frac{jk_o x_s x_d}{f_l}$$

7.5 From Eqs. (7.3.19) and (7.3.17),

$$H = |U_{fl}|^2/(2\eta_o) = |C_6(x_d, y_d)\mathbf{F}[U_t(x_s, y_s)]|^2/(2\eta_o)$$

$$= |\mathbf{F}[U_t(x_s, y_s)]|^2/(2\eta_o\lambda^2 f_l^2)$$

7.6 For $z_o = 0$,

$$C_6 = \frac{j}{\lambda f_l}\exp[-jk_o(nd_o + f_l)]\exp\left[-\frac{jk_o}{2f_l}(x_d^2 + y_d^2)\right]$$

Similar to Example 7.2, using (ξ, n) instead of (x_s, y_s), from Eq (7.3.19),

$$U_{fl}\left(\frac{x_d}{\lambda f_l}, \frac{y_d}{\lambda f_l}\right) = C_6 A \int_{-a}^{a}\int_{-b}^{b} \exp\left[-\frac{j2\pi}{\lambda f_l}(\xi x_d + \eta y_d)\right] d\xi\, d\eta$$

$$= C_6 A \left.\frac{\exp\left(-\frac{j2\pi x_d \xi}{\lambda f_l}\right)}{j2\pi x_d/\lambda f_l}\right|_{-a}^{a} \left.\frac{\exp\left(-\frac{j2\pi y_d \eta}{\lambda f_l}\right)}{j2\pi y_d/\lambda f_l}\right|_{-b}^{b}$$

$$= C_6 A\, ab\, \mathrm{sinc}(2\pi x_d a/\lambda f_l)\, \mathrm{sinc}(2\pi y_d b/\lambda f_l)$$

Therefore, $H = A_2(ab)^2\, \mathrm{sinc}^2(2\pi x_d a/\lambda f_l)\, \mathrm{sinc}^2\, (2\pi y_d\, b/\lambda f_l)/2\eta_o \lambda^2 f_l^2$ where ab is area of the aperture.

7.7 Aside from differences in phase, both C_1 and C_6 are proportional to λ^{-1}. Therefore, at the longer wavelength, amplitude and irradiance are one-half and one-quarter, respectively, of their values at the shorter wavelengths. In addition, the argument of the sinc function solution of Exercise 7.1 is $2\pi f_y b = y_d b/\lambda z$ (where $z = z_o$ for far-field lensless Fraunhofer diffraction and f_l for diffraction with a lens, respectively). Zeros occur when the argument is equal to multiples of π, that is, $2\pi y_d b/\lambda z = \pi, 2\pi, 3\pi, \ldots, n\pi$. At the shorter wavelength zeros occur when y_d is one-half the value of corresponding y_d for zeros at the longer wavelength. Therefore, the diffraction pattern "shrinks" with decreasing wavelength.

7.8 From Eq. (7.4.15),

$S_1 = dD/(1.22\lambda)$

For the eye, detected wavelength is $\lambda = \lambda_o/n_2 = \lambda_o/1.34$. Therefore,

$S_{1_e} = dDn_2/(1.22\lambda) = 1.5 \times 10^{11}(5 \times 10^{-3})(1.34)/(1.22\lambda)$

For the telescope, $D = 200$ in. $\times$ 2.54 cm·in.$^{-1}$ = 508 cm

$S_{1_t} = dD/1.22\lambda) = 1.5 \times 10^{11}(5.08)/1.22\lambda$

$$S_{1_e}/S_{1_t} = \frac{508}{5 \times 10^{-3}(1.34)} = 7.58 \times 10^2$$

The star pair can be detected with the telescope at a distance 758 times further away than with the unaided eye. If n_2 increases, this distance decreases since resolution of the human visual system improves.

7.9 (a) $\mathbf{F}\{U_t\} = C_6 A(8ab)\, \mathrm{sinc}(2\pi x_d 2a/\lambda f_l)\, \mathrm{sinc}(2\pi y_d b/\lambda f_l)$
$A(y_s) = \delta(y_s - 2b) + \delta(y_s + 2b)$

$\mathbf{F}[A(y_s)] = \exp[j2\pi(2b)y_d/\lambda f_l] + \exp[-j2\pi(2b)y_d/\lambda f_l] = 2\cos(4\pi b y_d/\lambda f_l)$

$\mathbf{F}\{u_{tn}\} = 16abAC_6\, \mathrm{sinc}(4\pi x_d a/\lambda f_l) \cdot \mathrm{sinc}(2\pi y_d b/\lambda f_l) \cdot \cos(4\pi b y_d/\lambda f_l)$

$$H\left(\frac{x_d}{\lambda f_l}, \frac{y_d}{\lambda f_l}\right) = \frac{128a^2b^2A^2}{\lambda^2 f_l^2 \eta_o} \operatorname{sinc}^2(4\pi x_d a/\lambda f_l)$$

$$\cdot \operatorname{sinc}^2(2\pi y_d b/\lambda f_l) \cdot \cos^2(4\pi b y_d/\lambda f_l)$$

(b) $\mathbf{F}\{U_t\} = C_6 A(\pi c^2)[2J_1(2\pi\rho c)/2\pi\rho c]$
where $\rho = (x_1^2 + y_d^2)^{1/2}/\lambda f_l$
and

$$A(x_s, y_s) = \exp(-j2\pi b y_d/\lambda f_l)[\exp(-j2\pi a x_d/\lambda f_l) + \exp(-j2\pi 2a x_d/f_l)]$$

$$= 2\exp[-j(2\pi/\lambda f_l)(3/2a x_d + b y_d)]\cos\pi a x_d/\lambda f_l$$

$$\mathbf{F}\{U_{tn}\} = 2C_6 A\pi c^2 A(x_s, y_s)\left[\frac{2J_1(2\pi\rho c)}{2\pi\rho c}\right]$$

$$H\left\{\frac{x_d}{\lambda f_l}, \frac{y_d}{\lambda f_l}\right\} = |\mathbf{F}\{U_{tn}\}|^2/2\eta_o$$

$$= \frac{2A^2\pi^2c^4}{\lambda^2 f_l^2 \eta_o}(\cos^2\pi a x_d/\lambda f_l)\left[\frac{2J_1(2\pi\rho c)}{2\pi\rho c}\right]^2$$

(c) $A(x_s, y_s) = \delta(x_s + a) + \delta(x_s - a)$
$\mathbf{F}\{A(x_s, y_s)\} = e^{j\pi a x_d/\lambda f_l} + e^{-\pi a x_d/\lambda f_l} = 2\cos\pi a x_d/\lambda f_l$
$\mathbf{F}\{U_t\}$ = same as in part (b).

$$\mathbf{F}\{U_{tn}\} = 2C_6 A\pi c^2(\cos\pi a x_d/\lambda f_l)\,\frac{2J_1(2\pi\rho c)}{2\pi\rho c}$$

$$H\left(\frac{x_d}{\lambda f_l}, \frac{y_d}{\lambda f_l}\right) = \frac{2A^2\pi^2c^4}{\lambda^2 f_l^2 \eta_o}\cos^2(\pi a x_d/\lambda f_l)\left[\frac{2J_1(2\pi\rho c)}{2\pi\rho c}\right]^2$$

Note that the irradiance distribution is identical to that in part (b), although CFA distributions differ because of phase differences. A somewhat overexposed example of this irradiance is shown below for $a >> c$.

(d)

$$A(x_s, y_s) = \sum_{n=0}^{3}\sum_{p=0}^{3}\delta(x_s - na, y_s - pb)$$

$$\mathbf{F}\{A(x_s, y_s)\} = \sum_{n=0}^{3}\sum_{p=0}^{3} \exp\left[-\frac{j2\pi}{\lambda f_l}(nx_d a + py_d b)\right]$$

$\mathbf{F}\{U_t\}$ = same as parts (b) and (c)

$$\mathbf{F}\{U_{tn}\} = C_6 A(\pi c^2)\left[\frac{2J_1(2\pi\rho c)}{2\pi\rho c}\right]\sum_{n=0}^{3}\sum_{p=0}^{3} \exp\left[-\frac{j2\pi(nx_d a + y_d b)}{\lambda f_l}\right]$$

Therefore,

$$H\left(\frac{x_d}{\lambda f_l}, \frac{y_d}{\lambda f_l}\right) = \frac{A^2\pi^2 c^4}{2\lambda^2 f_l \eta_o}\left[\frac{2J_1(2\pi\rho c)}{2\pi\rho c}\right]^2 \left|\sum_{n=0}^{3}\sum_{p=0}^{3} \exp\left[-\frac{j2\pi}{\lambda f_l}(nx_d a + py_d b)\right]\right|^2$$

A slightly overexposed photograph is shown below for $a = b >> c$. The diffraction pattern for a single circular aperture is sampled in both the x_d and y_d directions. The reason for the long exposure is to indicate what happens in the Fourier plane at distances relatively far from the origin where irradiance is weaker.

To understand the sampling, note the Fourier relationship

$$\sum_{n=-\infty}^{\infty} e^{j2\pi n f_x a} = \sum_{p=-\infty}^{\infty} \delta\left(f_x - \frac{n}{a}\right)$$

with regard to the aperture function $\mathbf{F}\{A(x_s, y_s)\}$.

The greater the density of aperture holes illuminated, the more ideal the sampling

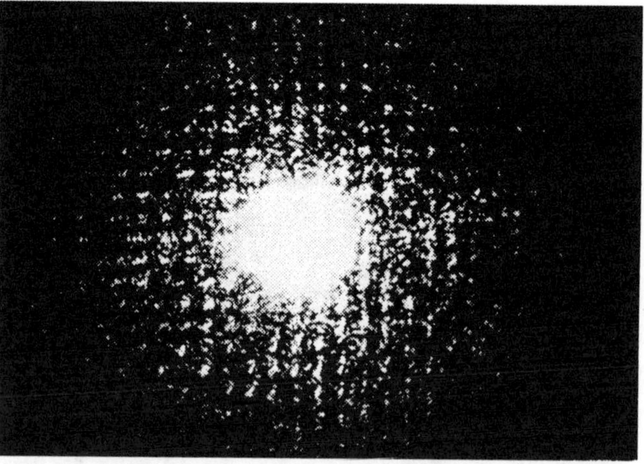

7.10 From Eq. (7.5.16), the maximum irradiance of primary maxima is proportional to N^2. Therefore, from Eq. (7.5.14), at half-maxima width, x_d must satisfy

$$\frac{2A^2a^2}{\eta_o\lambda^2 f_l}\frac{\sin^2[2\pi N(a + b)x_d/\lambda f_l]}{\sin^2[2\pi(a + b)x_d/\lambda f_l]} = \frac{N^2}{2}$$

7.11 As shown by Eq. (7.5.17), primary maxima occur when $x_{dm} = \pm m\pi/q = \pm mf_l\lambda/[2(a + b)]$. For a wavelength $\lambda + \Delta\lambda/2$, $x_{dm+} = \pm mf_l(\lambda + \Delta\lambda/2)/[2(a + b)]$. For a wavelength $\lambda - \Delta\lambda/2$, $x_{dm-} = \pm mf_l(\lambda - \Delta\lambda/2)/[2(a + b)]$. Therefore $x_{dm+} - x_{dm-} = \Delta x_{dm} = mf_l\Delta\lambda/[2(a + b)]$.

Minima occur when $(x_d)_{\min} = \pm m\pi/Nq = \pm mf_l\lambda/[2N(a + b)]$. Distances between adjacent minima are $(\Delta x_d)_{\min} = f_l\lambda[2N(a + b)]$. By the Rayleigh criterion, $\Delta x_{dm}/2$ is limited to one-half of $(\Delta x_d)_{\min}$ in either direction.

Therefore, $\Delta x_{dm} = (\Delta x_d)_{\min} = \pm mf_l\Delta\lambda/[2(a + b)] = f_l\lambda/[2N(a + b)]$. This yields $\lambda/\Delta\lambda = mN$. The greater the number of illuminated slits and the higher the order of interference, the better the wavelength resolution, that is, the narrower the linewidth passed by the monochromator. Significance of increase in m is that for a given wavelength the principal maximum is displaced further from the origin, thereby permitting more accuracy in wavelength measurement. Significance of increase in N is that linewidth of principal maximum narrows, thereby facilitating more exact determination of location of principal maximum.

7.12 Primary maxima occur at locations $x_d = \pm mf_l\lambda/[2(a + b)]$. We can see that x_d is proportional to the product $m\lambda$. Therefore, while the monochromator may "read" 900 nm for $m = 1$, second-order interference for $\lambda = 450$ nm, third-order interference for $\lambda = 300$ nm, etc., will all yield principal maxima at the same x_d location. Therefore, to be sure the measurement is indeed 900 nm, a filter rejecting 450-nm, 300-nm, etc., radiation should be placed either at monochromator input or output (preceding the detector, of course). High-pass wavelength filters are easily available with different wavelength cutoffs. Any cutoff between about 480 and 900 nm would do the trick. If the source is of narrow linewidth then ideally no radiation at 450 or 300 nm, etc., is input to the monochromator and none is detected at the output, in which case no filters are necessary.

7.13

$$2a = 2b = \frac{25.4 \times 10^{-3}\text{m}}{2(13{,}400)} = 0.95\ \mu\text{m} << \lambda$$

See assumption 1 concerning evaluation of Eqs. (7.1.12). The radiation is not transmitted through such narrow slits and therefore does not reach the observation plane since radiation behaves as if edges continue out about a distance λ beyond the actual physical edge.

7.14 Let $U_t(x_s, y_s)$ designate the original CFA image without the background irradiance b. Background CFA is $(2\eta_o b)^{1/2}$. Exiting plane P_1 is the CFA $U_t(x_s, y_s) + (2\eta_o b)^{1/2}$. Incident on plane P_2 is the CFA $C_6\mathbf{F}\{U_t(x_s, y_s)\} + C_6(2\eta_o b) \cdot \delta(x_d, y_d)$. Therefore, to get rid of the background irradiance one should block the origin in the Fourier transform plane P_2. In other words the transmission of the spatial filter $t(x_d, y_d)$ in Eq. (7.6.1) should be zero for $|x_d|, |y_d| < \epsilon$, and unity elsewhere. The smaller the value of ϵ, the less the distortion of the original image on plane P_3 after inverse transforming the CFA exiting plane P_2. Uniform background irradiance is commonly referred to as *dc noise*.

7.15 The Fourier transform of a rectangular grating as in Fig. 7.14 is described by Eq. (7.5.11). However, in view of Eq. (7.5.9) and (7.5.10), it can also be reexpressed in its original form as

$$\mathbf{F}\{U_{tn}(x_s)\} = C_6 A_a a(\mathrm{sinc} 2\pi x_d a/\lambda f_{l_2}) \cdot \sum_{n=0}^{N-1} \exp[-j2\pi(b + a)x_d n/\lambda f_{l_2}]$$

For large N, the sum on the right side may be represented by an equivalent sum of impulse functions:

$$\sum_{n=-\infty}^{\infty} \exp[-j2\pi(b + a)x_d n/\lambda f_{l_2}] = \sum_{n=-\infty}^{\infty} \delta\left(x_d - \frac{n\lambda f_{l_2}}{b + a}\right)$$

Therefore, for large N,

$$\mathbf{F}\{U_{tn}(x_s)\} \approx C_6 A_a a(\mathrm{sinc} 2\pi x_d a/\lambda f_{l_2}) \cdot \sum_{n=-\infty}^{\infty} \delta\left(x_d \frac{n\lambda f_{l_2}}{b + a}\right)$$

In this form the secondary maxima in Fig. 7.16 disappear, leaving only primary maxima, which narrow to become impulse functions. Therefore, what is required in plane P_2 of Fig. 7.17 is a bandpass filter passing only the components for $n = \pm 1$; that is,

$$t(x_d, y_d) = \begin{cases} 1 & 0 < x_d < \dfrac{2\lambda f_{l_2}}{b + a} \\ 0 & \text{elsewhere} \end{cases}$$

The CFA exiting such a filter is then

$$\mathbf{F}\{U_{tn}(x_s)\} \cdot t(x_d, y_d) = C_6 A_a a(\mathrm{sinc} 2\pi\ da/\lambda f_{l_2})$$
$$\cdot \{\delta[x_d + \lambda f_{l_2}/(b + a)] + \delta[x_d - \lambda f_{l_2}/(b + a)]\}$$

Inverse Fourier transform yields in plane P_3 of Fig. 7.17 convolution of a slit of width $2a$ at the origin with $2 \cos 2\pi x_3/(b + a)$. For very narrow slits, the cosine term is dominant since the sinc term in plane P_2 is almost a constant.

7.16 If the overall film transparency is placed in plane P_1 in Fig. 7.17 and illuminated, its Fourier transform exists in plane P_2. In plane P_1 we have a periodic series of lines $\Sigma_n e^{-j2\pi n d x_d/\lambda f_l}$. Similar to the treatment in Example 7.4, the vertical lines give rise in plane P_2 to a periodic sequence of primary maxima about the x_d axis separated by a distance $\pi/q = f_{l_2}\lambda/d$. Therefore, the lines can be removed by blocking these primary maxima in the Fourier transform plane P_2. The spatial filter should simply consist of opaque dots along the x_d axis beginning at the origin and spaced distances $f_{l_2}\lambda/d$ apart. Radii of dots are to be determined from the solution to Exercise 7.10.

7.17 If the CFA in plane P_1 is $U_t(x_s, y_s) + \cos(ay_s)$ then CFA in plane P_2 from the noise is

$$\mathbf{F}\{U_t(x_s, y_s)\} = \frac{1}{2}\,\delta\left(y_d - \frac{\lambda f_{l_2} a}{2\pi}\right) + \frac{1}{2}\,\delta\left(y_d + \frac{2\lambda f_{l_2} a}{2\pi}\right)$$

Therefore, the Fourier spectrum of the noise consists of impulses at $y_d = \pm\lambda f_{l_2} a/2\pi$ and can be filtered out by use of a spatial filter in plane P_2 that is transparent everywhere except at these two locations:

$$t(x_d, y_d) = \begin{cases} 0 & y_2 = \pm\dfrac{\lambda f_{l_2} a}{2\pi} \pm \epsilon \\ 1 & \text{elsewhere} \end{cases}$$

To prevent unnecessary distortion in plane P_3 of the signal $\mathbf{F}^{-1}\,\mathbf{F}\{U_t(x_s, y_s)\}$, ϵ in plane P_2 should be as small as possible.

7.18 In plane P_2, incident CFA is $\mathbf{F}\{\exp[j\alpha\phi_1(y_s)]\} \cong \mathbf{F}\{1 + j\alpha\phi_1(y_s)\} = \delta(y_d) + j\alpha\mathbf{F}\{\phi_1(y_s)\}$. Exiting plane P_2 is the CFA $t(y_d)[\delta(y_d) + j\alpha\mathbf{F}\{\phi_1(y_s)\}] \cong C_6 j\delta(y_d) + C_6 j\alpha\mathbf{F}\{\phi_1(y_s)\}$. In plane P_3, CFA is $C_7 j[1 + \alpha\mathbf{F}^{-1} \cdot \mathbf{F}\{\phi_1(y_s)\}]$. Irradiance in plane P_3 is $|1 + \alpha\phi_3|^2/2\eta_o \cong [1 + 2\alpha\phi_3(y_3)]/2\eta_o$ where $\phi_3(y_3) = \mathbf{F}^{-1}\mathbf{F}\{\phi_1(y_s)\}$. In this way phase information $\phi_3(y_3)$ is preserved. C_7 is defined in Eq. (7.6.6).

7.19 If $U_t(x_r, y_r) = A(1 + \cos 2\pi x_s/L) = A + (A/2)[\exp(2\pi x_s/L) + \exp(-2\pi x_s/L)]$, then incident on plane P_2 is the CFA $C_{6L_2} A\delta(x_d, y_d) + C_{6L_2}(A/2)[\delta(x_d - 2\pi\lambda f_{l_2}/L) + \delta(x_d + 2\pi\lambda f_{l_2}/L)$. If the filter passes radiation only for

$$x_d > \pi f_{l_2}/(L 2\pi/\lambda) = \frac{\lambda f_{l_2}}{2L} < \frac{2\pi\lambda f_{l_2}}{L}$$

then the dc term only is blocked by the filter, and exiting plane P_2 is the CFA

$$C_{6L_2}(A/2)[\delta(x_d - 2\pi\lambda f_{l_2}/L) + \delta(x_d + 2\pi\lambda f_{l_2}/L\}$$

This leads to CFA in plane P_3 equal to $C_7 A\cos(2\pi x_3/L)$ where C_7 is defined in Eq. (7.6.6).

7.20 Light reaching the film is that which passes through the slits in the diffraction grating. The grating causes a *sampled* rather than total image to be exposed on the film. The CFA reaching the film is then a *product* of the actual image CFA and the array function CFA described in Example 7.4 for diffraction grating. Multiplication in the spatial domain implies convolution in the Fourier domain. Therefore, if a positive transparency of the image multiplexed film $i(x_s, y_s)$ is placed in plane P_1 of the optical computer in Fig. 7.17, the Fourier transform in plane P_2 would be a convolution of the Fourier transform $\mathrm{I}(y_d, y_d)$ of $i(x_s, y_s)$ and the CFA in Eq. (7.5.12). The irradiance distribution in the Fourier plane P_2 would involve the irradiance in Eq. (7.5.14) and $|\mathbf{I}(x_d, y_d)|^2$. For example, assume the first image i_l is obtained when the grating slits are positioned horizontally. This image is sampled vertically. The Fourier transform of this image would then consist of spectra of $\mathbf{I}_1(x_d, y_d)$ positioned along the vertical (y_d) axis at the location of each principal maxima of Eq. (7.5.14), particularly since for small a $(\sin nqy_d/\sin qy_d)^2$ approaches a train of impulse functions which convolve $\mathbf{I}_1(x_d, y_d)$ to the spatial frequency location of each impulse function. If the second image (i_2) is obtained when the grating is rotated an angle θ clockwise, the direction of sampling is again perpendicular to the direction of the slits, and the Fourier transform of this sampled image in plane P_2 consists of spectra of $\mathbf{I}_2(x_d, y_d)$ along an axis at a clockwise angle θ with regard to the y_d axis. Similarly, if $i_3(x_s, y_s)$ is obtained when the grating is rotated another time through the angle θ, then in the Fourier domain the spectra $\mathbf{I}_3(x_d, y_d)$ are along a diagonal axis at an angle of 2θ clockwise from the y_d axis. (see the figure below). Thus many images can be multiplexed on the same piece of film, provided that for each image the diffraction grating has been rotated another angle θ, until almost 180° have been covered by grating rotation. From Exercise 7.11, the distance between centers of $\mathbf{I}_1$ spectra along the y_d axis is $f_{l_2}\lambda/[2(a + b)]$. This separation must be greater than the spatial-frequency bandwidth of each $\mathbf{I}_1$ spectrum so that no overlap between spectra occurs. This is in accordance with the sampling theorem and determines the required spatial sampling rate. If $2f_{r\,\max}$ designates a Fourier spatial-frequency bandwidth of $\mathbf{I}_1$ in the vertical direction from the origin, then $\Delta y_d = 2\lambda f_{l_2} f_{r\,\max}$ where Δy_d is the radius of each spectrum of $\mathbf{I}_1$. Therefore, $f_{l_2}\lambda/[2(a + b)] > 2f_{l_2}\lambda f_{r\,\max}$, or $4(a + b) > f_{r\,\max}^{-1}$. If L is the vertical dimension of the film, then $N = L/[2(a + b)] > 2Lf_{r\,\max}^{-1}$. In Eq. (8.5.8), a practical limit on $f_{r\,\max}$ is obtained as limited by lens diffraction. This provides a minimum value for N.

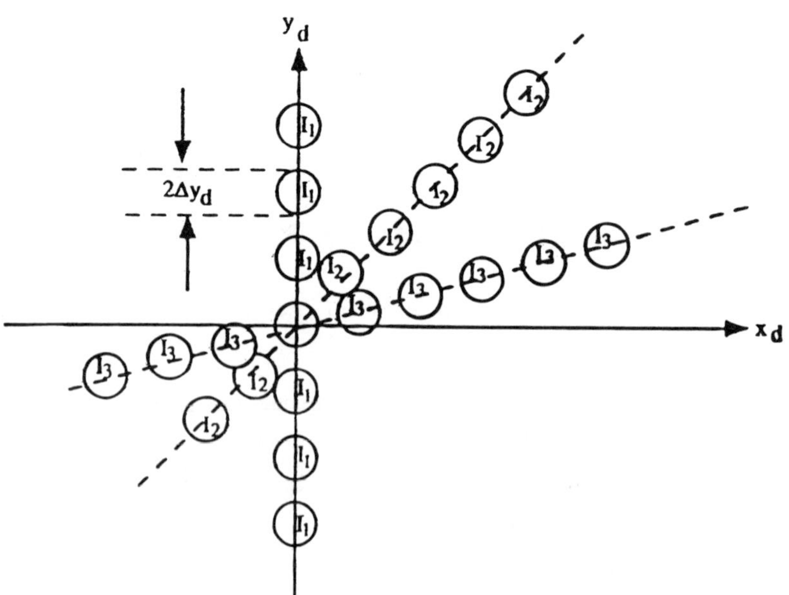

To demultiplex, simply cover all spectra in plane P_2 except the one whose original image is desired. That can be inversely Fourier transformed by lens L_3 in Fig. 7.17, and the original single image viewed in the output plane P_3.

7.21 When $h(x_4, y_4)$ is placed against L_4 rather than in its front focal plane, then from Eq. (7.3.17), since $z_o = 0$,

$$C_{6L4} = \frac{i}{\lambda f_{l4}} \exp[-jk_o(nd_o + f_l)] \exp \frac{-jk_o}{2f_l}(x_d^2 + y_d^2)]$$

and, as a result, the irradiance incident on the film in the Vander Lugt filter fabrication is described by Eq. (7.7.4), if

$$\mathbf{H}' = H \exp\left[-jk_o \frac{(x_d^2 + y_d^2)}{2f_{l_2}} + 2f_{l_4}\right]$$

is used in place of **H**.

Now, with regard to application in the optical computer, if P_1 in Fig. 7.17 is a distance z_o instead of f_{l_2} in front of lens L_2, then if $g(x_s, y_s)$ is the CFA signal exiting P_1, the CFA incident on the Vander Lugt filter is, from Eqs. (7.3.17) and (7.3.19),

$$U_{fl_2}(x_d, y_d, z_o + f_{l_2}) = C_6(x_d, y_d)\, \mathbf{G}(xd, y_d)$$

where

$$C_6(x_d, y_d, z_o) = \frac{j}{\lambda f_{l4}} \exp[-jk_o(nd_o + z_o + f_{l_2}) \exp\left[\frac{jk_o}{2f_{l_2}}\left(\frac{z_o}{f_{l_2}} - 1\right)(x_d^2 + y_d^2)\right]$$

When this is multiplied by the fourth term in the Vander Lugt filter transmittance, one obtains

$$C_6(x_d, y_d, z_o)\mathbf{G}(x_d, y_d) r_o C_{6L4}^*(\mathbf{H}')^* \exp(-2\pi\alpha y_d)$$

In order for this to be equal to the fourth term in Eq. (7.7.6), where $C_{6L_2} = (j/\lambda f_{l4}) \exp[-jk_o(nd_o + 2f_{l_2}])$, then

$$-z_o + \left(\frac{z_o}{f_{l_2}} - 1\right)\frac{(x_d^2 + y_d^2)}{2f_{l_2}} + \frac{(x_d^2 + y_d^2)}{2f_{l_2}} + f_{l4} = -2f_{l_2}$$

yielding

$$z_o = \frac{2f_{l_2}^2(f_{l4} - 2f_{l_2})}{x_d^2 + y_d^2 - 2f_{l_2}^2}$$

Now, since x_d and y_d are coordinates and z_o is distance between plane P_1 and lens L_2 and must therefore be constant, this solution is simply not feasible.

However, if $z_o = 0$, that is, transparency $g(x_s, y_s)$ in Fig. 7.17 is placed adjacent to lens L_2, then in order to obtain in plane P_2 the same fourth term as in Eq. (7.7.6), we have the inequality

$$-\left(\frac{x_d^2 + y_d^2}{2f_{l_2}}\right) + \frac{(x_d^2 + y_d^2)}{2f_{l_2}} + f_{l4} = -2f_{l_2}$$

or

$$f_{l4} = -2f_{l_2}$$

This solution is feasible. However, it is not even necessary since exp(−$jk_o f_{l4}$) is in any event a constant. Therefore, $z_o = 0$ is a feasible solution in order to obtain correlation between g and h despite the mistake in the Vander Lugt filter fabrication. To see this, multiply

$$C_6(x_d, y_d)\mathbf{G}(x_d, y_d)r_o C^*_{6L4}(\mathbf{H}')^* \exp(-j2\pi\alpha y_d) \text{ for } z_o = 0$$

The result is

$$\frac{j}{\lambda f_{l_2}} \exp\{-jk_o[nd_o + f_{l_2} - (x_d^2 + y_d^2)/2f_{l_2}]\}\mathbf{G}(x_d, y_d)r_o C^*_{6L4}H^*$$

$$\times \exp[jk_o(x_d^2 + y_d^2)/(2f_{l_2}) + 2f_{l4}] \exp(-j2\pi\alpha y_d)$$

$$= \frac{j}{\lambda f_{l_2}} \exp[-jk_o(nd_o + f_{l_2} + f_{l4} - 2\pi\alpha y_d]\mathbf{G}(x_d, y_d)r_o C^*_{6L4}\mathbf{H}^*$$

To within a constant, this is equal to the correlation term in Eq. (7.7.6)

7.22 From Eq. (7.9.36), $d_c = \bar{\lambda}R/(2\pi a)$. For the sun, $R = 93 \times 10^6$ miles, $a = 4 \times 10^5$ miles. Substituting, $d_c = 5.5 \times 10^{-7}(93 \times 10^6) = 2 \times 10^{-5}$ m. Even the sun exhibits partial coherence.

7.23 (a)

$$d_c = \lambda\bar{R}/(2\pi a) = \frac{0.63 \times 10^{-6}(1)}{4\pi(3 \times 10^{-3})} = 6.7 \times 10^{-5} \text{ m}$$

$$\rightarrow \text{coherence area } A_c = \frac{\pi}{4} d_c^2 = 3.5 \times 10^{-9} \text{ m}^2$$

(b)

$$d_c = \frac{0.63 \times 10^{-6}(1)}{\pi(3 \times 10^{-5})} = 6.7 \times 10^{-3} \text{ m} \rightarrow A_c = 3.5 \times 10^{-5} \text{ m}^2$$

The pinhole improves the coherence diameter because of the widened beam-width resulting from pinhole diffraction of beam.

Diffraction-limited beamwidth diameter r_d is aperture diameter D plus $2(1.22\bar{\lambda}R)/D$, where $R = 1$ m distance. Therefore, for part (a) $r_d = 3$ mm $+ 2(1.22\bar{\lambda})(R)/3$ mm $= 3.51$ mm $\approx 52.4d_c$. For part (b) $r_d = 30$ μm $+ 2[(1.22\bar{\lambda})(R)/30$ mm $= 5.13$ cm $\approx 7.66d_c$. Therefore, the pinhole improves the "coherence efficiency" of the beam since a greater portion of expanded beam is spatially coherent than of initial light beam.

7.24 From Eq. (7.9.35), $d_o = 1.22\bar{\lambda}R/2a = 1.22\,(5.89 \times 10^{-7})\cdot(0.5)/(5 \times 10^{-4}) =$ 0.72 mm. This yields the first zero of $|\gamma_{12}|$. Slit spacing should be just less than 0.72 mm.

8

8.1 Assume $A(\xi) = 1$ with no aperture limitations. Equation (8.1.10) becomes

$$\psi_i(x') = \int_{-\infty}^{\infty}\int_{-\infty}^{\infty} \psi_o(x) \exp\left[jk_o\xi\left(\frac{x}{s} + \frac{x'}{s'}\right)\right] d\xi\, dx$$

which, after integrating over ξ, becomes

$$\psi_i(x') = K_1 \int \psi_o(x)\delta\left(\frac{x}{\lambda s} + \frac{x'}{\lambda s'}\right) dx$$

$$= K_1\psi_o\left(-\frac{s}{s'}x'\right) = K_1\psi_o\left(-\frac{x'}{m}\right)$$

The image here is identical to the object but is inverted and is scaled with lateral magnification $m = s'/s$, as expected from geometrical optics. There is no diffraction blur.

8.2 From Eq. (8.5.2), assuming linearity with CFA, the impulse response for the obstructing aperture is subtracted from that of the clear aperture:

$$u(a) = K_1\pi\alpha_o^2\left[\frac{2J_1(\alpha_o a)}{\alpha_o a}\right] - K_1\pi(\epsilon\alpha_o)^2\left[\frac{2J_1(\epsilon\alpha_o a)}{\epsilon\alpha_o a}\right]$$

$$= K_1\pi\alpha_o^2\left\{\left[\frac{2J_1(\alpha_o a)}{\alpha_o a}\right] - \epsilon^2\left[\frac{2J_1(\epsilon\alpha_o a)}{\epsilon\alpha_o a}\right]\right\}$$

where α_o is defined in Eq. (8.5.1)

From L'Hopital's rule (see Section 7.4), CFA at the center where $a = 0$ is $K_1\pi\alpha_o^2(1 - \epsilon^2)$. Therefore,

$$\frac{s(a)}{s(0)} = \frac{1}{(1-\epsilon^2)}\left[\frac{2J_1(\alpha_o a)}{\alpha_o a} - \epsilon^2\frac{2J_1(\epsilon\alpha_o a)}{\epsilon\alpha_o a}\right]^2$$

8.3 The limiting aperture is the rectangle. From Exercise 7.6,

$$u(x', y') = K_2\beta\gamma\ \text{sinc}(\beta_o x')\sin(\gamma_o y')$$

where

$$\beta_o = \frac{2\pi a}{\lambda s'} \text{ and } \gamma_o = \frac{2\pi b}{\lambda s'}$$

Consequently $u(0, 0) = K_2\beta\gamma$. Therefore,

$$\frac{s(x', y')}{s(0, 0)} = [\text{sinc}(\beta_o x')\text{sinc}(\gamma_o y')]^2 \text{ (see Fig. 8.8a).}$$

8.4 From the solution of Exercise 7.9(c),

$$u(x', y') = 2K_2\pi\alpha_o^2 \cos(\beta_o a/2)\left[2\frac{J_1(\alpha_o a')}{\alpha_o a'}\right]$$

where

$\alpha_o = -\dfrac{\pi c^2}{\lambda s'}$, and a' is a radial coordinate equal to $[(x')^2 + (y')^2]^{1/2}$. $u(0, 0) = 2K_2\pi c^2$. Therefore,

$$\frac{s(x', y')}{s(0, 0)} = \left[\cos(\beta_o a/2)\ 2J_1(\alpha_o a')/(\alpha_o a')\right]^2$$

For the aperture of Exercise 7.9(b), CFA impulse response differs from the result here because of the multiplicative phase term. However, the normalized point spread function is identical to that above for aperture of Exercise 7.9(c).

8.5 For the rectangular aperture, using the figure below where $\beta_o = k_o(2a)/\lambda s'$ and $\gamma_o = k_o(2b)/\lambda s'$, the area of overlap is $[\gamma_o - (\omega_y - \gamma_o)][\beta_o - (\omega_x - \beta_o) = (2\gamma_o - \omega_y)(2\beta_o - \omega_x)$. Area of aperture is $4\beta_o\gamma_o$. Therefore,

$$\tau(\omega_x, \omega_y) = \begin{cases} (2\gamma_o - \omega_y)(2\beta_o - \omega_x)/4\beta_o\gamma_o & |\omega_x| \leq 2\beta_o,\ |\omega_y| \leq 2\gamma_o \\ 0 & \text{elsewhere} \end{cases}$$

or

$$\tau(\omega_x, \omega_y) = \begin{cases} |1 - \omega_y/2\gamma_o|\cdot|1 - \omega_x/2\beta_o| & |\omega_x| \leq 2\beta_o, |\omega_y| \leq 2\gamma_o \\ 0 & \text{elsewhere} \end{cases}$$

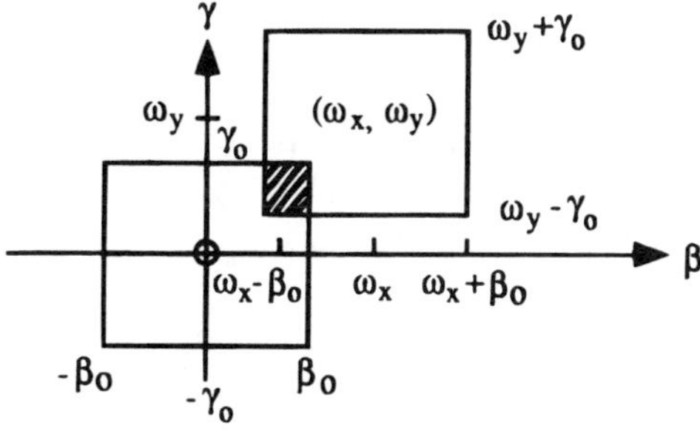

This result could have been achieved easily by applying Eq. (8.4.7) to both directions separately.

8.6 For the aperture of Exercise 7.9(c), the figure there is limiting aperture. Using the figure below, where

$$\alpha_o = -\frac{kc/2}{\lambda s'} \quad \text{and} \quad \beta_o = \frac{ka}{\lambda s'}$$

convolution yields the results shown for the ω_x direction, assuming $a > c/2$. In ω_y direction, each lobe is identical to OTF or $\tau(\omega)$ shown in Fig. 8.7.

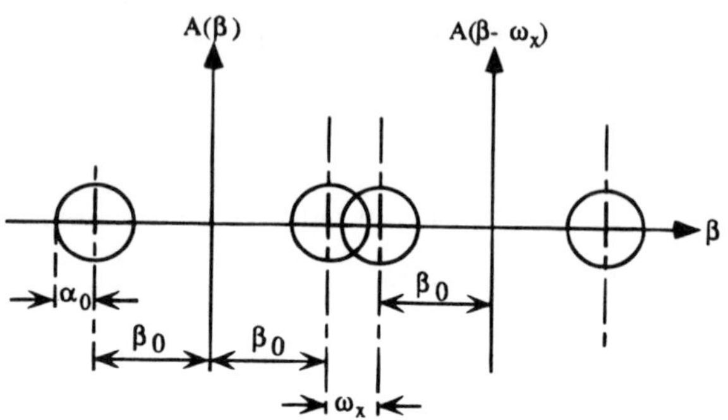

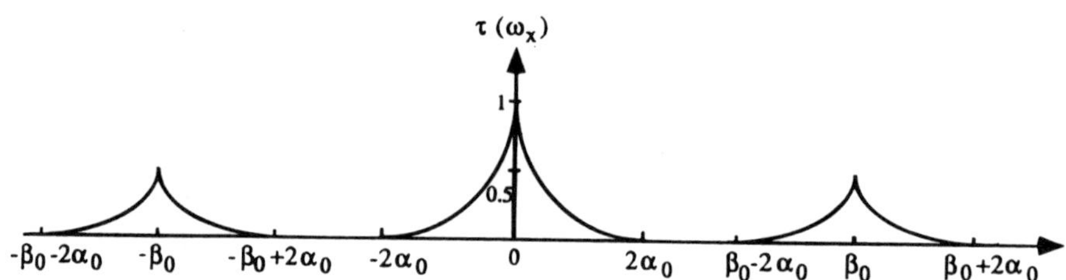

8.7 Because of the linearity of $u(x')$ with aperture function, the inner rectangular can be subtracted from the outer one, that is,

$$u(x') = 2K_1(\beta_{01}\gamma_{01}\,\text{sinc}\beta_{01}x'\,\text{sinc}\gamma_{01}y' - \beta_{02}\gamma_{02}\,\text{sinc}\beta_{02}x'\,\text{sinc}\gamma_{02}y')$$

where

$$\beta_{01} = 2k_0a/\lambda s', \quad \beta_{02} = k_0a/\lambda s' = \beta_{01}/2$$

$$\gamma_{01} = 2k_0b/\lambda s', \quad \gamma_{02} = \gamma_{01}/2$$

$$u(0) = 2K_1(\beta_{01}\gamma_{01} - \beta_{01}\gamma_{01}/4) = \frac{3}{2}\beta_{01}\gamma_{01}K_1$$

$$u(x')/u(0) = \frac{4}{3}\{\text{sinc}\beta_{01}x'\,\text{sinc}\gamma_{01},\, y' - (1/4)\text{sinc}(\beta_{01}x'/2)\text{sinc}(\gamma_{01}y'/2)\}$$

$$\frac{s(x)}{s(0)} = \left|\frac{u(x')}{u(0)}\right|^2$$

A sketch of $\tau(\omega_x)$ is shown below, based on Eq. (8.4.8). A similar sketch holds for $\tau(\omega_y)$ with cutoffs at $\omega_y = \pm 2\gamma_{01}$. If the stop in the middle increases in size, so too do the distortions of $\tau(\omega_x)$ and $\tau(\omega_y)$ from being straight lines as shown in Fig. 8.5. The result is a decrease in $\tau(\omega_x, \omega_y)$ at low spatial frequencies but an increase at high spatial frequencies.

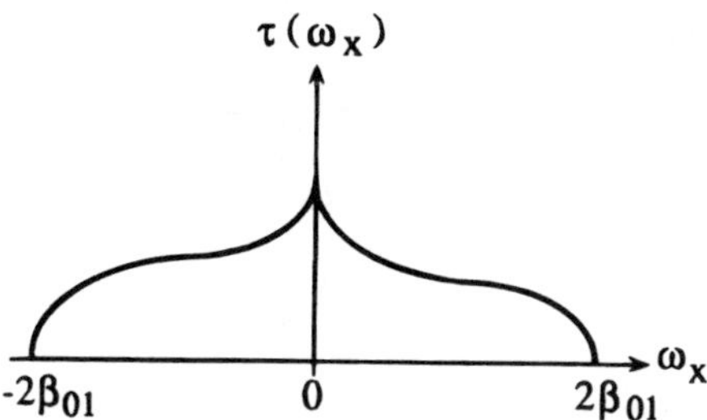

8.8 Applying Eq. (8.4.8), if the semicircular aperture is displaced in the horizontal direction, the overlap area is half that in Eq. (8.5.4). Since the aperture area is one-half that without the stop, the OTF in the horizontal direction is identical to Eq. (8.5.8), that is,

$$\tau(\omega_x) = \frac{2}{\pi}\left[\cos^{-1}\left(\frac{\omega_x}{2\alpha_0}\right) - \frac{\omega_x}{2\alpha_0}\sqrt{1 - \left(\frac{\omega_x}{2\alpha_0}\right)^2}\right]$$

where α_0 is described in Eq. (8.5.1). For vertical displacement of the aperture, the overlap and aperture areas are again identical to those without the field stop. However, while a horizontal displacement of at least $2\alpha_0$ is required to obtain zero overlap, a vertical displacement of only α_0 or more is required for zero overlap. Hence, $\cos\phi = \omega_y/\alpha_0$ and

$$\tau(\omega_y) = \frac{2}{\pi}\left[\cos^{-1}\left(\frac{\omega_y}{\alpha_0}\right) - \frac{\omega_y}{\alpha_0}\sqrt{1 - \left(\frac{\omega_y}{\alpha_0}\right)^2}\right]$$

The OTF has an elliptical rather than circular base. Spatial-frequency bandwidth in the y direction is from $-\alpha_0$ to $+\alpha_0$, while in the x direction it is from $-2\alpha_0$ to $+2\alpha_0$.

8.9 $A(\beta) = \delta(\beta + \beta_0) + \delta(\beta - \beta_0)$, $\beta_0 = k_o d/f_l$ where $s' = f_l$ because input radiation is plane wave.

(a) $u(x') = K_2 \int A(\beta) e^{-j\beta x'}\, d\beta = 2K_2 \cos\beta_0 x'$

(b) $s(x') = u(x')u^*(x') = 4K_2^2 \cos^2\beta_0 x' = 2K_2^2(1 + \cos\beta_0 x')$

(c) For incoherent imaging

$$H_i(x') = \int_{-\infty}^{\infty} H_0(x) s(x' - x)\, dx; \qquad H_0(x) = \begin{cases} 1 & |x| < D' \\ 0 & |x| > D' \end{cases}$$

$$H_i(x') = 2K_2^2 \int_{-D'}^{D'} (1)[1 + \cos 2\beta_0(x' - x)]\, dx$$

SOLUTIONS

$$= 2K_2^2\left[(2D') - \int_{x'+D'}^{x'-D'} \cos 2\beta_0 x_1\, dx_1\right], \qquad \text{where} \quad x_1 = x' - x$$

$$= 4K_2^2 D' + \frac{K_2^2}{\beta_0} \sin 2\beta_0 x_2 |_{x'-D'}^{x'+D'} = 4K_2^2 D'$$

$$+ K_2^2[\sin 2\beta_0(x' + D') - \sin 2\beta_0(x' - D')]/\beta_0$$

$$= 4K_2^2 D' + (2K_2^2/\beta_0) \sin 2\beta_0 D' \cos 2\beta_0 x'$$

$$= 4K_2^2 D' \,[(1 + \operatorname{sinc} 2\beta_0 D')\cos 2\beta_0 x']$$

The double-slit aperture gives rise to an interference pattern as in Young's experiment [Eq. (7.9.14)], thereby implying partial coherence does exist for "ordinary" light.

For *coherent* imaging

$$H_i(x') = \left|2K_2 \int_{-D'}^{D'} (1) \cos\beta_0(x' - x)\, dx\right|^2$$

$$= \left|2K_2 \int_{x'+D'}^{x'-D'} \cos\beta_0 u(-du)\right|^2, \qquad \text{where } u = x' - x$$

$$= \left|\frac{2K_2}{\beta_0} \sin\beta_0 u|_{x'-D'}^{x'+D'}\right|^2$$

$$= \left|\frac{2K_2}{\beta_0} [\sin\beta_0(x' + D') - \sin\beta_0(x' - D')]\right|^2$$

$$= \left|\frac{4K_L}{\beta_0} \sin\beta_0 D' \cos\beta_0 x'\right|^2 = 16K_2^2(b')^2 \operatorname{sinc}^2\beta_0 D' \cos^2\beta_0 x'$$

$$= 8K_2^2(D')^2 \operatorname{sinc}^2\beta_0 D'(1 + \cos 2\beta_0 x')$$

(d) From Eq. (7.9.16),

$$V = \frac{H_{i\,\max} - H_{i\,\min}}{H_{i\,\max} + H_{i\,\min}}$$

For incoherent imaging,

$$V = \frac{[4K_2^2 D'(1 + \operatorname{sinc} 2\beta_0 D')] - [4K_2^2 D'(1 - \operatorname{sinc} 2\beta_0 D')]}{4K_2^2 D'(1 + \operatorname{sinc} 2\beta_0 D' + 4K_2^2 D'(1 - \operatorname{sinc} 2\beta_0 D')}$$

$$= \operatorname{sinc} 2\beta_0 D' = \operatorname{sinc}(2k_0 dD'/f_i)$$

For coherent imaging,

$$V = \frac{8(K_2D')^2 \operatorname{sinc}^2\beta_0 D' - 0}{8(K_2D')^2 \operatorname{sinc}^2\beta_0 D' + 0} = 1$$

Better contrast (visibility) is obtained with coherent imaging as is to be expected for interferometry rather than with ordinary incoherent imaging.
(e) For incoherent imaging, $V = \operatorname{sinc}(2k_0dD'/f_l)$. A phase change of π in the argument of the sinc causes the visibility to change sign, that is, maxima and minima interchange:

$$\frac{4\pi(\Delta d)D'}{\lambda f_l} = \pi \rightarrow \Delta d = \frac{\lambda f_l}{(4D')}$$

Thus, changing the slit separation can be used to determine source diameter D'. This can be used in astronomy to determine star diameter.
(f) From Eq. (7.9.16), $V = |\gamma_{12}|$ if irradiance through both slits is equal.

9.1 Irradiance from the scene reaching the imager is proportional to the radiance of each object, which is equal to $H_s(\lambda)\rho(\lambda)/\pi$ assuming Lambertian objects, where $H_s(\lambda)$ is solar spectral irradiance. Therefore, at wavelength λ object scene MCF is

$$\mathrm{MCF}(\lambda) = \frac{[H_s(\lambda)/\pi][\rho_{\max}(\lambda) - \rho_{\min}(\lambda)]}{[H_s(\lambda/\pi[\rho_{\max}(\lambda) + \rho_{\min}(\lambda)]}$$

$$= \frac{\rho_{\max}(\lambda) - \rho_{\min}(\lambda)}{\rho_{\max}(\lambda) + \rho_{\min}(\lambda)}$$

9.2 (a) Since total thermal radiance is $N = \epsilon N_{bb}(T)$, total object scene MCF over essentially the whole spectrum is

$$\mathrm{MCO} = \frac{\epsilon_1 N_{bb}(T_1) - \epsilon_2 N_{bb}(T_2)}{\epsilon_1 N_{bb}(T_1) + \epsilon_2 N_{bb}(T_2)} = \frac{\epsilon_1 - \epsilon_2 N_{bb}(T_2)/N_{bb}(T_1)}{\epsilon + \epsilon_2 N_{bb}(T_2)/N_{bb}(T_1)}$$

$$= \frac{\epsilon_1 - \epsilon_2 (T_2/T_1)^4}{\epsilon_1 + \epsilon_2 (T_2/T_1)} = \frac{1 - (\epsilon_2/\epsilon_1)(T_2/T_1)^4}{1 + (\epsilon_2/\epsilon_1)(T_2/T_1)^4}$$

(b) Simply substitute $N_\lambda(\lambda, T_2)/N_\lambda(\lambda, T_1)$ in place of the Stephan-Boltzmann law in the above:

$$\mathrm{MCO}(\lambda) = (1 - a)/(1 + a)$$

where

$$a = \frac{\epsilon_2 (e^{h\nu/\lambda k T_1} - 1)}{\epsilon_1 (e^{h\nu/\lambda k T_2} - 1)}$$

9.3

$$\mathrm{LSF}(y') = \int_{-\infty}^{\infty} s(x', y')\, dx'$$

9.4 $H = \epsilon\sigma T^4 \rightarrow \Delta H = \sigma T^4 \Delta\epsilon + \bar{\epsilon}\sigma T^3 \Delta \mathrm{T}$
$= \sigma T^3 (T\Delta\epsilon + 4\bar{\epsilon}\Delta T)$
$= \sigma T^3 [(290(0.3) + 4(0.5)\Delta T] = \sigma T^3 (87 + 2\Delta T)$
As long as $\Delta T < 43.5$ K, thermal contrast derives primarily from emissivity differences. Therefore, since this condition on ΔT almost always holds for passive targets nighttime thermal contrast is very close to daytime thermal contrast for similar weather conditions [9.7]. This holds as long as $\Delta T <<$

$T_o\Delta\epsilon/4\bar{\epsilon}$, which is almost always the case for passive objects, where $T_o \approx$ 290 K. For active targets such as motors or engines, ΔT can be much larger.

9.5 From Fig. 9.4(a) at 900-nm wavelength, MCO $\approx (0.55 - 0.39)/(0.55 + 0.39) = 0.17$. At 500-nm wavelength MCO $\approx (0.34 - 0.28)/(0.34 + 0.28) = 0.097$. Note the contrast reversal at the two wavelengths. At the shorter wavelength, dry sand reflects more; whereas at the longer wavelength, concrete reflects more.

9.6

λ (μm)	2	5	8	12	20
MCO	0.91	0.86	0.84	0.83	0.82

The short-wavelength limit approaches unity if irradiance at shorter wavelengths is thermal rather than reflective. The long optical wavelength approximation limit approached is $|(1 - \epsilon_2/\epsilon_1)|/(1 + \epsilon_2/\epsilon_1) = 0.81$ since temperatures are similar. At 200-μm and 2-mm wavelengths, however, Eq. (9.6.2) with Eq. (9.6.3) substituted into it yields MCO $= 0.714$ and 0.974, respectively. Using Eq. (9.6.4) at 200-μm and 200-mm wavelengths, MCO $= 0.75$ and 0.974. Thus, as λ increases so too does the accuracy of the Taylor series approximation and Eq. (9.6.4) approaches unity. It is different from results of Eq. (9.6.7) because temperatures are close but not quite identical.

9.7 In Eq. (9.6.6), the argument of the exponents is almost always much less than unity. Therefore, using Taylor series expansions (first two terms only)

$$\text{MCO} \approx \frac{|1 - (1 + hc\Delta T/\lambda k R_1 T_2|}{1 + (1 + hc\Delta T/\lambda\lambda_k T_1 T_2} = \frac{hc\Delta T/\lambda k T_1 T_2}{2 + hc\Delta T/\lambda k T_1 T_2}$$

$$= \frac{\Delta T}{\Delta T + 2\lambda k T_1 T_2/hc}$$

9.8 The image of the illuminated pinhole should be less than the 10- × 10-μm pixel size. Thus magnification should be less than 0.5, implying

$$\frac{s'}{s} < 0.5 \rightarrow s > \frac{s'}{0.5} = \frac{12\text{ cm}}{0.5} = 24\text{ cm}$$

Now

$$\frac{1}{s} = \frac{1}{f_l} - \frac{1}{s'} \rightarrow s = \frac{s' f_l}{s' - f_l} = \frac{(12)(10)}{12 - 10} = \frac{120}{2} = 60\text{ cm}$$

This object distance satisfies the requirement $s > 24$ cm. At this distance image of pinhole diameter $= 20$ μm $(12/60) = 4$ μm $<$ pixel size, and image is real since $s > f_l$.

10.1 s' = image distance = $sf/(s - f) = 1000(50)/(1000 - 50) = 52.6$ mm
From Eq. (8.4.2),

$$\beta_{0x} = \frac{ka}{s'} = \frac{2\pi(11)}{5 \times 10^{-4}(52.6)} = 2630 \text{ rad}\cdot\text{mm}^{-1}$$

in the horizontal direction and

$$\beta_{0y} = -\frac{kb}{s'} = \frac{2\pi(18)}{5 \times 10^{-4}(52.6)} = 1911 \text{ rad}\cdot\text{mm}^{-1}$$

in the vertical direction. Therefore, in the horizontal direction,

$$\tau(\omega_{x\,\max}) = 0.02 = 1 - \omega_{x\,\max}/2\beta_{0x} \rightarrow \omega_{x\,\max} = 2\beta_{0x}\,(0.98)$$
$$\approx 2[2.630(0.98)] = 5155 \text{ rad}\cdot\text{mm}^{-1}$$

or $f_{x\,\max} = \omega_{x\,\max}/2\pi = 820$ cycles·mm^{-1}. In the vertical direction $f_{y\,\max} = f_{x\,\max}(8/11) = 597$ cycles·mm^{-1}. For detection, $n_{50} \approx 1$ and, from Eq. (10.2.4), $\Delta x \approx (f_{x\,\max}m)^{-1}$. Magnification $s'/s = 52.6/10^6 = 5.26 \times 10^{-5}$. Therefore,

$$\Delta x \approx [(820)(52.6 \times 10^{-6})]^{-1} = 23.2 \text{ mm}$$
$$\Delta y \approx [(597)(52.6 \times 10^{-6})]^{-1} = 31.8 \text{ mm}$$

10.2 The lens itself is the aperture and it is larger than the previous one. From Eq. (8.5.1),

$$\alpha_0 = \frac{kD}{2s'} = \frac{2\pi(25)}{2(5 \times 10^{-4})52.6} = 2986 \text{ rad}\cdot\text{mm}^{-1}$$

in any direction.

$$\tau(\omega_{r\,\max}) = 0.02 = \frac{2}{\pi}\left[\cos^{-1}\left(\frac{\omega_{r\,\max}}{2\alpha_0}\right) - \frac{\omega_{r\,\max}}{2\alpha_0}\sqrt{1 - \left(\frac{\omega_{r\,\max}}{2\alpha_0}\right)^2}\right]$$

Now, cutoff for an ideal circular aperture is the radian spatial frequency $2\alpha_o = 5973$ rad·mm^{-1}. A 2% threshold contrast implies $\omega_{r\,\max}$ slightly less, say about 5800 rad·mm^{-1}. This leads to $\tau = 5\%$ instead of 2%. Trial and error iterations lead to $\omega_{r\,\max} \approx 5580$ rad·mm^{-1} $\rightarrow f_{r\,\max} = 888$ cycles·mm^{-1}. ∴ $\Delta r \approx (2f_{r\,\max}m)^{-1} = [1776(5.26 \times 10^{-5})]^{-1} = 10.7$ mm from Eq. (10.2.4).

10.3 The TV tube is placed in the image plane. TV tube $f_{r\,\max}$ is $(N_{TV}/2)$ TV lines ÷ 11.8 mm. From Eq. (10.4.2) $x = 2$ m $= n_{50}(f_{r\,\max}m)^{-1}$ where $n_{50} = 4$

line pairs from the Johnson chart and $m = -f_l/s = 400$ mm/4 km $= 10^{-4}$ where f_l is focal length. Substituting, $2 = 4[N_{TV}/2)(0.0118)^{-1} \cdot 10^{-4}]^{-1}$ or $N_{TV} = 2(0.0236) \cdot 10^4 = 472$ TV lines.

10.4 From Eq. (10.4.2), $x = n_{50} \cdot (f_{r\,max} m)^{-1}$ where for identification in medium clutter $n_{50} = 6.4 \times 2 = 12.8$ line pairs. Optical magnification is $f_l/s = 0.5/5 \times 10^3 = 10^{-4}$. $f_{r\,max} = 250$ line pairs/11.8 mm. Therefore, $x_{identify} = 12.8\,(250/0.0118)^{-1}\,(10^4) = 6$ m;

$$\frac{n_{50recognize}}{n_{50identify}} = \frac{4}{6.4} \rightarrow x_{recognize} = \frac{4}{6.4} x_{identify} = 3.8 \text{ m}$$

For low clutter, object size resolved is reduced (improved) by 50%.

10.5 From Fig. 10.7, $n_{90}/n_{50} \approx 1.75$. Therefore, N_{TV} for 90% recognition probability $= 1.75\,(472) = 826$ TV lines.

10.6 $x_{identify} = (6 \text{ m})(1.75) = 10.6$ m. This is reduced by 50% to 5.3 m for low clutter and 90% identification probability.

10.7 $\tau(\omega_x) = 1 - |\omega_x|/2\beta_{0x}$. From Eq. (10.3.1),

$$N_{ex} = \int_{-\infty}^{\infty} |\tau(f_x)|^2\, df_x = 2\int_0^{\beta_{0x}/\pi} (1 - \pi f_x/\beta_{0x})^2\, df_x$$

$$= 2\int_0^{\beta_{0x}/\pi} [1 - (2\pi/\beta_{0x})f_x + (\pi/\beta_{0x})^2 f_x^2]\, df_x$$

$$= 2\left[f_x\Big|_0^{\beta_{0x}/\pi} - (\pi/\beta_{0x})f_x^2\Big|_0^{\beta_{0x}/\pi} + (\pi/\beta_{0x})^2 f_x^3/3\Big|_0^{\beta_{0x}/\pi}\right]$$

$$= 2\beta_{0x}[1/\pi - 1/\pi + 1/(3\pi)] = 2\beta_{0x}/(3\pi)$$

$$= 2[2630 \text{ rad}\cdot\text{mm}^{-1}/(3\pi \text{rad cycle}^{-1})]$$

$$= 558.1 \text{ cycles}\cdot\text{mm}^{-2}$$

$$N_{0y} = 2\int_0^{\beta_{0y}/\pi} (1 - \pi f_y/\beta_{0y})^2\, df_y = 2\beta_{0y}/(3\pi) = 2(1911)/(3\pi)$$

$$= 405.5 \text{ cycles}\cdot\text{mm}^{-1}$$

Note $N_{ey} < N_{ex}$ since $b < a$.

10.8

$$N_{ex} = \int_{-\infty}^{\infty} |\tau(f_x)|^2\, df_x$$

$\tau(f_x) = \text{sinc}2\pi f_x a$ as shown in Section 9.7.1. Here $a \equiv b = 10$ µm. From integral tables,

$$\int_0^{\infty} (\sin^2 x/x^2)\, dx = \pi/2$$

Let $x = 2\pi f_x a \rightarrow dx = 2\pi a\, df_x$ or

$$N_{ex} = N_{ey} = 2(2\pi a)^{-1} \int_0^{\pi} (\text{sinc} x)^2\, dx$$

where f_x and f_y are limited because of pixel size to $1/2a$ and $1/2b$ respectively. Therefore,

$$N_{ex} = N_{ey} = 2(2\pi a)^{-1} \int_0^{\pi} (\text{sinc} x)^2\, dx < 2(2\pi a)^{-1} \int_0^{\infty} (\text{sinc} x)^2\, dx$$

$$< 2(2\pi a)^{-1}(\pi/2) = \frac{1}{2a} = 10^5/2 = 50 \text{ cycles}\cdot\text{mm}^{-1}$$

Since contributions to the integral for frequencies beyond the first zero of the sinc^2 function are small, the above represents an upper limit approximation to N_{ex} and N_{ey}.

10.9 Since N_{ex} and N_{ey} in the previous exercise are much less than those in Exercise 10.7, it is clearly the CCD that limits resolution, rather than the optics. Another way of seeing this is to consider the spatial frequency cutoffs of both MTFs. For the CCD, $f_{x\text{ cutoff}} = f_{y\text{ cutoff}} = 1/2a = 1/2b = 1/20\ \mu\text{m} = 50$ cycles·mm^{-1} in each of the positive and negative directions, or 100 cycles·mm^{-1} altogether in each direction. (This is equal to 2 Ne). For the lens and rectangular aperture combination, spatial frequency cutoffs are at $\pm 2\beta_{ox}$ and $\pm 2\beta_{oy}$ or at $\pm 2(2630)/2\pi$ and $\pm 2(1911)/2\pi$ cycles·mm^{-1} in each of the plus-and-minus x and y directions. Clearly, the optics cutoffs are at spatial frequencies $>>50$ cycles·mm^{-1}, which is the CCD cutoff.

10.10 From the previous exercise, optics MTF can be neglected. System MTF is essentially that of the CCD, and $\tau(f_x) = \text{sinc}(2\pi f_x a)$. Assuming threshold contrast of 2%, $\tau(f_{x\text{ max}}) = 0.02 = \text{sinc}(2\pi f_x a)$ where $a = 10\ \mu\text{m} = 0.01$ mm. Since $f_{x\text{ max}} < f_{x\text{ cutoff}} = 50$ cycles·mm^{-1}, let us try $f_{x\text{ max}} \approx 47$ cycles·mm^{-1}. This leads to $\tau(f_{x\text{ max}}) = 0.063$ instead of 0.02. Trial and error leads us to $f_{x\text{ max}} = f_{y\text{ max}} \approx 48.95$ cycles·mm^{-1}. Substituting into Eq. (10.2.4), $\Delta x \approx (2f_{x\text{ max}} \cdot m)^{-1} = [2(48.95)(5.26 \times 10^{-5})]^{-1} = 194$ mm, which is much worse than the diffraction-limited resolution.

10.11 As shown above, the limiting resolution derives from the CCD rather than from the optics. This is even more so in the present exercise since lens diameter is twice that in the previous exercise. Although focal length affects optics MTF, its primary effect on resolution derives from its effect on magnification. At kilometer-long distances, $s' \approx f_l$ and $m = f_l/s$, where f_l is focal length, and s and s' are object and image distances, respectively. From Eq. (10.4.5), $x \approx 2n_{50}(f_{r\text{ max}} m)^{-1} = 2n_{50}(f_{r\text{ max}} f_l/s)^{-1}$. Substituting, using minimum object dimension and the Johnson chart value of $n_{50} = 2(4) = 8$ for recognition through medium clutter, we have

$$4 \geq 16(48{,}950)^{-1}(f_l)^{-1}(2000) \qquad \text{where } f_{x\,\max} = 48{,}950 \text{ cycles}\cdot\text{m}^{-1}$$

from the previous exercise and derives from CCD limiting resolution. Accordingly, $f_l \geq 163$ mm. For 90% probability, $n \approx 1.75n_{50} = 14$ line pairs instead of 8. Therefore, $f_l \geq 288.8$ mm.

10.12 From Eq. (10.5.4a), $P(t) = 50\% = P_\infty\{1 - \exp(-t/6.8n_{50}/n)\}\}$ where t is 8 s. However, it is more convenient to use the approximation $\tau \approx 3.4/P_\infty$ since this yields

$$0.5 = P_\infty - P_\infty \exp(-8P_\infty/3.4)$$

which is limited to one variable, a graphical solution to which converges to $P_\infty \approx 0.64$. From Fig. 10.7, this value of P_∞ corresponds to $n = 1.17n_{50} = 1.17(2) = 2.34$ line pairs.

Equation (10.4.5) can be used now to yield $x = 4 \leq 2n(f_{x\,\max}f_l)^{-1}\cdot s = 2(2.34)(48{,}950)^{-1}f_l^{-1}$ (2000), the result of which is $f_l \geq 48$ mm. For this case, from Eq. (10.5.10) glimpse probability is about $0.044n/n_{50} = 0.044(1.17) = 0.05$.

10.13 From Fig. 10.7, $n = 1.75n_{50}$ for 90% probabilities = 1.75(6.4) for 90% identification probability. This product yields 11.2 line pairs. Therefore, using Eq. (10.4.5), $x = 2 \leq 2(11.2\cdot)\,[(48{,}950)]^{-1}\,(f_l)^{-1}(5000) \rightarrow f_l \geq 1144$ mm. Expected acquisition time is $\tau = 6.8n/n_{50} = 6.8\,(1.75) = 11.9$ s.

10.14 In Exercise 10.11, $n/n_{50} = 1 \rightarrow \tau \approx 6.8$ s for 50% probability; for increased magnification and focal length, $n/n_{50} = 1.75 \rightarrow \tau = 6.8/1.75 = 3.9$ s for 90% probability. In Exercise 10.12, $n/n_{50} \approx 1.17 \rightarrow \tau = 6.8/1.17 = 5.8$ s. From Eq. (10.5.4b) this corresponds to ≈48% acquisition probability during 8 s.

10.15 From Eq. (10.4.5), $x = 4 = 2n(f_{r\,\max}m)^{-1}$ where again $f_{r\,\max} = 48.95$ cycles·mm^{-1} since the CCD is still limiting resolution, particularly with such a large diameter objective. [For the telescope, $f_c = 2\alpha_o/2\pi = kD/f_l = 2(80)/(5 \times 10^{-4})(1200) = 266.7$ cycles·mm^{-1} >> $f_{x\,\text{cutoff}}$ for the CCD, which is 50 cycles·mm^{-1}.] Using the above value for $f_{r\,\max}$, and considering that $m = 1.2/8000 = 1.5 \times 10^{-4}$, $n = 2(48{,}950)(1.5 \times 10^{-4}) = 14.7$ line pairs. For medium clutter recognition, $n_{50} = 6.4 \times 2 = 12.8$. Therefore, $n/n_{50} = 1.15$.

From Fig. 10.7, $P_\infty^* = 0.63$. For detection, however, $n/n_{50} = 14.7/(1 \times 2) = 7.35$, and $\tau = 6.8/7.35 = 0.93$ s. Substituting into Eq. (10.5.4b), $P(t) = 0.63(1 - e^{-5/0.93}) = 0.63$. Expected acquisition time $\tau = 0.93$ s, and $P_g = 0.044(n/n_{50}) = 0.044(7.35) = 0.32$. Therefore, from Eq. (10.5.9b), $t_g = t_f/P_g = 0.3/0.32 = 0.93 \text{ s} \approx \tau$.

10.16 **(a)** n_{FOV} = FOR/FOV = 98.3 mrad/(11.8/1200) = 10 FOV.

(b) From Eq. (10.5.11), $P_s = 1 - e^{-t_o/\tau} = 1 - e^{-1.8/5.8} = 0.27$ using solution 10.14 for τ.

SOLUTIONS

(c) From Eq. (10.5.12), $P_5 = P_\infty[1 - (1 - P_s)^5] = 0.63[1 - (1 - 0.27)^5] = 0.50$.

(d) From Eq. (10.5.13), $t_1 = t_o n_{FOV}[(2 - P_s)/2P_s] = (1.8)(10)[(2 - 0.27)/0.54] = 57.7$ s.

Note that $t_1 >> n_{FOV} t_o = t_s = 18$ s. The approximation $t_1 \approx n_{FOV}\tau = 10(5.8) = 58$ s is then valid.

(e) From Eq. (10.5.14), $P(t) = 0.63(1 - \exp(-15/57.7) = 0.14$ for 15 s and $P(t) = 0.63(1 - \exp(-70/57.7) = .44$ for 70 s. Note that the latter is less than $p_5 = 0.50$ in part (3) because five complete scans require $5t_s = 5n_{FOV}t_o = 5(10)(1.8) = 90$ s, which is greater than the 70 s here.

10.17 If n_{FOV} is large, then $t_1 \approx t_o n_{FOV}/P_s = t_o n_{FOV}/[1 - \exp(-t_o/(\tau)] \approx t_o n_{FOV}/[1 -(1 - t_o/\tau + \cdots)] \approx t_o n_{FOV}/(t_o/\tau) = n_{FOV}\tau$ provided $t_o << \tau$.

10.18 Using Eq. (10.6.2), $xy = 400 \leq 4n_x n_y (f_{x\ max}\ f_{y\ max})^{-1}\ m^{-2} = (4)16.5[(48{,}950)(245/0.0088)]^{-1}\cdot(f_l/10^4)^{-2} \rightarrow f_l \geq 110$ mm. By Johnson's criteria, the vertical dimension is the shorter or critical dimension. Accordingly, from Eq. (10.4.7), $y = 2n_{y50}\ (245/0.0088)^{-1}(f_l/10^4)^{-1}$ where $f_l = 110$ mm and $n_y = 4$ line pairs. The right side indicates an object size of 26.1 m would be recognized with this focal length by the Johnson criteria. This is much greater than the ship height of 8 m. Hence, the focal length would have to be increased above 110 mm to meet the 50% probability Johnson criteria. The focal length would have to be at least 110 mm (26.1/8) = 359 mm. Note that here $f_{y\ max}$ is evaluated according to raster scan (245 pixel pairs over 8.8 mm) since this (27.84 line pairs·mm^{-1}) is much less than that obtained from pixel size (48.95 line pairs·mm^{-1}).

10.19 (a) The parabolic index fiber serves as a lens as shown in Section 2.10. In order for the input picture to fill up the entire field of view,

$$\frac{x/2}{s} = \tan\theta = \frac{2.5}{s}\ \text{cm}$$

where θ is the half-angle of acceptance to the fiber. By definition, numerical aperture = 0.3 = (1) sinθ. As a result, $\theta = 17.45°$ and $s = 7.95$ cm. (b) Assuming the angular cone of divergence at fiber output is equal to that at input, the lateral magnification is $m = s'/s = x'/x = 0.3\ \text{cm}/5\ \text{cm} = 0.06$. Therefore, $s' = 0.6\ s = 0.48$ cm. The fiber exit should be this distance from the eye. The same result can be obtained from $x'/(2s') = \tan\theta = 0.15\ \text{cm}/s'$ where tan (17.45°) = 0.3145 as in part (a); (c) $1/s + 1/s' = 1/f_l \rightarrow$ focal length = 0.45 cm; (d) spatial frequency bandwidth required is 130 lp/0.3 cm = 430 cycles·cm^{-1}. This must be less than the diffraction-limited bandwidth $2\alpha_o$ for a circular aperture fiber of diameter D, where

$$2\alpha_o = D/(\lambda s') \geq f_{r\ max} = 430\ \text{cycles}\cdot\text{cm}^{-1}$$

Consequently,

$$D \geq \lambda s' f_{r\ max} = (0.55\ \mu\text{m})(0.45\ \text{cm})(430\ \text{cm}^{-1}) = 0.01\ \text{cm}$$

10.20 In example 10.3, MCI = MCO · M_B = 0.229. We assume again a required threshold contrast of 0.02. Overall modulation contrast in the image plane is MCI = (MCO)M_B $|\tau(f_x)|$ = 0.229 sinc$2\pi f_x a$. Therefore, MCI = 0.02 at the spatial frequency where sinc $2\pi f_{x\,\max} a$ = 0.02/0.229 = 0.0873. Because of the reduced image contrast stemming from reduced object plane contrast, $f_{x\,\max}$ is reduced to 45.95 cycles·mm^{-1}. Substituting into Eq. (10.2.4), $\Delta x = (2 f_{x\,\max} m)^{-1} = [2(45.95)(5.26 \times 10^{-5})]^{-1} = 207$ mm. Because the sinc function falls off quite rapidly as it approaches zero, the resolution impairment as compared to Exercise 10.10 stemming from reduced object plane contrast is not too severe in this exercise.

11.1 From Eq. (11.1.16),

$$\mathrm{SNR}_D = \frac{2^{3/2}(\mathrm{MCO})(a/A)^{1/2} i_{av} G_I G_s (B\tau_i)^{1/2}(\mathrm{MTF})_{sT}}{\{[2qi_{av}B + (i_{av}\sigma_\eta/\overline{\eta})^2]G_I^2 G_s^2 + 2qi_{spd}G_s^2 B\}^{1/2}}$$

As irradiance increases, so too does i_{av}. Therefore, the spatial noise is dominant in the denominator and

$$\mathrm{SNR}_D \approx 2^{3/2}(\mathrm{MCO})(a/A)^{1/2}(B\tau_i)^{1/2}(\mathrm{MTF})_{sT}\overline{\eta}/\sigma_\eta$$

In this case SNR_D is proportional to $\overline{\eta}/\sigma_\eta$. Improvements in uniformity are desired to improve signal-to-noise ratio.

11.2 (a) From Eq. (11.1.14b),

$$\mathrm{SNR}_{IT} = \frac{2^{3/2}(\mathrm{MCO})(a/A)^{1/2} i_{av}(B\tau_i)^{1/2}(\mathrm{MTF})_s}{[2qB(i_{av} + i_{spd})]^{1/2}}$$

$$= \frac{2(\mathrm{MCO}) i_{av}(a/A)^{1/2}(\mathrm{MTF})_s}{(\mathrm{FR})^{1/2}[q(i_{av} + i_{spd})]^{1/2}}$$

where FR is frame rate. Here, $i_{av} << i_{spd}$ and $(\mathrm{MTF})_s = M_B \cdot M_o$ where M_o is optics MTF and

$$(\mathrm{MCO})M_B = \mathrm{MCI} = \frac{\rho_t}{\rho_t + 2H_A/(\tau' H_s)}$$

from Eq. (10.7.2), where $\tau' = \exp(-3) = 0.05$ since optical depth of atmosphere is three. Substituting,

$$\mathrm{MCI} = \frac{0.7}{0.7 + 2(100)/[(0.05)(1000)]} = 0.15$$

and $i_{av} = (i_T + i_b)/2 = \Re\, A[H_T(a/A) + H_b]2$ where $\Re$ is responsivity. Although H_T and H_b are target and background irradiances incident on receiver optics, currents are determined by irradiance in image plane. From Eq. (3.4.8), $H_T = \tau' W/(4F^2) = \tau' \rho_t Hs/(4F^2) = 0.05(0.7)1000/[4(64)] = 0.14\ \mathrm{W \cdot m^{-2}}$.

Path luminance background passes through the same optics and, assuming $\tau' \approx 1$ for H_b near the receiver, $H_b \approx 100/[4(64)] = 0.39\ \mathrm{W \cdot m^{-2}}$. From the above,

$$i_{av} = 0.1(11.8 \times 10^{-3})(8.8 \times 10^{-3})[0.14(1/10) + 0.39]/2$$

$$= 2.1\ \mu\text{A} << i_{spd}$$

In view of the above,

$$\text{SNR}_{IT} = \frac{2(0.15)(2.1 \times 10^{-6})(0.1)^{1/2} M_o(f_r)}{(25)^{1/2}[1.6 \times 10^{-19}(3.002 \times 10^{-3})]^{1/2}} = 1.82 \times 10^3 M_o(f_r)$$

11.3 (a) From Eq. (11.1.16),

$$\text{SNR}_D = \frac{2^{3/2}(\text{MCO})(a/A)^{1/2} G_I G_s i_{av} (B\tau_i)^{1/2} (\text{MTF})_{ST}}{(2qi_{av} G_I^2 G_s^2 B + 2qi_{spd} G_s^2 B)^{1/2}}$$

$$\approx \frac{2\text{MCI}(a/A)^{1/2} G_I i_{av} M_o(f_r) M_{ID}(f_r)}{(FRqi_{spd})^{1/2}}$$

$$= \frac{2(0.15)(0.1)^{1/2}(1)(2.1 \times 10^{-6}) M_o(f_r) M_{ID}(f_r)}{25^{1/2}[1.6 \times 10^{-19}(3 \times 10^{-3})]^{1/2}}$$

$$= 1.82 \times 10^3 M_o(f_r) M_{ID}(f_r)$$

where $M_{ID}(f_r)$ is combined imager-signal-processing-display MTF.

(b) The above results indicate that SNR_D is reduced to threshold values as in Table 11.1 only at spatial frequencies so high that $M_o(f_r) \cdot M_{ID}(f_r)$ is on the order of 0.2 or 0.3%. Such low MTF values are much less than required contrast thresholds for the human visual system at high spatial frequencies, so that $f_{r\,\max}$ is determined by required contrast threshold rather than by SNR_D threshold. Therefore, imaging is contrast-limited.

11.4 (a) If $G_I = 0.005$, then

$$\text{SNR}_D = \frac{2(0.15)(0.1)^{1/2}(0.005)(2.1 \times 10^{-6}) M_o(f_r) M_{ID}(f_r)}{25^{1/2}[1.6 \times 10^{-19}(3 \times 10^{-3})]^{1/2}}$$

$$= 9.1 M_o(f_r) \cdot M_{ID}(f_r)$$

Therefore, for threshold SNR_D contrast levels on the order of 2.8 to 5.8, $M_o(f_r) M_{ID}(f_r)_I$ must be between 31% and 64%. This can be expected to be well above required threshold contrast values, so that $f_{r\,\max}$ is determined now by SNR threshold rather than by contrast threshold.

(b) If $G_I = 0.001$, then $\text{SNR}_D = 1.82 M_o(f_r) M_{ID}(f_r)$. Since required SNR_D thresholds are on the order of 2.8 to 5.8, then at all spatial frequencies, including dc, SNR_D is below required threshold and image quality is so poor that nothing can be "detected." Imaging is very noise limited.

11.5 (a) From the solution to Exercise 11.4(a), using Fig. 11.1,

$$\text{SNR}_D = 9.1 M_o(f_r) \cdot M_{ID}(f_r)$$

Using Table 11.1 for 50% probability of detection, $2.8 = 9.1 \exp(-0.04 f_r)$.

S O L U T I O N S

$$\therefore f_{r\,\max} = -\frac{1}{0.04}\ln\left(\frac{2.8}{9.1}\right) = 29.5\ \text{cycles}\cdot\text{mm}^{-1}$$

(b) For contrast-limited imaging with 2% threshold contrast

$$0.02 = \exp(0.04 f_r)$$

$$f_r = -(1/0.04)\ln(0.02) = 97.8\ \text{cycles}\cdot\text{mm}^{-1}$$

(c) Comparison of parts (a) and (b) indicates the system is noise limited and, from part (a),

$$\Delta r' = (2 \cdot f_{r\,\max})^{-1} = [2(29.5)]^{-1}\ \text{mm} = 0.017\ \text{mm}$$

This refers to the small imager (sensor) plane. On a 14- or 20-in. monitor this detail would be much larger and easily detectable if SNR_D were sufficiently high.

(d) Assuming a rectangular target, the spatial extent of target image in x direction $\approx 0.1W_x$ and spatial extent of target in y direction $\approx 0.1V_y$. For identification through uniform clutter, Table 11.1 indicates 13 equivalent bars are required over the target's critical dimension. Therefore, $N_{sf} \approx 13/01 = 130$ bars/critical dimension. Extrapolating from data for $N_{sf} = 100$ and 300, SNR_D threshold ≈ 5.5 for $N_{sf} = 130$. From part (a), $5.5 = 9.1\exp(0.04f_r) \rightarrow f_{r\,\max} = 12.6\ \text{cycles}\cdot\text{mm}^{-1}$. Since this refers to each equivalent bar, $\Delta r' = 13 f_{r\,\max}^{-1} = 1.03$ mm.

11.6 According to Table 11.1 and Fig. 11.3, $\text{SNR}_{DT} = (3.6)(1.30) = 4.68$. Using Eq. (11.1.25), $(\log \text{SNR}_D - \log 3.6)/0.198 = 1.282$. Accordingly $\log \text{SNR}_D = 0.810$ and $\text{SNR}_D = 6.46 > 4.68$.

11.7 From Eq. (11.2.10),

$$z = [\alpha(\lambda, T) + \beta_{sys} n/2\Delta x]^{-1} \ln\left(\frac{\Delta T\sqrt{\alpha_T/7}}{\text{SNR}_N \text{MRT}_o}\right)$$

From Section 10.6, for thermal image identification $n = 8$. Therefore, $\alpha_T = 48{:}1$.

$$8.2\ \text{km} = [0.5\ \text{km}^{-1} + \beta_{sys}(8/3)]^{-1} 5.31$$

$$0.50\ \text{km}^{-1} + (8/3)\, m^{-1}\beta_{sys} = (5.86/8.2)\ \text{km}^{-1}$$

$$\beta_{sys} = (3/8)(0.71 - 0.50)\ \text{km}^{-1} = 0.08/\text{mrad}\cdot\text{cycle}^{-1}$$

Therefore, focal length and magnification must be increased by a factor of $0.63/0.08 = 7.84$ for the same thermal imager.

11.8 If viewed from the front, then for recognition the tank dimensions must be filled by at least 8 bars, where each is of aspect ratio 8:1. Therefore, $\alpha_T = 8$, and

$$z = [\alpha(\lambda, T) + \beta_{sys} n/\Delta x]^{-1} \ln\left[\frac{\Delta T\sqrt{\alpha_T/7}}{(\mathrm{SNR}_N)\mathrm{MRT}_o}\right]$$

$$= 0.75 \text{ km} \ln\left[\frac{2(\sqrt{8/7})}{(1.5)(0.01)}\right] = 0.75 \text{ km } (4.96) = 3.72 \text{ km}$$

If a focal plane array is used $z = 1.38$ km.

11.9 Since SNR_N of 1.5 is required, P^*_∞ is given as being 90% for recognition. From Eq. (10.5.3), $\tau \propto (n/n_{50})^{-1}$. From the Johnson chart, n/n_{50} for detection is one-quarter that required for recognition. Furthermore, from Eq. (10.5.3), we also see that $\tau \propto P_\infty^{-1}$. Hence, for detection $\tau \approx 3.4/P_\infty \approx (1/4)(3.4/P^*_\infty) = (1/4)(3.4/0.9) = 0.94s$. (Note that the slope of Fig. 11.5 is almost linear over this region of probabilities.) From Eq. (10.5.4b), $P(2s) = P_\infty[1 - \exp(-t/\tau)] = 0.9[1 - \exp(-2/0.94)] = 0.79$.

11.10 From Exercise 11.9, $\alpha_T = 8$. From Eq. (11.2.9),

$$\mathrm{SNR}_N(z, f_{a\,\max}) = \frac{\Delta T\sqrt{7/\alpha_T}\exp[-\alpha(\lambda, T)z]}{\mathrm{MRT}_o \exp(\beta_{sys} nz/x)}$$

$$= \frac{2\sqrt{7/8}\exp[-(0.5)3]}{0.01\exp[0.63(4)(3)/3]} = 3.14$$

From Fig. 11.5, $P_\infty \approx 1$. From Eq. (10.5.3) and the analysis in the previous solution, $\tau \approx (1/4)3.4 = 0.85s$. Therefore, $P(4s) = P_\infty[1 - \exp(-\tau/t)] = 1[1 - \exp - (4/0.85)] = 0.99$.

SOLUTIONS

12

12.1 The planar angle subtended by the letters is 1 mm/2.5 × 10^5 mm = 4 × 10^{-6} rad. Now, 1 minute of arc is 2.91 × 10^{-4} rad. Normal visual acuity requires a resolution angle five times larger, or 1.455 × 10^{-3} rad. Therefore, angular magnification required is 1.455 × 10^{-3}/4 × 10^{-6} = 364.

12.2 The J curve in Fig. 12.17 indicates minimum contrast threshold is at around 2 cycles·deg^{-1}. From Example 12.1, $\Delta\alpha' = (2f_{a\,\max})^{-1} = (2.2\ \text{cycles·deg}^{-1})^{-1}$ = 0.25 deg = 4.36 mrad. Best magnification would be 4.36 × 10^{-3}/4 × 10^{-6} = 1091.

12.3 Normal photopic visual acuity is based on human visual system MTF, which peaks at about 6 cycles·deg^{-1} as shown in Fig. 12.13. Contrast threshold required is minimum at about 2 cycles·deg^{-1} as shown in Fig. 12.17. Therefore, the above two magnification solutions differ by about a factor of 3. If the object is of high contrast and there is little path luminance, then 364 magnification is better. For high path luminance or small system MTF bandwidth, contrast is degraded and 1091 magnification would be required.

12.4 From Section 2.12 the planar field of view is approximately 1 in./1 m = 22.5 mrad. The planar angle subtended by the object is 2 m/5 km = 0.4 mrad. Therefore, the object fills 0.4/22.5 = 1.78% of the field of view in either of its two orientation directions. A 12- × 9-in. display would then allow the scene to occupy a 9-in. diameter circle. The target image would occupy 1.78% of this or 0.16 in. or 0.4 cm in either of its orientation directions. If z is the viewer distance then, using the nomenclature of Example 12.1,

$$\Delta\alpha' = 0.4\ \text{cm}/z \text{ and } f_a = (0.8\ \text{cm}/z)^{-1} = z/(0.8\ \text{cm})\ \text{cycles·rad}^{-1}$$

From Fig. 12.17, this should be equal to about (2 cycles·deg^{-1} · 180/π) = 114.6 cycles·rad^{-1}. Therefore, z = (0.8 cm)(114.6 cycles·rad^{-1}) = 91.7 cm.

For high-contrast viewing, f_a should be equal to about 6 cycles·deg^{-1} or 343.8 cycles·rad^{-1}. Therefore, z = 0.8 cm (343.8 cycles·rad^{-1}) = 2.75 m, or about a distance three times larger than that for low contrast.

12.5 Since −1 diopters are appropriate when a person cannot see clearly beyond 1 m, then −5 diopters are required if he cannot see clearly beyond 20 cm.

13

13.1 Since resolution is twice as good in the vertical direction and somewhat more in the horizontal direction, then to see such small detail observers should be on the order of half the distance or slightly less than that for HDTV displays as compared to conventional TV displays.

13.2 In the horizontal direction, pixel pitch is $p = 11.8$ mm/512 pixels $= 23$ μm. Hence pixel dimensions of 10 and 20 μm can exist, but 30 μm cannot. The Nyquist spatial-frequency cutoff is $f_y = (2p)^{-1} = 21.7$ cycles·mm^{-1}. Pixel MTF is $\mathrm{sinc}(2\pi f_x a)$, where $a = 5$ and 10 μm for 10- and 20-μm-wide pixels, respectively. Pixel MTF reaches zero at spatial frequencies $(2a)^{-1}$ $= 100$ cycles·mm^{-1} and 50 cycles·mm^{-1} for 5- and 10-μm-wide pixels, respectively. Hence, resolution is limited by Nyquist criterion to the spatial frequency of 21.7 cycles·μm^{-1} for *both* size pixels. The smallest detail that can be resolved is $2p = 46$ μm, which is much larger than pixel size. However, contrast of this point spread function detail is better for the smaller pixel array than for the larger since the pixel MTF at this Nyquist spatial frequency is 0.92 for the smaller pixel and 0.72 for the larger pixel array.

13.3 If ϕ_{ij} represents the number of photons incident on pixel ij and if η_{ij} is quantum efficiency of that pixel, then the variance in the number of photogenerated electrons across the array per scene of N pixels is, assuming Poisson photon statistics,

$$\sigma_s^2 = \sum_{ij} = \frac{[\eta_{ij}\phi_{ij} - \langle\eta\phi\rangle]^2}{N} = \langle(\eta\phi)^2\rangle - \langle\eta\phi\rangle^2$$
$$\approx \langle\phi\rangle^2[\langle\eta^2\rangle - \langle\eta\rangle^2] = \bar{\phi}^2\sigma_\eta^2$$

where σ_η is quantum efficiency variance across the array. Temporal noise is given by

$$\sigma_T^2 = \bar{\eta}\sigma_\phi = \bar{\eta}\bar{\phi}$$

where σ_ϕ is the variance in photon number at a given pixel and spatial stationarity is assumed. Hence, total noise is

$$\sigma^2 = \sigma_T^2 + \sigma_S^2 = \bar{\eta}\bar{\phi} + (\bar{\phi})^2\sigma_\eta^2 = \bar{\eta}\bar{\phi} + (\bar{\eta})^2(\bar{\phi})^2U^2$$

where $U = \sigma_\eta/\bar{\eta}$ is pixel nonuniformity. Therefore,

$$\mathrm{SNR} = \frac{\eta\phi}{[\bar{\eta}\bar{\phi} + (\bar{\eta})^2(\bar{\phi})^2U^2]^{1/2}}$$

where the signal $\eta\phi$ is the number of photoelectrons in a detector site. For smaller $\eta\phi$, temporal noise dominates and SNR $\propto \sqrt{\eta}$. For large $\eta\phi$, SNR $\propto U^{-1}$. For more details see Ref. [11.3].

13.4 If $L_{diff} = 75$ μm and $L_{depl} = 5$ μm, then from Eq. (13.2.2) $L = 75\ \mu\text{m}/[1 + 4\pi^2(0.75)^2f^2]^{1/2}$ where f is spatial frequency in cycles·mm^{-1}. Therefore, $L = 75\ \mu\text{m}/[1 + 22.2f^2]^{1/2}$. For 1.5-eV photons (820-nm wavelength), $\alpha = 10^3\ \text{cm}^{-1} = 0.1\ \mu\text{m}^{-1}$ and $\alpha L_{depl} = 0.5$. Consequently, from Eq. (13.2.1), for 820-nm wavelength

$$\text{MTF}_{diff} = \frac{1 - \{\exp(-0.5)/[1 + 7.5/(1 + 22.2f^2)^{1/2}]\}}{1 - \{\exp(-0.5)/[1 + 7.5]}$$

$$= \frac{1 - \{0.6/[1 + 7.5/(1 + 22.2f^2)^{1/2}]\}}{1 - 0.071}$$

$$= 1.077 - 0.646/[1 + 7.5/(1 + 22.2f^2)^{1/2}]$$

When $f = 0$, $\text{MTF}_{diff} = 1$. When $f = f_N = 21.7$ cycles·mm^{-1}, $\text{MTF}_{diff} = 0.60$. Therefore, the MTF values at the Nyquist spatial frequency are reduced from 0.92 to 0.55 and from 0.72 to 0.43 for the smaller and larger pixel arrays, respectively. For 3-eV photons (410-nm wavelength), $\alpha = 7 \times 10^4$ cm^{-1} = 7 μm^{-1} and $L_{depl} = 35$. Therefore, again using Eq. (13.2.1),

$$\text{MTF}_{diff} = \frac{1 - \{\exp(-35)/[1 + 525/(1 + 22.2f^2)^{1/2}]\}}{1 - [\exp(-35)/(1 + 525)]}$$

$$\approx 1 - \{6.3 \times 10^{-16}/[1 + 525/(1 + 22.2f^2)^{1/2}]\}$$

When $f = 0$, $\text{MTF}_{diff} = 1$. When $f = f_N = 21.7$ cycles·mm^{-1}, $\text{MTF}_{diff} = 1 - 1.03 \times 10^{-16} \approx 1$. Hence, at the shorter wavelength $\text{MTF}_{diff} \approx 1$ and CCD MTF is essentially pixel MTF. It is unaffected by diffusion and the MTF values remain 0.92 and 0.72 at the Nyquist frequency for the smaller and larger pixel arrays, respectively. Image quality is noticeably better at the shorter wavelength.

Note that for FT CCDs, there is little dead space.

14.1 For exposures shorter than $t_{e\ \max}$, resolution is limited by the film. For exposures longer than $t_{e\ \max}$ resolution is limited by image motion. A required contrast threshold of at least 3% for image motion will be assumed. From Eq. (14.3.3) $M_t(f) = \text{sinc}\pi(fvt_e) = 0.03$. This implies $\pi fvt_e = 3.05$ or $t_e = 3.05/\pi fv = 3.05/\pi(50)(4) = 4.85$ ms where the limiting spatial frequency of the film is utilized. Exposures longer than 4.85 ms will result in blur deriving from image motion rather than from the quality of the film.

14.2 The image irradiance is assumed to be of the form

$$x' = x_o + D\sin\omega t$$

Blur radius is $2D$, and does not increase with exposure time, as long as $t_e \geq T_o$. As in Example 14.1, we are interested in the sine wave response. Hence, we assume an image

$$i(x') = B_o + B_m \cos 2\pi fx'$$

Substituting for x', the vibrated image irradiance pattern is

$$\begin{aligned} i(x, t) &= B_o + B_m \cos 2\pi f(x_o + D\sin\omega t) \\ &= B_o + B_m \cos 2\pi fx_o \cos 2\pi f(D\sin\omega t) \\ &+ B_m \sin 2\pi fx_o \sin 2\pi f(D\sin\omega t) \end{aligned}$$

Again the mean intensity over the exposure period t_e may be computed from the integral:

$$\begin{aligned} \bar{i} = B_o &+ \frac{(B_m \cos 2\pi fx_o)}{t_e} \int_0^{t_e} \cos 2\pi f(D\sin\omega t)\, dt \\ &+ \frac{(B_m \sin 2\pi fx_o)}{t_e} \int_0^{t_e} \cos 2\pi f(D\sin\omega t)\, dt \end{aligned}$$

If there are a number of cycles of vibration occurring within the exposure period, the neglect of a portion of a cycle would have no effect on blur radius. If only the case of exposures containing an integral number of vibration cycles were to be considered, blur radius would still be the same. In this case the second integral, above, has the value zero since the function is odd, being symmetrically positive and negative. The rest of the equation may be written:

$$\bar{i} = B_o + B_m \cos 2\pi f x_o \left[\frac{1}{t_e} \int_0^{t_e} \cos 2\pi f (D \sin \omega t)\, dt \right]$$

The bracketed expression above is the transfer function associated with simple harmonic motion. Since by definition $T_0 = 2\pi/\omega$, then because of the simplifying assumption the exposure contains an integral number of cycles,

$$t_e = nT_o = \frac{2\pi n}{\omega}$$

This value of t_e is substituted into the integral above and a change of integration to $y = \omega t$ is made. The result is recognized as a Bessel function. Thus:

$$\frac{1}{2\pi n} \int_0^{2\pi n} \cos(2\pi f D \sin y)\, dy = J_0(2\pi f D)$$

When this result is substituted into the expression for image irradiance distribution, $\bar{i}$ then becomes

$$\bar{i} = B_o + B_m J_0(2\pi f D) \cos 2\pi f x_o$$

The modulation contrast of the static image is simply

$$\text{MCO} = \frac{(B_o + B_m) - (B_o - B_m)}{(B_o + B_m) + (B_o - B_m)} = \frac{B_m}{B_o}$$

as in Example 14.1. However, the modulation contrast of the vibrated image is

$$\text{MCI} = \frac{B_m}{B_o} [J_0(2\pi f D)]$$

The modulation transfer function is therefore the ratio of the two, or

$$M_s(f) = \frac{\text{MCI}}{\text{MCO}} = \frac{B_m}{B_o} [J_0(2\pi f D)] \div \frac{B_m}{B_o} = J_0(2\pi f D)$$

as in Eq. (14.4.6).

14.3 (a) From Fig. 14.7, in order for $M_s(f)$ in Eq. (14.4.6) to be greater than 0.1, $J_0(\pi f f_l \theta_{pp}) > 0.1$, which implies $f f_l \theta_{pp} \approx 0.7$, or

$$f_{r\,\max} = 0.7/(10^{-4}\ \text{rad}) f_l = 7000/f_l$$

where $f_{r\,\max}$ here refers to vibrations only. The number of TV line pairs is limited effectively by vibrational blur from a hardware maximum of 700 to $(f_{TV\,r\,\max})$ (11) mm = $(7000/f_l)$(11 mm). On the other hand, with regard to the TV system, $f_{TV\,r\,\max}$ = 700 line pairs/11 mm. Equating both limiting spatial frequencies yields the largest focal length for which blur by both the TV system and vibrations are equal. This occurs when

$$7000/f_l = 700/11$$

where f_l is in millimeters, just as is the effective camera dimension 11 mm. Accordingly, optimum focal length is $f_{l\,\text{opt}} = 110$ mm. If the zoom lens focal length is longer, the higher magnification also magnifies the image vibrations, thus causing this motion and resulting blur to be more noticeable. If the focal length is shorter, image resolution is limited by the TV system and vibrations are not noticeable because of the small image of the target.

(b) From similar triangles connecting object and image planes, angular resolution limit is, from Sections 10.2 and 10.4,

$$\frac{\Delta x}{s} = \begin{cases} -\dfrac{\Delta x'}{s'} = \dfrac{1}{2f_{r\,\max}s'} = 2(7000/f_l)s' = \dfrac{1}{14{,}000}, & f_l \geq 110 \text{ mm} \\ \quad = \dfrac{1}{2(700/11)s'} \approx \dfrac{11}{1400 f_l}, & f_l \leq 110 \text{ mm} \end{cases}$$

where f_l is in millimeters. For $f_l = 110$ mm, angular resolution limit is $\Delta x/s = 1/14{,}000 = 7.14 \times 10^{-5}$ radians. Angular resolution limit is depicted graphically on the next page. Any improvement in resolution at longer focal lengths is prevented by vibrational blur, which becomes more and more apparent with increased magnification.

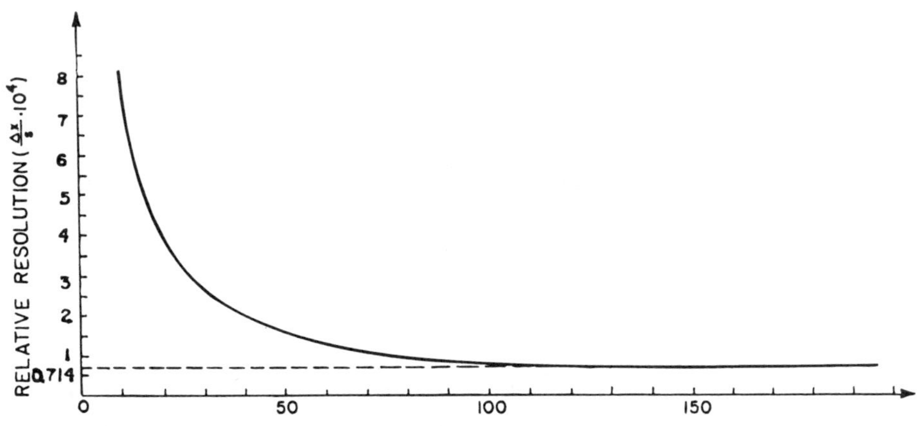

(c) Field of view decreases with increased focal length, as shown in Section 2.12. Hence, if one uses a focal length longer than 110 mm, not only does one see no smaller detail than for 110 mm, but field of view is decreased.

14.4 Largest blur extent $d_{\max}$ is obtained for exposures where $[x(t) = 0]$. These are centered between two peaks of the sine wave motion. Assume that $x(t) = \cos(2\pi t/T_o)$ and $t_s = T_o/4 - t_e/2$ (starting point for the largest blur). Hence, from Eq. (14.4.12b),

$$\mathrm{MTF}(f) = \left| \frac{1}{\pi} \int_{x(T_o/4-t_e/2)}^{x(T_o/4+t_e/2)} \frac{\exp(-j2\pi f x)}{\sqrt{D^2 - x^2}}\, dx \right|$$

By changing the variable of integration from space to time,

$$x = D \cos\omega_o t$$

$$dx = -\omega_o D \sin(\omega_o t)\, dt$$

and

$$\mathrm{MTF}(f) = \left| \frac{-\omega_o}{\pi} \int_{T_o/4-t_e/2}^{T_o/4+t_e/2} \exp[-j2\pi f D \cos(\omega_o t)]\, dt \right|$$

$$= \left| -\frac{\omega_o}{\pi} \int_{T_o/4-t_e/2}^{T_o/4+t_e/2} [\cos(2\pi f D) \cos(\omega_o t) - j \sin(2\pi f D) \cos(\omega_o t)] \right| dt$$

In this case the second integral, above, has the value zero since the function is odd, being symmetrically positive and negative. From trigonometry $\cos[\omega_o(t)] = \sin[\pi/2 - (\omega_o t)]$. When the argument of the sine wave approaches zero (as in the case for $T_o/4 - t_e/2 < t < T_o/4 + t_e/2$ and t_e is of very low value) then $\sin(\pi/2 - \omega_o) \cong \pi/2 - \omega_o t$. Therefore,

$$\mathrm{MTF}(f) = \left| \frac{-\omega_o}{\pi} \int_{T_o/4-t_e/2}^{T_o/4+t_e/2} \exp[-j2\pi f D \cos(\omega_o t)]\, dt \right|$$

$$= \frac{1}{\pi} \cdot \frac{1}{2\pi f D} \cdot \sin\left[2\pi f D \left(\frac{\pi}{2} - \omega_o t \right) \right] \Bigg|_{T_o/4-t_e/2}^{T_o/4+t_e/2}$$

$$= \frac{1}{\pi} \cdot \frac{1}{\pi f D} \sin(\pi f D \omega_o t_e)$$

Let $D\omega_o t_e$ be designated as blur extent d. Then the above reduces to

$$\mathrm{MTF}(f) = \frac{t_e \omega_o}{\pi} \frac{\sin(\pi f d)}{(\pi f d)} = 2\left(\frac{t_e}{T_o}\right) \mathrm{sinc}(\pi f d)$$

After normalizing this function by its maximum value the resultant MTF is the same as for linear motion MTF [Eq. (14.3.3)]. This is to be expected since at $x(t) = 0$ the velocity is uniform for a short exposure particularly as t_e approaches zero. The factor $2t_e/T_o$ is the area under the $f_x(x)$ curve. If the LSF or $f_x(x)$ curve were normalized in order to give unity area, then maximum value of MTF is unity. The above holds for maximum blur.

14.5 The minimum blur occurs when the exposure is centered at peak or at the bottom of the sine wave vibration curve, as shown in Fig. 14.5. In this case the calculation of the MTF is very simple if it is assumed that $t_e << T_o$. The starting point now is $t_s = T_o/2 - t_e/2$ and $x^2 = D^2$ around $T_o/2$. As t_e approaches zero, the image motion and vibration LSF in the first equation in the previous solution approaches a delta function since blur here is zero, and the image motion and vibration MTF for this case then approaches a constant value of unity.

SOLUTIONS

16

16.1 Even though the 5/3 power law dependence is dropped for plane wave propagation, the height profiles for C_n^2 still exist. Hence, for downward viewing,

$$M_{TL}(f_a) = \exp\left[-57.53 f_a^{5/3} \lambda^{-1/3} \sec\theta \int_0^L C_{no}^2 z^{-4/3}\, dz\right]$$

where $z = 0$ at the surface and 2×10^5 cm at $z = L$.

Note that if $z = 0$, the integrand approaches infinity. Hence, consider a propagation channel from 1 cm to $L = 2 \times 10^5$ cm. In this case,

$$\begin{aligned} M_{TL}(f_a) &= \exp\left[-57.53 f_a^{5/3} \lambda^{-1/3} \sec\theta\, C_{no}^2 \int_1^L z^{-4/3}\, dz\right] \\ &= \exp[-57.53 f_a^{5/3} \lambda^{-1/3} \sec\theta\, C_{no}^2 (-3)(L^{-1/3} - 1^{-1/3})] \\ &= \exp[-(3) 57.53 f_a^{5/3} (5 \times 10^{-5})^{-1/3} (2)\, 9 \times 10^{-14} (0.9829)] \\ &= \exp(-8.29 \times 10^{-10} f_a^{5/3}) \end{aligned}$$

The above exponent is 230 times larger than that in Eq. (16.4.2). It predicts turbulence degradation much more severe than that which is actually expected to take place. Hence, the plane wave approximation is much too severe.

For upward imaging, since the turbulence is again inhomogeneous we must consider the dependence of C_n^2 with elevation. Using the height profile in Eq. (16.4.4), we have

$$M_{TL}(f_a) = \exp\left\{-57.53 f_a^{5/3} \lambda^{-1/3} \sec\theta\, C_{no}^2 \int_0^L [(z + 1)/L]^{4/3}\, dz\right\}$$

where $z = 0$ at the object plane which is at elevation L. This simplifies to

$$M_{TL}(f_a) = \exp\left[-57.53 f_a^{5/3} \lambda^{-1/3} \sec\theta\, C_{no}^2 L^{-3/4} \int_1^{L+1} x^{4/3}\, dx\right]$$

$$= \exp\left\{-57.53 f_a^{5/3} \lambda^{-1/3} \sec\theta\, C_{no}^2 L^{-3/4} \left(\frac{3}{7}\right) [(L + 1)^{4/3} - 1^{4/3}]\right\}$$

$$\approx \exp(-1.49 \times 10^{-7} f_a^{5/3})$$

The exponent here is two orders of magnitude smaller than that in Eq. (16.4.5). Hence, the plane wave approximation would severely underestimate turbulence degradation for upward imaging.

16.2 r_o and $\rho_{1/e}$ can be calculated directly from Eqs. (16.6.9) and (16.6.10). However, since MTF expressions have already been calculated it is faster to use Eqs. (16.4.2) and (16.4.5) together with Eqs. (16.6.7) and (16.6.8).

For downward viewing

$$f_{a\,1/e} = (3.60 \times 10^{-12})^{-3/5} = 7.35 \times 10^6 \text{ cycles}\cdot\text{rad}^{-1}$$

$$\rho_{1/e} = \lambda f_{a\,1/e} = (5 \times 10^{-5}\text{ cm})(7.35 \times 10^6 \text{ cycles}\cdot\text{rad}^{-1}) = 367 \text{ cm}$$

$$r_o = 2.1\rho_{1/e} = 772 \text{ cm}$$

This represents excellent viewing conditions. For upward viewing,

$$f_{a\,1/e} = (1.4 \times 10^{-5})^{-3/5} = 815 \text{ cycles}\cdot\text{rad}^{-1}$$

$$\rho_{1/e} = \lambda f_{a\,1/e} = 0.04 \text{ cm}$$

$$r_o = 2.1\rho_{1/e} = 0.9 \text{ mm}$$

This value of r_o is very small and reflects the relatively high value of C_n^{-2} for the strong turbulence used here.

16.3 From Eq. (16.6.4b),

$$r_o \approx 0.185[\lambda^2/(3/8)(C_n^2 L)^{3/5}$$

$$= 0.185\{(6.3 \times 10^{-5})^2/[(3/8)(1.3 \times 10^{-15})(5 \times 10^5 \text{ cm})]\}^{3/5}$$

$$= 0.39 \text{ cm}$$

Image quality is almost as bad as when imaging upward even though the turbulence is weaker here.

16.4 Equation (16.2.19) indicates that for a 60-cm focal length, turbulence limits the resolution angle for detection to $(2f_{a\,\max})^{-1} = 3.45 \times 10^{-5}$ rad. For recognition, $n_{50} \approx 4$ line pairs from the Johnson chart and $f_{r\,\max} = f_{a\,\max}/f_l = 1.45 \times 10^4/60 = 241.7$ cycles$\cdot$cm^{-1}. From Eq. (10.4.2), $|x| = n_{50}\,(f_{\max}\text{ m})^{-1} = 4[(2.42 \times 10^2)\cdot(60/5 \times 10^5)]^{-1} = 138$ cm $= 1.38$ m. For a 50-cm focal length, the image resolution is limited by the TV to $\Delta x/s = 0.002/f_l$ for 50% probability of detection, where f_l is in centimeters. However, for the vidicon system $f_{r\,\max} = 250$ lp$\cdot$cm^{-1}. Therefore, $|x| = 4[(250)(50/5 \times 50^5)]^{-1} = 160$ cm $= 1.6$ m $>$ asymptotic limit of 1.38 m which exists for all focal lengths $>$ 57.9 cm. For focal lengths $<$ 57.9 cm, $|x|$ increases as f_l decreases.

16.5 In this case, Eq. (16.4.2) becomes $M_{TL} = \exp(-3.60 \times 10^{-10} f_a^{5/3})$. At the relatively high spatial frequency of 50 cycles$\cdot$mrad^{-1}, M_{TL} here equals $\exp(-0.0245) = 0.98$. Hence, even here turbulence is not a serious factor for downward viewing.

17

17.1 For $f_l = 60$ cm, which is greater than the critical focal length of 47.1 cm, resolution angle $\Delta x/s$ equals 4.24×10^{-5} rad. Now, $f_{a\ max}$ equals the reciprocal of twice this resolution angle. For recognition, $n_{50} \approx 4$ line pairs from the Johnson chart. Hence, for a 5-km distance, $x = (5 \times 10^3)(2)4\ (4.24 \times 10^{-5}) = 1.70$ m. Hence, consideration of aerosol MTF implies recognizable object detail is more than 1.23 times larger than that required in Exercise 16.4. For 90% probability, $n_{90} = 1.75 n_{50}$. Therefore, $x = 1.75\ (1.70) = 3.0$ m.

17.2 Two pixels are needed to represent 1 line pair. Hence, maximum angular spatial frequency is 70 line pairs over 2.5° (or 43.6 mrads), or 1604 line pairs or cycles·rad^{-1}. This is the hardware limitation. Long-exposure turbulence MTF is $M_{TL} = \exp[-(3/8)57.53\ f_a^{5/3}\ \lambda^{-1/3}\ C_n^2\ L] = \exp[-(3/8)57.53\ f_a^{5/3}\ (10^{-5})^{-1/3} \cdot 10^{-13} \cdot 2 \times 10^3] = \exp(-2 \times 10^{-7}\ f_a^{5/3})$. At $f_{a\ max} = 1604$ cycles·rad^{-1}, $M_{TL} = e^{-0.117} = 0.96$. Short-exposure turbulence MTF would be even higher. Hence, within the spatial-frequency limitations of the equipment, turbulence is not a significant problem at this distance.

With regard to hardware MTF, it was shown in Section 9.7.1 that pixel MTF equals $\text{sinc}(2\pi f_r a)$, where a is one-half the pixel dimension, which is unknown here. Since $f_r = f_a \cdot f_l$, pixel MTF can also be written as $\text{sinc}(2\pi f_a a/f_l)$, where a/f_l equals one-half the field of view seen by each pixel. (Field of view seen by each pixel is called instantaneous field of view, or IFOV). Since there are 140 pixels over 43.6 mrad, each pixel sees 43.6 mrad/140 = 0.31 mrad = $(2f_{a\ max})^{-1}$, since each pixel represents one-half of a line pair. The angle a/f_l is one-half of this, since a is one-half the pixel dimension. Consequently, $a/f_l = (4f_{a\ max})^{-1}$. At the angular spatial frequency $f_{a\ max}$, pixel MTF equals $\text{sinc}(\pi/2) = 2/\pi = 0.64$. Thus, hardware resolution is much more of a limitation than turbulence in this example, despite the high value of C_n^2.

17.3 In this case of $f_{a\ max}$ = 256 line pairs/43.6 mrad = 5.87 line pairs or cycles·mrad^{-1}, which is 3.66 times better than the previous system. In this case $M_{TL} = \exp(-2 \times 10^{-7} f_a^{5/3}) = e^{-1.02} = 0.68$ at this value of $f_{a\ max}$. Here turbulence is a more serious problem because of the higher spatial frequencies. It still is not necessarily as dominant a problem as in the visible because of the longer infrared wavelength.

17.4 The high spatial-frequency practical aerosol MTF asymptote is, using Eq. (17.2.1), equal to $\exp[-(A_a + S_a^*)L] = \exp[-(0.7)(2)] = \exp(-1.4) = 0.25$. This is much more serious degradation than that imposed by turbulence, and even more serious than that imposed by pixel size.

17.5 For this equipment, practical aerosol MTF becomes $\exp[-0.2(2) - 0.3(2)(f_a/2000)^2]$ from Eq. (17.2.1) for a 2-km path length, assuming $f_a < f_{ac}^*$. Otherwise, it equals $\exp[-(0.2 + 0.3)2] = \exp[-1] = 0.37$. At the equipment high spatial-frequency cutoff of 5.87 line pairs or cycles·mrad^{-1} (see Solution 17.3), turbulence and aerosol MTF values are almost similar, but aerosol MTF is still more dominant. [M_{TL} ($f_{a\ \max}$) has been calculated in Solution 17.3.] However, at lower angular spatial frequencies, aerosol MTF is also much lower than turbulence MTF and therefore dominant too. For example, at $f_a = f^*_{ac} = 2$ cycles·mrad^{-1}, aerosol MTF equals $\exp(-1) = 0.37$, while M_{TL} equals $\exp[-2 \times 10^{-7}\ (2000)^{5/3}] = \exp(-0.17) = 0.94$. For short exposures atmospheric MTF is dominated even more by aerosol MTF, since turbulence MTF increases.

Index

Norman S. Kopeika received the BS, MS, and PhD degrees in Electrical Engineering from the University of Pennsylvania, Philadelphia, in 1966, 1968, and 1972, respectively. His PhD dissertation, supported by a NASA Fellowship, dealt with detection of millimeter waves by glow discharge plasmas and the utilization of such devices for detection and recording of millimeter wave holograms. In 1973 he joined the Department of Electrical Engineering, Ben-Gurion University of the Negev, Beer-Sheva, Israel, as a Lecturer. He was appointed professor in 1988 and incumbent of the Reuben and Frances Feinberg Chair in Electro-Optics in 1994. From 1989 to 1993 he served two terms as department chair. From 1978 to 1979 he was a visiting associate professor in the Department of Electrical Engineering, University of Delaware, Newark. His research interests include atmospheric optics, target acquisition, effects of surface phenomena on optoelectronic device properties, optical communication, electronic properties of plasmas, laser breakdown of gases, the optogalvanic effect, electromagnetic (EM) wave-plasma interaction in various portions of the EM spectrum, and utilization of such phenomena in EM wave detectors and photopreionization lasers. He has published over 120 reviewed journal papers and over 90 conference papers in the above areas. He was particularly active in research of time response and impedance properties of plasmas and authored a general unified theory to explain EM wave-plasma interactions all across the electromagnetic spectrum. His earlier published work on the optogalvanic effect preceded the naming of the effect. He has done extensive work on the wavelength and weather dependences of image resolution through the open atmosphere, particularly with regard to spatial coherence degradation and spatial frequency dependence resulting from forward light scattering by relatively large airborne particulates. He and his students have developed methods to predict atmospheric modulation transfer function (MTF), including both turbulence and aerosol MTF components, according to weather, and to use that information in image restoration so as to deblur such effects. Also, methods to calculate numerically in real time optical transfer function for any type of image motion and vibration have been developed, and these too have also been used in image restoration. He contributed actively toward the development of postfabrication techniques for improvement of photodiode responsivity and uniformity and for wavelength-tuning of semiconductor light sources by external means. Recently he has been involved in adaptive techniques for satellite optical communication networks and effects of the atmosphere and image motion and vibration on target acquisition. Dr. Kopeika is a senior member of the IEEE, and a member of SPIE, the Optical Society of America, and the Laser and Electrooptics Society of Israel.